Introduction to
Nanoscience

Introduction to
Nanoscience

Gabor L. **Joydeep** **Harry F.** **Anil K.**
Hornyak • **Dutta** • **Tibbals** • **Rao**

CRC Press
Taylor & Francis Group
Boca Raton London New York

CRC Press is an imprint of the
Taylor & Francis Group, an **informa** business

CRC Press
Taylor & Francis Group
6000 Broken Sound Parkway NW, Suite 300
Boca Raton, FL 33487-2742

First issued in hardback 2019

© 2008 by Taylor & Francis Group, LLC
CRC Press is an imprint of Taylor & Francis Group, an Informa business

No claim to original U.S. Government works

ISBN-13: 978-1-4200-4805-6 (hbk)

Library of Congress Cataloging-in-Publication Data

Introduction to nanoscience / Gabor Louis Hornyak ... [et al.].
    p. cm.
  Includes bibliographical references and index.
  ISBN 978-1-4200-4805-6 (alk. paper)
  1. Nanostructures. 2. Nanoscience. 3. Nanostructured materials. 4. Nanotechnology. I. Hornyak, Gabor Louis.

QC176.8.N35I685 2008
620'.5--dc22   2008013710

Visit the Taylor & Francis Web site at
http://www.taylorandfrancis.com

and the CRC Press Web site at
http://www.crcpress.com

I would like to dedicate this book to my partner, musician, fellow bird watcher, hiker, Scrabbler, and best friend Kathleen Ann Chiras of Evergreen, Colorado. Her support went far beyond reason and her unwavering belief in me was my source of inspiration and energy.

$\mathcal{C}$ONTENTS

SECTION 5:   NATURAL AND BIONANOSCIENCE   693

Chapter 13   NATURAL NANOMATERIALS   695

PREFACE

"There is plenty of room at the bottom." This bold and prophetic statement from Nobel laureate Richard Feynman back in the late 1950s at Cal Tech launched the *Nano Age* and predicted, quite accurately, the explosion in nanoscience and nanotechnology that we have witnessed over the past decade and a half. Actually, there now is so much information available about this burgeoning new field that it must be, in no uncertain terms, a formidable task for anyone trying to keep pace—let alone to write a text. The challenge, however, has not discouraged those of us wishing to develop pedagogical tools that make the science and technology of the really small understandable yet challenging and rewarding to the motivated student.

Pedagogy versus "Introductionology." A few excellent introductory books that function as teaching tools have been published over recent years (e.g., those by Poole, Cao, and Kelsall). Common threads in these books contain the following characteristics: (1) they are written by one or a few authors (not editors), (2) the writing style is uniform throughout the book (e.g., continuity of theme and message), and (3) complex properties and phenomena are actually explained. This is in stark contrast to numerous introduction to nanotechnology books that consist of contributed works compiled by one or more editors. We do not mean to imply that books of such genres are not informative, interesting, and invaluable in many other ways because they most certainly are. They are simply not pedagogical in the sense of the specific meaning of the word. An evolutionary process is in place, and the timing is ideal to publish relatively comprehensive nanoscience and nanotechnology pedagogical tools designed specifically to support a college curriculum at the junior–senior level.

How does this text differ from other texts that have pedagogical objectives? To begin with, *Introduction to Nanoscience* is based on one fundamental precept: that only *science* is presented and discussed without any emphasis on *technology*. Science usually precedes technology. Technology, after all, is the application of science. In the purest sense, science is the study of Nature (with a capital "N"). It can be argued that the study of Nature is the study of Nature's technology (its nanotechnology)—the application of its science (its nanoscience). We are therefore caught up in a minor semantic chicken-and-egg dilemma in that our nanoscience is based on Nature's nanotechnology and that our nanotechnology is based on our nanoscience—got it?

Nature's Nanoscience → Nature's Nanotechnology → Our Nanoscience
     → Our Nanotechnology

We shall therefore, in this text, commit ourselves to studying the *science of the small* without indulging in its technology (applications, devices, and technological materials). This is easier said than done because some mention of technology is unavoidable. However, we will do our best to maintain this apparent distinction.

Teaching Nanoscience and Nanotechnology. There are three fundamental issues that confront the potential nano text author (or nano class instructor):

1. How does one write a college textbook (or teach a college course) that covers such a vast range of topics that over the past 10 years has undergone an explosion in information and communication (press releases, papers, journals, books, sessions and conferences, etc.)?
2. How does one organize a college text (or college course) that is highly interdisciplinary in nature—a text (course) that must plausibly integrate all scientific and engineering disciplines in some meaningful way?
3. How does one write a textbook (or teach a course) that addresses the individual needs of biologists, chemists, physicists, and engineers—all in one class?

From another perspective, how does one write a text (or teach a course) about nanophysics, nanochemistry, and nanoengineering to biologists or about nanobiology, nanochemistry, and nanoengineering to physicists—and all the other possible permutations?

Nanoscience (and nanotechnology) can be taught within the traditional monolithic academic framework, independent of traditional disciplines akin to materials science or embedded within traditional academic disciplines of physics, chemistry, biology, and engineering. There are proponents of both schools of thought. Nanoscience (and nanotechnology) can be taught to just chemists, just physicists, just biologists, or just engineers *or* to mixed classes with students from all the aforementioned academic disciplines. How much nanobiology should a physicist know? How much nanochemistry should an engineer absorb? How much nanophysics should a biologist learn? With how much nanomechanics should any future scientist or engineer become familiar? Is it feasible to teach nanoscience to a class that contains a mixture of physicists, engineers, chemists, and biologists or is it better to teach specifically oriented nanoscience to students of one traditional academic discipline?

The physicist will continue to work side by side with the biologist, the engineer side by side with the chemist, the biologist side by side with the engineer, and so on and so forth. Buildings that accommodate nano and biotechnology are specifically designed to accommodate such interdisciplinary overlap. The handling of intellectual property will require patent attorneys with an interdisciplinary background. Manufacturing practice, more than ever, will require engineers and technicians to acquire knowledge beyond their initial area of training.

Nanoscience as a Stand-alone Course. Nanoscience can be taught as a separate stand-alone course at the upper division level—once again, akin to materials science. In this way, *Introduction to Nanoscience* can serve as a stand-alone textbook. In an ideal world, we believe that before such a stand-alone nanoscience course, fundamental nanoscience topics should be embedded within general chemistry, general physics, general biology, and engineering courses. It should not be difficult to integrate selected (and relevant) nanoscience topics within the syllabus of a general biology, general chemistry, or general physics course. For example, the concepts of collective and specific surface area, surface energy, cluster orbitals, intermolecular interactions, and supramolecular chemistry are amenable for facilitative incorporation into freshman and sophomore level courses. Some mention of nanoscience and nanotechnology should be included in advanced courses as well: physical chemistry (quantum mechanics and thermodynamics), classical mechanics, solid state physics, materials science, molecular biology, biochemistry, genetics, spectroscopy, environmental biology, and so forth. Materials science texts in particular should add a few nanoscience chapters or use *Introduction to Nanoscience* as a *supplemental text*.

A One-Semester Course. *Introduction to Nanoscience* is designed as the foundation of a one-semester course. Supplementary materials, depending on the academic disciplinary emphasis, are always recommended. *Introduction to Nanoscience* should provide enough diversity with regard to topics to provide the potential student a sound basis in nanoscience. *Introduction to Nanoscience* should provide enough specifics to expand the knowledge of the student beyond that encountered in fundamental core course curriculum. "Introduction to Nanoscience (the course)" would be most effective if offered as an upper-division elective in chemistry, physics, biology, or engineering programs. Ideally, if the university already has an interdisciplinary program in place, such as a materials science program, a course in nanoscience based on this text should help prepare the student for things to come.

Textbook Strategy. The text is divided into five generic divisions: (1) perspectives, (2) tools, (3) physics, (4) chemistry, and (5) biology. The "Perspectives" division consists of two chapters: chapter 1, "Introduction," in which definitions, the historical perspective, and an overview of *nano* are presented, and chapter 2, "Societal Implications of Nano," in which issues confronting nanoscience and nanotechnology that exist outside the laboratory are presented and discussed. Our goal is to educate students about the potential ramifications of their future work.

The "Nanotools" division consists of two chapters: chapter 3, "Characterization Methods," and chapter 4, "Fabrication Methods." We believe that placement of these chapters up front in the text is a good idea because (1) they serve as a link to previous knowledge attained by the student from other courses, thereby providing perspective in addition to a comfort zone; and (2) characterization and fabrication methods are referred to specifically and frequently in subsequent chapters of the text and familiarization with them early is a strategically prudent course.

When describing the relationships between and among scientific disciplines, a logical case is made for using the "level of complexity" of the discipline as a parameter. For example, physics is fundamental to all of chemistry and all of biology, and chemistry is fundamental to all of biology:

$$Physics \rightarrow Chemistry \rightarrow Biology$$

Starting with Schrödinger's equation, we are able to describe the hydrogen atom quite adequately. However, from first principles, it becomes exceedingly more difficult to describe larger entities as we proceed from physics into chemistry and on into biology. The physics division (chapters 5–8) is concerned primarily with properties and phenomena with little mention of chemistry and biology. The chemistry division (chapters 9–12) focuses on synthesis and modification and prepares concepts and mechanisms that become relevant in the biological division of the book. The biological division (chapters 13 and 14) is centered on the properties, phenomena, synthesis, and modification of living materials and materials derived from them.

The "Physics: Properties and Phenomena" division comprises four chapters. Chapter 5 addresses properties and phenomena associated with the surface (collective and specific), size, shape, composition, and orientation. Chapter 6 is focused on energetic considerations associated with nanomaterials and mechanisms that serve to mitigate the excess energy associated with nanosurfaces. Chapter 7, "The Material Continuum," examines how macroscopic materials are related to micron-sized and nano-sized materials. Does a smooth continuum curve accurately address the transition from the classical domain to that of nanomaterials and subsequently to atoms and molecules of the quantum domain? Zero-, one-, and two-dimensional nanomaterials are introduced. Chapter 8 delves into a new field of thermodynamics called (not surprisingly, but most certainly arguably) nanothermodynamics.

The "Chemistry: Synthesis and Modification" division has four chapters. Chapter 9 introduces carbon-based nanomaterials that serve as examples of zero-, one-, and two-dimensional nanomaterials. Chapter 10 delves into the types of chemical bonding available to nanomaterials—an area with which most students are already quite familiar. Chapter 11, "Suparmolecular Chemistry" is very chemical in its presentation. Chapter 12 covers more synthesis and chemical modification of nanomaterials. Template synthesis in particular is an important route to forming nanomaterials.

The "Natural and Bionanoscience" division has two chapters: chapter 13, "Natural Nanomaterials," and chapter 14, "Biomolecular Nanoscience." Natural nanomaterials are all around us, and they offer insights into how we should go about making our synthetic nanomaterials and devices. The field of biomimetics is dedicated to copying natural materials. Biochemical nanoscience is most familiar to students that have taken already biochemistry. Understanding that nature is the master of nanotechnology is illustrated time and time again by biochemical machinery involved in metabolism and structure. With this last division, we wrap up *Introduction to Nanoscience*.

Development of the Student. The ultimate goal for those of us that teach and the ultimate goal of this textbook is the development of the student. Nanotechnology is the next industrial revolution. It is an interdisciplinary converging technology that brings together aspects of hitherto unrelated fields of biology and engineering, among many others. It is a disruptive technology in that it will change the way we do business. It may well be disruptive to the way we traditionally teach science and engineering. It is a technology that is horizontal, much like materials science, that spans every industrial sector, including energy, aerospace, medicine, construction, transportation, information technology, biotechnology, textiles, defense, electronics, recreation equipment, and many others. In other words, nanotechnology will affect every product made—eventually.

The societal implications of nanoscience and technology—everything "outside the science and technology"—must be also taken into account and discussed. The underlying goal of this text is to help prepare the student for a career in nanoscience and technology. The twenty-first century graduate needs to be prepared, flexible, adaptable, and ready for the challenges of an uncertain world, one that changes dramatically and with increasing frequency. It is our obligation as educators to ensure that students, citizens, cities, states, and nations are ready to undertake the challenges that the future springs upon us. As U.S. Commerce Department Undersecretary Phil Bond poignantly stated, "The wealth and security of nations will depend on who can commercialize nano" and, we add, those that teach, learn, understand, and develop nanoscience.

Links to Other Books. We specified earlier that *Introduction to Nanoscience* is focused primarily on the science of nano. Logically, a sister volume that follows must focus on the technology (the applications, devices, advanced materials, etc.) of nano. This textbook is titled *Fundamentals of Nanotechnology* (ISBN 978-1-4200-4803-2) and should be utilized in a complementary semester after completion of *Introduction to Nanoscience*. A combined volume is also available if professors wish to develop a complete one-year curriculum in nano. The title of the combined volume is *Introduction to Nanoscience and Nanotechnology* (ISBN 978-1-4200-4779-0). The two volumes should offer relatively complete coverage of the topic and provide the student with the tools necessary to proceed in the field.

Online Textbook Community Information for Introduction to Nanoscience. An online community has been created for you, located at www.nanoscienceworks.org.

Become a member (membership is free) and receive www.nanoscienceworks.org/textbookcommunity/introtonanoscience at no additional cost. This online community provides:

- Study problems and guide, solutions to selected problems, and additional problem sets not found in the book
- Sample exams
- Q & A from the author and the author's blog
- Lectures and PowerPoints
- NanoNews—a summary of the most recent developments
- National Nanotechnology Initiative developments
- Conference notification

This site will contain homework and classroom learning, as well as information for the individual learner. Please refer to www.nanoscienceworks.org/textbookcommunity/introtonanoscience for all resources regarding this highly anticipated textbook.

$\mathscr{A}$CKNOWLEDGMENTS

Many individuals, organizations, and companies from the public sector, academia, industry, and government donated valuable resources, advice, and encouragement to the writing of this text.

First and foremost, we humbly pay tribute to the late Nobel laureate **Richard Everett Smalley** who, in addition to his scientific contributions, was a great human being who cared about the planet and the future of our children. His motto, "Be a scientist, save the world," is surely heartfelt by us all. I had the pleasure of meeting him at the National Renewable Energy Laboratory in 2004, where he spent a few hours with us following his presentation to address and discuss our questions. He always answered e-mail despite his numerous speaking engagements, faculty responsibilities at Rice University, numerous consultations with the U.S. government, and, sadly, diminishing health. Thank you, Professor Smalley, for inspiring all of us.

Several wonderful companies provided images, written contributions, and consultation on behalf of the text. Special thanks go to **Karen Gertz**, marketing communications program manager, *Veeco Instruments, Inc.*; **Jeff R. Christiansen**, managing director, visual science and technology, *Halff Associates, Inc.*, whose finite element analysis depiction of a butterfly made of nanocomponents is awe inspiring; **Dr. Martin Thomas**, director of business development, *Quantachrome Instruments*, one of the leaders in BET analysis and equipment; **Dr. Tapani Viitala**, *KSV Instruments, Ltd.*, Finland, who provided technical insight into numerous instruments involved in surface tension, Langmuir–Blodgett instruments, and quartz crystal microbalance analysis and was so generous with his and his company's research and development; **Karen Vey**, *IBM-Zurich*, Switzerland; **Jenny Hunter**, communications manger, and **Dr. Donald Eigler**, *IBM-Alamden*

Research Center, whose image made of xenon atoms placed a face on nanotechnology and introduced it to the world; **Marc Pisnard**, Marketing, *Babolat USA*, showed us how nanotechnology has made it into recreation endeavors—the tennis racquet that I need to acquire soon; **James Hodge**, market media specialist, *JEOL USA, Inc.*, one of the leading manufacturers of scanning and transmission microscopes in all the world; **Dr. Tom Campbell**, senior research scientist, Advanced Programs, *ADA Technologies*, Littleton, Colorado, and good friend who contributed to the discussion about characterization methods; **Dr. Patrick M. Boucher**, intellectual property attorney, *Townsend & Townsend & Crew*, Denver, Colorado, who lent us his expertise concerning the impacts of nanotechnology on the world of intellectual property, regulation, and technology transfer and placed those thoughts into print in chapter 2; **Dr. Jo Anne Shatkin**, managing director, *CLF Ventures* and formerly of the *Cadmus Group*, who shared her expertise and knowledge about environmental impacts of nanotechnology in chapter 2.

Museums and foundations contributed in many ways to the work. **Steven Kruse**, general assistant, Department of History and Philosophy of Science, *Whipple Museum of History and Science*, United Kingdom; the **British Museum**'s contribution of the Lycurgus cup images, which is greatly appreciated; **Jonna Petterson**, public relations manager, *The Nobel Foundation*, Stockholm, Sweden; **Anne Hays**, of the *Melanie Jackson Agency*, LLC; **Linda Bustos**, *Cal Tech*, *public relations, photo/media archives*; and **Jade Boyd**, associate director and science editor, Office of News and Media Relations, *Rice University*, Houston, Texas.

National Laboratories and other organizations contributed valuable resources and information. **Dr. Alex Zunger**, fellow, *National Renewable Energy Laboratory*, Golden, Colorado, one of the world's leading physicists in the field of quantum dot theory; **Donald Freeburn**, general engineer, Office of Basic Energy Sciences, *U.S. Department of Energy*; **Dr. Meyya Meyyappan**, chief scientist for exploration technology, Center for Nanotechnology, *NASA Ames*, for his generous offer of information and images; **Mayuree Vathanakuljarus** and **Nares Damrongchai**, executive director, *APEC Center for Technology Foresight*, Bangkok, Thailand; **Michael Laine**, director, *Liftport Group*, in Washington state, whose ideas and vision have inspired a new generation of dreamers; **Kim Jones**, Electron Microscopy, *DOE National Renewable Energy Laboratory*, Golden, Colorado; **Dr. Alexandr Noy**, *Lawrence Livermore National Laboratory* in California; **Chris Phoenix**, *Center for Responsible Nanotechnology*, who contributed significant assistance in the areas of scaling laws; **Sunandan** and others at the *Asian Institute of Technology*, Bangkok, Thailand. **Dr. Tom Frey**, the executive director of the futurist think tank the *daVinci Institute* in Lafayette, Colorado, gave us a peak at the future of nanotechnology in chapter 2.

The academic world, of course, is where nano lives and thrives. First and foremost, I would like to thank **Professor Dr. Günter Schmid** of the Department of Inorganic Chemistry at the *University of Essen* (now in conjunction with the *University of Duisburg*, Germany) for his donation of numerous images of gold-55 clusters, chapter problems, and procedures; and **Professor George Whitesides**, Department of Chemistry, *Harvard University*, Cambridge, Massachusetts, for his generosity and his immense contribution to the field of nanoscience. Special thanks are extended to **Dr. Deb Bennett-Woods**, director and associate professor, Department of Health Care and Ethics, Rueckert-Hartmen School for Health

Professionals, *Regis University*, Denver, Colorado, an expert in bioethics and now nanoethics, who provided an overview of the ethical implications that will confront (and already confront) nanotechnology, for her dedicated effort to compile and edit chapter 2—we thank her wholeheartedly for her tireless effort; and to **Dr. Susanna Hornig Priest**, formerly of the University of South Carolina, now a professor in the Hank Greenspan School of Journalism and Media Studies at the *University of Nevada-Las Vegas*, for her contribution on the public perception of nanotechnology; and, of course, to **Dr. Peter Vukusic** of the School of Physics at the *University of Exeter*, United Kingdom, who provided the cover image for the text as well as fascinating insights into the nanostructure of butterflies, photonics, and the physics of nature at the nanoscale.

Many others from academia have made welcome contributions to the text. I would like to extend special acknowledgment to **Dr. Trent H. Galow**, Department of Chemistry, *University of Edinburgh*, United Kingdom, for his clear and well-developed presentations; **Dr. Giacinto Scoles**, professor of biophysics, *Scuola Internazionale Superiore di Studi Avanzati*, Italy, and professor of science, Department of Chemistry, *Princeton University*, New Jersey; **Dr. Zhiqun Lin**, assistant professor, Materials and Science and Engineering Department, *Iowa State University*, Ames, for his work on quantumn dots; **Chanchana Thanachayanont** of the *Asian Institute of Technology* in Bangkok for her contribution to the section on electron microscopy; **Professor Dr. Klaus Müllen**, *Max Planck Institute for Polymer Research*, Mainz, Germany; **Dr. Hong-Bo Sun**, professor in electronics, *Jilin University*, China, and professor in photonics, *Osaka University*, Japan; **Dr. Maziar S. Yaghmaee**, *University of Miscolc*, Hungary, for his insights and contributions to the nanoscale energy chapter of the text; **Nicholas J. Turro**, Department of Chemistry, *Columbia University*, New York, one of the world leaders in supra-molecular chemistry; **Joel Olson**, assistant professor, Chemistry Department, *Florida Institute of Technology*, Melbourne, Florida; **Ralph Merkle**, professor, *Georgia Tech College of Computing*, Atlanta; **Luigi Cristofolini**, professor, Department of Physics, *University of Parma*, Italy; **Linda Schadler**, professor, materials science and engineering, *Rensselaer Polytechnic Institute*, Troy, New York; **Kostay Novoselov**, School of Physics and Astronomy, *University of Manchester*, United Kingdom; **Glen P. Miller** of the Department of Chemistry at the *University of New Hampshire*, whose leading-edge research in physical organic chemistries and nanotechnologies is making huge contributions to nanomanufacturing.

Several students have assisted us during the writing. We would like to extend special thanks to **Raechelle D'Sa**, PhD candiate, Nanotechnology and Integrated Bio-Engineering Centre (NIBEC), *University of Ulster at Jordanstown*, Northern Ireland, for her section on biodegradable polymers and drug delivery systems; **Clifton Oertli**, Department of Electrical Engineering, *Colorado School of Mines*, Golden, Colorado, for his help with the numerous mathematical problems; **Kun Tae Pak**, Department of Chemistry, *University of Colorado at Denver*, for his support in many ways; and **Devi Rao** (the daughter of our coauthor Dr. Anil K. Rao), a freshman at the Colorado School of Mines in Golden, Colorado, who provided us with research articles and other helpful services.

I would like to extend special thanks to the people at Taylor & Francis who not only provided the opportunity for making this work possible but also had to endure the bumps in the road—some small and some rather large—on the way to completion of the text. First and foremost, we are indebted and especially grateful to **Nora Konopka**, publisher, Engineering and Environmental Sciences Division, *CRC Press*, Boca Raton, Florida, who has been a staunch ally through thick and thin—although her patience was stretched to the absolute limit—and **Ashley Gasque**, editorial assistant; **Jessica Vakili**, project coordinator, *Taylor & Francis Group*, New York; **Kacey C. Williams**, content and community director, *Taylor & Francis Group, CRC Press*, Nashville, Tennessee; and production experts **Glenon C. Butler**, project editor, *Taylor & Francis Group, CRC, Routledge, LEA*, Boca Raton, Florida; **Randy Burling**, production manager, *Taylor & Francis Group, CRC, Routledge, LEA*, Boca Raton, Florida; and **Anitha Johny**, account manager, *SPi*, Pondicherry, India, who coordinated the copy editing and other formatting issues related to the text.

Graphics are an important aspect of any textbook. Our renditions and images were placed in the expert, capable, and creative hands of **Mike Hamers**, *Lightspeed Commercial Arts*, in Niwot, Colorado. We were also blessed to have **Anil K. Rao**, one of our coauthors and professor in the Department of Biology, *Metropolitan State College of Denver*, contribute numerous images throughout the text.

Gabor L. Hornyak

I would like to acknowledge support, advice, moral support, and encouragement, as well as helpful suggestions, from **Cynthia Gillean**, science curriculum director, *Highland Park Independent School District*; and suggestions and background material sources from **Dr. Shou Tang**, MD, *University of Texas Southwestern*, **Dr. J. C. Chiao**, *University of Texas Arlington*, and others on the faculties of *University of Texas Southwestern, University of Texas Dallas, University of Texas Arlington; NASA; Rice University; Baylor University*; and *Sandia Laboratories*. Special thanks to **Joe Dell Gillean** for assistance in images for natural materials.

Special thanks to **Louis Hornyak** for helpful and expert editorial guidance, and to the staff of *Taylor & Francis* for expert editorial support. Other thanks and acknowledgments are given throughout the chapters to those who provided materials and permissions and expert illustration work. Also, special thanks to **Kathleen Chiras** for hosting editorial meetings, and to the *Milford Founders Trust* for hosting writing getaways in Vail, Colorado.

Special thanks to the late **Dr. Smalley** and his staff for their time during my visits to *Rice University*, and **Jim Von Ehr** and the staff at *Zyvex* during visits and seminars at their facilities.

Harry F. Tibbals

I would like to take this opportunity to thank several people for their suggestions, comments, and encouragement during the preparation of the manuscript—notably, **Heinrich Hofmann** of *Ecole Polytechnique Federale de Lausanne*, Switzerland; **Jons Hilborn** of *Uppsala University*, Sweden; **Mamoun Muhammed** of the *Royal Institute of Technology*, Sweden; **Sanjay Mathur**, Germany; **Frederico Rosei** of *INRS*, Canada; **Ioan Marinescu** of the *University of Toledo*, Ohio; and **Vivek Subramanian** of *University of California Berkeley*. It would not have been possible for me to contribute to this work without the active support of members of the *NANOTEC Center of Excellence in Nanotechnology* at AIT supported by the National Nanotechnology Center (NSTDA) of the Thai Ministry of Science and Technology (MOST). I would like to thank **Wiwut Tanthapanichakoon**, director of NANOTEC, for his kind support. Special thanks to **Said Irandoust** and **Worsak Nakulchai** for constant encouragement and to all my graduate students who supported me during the writing—notably, **Abhilash**, **Hemant**, **Rungrot**, **Jack**, **Tanmay**, **Botay**, and **Ann.** I acknowledge the dedication of **Louis Hornyak**, who, with his ready suggestions and advice, was the driving force to complete this manuscript on such a short notice. Finally, I thank my loving wife, **Sonali**, and our two beautiful children, **Joya and Jojo**, for their patience and support and for not complaining much in spite of missing me most of the time during the period I have been working on this book.

Joydeep Dutta

AUTHORS

Joydeep Dutta was born on May 5, 1964, and is currently director of the Center of Excellence in Nanotechnology and an associate professor in microelectronics at the Asian Institute of Technology (AIT), Bangkok, Thailand, whose faculty he joined in April 2003. He received his PhD in 1990 from the Indian Association for the Cultivation of Science, India. In 1991 and 1992, he did postdoctoral work at the electrotechnical laboratory (Japan) and at Ecole Polytechnique (France) before moving to Switzerland in 1993, where he was associated with the Swiss Federal Institute of Technology, Lausanne (EPFL), until 2003. From 1997 to 2001, Dr. Dutta worked in technical and managerial qualities in high-technology industries in Switzerland before returning to academia in 2002. He has been a member of the board of two companies working in high-technology electronics and environmental consulting respectively.

Dr. Dutta has been teaching courses in Microelectronics Fabrication Technology, Nanomaterials and Nanotechnology, Optoelectronic Materials and Devices, Failure Analysis of Devices, and Emerging Technologies at AIT. He has also taught Nanomaterials (since 1997) at EPFL, and in Uppsala University (2003–2005) and Royal Institute of Technology (2005–present), both in Sweden.

Dr. Dutta is a fellow of the Institute of Nanotechnology (IoN) and the Society of Nanoscience and Nanotechnology (SNN) and a member of several professional bodies, including the Institute of Electrical and Electronics Engineers (IEEE), Materials Research Society (MRS), Society of Industry Leaders, Gerson Lehrman Group Council, the NanoTechnology Group Inc., and the Science Advisory Board—all in the United States; the Asia-Pacific Nanotechnology Forum (APNF), Australia; and the U.K. Futurists Network, United Kingdom, among others. He has reviewed projects of various scientific organizations of different countries (lately in Sweden and Ireland) and has organized a few international conferences and served as a member in several others.

Dr. Dutta's broad research interests include nanomaterials in nanotechnology, self-organization, microelectronic devices, and nanoparticles and their applications in electronics and biology. He recently completed a textbook on nanoscience, and he has over 100 research publications, over 350 citations (Scopus), five chapters in science and technology reference books, three patents (five ongoing applications), and has delivered more than 50 invited lectures.

Gabor L. Hornyak has an interdisciplinary background in biology, chemistry, and physics. He received his PhD in chemistry from Colorado State University in 1997; BS and MS degrees in biology (human genetics) from the University of

Colorado at Denver in 1976 and 1981, respectively; and a BS degree in chemistry from the University of California at San Diego in 1990. Dr. Hornyak worked in the aerospace industry as a senior scientist from 1978 to 1990 in San Diego at the Convair Division of General Dynamics in coatings, adhesives, and corrosion. From 1997 to 2002, he worked at the National Renewable Energy Laboratory on development of chemical vapor deposition synthesis of carbon nanotubes. He has over 15 years of experience in nanotechnology research and development. Dr. Hornyak played a major role in bringing awareness about the promise of nanotechnology to the citizens and institutions Colorado and the surrounding region (2003–2005). He is the editor of the "Perspectives in Nanotechnology" series—a group of books dedicated to bringing topics to the general public that address issues about nanotechnology that are outside the laboratory and production line.

Anil K. Rao completed his PhD in comparative endocrinology at the University of Denver in 1986. He then continued as a postdoctoral student at the National Jewish Center for Immunology and Respiratory Medicine, where he studied fetal pulmonary electrophysiology and helped developed a lung tissue culture technique. After 3 years, Dr. Rao went on to a teaching position in the Department of Biology at the Metropolitan State College of Denver, where he is presently a tenured, full professor of biology. His teaching duties cover, but are not limited to, human anatomy and physiology, histology, and topics in advances in biology. Dr. Rao's personal interests include three- and two-dimensional computer graphics, scientific illustration, science education, and nanotechnology.

H. F. Tibbals has served as director of the Bioinstrumentation Resource Center for the University of Texas Medical Center since 1997, where he is responsible for providing engineering support to clinical and basic biomedical life science researchers. His funded research includes development of pressure and electrochemical sensors for medical applications, testing and evaluation of life support systems for NASA use in space flight and extravehicular activity, and development of technology for Alcon Research Ltd. for diagnosis of diseases of the eye. Dr. Tibbals's work involves consultation and team leading on a wide variety of analytical, materials, and systems technology in support of medicine and biomedical research. He is frequently called upon to advise on risks and cost benefits for technology decisions.

Prior to joining UTSW, Dr. Tibbals was product line manager, Digital Cardiology Products for Jamieson Kodak, working in Dallas, Texas; Rochester, New York; and at hospital cardiac catheterization laboratories across the United States and in Europe. He was responsible for developing and succeeding in gaining FDA approval for the first fully digital imaging systems capable of showing the living human heart with medical radiology standards of precision and accuracy. He also consulted for the development of patient anesthesiology systems. As president of Biodigital Technologies, Inc. and board member and consultant to Martingale Research Corporation from 1989 to 1995, he led teams in the development of real-time systems for the identification of bacterial and viral disease agents, and systems for monitoring, analyzing, and reducing environmental hazards, From 1988 to 1991, Dr. Tibbals was product line manager on contract to Inmos and later SGS Thompson for digital signal processing products and applications. Clients for product and development management projects included NEC, Teledyne, Marathon Energy Systems, Coors, Bank One, Shelby Technologies, Innovative Systems SA, Optical Publishing Inc., Gentech, Colorado Medical Physics, and others. He was a consultant and project manager for the development of a production and distribution control system for the world's largest nitrogen fertilizer complex at BASF Ludvigshafen.

For most of the 1980s, Dr. Tibbals worked for Rockwell International on trusted and secure systems in government and private areas, serving as principal investigator, project engineer, and systems engineer on projects in the United States and around the world. He was product planner and Rockwell representative on standards committees in the development of the first Rockwell Digital Facsimile Modem systems, which achieved more than 90% market share over a sustained period following its introduction. From 1983 to 1985, he was a principal design engineer for Mostek Systems Technology, heading work on standards, design, and introduction of the VMEbus product line and the development of applications that became dominant in telecommunications, process control, high-end workstations, and scientific and medical instrumentation.

During the 1970s, Dr. Tibbals served on the academic and research staff of Glasgow and Durham Universities in his United Kingdom, where he worked with the Edinburgh Regional Computing Centre, the Digital Cartography Unit, and the Glycoprotein Research Laboratory. He also taught for the Open University and Jordanhill College. He held visiting research and teaching positions at Bogazici University and the University of North Texas, working on instrumentation and systems for environmental monitoring and remediation.

Dr. Tibbals earned his BS degree in chemistry and mathematics at Baylor University in 1965, where he held scholarships in chemistry, English, and Old Testament studies, and was an undergraduate research fellow in electrochemistry. He was awarded his PhD degree from the University of Houston in 1970 for theoretical and experimental research in nonequilibrium statistical mechanics and kinetics of ion–molecule reactions. He won an SRC postdoctoral fellowship in physical silicon chemistry at the University of Leicester from 1970 to 1972. He has published a number of refereed scientific and technical papers; has received grants and study awards in computer systems architecture, man–machine interfaces, and applications of computers in chemistry; and holds two patents. In 1990, he was awarded a grant from the National Center for Manufacturing Science to visit key technology centers in Japan and study applications of advanced signal processing technology, including fuzzy logic and neural networks. He received a grant from Rockwell to organize a series of three symposia on networks in brain and computer architecture from 1986 to 1988. He served as adjunct professor in the University of Texas at Dallas School of Human Development, and he was a member of the advisory board for the University of Texas at Arlington Advanced Automation and Robotics Center and for the Rutgers Center for Advanced Information Processing.

Dr. Tibbals has served the IEEE Dallas CN group as treasurer, program chair, and board member. He is a member of the American Chemical Society, the Biophysical Society, the Royal Society of Chemistry (chartered chemist), The American Association for the Advancement of Science, The British Computer Society, and the Sigma Xi Society for Scientific Research. He has served on the boards of the Audubon Society Prairie and Cross Timbers Chapter and the Dallas

Nature Center. He has served as an advisor to government bodies and companies of information systems and technology for environmental monitoring and improvements. He served on the U.K. Science and Religion Forum and the Commission on Caring for Creation, and he worked with George Dragus and Nicholas Madden on the generation of computerized concordances for early Greek Christian patriarchal writings.

Section 1

Perspectives

INTRODUCTION

I would like to describe a field, in which little has been done, but in which an enormous amount can be done in principle. This field is not quite the same as the others in that it will not tell us much of fundamental physics (in the sense of, 'What are the strange particles?') but it is more like solid-state physics in the sense that it might tell us much of great interest about the strange phenomena that occur in complex situations. Furthermore, a point that is most important is that it would have an enormous number of technical applications. What I want to talk about is the problem of manipulating and controlling things on a small scale.

RICHARD FEYNMAN, CALTECH, 1959 [FIG. 1.0],
"There's Plenty of Room at the Bottom"

Chapter 1

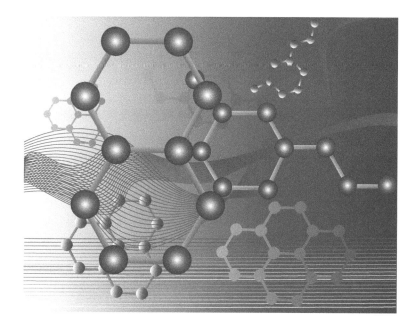

FIG. 1.0	*Nobel Prize laureate Richard Feynman at the blackboard.*

Source: Photograph of Richard Feynman used by permission of Melanie Jackson Agency, LLC. and Caltech Public Relations.

THREADS

"Threads" sections provide perspectives, reinforce general themes and at times serve as a navigational aid—what to expect, where we have been, and what our next objective is. The "Threads" sections provide the site map for the text, a short forum within which we discuss the order of subject matter and emphasize unifying themes. Here we address a fundamental pedagogical theme—specifically, that *Introduction to Nanoscience* is about the science of nanomaterials and not about the technology. The technological aspects of nanoscale materials are covered in our sister volumes "Fundamentals of Nanotechnology" and the combination text, *Introduction to Nanoscience and Nanotechnology.*

THREADS (CONTD.)

Another unifying theme of the book involves definitions and boundaries. In particular, we construct boundaries only to flagrantly disregard them later. The distinction between science and technology is the first in a series of boundaries that suffers from such apparent nebulosity. We firmly believe that all things converge, forming a continuum. The theme of the integration of everything is a fundamental tenet of nanoscience and nanotechnology. As a consequence, we feel that distinctions between and among things are presented solely for the purpose of perspective (and convenience). Lastly, we attempt to weave the wonders of the natural world into our discussions as often as possible. Nature, the ultimate nanotechnologist, has much to share.

Introduction to Nanoscience consists of five divisions: I. Perspectives, II. Nanotools, III. Physics: Properties and Phenomena, IV. Chemistry: Synthesis and Modification, and V. Natural and Bionanoscience. Chapters 1 and 2 comprise the perspectives division of the text. Emphasis on our historical scientific heritage is a major theme throughout this book. Chapter 2 delves into societal implications of nanoscience and nanotechnology. Its placement early in the book is strategic because we hope students *always* keep in mind the multifaceted ramifications of past, present, and, hopefully, their own future research. Characterization and fabrication of nanomaterials are the topics of

chapters 3 and 4, respectively, and they comprise the nanotools division of the text. The placement of these chapters early in this text is pedagogically motivated. We present images of nanomaterials, analytical data, and synthesis methods throughout the text. Why not understand early on how images and spectra of nanomaterials are acquired? And why not understand early on how nanomaterials are fabricated? In most cases, characterization methods and fabrication procedures predate the formal development of nanotech. The remainder of the text is divided into chapters about properties and phenomena (e.g., physics), synthesis and chemical modification (e.g., chemistry), and, lastly, two chapters that review the remarkable nanomaterials found in the natural world around us (natural nanomaterials and biochemistry).

In the seventeenth century, Sir Isaac Newton provided a fundamental thread that is quite relevant to nanoscience: "If I have seen farther, it is by standing on the shoulders of giants." Newton is credited for making this statement in a letter to Robert Hooke in which he refers to Galileo and Kepler. In the same vein, we all must pay tribute to those who have come before us—those who established the foundation for what we have today—for the benefit of the scientist, technologist, and those properly enlightened with a strong social consciousness.

1.0 NANOSCIENCE AND NANOTECHNOLOGY— THE DISTINCTION

We have stated clearly in "Threads" that the focus of *Introduction to Nanoscience* is on the science, not the technology. However, especially in the case of "nano," it is often difficult to distinguish between the two. Science involves theory and experiment. Technology involves development, applications, and commercial implications. Both feed on each other (**Fig. 1.1**). We will do our best to adhere to "pure science" without the "applied" component in this text. By necessity, the historical perspectives of both nanoscience and nanotechnology will be presented together, intermingled. Our first boundary therefore already is at risk of "fuzzification."

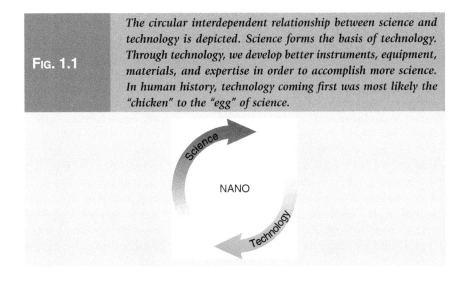

FIG. 1.1 *The circular interdependent relationship between science and technology is depicted. Science forms the basis of technology. Through technology, we develop better instruments, equipment, materials, and expertise in order to accomplish more science. In human history, technology coming first was most likely the "chicken" to the "egg" of science.*

Before we begin to study nanoscience, all of us sincerely hope that you, the student, become inspired by what is to unfold. Pure science is, after all, about truth and the process of digging for the truth—idealistically a good thing for humanity. Democritus is known for saying, "I would rather discover one scientific fact than become king." We unilaterally support Democritus's position. We hope that you are inspired and seriously consider and then pursue a career in nanoscience and technology. Or, we hope that you become involved in a field that is affiliated with nanoscience and nanotechnology outside the laboratory or manufacturing environment. This lofty objective is an underlying goal of this book (both subliminally and overtly stated) with the intent of creating more scientists in a world where technological breakthroughs are needed to help humanity face the grand challenges of the future [1]. The late Nobel laureate Richard Smalley (**Fig. 1.2**) stated directly and sincerely, "Be a scientist—Save the world."

1.0.1 Requisite Definitions

Science requires measurement. Without measurement there is no dissemination of theory and experiment. Civilization requires language. Without language we cannot communicate. Measurement is the language of science. Nanoscience implies a scale of measurement that exists at the level of the nanometer. We begin our journey, then, by submitting fundamental definitions that together yield a sense of the nature of nanoscience and nanotechnology.

The etymological derivation of the prefix *nano* can be traced back to the Latin *nanus* and further back to the Greek root *nan(n)os*, which means dwarf or "little old man." *Nanos* is the foundation of words such as *nana*, *nanny* (aunt), *nannus* (uncle), and even *nun* [2]. According to dictionary definitions, it is "specialized in certain meanings to mean one-billionth" [3] and has evolved to indicate "extreme smallness" [2]. Nanotechnology therefore is a technology defined by the nanometer. One nanometer is equivalent to one-billionth of a meter (10^{-9} m).

| FIG. 1.2 | *Nobel laureate and professor Richard E. Smalley of Rice University was a strong proponent of energy development through nano-technology. He was very passionate about students making a difference in a world that needs technology to help mitigate the challenges that face humanity in the twenty-first century.* |

Source: Image courtesy of Rice University Office of Media Relations and Information.

EXAMPLE 1.1 Nanometer Games

(a) How many carbon nanotubes 1 nm in diameter can be tightly packed into a cylinder defined by a human hair 100 μm in diameter? Assume packing is done parallel to the long axis and that packing efficiency is not a concern.

(b) Assume that a cubic-shaped transistor in a computer chip has volume of 10 nm³. How many would fit into a 5-mL drop of water? If currently one billion transistors are fabricated every second, how much time in years is required to manufacture this number of transistors?

(c) Denver is called the Mile-High City. What is its height in nanometers? How far away is our Sun in nanometers?

Solutions:

(a) *The area of the cross-section of a 100-μm diameter hair is $A = \pi r^2 = 7.85 \times 10^3$ μm². The cross-section area of a nanotube is 7.85×10^{-1} nm². The number of tubes that can be packed into this area is 7.85×10^3 μm² ÷ 7.85×10^{-1} nm² $= 1.00 \times 10^{10}$ tubes or about 10 billion tubes.*

(b) *5 mL = 5 cm³. 5 cm³ ÷ (1 transistor area · 10 nm⁻³) $= 5 \times 10^{20}$ transistors. If one billion are made every second today, it would take approximately 16,000 years to make this many transistors. It is safe to assume that this number is greater than that of all the transistors ever made.*

(c) *Denver is at an altitude of 5280 ft. This is equal to 1.61×10^5 cm, 1.61×10^{12} nm, or approximately 1.6×10^{12} nm. The sun is 92.9×10^6 miles or 1.496×10^8 km from Earth. This is 1.496×10^{20} nm—surely an astronomical number.*

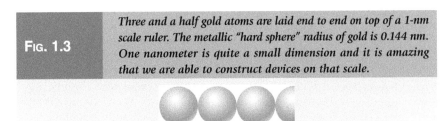

FIG. 1.3	*Three and a half gold atoms are laid end to end on top of a 1-nm scale ruler. The metallic "hard sphere" radius of gold is 0.144 nm. One nanometer is quite a small dimension and it is amazing that we are able to construct devices on that scale.*

One nanometer

When considering covalent "hard sphere" radii of atoms, 3.5 gold atoms (covalent radius equal to 0.144 nm) placed in a row equal 1 nm [4] (**Fig. 1.3**). A human hair ranges in size from 50 to 100 μm. One nanometer is approximately 1/50,000–1/100,000 as thick as a human hair.

Science (from Latin *scientia*, "knowledge," based on the Greek *skhizein*, "to split, rend, cleave, to separate one thing from another, to distinguish") is defined as "knowledge or a system of knowledge addressing general truths" or the "operation of general laws especially as obtained and tested through scientific method" [3]. Technology (from the Greek *techno*, "art, skill, craft, method, system," based on *tek*, "indicating shape, make" + *logia*, "a study of") is the "practical application of knowledge (science), especially in a particular area" [3]. It is clear to see that science and technology are integrally linked. We will do our best to keep them separate.

1.0.2   Government Line

In February 2000, the National Science and Technology Council (NSTC) Committee on Technology, Subcommittee on Nanoscale Science, Engineering and Technology (NSET) derived the following definition for nanotechnology:

> Research and technology development at the atomic, molecular or macromolecular levels, in the length scale of approximately 1–100 nanometer range, to provide a fundamental understanding of phenomena and materials at the nanoscale and to create and use structures, devices and systems that have novel properties and functions because of their small and/or intermediate size. The novel and differentiating properties and functions are developed at a critical length scale of matter typically under 100 nm. Nanotechnology research and development includes manipulation under control of the nanoscale structures and their integration into larger material components, systems and architectures. Within these larger scale assemblies, the control and construction of their structures and components remains at the nanometer scale. In some particular cases, the critical length scale for novel properties and phenomena may be under 1 nm (e.g., manipulation of atoms at ~0.1 nm) or be larger than 100 nm (e.g., nanoparticle reinforced polymers have the unique feature at ~200–300 nm as a function of the local bridges or bonds between the nanoparticles and the polymer).

A rather cumbersome definition to be sure, but if you are in the process of applying for grant money from the federal government, abiding by these guidelines is a prudent choice. Strict definitions such as these are necessary to filter unwanted solicitations. As with any new technology that is hot and trendy, blatant exploitation of the word is expected. Please recall the buzzword *turbo*. Enter the *nano-pretenders*, those who stretch the concept of nanotechnology beyond any reasonable boundary for the purpose of exploitation. Nano-pretenders are generally business folk or academics that incorporate the prefix nano into a product, company name, or proposal but in reality have nothing or very little to do with it [5]. In fairness to all parties, however, the nebulous nature of nano makes a formal definition rather difficult.

Although this volume is dedicated to nanoscience and not nanotechnology per se, both areas will be addressed early on and used at times interchangeably. Students are encouraged to modify all terms and definitions to suit themselves. Definitions, after all, are basically useful tools that provide a good starting point—a frame of reference. The following obligatory definitions of the nanoscale, nanoscience, and nanotechnology are given to provide a perspective and to lay the foundation for a succinct, meaningful, and luminous 5-min. elevator pitch.

1.0.3 Working Definitions

What is nanotechnology? Nanoscience? **Figure 1.4** illustrates some of the complexity involved in generating a strict definition of such simple words.

FIG. 1.4 *What exactly is nanotechnology? Nanoscience? This schematic illustrates the multiple dimensions, definitions, permutations, and indications of nano above and beyond its relationship to size. The definition of* microtechnology *never underwent such scrutiny or confusion. The definition of* materials science, *when it first appeared on the technological stage several decades ago as a formal academic discipline, was never burdened with such nebulosity or, for that matter, promise.*

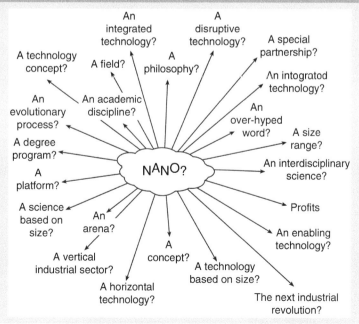

Nanotechnology and nanoscience have different meanings to different people. In the ultimate analysis, a formal definition of nano is not required, but we need to communicate what we discover and manufacture in the language of science.

We begin by defining the nanoscale. By necessity, we must include a broader perspective with regard to size than that given by NSET. Micron-sized particles and those in the submicron realm are generally considered to be bulk materials. In other words, the physical properties of micron-sized materials resemble those of bulk materials; for example, they possess *continuous* (macroscopic) physical properties. It is only when particles assume proportions smaller than ca. 10 nm that the principles of classical physics begin to wobble a bit. We then define the nanoscale to include (1) the size domain less than ca. 10 nm, where materials demonstrate remarkable properties and phenomena apart from the bulk form of that material; (2) larger materials (colloids, thin films) with properties and phenomena between the thresholds of classical physics and the quantum domain; and (3) large molecules that are less than 1 nm in size—and why not include materials that are less than 1 μm in some dimension? We propose the following definition of the nanoscale as it applies to nanoscale science: "The nanoscale, based on the nanometer (nm) or one-billionth of a meter, exists specifically between 1 and 100 nm. In the general sense, materials with at least one dimension below one micron but greater than one nanometer are nanoscale materials."

Nanoscience is the study of nanoscale materials. It is the study of the remarkable properties and phenomena associated with nanoscale materials. Nanoscience is the study of properties and phenomena of materials that are a function of size and size alone. "Nanoscience is the study of nanoscale material—materials that exhibit remarkable properties, functionality, and phenomena due to the influence of small dimensions."

Nanotechnology is the natural progression of technology miniaturization from the bulk macroscopic world (e.g., the plow) to millimeter-sized objects (e.g., the first transistor) to micron dimensions (e.g., integrated circuits), and, finally, into the nanoworld (e.g., the quantum dot). The definition of nanotechnology connotes industry, products, and commerce. "Nanotechnology, based on the manipulation, control, and integration of atoms and molecules to form materials, structures, components, devices, and systems at the nanoscale, is the application of nanoscience, especially to industrial and commercial objectives."

Is nano in general (the combination of nanoscience + nanotechnology) a *platform*, an *arena*, a *field*, or a *frame of reference*? Is nanoscience purely an academic enterprise? And is nanotechnology to be viewed purely from the perspective of applications? Such superficial boundaries have not stopped universities worldwide from teaching nanoscience and nanotechnology or offering special degree programs in either. Nor has it stopped businesses from creating nanotechnology divisions within their companies. These programs and divisions without question are all perfectly valid and grounded in reality. On the flip side, others believe that nano should be incorporated into traditional compartmentalized academic departments like physics, engineering, materials, and chemistry courses without the need to create radically new curriculum or degree programs—once again, a valid position.

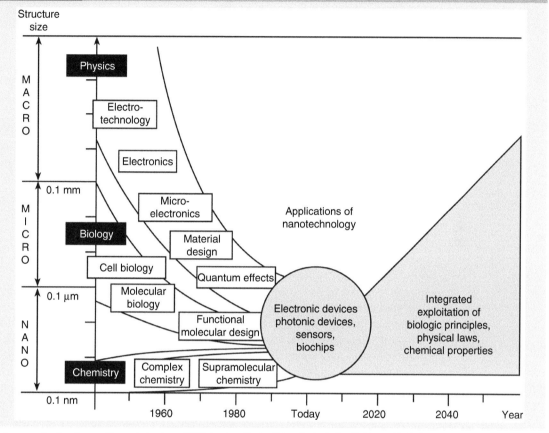

| FIG. 1.5 | *Size scale of nanoscience divisions of physics, chemistry, and biology are expanded to include associated technologies. The graph clearly places the juxtaposition in today's time. From now on, manufacture of very small to very large integrated devices will be made possible by advances in nanoscience and nanotechnology.* |

Source: Graph redrawn with permission from VDI-Technology Center, Future Technologies Division—APEC Center for Technology Foresight, Thailand.

The academic platform commonly known as *materials science* (*and engineering*) comes closest to resembling nanoscience. Although both are inherently interdisciplinary in nature, nanoscience transcends the boundaries of materials science by adding biology and biochemistry into the mix. In addition, materials science, unlike nanotechnology, was never labeled as "the next Industrial Revolution" and societal implications were not a serious part of the materials science equation. A comprehensive graphic relating science, technology, size and a timeline is shown in **Figure 1.5.**

Industrial sectors such as aerospace, biotech, energy, transportation, health care, telecommunications, and information are considered to be vertically oriented (**Fig. 1.6**). Since all industrial sectors depend on materials and devices made of atoms and molecules, by default they can all be improved by application of nanomaterials and nanotechnology. There is no argument with this line

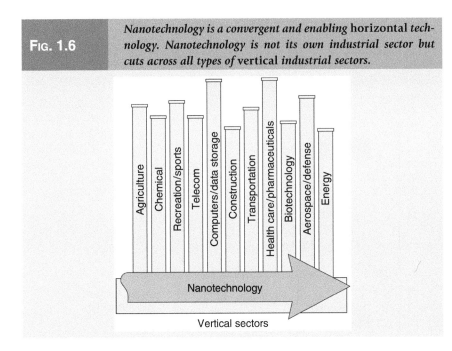

FIG. 1.6 *Nanotechnology is a convergent and enabling horizontal technology. Nanotechnology is not its own industrial sector but cuts across all types of vertical industrial sectors.*

of reasoning. Nanotechnology is, therefore, a *horizontal* and *enabling* technology that will potentially impact ALL industries. The word *convergent* is quite frequently associated with nano. Convergence is the coming together from different directions of previously equal but independent areas of science and technology. The integration of DNA with silicon is an example of convergence of biology with inorganic chemistry. We add, then, another definition to the previous ones: "Nanotechnology is a horizontal-enabling convergent technology that cuts across all vertical industrial sectors, while nanoscience is a horizontal-integrating interdisciplinary science that cuts across all vertical science and engineering disciplines."

Nanotechnology has the potential to impact all products manufactured now and in the future. It has the potential to change the way we all live. This makes nanotechnology a *disruptive* technology. The iron sword of the Hittites, gunpowder, atom bomb, automobile, telephone, penicillin, and computer are examples of disruptive technologies that molded history and changed our lives. These examples, however, are a list of materials and devices rather than a wave of a new type of technology. In this sense, nanotechnology is not a specific entity but rather a generalized form—one that has the potential to disrupt. Nanotechnology, unlike the "dot.com" businesses, requires PhD expertise, significant capital, and a new generation of partnerships to "make it go."

Nanotechnology is a disruptive technology with a high barrier of entry that will impact the development of materials and devices. Nanotechnology will require that a new genre of partnership be formed among and between business, academe, and government. It will devote study and effort to potential societal implications. Nanotechnology is predicted to significantly impact the wealth and security of nations. Nanotechnology is the next Industrial Revolution.

Societal implications of technology are important now more so than ever. The relationship of nanoscience, nanotechnology, and society will be discussed in more detail in chapter 2. Nanotechnology is expected to exert widespread societal impact. "Nanotechnology is considered, more so than most other technologies, to have great impact on all aspects of culture and society."

We have listed some definitions, platitudes, and phraseology concerning nanoscience and nanotechnology. We have attempted to define the nebulous, the transitory, the ephemeral, and, perhaps, the concrete. Nanotechnology is everything in the basest sense but it is also unique. Nanotechnology has been around for quite some time—thousands of years in our synthetic civilization and billions of years if nature is included. The onus is now upon you, the student, to formulate an impression of this exciting and important field, platform, technology, or business—or word. A Japanese production engineer devoted to accuracy and precision, Norio Taniguchi of the Tokyo Science University, introduced the term *nanotechnology* in 1974. His statement is recorded in the *Proceedings of the International Conference of Production Engineering, On the Basic Concept of NanoTechnology*, 1976. Thus did the word begin.

1.1 HISTORICAL PERSPECTIVES

Historical transcription of nanotechnology or any science comprises individuals, records, ideas, experiments, tools, applications, and everything outside the science and technology. Historical accounts are a blend of many things. The word *history* itself (from the Greek *historia*, meaning "record, account") is based on the root *histor*, or "wise man." An individual able to grasp all the knowledge of the ancient world was referred to as a *polyhistor*. There are no *polyhistors* of nanotechnology today. As a case in point, a team of contributors was required to compile this textbook. Nanoscience was a part of our heritage well before the formal designation of the concept of nano arrived on the scene. We are at a crossroads, a juxtaposition between the past and present. We ask that you absorb the following historical perspective so that you will be able to stand on the shoulders of the giants and take nanoscience to the next level.

1.1.1 *Concept of Atomism*

The earliest roots of recorded atomism date back to India 2600 years ago. The *Vaiseshika* philosophical school founded by the Hindu sage Kanada proposed the first record of atomistic theory. The philosophers of the school described atoms as eternal, indivisible, infinitesimal, and ultimate parts of matter [6]. According to the school, if matter could be subdivided infinitesimally, "there would be no difference between a mustard seed and the Meru Mountain" [6]. The size of an atom, from the Buddha biography *Lalitavistara* 2200 years ago, was estimated to be 10^{-10} m [6].

In the West, Anaximander in the sixth century B.C. speculated that the basic element of the universe was an unobservable infinite substance referred to as *apeiron* [7]. Leucippus, a Greek philosopher of the fifth and fourth centuries B.C., invented the concept of atomism (from the Greek *atomos*, uncut: "not" + *tomos*, "cut") [8]. Specifically, he stated that matter is comprised of invisible particles: "imperceptible, individual particles that differ only in shape and position." Democritus (ca. 430 B.C.), Leucippus's student, further developed the atomistic concept and extended the concept of the *apeiron* of Anaximander. Democritus is known for the following atomic maxims [7]:

> Nothing exists except atoms and empty space; everything else is opinion.
> Atoms are indivisible ... eternal, unchangeable, indestructible.
> Atoms differ from each other physically, and in this difference is found
>   an explanation for the properties of various substances.

About 100 years later, Epicurus kept Democritus's principle of atomism alive [7].

With the lack of formal experimental methodology, Aristotle and philosophers who followed him adopted other equally intuitive but unfortunately misleading ideas. During the sixth and fifth centuries B.C., Thales and then Anaximenes, respectively, believed that the fundamental elements of the universe are *water* and *air*. Later, Xenophanes and Heraclitus added the elements of *fire* and *earth* to the mix. Empedocles (fifth century B.C.) merged all four and stated that all things are made up of combinations of *earth, air, water,* and *fire* [7]. Certainly this must be the first time in recorded history that a convergence of scientific principles took place—the first grand unification theory! Aristotle adopted and then embellished Empedocles's four-element concept by adding a fifth element, *aether*, described as the ingredient of the heavens [7]. Unfortunately, Democritus's intuitive concept was not to reemerge for two millennia.

In the first century, the Greek engineer Hero postulated that air was compressible and therefore comprised of individual particles separated by space [7]. Unfortunately for Hero, the Aristotelians, who did not advocate experimentation, beat his idea down. It was not for another 1500 years that the principles of Aristotelian thought were finally challenged. Robert Boyle wrote the *Sceptical Chymist* in the 1660s and reintroduced the corpuscular concept of matter [9] (**Fig. 1.7**): "minute masses or clusters that were not easily dissipable into such particles that composed them."

Because Boyle was privately tutored, he was not influenced by the "didactic Aristotelianism that still victimized most universities" [7]. Boyle reaffirmed Hero's observation that air was compressible: "It must be composed of discrete particles separated by a void." Democritus's concept of atomism emerged once again, but this time it was to prevail.

John Dalton and his "law of multiple proportions" in 1805 reaffirmed that elements are made of atoms [10]. He also stated that atoms of one element are identical to each other and different from atoms of other elements, and that they are capable of combining with other atoms of other elements to form compounds that exist in atomic proportions [10,11]. Lastly, in his treatise he asserted that atoms cannot be divided into smaller particles or destroyed in chemical reactions [11]. The unit of the dalton, represented by "Da," is equivalent to one atomic mass unit. Thus, arguably, began the rise of modern chemistry [10,11].

1.1.2 Colored Glasses

Metal colloids (from the Greek, *kolla*, meaning "glue") provide the best examples of nanotechnology throughout ancient, medieval, and modern times. Metal colloids are the "finely divided" components of catalysts, colored glasses, dyes,

| FIG. 1.7 | *In his work the* Skeptical Chymist, *Robert Boyle promotes the* **corpuscular theory** *of matter, thereby laying the foundation for the modern interpretation of the elements.* |

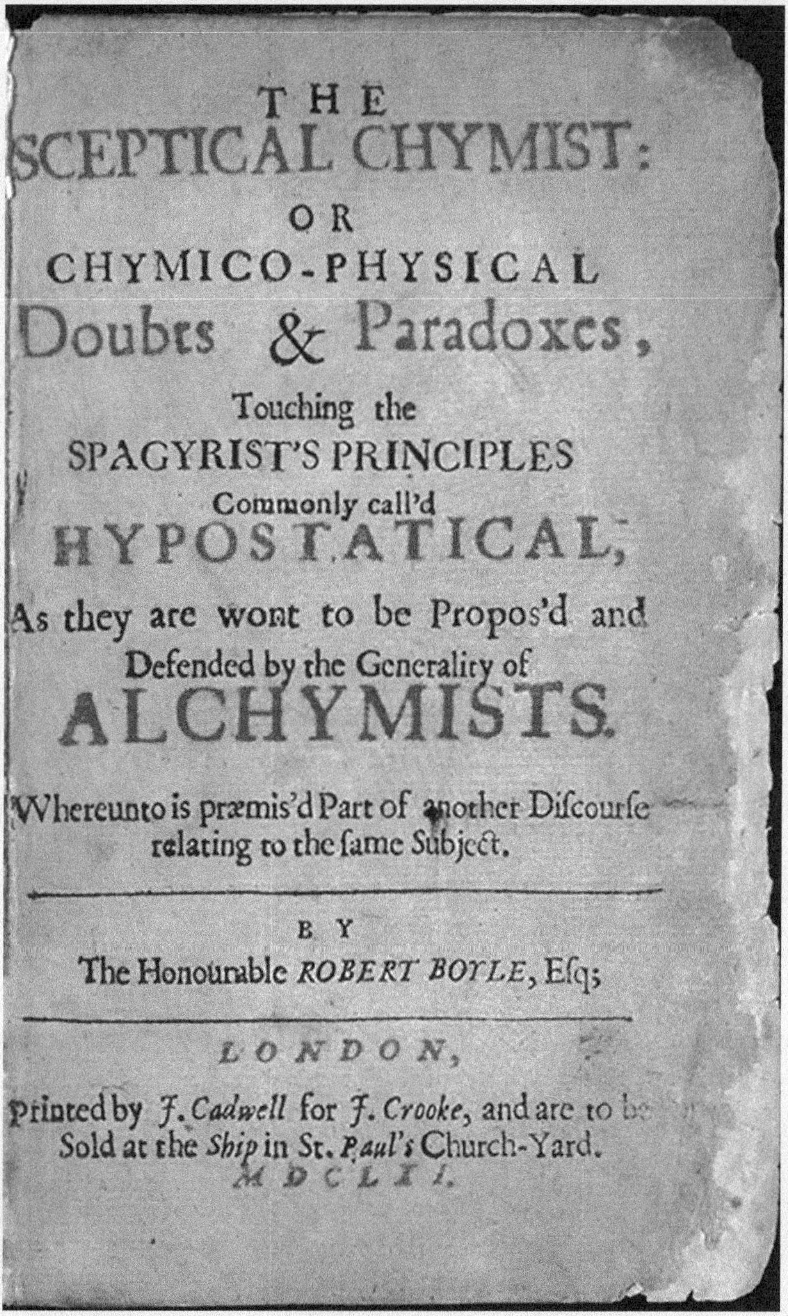

THE
SCEPTICAL CHYMIST:
OR
CHYMICO-PHYSICAL
Doubts & Paradoxes,

Touching the
SPAGYRIST'S PRINCIPLES
Commonly call'd
HYPOSTATICAL,

As they are wont to be Propos'd and
Defended by the Generality of
ALCHYMISTS.

Whereunto is præmis'd Part of another Discourse
relating to the same Subject.

BY
The Honourable *ROBERT BOYLE*, Esq;

LONDON,

Printed by *J. Cadwell* for *J. Crooke*, and are to be
Sold at the *Ship* in St. *Paul's* Church-Yard.
MDCLII.

and photographic materials, and, eventually, the progenitors of quantum dots. Metal implements fabricated during the Copper, Bronze, and Iron Ages surely were composed of micro- and nanosized metal particles. The same can be said for ceramic materials, dyes, and potions that were developed throughout history. The development of semiconducting colloidal materials also occurred well before the formal embodiment of the *Nano Age*.

A definitive and remarkable piece of Roman glasswork, dating to the fifth century, clearly demonstrates one of the grandest examples of nanotechnology in the ancient world. This magnificent cup, housed in the British Museum, depicts King Lycurgus being dragged into the underworld by Ambrosia. When illuminated from the outside, the cup appears green (**Fig. 1.8a**). When illuminated from the inside, the cup appears crimson red except for the King, who looks purple

FIG. 1.8 *Two versions of the famous Lycurgus Cup, displayed in the British Museum, are shown in this figure. (a) The green appearance is produced by reflection. (b) Red and purple colors are due to transmitted light. The apparent dichroism is due to the interaction of light with gold–silver and copper nanometals and alloys embedded within the soda glass matrix of the cup. King Lycurgus is shown being strangled by the maenad Ambrosia, who was changed into a vine by the goddess Diana. The sad state of King Lycurgus, a Thracian king in 800 B.C., was a consequence of his assault on the god Dionysius and Ambrosia.*

(a)   (b)

Source: Images reprinted with permission from British Museum Images.

(**Fig. 1.8b**). It was not until 1990 that the specific cause of this color was uncovered after shards of the cup were analyzed with a scanning electron microscope (SEM). Scientists found that the splendid dichroism was due to the presence of nanosized particles of silver (66.2%), gold (31.2%), and copper (2.6%), up to 100 nm in size, that were embedded in the glass matrix [12,13]. The majority of nanometals range from 20 to 40 nm in size. The red color is due to absorption of light by gold at ca. 520 nm. The purple color following absorption is due to larger nanoparticles. The green is attributed to scattering by large silver colloids >40 nm in size. The glass is 73.5% silica, 14.0% Na_2O, 6% lime, and 0.9% K_2O and also plays a role in the optical response of the nanometal-insulator composite glass. The Lycurgus Cup is actually a nanocomposite material.

Following the Romans, medieval craftsman also exploited the addition of metallic constituents to glass to create beautiful stained glass windows. Johann Künckel in the mid-1600s created beautiful ruby-colored glasses from a method to infuse gold into glass [14]. The "purple of Cassius" is a colloidal mixture of gold nanoparticles and tin dioxide in glass [15]. In 1718, Hans Heicher published a complete summary of gold's medicinal uses. In it he discussed how the shelf life of potable gold solutions is stabilized by the addition of boiled starch—perhaps an example of the first ligand-stabilized gold colloids to be synthesized [15,16]. Ancient Chinese cultures are well known for producing beautiful colored glass [17,18]. Chinese porcelain known as *famille rose* contains gold nanoparticles that are 20–60 nm in size. Alchemists have known of colloidal suspensions and potential applications, without providing disciplined experimentation or any intrinsic explanation of them.

Francesco Selmi in 1845 published the first systematic studies of inorganic colloids, particularly those made of silver chloride. He found that a detectable change in the temperature of a solution was absent following coagulation and concluded that such particles could not possibly exist in a state of molecular dispersion [19]. Thomas Graham in 1864, the developer of "Graham's law of diffusion" and considered to be the father of colloid chemistry, discovered that low diffusivity behavior exhibited by colloid solution resulted from their large size (nanometer to micron range) in relation to "ordinary molecules" [20]. Therefore, without having intimate knowledge of nanoparticles, nineteenth century scientists were able to predict the existence of colloids.

It was Michael Faraday (**Fig. 1.9**) in 1857 that first conducted rigorous scientific study concerning the preparation and properties of colloids—those made of gold in particular [21]. The purple color slide he presented at the lecture is shown in **Figure 1.10.** Michael Faraday, in his lecture at the Royal Society of London in 1857, stated that "Gold is reduced in exceedingly fine particles, which becoming diffused, produce a beautiful ruby fluid … the various preparations of gold, whether ruby, green, violet or blue … consist of that substance in a metallic divided state."

Not only did Faraday postulate correctly about the physical state of colloids but he also found that the addition of salts transformed ruby-colored gold solutions into blue ones. He demonstrated that such color transformations were prevented by the addition of organic materials such as gelatin. From these experiments and others, Faraday indirectly contributed to organic monolayer chemistry. He also was the first to record the relationship of metal colloids, their surrounding media, and the resulting optical properties.

FIG. 1.9	*A print from the* Illustrated London News *of Michael Faraday at a lecture in 1856. Prince Albert is at the front in the center of the audience.*

Source: Image reprinted with permission from the Whipple Museum of the History of Science, Cambridge University.

In the early twentieth century, Gustav Mie presented his *Mie theory,* a mathematical treatment of light scattering that describes the relationship between metal colloid size and optical properties of solutions containing them [22]. Mie was one of the first to mathematically demonstrate the link between the optical properties of spherical colloids and particle size—one of the prime movers of

FIG. 1.10	*The slide depicting Faraday's gold sol was used in his lecture on nanogold, titled* On the Relation of Gold to Light, *on June 12, 1856. The slide has a reddish-purple color, indicating particles on the order of 20–40 nm in diameter on the average.*

Source: Image reprinted with permission from the Whipple Museum of the History of Science, Cambridge University.

early nanotechnology. In the early 1920s, J. C. Maxwell-Garnett, an outstanding woman scientist who hid her identity from a male-dominated science community, submitted two fundamental papers about the optical properties of metallic colloid solutions. She developed one of the first *effective medium* theories. Effective medium theory relates particle size, shape, orientation, composition, and the dielectric function of the host medium as factors contributing to the optical response of the resulting composite [23,24]. Colored glasses served as one of the first examples of a synthetic nanocomposite material.

1.1.3 Photography

Images were first observed in the *camera obscura*, a phenomenon clearly described by Leonardo da Vinci in 1490 and by Chinese philosopher Mo-Ti (Mo-Tsu) in the fifth century B.C. [26]. Generating an image was one thing; however, recording that image was quite another. Thomas Wedgewood in the early 1800s compiled the observations and work accomplished by numerous scientists throughout the ages to establish the beginning of modern photographic science [25]. His effort demonstrated the first union of interdisciplinary nanoscience followed by a nanotechnology (or was it the other way around?).

Robert Boyle reported that silver chloride blackened upon exposure to the atmosphere; he mistakenly believed that it was exposure to air and not light that caused the reaction. Chemists understood only later that light is able to interact with specific substances. The darkening reaction silver salts undergo when exposed to sunlight has been known since 1585 [25]. In 1727, J. H. Schulze discovered that silver nitrate, when mixed with chalk and placed under stenciled letters, darkened when exposed to light. We should also wonder, as potential nanotechnologists, whether the chalk used in Schulze's experiment was in the form of micro- or nanosized particles. If true, then this would be another example of a human-made nanocomposite. Sir Humphry Davy suggested that silver chloride, a substance more sensitive to light, be used instead [25]. Unbeknownst to the researchers of the time, silver chloride decomposes and is transformed into finely divided silver particles after exposure to light.

In 1839, French chemist Louis J. M. Daguerre was able to piece many aspects of the photographic puzzle together and created the first commercially viable print, the daguerreotype [25]. Several steps were involved in making a daguerreotype print. First, a layer of photosensitive silver iodide was exposed to light. Development of the layer was accomplished over a cup of mercury at 75°C, creating an amalgam. The latent image thus formed was fixed by immersion into a hyposulfite solution. In 1840, William F. Talbot developed a means to coat paper with silver chloride to create the first negative image. This procedure was the earliest way to reproduce images [25]. Ironically, the foundation of photography was based on chemical nanotechnology. We now capture our images with a new form of nanotechnology, as digital nanotechnology displaced the Polaroid camera in quite a disruptive manner.

1.1.4 Catalysis

The ancient Babylonians 5000 years ago stored soap-like materials in clay pots. Writing deciphered on the pots indicated that these materials mere made by boiling fatty substances along with ash. The Egyptians combined animal and vegetable oils with alkaline salts to form soap-like materials for medicinal purposes as well as for cleansing. The word *soap* comes from the Roman period; Mt. Sapo was the site of altar sacrifices where animals were slaughtered and burned. Melted animal fat (tallow) and ashes were washed down into the river. Servants discovered that soapy materials embedded in clay sediments along the riverbank downstream from the mountain aided in washing clothes. The Gauls during the time of the Roman Empire over 2000 years ago are formally credited with inventing soap [27]. Soap is the product of a catalytic process. When boiled in the presence of wood ash (the alkali catalyst), animal fat (the substrate) yields soap and glycerin. The formal investigation of catalytic processes, however, did not occur for another 1400 years. Thus, over the course of several millennia, both catalysts and surfactants were discovered; both materials play a major role in modern nanoscience.

Near the turn of the eighteenth century, Gottlieb Kirschof demonstrated the first controlled catalytic reaction (e.g., the acid hydrolysis of starch to produce glucose [28]). In 1836, J. J. Berzelius coined the word *catalysis* (from the Greek *katalusis*, meaning "dissolution," from *kata*, "down," + *lysis*, "a loosening") to describe the process. Catalysis is the acceleration of a chemical process by a substance known as a catalyst—a material not consumed in the reaction and, ideally, reusable [28]. Heterogeneous catalytic reactions consist of a solid phase catalyst and substrates that exist in liquid or gaseous states. Heterogeneous catalysis is the type of catalysis most important to nanotechnology. The solid phase catalytic particles are usually micro- to nanosized metals made of iron, nickel, or cobalt. Substrates are usually carbon-containing compounds.

Humphry Davy in the early 1800s observed that a heated platinum wire in the presence of air produced water. This is one of the first experiments to demonstrate heterogeneous catalysis [27]. J. J. Berzelius also observed that platinum accelerated chemical reactions without being consumed [27]. Nobel laureate Paul Sabatier discovered the process of catalytic hydrogenation in the early 1900s. He ascertained that catalysts were composed of "finely divided" particles [29,30]—"Finely-divided metal hydrogenation catalysts subsequently formed the bases of the margarine, oil hydrogenation, and synthetic methanol industries."

Heterogeneous catalysis made a significant impact on the chemical and petroleum industries in the mid-1900s. With synergistic breakthroughs in the automobile, oil, and plastic industries to follow, heterogeneous catalyst technology secured a prominent place in the development of nanomaterials to come. We can therefore conclude, with much confidence, that a minitechnological revolution was already intact, thriving, and based on nanotechnology well before the word *nano* became popular.

Interestingly, a pesky by-product of the catalytic decomposition of hydrocarbons also became an unintended contributor to nanotechnology [31]. Carbon fibers, multiwalled herringbone carbon tubes, turbostratic graphite, and amorphous carbons were all considered undesirable by-products of the catalytic decomposition of hydrocarbons. These *soots* typically poisoned the metal catalyst and terminated the primary chemical reaction [31]. How many times have single-walled carbon nanotubes (SWNTs) unknowingly been produced in catalytic decomposition reactions? There is no recorded evidence of nanotubes until the advent of the electron microscope in the 1930s, and in those days the resolution was not good enough to observe a single-walled carbon nanotube. Ironically, the catalytic production of single-walled carbon nanotubes is now a goal and not an undesired byproduct.

Along with carbon nanotubes, the study of catalysis has expanded our knowledge of nanoscience. Properties associated with nanomaterials, such as high surface area, high surface energy, and unique molecular structure, are better understood. We can now, for example, use nanoscopic gold in catalysis. Gold typically is not considered to have catalytic function.

1.1.5 *Integrated Circuits and Chips*

The drive to minimize electronic components contributed to the onset of the *Nano Age*. J. Bardeen, W. Shockley, and W. H. Brattain invented the transistor on December 23, 1947, and ended the reign of the bulky, heat-generating vacuum tube in electronic amplifiers and other devices. Brattain et al. described their first transistor in their journal [32]:

> Two hair-thin wires touching a pinhead of a solid semi-conductive material soldered to a metal base are the principal parts of the Transistor. These are enclosed in a simple, metal cylinder not larger than a shoelace tip. More than a hundred of them can easily be held in the palm of the hand.

One hundred can be held in the palm of the hand! Amazing! In that time, it was most certainly an amazing accomplishment. Bardeen and his teammates were awarded the Nobel Prize in physics in 1956 for their discovery. Bardeen was awarded another Nobel Prize in physics later for his work on superconductivity.

The first silicon transistor was created in 1951 and the first integrated circuit (IC) was made in 1958. Integrated circuits have shrunk in size over the past 50 years to the level where they certainly qualify as nanotechnology under the definition promoted by NSET that was given earlier. The Intel Corporation marketed a transistor in 2003 that was smaller than the human influenza virus (80 vs. 100 nm, respectively) (**Fig. 1.11**) [33]. Advances in reducing resolving power of lithographic techniques have made it likely that components will become smaller. Lithography will be discussed in more detail in chapter 4, "Fabrication Methods."

Gordon E. Moore helped found Intel (*Int*egrated *El*ectronics Company) in 1968. In 1965, he derived, with input from Douglas Engelbart of the Stanford Research Institute, "Moore's law" [34]. A version of Moore's law is graphically depicted in **Figure 1.12** and is generally summarized: "At our rate of technological development, the complexity of an integrated circuit, with regard to minimum component cost, will double every eighteen months."

The current interpretation of Moore's law states that data density will double every 18 months. In 1965, 30 transistors populated the chip; in 1971, 2000 populated it and ca. 40 million do so today. Reality tracked fairly well with this prediction except that the doubling actually occurred every 2 years rather than every 18 months. How will nanotechnology influence Moore's law? According to the National Nanotechnology Initiative, transistors must be "scaled down" to at least 9 nm. In order to keep pace with Moore's law, downsizing to a 9-nm transistor would result in billions of transistors on a chip by 2016.

FIG. 1.11

(a) A vintage 2003 transistor fabricated by the Intel Corporation is compared to the human influenza virus. We are now able to make devices smaller than one of the smallest "complete" biological structures. (b) The decreasing trend in transistor size is shown. By 2017, transistors under 10 nm in size are expected to be components in chips.

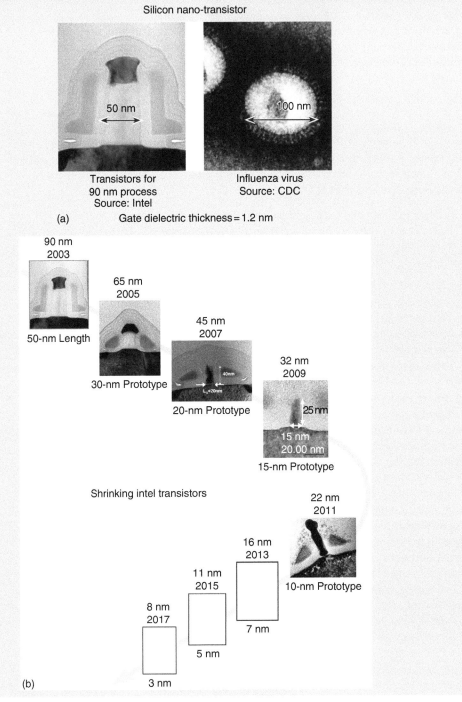

Silicon nano-transistor

50 nm

100 nm

Transistors for
90 nm process
Source: Intel

Influenza virus
Source: CDC

(a)   Gate dielectric thickness = 1.2 nm

90 nm
2003

50-nm Length

65 nm
2005

30-nm Prototype

45 nm
2007

20-nm Prototype

32 nm
2009

40nm

L=20nm

25 nm

15 nm
20.00 nm

15-nm Prototype

Shrinking intel transistors

22 nm
2011

16 nm
2013

11 nm
2015

10-nm Prototype

8 nm
2017

7 nm

5 nm

(b)   3 nm

Source: Images reprinted with permission of George Thompson, Intel Corporation.

| FIG. 1.12 | *An adaptation of Moore's law shows how computing power has increased since the onset of the first mechanical computing devices.* |

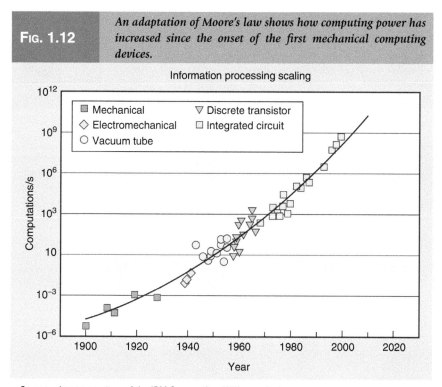

Source: Image courtesy of the IBM Corporation. With permission.

1.1.6 *Microelectromechanical Systems*

Microelectromechanical systems (MEMS) are devices that contain micron-sized components. MEMS devices are fabricated by conventional integrated circuit batch processing techniques. With the onset of ICs and chips and the development of lithography, MEMS production has burgeoned over the past 20 years. Development of micromachined silicon processes led to the development of electronic comb actuators and micropositioning disc drive heads. The automotive and defense industries in particular exploited MEMS technology to produce advanced sensors and actuators for fuel pressure, tire pressure, air conditioner compressors, and exhaust gas sensors. The accelerometer used in airbags is a MEMS device. MEMS devices have long since been applied to medical, environmental, aerospace, electronics, and other industries.

A relatively new area of technology, called NEMS (nanoelectromechanical systems), is an example of the evolutionary trend to make smaller and smaller components. A compilation of MEMS and microlithographically formed devices is shown in **Figure 1.13a–g.**

1.2   ADVANCED MATERIALS

Examples of nanomaterials are evident throughout the twentieth century. The oxide layer formed on anodized aluminum, for example, is a nanostructured

FIG. 1.13   *Images of microelectromechanical system (MEMS) devices are shown in the figure. (a) Drive mechanism with dust mite on tiny gears, (b) Close-up of drive-gear hub of a micro-engine, (c) Single-piston microsteam engine, (d) Indexing motor, (e) Grain of pollen and red blood cells on a drive-gear linkage of a microengine, (f) Torsional ratcheting actuator that uses rotationally vibrating (oscillating) inner frame to ratchet surrounding gear, and (g) Six-gear chain capable of 250,000 rpm.*

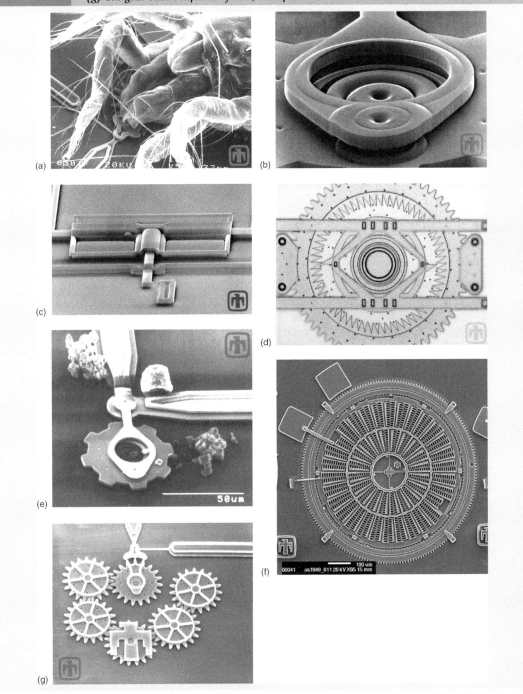

Source: Images courtesy of Sandia National Laboratories, SUMMiT Technologies, www.mems.sandia.gov. With permission.

FIG. 1.14	*Atomic force microscope (AFM) image of a porous alumina membrane surface is shown. The pores are ca. 50 nm in diameter. The film was processed in a 10 wt.% oxalic acid solution at 0°C under 30 V conditions. The hexagonally packed pore channel structure is clearly depicted. Alumina membranes such as this form perfect templates for the synthesis of nanomaterials.*

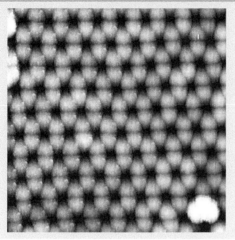

Source: Image courtesy of the National Renewable Energy Laboratory in Golden, Colorado. With permission.

material. The process of anodizing was patented in the United Kingdom in 1926 and then acquired by Alcoa in the early 1930s. Nearly every aluminum panel or part that is used in industry applications has an anodized finish. Anodization is a simple process that provides a protective oxide layer on aluminum to prevent corrosion. Little did the developers know at the time that the anodic layer was composed of hexagonally close-packed channels with diameter ranging from 10 to 250 nm or greater. The aspect ratio (length to diameter) of the channels ranges from 1 to as high as 10,000. Anodically formed porous alumina membranes (**Fig. 1.14**) are used extensively in template synthesis of nanomaterials. Anodic films are discussed more fully in chapter 12 [35].

Other familiar examples include nanoparticles that are found in the rubber component of automobile tires, titanium dioxide pigment found in white paint, chip components in computers, large synthetic biomolecules, polymer monomeric units, nanoparticle slurries for polishing, molecular sieves, ceramic cells and structure, and many more. The recreation industry has already taken advantage of nanomaterials. Multiwalled carbon nanotubes are routinely used as filler material in polymer composites. Single-walled carbon nanotubes (**Fig. 1.15**) can be woven into ropes that exhibit strength greater than that of silk. An image of a *Babolat* tennis racquet reinforced with carbon nanotubes is shown in **Figure 1.16**.

1.2.1 Thin Films

Chemists state, perhaps somewhat ostentatiously, that the practice of nanotechnology originated in chemistry labs long before nanotechnologists claimed to

FIG. 1.15	*Single-walled carbon nanotubes (SWNTs) are arranged in a hypothetical three-dimensional superstructure. Materials reinforced with carbon nanotubes promise to be the strongest known to science.*

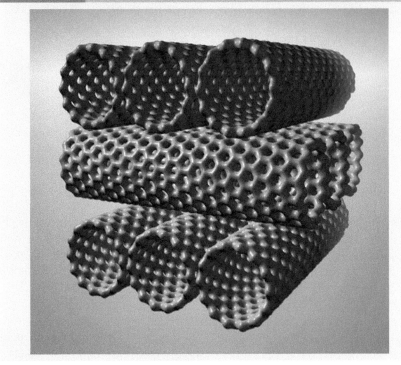

Source: Image courtesy of Anil K. Rao, Department of Biology, the Metropolitan State University, Denver, Colorado. With permission.

be nanochemists. Chemists in general manipulate large numbers of atoms and molecules from the "bottom up" to form bulk materials without much consideration of the nanoscale. A major exception to this rule is perhaps found in colloid chemistry, where nucleation and precipitation mechanisms are most certainly nanoscale phenomena. The mainstream chemist designs and directs reactions that require the rupture and formation of strong chemical bonds under kinetic control. The nanochemist, on the other hand, is more likely to conduct synthesis of nanomaterials under thermodynamic control with the formation and dissolution of weaker chemical bonds. Under thermodynamic control, weak bonds are made and broken under mild conditions. More discussion about the distinction between the two types of reactions is presented in chapters 10 through 12.

Thin film technology is a nanotechnology that has been practiced for millennia [36]: "Thin film technology is simultaneously one of the oldest arts and one of the newest sciences. Involvement with thin films dates to the metal ages of antiquity." The ancient Egyptians were capable of producing gold leaf as thin as 3 μm or less nearly 4000 years ago. This was accomplished by physical means such as rolling and beating. Inca goldsmiths practiced "depletion gilding," a process whereby gold is enriched at the surface, over 2000 years ago. The gold leaf

FIG. 1.16	*Recreational applications of nanomaterials include the Babolat tennis racquet depicted in the figure. The racquet's structure at the throat of the frame is reinforced with nanocarbon technology that affords the player better power, control, and stability. "NS" stands for nanostrength technology.*

industry still thrives today and, by means of "machine beating," foils 50 nm thick or less can be attained [36].

Benjamin Franklin in 1773 recorded the unique relationship between oil and water after noticing that the wakes of ships' waters were calmed after cooks dumped cooking grease overboard, as was the custom [37]. John W. Strutt,

FIG. 1.17

Lord Rayleigh (a.k.a. John William Strutt) was the third Baron Rayleigh. He also happened to be one of the leading scientists of his day. He was capable of "understanding everything just a little more deeply than anyone else."

Source: Image courtesy of the Nobel Foundation, NobelPrize.org. With permission.

better know as Lord Rayleigh (**Fig. 1.17**), in 1891 studied the surface tension of contaminated waters. He also investigated and described the scattering of light, a fundamental phenomenon responsible for Raman spectroscopy. In 1924, Lecompte du Nuoy, inventor of the de Nuoy ring surface tensiometer, estimated Avogadro's number to be 6.004×10^{23} by application of a thin film of sodium oleate onto water [38]. The amphiphilic molecules formed an organic monolayer on the surface of the water.

One of the greatest influences behind surface chemistry (monolayers in particular) was Irving Langmuir, who, along with Katherine Blodgett, developed Langmuir–Blodgett technology. Langmuir was the first to study monomolecular films and for his effort was awarded the Nobel Prize in chemistry in 1932. While working at General Electric Research Laboratory in Schenectady, New York,

Katherine Blodgett, the first female scientist to be hired by GE, developed the first "invisible glass" in 1938, an application of nanometer thin films. The glass was invisible because it did not reflect light due to the interference effect produced by the 44-molecule-thick film. She also developed a way of measuring thin films by means of interference colors generated by varying the thickness of stearic acid monolayers [39].

In 1980, Jacob Sagiv of the Weizman Institute in Israel found out that octadecyltrichlorosilane would react spontaneously with a glass surface. In 1983, David Allara of Bell Labs found that alkanes containing thiol groups would self-assemble on a gold surface into a monolayer. George Whitesides of Harvard (**Fig. 1.18**) also

Fɪɢ. 1.18 *George M. Whitesides was the Mallinckrodt Professor of Chemistry from 1982 to 2004 at Harvard University. He is currently the Woodford L. and Ann A. Flowers University Professor. Dr. Whitesides made significant contributions to nanoscience in general but specifically in the field of molecular self-assembly. The underlying theme of his lab is to* **fundamentally change the paradigms of science.**

Source: Image courtesy of George Whitesides, Harvard University. With permission.

contributed significantly to the understanding and development of monolayer science and technology. Colin D. Bains, a coauthor of Whitesides's landmark paper in 1989 [40–42], found that "stable surfaces exposing a single organic functional group could be created by adsorption of thiols on gold—for a dozen different groups, from non-polar methyl and CF_3 groups to halides, nitriles and carboxylic acids ... and how these surfaces could be used to predict, tailor, and control wettability."

Thin film chemistry is a central component of nanoscience and nanotechnology and represents a class of fabrication known as "bottom-up" synthesis. Thin films are characterized as two-dimensional materials and their chemistry is introduced in chapter 12; they are also classified as two-dimensional (2-D) quantum structures—a material with length and width but without any apparent thickness (chapter 10).

1.2.2 Fullerenes and Carbon Nanotubes

In 1985, R. E. Smalley, H. W. Kroto, and R. F. Curl discovered a third major form of elemental carbon: the *buckyball*, a molecule consisting of 60 atoms of carbon, C_{60}, assembled in the form of a soccer ball pattern [43]. This new allotrope of carbon, formally called *Buckminsterfullerene*, was named after R. Buckminster Fuller, the architect best known for the geodesic dome (**Fig. 1.19**). Buckyballs were discovered somewhat serendipitously after vaporization of a carbon target with a high-powered laser and then channeling the products into a mass spectrometer (MS). The MS signature of C_{60} and other fullerenes was unmistakable. The primary goal of the Kroto and Smalley group was to create long chain carbons similar to those detected around carbon stars. Little did they know at the time that a Nobel Prize was in the offing or that their discovery would change the face of nanotechnology.

| **FIG. 1.19** | *Buckminster Fuller was an architect known for his geodesic dome design. Carbon C_{60} molecules are called buckminster fullerenes in his honor.* |

Source: Image courtesy of the United States Postal Service. With permission.

If one slices a buckyball in half and then extrudes a tube from the hemispherical structure (resembling chicken wire), a single-walled nanotube (a buckytube or SWNT) is formed. Depending on which symmetry axis of the fullerene is chosen—whether projecting from the pentagon apex or the hexagon apex of a C_{60} molecule—different types of SWNTs are formed. Sumio Iijima, of NEC Corporation, reported in the journal *Nature* about his experimental validation of multiwalled carbon nanotubes (MWNTs) [44]:

> Here I report the preparation of a new type of finite carbon structure consisting of needle-like tubes. Produced using an arc-discharge method.... The formation of these needles, ranging from a few to a few tens of nanometers in diameter, suggests that engineering of carbon structures should be possible on scales greater than those relevant to the fullerenes.

In 1993, Iijima described the synthesis of SWNTs [45]. The discovery of carbon nanotubes, just like with fullerenes, was not new. J. R. Rostrup-Nielsen in 1972 reported thermodynamic parameters of carbon fibers from his studies on carbon fibers formed by catalytic chemical vapor deposition reactions [31]. Roger Bacon, a Union Carbide scientist in 1960, determined that samples from a chemical vapor deposition process yielded certain types of MWNTs [46]. Later, in 1979, MWNTs were produced in carbon arc discharge experiments. Theoretical work describing the mechanisms of carbon nanotube formation and structure soon followed [47].

SWNTs are called *quasi*-one-dimensional (1-D) quantum structures or, more accurately, pseudoquantum wires. One-dimensional structures have length but no height or width—a line consisting of infinitely small points (chapter 7).

1.2.3 *Quantum Dots*

Quantum dots, known as artificial atoms, are called zero-dimensional materials (0-D) (chapter 7). Electrons in a zero-dimensional material are "confined" in all three physical dimensions: the x, y, and z. Quantum dots are quite small, usually less than 10 nm. Quantum dots exhibit properties that are intermediate between the bulk and atoms or molecules. A metal cluster, a type of quantum dot, has characteristic "lowest occupied cluster orbitals" (LOCOs) and "highest unoccupied cluster orbitals" (HUCOs). The electronic properties of colloids, on the other hand, resemble those of the bulk material. Quantum dots are usually smaller than 10 nm and exhibit relatively monodisperse size distribution. Colloids are generally polydisperse with respect to size and shape and are much larger [48,49].

The history of the quantum dot takes us back to the *Stone Ages*, when early researchers developed colloidal pigment elements for wall art. The ancient Egyptians were known to add colloidal constituents into inks to give color and consistency. Medieval artists and craftsmen utilized metal colloids in glasswork. The beautiful stained glasses found in cathedrals certainly were able to impart a sense of other-worldliness due to nanotechnology. Chromium and iron nanoparticles were responsible for Maya blue pigment found in Mayan artifacts and art at the Chichén Itzá site in the Yucatan Peninsula [50].

As we know, the utilization of colloidal metals is well represented in the *Nano Age*. One of the leading nanopioneers in this field is Professor Günter Schmid of the University of Essen (now Duisburg-Essen) in Germany. The development of ligand-stabilized Au-55 quantum dots, transition metal clusters, and bimetallic catalysts offers examples of some of the important developments to emerge from his laboratory [51,52]. Applications of clusters include catalysis, bio-imaging, nano-electronic circuitry [53], and information storage. Both metal and semiconductor clusters have been inserted into porous alumina membranes [54] or secured otherwise into two-dimensional arrays [53]. Ligand-stabilized gold quantum dots comprising 5 to as many as 55 or more gold atoms have been produced experimentally [55]. More discussion will be presented about Au-55 in chapter 12.

Semiconductor quantum dots also possess a rich history. Victorian period glasses that appear red, red-orange, or orange were found to contain nano-crystallites of Zn or Cd sulfides and selenides [56]. Modern era investigations into quantum dots began in earnest in the early 1960s and continued more extensively in the 1970s and 1980s. In 1960, W. D. Lawson and his colleagues wrote a paper titled "Influence of Crystal Size on the Spectral Response Limit of Evaporated PbTe and PbSe Photoconductive Cells" [57]. In the 1970s and 1980s at Bell Labs and other places, A. Efros, L. Brus, M. Bawendi, and P. Alivisatos contributed extensively to the understanding of the role that particle size plays in the manifestation of physical properties [58–62]. In 1982, A. I. Ekimov's paper titled "Quantum Size Effect in Three-Dimensional Microscopic Semiconductor Crystals" describes the importance of size in determining quantum dot properties [62]. In those days, the word *nano* was used sparingly or not at all and most likely in the context of a measurement. An image of CdSe quantum dots synthesized in the laboratory of Professor Zhiqun Lin at Iowa State University is shown in **Figure 1.20**. The vivid fluorescent colors are dependent on the size of the quantum dots in solution.

The drive to develop solar cells also contributed to the development of quantum dots. The Solar Energy Research Institute (SERI, name changed to the National Renewable Energy Laboratory, NREL) was formed in 1978 under the Carter administration. NREL scientists like Arthur Nozik are considered to be pioneers in applications of quantum dots to the field of solar conversion [59]. NREL physicist Alex Zunger (**Fig. 1.21**) contributed significantly to the theoretical understanding of the quantum machinery responsible for the behavior of quantum dots. Michael Grätzel of the Swiss Federal Institute of Technology in Lausanne developed a solar cell based on TiO_2 nanoparticles [63].

1.2.4   Other Advanced Materials

Ferrofluids. In 1960, NASA researchers discovered that ferrofluids could be used in the attitude control devices of spacecraft. A ferrofluid is a liquid containing magnetic nanoparticles that reacts strongly to a magnetic field. Ferrofluids lose their magnetic properties once the field is discontinued. Nanoparticles, however, tend to aggregate (chapter 6). Ligand stabilization of nanomagnetic particles made of iron was prevented from aggregation in polar or nonpolar liquids. Ferrofluids have numerous applications in the electronic, defense, aerospace, metrology, medicine, and automotive industries.

Zeolites. Charles Plank and Edward Rosinski invented a means of converting petroleum base into gasoline by the use of inorganic catalytic

| FIG. 1.20 | *The wavelength of quantum resonance fluorescence emission of cadmium selenide quantum dots is related to the size of the nanoparticle. The emission of longer wavelengths is indicative of larger size.* |

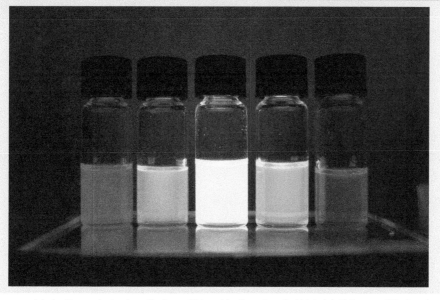

Source: Image reprinted with permission from Professor Zhiqun Lin, Department of Materials Science and Engineering, Iowa State University.

materials called zeolites. Zeolites have been used for decades to filter water out of nonpolar liquids and to separate gases.

DNA–metal inorganic structures. In 1996, C. Mirkin and R. Letsinger of Northwestern University in Chicago attached DNA to gold nanoparticles. This breakthrough helped bring on a new generation of bio-inorganic nanostructured materials. The convergence of biology, chemistry, physics, and engineering is apparent in the physical expression of these advanced materials.

1.3 TOOLS OF NANO

Tools (from Old English *tol*, meaning "instrument, implement, prepare") are physical devices that are used to carry out a particular function. The tools of nanotech are the instruments that measure, observe, and manipulate atoms and molecules. Without such tools, there would be no nanoscience, no nanotechnology. The tools of nano are at the heart of nanoscience and nanotechnology. The development of instruments and devices runs parallel to the history of nano and will mold the shape of things to come and perhaps be molded in return.

G. Binnig and H. Rohrer discovered the scanning tunneling microscope (STM) and the atomic force microscope (AFM) in 1982 and 1986, respectively [64,65]. In 1989, courtesy of Donald Eigler and his group at IBM Almaden in

Source: Image courtesy of Alex Zunger, National Renewable Energy Laboratory, Golden, Colorado. With permission.

California, 35 xenon atoms were moved about on a nickel (110) surface at low temperature by a homemade STM to spell out the "IBM" logo (**Fig. 1.22**) [66,67]. The "control, manipulation, and integration" of xenon atoms, according to our definition, has been accomplished in spectacular fashion, and the *Nano Age* was formally launched about 30 years after Richard Feynman's bold prediction in 1959 [68]. *Time Magazine* featured the IBM accomplishment in its April 16, 1990, issue [69]. In 2005, in an interview with *Nanooze* (a kids' science

FIG. 1.22 *Thirty-five xenon atoms on a nickel (110) surface at ultralow temperature were placed to spell "IBM" with the aid of an STM by Donald Eigler and his group at the IBM Almaden Research Center. The actual writing took 22 hours to complete. The image was published in* Time Magazine *in 1990 and formally ushered in the* Nano Age.

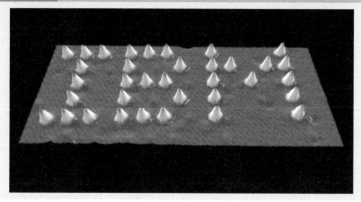

Source: Image reprinted with permission from IBM Research, Almaden Research Center.

magazine sponsored by the National Nanotechnology Infrastructure Network [NNIN]), Donald Eigler said,

> It all depends upon what you call nanotechnology. Rather than get everyone confused about what is or what is not nanotechnology, let's just think about the role that nanometer-scale structures play in our lives and are likely to play in the future. From the transistors in the computer on which I am composing this sentence, to the coatings on the window in front of me, to the drugs and chemicals that are already a part of everyday life, nanometer-scale structures are crucially important. As time goes on, our ability to engineer and fabricate new and useful nanometer-scale structures will only increase. Nanotech will have a profound impact on our lives.

1.3.1 Electron Microscopy

In 1931, Max Knoll and Ernst Ruska invented the transmission electron microscope (TEM), nearly 34 years after the discovery of the electron by J. J. Thomson. M. von Ardenna built the first commercial SEM in 1938, followed soon by the first commercially available TEM, built by Siemens in 1939. The principle of electron beam methods is simple. In order to observe objects with smaller dimensions, the wavelength of the probing source must also acquire comparably smaller dimensions. Erwin Müller, a professor in the Penn State University Physics Department, invented the field-ion EM. He is the first person known to see atoms. The first scientists however to fully exploit electron beam methods were biologists, an interesting irony. Since the resolution of the early scopes was not at the atomic level, it makes sense that relatively large biological structures were first in line to be described by TEM and SEM.

In 1951, x-ray spectroscopic capability was added to TEM and SEM systems with the purpose of conducting in situ elemental analysis. EDS (or EDX, energy dispersive x-ray spectroscopy) systems are included in virtually all EM systems today. In 1970, a high-resolution TEM (HR-TEM) achieved 4-Å resolution. In 2004, Stephen Pennycock of the Oak Ridge National Laboratory (ORNL) achieved 0.6-Å resolution, *transcending the* 1-Å *atomic resolution barrier in the TEM* [70]. Welcome to the *Pico Age*.

Single instruments capable of executing numerous analytical tasks represent another type of convergence. Over the past few decades, development of instruments capable of multiple analyses proved to be a powerful combination in the quest to unravel the mysteries of the nano dominion.

1.3.2 Atomic Probe Microscopes

In 1958, Leo Esaki, a Sony physicist, discovered the phenomenon of electron tunneling and was the first to comment on the special behavior of nanosized materials. His name is attached to the *tunneling diode* and he shared the Nobel Prize in physics in 1973. As a result of Esaki's discovery, another phenomenal research tool was developed—one related in principle to the SEM (e.g., electron emission), but based on the principle of electron tunneling. Gerd Binnig and his colleague Heinrich Rohrer at IBM Zürich introduced a new kind of microscope that is not dependent on wavelength [64].

The STM is capable of creating a topological map of a conductive surface by providing a picture of electron density. The tunneling current is applied under a preset bias between the probe tip and specimen surface. Electrons "jump" from the tip to the surface similar to the way electrons are emitted from the gun of an SEM. Because the size of the analytical current is exponentially proportional to the distance that the tip is from the surface, the topology is mapped by rastering the probe over the surface.

Binnig, Quate, and Gerber invented the atomic force microscope (AFM) 4 years later for the specific purpose of investigating surfaces on the atomic scale that are not conducting [65]. The AFM is a combination of an "STM with a stylus profilometer" [65]. The AFM also is equipped with an atomically sharpened probe tip. The tip is attached to a shiny cantilever that reflects laser light into a detector as the probe is moved across the surface (rastered). Differences in the reflection are recorded in a split photodiode and recorded as changes in topography. Use of the STM is limited to electrically conducting samples. Therefore, the atomic force microscope was developed to image insulating samples. The ability to see atoms and molecules with a device that is purely mechanical is simply amazing. Binnig, Rohrer, and Ruska received the Nobel Prize in physics in 1986 for these incredible inventions—amazing tools that are tailor made for investigating nanomaterials [71].

1.3.3 X-Ray Spectroscopy

German scientist Wilhelm C. Röntgen (**Fig. 1.23**) discovered the x-ray in 1895. He discovered that paper covered with barium platinocyanide fluoresced when placed in the pathway of the radiation emitted from a cathode ray discharge tube [72]. He theorized that x-rays were produced by the impact of the "cathode

| FIG. 1.23 | *Wilhelm Conrad Röntgen was awarded the Nobel Prize in physics in 1901 for discovering x-rays in 1895 and their application. One of the most useful tools of nanoscience is x-ray diffraction; its roots were laid down at the time of Röntgen.* |

Source: Image courtesy of the Nobel Foundation, NobelPrize.org. With permission.

rays" on an object [72]. Max von Laue demonstrated later that x-rays are electromagnetic in nature but with extremely short wavelength [73]. Röntgen's and von Laue's contributions led eventually to the development of x-ray diffraction (XRD). Because the wavelength of x-rays is on the order of distance between crystal planes, XRD is a useful tool in analyzing three-dimensional crystal structure of materials.

The father and son team of William and Lawrence Bragg contributed significantly to this new branch of science [74]. Bragg's law developed instrumental methods to analyze crystal structure based on XRD. The Bragg law (chapters 3 and 5) is the fundamental equation relating interplanar crystal spacing to the x-ray wavelength. XRD methods were expanded to include analysis of powders—another significant contribution to the emerging areas of nanoscience and engineering.

1.3.4   *Surface Enhanced Raman Spectroscopy*

In 1930, Sir Chandrasekhara Venkata Raman was awarded the Nobel Prize in physics for his discovery of Raman spectroscopy. In 1974, Martin Fleischman et al. reported that metal nanoparticles enhanced the Raman signal of pyridine adsorbed to a roughened silver electrode [75] and erroneously attributed the phenomenon to increased surface area (roughness) alone [75]. In 1977, Richard P. Van Duyne of Northwestern University is credited with explaining the mechanism of surface enhanced Raman spectroscopy (SERS). He found that enhanced Raman signal of the adsorbed analyte species was due to chemical and electromagnetic factors in addition to signal enhancement from increased surface area. Martin Moskovits of the University of Toronto explained the phenomenon in detail in 1985 [76].

The electromagnetic enhancement of SERS is due to induced dipolar and higher order resonance of the surface plasmon on nanometal particles or surface facets. Chemical enhancement, on the other hand, occurs when adsorbed compounds have inherent charge transfer capability (e.g., molecules with lone pairs or π-clouds such as pyridine show the strongest SERS signals). Five or more orders of enhancement using SERS substrates have been reported. Silver is one of the best substrates for SERS use.

SERS techniques have evolved sufficiently to push the detection limit past the attomole (10^{-18} mol) level. This is because SERS is a nanotool that relies on nanoparticles to generate a signal. Specifically, the SERS signal is mediated by the characteristics of the SERS substrate: particle size, shape, and orientation and its composition. The scattering cross-section characteristic of traditional Raman spectroscopy precluded its use for single molecule detection. It is only with the advent of SERS that single molecule Raman detection (SERS-SMD or smSERS) is possible. Single molecule detection of rhodamine-6G on silver nanoparticles was verified by SERS in 2002 [77]. The scattering cross-section of rhodamine-6G was found to be $\sigma = 2 \times 10^{-14}$ cm^2, the largest to date. Although there are skeptics [78], in 2006, Le Ru et al. showed that single molecule signals are very common in SERS [79]. The power of nano!

1.3.5   *Lithography*

Lithography (from the Greek *lithos*, "stone" + *graphien*, "to write"), a term coined in 1813, has its roots in printing technology. The first lithography was observed in cave etchings dating to the Neanderthal period. Clay Sumerian tablets revealed the earliest evidence of writing. Woodblock printing was developed in China in A.D. sixth century and, in 1040, Pi Sheng invented movable type. Much later, the Koreans also developed movable metal type printing in the thirteenth century. Johannes Gutenburg introduced the printing process to Europeans in 1440. Lithography, however, was not to emerge on the scene for another 300 years.

In 1798 Alois Senefelder of Bavaria invented the modern lithographic process, and in 1819 he wrote a treatise on the subject [80]. Senefelder desperately needed a new method to publish a play he wrote after falling in debt due to problems associated with a "conventional" publisher. He used hydrophobic acid-resistant ink as the *resist* (the positive image) on smooth, finely grained Solnhofen limestone [80]. When the etched plate was immersed in an ink–water mixture, the ink adhered to the positive hydrophobic image and the water adhered to the negative hydrophilic component. He was then able to print from the flat surface of a stone; this was called a *planographic* process. He named it *stone printing* (i.e., lithography).

The photolithography process, a descendent of lithography, is initiated by shining light through a photomask onto a photoresist material. The photoresist material is chemically altered by exposure to the light. Application of a subsequent chemical step etches away the exposed image area (positive resist) or the area around the image (negative resist), depending on the nature of the substrate. Following the etching process, either a positive or a negative of the photomask pattern is produced. Integrated circuits are still produced this way.

Photolithographic processes are limited by the wavelength of the light used. Extreme ultraviolet (EUV) light can be used (wavelength equal to ca. 14 nm), but there are several engineering issues that need to be resolved before its use becomes widespread. There are many more types of lithography, and several will be discussed in more detail in chapter 4.

In 1999, Chad Mirkin of Northwestern University developed the first dip-pen nanolithography system. Nanolithography is lithography at the nanoscale and is capable of producing nanofacets and nanostructures 100 nm or less in size. J. Lyding of the University of Illinois Urbana-Champaign and M. Hersam of Northwestern University developed a variation of lithography called feedback-control lithography (FCL) in 2000. In FCL, an STM is used to build nanostructures on hydrogen-passivated silicon substrates. With the FCL technique, the study of the chemistry of single atoms is possible.

1.3.6 Computer Modeling and Simulation

A valuable nanotool is computer modeling and simulation of atomic and molecular structures, processes, and devices. Although we will not delve into computer modeling and simulation of nanomaterial properties and phenomena in this text, several methods have been developed to probe the behavior of small materials. Computer capability at this time, however, is still limited by computational expense in terms of time and funds to ~100 atoms and timescales on the order of picoseconds.

A versatile and accurate method is molecular dynamic reactive force field simulations based on first principles. Bond order, bond distance, bond energy, and molecular environment dependent charge distribution are examples of parameters that are evaluated by molecular simulation methods. Ab initio Monte Carlo and finite element analysis are also very popular computer simulation methods. Such techniques are analogous to CAD (computer-aided design) programs used by engineers, but operate instead at the nanoscale. Since the design of nanomaterials is usually a costly undertaking, conducting computer modeling beforehand saves time and resources a priori. The nano-engineer–chemist of the future needs to forge good working partnerships with experts in computer modeling.

Phenomena such as hydrogen adsorption on carbon nanotubes, the lotus effect on hydrophobic surfaces, flow through nanochannels, thermal conductivity of carbon nanotubes, diffusion of carbon through (or on the surface of) catalyst particles, light interaction with photonic band gap structures, nanomechanics, and quantum dot optical response all have been modeled with computer simulation techniques. Biological modeling is also routinely done. For example, computer models have simulated the transport of ions via ion channels in membranes [81]. The National Nanotechnology Initiative (NNI) is vigorously funding theory, modeling, and simulation investigations [81].

1.3.7 *Molecular Electronics*

In 1974, researchers from Northwestern University (Mark A. Ratner) and IBM (Ari Aviram) proposed that electronic circuits be built from the bottom up. Ratner is considered to be the "father of molecular-scale electronics." In 1987, Dmitri Averin and Konstantin Likharev introduced the concept of the "single-electron tunneling transistor" (SET). In 1989, Bell Lab researchers T. Fulton and G. Dolan assembled such a device. The control of the movement of single electrons by a nanodevice was achieved!

1.4 NATURE'S TAKE ON NANO AND THE ADVENT OF MOLECULAR BIOLOGY

Nature was the master of nanotechnology well before the arrival of the human version (or vision) of nanotechnology. The question to ask nature is, "Why do so many structures and functions exist at the nanoscale?" Nature, of course, does not yield secrets easily. The fundamental building blocks of living things are synthesized from the bottom up. Some remain in the form of individual nano-materials (e.g., enzymes) and others form hierarchical structures composed of nanomaterials (e.g., connective tissue). The common thread is that all living things without exception are composed of nanomaterials.

The nanoscale represents the size domain above independent atoms and molecules. It also represents the smallest size domain in which there is evidence of functionality. Atoms and molecules per se do not possess functionality; they react and interact according to inherent physical laws, proximity, and probability. It is at the nanoscale, however, where the first examples of function are detected.

1.4.1 *Macroscopic Expressions of Natural Nanomaterials*

All living matter is composed of nanomaterials that were assembled from the bottom up. The level of complexity (or size) depends on where you look. Following is an example of a hierarchy proceeding from nanoscale materials to assemblies of nanoscale materials. The selected hierarchy starts with the light-absorbing chlorophyll molecule found in plants:

Chlorophyll → Thylakoid → Granum → Chloroplast → Cell → Leaf

Within the chloroplast, the plant cell, and larger structures like leaves, there is a convergence of many types of nanomaterials, structures, and functions (**Fig. 1.24**). Hierarchies such as this can be constructed for any naturally occurring biological material and system, whether structural, metabolic, or even behavioral (e.g., bird feathers). The pads on a gecko's feet, the color of peacock

FIG. 1.24	*A chloroplast is a plastid with a lens-shaped structure about 5 µm in length. It has an inner and an outer membrane. Chlorophyll resides in stacks of* thylakoid *membranes called* grana. *The structure is interconnected with a network of* lamellae.

Source: Image courtesy of Anil K. Rao, Department of Biology, the Metropolitan State University, Denver, Colorado. With permission.

feathers and butterfly wings, the structure of a moth's eye, the silica-based infrastructure of diatoms, the structure of abalone shells, the bone structure of parrot fish teeth, the nanostructured facets of the lotus plant leaf surface, and spider silk comprise a short list of naturally occurring structures composed of nanostructured materials. Natural nanotechnology will be discussed further in chapter 13.

1.4.2   *Cell Biology*

With the advent of the microscope, cell biologists were the first to investigate micron-sized living things and structures. They also were aware of smaller but irresolvable structures like cellular organelles. The tool of the microworld in the eighteenth and nineteenth centuries was the optical microscope. Dating back to ancient Rome, magnifying lenses are mentioned in the works of Seneca and Pliny the Elder. Spectacles were invented in the thirteenth century (named "lenses" due to their similarity to lentil seeds). Zaccharias and Hans Janssen of Holland in 1590 made the first rudimentary microscope. Galileo in 1609 seized upon that invention and created the first telescope [7]. Robert Hooke invented the compound microscope and in 1665 described microscopic organisms and fossils [7,82]. Hooke is also credited with using the word *cell* (from the Latin *cellulae,* meaning "little rooms") to describe the structure of cork (at 30× magnification). Matthias Schleiden (plant cells) and Theodor Schwann (animal cells) proposed the cell doctrine in 1839 and it remains as the foundation of biology to this day [7].

The "father of microscopy" was Dutchman Anton van Leeuwenhoek [7]. In 1677, with the aid of a microscope capable of 300× magnification, he described entities that dwell at the micron level of measurement: protozoans, red blood cells, and even bacteria. After a 100-year hiatus between major biological break-throughs, biologists got back on track as Scottish botanist Robert Brown in 1833 coined the word *nucleus* (from the Latin *nucleus*, "kernel," and from the Greek "little nut") after observing cells in the "cellular juice of the orchid" [7]. He is also credited with the first experimental observation of *atomism* after describing Brownian motion.

The discovery of organelles within the cell was documented in the 1800s. Mitochondria and chloroplasts are micron-sized organelles that are composed of nanostructured materials. In 1817, French chemist P. J. Pelletier isolated a nanosized molecule called chlorophyll (from the Greek *khloros*, meaning "pale green" + *phyllon*, indicating "leaf") [7]. Although Julius von Sachs linked the molecule chlorophyll to chloroplasts (an organelle of submicron to micron size) in 1865, Nehemiah Grew is credited with their discovery in 1682. Rudolf Albert von Kolliker described the mitochondrion (from Greek *mitos*, meaning "thread" + *khondrion*, indicating "little granule") in animal cells in 1857 due to improved resolution of optical microscopes. German microbiologist Carl Benda in 1901 named the tiny structures mitochondria. The cellular biologists of the seventeenth through nineteenth centuries went on to investigate and catalogue numerous other intricate microstructures and organelles found in cells. Better microscope resolution would accelerate the study of smaller and smaller components beyond the level of the organelle after the invention of the TEM.

1.4.3 Molecular Biology and Genetics

The area of molecular biology is divided into two parts. One part is biochemical and involves the study and characterization of metabolic processes and synthe-ses. The study of metabolic pathways fell into the domain of chemists and later biochemists. The other part is the study and characterization of biological nano-materials like DNA. Description of nanostructured materials in cells became part of the domain of the microscopists. Molecular biology represents the con-vergence of genetics, biology, and chemistry.

In the 1870s, Walther Flemming discovered chromatin after staining cells with synthetic dyes. He noticed them while observing a cellular division pro-cess. Heinrich Wilhelm Waldeyer-Hartz suggested the name *chromosome* (from the Greek *khroma*, "color" + *soma*, "body"). In 1910, Thomas Hunt Morgan linked genetic material to chromosomes of *Drosophila*, the fruit fly. Hermann J. Müller in 1926 postulated that the gene is the basis of life [83]. Unraveling the structure of deoxyribose nucleic acid (DNA) by Watson, Crick, and Franklin in the early 1950s demonstrated that nanomaterials are fundamental to life. Thus was established the *central dogma of biology*—that all life depends on DNA. The recently completed mission of the Human Genome Project (HGP) was to generate a map of all the nucleotides of genes in the human genome.

Diastase, the enzyme responsible for converting starch into sugar, was the first enzyme to be isolated. In 1833, a French chemist, Anselme Payen, purified diastase from a malt extract [7]. The suffix "ase" henceforth has indicated that the substance is an enzyme and performs some metabolic function. He also isolated cellulose from wood, another material that is made of nanosized building blocks, and is credited with introducing the word *ferment* into the lexicon of the early biologists. In 1878, Wilhelm Kühne, a German physiologist, suggested the word *enzyme* (from the Greek word meaning "in yeast," *enzymos,* "leavened") be applied to those substances that "brought about chemical reactions associated with life" [7]. In 1836, Theodor Schwann, who established the *cell theory* along with Matthias Schleiden, isolated pepsin, the first enzyme derived from animal tissue [7]. Biochemist James B. Sumner was the first to crystallize an enzyme, urease, from the extract of black beans in 1926. He was also the first to suggest that enzymes were made of proteins [7].

After he won the Nobel Prize in chemistry in 1902 for his research in purine and sugar chemistry, the great German chemist Emil Fischer (**Fig. 1.25**) joined

| **FIG. 1.25** | *Hermann Emil Fischer was awarded the Nobel Prize in chemistry in 1902 for his work with purines and sugars. He is also the first to propose the* lock and key *hypothesis to describe the relationship between an enzyme (the lock) and its substrate (the key).* |

Source: Image courtesy of NobelPrize.org. With permission.

the carboxyl and amino ends of two amino acids together and synthesized the first peptide [7]. Molecular biology became a formal discipline in the 1920s. Warren Weaver, of the Natural Sciences Division of the Rockefeller Foundation, said in 1938, "And gradually there is coming into being a new branch of science—molecular biology—which is beginning to uncover many secrets concerning the ultimate units of the living cell…in which delicate modern techniques are being used to investigate ever more minute details of certain life processes."

The development of a protein synthesizing machine by Bruce Merrifield in 1971 was the first attempt to manufacture proteins in a stepwise synthetic process. The process involved making and breaking strong bonds and therefore required the use of strong acids such as trifluoroacetic acid. Merrifield and other protonanotechnologists were able to anchor amino acids to a board and fabricate protein structures, thus laying down the groundwork for prototypes of "lab-on-a-chip" sensors.

1.5   The Nano Perspective

We have discussed some history, introduced some of the scientists involved, and made mention of some of the significant developments involved in making nanoscience and nanotechnology what they are today. Now we need to delve into the nanoscale itself and gain a feel for the physical perspective. **Figure 1.26** depicts the relative scale of things, depicting images from nature and comparing them to synthetic materials [81]. The electromagnetic radiation spectrum is also shown in order to provide the perspective.

1.5.1   Integration of Everything

We all understand that the major academic disciplines of physics, chemistry, biology, and engineering serve as convenient boundaries to help us make sense of the universe. From the eyes of nature, we can only imagine that the universe is seamless, without demarcation. Intuitively, we sense that all things are connected in some way—that nothing exists independently. The concept of the continuum is referred to often in this text. However, it is difficult to think in terms of "continuum"; otherwise, we would become the polyhistors who know everything. We need boundaries, definitions, and guidelines to sort out our universe. We need to parse data. To cope with the continua abounding in the universe, we invented the major scientific disciplines of mathematics, physics, chemistry, and biology.

Disciplines are created from the top down. For example, chemistry spawned numerous subdisciplines such as organic chemistry, inorganic chemistry, physical chemistry and, later, colloid chemistry and surface chemistry. The list is imposing if we attempt to distinguish more specialized fields such as thin film chemistry, supramolecular chemistry, and, yes, nanoscale chemistry. However, we found that it was not enough to create more specific sub-subdisciplines and sub-sub-subdisciplines complete with journals and conference symposia. As we learned more and more, we needed to cross borders. Hence, subjects such as chemical physics, biophysics, biochemistry, and, yes, even biophysical chemistry emerged.

FIG. 1.26 *The scale of natural things compared to the electromagnetic spectrum are shown in (a), and in (b), synthetic materials of diminishing size are dipicted.*

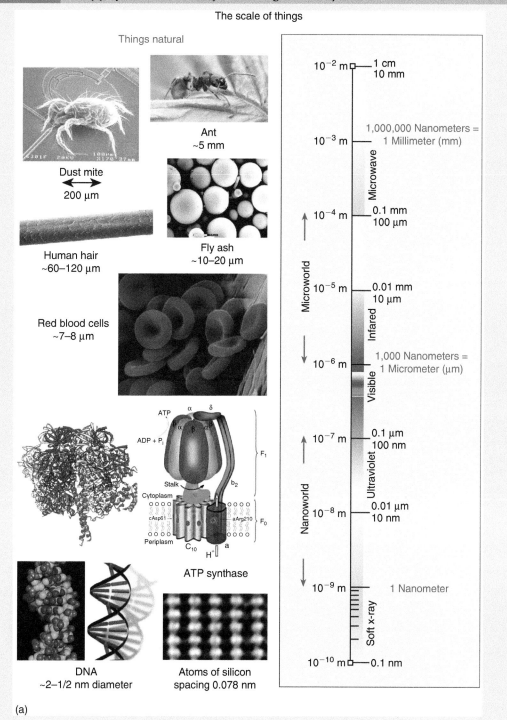

The scale of things

Things natural

Dust mite
200 μm

Ant
~5 mm

Human hair
~60–120 μm

Fly ash
~10–20 μm

Red blood cells
~7–8 μm

ATP
ADP + Pᵢ
Stalk
Cytoplasm
cAsp61
Periplasm
C₁₀
H⁺
a
F₁
F₀

ATP synthase

DNA
~2–1/2 nm diameter

Atoms of silicon
spacing 0.078 nm

10^{-2} m — 1 cm / 10 mm

10^{-3} m — 1,000,000 Nanometers = 1 Millimeter (mm)

Microwave

10^{-4} m — 0.1 mm / 100 μm

Microworld

10^{-5} m — 0.01 mm / 10 μm

Infared

10^{-6} m — 1,000 Nanometers = 1 Micrometer (μm)

Visible

10^{-7} m — 0.1 μm / 100 nm

Nanoworld

Ultraviolet

10^{-8} m — 0.01 μm / 10 nm

10^{-9} m — 1 Nanometer

Soft x-ray

10^{-10} m — 0.1 nm

(a)

(continued)

**Fig. 1.26
(Contd.)**

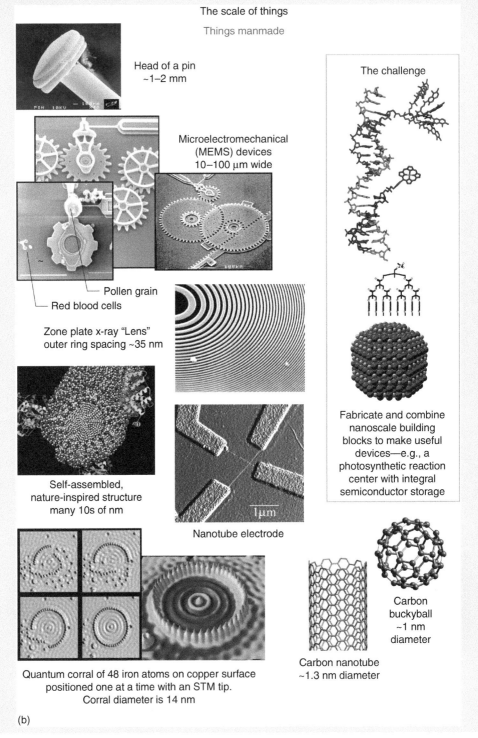

The scale of things

Things manmade

Head of a pin
~1–2 mm

Microelectromechanical
(MEMS) devices
10–100 μm wide

Pollen grain
Red blood cells

Zone plate x-ray "Lens"
outer ring spacing ~35 nm

Self-assembled,
nature-inspired structure
many 10s of nm

Nanotube electrode

Quantum corral of 48 iron atoms on copper surface
positioned one at a time with an STM tip.
Corral diameter is 14 nm

The challenge

Fabricate and combine
nanoscale building
blocks to make useful
devices—e.g., a
photosynthetic reaction
center with integral
semiconductor storage

Carbon
buckyball
~1 nm
diameter

Carbon nanotube
~1.3 nm diameter

(b)

Source: Image reprinted with permission from the Office of Basic Energy Sciences, Office of Science, U.S. Department of Energy.

Since students are not in the habit of reading the preface, we are forced to reiterate some of the salient aspects mentioned there. *Physics* (from the Greek *physikos*, "natural," or *physis*, "nature") is the study of the natural world: the "science that deals with matter and energy in terms of motion and force" [2,3]. The discipline of physics, in the context of this text, is concerned mostly with *properties* and *phenomena* of nanomaterials and is therefore presented first. Properties and phenomena include conductivity, semiconductivity, magnetism, optical response, band gap, heat conduction, quantum phenomena, surface energy, and interaction with electromagnetic radiation and forces in general. Chapters 5–8 comprise the "Properties and Phenomena" (physics) division of the book.

Chemistry (from the Medieval Latin *alchymia*, "the art of transmutation," from the Greek *khenia*, "the art of transmutation as practiced by the Egyptians," and from the Egyptian *Kh'mi*, meaning Egypt) has been in existence for more than four millennia. Chemistry is the science that "investigates the composition, properties and changes of composition, structure, properties and reactions of substances" [2,3]. Within the disciplinary boundaries of this text, chemistry is concerned with *synthesis* of nanomaterials and their modification by chemical methods and is therefore presented after physics. Chemistry also represents the next level of complexity above physics (from a first-principles computational point of view). The content of chapters 9 through 12 comprises the "Chemistry: Synthesis and Modification" division of the text.

Biology (from the Greek *bios*, "life" + *logia*, meaning "study of") is the science of life and life processes, including the study of structure, function, growth, origin, evolution, and distribution of living organisms [2,3]. Biology is where nano begins and is optimized by nature. Within the disciplinary boundaries of this text, the discipline of biology as it applies to nanomaterials is concerned mostly with *describing* and *understanding* natural nanomaterials and is presented after physics and chemistry. Chapters 13 and 14 are in the "Natural and Bionanoscience" division of the book, dedicated to the nanoscience associated with biological and inorganic natural phenomenon. Biology also represents the next level of complexity above chemistry (from a first-principles computational point of view).

Engineering (from O.Fr. *engine*, meaning "skill, cleverness, or war machine" based on the L. *ingenium*, "inborn qualities, talent," *gen* meaning "to produce") is an applied science. The objectives of engineering (chemical, mechanical, and electrical) are to understand the properties of materials and then design devices, circuits, and structures to solve a problem. Engineering is the synthetic analog of biology if materials, structures, and devices are manufactured from the bottom up. Materials science is a subject usually taught in engineering schools. Nanoengineering (e.g. nanomechanics, nanofluidics, nanotribology, nanoelectronics, and the recent fusion of engineering, computer science, and biology via nanotechnology) does place a remarkable spin on the interdisciplinary perspective. Engineering topics are not emphasized in this text.

One more philosophical topic will be discussed and it concerns the convergence of ideas. We have been driven to devise "grand unification theories" ever since earth, air, water, and fire were first merged over 2000 years ago. We know that all academic disciplines are inherently integrated—within an *integration of everything* scenario. We are very interested in the outcome of nanoscopic, microscopic, and macroscopic expressions of nanomaterials from both the synthetic and natural points of view. In many ways we are integrating, unifying, and converging by way of nanoscience and technology. Nanoscience and technology, unlike anything we have seen before, have already raised the bar of interdisciplinary collaboration—proceeding along the path of integrating all scientific and engineering disciplines.

1.5.2 *Scale of Things and Timescales*

How does one acquire a feel for the nanoscale? One nanometer is $10 \times 10 \times 10 \times 10 \times 10 \times 10 \times 10 \times 10 \times 10$ times smaller than a meter. The distance between cones in the human eye is ca. 2 µm. It is the distance between those cones that limits resolution in the human eye. Robert Hooke in 1673 estimated the resolving power of the human eye to be 70 µm (or 70,000 nm) at the near view limit. This is similar in size to the thickness of a human hair—quite a bit above the nanoscale. Although we cannot see nanoscale materials, we do see the expression of the collective sum of nanomaterials in a macroscopic structure.

Anodized alumina films, ranging in thickness from a few to 20 µm and above, can be easily seen upon close inspection without a microscope, but they appear more like a continuous film without thickness. In order to view the pores of the alumina film, the aid of an electron microscope is required. Light microscopes help to reduce the optical resolution limit to ca. 0.100 µm (or 100 nm), the size of a small bacterium. Electron microscopes, of course, allow us to see atoms 0.0001 µm (or 0.1 nm) and below 1 pm if needed. We are able to see nanomaterials if they are in the form of a powder or in the form of a black soot. We are not able, however, to resolve the detail of the mass without the aid of an electron or atomic force microscope.

The nanoscale is all about size. You will read over and over again in this text how size affects properties and phenomena. The nanoscale is also about time. Emppu Salonen at the Helsinki University of Technology correlated timescales with size. We have added a few extra examples. In terms of seconds, humans live for 2.52×10^9 s (ca. 80 y); red blood cells live for about 1.1×10^7 s (ca. 130 days); *Escherichia coli* can live for $60 – 8.6 \times 10^5$ s (ca. 10 days); it takes protein about 50–100 s to fold; nanocolloid diffusion happens in 10^{-6} s; surfactant dynamics in 10^{-10} s; molecular collisions in 10^{-12} s; and atomic vibrations every 10^{-14} s in a molecule. The smaller something gets, the faster it reacts and the higher its voice becomes.

1.5.3 *Grand Challenges Facing Nanoscience and Nanotechnology*

The atomic abacus shown in **Figure 1.27** is an example of manipulation and control of atoms at the nanoscale to form a simple computing system, the abacus. The *beads* consist of C_{60}—otherwise known as a buckyballs—placed in rows of 10 on the steps (*rails*) of a copper substrate [84]. The abacus is manipulated by an STM *finger* probe tip in order to perform simple calculations. The abacus can be operated at room temperature. James K. Gimzewski, leader of the nanoscience project at the Zurich Research Laboratory, states:

> We have made significant progress in handling objects and creating functional
> units on the nanometer scale at room temperature. Our work demonstrates a

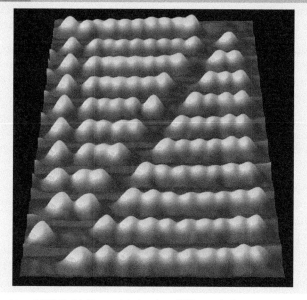

FIG. 1.27 *A nanoscale abacus created in the IBM-Zurich laboratory by Cuberes et al. The beads are actually C_{60} molecules. The rails along which the beads are moved are steps in the copper substrate. Manipulation (calculation) is accomplished with the tip of an STM.*

Source: Courtesy of IBM Zurich Research Laboratory. With permission.

further step in the new and fascinating field of "nano-engineering," where solid-state physics and chemistry merge. We may be able to assemble more complex structures from the bottom up, as nature does, molecule by molecule, and thus break ground for entirely new fabrication technologies with a broad range of applications.

But how practical is a manufacturing process based on this kind of "building one-atom-at-a-time" process? Building nanoscale devices one atom at a time by STM is simply not a reasonable option in today's mass production economy. There needs to be a way to manufacture such devices on a mass scale, simultaneously, and preferably from the "bottom up."

It is also quite straightforward, for example, to fabricate SWNTs. A furnace, some catalyst particles, an inert atmosphere, and a carbon gas source like methane are all that is required. Manufacturing SWNTs that are pure, of the same diameter, of the same orientation, and of the same chirality (chapter 9) is quite another matter. It is somewhat easier to manufacture raw nanomaterials, but it is another matter to manufacture devices that are all the same on a mass scale. George Whitesides of Harvard summarizes some of the challenges that nano will have to address in the next decade [84,85]:

> The age of nanofabrication is here, and the age of nanoscience has dawned, but the age of nanotechnology—finding practical uses for nanostructures—has not really started yet.

Many interesting problems plague the fabrication of nanodevices with moving parts. A crucial one is friction and sticking .... Because small devices have very large ratios of surface to volume, surface effects—both good and bad—become much more important for them than for large devices .... We will undoubtedly progress toward more complex micromachines and nanomachines modeled on human-scale machines, but we have a long path to travel before we can produce nanomechanical devices in quantity for any practical purpose. Nor is there any reason to assume that nanomachines must resemble human-scale machines.

Within our lifetime, we have seen how technology is able to change the way we live. Nanotechnology, with all of its hype—ranging from eternal life to great wealth—is supposed to change the way we live in revolutionary fashion [86]—or will it? George Whitesides continues:

None of these things will happen overnight. It is likely that it will take decades ... and nanotechnology will be seen for what it is: another of the wonderful portfolio of technological tools we have available. Not a revolution but an evolution, and one that we don't notice but we take on as another step forward.

Meyya Meyyappan, the director of NASA Ames Nanotechnology Center, states the following about nanosience and its relation to nanotechnology [87]: "lots of nanoscience, very little nanotechnology at this time."

According to Michael Roukes, realistic nanomanufacturing is a long way off. There are two basic types of challenges facing the development of nanotechnology [85]. Communication between the macroworld and the nanoworld is *challenge no. 1*. Starting with the *Heisenberg uncertainty principle*, measuring a system perturbs it. Therefore, any link to a macroscopic detector will change the nanosystem from the ideal state [88]. An interesting permutation of Roukes's proposition is the following: When nanosystems are measured, would the measurement itself serve to stabilize nanocomponents and thereby lose energy (reduce or remove unique properties or phenomena)? *Challenge no.2* involves the nanosurface [88]. The development of an ideal nanocrystal with ultrahigh purity is intended to operate perfectly in ultrahigh vacuum. In any other environment, contact with contaminants would reduce its properties [88].

Some outstanding challenges facing the practical development of nanotechnology are summarized in **Figure 1.28** [85–89].

1.5.4   *Next Industrial Revolution*

Nanotechnology (as well as nanoscience) has indeed taken center stage in the arena of academic science and is making significant headway into numerous industrial sectors. The number of published scientific literature papers has burgeoned over the past 10 years. The concomitant creation of numerous new journals dedicated specifically to nanoscience and technology soon followed. Hardly any technical conference, whether engineering, optics, chemistry, physics, or biochemistry, is without a symposium on nanoscience and technology. According to government sources, nanotechnology-enhanced products are estimated to contribute \$1 trillion to the global GDP and create over two million jobs by 2015 [90].

While nanotechnology is in the "pre-competitive" stage (meaning its applied use is limited), nanoparticles are being used in a number of industries ... electronic,

FIG. 1.28 *The challenges facing nanoscience and nanotechnology are depicted schematically. Reproducibility, reliability, and facilitative upscale are just a few of the challenges shown here. Can you think of any more?*

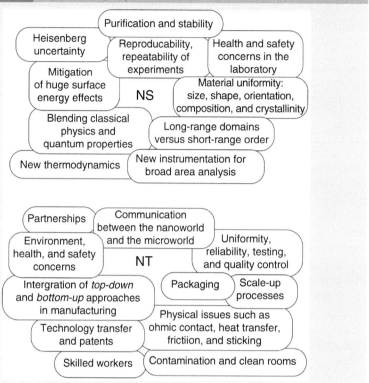

magnetic and optoelectronic, biomedical, pharmaceutical, cosmetic, energy, catalytic and materials applications. Areas producing the greatest revenue for nanoparticles reportedly are chemical–mechanical polishing, magnetic recording tapes, sunscreens, automotive catalytic supports, biolabeling, electroconducting coatings and optical fibers [90].

Basic nanoscience research has been in existence for over 20 years. Since then, many products have made their way into the marketplace. In 2003, for example, Intel introduced the sub-100-nm transistor, smaller than the human influenza virus. Medical diagnostics has exploited the remarkable optical properties of quantum dots. Investigations are taking place now that link quantum dots with molecular recognition systems. Of course, the reliance on nanocatalysts by the oil and gas industries continues unabated. Display technology relies heavily on nanostructured polymer films. Organic light emitting diodes (OLEDs) are being produced with "brighter images, lighter weight, less power consumption and wider viewing angles" [90]. Water purification systems enhanced by carbon nanotube nanotechnology offer inexpensive solutions to filter viruses, bacteria, particulates, and unwanted chemicals. Other products include wear- and chemical-resistant coatings, stain-resistant clothing, inks, dental bonding agents, long-lasting tennis balls, smart golf balls, tennis racquets, cutting tools, special coatings for glasses, waterproof textiles, and catalytic converters.

Be on the lookout for converging technologies, collectively referred to as *nano-bio-info-cogno* (a.k.a. NBIC) [91]. Converging NBIC technologies are collectively called *metatechnology* [88]. The interrelationships are characterized by synergy; development in one of the areas forms a base for developments in the other [91]. There is an apparent "fast spiral-growing positive feedback loop" [91]. With the way information is spread worldwide in milliseconds, this should not be much of a surprise.

1.6 Concluding Remarks

What role did nanoscale materials play in biological development? For example, one would ask at what size scale life first appeared. Would it be, perchance, at the nanoscale? Secondly, were functional nanomaterials the gatekeepers of the threshold between single- and multicellular organisms? Is mitosis itself a process driven by surface area considerations? Large prganisms have a small external surface-to-volume ratio but a huge internal surface-to-volume ratio if all components are summed. Life is all about energy and entropy. For some reason, living things have been able to exploit the sun's energy to form systems that are not clearly favored by entropy (more about this topic in chapter 8). Where do entropic considerations fit into life processes and what role does nano play there?

We have placed a picture of a *Morpho* butterfly on the cover of this text. Its phenotypic expression is a direct result of nanoscale structure and function. Nature has accomplished nano like nobody's business, and we hope you learn much from this great master. We call the study of nature *nanoscience*. From nature's perspective, it is *nanotechnology*.

Due to the accelerated rate of technological development, societal implications of nanoscience and nanotechnology must be considered before, during, and after you write your first research paper. With instant worldwide communication, rapid transport of people and materiel around the globe, and the "accelerated rate of technological development," we can ill afford to neglect consideration of both the fruits or lemons of our nanolabor! As you read and, hopefully, study this text, ask the question, "What is the potential of these things I read about?" Then, as you conduct your own research, ask the question, "What is the potential of these things I am creating?" It all begins with you; where it ends should not be a guessing game.

Acknowledgments

We would like to acknowledge the National Nanotechnology Initiative and its director Mihail C. Rocco. He has worked relentlessly for the past several years to make nanotechnology relevant. His emphasis on societal implications in particular has placed such issues at the forefront of an emerging technology like never before. For more information about the NNI, please refer to www.nano.gov.

References

1. R. E. Smalley, Our energy challenge, presentation, *27th Illinois Jr. Science & Humanities Symposium* (2005).
2. *The American heritage dictionary of the English language*, Houghton Mifflin, New York (1969).
3. *Webster's encyclopedic unabridged dictionary of the English language*, Random House, New York (1994).
4. *CRC handbook of chemistry and physics*, 66th ed., CRC Press, Boca Raton, FL (1985).
5. P. McFedries, *IEEE Spectrum Online*/spectrum.ieee.org/apr05/110 (2005).
6. Physics, In *Mechanics of solids*, National Council of Educational Research and Training, India, http://ncert.nic.in/textbooks/, chap. 9 (2005).
7. I. Asimov, *Asimov's biographical encyclopedia of science and technology*, Doubleday & Company, Inc., New York (1982).
8. J. Barnes, Reason and necessity in Leucippus, *Proceedings of the First International Congress Democritus, Xanthi*, 141–158 (1984).
9. R. Boyle, *Sceptical chymist*, Henry Hall, London (1661).
10. J. Dalton, J. Gay-Lussac, and A. Avogadro, *Foundation of the molecular theory: Comprising papers and extracts*, The University of Chicago Press, Chicago, 5–7 (1906).
11. H. E. Roscoe, *John Dalton and the rise of modern chemistry*, MacMillan & Co., New York (1895).
12. D. J. Barber and I. C. Freestone, An investigation of the origin of the color of the Lycurgus cup by analytical transmission electron microscopy, *Archaeometry*, 32, 33–45 (1990).
13. F. E. Wagner, S. Haslbeck, L. Stievano, S. Calogero, Q. A. Pankhurst, and K.-P. Martinek, Before striking gold in gold-ruby glass, *Nature*, 407, 691–692 (2000).
14. F. Mehlman, *Phaidon guide to glass*, Prentice Hall, Englewood Cliffs, NJ (1983).
15. M.-C. Daniel and D. Astruc, Gold nanoparticles: Assembly, supramolecular chemistry, quantum-size-related properties, and applications toward biology, catalysis and nanotechnology, *Chemistry Reviews*, 104, 293–346 (2004).
16. H. H. Heicher, *Aurum Potabile oder Gold Tinstur*, J. Herbord Klossen, Breslau, Lepzig (1718).
17. J. Ayers. In *Ceramics of the world: From 4000 BC to the present*, L. Camusso, and S. Bortone, eds., Abrams, New York, p. 284 (1992).
18. H. Zhao and Y. Ning, Techniques used for the preparation and application of gold powder in ancient China, *Gold Bulletin*, 33, 103 (2000).
19. *Encyclopedia Britannica*, Colloid, Vol. 6, Encyclopedia Britannica, Inc. (1970).
20. T. Graham, On the properties of silicic acid and other analogous colloidal substances, *Journal of the Chemical Society of London*, 17, 318–323 (1864).
21. M. Faraday, Experimental relations of gold (and other metals) to light; the Bakerian lecture, *Philosophical Transactions of the Royal Society of London*, 147, 145–181 (1857).
22. G. Mie, Beiträge zur Optik trüber Medien speziell kolloidaler Metallösungen, Leipzig, *Annals of Physics*, 25, 377–445 (1908).
23. J. C. Maxwell-Garnett, Colours in metal glasses and metal films, *Philosophical Transactions of the Royal Society A*, 203, 385–420 (1904).
24. J. C. Maxwell-Garnett, Colours in metal glasses, in metallic films and in metallic solutions—II, *Proceedings of the Royal Society of London*, 76, 370–373 (1905).
25. *Encyclopedia Britannica*, Photography, Vol. 17, Encyclopedia Britannica, Inc. (1970).
26. photography.about.com/library/weekly/aa010702a.htm
27. http://catalyse.univ-lyon1.fr/0catal.htm
28. *Encyclopedia Britannica*, Catalysis, Vol. 5, Encyclopedia Britannica, Inc. (1970).
29. P. Sebatier, *La catalyse en chimie orgarnique* (first published in 1913, with a second edition in 1920; English translation by E. E. Reid) (1923).

30. P. Sabatier, nobelprize.org/chemistry/laureates/1912/sabatier-bio.html

31. J. R. Rostrup-Nielsen, Equilibria decomposition reactions of carbon monoxide and methane over nickel catalysts, *Catalysis*, 27, 343–356 (1972).

32. W. H. Brattain, Laboratory notebook of December 24, 1947, Press Release, Bell Telephone Laboratories, New York (1948).

33. G. Thompson, Silicon nanoelectronics, INTEL, presentation, Colorado Nanotechnology Summit, Boulder, CO, May (2004).

34. G. E. Moore, Cramming more components onto an integrated circuit, *Electronics*, 38, April 19 (1965).

35. G. L. Hornyak, *Characterization and optical theory of nanometal—Porous alumina composite membranes*, Ph.D. dissertation, Colorado State University (1997).

36. M. Ohring, *Materials science of thin films*, 2nd ed., Academic Press, London (2002).

37. B. Franklin, Letter from Benjamin Franklin to William Brownrigg (1773) www.njsas.org/projects/atoms/monolayers/hist.php

38. P. Becker, History and progress in the accurate determination of the Avogadro constant, *Reports on Progress in Physics*, 64, 1945–2008 (2001).

39. edisonexploratorium.org/bio/blodgett.htm

40. C. D. Bain, E. B. Troughton, Y.-T. Tao, J. Evall, G. M. Whitesides, and R. G. Nuzzo, Formation of monolayer films by the spontaneous assembly of organic thiols from solution onto gold, *Journal of the American Chemical Society*, 111, 321–335 (1989).

41. C. D. Bain and G. M. Whitesides, Formation of monolayers by the coadsorption of thiols on gold: Variation in the head group, tail group and solvent, *Journal of the American Chemical Society*, 111, 7164–7175 (1989).

42. C. D. Bain and G. M. Whitesides, Modeling organic surfaces with self-assembled monolayers, *Angewandte Chemie International Edition, England*, 28, 506 (1989).

43. H. W. Kroto, J. R. Heath, S. C. Obrien, R. F. Curl, and R. E. Smalley, C-60—Buckminsterfullerene, *Nature*, 318, 162–163 (1985).

44. S. Iijima, Helical microtubules of graphitic carbon, *Nature*, 354, 56 (1991).

45. S. Iijima and T. Ichihashi, Single shell carbon nanotubes of one nanometer diameter, *Nature*, 363, 603 (1993).

46. R. Bacon, Growth, structure, and properties of graphite whiskers, *Journal of Applied Physics*, 31, 283–290 (1960).

47. G. G. Tibbetts, *Journal of Crystal Growth*, 66, 632 (1984).

48. R. G. Finke, Transition—Metal nanoclusters. In *Metal nanoparticles: Synthesis, characterization and applications*, chapter 2, D. L. Feldheim, and C. A. Foss, eds., Marcel Dekker, Inc., New York (2002).

49. D. L. Feldheim and C. A. Foss, *Metal nanoparticles: Synthesis, characterization and applications*, Marcel Dekker, Inc., New York (2002).

50. M. Jose-Yacaman, L. Rendon, J. Arenas, and M.C.S. Puche, Maya blue paint: An ancient nanostructured material, *Science*, 273, 223–225 (1996).

51. G. Schmid, Large clusters and colloids: Metals in the embryonic state, *Chemical Reviews*, 92, 1709–1727 (1992).

52. G. Schmid, ed., *Clusters and colloids: From theory to applications*, VCH, New York (1994).

53. G. Schmid and G. L. Hornyak, Metal clusters: New perspectives in future nanoelectronics, *Current Opinion in Solid State Materials Science*, 2, 204 (1997).

54. G. L. Hornyak, M. Kroell, R. Pugin, T. Sawitowski, G. Schmid, J-O. Bovin, H. Hofmeister, and S. Hopfe, Gold clusters and colloids in alumina membranes, *Chem. Eur. J.*, 3, 1951 (1997).

55. J. Zheng, C. Zhang, and R. M. Dickson, Highly fluorescent, water-soluble, size-tunable gold quantum dots, *Physics Review Letters*, 93, 077402 (2004).

56. www.qdots.com (Quantum Dot Corporation)

57. W. D. Lawson, F. A. Smith, and A. S. Young, Influence of crystal size on the spectral response limit of evaporated PbTe and PbSe photoconductive cells, *Journal of the Electrochemical Society*, 107, 206–210 (1960).

58. B. I. Shklovski and A. l. Efros, Interband absorption of light in strongly doped semiconductors, *Soviet Physics Journal of Experimental and Theoretical Physics*, 32, 733–738 (1971).

59. A. Nozik, Photoelectrochemistry applications to solar energy conversion, *Annual Review of Physical Chemistry*, 29, 189–222 (1978).

60. A. I. Ekimov and A. A. Onuschenko, Quantum size effect in three-dimensional microscopic semiconductor crystals, *Journal of Experimental and Theoretical Physics Letters*, 34, 346–349 (1982).

61. L. E. Brus, A simple model for the ionization potential, electron affinity, and aqueous redox potentials of small semiconductor crystallites, *Journal of Chemical Physics*, 79, 5566–5571 (1983).

62. A. I. Ekimov and A. I. Efros, Quantum size effect in semiconductor microcrystals, *Solid State Communications*, 56, 921–924 (1984).

63. M. Graetzel, Artificial photosynthesis: Water cleavage into hydrogen and oxygen by visible light, *Accounts of Chemical Research*, 14, 376–384 (1981).

64. G. Binnig, H. Rohrer, Ch. Gerber, and E. Weibel, Surface studies by scanning tunneling microscope, *Physics Review Letters*, 49, 57 (1982).

65. G. Binnig, C. F. Quate, and Ch. Gerber, Atomic force microscope, *Physics Review Letters*, 56, 930 (1986).

66. D. M. Eigler and E. K. Schwiezer, Positioning single atoms with a scanning tunneling microscope, *Nature*, 344, 524–526 (1990).

67. D. M. Eigler, C. P. Lutz, and W. E. Rudge, An atomic switch realized with the scanning tunneling microscope, *Nature*, 352, 600 (1991).

68. R. P. Feynman, There's plenty of room at the bottom: An invitation to enter a new field of physics, Annual Meeting, American Physical Society/California Institute of Technology, December 29, 1959, California Institute of Technology, *Engineering and Science Magazine*, 23, 22, (1960).

69. *Time Magazine*, 135, 16 (April 16, 1990).

70. M. A. O'Keefe, L. F. Allard, S. J. Pennycock, and D. A. Bloom, Transcending the one-ångstrom atomic resolution barrier in the TEM, *Microscopy Microanalysis*, 12, 162–163 (2006).

71. G. L. Hornyak, S. Peschel, T. Sawitowski, and G. Schmid, TEM, STM and AFM as tools to study clusters and colloids, *Micron*, 29, 183 (1998).

72. W. C. Röntgen, Biography, NobelPrize.org

73. M. von Laue, Biography, NobelPrize.org

74. William Bragg and Lawrence Bragg, Biography, NobelPrize.org

75. M. Fleischman, P. J. Hendra, and A. J. McQuillan, Raman spectra of pyridine. Adsorbed at a silver electrode, *Chemical Physics Letters*, 26, 163 (1974).

76. M. Moskovits, Surface-enhanced spectroscopy, *Review of Modern Physics*, 57, 783–826 (1985).

77. K. A. Bosnick, J. Jiang, and L. E. Brus, Fluctuations in local symmetry in single-molecule rhodamine 6g Raman scattering on silver nanocrystal aggregates, *Journal of Physical Chemistry B*, 106, 8096–8099 (2002).

78. K. Kneipp, G. R. Harrison, S. R. Emory, and S. Nie, Single molecule Raman spectroscopy: Fact or fiction? *Chimia*, 53, 35–37 (1999).

79. E. C. Le Ru, M. Meyer, and P. G. Etchegoin, Proof of single-molecule sensitivity in surface enhanced Raman scattering (SERS) by means of a two-analyte technique, *Journal of Physical Chemistry B*, 110, 1944–1948 (2006).

80. A. Senefelder, *Vollstandiges Lehrbuch der Steindruckery* (1819) (www.lib.edel.edu/ud)

81. Theory, modeling and simulation, research and development supporting the next Industrial Revolution, Supplement to the President's FY 2004 Budget National Nanotechnology Initiative (2004).

82. R. Hooke, *Micrographia: Some physiological descriptions of minute bodies made by magnifying glasses with observations and inquiries thereupon*, The Royal Society of London, J. Martyn and J. Allestry (1665).

83. J. H. Müller, The gene as the basis of life, *Proceedings of the International Congress of Plant Science*, 1, 897–921 (1926).

84. M. T. Cuberes, R. R. Schlitter, and J. K. Gimzewski, Room temperature—Repositioning of individual C60 molecules at Cu steps: Operation of a molecular counting device, *Applied Physics Letters*, 69, 3016–3018 (1996).

85. G. M. Whitesides, The art of building small. In *Understanding nanotechnology*, Scientific American, Time-Warner Book Group, New York (2002).

86. G. M. Whitesides, The once and future nanomachine, *Scientific American*, 285, 78–83 (September 2001).

87. P. Binks, The challenges facing nanotechnology. Interviewed by R. Williams, *Ockham's Razor*, www.abc.net.au/rn/science/ockham/stories/s1304778.htm (2005).

88. M. Meyyappan, Nanotechnology: The next frontier, presentation, NASA Ames Research Center, Moffett Field, CA.

89. M. Roukes. Plenty of room indeed. In *Understanding nanotechnology*, Scientific American, Time-Warner Books (2002).

90. G. Cao, *Nanostructures & nanomaterials: Synthesis, properties and applications*, Imperial College Press, Singapore (2004).

91. National Nanotechnology Initiative, nano.gov

Problems

1.1 Scientists have made deductions indicating the presence of nanoparticles without actually having seen them. The experiments of Graham and Faraday have done as much. What other examples of such experiments (other than those mentioned in the text) can you think of that imply the existence of nanoparticles before the advent of the electron and atomic microscopes?

1.2 There are numerous examples of nano-material-based phenomena all around us, from both technology and nature. Colored materials, for example, have a history of participation of nanomaterials. Can you figure out the basics of the color-generating component of the following (feel free to use the Internet as a research tool)? Not all are based on nanophenomena.

a. Colored titanium jewelry
b. Colored anodized aluminum siding for skyscrapers
c. White/green gold leaf
d. Origin of the color of ruby-colored glass vases in your grandmother's dining room cabinet? What is the origin of the color for the blue-colored glass vases?
e. Colors of bubbles derived from soapy films
f. White-colored titanium dioxide particles
g. Color imparted to a male peacock's feathers
h. White of the edelweiss (*Leontopodium nivale*), an alpine flower that lives at high altitudes, up to 3000 m/10,000 ft, where UV radiation is strong
i. Color of the blue *Morpho* butterfly found in Central and South America
j. Color of emerald, sapphire, and ruby minerals
k. Color of green leaves of plants
l. Color of the sky at noon and at sunset
m. Color of opaline materials like pearls
n. Color of flames

How many of these are based on nanophenomena?

1.3 Without being able to see enzymes, how did chemists and biologist know that enzymes existed and that they performed metabolic functions? How were enzymes discovered?

1.4 What is the nature of the moiety that requires "fixing" in a photographic process? Why is fixing required?

1.5 Why would a nanobrick type of layered structure in the shell of the abalone aid in its survival? The bricks are fused laterally into the shell and there is a biological polymeric adhesive between layers.

1.6 Using only your chemical or physical intuition, provide an empirical explanation for Faraday's observation about the different colors that solutions containing gold particles assume "whether ruby, green, violet or blue." Why would the addition of gelatin or salts prevent the change in color of the colloid solutions?

1.7 What would happen to metabolism in general if all enzymes would suddenly acquire micron-sized dimensions? How would you mitigate the consequences?

1.8 Do you consider nanotechnology to be the next industrial revolution or is it simply a natural progression towards making things smaller?

1.9 What are your views about how nanoscience–nanotechnology should be taught? Should there be degree programs that are anchored in nanoscience and/or technology? Or should the subjects be included in core courses?

1.10 The digital camera all but displaced the photographic film: one nanotechnology displacing another. How many other disruptive technologies have occurred over the years that you are able to recall?

1.11 More nanogames:

    a. How many nanotetrahedra can be packed into 1 m³? Assume that the all sides of the tetrahedra are 10 nm.

    b. How many C_{60} would it take to fill a 10-μm long single-walled carbon nanotube of 1.5 nm diameter?

    c. Assuming van der Waals radii, how many hydrogen atoms can be squeezed into a length of 1 nm?

    d. How many yoctameter-sized cubes can fit into a cubic nanometer?

1.12 What wavelength of probing radiation is required to inspect objects or phenomena on the order of the size of

    a. Airplanes
    b. Paramecium
    c. Gold clusters (~2 nm diameter)
    d. Humans
    e. Rotation of water molecules
    f. Chemical bonds
    g. Protons
    h. Crystal planes
    i. Viruses
    j. Chromosomes

1.13 Do you believe that nanobots able to clean plaque from your artery walls will ever be developed?

1.14 Nature's incredible use of nanomaterials and bottom-up synthesis is exhibited all around and within us. Give an example of a process, starting with molecules and ending with a macroscopic structure (e.g., tendon, bone, muscle). Describe the levels of hierarchy.

1.15 Why are nanomaterials inherently unstable? Why are they stable in living systems?

1.16 Which group—chemists, engineers, physicists, or biologists—do you think made the greatest contribution to nanoscience?

1.17 The space elevator project is an effort that intends to place an elevator system grounded on Earth into geosynchronous orbit. Carbon nanotubes are the only material strong enough to make this a reality. What is it about carbon nanotubes that make people think this way? Do you think a space elevator is possible? What are some potential issues?

1.18 Give a 5-min. space elevator speech about nanotechnology.

1.19 How would you teach a course in nanoscience and/or nanotechnology? Is this an important concern?

1.20 List some characteristics of nanomaterials that may be considered to be phenomenal.

1.21 Propose another scenario for the invention of soap.

1.22 Repeat example 1.1 but assume tubes have two-dimensional coordination number of 4. Repeat again with coordination number of 6. What is the number of tubes able to be packed in the hair cylinder in these cases?

1.23 What is grey goo? Green goo?

1.24 What one individual do you think made the greatest contribution to nanoscience?

1.25 List as many women scientists as possible involved in nano in the past.

1.26 How are nanoscience and nanotechnology to be defined? (Question courtesy of Günter Schmid, Uni-Essen.)

SOCIETAL IMPLICATIONS OF NANO

Addressing societal and ethical issues is the right thing to do *and* the necessary thing to do. *It is* the right thing to do *because as ethically responsible leaders we must ensure that technology advances human well-being and does not detract from it. It is* the necessary thing to do *because it is essential for speeding technology adoption, broadening the economic and societal benefits, and accelerating and increasing our return on investment.*

PHILIP J. BOND
**Former undersecretary for technology,
U.S. Department of Commerce (2003) [1]**

*C*hapter 2

FIG. 2.0	*Former undersecretary of commerce for technology, Technology Administration, Phillip J. Bond. He began his career at the U.S. Department of Commerce on October 30, 2001, and recognized readily the importance of nanotechnology to the prosperity and security of this country.*

Source: U.S. Department of Commerce, Technology Administration, www.technology.gov.

THREADS

Definitions, the historical perspective, and the challenges facing nanotechnology from the scientific and manufacturing points of view have been briefly introduced in chapter 1. We now delve into topics that address formidable challenges other than those posed by pure science and technology. In this chapter, we grapple with the complex societal issues that inevitably accompany any new technology, focusing on those that are unique to nano. The world has changed much since the straightforward era of hunting and gathering when choices were ruled by day-to-day-survival. Now, with billions of people living on the planet and communication occurring instantly and globally, we need to be prepared for the ramifications of a new type of technology that is already changing our way of life. In this chapter, we

are fortunate to compile the opinions of five experts that present, discuss, and summarize a variety of issues that face the advent of nanotechnology.

The mission of this chapter is awareness. We hope that message stays with you while you read this text—to reflect, contemplate, and formulate and to look into the past, assimilate the present, and project into the future. Societal implications are broad, complex, and highly integrated. For example, specialized groups already protest the potential of nano. Please supplement information gleaned from this chapter with your own outside reading. Stay informed! As the butterfly pictured on the cover of this text flutters its wings, what are the consequences of that action elsewhere? Interestingly, our choices are still ruled by survival.

2.0 INTRODUCTION TO SOCIETAL ISSUES

This chapter is intended to provide a brief introduction to a broad array of issues that arise as a result of, or in response to, new and potentially disruptive technologies such as nanotechnology and the various other technologies it may enable. Science and technology do not exist in a vacuum separate and apart from the society into which they are introduced. Rather, science and technology influence key dimensions of the larger society, including law, politics, economics and business, public health and safety, national security, education, and popular culture. Science and technology have long challenged and contributed to the evolution of our Western philosophical worldview, altering cultural values, attitudes, beliefs, and practices along the way. Finally, science and technology have become strongly embedded in our assessment of the human condition and in our vision and aspirations for the human future.

As noted in chapter 1, nanoscience and nanotechnology hold the potential for rapid and dramatic changes in our technological capabilities, raising significant practical and moral opportunities and concerns. In this chapter, we focus on the particular dimensions of ethics, law, environment, public perception, and technological forecasting to illustrate the complex relationship between advances in nanotechnology and the larger society. Roco and Bainbridge of the National Nanotechnology Initiative (NNI) stated in 2001:

> Over the next 10–20 years, nanotechnology will fundamentally transform science, technology, and society. However, to take full advantage of opportunities, the entire scientific and technology community must set broad goals; creatively envision the possibilities for meeting societal needs; and involve all participants, including the general public, in exploiting them [2].

Great opportunities are generally accompanied by great challenges and such is the case with nanoscience and nanotechnology. Much has been written about the tremendous potential of nanotechnology; however, barriers exist that go beyond basic science and engineering. The magnitude of scientific advance predicted by Roco and Bainbridge also portends the emergence of a wide range of ethical and societal implications including challenges related to basic moral questions, legal and regulatory responses, environmental concerns, national and global economic and political impacts, public perception, and planning for a range of future scenarios.

2.0.1 Societal Implications—The Background

To understand the current context of societal implications research, one must travel back in time to the launch of the Human Genome Project (HGP) in October of 1990 [3]. The HGP was an effort to identify and map all of the human genes, as well as determine the complete sequence of human DNA. Once available, the fully mapped human genome would provide rich new ground for biological study and potential medical advances. However, the project was controversial on many fronts. First, there were many *naysayers* who believed the project was not feasible and amounted to an enormous waste of time and resources. Other critics were concerned with how genetic knowledge would be used, how it might affect human self-perception, and what potential harms could come from the ability to manipulate and control human genetics at its most basic level. In fact, many of the same general societal implications of the HGP pertain to nanotechnology, including matters related to privacy, intellectual property and patent questions, health and environmental concerns, and moral dilemmas related to human bioengineering and commercialization.

In response to these critics, the project identified a key research area that came to be known as ELSI, or ethical, legal, and social issues research. The U.S. Department of Energy (DOE) and the National Institutes of Health (NIH) devoted between 3% and 5% of their total annual HGP budget to the study of ethical, legal, and social issues associated with the availability of genetic information. The ELSI model is now applied around the world with respect to a variety of scientific endeavors. Research continues and the ELSI Website of the HGP has become a clearinghouse for developments in these areas.

Similar to the precedent set by the HGP, and in anticipation of the enormous potential of nanoscience and nanotechnology to impact virtually every realm of human existence, the National Science Foundation (NSF) sponsored a workshop in 2001 from which a report was published. The report, *Societal Implications of Nanoscience and Nanotechnology* [2], is a collection of expert opinions and recommendations on how best to "(a) accelerate the beneficial use of nanotechnology while diminishing the risks, (b) improve research and education, and (c) guide the contributions of key organizations."

The first recommendation reads as follows:

> Make support for social and economic research studies on nanotechnology a high priority. Include social science research on the societal implications in the nanotechnology research centers, and consider creation of a distributed research center for social and economic research. Build openness, disclosures, and public participation into the process of developing nanotechnology research and development program direction.

Additional recommendations address the need to inform, educate, and involve the public; create an infrastructure for interdisciplinary evaluation of the scientific, technological, and social impacts in the short, medium, and long terms; and educate a workforce.

In 2005, the President's Council of Advisors on Science and Technology (PCAST) released the first major progress report on the NNI, entitled *The National Nanotechnology Initiative at Five Years: Assessment and Recommendations of the National Nanotechnology Advisory Panel (President's Council of Advisors on Science and Technology, 2005)* [4]. With respect to societal concerns, 8% of the total NNI budget for 2006, or about $82 million, was requested for societal implication activities, sometimes referred to as SEIN (societal and ethical implications of nanotechnology). Approximately $38.5 million was earmarked for programs working on environmental, health, and safety research and development. The remaining $42.6 million was split between education-related activities targeted at workforce development, as well as public understanding and acceptance, and research on the broad implications of nanotechnology for society, including economic impacts, barriers to adoption, and ethical issues, particularly as related to research priorities [5]. All NNI-funded nanotechnology research must have a societal dimensions component that addresses one or more of the preceding categories.

Presently, the societal dimensions program component area (PCA) of the NNI encourages and funds research initiatives in three broad areas: (1) environmental, health, and safety (EHS) impacts of nanotechnology development and risk assessment of such impacts; (2) education-related activities, such as the development of materials for schools and universities as well as public outreach; and (3) identification and quantification of the broad implications for society. In this last category are included social, economic, workforce, educational, ethical, and legal implications (please consult the National Nanotechnology Initiative website at nano.gov) [6].

In 2007, there was a shift in funding, with approximately $44 million earmarked for EHS programs and $38 million for the combined areas of education and all other societal implications. Among SEIN researchers and other observers, the EHS allocation is somewhat controversial. For example, one might expect environmental, health, and safety concerns to be part of the basic cost of research and development funding as it is in general industry. Critics suggest this assignment of EHS to societal implications effectively reduces the budget for all other societal implications research to far less than what is needed, making SEIN research commitments largely symbolic, geared more towards managing public perception and support than minimizing negative impacts [7].

Regardless of motive, this unprecedented commitment to assessing the societal impacts of new technology does illustrate the breadth and complexity of the issues. At the time of this writing, an infrastructure for interdisciplinary collaboration is beginning to emerge. In reality, there is not much precedent for the idea that scientists and engineers should be working collaboratively with the social sciences and humanities to proactively assess societal impacts and ethical implications. However, National Science Foundation grants have funded large centers at Harvard University, University of South Carolina, Arizona State University, and the University of California at Santa Barbara in an effort to encourage collaboration and coordination of ongoing NSF-funded research initiatives across the country. The overall approach represents a fundamental change in the culture of science and technology, as well as its relationship to both policy making and public dialogue. As scientists and engineers, you will increasingly find yourselves collaborating with ethicists, social scientists, and others to address these larger implications of your work.

2.0.2 Breadth of Societal Implications

As discussed in earlier sections, the range of societal implications is quite broad, reaching into many, if not all, areas of society (**Fig. 2.1**). Berube attempts to define SEIN research as follows [7]:

> SEIN encapsulates research by toxicologists, ethicists, and futurists into a range of issues from environmental impacts, regulatory regimes, workplace and economic dislocations, bionanotechnology convergence, and transhumanism and posthumanism. It involves experts from philosophy, communication studies, law, and political science, as well as fiction studies and art. It also includes the less expert as well as the self-proclaimed technophiles, the social critic, and the crank.

FIG. 2.1 *Societal dimensions of nanotechnology. A single-walled carbon nanotube frames many aspects of nanotechnology within its hexagonal-carbon superstructure.*

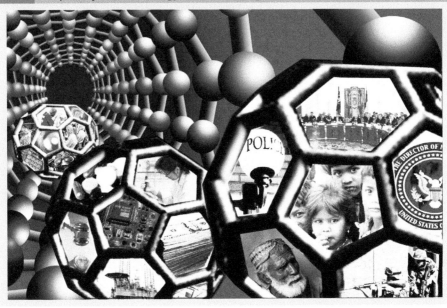

Source: Image reprinted with permission from R. M. Bennett-Woods.

Due to the sheer range of issues, there is no easy way to categorize and describe societal implications. Many, if not all, industries are anticipated to experience significant change in response to discoveries and applications of nanoscience. However, to fully understand the true extent of societal implications, one must also recognize connections branching into nearly every corner of the public sphere; general categories of societal impacts that have been raised as potential concerns range from safety and environmental impacts to workforce and global economic disruptions to controversial applications in medicine.

Any approach to societal implications research is further complicated by the dynamic complexity inherent in any societal shift of the magnitude suggested by nanoscience and its related technological offspring. One example of dynamic complexity is the potential for a relatively small change in one industry to become a large change in another industry, or for changes in the economic sphere to escalate rapidly into the political sphere. For example, nanoscience advances in the biotechnology industry have the potential to alter the very foundation of how medicine is practiced within the health care industry, affecting both the general biotechnology sector and the health care sector. Furthermore, disruptive changes in the health care sector that affect cost or access to services will intensify existing political pressures to reform the entire health care system.

To further illustrate the complex relationships among nano-enabled technological advances, nested within the relationship between the biotechnology and health care industries is the Massachusetts Institute of Technology *MIT Institute for Soldier Nanotechnologies*, which is supporting a number of projects that employ elements of biotechnology and medicine to "enhance soldier survivability" [8]. The combined goals of various projects include improving battlefield triage and the treatment of injuries while still on the battlefield, detection of biohazards, and even augmentation of a soldier's strength and physical agility. Plans are already underway to allow technologies developed for military use on the battlefield to find their way into commercial application within the health care industry and others. At least some of these innovations will likely spark controversy insofar as they raise issues of privacy, cost, access, and human enhancement. Consider, for example, the impact on the sports industry if military applications originally intended to protect soldiers on the battlefield become available as a means to improve endurance or mask injury in sports competitions.

The important point here is to emphasize the complex, highly integrated, and unpredictable nature of societal implications. Some advances will be relatively unproblematic while others will raise a range of concerns that may be difficult to sort out before questionable applications have already made their way into the marketplace. In general, societal implications of nanoscience in most industry sectors raise at least some potential concerns related to health and safety as well as many ethical dilemmas, legal and regulatory issues, barriers to public acceptance, workforce disruptions, education and training needs, and public policy questions.

2.0.3 Meet the Experts

Addressing the full breadth of SEIN is clearly beyond the scope of this introductory chapter. However, we will briefly focus on five specific dimensions of societal impact: ethics, law, environment, public perception, and future forecasting. Experts from these areas will provide an overview of issues that confront scientists, engineers, corporate leaders, policy makers, and community stakeholders involved in research and development of nanotechnology and its potential outcomes.

In the section entitled "Ethical Implications," Professor Deb Bennett-Woods of Regis University (**Fig. 2.2**) introduces a set of concepts from the field of ethics that provide a framework for broadly considering the moral dilemmas raised by nanotechnology in all other dimensions. She suggests that questions based on these concepts can become the basis for collaborative and informed dialogue among scientists, engineers, policy makers, business leaders, and community stakeholders.

"Legal Implications," by intellectual-property attorney Dr. Patrick Boucher (**Fig. 2.3**), a patent attorney for a technology law firm that handles intellectual property, explore the complex legal issues posed by a truly interdisciplinary science and technology [9]. This section takes the approach of illustrating the impact of two fundamentally different approaches to legal doctrine on the development and use of nanoscience. The role of intellectual property in providing inducements to nanoscience developers to share their results for the benefit of all is contrasted with various liability doctrines that attempt to restrain those same developers from prematurely allowing those results to be used without adequate consideration of safety issues. The author suggests a natural legal tension affecting the development of nanoscience that, between factors favoring rapid development and factors favoring a more cautious approach, is reflected in the way these factors interact with existing legal doctrine.

FIG. 2.2

Dr. Bennett-Woods is the chair of the Department of Health Care Ethics and the director of the Center for Ethics and Leadership in the Health Professions at Regis University in Denver, Colorado. She is also a fellow of the Center on Nanotechnology and Society at the Chicago-Kent College of Law. She is the author of Nanotechnology: Ethics and Society *(CRC Press, 2008). Her particular interests are with ethical issues arising from emerging technologies in health care and the convergence of technologies such as nanotechnology, biotechnology, artificial intelligence, cognitive science, and robotics.*

Source: Image reprinted with permission from R. M. Bennett-Woods.

Dr. Jo Anne Shatkin (**Fig. 2.4**), of CLF Ventures, Inc., in the section entitled "Environmental Implications," provides an overview of the process and importance of risk assessment as it relates to nanotechnology. Potential environmental impacts and important questions regarding nanotoxicology are explored.

One specific area of focus for proponents of the NNI has been public understanding and acceptance of nanotechnology, particularly in the face of media hype and science fiction accounts of *gray goo* and *nanorobots* run amuck. In the section entitled "Public Perception," Professor Susanna Priest (**Fig. 2.5**) of the University of Nevada, Las Vegas addresses what the public really thinks

FIG. 2.3 *Dr. Patrick M. Boucher holds a PhD in physics (Queen's University, Canada) and a J.D. (Touro College, United States). His technical publications have been in the areas of condensed-matter, nuclear, and astrophysics. For several years, he was associated with Physical Review B, which publishes much of the world's technical nanotechnology research and where he acted as associate editor and managed the journal's scientific editorial staff. He is currently a patent attorney practicing in Denver, Colorado, as a partner of Townsend and Townsend and Crew LLP.*

and what, if any, barriers exist to widespread support and adoption of nano-related products.

Finally, in the last section, "The Future of Nanotechnology," futurist Dr. Thomas Frey (**Fig. 2.6**) of the DaVinci Institute in Boulder, Colorado, introduces his brand of forecasting and introduces us to 10 endpoints of nanotechnology that will influence the society of the future.

FIG. 2.4

Dr. Jo Anne Shatkin is managing director of CLF Ventures, a nonprofit group that helps organizations develop complex projects with demonstrable environmental and economic benefits. She provides technical and strategic expertise on projects and is a recognized expert in strategic environmental initiatives, human health risk assessment, technical communication, and environmental aspects of nanotechnology. Before joining CLF Ventures, she managed a human health risk assessment and nanotechnology practice for the Cadmus Group. She brings a wealth of experience in research and application of quantitative human health risk assessment for site redevelopment and remediation, drinking water and air quality, and environmental evaluations of emerging contaminants. Her specialty is the application and communication of innovative science-informed analysis to address complex emerging issues affecting businesses and communities. She has been an active member of the Society for Risk Analysis since 1989, and recently founded the Emerging Nanoscale Materials Specialty Group of the Society for Risk Analysis.

2.0.4 The Nano Perspective

Clearly, the implications for nanoscience and nanotechnology go well beyond the boundaries of the science itself. Renowned science fiction author Asimov once said [10] that "the saddest aspect of life right now is that science gathers

FIG. 2.5

Professor Susanna Hornig Priest is professor in the Hank Greenspun School of Journalism and Media Studies at the University of Nevada, Las Vegas, and is the editor of the journal Science Communication. Previously she served as director of research for the College of Mass Communications and Information Studies at the University of South Carolina and as director of the science journalism master's program at Texas A&M University. Her own research focuses on public and media responses to controversial issues in science and technology, a subject on which she has published many dozens of research articles, book chapters, and unpublished reports. She is also the author of two research methods textbooks for media studies students.

knowledge faster than society gathers wisdom." Take a moment to reflect on what this means for you, as a student of nanoscience and nanotechnology. What wisdom is needed to ensure that scientific advances uniformly guarantee progress and improvements in the human condition rather than making existing social problem worse or creating new ones? And, of equal importance, what is

FIG. 2.6

Dr. Thomas Frey is the executive director and senior futurist at the DaVinci Institute. He is currently Google's top-rated futurist speaker, working with CEOs and executives at top companies from around the world. His talks are known for continually pushing the envelope of understanding, creating fascinating images of the world to come. Before launching the DaVinci Institute, Tom spent 15 years at IBM as an engineer and designer where he received over 270 awards, more than any other IBM engineer.

your responsibility as an agent of scientific and technological change for ensuring that the broader social impacts of your work have been considered? Keep this in mind as you explore the thoughts of various SEIN experts in the following sections.

2.1 ETHICAL IMPLICATIONS

The term *nanoethics* has begun to appear in the SEIN literature. Whether this term, which translates to *ethics of the extremely small*, catches on or not remains to be seen. However, the broad discipline of ethics does provide a unique and useful lens for framing the diverse range of potential societal effects of nanotechnology with a common language. As such, it is a good starting point for the discussion of societal impacts.

2.1.1 Ethics in the Context of Research and Applied Science

Ethics can be loosely defined as the general study of morals and moral choices. The field encompasses an array of theories, concepts, and methods for conducting moral analysis and providing insight into moral dilemmas. Ethical problems arise when it is not clear what action is the right one to take in particular circumstance. The classic definition of a moral or ethical dilemma is a situation in which *doing something right* also results in *doing something wrong*.

Perhaps the best analogy to the ethical dimensions of nanotechnology can be drawn from the realm of human-subject research in medicine. Modern medicine has come a long way from its primitive roots in which patients were subjected to largely untested, at least by scientific standards, folk remedies, bloodletting, and various cultural rituals. The introduction of modern sanitation, nutrition, and antibiotics has largely shifted the focus of medicine in the developed world from the control of infectious disease to complex management of acute trauma and noninfectious diseases such as diabetes and cancer. However, nearly every major advance in medical science has come as a direct result of research on living human subjects. Such research can often be risky and most subjects never directly benefit from their participation. Abuses in human-subject research, such as the infamous Tuskegee study, in which 400 African American men with untreated syphilis were observed for 40 years in order to document the natural progression of the disease, have led to a set of commonly accepted procedural standards based on ethical principles that safeguard the rights of patients.

Three such principles are formally established in a document entitled the *Belmont Report* produced by the National Commission for the Protection of Human Subjects of Biomedical and Behavioral Research in 1979 [11]. The first principle, *respect for persons*, is the basis for the concept of patient autonomy (self-determination or self-governance) and requires guidelines for obtaining informed consent and the protection of vulnerable populations such as children. The second principle, *beneficence*, obligates researchers to avoid needless harm by minimizing any inherent risks while also attempting to maximize the potential benefits to subjects. *Justice*, the third principle, examines more broadly how likely the benefits and burdens of the research are to be fairly distributed. Using the Tuskegee study as an example, all three principles were violated. The subjects were misled into believing they were being treated and so never gave informed consent; they were needlessly harmed since a cure was known and available; and they bore the full burden of the research with no benefits to themselves or even to science since the stages of syphilis were relatively well documented at the time of the study.

Fascinating, but how does this relate to the societal implications of nanotechnology? Commentators in nanotechnology have actually drawn an analogy between the potential impacts of nanotechnology and human-subject research [12,13]. For example, when exploring the societal implications of nanotechnology, Sarewitz and Woodhouse [12] describe the "unfolding revolution as a grand experiment—a clinical trial—that technologists are conducting on society." The direct implication here is that the considerations and review processes used for clinical trials in human-subject research may also be applicable to the various uncertainties posed by nanotechnology.

Naturally, there are no formal mechanisms in place to accomplish the sort of review and oversight mandated in human-subject research. However, there are key decision points in the continuum of science and technology research and development at which ethical principles can be applied [14]. For example, should we fund new or continued research and development? With increasingly constrained budgets for research and development, we may be obligated to prioritize some research initiatives over others on the basis of a greater good. Should we transfer knowledge developed for one purpose in one industry sector to another? A common example is the transfer of technologies originally developed for military purposes into the private sector. The navigation system in your new car has its origins in military applications of global positioning technology. This raises a legitimate question of fairness and who should benefit from research conducted at the public expense. Finally, do we regulate? This question is often posed only after serious environmental, health, and safety effects have been observed in a product already on the market. Several recent and widely publicized cases of drugs being withdrawn from the market due to serious safety concerns are examples of the value of regulatory assessment and oversight in promoting consumer safety and informed consent in the face of corporate interests.

Although the principles outlined in the Belmont Report are narrowly directed at protecting the individual human subject, they can be expanded to consider broader societal impacts, providing a framework within which ethical analysis and dialogue can occur. The principle of respect for persons can be expanded to the principle of respect for communities, while the principles of beneficence and justice can likewise be respectively broadened to incorporate the common good and a broader conception of social justice [14]. Let us explore each of these individually.

2.1.2   *Principle of Respect for Communities*

Just as the principle of respect for persons requires that each person be treated with dignity and as an autonomous agent capable of informed decision making in his or her own best interests, so too communities have the ability to act as autonomous, self-governing agents. Naturally, this is much easier said than done, as noted by Sarewitz when he characterizes the "compulsory nature of technological assimilation" as a central cause of social change. In other words, most technology has a tendency to be widely introduced and accepted long before most of us have given much thought to any potential downsides [15].

There are many barriers to a communal informed consent. First, there are few effective mechanisms for timely public engagement in the development and introduction of new technologies. And, even if there were effective forums for public dialogue, informed consent requires that citizens have the information needed to conduct a reasoned deliberation. In a rapidly paced and highly competitive marketplace, the interests of researchers and corporations are not always well served by being too forthcoming about the technical nature of their developments and, in particular, any untoward hazards. Again, even if the information were available, communities marginalized by poverty and illiteracy are not likely to receive it or take advantage of public forums to express their views. Nonetheless, there have been examples of public backlash against the introduction of new technologies, and the NNI's strong emphasis on research initiatives related to public education and perceptions is directly intended to help manage the issue of community consent. Consider the recent rejection of genetically modified organisms (GMOs) by the European Union.

In a nano-related example, let us examine the question of privacy. The development of sophisticated inventory tracking devices, nanosensors for military surveillance, neural implants, and nanobiosensors for detecting biological and environmental threats has raised concerns about individual privacy and related civil liberties [16–18]. The principle of respect for communities would raise a number of basic questions when considering the development of these devices:

- Do potential violations of privacy undermine or enhance human dignity?
- Do these devices have the potential to violate fundamental human rights, including privacy, freedom of conscience, and other basic liberties?
- Are there sufficient information and an effective mechanism available to the public regarding the potential outcomes so that informed and meaningful community consent can occur?
- Are there certain populations that are more vulnerable or under-represented than others, and are there sufficient protections to ensure that vulnerable populations within larger communities have a meaningful voice in the consent process?
- Are there barriers to consent being competent and voluntary?

By periodically posing these questions at those key decision points during the nanotechnology (NT) development process and targeting the most questionable applications for more careful scrutiny and planning, potential problems can more effectively be anticipated and solutions developed or products abandoned. On the other hand, failure to address these issues effectively can lead to widespread public rejection of such technologies, including those that offer great potential benefits to the society as a whole.

2.1.3 *Principle of the Common Good*

The principle of the common good obligates us to act in ways that respect shared values and promote the common good of communities. On one level, this involves an obligation to promote the well-being of individuals and groups specifically by avoiding harm, minimizing risks, and maximizing benefits. On another level, the concept of a common good is a bit more complicated, but basically encourages actions that respect those shared values that allow us to live effectively in community. In the communitarian traditions of political science and ethics, primary attention is given to a balance between individual liberty and responsibility towards the community, including obligations of sustainability and justice between generations.

Defining the common good is very difficult to pin down in the larger scope of society. The first complicating factor is that, in a complex society, we function as many smaller communities nested within larger ones, often with competing interests and differing values. Second, in order to avoid *harms* you need to know what those *harms* are likely to be—something that is only a matter of informed speculation in the case of emerging technologies. If a technology results in serious environmental degradation or exposes certain populations to a high level of risk, you could argue that the common good has been violated. However, it is not always clear how to identify the greater harm when the impacts are more subtle or complex. How much unintended environmental degradation is acceptable if food production is greatly enhanced? Technology is often referred to as a two edged sword that results in opportunities or solves problems for some, while creating barriers or other problems along the way.

To complicate matters further, when applying this principle, you also need to agree on how to define and weigh various societal benefits with harms, not just in the short term but also in the long term. We often see this dynamic when we aggressively exploit natural resources, such as oil and gas, in the short term for commercial gain and convenience, leaving environmental degradation and limited resources for future generations to confront. And, again, we are faced with the question of how to accommodate vulnerable populations who are not likely to benefit or are actively harmed by the technology in question.

Moving beyond benefits and harms to the context of shared values is even more challenging. What are the shared values and best interests of society and what responsibilities do we owe one another? Where do we draw the line between the ideals of community and individual liberty? In other words, where do my liberties stop and the interests of the larger community kick in? A good example of a nano-related dilemma that may pit the interests of the individual against the common good of the community involves projected shifts in manufacturing and the workforce.

As noted earlier, a primary research focus of the NNI is workforce preparation. Nano-driven advances in business and industry will require a highly trained and specialized workforce that does not currently exist. While this opens the door of opportunity for a segment of the workforce with the means to retool itself, large numbers of manufacturing jobs are likely to be rendered obsolete or displaced [19]. Furthermore, the competition for global dominance in nanotechnology may continue the current trend of any new jobs moving offshore, depending on which countries take the scientific and technological lead in nano-innovation. As corporations position themselves to compete in the global marketplace, what obligations, if any, do they have to the local communities disrupted by rapid shifts in the job market and economic uncertainties? What constitutes an acceptable balance between corporate well-being and individual well-being? In a similar vein, the extraordinary investment in nanoresearch has, by necessity, reduced funding available for unrelated research interests, educational initiatives, and other funding priorities. When faced with scarce resources, every funding decision is a pragmatic choice, to be sure; however, every funding decision is also a statement of what the funding source values and this may or may not align well with the values of the various community stakeholders.

When considering the principle of the common good with respect to a particular issue, several additional questions can be posed:

- How are the values and priorities of communities represented and served?
- How might the values and priorities of communities be violated or undermined?
- What are the potential short-, medium-, and long-term benefits and burdens for individuals and communities?
- What are the most likely outcomes: positive, negative, or neutral?
- What unintended outcomes can be anticipated? What is the level of risk?
- What limitations or safeguards are prudent to prevent negative outcomes?

The principle of the common good essentially boils down to holding all members of the community responsible for the well-being of the larger society and the interests of individual community members. In the face of rapid or unprecedented technological innovation, effective consideration of the common good can allow for a mutually beneficial transition for all members of the larger society.

2.1.4 Principle of Social Justice

The final principle is that of social justice, obligating us to act in ways that maximize the just distribution of benefits and burdens within and among communities. Perhaps more than the first two principles, this principle calls us to think globally. The nature of major technological innovations is that they initially become available largely to those who can afford them. A secondary effect is that related outcomes, such as environmental degradation or political and economic disruptions, tend to result in the least advantaged segments of the global society experiencing the highest level of burden and risk, while receiving little or no benefit. The principle of justice asks us to always give primary consideration to alleviating the suffering of the most vulnerable among us. The sustainability movement provides an obvious example of just such an opportunity.

At least nine biotechnologies related to medicine and the environment have been identified as having the potential to greatly impact health in the developing world [20]. However, there are clear barriers to those technologies becoming available. One barrier is a matter of research priority and investment. The risk profiles of developed and developing countries differ significantly [21]. If research priorities allocate the majority of nanotechnology research and development towards the diseases of the developed world, health disparities may actually increase. A second barrier involves basic market forces and the issue of access. Most companies engaged in NT development seek a solid return on their investment. Increasingly, the partnerships between research universities and the private sector that have been promoted by the NNI have resulted in patenting and restrictive use licensing that directly blocks access to technologies for humanitarian purposes [22].

Assessing the aspect of social justice requires posing one additional set of questions:

- How are benefits and burdens balanced within and between communities?
- How might current social, economic, and political boundaries be enhanced or disrupted?
- How will those communities that are harmed be compensated?
- Does potential lack of access constitute a basic violation of human rights?
- Who is accountable for the fair distribution of benefits?

Taken together, these three principles strive to balance concerns at the levels of the individual, the local community, and the global society.

2.1.5   *You as Moral Agent*

A final word about ethical implications and you as a moral agent needs to be addressed. Individuals all wear multiple hats when it comes to ethical obligations. In the language of ethics, this is termed *conflicting loyalties* or *conflicts of fidelity*. In any particular situation you may find yourself needing to be loyal to your employer, your colleagues, your profession, your family, your community, and yourself. These multiple loyalties often challenge us to consider what we do, and the manner in which we do it, in light of each different stakeholder. For example, the immediate interests of the company you work for may conflict with what you believe to be legitimate concerns regarding public safety. In another example, public funding priorities may violate your sense of how resources should be allocated to serve the greater good. On the flip side, your personal financial well-being could be at stake if funding priorities shift from one direction to another.

We do not have good systems and processes for dealing with such conflicts in modern organizations and social institutions. However, the discomfort you feel when such conflicts arise is evidence that scientists and engineers do have a role and a responsibility to take the societal implications of their work seriously. Using the language and concepts of ethics in a productive and collaborative dialogue, the evolving opportunities of nanotechnology can be realized while also minimizing the inevitable disruptions and harms of technological advancements.

2.2   LEGAL IMPLICATIONS

The previous section explored some factors that bear on how ethics apply to nanoscience and nanotechnology. Because one of the goals of the law is to provide a framework for promoting morally defensible choices, we now turn to the complex intersection of nanoscience, nanotechnology, and the law. The topics of intellectual property and civil liability are used to introduce the reader to some of the legal challenges raised by nanoscience and technology.

2.2.1   Interaction of Law and Nanoscience

Fundamentally, the purpose of law is to control the behavior of people. No matter what the philosophical underpinnings of different governments in the world might be or what different views those philosophical underpinnings represent about the nature of individual liberty, they all agree there is a need to control aspects of the behavior of those within their borders. Such control can be effected in positive or negative ways—that is, by providing inducements that encourage behavior viewed as socially beneficial or by providing punishments that discourage undesirable behavior. As such, the law insinuates itself into virtually every aspect of human activity, from governing the way children are identified and treated at the moment of their birth to influencing the way they interact with others during their lives to resolving their affairs after death. Nanoscience is one of those rare disciplines that may have the potential to enjoy a similar breadth.

In considering the various ways nanoscience and the law may intersect, it is important to keep in mind that these intersections arise even when laws have not been tailored specifically to address nanoscience. The history of legal doctrine recognizes that human society is constantly in flux and has developed flexibility that aims to have it applied to new circumstances and to new technologies. In this way, legal doctrine attempts to recognize that while circumstances change over time, there are certain constants about the nature of human beings that permit general principles to be applied in circumscribing limits on their behavior.

To be sure, there will be specific issues raised by nanotechnology and nanoscience that will be addressed with targeted legislation. For the most part, this will happen when a perception develops that the more generic legal principles, even when faithfully applied, are not leading to the most desirable result. But in the broad scheme, these targeted laws will be of relatively minor importance in the way nanoscience and the law interact. The vast majority of legal issues that involve nanoscience will draw on legal doctrines that have more general application.

To illustrate how this happens, this section uses two examples of legal frameworks that have been designed with flexibility in mind: intellectual property and civil liability. The first is an example of a set of laws that intends to encourage positive behavior—namely, to provide an inducement for creative individuals to develop and share their creations with the public so that society as a whole may advance more quickly. The second is an example of a set of laws intended to discourage negative behavior by holding those who engage in activities that harm others responsible for their actions. Neither of these doctrines was developed with nanoscience in mind—indeed, they have pedigrees that span centuries before the discipline was even conceived—and yet their principles find ready application to issues of relevance to nanoscience today.

2.2.2   Intellectual Property

Intellectual-property law is a set of legal doctrines that governs the ownership of inventions and creations. It has a number of subcategories—patents, copyrights, trademarks, and trade secrets—that are directed to more specific aspects, with each of these subcategories providing different types of protections to inventors and creators. While those involved in the development of nanoscience have the potential to benefit from any of these doctrines, by far the most important is the ability to obtain patents for inventions in the fields of nanoscience and nanotechnology.

The current patent system has aspects of its origins in laws passed by the city–state of Venice in the fifteenth century. There is a natural tendency among those who invent something useful to want to control the essence of their discovery and thereby to derive a commercial advantage from having exclusive access to it. But such secrecy does little in allowing the discovery to benefit society as a whole. What Venice did, in its years as a thriving center of Middle Ages commerce, was to experiment with a system that would provide a state-sanctioned monopoly on use of an invention in exchange for having the inventor disclose it to the public. The public knowledge that resulted would enable others to make improvements on the invention, and thereby accelerate the pace of innovation. And the inventor would be the beneficiary of certain exclusive rights backed by the power of government. Venice initially experimented with patents in the area of silk-making, but soon expanded to providing protection for other types of devices, including patents for flour mills, cook stoves for dye shops, printing methods, and mills. Because of Venice's commercial importance at the time, its system had the effect of inducing even outsiders to disclose their inventions in exchange for exclusive rights within the city–state, proving the value of the system in putting useful information into the public domain.

The patent system today is predicated on the same basic philosophy. In exchange for disclosure of an idea, the government grants a right to an inventor to exclude others from making, using, or selling that invention unless they have her permission. The right exists for a limited period of time—in almost every country it is 20 years from the date the application is filed with the Patent Office—after which anyone is free to use the invention. One important aspect of this bargain between the inventor and the government is that this right can be exercised even against those who independently develop the same invention. This makes patents especially important in areas like nanoscience, where many researchers are working actively and competitively on developing similar ideas. The first to discover and patent inventions that result from them will have a right of exclusion that can be exercised against all of the others.

The same is not true of trade secrets, which are a form of intellectual property that is in many ways the antithesis of patents. As its name suggests, a trade secret is some kind of information that is withheld from the public. Nothing in the patent system obligates disclosure by an inventor; indeed, when a researcher creates an invention, he still has the choice of disclosing it to the world in exchange for a patent or concealing it to be used as a trade secret. The risk of concealment is that there is no protection provided for independent discovery of a trade secret. If another researcher develops the same idea, there will be no way for the first inventor to interfere with its use by the second inventor and any commercial advantage from being the first to make the discovery will be largely lost. The most significant benefit of a trade secret is that there is no time limit on it. While patents expire in 20 years, trade secrets can be maintained indefinitely, and there are commercial examples of trade secrets that have been maintained for close to a century.

Decisions of whether to patent inventions or to maintain them as trade secrets therefore hinge to a great extent on the likelihood that the invention will be independently developed and on the ease with which its secrecy can be maintained. There is no general prohibition on using reverse-engineering techniques to try to discover how some product that incorporates nanoscience is made. Indeed, the use of such techniques is mostly encouraged as a matter of legal policy since the result is a greater contribution to the public knowledge. Instead, the protection the law affords to those who decide to maintain information as a trade secret is to provide punishments that can be imposed when they are misappropriated. The most dramatic examples of trade-secret theft are sometimes glamorously described as *economic espionage*, but more mundane forms of theft can have similarly negative financial impacts.

Consider, for example, a nanoscience researcher who develops a method of fabricating doped nanotubes that can be easily formed into superconducting wires. In this example, the method uses a catalyst during a reaction that was only accidentally discovered and comes as a complete shock to the scientist who noted it. Should the method be patented or kept secret? The answer is not clear and the decision is a strategic one that requires the exercise of some judgment. Relevant considerations include:

How likely is it that another researcher will stumble on the same discovery?

How difficult will it be to maintain control over the information in production and sales environments?

Is it possible to analyze the superconducting wires and detect use of the catalyst?

The last of these questions is especially important because it relates not only to how easy it is for others to discover the secret but also to how difficult it would be to detect whether there was any actual infringement of a patent.

The other two types of intellectual properties are probably of less direct relevance to nanoscience. Copyright law does have the potential to apply to certain kinds of very small creations—for example, sculptural works created using methods common in nanotechnology. For instance, the Osaka Bull was carved out of resin in 2001 by a group of researchers at Osaka University using a two-photon micropolymerization technique (**Fig. 2.7**).

While one might quibble about whether the bull itself qualifies as "nano," given that its size is on the order of microns, it illustrates the point that creative works are sometimes made using nanoscience and these are entitled to copyright protection. In some ways, an example like the Osaka bull is little more than an isolated curiosity. More practical application of copyright law to nanoscience may arise from ideas currently being formulated for using chemical structures in information processing to produce a *nanocomputer*. The sequence of such chemical structures could easily qualify as a work entitled to copyright protection in much the same way that conventional source code is protected by copyright today. Similar considerations apply to "quantum computing" in which quantum states of particles are used in defining bit states and assembled to produce a form of computational program.

| FIG. 2.7 | *The Osaka Bull was formed by a two-photon polymerization technique. The scale bar is 2 μm. Parts a through f represent different perspectives of the bull.* |

Source: Kawata, S. et al., *Nature*, 412, 697–698, 2001. Image courtesy of the Nature Publication Group. With permission.

Trademarks, the last form of intellectual property, are likely to find application in nanoscience in only the most peripheral way. Trademarks are intended to identify the source of goods or services. Thus, the names of nanotechnology companies could be protected as trademarks to identify the source of the products they produce. And while it is conceivable structures could be included on nanotechnology products as a source identifier, this is relatively unlikely to be done.

2.2.3 Civil Liability Issues

As the discoveries of nanoscience become commercialized, so does the potential for them to act as a source of liability. As discussed in the next section, there are already certain specific concerns about the impact that nanosized structures may have on the environment or human beings, and it is certain that a greater understanding of those impacts will be developed in the future. What the law of liability seeks is a sensible apportionment of financial responsibility when harms occur. For instance, if it turns out that a certain product incorporating nanoscience causes harm, how are the financial and other consequences of the harm to be assigned?

Modern product liability law is based on the idea that there is liability whenever a party introduces a *defective* product into the marketplace. There are three recognized types of defects, the most important of which is a *design defect*. It is not easy to identify when the design of some product is defective since any design necessarily attempts to balance a number of different factors. While the safety of a product is an important factor, economic considerations, the appeal of the product to consumers, the durability of the product, and other factors also bear on the overall design selection. Nonetheless, some objective criteria are needed to evaluate when designs are defective. Two tests have been developed.

The first, the *consumer expectations* test, attempts to discern what levels of safety reasonable consumers anticipate a particular product will have. If the safety level the product provides is incommensurate with what consumers reasonably expect and someone suffers harm as a result of the design, the designer or seller is usually liable for that harm. Difficulties arise in applying this test to *complex designs*—a categorization that applies to many nanotechnology products because of the inherent difficulty for consumers to devise reasonable expectations about the nature of the design. What is more frequently applied to these types of designs is the *risk–utility* test, which seeks to balance the safety risk that a product design has against its utility. This test recognizes more explicitly that making a trade-off with safety risks may often be legitimate to achieve desired functionality. Under this test, a design is defective if the cost of avoiding particular risks is greater than the safety benefit that would result from such avoidance. Particularly relevant in performing such a balancing is whether there are alternative designs for the product that would not entail the same level of risk.

The flexibility of this test can be illustrated by considering the wide variety of products that might incorporate nanoscience and nanotechnology. At one end of the spectrum are consumer products that are entirely traditional except for being augmented using nanoscience to provide some particular benefit. Examples include such things as incorporating nanoparticles in fabrics to enhance resistance to stains or in children's toys to provide antibacterial properties. Most consumers expect such products to be just as safe as their conventional versions, and these conventional versions serve as examples of alternative designs. Even very modest increases in safety risks attributable to the nanotech structures are likely to be seen as outweighing the utility that those structures provide. For instance, improved resistance to staining of fabrics is likely to be seen as having insufficient utility if the risk of illness from wearing the fabrics is even only slightly elevated by the use of nanoparticles. When people are harmed as a result of incorporating nanoscience into consumer products, the producers of those products should expect to have little latitude in avoiding liability.

The same is not true of other types of products that might use nanoscience, with various military applications perhaps illustrating the opposite end of the spectrum. For instance, a very high degree of risk is likely to be acceptable for products that incorporate nano-energetic materials that are intended as explosive devices. The fact that a product is intended to cause harm or to be used in settings where injuries are common is relevant both to consumer expectations and to the balancing of risk and utility.

Where the real issue arises with these kinds of products is under circumstances where the *nature* of the risk is changed by the nanotech structures. When they present a risk that is not conventionally associated with the product, consumer expectations may be quite different from the actual nature of the risk. In such cases, the utility of the nanotech structures when applying the risk–utility test is better considered to be the difference in utility of the product with and without those structures. It is this incremental utility that should be weighed against the excess risk presented by the incorporation of nanotechnology. For example, consider the use of nano-energetic materials for the demolition of structures. There is a certain well-known risk of injury attributable to such demolitions. The mere fact that the demolition is performed with a nano-energetic material would not by itself change the calculus by which liability was determined. It is when the use of such materials produces a new form of risk, such as when the release and inhalation of nanoparticles at demolition sites causes new types of harms, that the evaluation of liability changes. The benefit of using nano-energetic materials as compared with conventional explosives should then be weighed against this new risk. With the test properly applied this way, there is a strong likelihood of liability whenever nanotech structures result in harms that are not well foreseen.

It is worth noting that the law affords special treatment to a class of products described as *unavoidably unsafe*. The doctrine has traditionally found its most effective application in medical contexts and many of the various nanoscience products being developed for medical uses are likely to fall in this category. The doctrine embraces a relatively extreme form of the risk–utility test—that there are certain products that have legitimate uses but are simply of a nature that makes them inherently unsafe. When this is the case and the user of the product is adequately informed about the potential risks, liability for harms caused by the product can be completely avoided.

It is not only in the context of unavoidably unsafe products that the need to inform users about risks arises. Indeed, the principal way to limit liability is to be candid about the nature of all known risks in the form of warnings. A consumer who is adequately warned about the risks associated with a product who still proceeds to use it is generally considered to have assumed the risk and its potential consequences. This illustrates the second type of product defect that is recognized: a warning defect. Manufacturers of products are expected to provide *adequate* warnings of *foreseeable* risks. For a warning to be adequate, it must be clear and specific, and communicate some understanding of the degree of risk to the consumer.

The final recognized product defect—a manufacturing defect—arises when there is a deviation from the standard product design in a particular product. The issues surrounding these kinds of defects are much more straightforward than those related to design defects: If the deviation from the product design resulted in a harm, there is liability, irrespective of almost any other factor that can be conceived.

Each of these liability theories for products is a form of *strict* liability, meaning there is no need to prove any sort of *fault* in order to establish liability. All that is needed is to prove the defect existed and a plaintiff was harmed as a result. There are other theories of liability in which it *is* necessary to establish a form of fault, and while these theories do have potential application to products, they are more usually applied in other contexts. This is specifically true of negligence, which is a legal doctrine that applies when one party owes a duty of care to another party. It may find particular application in medical-treatment settings that use nanotechnology.

Medical malpractice is essentially a specialized form of negligence that considers breaches of the duty owed by physicians or other medical practitioners to their patients. To be liable for medical malpractice, the physician must not only cause harm to a patient, but must also have done so in a way that breaches the duty towards the patient. He must have acted in a manner that is unreasonable according to the standards of the profession. Would a reasonable physician have prescribed the nanoscience treatment for this condition? Would a reasonable physician have administered the treatment in a similar way? If the answers to questions like these are "yes," then there will be no liability for harms caused by the treatment because there has been no breach of the duty. The law of negligence demands only that people behave reasonably; it does not demand infallibility of anyone.

2.2.4 *Evaluation*

These brief examples have been intended to give a flavor of how legal issues may bear on nanotechnology in both positive and negative ways. The rights of intellectual property encourage nanoscience researchers to disclose their results to the public so that all may benefit from developments that are made. The law of liability acts conversely to cause those who introduce nanoscience products or treatments to pause before doing so. The potential to be held accountable for the harms the products cause is intended to ensure that adequate testing is performed, the public is apprised of any risks, and the products are going to be introduced into the market in a responsible way.

There are many other ways in which nanotechnology might be affected by different laws. But ultimately all of those laws will act in one of these two ways. They will either provide an inducement for positive behavior or will provide a penalty to discourage negative behavior. It is in this way that the law acts symbiotically with nanoscience in promoting societal values.

2.3 ENVIRONMENTAL IMPLICATIONS

Another aspect of liability is how nanomaterials impact the environment and health. We need not look far into the past to realize the significance of releasing toxic materials into the environment, thus potentially harming people who work with them, use them in consumer products, or are exposed through other environmental pathways such as in their drinking water or food supply. Those who develop these materials and use them in technological applications bear responsibility to ensure they do not harm health or the environment. Nanoscientists and engineers are developing novel materials that did not previously exist, and are using them for broad applications in many economic sectors: medicine, energy, industrial applications, materials science, engineering, electronics, cosmetics, and food science.

The unique behavior of materials 100 nm and below is what makes them attractive for developing new applications. The size raises questions, however, about whether the unique behavior also affects biological systems (people and the environment) differently than other materials. Deoxyribonucleic acid (DNA), the basic building block of life, is made of molecules that are nanosized, and can be called a nanoscale material, since the width of DNA is about 2–3 nm. Because it is not engineered, DNA may not fall into everyone's definition of nanoscale materials; however, a key point is that other nanoscale materials are in the same size range, so they can react with biological systems in ways that larger particles do not necessarily.

Environmental concerns about nanotechnology and nanoscale materials include the use of nanoscale materials in applications likely to result in releases to the environment, and the release of materials throughout the life cycle of a material, from its manufacturing, distribution, use and reuse, recycling, or disposal. Consider sunscreen as an example. First, one must fabricate the nanoscale particles for the sunscreen, which include nanoscale titanium dioxide ($nTiO_2$) and nanoscale zinc oxide ($nZnO$). The $nTiO_2$ and $nZnO$ are packaged and shipped to the sunscreen manufacturer. When the sunscreen is sold, the consumer applies it directly to his or her skin, then swims or sweats off the sunscreen onto a towel or showers off the sunscreen. In each case, the $nTiO_2$ and $nZnO$ are released to water, a pool, lake, river, or the ocean while swimming, or to wastewater from bathing or laundry. When the container of sunscreen is empty, it either enters municipal solid waste for land or incineration disposal or is recycled. Each of these steps potentially releases the nanoparticles to the environment.

The breadth of potential applications raises questions about the safety of nanomaterials, and whether they have the potential to adversely affect health. C_{60} fullerenes are spherical molecules with a diameter of about 1–2 nm and have been reported to behave as antioxidants, scavenging radical oxygen molecules. Hydroxyl radicals have been associated with aging and stress, and antioxidants are hot market items for skin creams and *nutraceuticals*. Fullerenes are in at least six skin creams currently on the market [26]. C_{60} also appears to have catalytic effects on cell membranes reducing cell viability, and is being investigated as an antibacterial additive for disinfectants. Some early studies on the behavior of C_{60} raise concerns about whether it might cause unintended effects.

In a hospital environment, it is very important to keep surfaces free from contamination by bacteria, and many cleaning and disinfection products are used for washing equipment, hands, floors, and surfaces to help prevent the spread of germs. Using a product containing C_{60} as a disinfectant might mean it would be sprayed, wiped, poured into buckets and on floors, and washed down drains. Where could the C_{60} end up?

Researchers are now detecting bacteriocidal chemicals such as triclosan, commonly found in antimicrobial soaps and cleaning products, in rivers and other water bodies. Populations of bacteria that are resistant to antibiotics are also being found in the environment. Antimicrobial resistance is a big problem, since bacteria are no longer susceptible to the treatments we have for them. Could C_{60} cause antimicrobial resistance in the environment? What other unintended effects could a substance that catalyzes cell membranes cause?

There are hundreds of studies that have measured the effects of nanoparticle exposure on health, but they often suffer from methodological concerns and raise more questions than they answer. Using nanomaterials includes some potential for exposure of those developing the products and those using them to come into contact with nanoparticles. This section defines the field of nano-toxicology, introduces the concepts of environmental health and safety, and addresses larger concerns about environmental aspects of nanoscience.

2.3.1 Nanotoxicology

All materials pose some threat to health. Even water, the sustenance of life, can be toxic. When too much water is consumed, it causes an electrolyte imbalance leading to hypotonia, a condition in which a salt imbalance leads quickly to death. Thus, a key aspect to understanding the toxicity of materials is to know how much of a substance—the dose—can cause harm. Certainly we will not stop drinking water because too much water will kill us. Instead, it is reasonable to assume we will drink adequate, but not excessive, amounts of water. This concept, introduced by Paracelsus in the fourteenth century, forms the foundation of modern toxicology: "The dose makes the poison."

This is the rationale behind alchemy, as well as behind pharmacology, or the use of drugs to treat medical conditions. At some dose level, an effect occurs. In pharmacology, the effect is medicinal to treat some disorder. In toxicology, the effect is often an adverse one; that is, a substance causes toxicity to a cell or an organ system. We all know drugs can have side effects, so one area of toxicology examines the unintended effects of medicines. There are indeed many substances required for health that cause toxicity if taken in excess, including proteins, enzymes, and vitamins. When our bodies are introduced to substances at the wrong dose, or unintentionally, the exposure can result in toxicity.

There are many ways to study the effects of substances in the body. Commonly, substances are tested at different doses in laboratory animals to determine whether adverse effects occur, the types of effects, and the doses at which they occur. While mice and rats are not people, they are mammals and can indicate effects that may occur in people. Alternatives to animal testing include the use of cell culture assays to observe effects at the cellular level. These in vitro tests can be informative, but do not mimic the dynamics of human physiology and thus cannot completely replace animal testing.

More recent developments in toxicology include conducting assays using genetic material and proteins to evaluate specific biological interactions. Termed -*omics*, studies including genomic, proteomic, and metabolonomic assays help scientists understand the fundamental mechanisms by which substances cause toxicity. A genomic test examines how substances interact with genetic material, or DNA. Some substances cause changes to DNA that can lead to the formation of diseases such as cancer. Similarly, metabolonomics observed whether or how substances affect metabolism and can predict toxic interactions in the body. With tools such as micro- and nanoarrays, many tests can be conducted simultaneously. Eventually, there will be databases describing all types of toxicological mechanisms, and we may be able to use these data to predict the toxicology of new nanoscale materials based on their interactions with specific biological molecules.

It is important to keep in mind that we are all a bit different, and some of us are more sensitive than others. One person may be unaffected by an exposure to a substance while another person may suffer an adverse effect. One example of this is allergic reactions. In some cases, genetic differences can be linked to why some people are more sensitive than others to the effects of particular exposures. In other cases, it is unclear why some, but not all, exposed people have a particular reaction.

Nanotoxicology is a new field that evaluates the toxicity of nanoscale materials on the health of organisms. It is the application of toxicology to nanoscience. The term was coined by Oberdörster, Oberdörster, and Oberdörster in 2005 and defined as the "science of engineered nanodevices and nanostructures that deals with their effects in living organisms" [23]. The field of nanotoxicology explores the effects of exposure to nanomaterials.

There are few data available on the toxicology of nanomaterials. Type the search term *nanotoxicology* into the database of environmental health and safety (EHS; work on nanomaterials developed and maintained by the International Council on Nanotechnology [ICON]). How many papers appear in comparison to a search in the ICON EHS database with the term *toxicology*? Most of these studies fall in a few categories: simulations of particles inhaled into the lungs, cellular assays, and short-term ingestion exposure tests. Many are studies of ultrafine articles that are not engineered and are more diverse than engineered particles with a range of sizes and different levels of contamination. Each of these provides different types of information. But with nanoscale materials, the results of the studies are not always easy to interpret. One of the complexities is that some particles behave differently when they are nanoscale than when they are larger. This may be because of increased surface area of the particle, so doses equivalent on the basis of mass are very different on the basis of area.

Alternative explanations for effects from exposure to nanoparticles relate to particle surface charge or shape, among others. Toxicology has not traditionally addressed these complexities about dose, leaving considerable uncertainty about what to measure in studies. For example, what are the key parameters to describe for nanotubes, nanohorns, and aggregate particles? Because of this and similar

---

EXAMPLE 2.1   *How to Define the Toxic Dose*

A study by the National Institute for Occupational Safety and Health found that the lungs of mice exposed to single-walled carbon nanotubes (SWNTs) formed "unusual fibrotic responses" compared to mice exposed to other types of carbon particles—specifically, an amorphous particle called carbon black. This response was seen in mice exposed to a dose of 10–40 μg/mouse aspirated into the lungs. The authors conclude that SWNTs are more toxic than carbon black. However, if compared on the basis of surface area instead of mass, 40 μg/mouse of SWNTs is estimated to have a surface area of 1040 m^2/g, while the surface area of the carbon black is 254 m^2/g [24]. In the study, the surface area was determined by Brunauer, Emmett, and Teller (BET) analysis, and diameter was measured by transmission electron microscopy (TEM).

On the basis of surface area, the mice receiving SWNTs received four times as much exposure to carbon particles in their lungs. So, on the basis of surface area, the results showing SWNTs are three to five times more toxic to lungs were actually done at doses four times higher! While this is not definitive evidence that nanotubes are not more toxic to the lung than other carbon nanoparticles, on the basis of surface area, SWNTs may not be more toxic than carbon black when inhaled in the lung.

Another complicating factor in this study is that the SWNTs were found to contain about 0.23% iron, which some have suggested may have contributed to the toxic responses in the lung [25]. Thus, it is not clear whether carbon nanotubes are more toxic than other shapes or sizes of carbon particles, and toxicology research now must consider new factors in the assessment of toxicity.

issues, almost every conference or meeting on nanotechnology includes a discussion of EHS risks. This uncertainty affects decisions about how to work safely with nanoscale materials, since we do not yet understand whether materials can cause toxicity or, if they do, how much is required for an effect to occur. There are many questions about the toxicology of nanoscale materials, and the questions suggest we take a broader look at the potential EHS risks in a larger context.

2.3.2   *Nanotechnology Risk Assessment*

In section 2.1.3, we asked, "What is the level of risk?" as a question about the nature of a technology. In this section, we discuss environmental risk assessment, a decision-making tool used to analyze and help make decisions about substances and technologies. Risk assessment may be defined as

> a process intended to calculate or estimate the risk to a given target organism, system or (sub)population, including the identification of attendant uncertainties, following exposure to a particular agent, taking into account the inherent characteristics of the agent of concern as well as the characteristics of the specific target system. [27]

More simply, risk assessment allows the estimation of health and environmental impacts from exposure to a substance.

Governmental and private organizations all over the world use risk assessment for environmental and public health decision making for risk management. In the United States, risk assessment is used to understand risks and make management decisions regarding hazardous waste site cleanup, closing municipal solid-waste landfills, setting standards and managing safe drinking water, chemicals' regulation, allowing additives to food and food packaging, and food safety regulations. The European Commission passed legislation in December 2006 to require risk assessments for all chemicals used in commerce in the European Union (Registration Evaluation and Authorization for Chemicals). Companies often perform risk assessments on the substances they manufacture or the products they sell.

You likely conduct risk assessments, too. Is it safe to drink the water from the tap? Is the food in the refrigerator still safe to eat, or has it gone bad? Are you adequately protecting yourself from exposure to materials you handle in the lab? Think of something important for you to succeed in. Will you go for it, or not even try for fear of failing? On an individual basis, we might have different answers to some of these questions. You might be unconcerned about your drinking water, but your classmate may insist on bottled water. If there were an absolute answer, we would not need to conduct an assessment of risk. However, in each case, there is uncertainty that requires some data for our decisions and also means we have to interpret the data, using our judgments. The uncertainty and our individual tolerance for risk affect how we judge the data and the conclusions we reach. The combination of science and judgment constitutes the assessment of risk. Conducting formal risk assessments allows us to apply a process for risk assessment to determine the relative level of risk and judge whether the risk is acceptable or not.

There are four basic steps in a risk assessment. First, we must define the problem or the hazard. Different international frameworks use different terminology for this step, referring to it as problem formulation, hazard characterization, or hazard identification. The hazard identification defines how to conduct the remainder of the assessment. It defines the questions the risk assessment will ask and answer. Hazard identification questions for nanomaterials may not differ from those for other substances; however, the necessary measurements are not widely available.

The second step is to develop an exposure assessment. Exposure assessment considers who might be exposed to the agent and defines the circumstances of those exposures. In human health risk assessment, receptors are those who may be exposed under different scenarios, such as workers in an occupational environment who manufacture a chemical, users of a product, and others who may have incidental exposure occurring as a result of manufacturing, use, or disposal of a product. In ecological risk assessment, the receptors may be specific species, populations, or entire ecosystems.

Substances behave differently in the environment, so it is important to understand whether exposure is likely to occur by a particular pathway. For example, a substance that is a solid would have to somehow be released in the air that we breathe; otherwise, we would not evaluate an inhalation pathway for that substance. For nanoscale materials, it is not clear how to measure exposure as it relates to toxicity (as a concentration, by surface area, or reactive surface area), and there are few analytical techniques currently available that quantitatively measure substances at the nanoscale.

The next step in a risk assessment is to look at how or whether the chemical causes harm. The dose–response assessment identifies the nature of a substance's toxicity by different exposure routes. This assessment relies on data from toxicology studies and, in some cases, from epidemiology studies in populations. The dose–response assessment identifies the effects at different doses and the lowest levels where effects or no effects have been observed. These effect levels become the basis for comparing to the exposure levels. One widely observed effect from exposure to nanoparticles is inflammation, associated with the development of many diseases. It is presently unclear whether the chemical composition, size, shape, or surface characteristics of particles affect the toxicity of nanoscale materials.

The risk characterization step brings together the exposure and dose–response assessments. This is the process of comparing exposure levels to effect levels to see whether the exposures that could occur under different scenarios are significant. The risk characterization also looks at the risks in context of regulations that define how much exposure is allowed under different circumstances and makes comparisons with other types of risks that help to inform how the risks are managed.

Generally, there is a lot of uncertainty associated with this comparison, and the risk characterization step considers these uncertainties. One example of an uncertainty is that we often cannot measure exposure exactly, so we make estimates that rely on assumptions (i.e., assuming a person drinks 2 liters (L; about a gallon) of water every day. Not all of us drink exactly 2 L of water per day; some of us drink 1 L, some drink 3 L, and others drink little if any water. This adds uncertainty to risk estimates because, when we assume some exposure occurs when a person ingests 2 L of water per day, we have simplified reality.

Despite uncertainty, risk assessment is a valuable technique for estimating the potential health and environmental impacts of nanoscale materials used in nanoscience and technology. Even when all of the necessary information is not available, risk assessment can still be helpful to make estimates of potential risk to set research agendas, or make safety decisions about working with or using materials.

2.3.3   *Environmental Aspects of Nanotechnology*

Many of the current applications of nanotechnology are in consumer products. Maybe some of your personal consumer products use nanotechnology (e.g., an MP3 player or the coating on your cell phone); your laptop screen may use nanotechnology for a stronger, scratch-resistant coating or a more efficient and higher resolution display. You could be wearing antimicrobial socks or static-free pants. Your toilet may have a self-cleaning surface (or perhaps you wish it did). As a student, you may be working with nanomaterials in a laboratory.

How do any of these applications of nanoscience affect the environment? If nanoscale materials are part of a coating on an electronic device, how could the nanomaterials enter the larger environment? When you use your cell phone, how could nanomaterials in the coating of the phone affect the environment? Generally, *exposure* must occur for the environment to be affected. Using a cell phone with a coating including nanomaterials does not result in a release of nanomaterial to the environment, so it is unlikely that it could have an effect.

Working in a laboratory, you might be handling nanoscale materials. You might be making new materials, or testing the properties of a material. Do you touch the materials with your hands? Could the surfaces of your lab have some nanomaterials on them? How could these get into your body? Do you ever put your hands near your mouth or eyes? Is it possible that nanoscale materials are airborne and could be inhaled? When an experiment is finished, where do the materials go? Do they get put in the trash or washed down the drain? If you are working in a clean room or a hood, are particles filtered by an air-circulating system? Are they part of the filter?

These questions are an informal way of conducting an exposure assessment. How materials are used and how they are disposed of influence how and whether they can affect the environment. It is important to consider the entire life cycle of a material to understand the potential for impacts to the environment. What is the life cycle? Some refer to it as "cradle to grave" or "cradle to cradle." Considering the potential for effects throughout the life cycle is an important step in generating new materials.

These questions about whether and how much nanoscale materials associated with nanoscience and nanotechnology enter the environment remain unanswered at the present. It is also unclear whether small amounts released make a difference in terms of health or environmental impact. There are more questions than answers about the fate of nanoscale materials in the environment. Will they enter the air we breathe? Will they contaminate our rivers and our drinking-water supplies? Will they leak out of landfills and enter unexpected places in the environment—for instance, the water we use to irrigate our food crops? Without the answers to these questions, we can only speculate about the behavior of nanoscale materials in the environment. We know nanotechnology is used and useful because of the unique properties at the nanoscale, so we can also speculate that the environmental transport and fate may also be unique from material to material. Considering the life cycle of materials offers an approach to think through the consequences of using nanoscale materials, in the context of their life cycle, both for health and safety and for society as whole.

Whether working in a research lab or for a manufacturer, or consuming products containing nanoscale materials, individual actions will affect whether and how much of the material enters the environment. It is not clear today what the potential impacts are from nanoscale materials in the air, water, and soil. It is not understood whether nanomaterials might enter the food web and become part of what we eat and drink, or whether they can affect our forests and coral reefs and air quality. There is much research to be done to understand the potential impacts. But your actions as a scientist, an inventor, and a consumer make a difference in terms of the extent of impact that occurs.

2.4   PUBLIC PERCEPTION

Not wanting nanotechnology to be subject to the same sorts of public criticisms that have emerged for controversial forms of biotechnology (such as genetically modified foods, cloning, and stem cell research), considerable effort and expense is being invested in the United States and in Europe in trying to engage the general public early on in considering nanotechnology's benefits and risks. The hope is that later reactions will be better informed if public discussion starts at an earlier point, sometimes referred to as *upstream engagement*.

Ideally, encouraging people to think and talk about emerging technologies at an early stage will also help the futures of those technologies be shaped (in turn) by popular values and priorities. After all, much of the research on these technologies is paid for by tax dollars, so if the public is not "on board" with the directions in which this research is taking us, the odds of negative reactions inevitably rise.

2.4.1   *Factors Influencing Public Perception*

Many scientists and policy makers alike assume that popular reactions to new technologies are based on superstitions and exaggerated fears when they ought to be based on science. In fact, research has shown that scientific knowledge may not make as much difference as other factors, such as social values (whether the technologies appear to achieve goals that ordinary people consider important) and whether people trust business and government to do the right thing.

Of course, nano is a scale, not a thing! Determining what people think nanotechnology is, let alone whether they think it is good or bad, is not really an easy matter. Focus groups with adults across the United States and Canada, as well as interviews with students at the University of South Carolina in Columbia, have established that, at present, most nonscientists have very little notion of what nanotechnology might be or how it might work. While experts may worry that ordinary people will become unreasonably fearful because of the influence of images of *gray goo* or runaway *nanobots,* so far there is little evidence this is taking place.

Instead, people tend to react to nanotechnology according to their value systems and their experiences with previous technologies. This is not to say that everyone who is opposed to (say) stem cell research or nuclear power is opposed to all other forms of new technology. Rather, people tend to draw from their experience to determine whether the regulatory system for technology is adequate, whether environmental impacts will be difficult to control, whether scientists are conscious of their social responsibilities, and so on.

2.4.2   *Nano and Public Opinion Polls*

Public opinion polls, while not perfect, are one way we are accustomed to *taking the public pulse* in modern societies. One of the reasons polls are tricky in the very early stages of the development of a technology is because when people have little information, they are very susceptible to small differences in question wording or question order. For nanotechnology, for example, what may seem like minor differences in definitions or the types of examples offered can be quite influential. Also, people may answer questions about their *opinions* even when, in truth, they have not had much opportunity to consider the topic. In fact, some observers would say these people do not really have opinions at all!

Nevertheless, polls are a useful indicator of both initial public reactions and general trends in those reactions. We have developed relatively reliable procedures for conducting polls, even though declining rates of response and shifts away from traditional land-line telephones in favor of cell phones have caused a lot of concern.

One of the first American surveys of public opinion about nanotechnology was conducted over the Internet with over 3900 respondents [28]. As is the case for many Internet surveys, this study was not actually a random or probability sample representative of the U.S. population as a whole. People who chose to answer this survey were likely to have been volunteers and may have been more interested in nanotechnology and more knowledgeable about science and technology generally than the average person. Nevertheless, the results were consistent with other research on popular reactions to technologies in several ways. Women were less optimistic about the risks (a common pattern), whereas people who are generally protechnology were more likely to react favorably.

In this survey, 57.5% of those responding felt that *human beings will benefit greatly* from nanotechnology. This high percentage must be judged in the context of the fact that volunteers in such cases are likely to include a higher proportion of those who are enthusiastic about the technology in question, whatever it might be. However, other surveys also indicate a generally positive public reaction.

In what was probably the first published U.S. phone survey of nanotechnology opinions designed to be representative of the general public rather than composed of volunteers that was conducted a few years later, similar optimism was visible [29]. In this survey of just over 1500 people, respondents felt that the benefits of nanotechnology would outweigh the risks, even though 51.8% had heard "nothing" about nanotechnology! And 80.3% were "not worried" about developments in this area.

In other words, most Americans, in their initial reaction, were not especially concerned about nanotechnology. Treating human disease was identified in this study as the most important of five benefits, clearly reflecting participants' underlying values as well as their overall positive expectations from science and technology. In our proscience, protechnology culture and in the absence of specific evidence that a technology represents cause for concern, this may not be surprising, but it is different from the reactions to some other major technologies we have confronted in recent years, especially biotechnologies. And it is also different from reactions in other parts of the world, to a certain extent.

According to a comparative analysis using data from the "Eurobarometer" (a periodic survey based on in-person interviews with individuals from across Europe—15,000 of them in this particular case) and a phone survey of 850 people in the United States, fully 50% of U.S. respondents believed nanotechnology would improve the quality of life in their country over the following 20 years [30]. Only 29% of Europeans agreed, most likely reflecting cultural differences in technology-related optimism.

However, only small numbers in either location responded that they felt nanotechnology would actually have a negative effect (4% in the United States and 6% in Europe). And in a slightly later U.S. survey of just over 700 people, 74% predicted nanotechnology would help in the area of detection and treatment of human diseases, 64% felt it would help increase national security, and 62% thought it would help in environmental cleanup [31]. A subsequent survey in January of 2005 found 46% of those in the United States foreseeing a positive impact [32]. Nevertheless, only about half the U.S. population seemed to believe that nanotechnology's developers shared their values.

Further, the North American focus group research discussed earlier did suggest some concerns that seemed to "resonate" with the general public [33]. In other words, when small-group discussions turned to certain topics, concerns seemed to be sparked. These topics included issues of the adequacy of control and regulation by government agencies, and also questions of negative environmental and health effects. In addition, more diffuse concerns emerged regarding social disruption, workforce impact, and distribution of benefits. If there are going to be new cures for cancer, for example, people seem to want to know if their insurance will cover these adequately or if these treatments will only be available to the economically well off. They also want to know if some people will lose their jobs or possibly go out of business as a result of technology's evolution.

Why do nanotechnology and biotechnology seem to elicit such different public reactions? Apparently, biotechnology—especially the manipulation of genetic material through recombinant DNA techniques—produces reactions and concerns that are less immediately obvious for nanotechnology. Biotechnology can challenge our cultural notions of individuality (through cloning) and species identification (through genetic engineering that combines genes from different kinds of plants and animals). Stem cell research causes some people pause to reflect on what it means to use human embryos in the laboratory, even for good purposes. Or, as some commentators put it, the *yuck factor* for biotechnology is much closer to the surface than for nanotechnology.

2.4.3 A Call for Two-Way Communication

What does all this mean for the science behind nanotechnology, and what do scientists need to do about public opinion? In some ways, nanotechnology has become a giant experiment in new ways of introducing new technologies to society. Because of past difficulties, much more attention is being paid to how the public will react. Increasingly, scientists feel an obligation to become part of the solution here rather than just part of the problem—to help educate the public in the hopes that popular concerns will not mushroom out of proportion. However, communication between scientists and the public—whether it takes place through public meetings, at events at universities or science museums, or through stories in the mass media—will be most meaningful if it is two-way. At the same time scientists are striving to educate the public, they should be striving to understand the public's actual concerns. Ideally, in a democratic society, science policy—the ways we choose to invest in research and development, and the approaches we take to regulation—will evolve in ways that reflect public values.

2.5 FUTURE OF NANOTECHNOLOGY

Professor Kip Thorne of Caltech is a theoretical physicist, known for his visionary understanding of the universe, who was named the 2004 California Scientist of the year. He asks a very probing question: "In an infinitely advanced universe, what things will be possible and what things won't?" The limitations of both physics and humanity are a subject that is not well understood. The future of nanotechnology, as described in this chapter, will be framed by the physical limitations to which Dr. Thorne is referring. Currently some are in the process of being understood and others only imagined.

The study of the future is commonly referred to as a soft science because it often lacks concrete data points around which decisions can be made. However, the most beneficial aspect of the future remains that it is largely unknowable. A known future is demoralizing. Inside the mysteries of a future that we cannot know lie the hope and inspiration that drive humanity forward.

On the following pages we will discuss some of the tools that give us clues about the world ahead, and apply these tools to nanotechnology and its surrounding developments. The tools that will be discussed include cycles and patterns, trend forecasting, attractionary futuristics, and Maximum Freud.

2.5.1 Cycles and Patterns

Edward R. Dewey (1895–1978) was an economist who studied cycles and patterns in economics and other fields. Dewey first became interested in cycles while he was chief economic analyst of the Department of Commerce in the 1930s. President Hoover wanted to know what caused the great depression. Dewey found that economists could give no consistent answers on the cause of the depression and he lost faith in the existing economic methods. He then shifted his study to how business behavior occurs, rather than why, and in doing so was able to map business patterns and form a more reliable understanding of the future than economists were able to give.

Dewey devoted his life to the study of cycles, claiming that "everything that has been studied has been found to have cycles present." He carried out extensive studies of cyclicity in economic, geological, biological, sociology, and physical sciences and other disciplines. So far, more than 500 different phenomena in 36 different areas of knowledge have been found to fluctuate in rhythmic cycles. The study of nanotechnology is tied to many different cycles, ranging from harmonic oscillation to circadian rhythms to absorption spectroscopy to standard business cycles.

2.5.2   *Trend Forecasting*

Similar in some respects to the study of cycles, trend forecasting is an exercise in pattern recognition where a researcher is able to create a high probability trend line, using historic data points, to determine a future event. Commonly adopted methods of trend forecasting include the Delphi method (averaging the opinions of experts), forecast by analogy, growth curves, and extrapolation. Normative methods of technology forecasting such as the relevance trees, morphological models, and mission flow diagrams are also commonly used.

While this list of methodologies makes it sound like trend forecasting has been reduced to some very scientific processes, it really boils down to either rational and explicit methods using hard data or intuitive methods using personal opinion. Moore's law with its prediction of "doubling of the number of transistors on integrated circuits every 18 months" and Metcalf's law, which states that "the value of a telecommunications network is proportional to the square of the number of users of the system" are two examples of how the study of trends can be used to predict future events.

2.5.3   *Attractionary Futuristics*

An *attractor* is a future event that humanity is being drawn towards. Future events such as putting a man on Mars, finding a cure for cancer, developing a mass energy storage system, or inventing a flying car are all well-known endpoints that have become staples of our culture, and a common theme in media as well as the global conversation. In our daily lives we often discuss these attractors without any awareness that the mere discussion of the topic reinforces its role as an attractor.

A new science that I am involved in developing at the Colorado-based DaVinci Institute, *attractionary futuristics*, is the science of attractors where the effect that a known future event is having on present-day research is being studied. Research is being focused on identifying known attractors, the creation of new attractors, categories of attractors, range of influence, intensity of the attraction, and the directional vectors of these forces. An example of an attractor in the nanotech field is the creation of the ultimate small memory storage particle and the effect this is having on high tech.

Past visionaries have given us the visions we hold today of the future. Sometimes these visions come in the form of illustrations or artwork and at other times in movies or video clips, but very often they start as nothing more than ideas in literature. As an example, Leonardo da Vinci dedicated over 500 drawings and 35,000 words to the concept of flying. He used the tools at his disposal to convey the idea that flying would someday become both viable and practical. His ideas managed to survive the centuries and eventually came to life, first in the form of Joseph and Jacques Etienne Montgolfier's hot air balloon, and later with the Wright brothers' flying machine in 1903 at Kitty Hawk, North Carolina. The da Vinci drawings served as a source of inspiration and as a blueprint of sorts for making it happen.

Other notable visionaries who have influenced our thoughts on the future include Jules Verne's visions of submarines and space travel, Gene Roddenberry's visions of cell phones and teleportation, Arthur C. Clark's visions of talking computers and the space elevator, Philip K. Dick's visions of flying cars and time travel, and George Lucas's visions of robots and space travel.

In 1986, the field of nanotechnology was jolted into existence with K. Eric Drexler's book, *Engines of Creation,* and his visions of what may or may not be possible once we have the ability to work with nanoscale materials. While some of his visions for molecular-scale factories and self-assembling machines remain the source of much discussion and debate, his visions served as a significant turning point for scientific research and commercial development.

Our visions of the future determine our actions today. Very often the images of the future that we hold in our heads have a profound impact on the ways we lead our lives and the decisions we make at work. For this reason it becomes imperative that we continue to improve our visions of the future.

2.5.4 Maximum Freud

Maximum Freud is a technology life-cycle tool that can be used to predict technology endpoints. The topic is best explained with the story in example 2.2.

Clearly this period of time was the end of an era. It was the end of the slide rule era and the beginning of the calculator era. As a society we have not seen

---

EXAMPLE 2.2	*The Slide Rule*

In 1972, I was a young engineering student at South Dakota State University in Brookings, South Dakota. One of the first courses I was required to take was a short course on slide rules. For those of you who do not know what a slide rule is, first came the abacus, then came the slide rule, and then came the calculator.

This was a time when the real "cool geeks" on campus walked around proudly displaying their black carrying cases for their slide rule attached to their belts. "Brainiacs on parade," as some described them, used their slide rules as a symbol to tell the world how smart they were.

Early calculators were first making their way into stores around 1970, and in 1972 they were still rather expensive. I remember arguing with my teacher about whether or not the slide rule course was necessary and his response was a dismissive "all engineers need to know how to use the slide rule."

Of course his thinking was wrong. Even though I completed the course with flying colors, I have never used a slide rule in my engineering work. Engineers at Hewlett Packard and Texas Instruments who were working on next-generation calculators at the time would have laughed at my teacher's assertion that slide rules were always going to be the centerpiece of the engineer's tool chest.

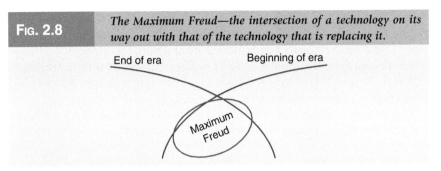

FIG. 2.8   *The Maximum Freud—the intersection of a technology on its way out with that of the technology that is replacing it.*

End of era   Beginning of era

Maximum Freud

Source: Courtesy of the DaVinci Institute. With permission.

the end of too many eras, but we are on the verge of experiencing many things disappearing in the near future. Most will not be as cleanly defined as the slide rule being replaced by the calculator. Often the soon-to-be-obsolete technology will be replaced by two or three other technologies.

As I sketched out the simple diagram showing the end of one era and the beginning of another, the point where the two eras overlapped caught my attention. This period of time is important to isolate because of the extreme dynamics happening there. It also occurred to me that we did not have a name for this intersection of technology, this collision of business forces. So I came up with the name *Maximum Freud* (**Fig. 2.8**). Yes, it is a rather wacky name, but useful in its description.

As technologies approach Maximum Freud, this is the period when industry players have to spend their time on the Freudian couch to understand what is going on. This is a period of extreme chaos, and also a period of extreme opportunity. But here is the most important part:

> All technologies end. Every technology that we use today will someday go away and be replaced by something else. Every technology will approach its own period of Maximum Freud. So, from the standpoint of making bold predictions, the imminent demise of many of our technologies is a certainty. [34]

Here are just a few examples of technologies that are currently approaching Maximum Freud:

- Checking industry—This is already in decline; the end of the hand-written check is drawing near.
- Space shuttle—This Model T of the space age is long overdue to be replaced by an efficient, low-prep craft that makes space accessible to the common man.
- Sign language—Advances in cochlear implant technology will soon make the need for visual person-to-person sign language unnecessary.
- Fax machine—Museum curators are already dusting off a spot for this once staple of the business world. Already in its twilight, the remaining days of the fax machine are numbered.
- Traditional AM–FM radio—As wireless technologies become more ubiquitous, aspiring radio station owners will find many new low-barrier entry points for commercial broadcast without the need for assigned radio frequency spectrum.

A good example in the nanotech field is the coming demise of the diamond mining industry. Artificial diamonds are on the verge of being created faster, more cheaply, and better than diamonds that nature produces; consequently, the labor-intensive mining industry will soon enter into the chaos of Maximum Freud.

The Maximum Freud concept is a tool we can use to determine when the technologies that we use today will be gone tomorrow, and in the process speculate on their replacements. We live in a very fluid, changing world and each step we take towards the future will enable us to experience life in a new and different way, and nanotechnology will be one of the key agents of change.

2.5.5 Nanotechnology End Points

Tying together this discussion of the future of nanotechnology, we will take a brief look at 10 of the possible end goals, or *attractors,* that are driving the industry. Each of these items can best be described as a disruptive technology with far-ranging business potential:

1. *The ultimate small storage particle.* Eventually Moore's law will end and we will reach a point where we have created the smallest possible particle on which we can store data. Once it is realized that we have created the smallest possible storage particle, the industry will begin to establish standards for information storage, so researchers 200 years in the future will be able to access the data.
2. *The ultimate small motor.* Ultratiny robots will be driven by ultratiny motors—so how small can we make a motor? The Foresight Institute is currently offering a $250,000 Feynman Prize to the person or team that creates a motor no more than 100 nm wide in any direction that is capable of moving atoms around.
3. *The ultimate small flying machine.* The U.S. Department of Defense has funded many projects related to testing the limits of aircraft miniaturization. So far they have been unable to achieve their goal of creating an invisible spy craft that can be flown into an enemy camp without detection. The ultimate small flying machine will, of course, have many other uses.
4. *The ultimate small decision-making (smart) machine.* How small can we scale intelligence and how do we go about building that intelligence into nanoscale machines? The Foresight Institute is currently offering a $250,000 Feynman Prize to the person or team that creates a machine less than 50 nm wide that is capable of adding numbers.
5. *Binary power.* Binary power is a technology where two otherwise harmless beams of energy intersect at a point in space and create power. When imagining systems for powering nanoscale machines with molecule and atom-specific precision, the idea of binary power begins to make sense. This type of power will be very useful for powering nanomotors and nanoscale flying machines.
6. *Molecular alteration.* Much like the medieval dream of being able to turn lead into gold, molecular alteration brings with it the promise of turning inferior grade materials into superior grade materials, useless substances into valuable substances, and problem cells into solution cells.

7. *Robotic nanosubmarine for repairing the human body.* As the ultimate medical device that can be driven to problem areas of the body, the nanosubmarine will come complete with vision tools, cellular diagnostic equipment, and arms, legs, and lasers for effecting change to problem areas of the body.

8. *The ultimate indestructible materials.* Virtually every material known to mankind enters a deterioration process as soon as it is created. The dreams of being able to build a house that lasts forever or roads that never show signs of wear all start with the creation of an indestructible material. However, a truly indestructible material will result in landfills filled with indestructible trash. So the ultimate indestructible material will have a disintegration feature that can be triggered at the end of product life.

9. *Self-repairing cells* (self-healing body tissue). Much like the lizard that loses a leg and is able to grow it back, the idea of creating regenerative cells brings with it the promise of eternal youth and the ability to live forever.

10. *The trifecta particle.* If you can imagine the most useful nanoscale material and the properties it will have, you will have imagined what we call the *trifecta particle*. This particle will have the attributes of being designable, controllable, and intelligent. Computer programming is all about architecting the flow of electrons. In the future, nanotechnology will be all about architecting the flow of matter, and these are the particles that will make that possible.

This list of end goals or "attractors" is not intended to be an exhaustive list of all the possibilities. Rather, its purpose is to stimulate thinking and stretch the imagination. Our visions of the future determine our actions today, and these examples are the first step towards improving your vision.

Although by no means exhaustive, the purpose of this chapter was to help you recognize the tremendous implications nanoscience and technology have for influencing the larger society now and in the future. Proponents of nanotechnology laud the potential for advances at the nanoscale to increase energy efficiency, improve human health, mitigate environmental degradation, and create new economic opportunities. Skeptics are less optimistic and warn that we need to proceed more cautiously to avoid the untoward consequences that inevitably accompany such revolutionary advances. Keep these various issues in mind as you continue through the course. If nanoscience and technology represent *the next big thing*, and broad societal impacts are inevitable, then all of society benefits from thoughtful awareness and responsible action on the part of scientists and engineers.

Acknowledgment

We are extremely grateful to Professor Deb Bennett-Woods of Regis University in Denver, Colorado, for donating her time and effort to organizing and editing this chapter. She also wrote section 2.1, the portion of the chapter that addressed the ethical implications of nanotechnology.

References

1. P. J. Bond, Preparing the path for nanotechnology. In M. C. Roco and W. S Bainbridge, eds., *Nanotechnology: Societal implications—Maximizing benefits for humanity;* Report of the National Nanotechnology Initiative Workshop, Arlington, VA, 16–21, December 2–3 (2003).

2. M. C. Roco and W. C. Bainbridge, eds., *Societal implications of nanoscience and nanotechnology: NSET Workshop Report*, National Science Foundation, 12 (2001).

3. Human Genome Project, U.S. Department of Energy, www.ornl.gov/sci/techresources/HumanGenomne/elsi.shtml (2006).

4. *The national nanotechnology initiative at five years: Assessment and recommendations of the national nanotechnology advisory panel*, President's Council of Advisors on Science and Technology, The National Nanotechnology Initiative (2005).

5. *The National Nanotechnology Initiative: Research and development leading to a revolution in technology and industry, supplement to the President's FY2006 budget*, Nanoscale Science, Engineering, and Technology Subcommittee, Committee on Technology, and the National Science and Technology Council, March (2005).

6. *Societal dimensions*, National Nanotechnology Initiative, www.nano.gov/html/society/home_society.html

7. D. M. Berube, *Nano-hype: The truth behind the nanotechnology buzz*, Prometheus Books, New York (2006).

8. *MIT Institute for Soldier Nanotechnologies*, Massachusetts Institute of Technology, December 24 (2006) web.mit.edu/ISN/resedarch/team02/index.html

9. *Ethical, legal and other societal issues*, National Nanotechnology Initiative, December 15 (2006), www.nano.gov/html/society/ELSI.html

10. J. A. Shulmam and I. Asimov, *Isaac Asimov's book of science and nature quotations*, Grove Press, New York (1988).

11. *The Belmont Report: Ethical principles and guidelines for the protection of human subjects of research*, The National Commission for the Protection of Human Subjects of Biomedical and Behavioral Research, April 18 (1979). www.ohsr.od.nih.gov/guidelines/belmont.html

12. D. Sarewitz and E. Woodhouse, Small is powerful. In A. Lightman, D. Sarewitz, and C. Desser, eds., *Living with the genie: Essays on technology and the quest for human mastery*, Island Press, Washington, D.C., 63–83 (2003).

13. D. Bennett-Woods and E. Fisher, Nanotechnology and the IRB: A new paradigm for analysis and dialogue, *Proceedings of the Joint Meeting of the European Association for the Study of Science and Technology and Society for Social Studies of Science.* Paris, France (2004), www.csi.ensmp.fr/csi/4S/index.php

14. D. Bennett-Woods, Integrating ethical considerations into funding decisions for emerging technologies, *Journal of Nanotechnology Law & Business*, 4 (2007).

15. D. Sarewitz, Science and happiness. In A. Lightman, D. Sarewitz, and C. Desser, eds., *Living with the genie: Essays on technology and the quest for human mastery*, Island Press, Washington, D.C., 181–200 (2003).

16. B. Gordijn, Converging NBIC technologies for improving human performance, *Journal of Law, Medicine & Ethics*, 34, 726 (2006).

17. R. Rodrigues, The implications of high-rate nanomanufacturing on society and personal privacy, *Bulletin of Science, Technology & Society*, 26, 38 (2006).

18. E. Gutierrez, *Privacy implications of nanotechnology*, Electronic Privacy Information Center, www.epic.org/privacy/nano

19. M. C. Roco and W. S. Bainbridge, eds., *Nanotechnology: Societal implications—Maximizing benefits for humanity*, Theme 10: Education and Human Resource Development, Report of the National Nanotechnology Initiative Workshop, 88, Arlington, VA, December 2–3 (2003).

20. A. S. Daar, H. Thorsteinsdóttir, D. K. Martin, A. C. Smith, S. Nast, and P. A. Singer, Top ten biotechnologies for improving health in developing countries, *Nature Genetics*, 32, 229–232 (2002).

21. *Reducing risks, promoting healthy life*, World Health Organization, The World Health Report 2002. Published report, World Health Organization, Geneva (2002), www.who.int/whr/2002/en

22. *Nanotechnology and the poor: Opportunities and risks*, Meridian Institute, published report, Meridian Institute (2005), www.meridian-nano.org/gdnp/paper.php

23. G. Oberdörster, E. Oberdörster, and J. Oberdörster, Nanotoxicology: An emerging discipline evolving from studies of ultra-fine particles, *Environmental Health Perspectives*. Online, July (2005).

24. J. B. Mangum, E. A. Turpin, A. Antao-Menezes, M. F. Cesta, E. Bermudez, and J. C. Bonner, Single-walled carbon nanotube (SWCNT)-induced interstitial fibrosis in the lungs of rats is associated with increased levels of PDGF, mRNA, and the formation of unique intercellular carbon structures that bridge alveolar macrophages in situ, *Particle Fibre Toxicology*, 3, 15 (2006).

25. A. A. Shvedova, E. R. Kisin, R. Mercer, A. R. Murray, V. J. Johnson, A. I. Potapovich, Y. Y. Tyurina, O. Gorelik, S. Arepalli, and D. Schwegler-Berry, Unusual inflammatory and fibrogenic pulmonary responses to single walled carbon nanotubes in mice. *American Journal of Physiology—Lung Cellular and Molecular Physiology*, 289, L698–L708 (2005).

26. Woodrow Wilson International Center for Scholars. *Project on emerging nanotechnologies. A nanotechnology consumer products inventory.* Retrieved April 29 (2007), www.nanotechproject.org/

27. IPCS risk assessment terminology. Part 1: IPCS/OECD Key generic terms used in chemical hazard/risk assessment. Part 2: OPCS Glossary of key exposure assessment terminology; World Health Organization (WHO), January 14 (2007). www.who.int/ipcs/methods/harmonization/areas/ipcsterminologyparts1and2.pdf

28. W. S. Bainbridge, Public attitudes toward nanotechnology, *Journal of Nano Research*, 4, 561–570 (2002).

29. M. D. Cobb and J. Macoubrie, Public perceptions about nanotechnology: Risks, benefits and trust, *Journal of Nanoparticle Research*, 6, 395–405 (2004).

30. G. Gaskell, T. T. Eyck, J. Jackson, and G. Veltri, Imagining nanotechnology: Cultural support for technological innovation in Europe and the United States, *Public Understanding of Science*, 14, 81–90 (2005).

31. D. A. Scheufele and B. V. Lewenstein, The public and nanotechnology: How citizens make sense of emerging technologies, *Journal of Nanoparticle Research*, 7, 659–667 (2005).

32. S. H. Priest, The North American opinion climate for nanotechnology and its products: Opportunities and challenges, *Journal of Nanoparticle Research*, 8, 563–568 (2006).

33. S. H. Priest and H. Fussell, *Nanotechnology: Constructing the public and public constructions.* Paper presented at the Annual Meeting, Association for Education in Journalism and Mass Communication, San Francisco, CA, August (2006).

34. T. Frey, Collision course: Comings and goings of technologies marked by futurist's "Maximum Freud," *Rocky Mountain News*, July 1 (2005).

35. W. Joy, Why the future doesn't need us, *Wired*, 8.04 (2000) www.wired.com/wired/archive/8.04/joy.html?pg=1&topic=&topic_set=

36. D. B. Hughes, Let the nanotech wars begin! KurzweilAI.net, Dec. 14 (2003) www.kurzweilai.net/news/frame.html?main=/news/news_single.html?id%3D2748

37. R. E. Smalley, Of chemistry, love and nanobots—How soon will we see the nanometer-scale robots envisaged by K. Eric Drexler and other molecular nanotechnologists? The simple answer is never, *Scientific American*, 68–69, September (2001).

38. Public policy, definition, www.bitpipe.com/tlist/Public-Policy.html (2007).

39. M. Crichton, *Prey*, Avon Books, Harper-Collins Publishers, New York (2002).

Problems

Introduction to Societal Issues

2.1 Locate an article on a new development in nanoscience or nanotechnology and list the possible societal implications of such a development.

2.2 Using the NNI Website as a resource, review current activities supported by the National Science Foundation. Briefly describe what you believe to be the more important areas of focus. Imagine that $40 million has already been allocated for environmental, health, and safety research and you have been given the job of authorizing the remaining $40 million in miscellaneous SEIN research funds for the coming year. How might you prioritize the allocation of funds and what would be your justification?

Ethical Implications

2.3 Write a paragraph describing the ethical responsibilities of scientists and engineers in nanotechnology.

2.4 One of the most controversial areas of nanotechnology relates to the convergence of nanotechnology, biotechnology, artificial intelligence, and cognitive science, among others. Developments in these areas might not only help restore people who are ill or have been injured to normal functioning, but also may be used to dramatically enhance human performance and perhaps extend the normal human life span. Using the ethics section for reference, pose a list of ethical questions that might be used to consider the societal implications of the use of nanotechnology to enable human enhancement and life extension.

Legal Implications

2.5 Mary is a graduate student working under the supervision of Dr. Brilliant at Prestigious University. The research Mary conducts on the structures of nanotubes leads to a remarkable new process that can produce very long nanotubes with extraordinary efficiency. The most innovative aspects of developing the process are due to Mary, but Dr. Brilliant has consistently overseen how the research was conducted and offered helpful suggestions at many key points in the development. Mary and Dr. Brilliant patent the process and sell it for a very tidy sum to NanoCorp, which becomes the exclusive user of the process. NanoCorp enjoys impressive profits and begins to dominate the nanotube industry as a direct result of using Mary's process. It sells its products in the United States through an exclusive relationship with a single retailer, Bucky Inc. There are no U.S. regulations that restrict it in any way from adopting Mary's process in the manner it deems most useful.

Five years later, physicians begin to report a previously unknown respiratory illness in patients employed by Nano-Corp, specifically those who work most actively on implementing Mary's process. Over the next several years, research identifies a biological mechanism that conclusively demonstrates a causal relationship between Mary's process and this illness. Over that same period of time, mortality rates begin to increase among NanoCorp workers and a few reports start end consumers suffering from similar afflictions to appear. Those reports also increase and, after another 10 years, it becomes clear that tens of thousands of consumers have either died or suffered a great loss in the enjoyment of their lives as a result of the process Mary developed. Calls for regulation begin, with many public commentators chastising the U.S. government for taking too little action too late.

Work with a partner. One of you should identify the strongest arguments you can devise for finding each of the following parties liable for the injuries suffered by NanoCorp workers and by consumers:

a. Mary
b. Dr. Brilliant
c. Prestigious University
d. NanoCorp
e. Bucky Inc.
f. The United States

The other of you should respond to those arguments with the best defenses you can devise. In each instance, who do think has the stronger argument? Are there any parties for whom you would like to have additional facts? Which parties and which facts?

After completing the exercise, research the history of asbestos litigation in the United States, particularly the scope of its economic impact. What parallels exist with the hypothetical exercise? What lessons does asbestos litigation provide for those involved in the development of nanotechnology? Summarize your conclusions in a short essay.

2.6 The implementation of the patent system in the United States is frequently criticized for permitting too many obvious inventions to be patented. In some ways, this is reminiscent of an exercise performed several years ago by a major physics journal in which many readers thought too many of the published papers to be of only marginal interest; further investigation found general agreement that about 10–15% of the papers would not be missed if they had never been published, but almost no agreement on which papers were in that group. You are also likely to find that many of the technical ideas discussed in this book will present you with conceptual challenges, but when you review them years from now they will no longer seem at all difficult and the struggles you endured to master them will seem almost quaint; you have probably had similar experiences in the past.

With these issues in mind, explain how you would define an *objective* test for obviousness of a patent application. What are the key factors that indicate something was obvious at the time it was "invented"? Is your test effective at putting the person applying it in the position the inventor found himself in? What safeguards does your test have to prevent a hindsight view from being applied unfairly? Is your proposed test dependent on the technology involved—that is, would you apply the same test to judge the obviousness of a nanotechnology invention as you would to judge the obviousness of a new type of pillow or a new design for a shoehorn? Do you think it is more sensible to have a single objective test applicable across all technology or different tests for different technologies?

2.7 One of the aspects of nanoscience that is commonly lauded is its relevance to an unusually wide range of human activities. When considering regulation of nanotechnology, there are then (at least) two approaches that could be adopted by governments: (1) form a centralized agency that regulates all aspects of nanoscience and nanotechnology or (2) allow existing agencies that are focused on particular areas to regulate nanoscience and nanotechnology as they apply to their own areas. Which approach do you think is better? Defend your view fully.

Environmental Implications

2.8 Let us consider the potential for effects using nanoscale titanium dioxide (nano-TiO_2) in a coating for your cell phone. Here is a brief description of the life cycle of the nano-TiO_2. Titanium dioxide is mined and shipped to a raw material manufacturer. A chorination process converts the raw mineral to liquid titanium tetrachloride, a toxic and unstable substance. The nano-TiO_2 is manufactured from the titanium tetrachloride at high temperature with the addition of water. A coating company buys the nano-TiO_2 and it is shipped in a drum in powder form to the company, which mixes it with a polymer coating. The coating company packages the mixture and ships it to the cell phone housing manufacturer, where it is applied as a coating in the fabrication process and shipped to a company that manufactures your phone. Your phone is shipped to a retailer, where you buy it, and then use it for a period of time. Suddenly, it is time for a new phone, so you dispose of the old phone. If you toss it in the trash, a waste hauler collects your phone, and delivers it to a waste facility—either a landfill, where it enters a pile of trash, or an incinerator, where it is burned. Identify the steps in this scenario in which the nano-TiO_2 in your phone could enter the environment.

2.9 Now, using the example in problem 2.8, let us consider the possibility that when you replace your phone, you send it to an electronics recycling program. Instead of entering the trash, the phone is sent to a company that dismantles the phone and separates it to the original components, from which the metallic components are extracted. The nano-TiO_2 is recovered and placed in a container and shipped to a coating manufacturer. In this case, the TiO_2 is much less likely to enter the environment. Should recycling of such electronics be mandatory? How would this be regulated and enforced? Who is ultimately responsible for the consequences of allowing these materials to enter the environment?

2.10 Finally, again using the example of problem 2.8, let us say you work for a company that makes consumer electronic devices. The company decides to put a coating on the surface of a cell phone to increase its scratch resistance, and protect the underlying paint from fading in sunlight using nanoscale titanium dioxide. You are the scientist in charge of testing the coating. Will your work bring you into contact with the nano-TiO_2? If so, how will it?

When you are finished with your testing, you must dispose of the test material. Your safety officer instructs you to put the test material in the trash. Where does it go then? Your testing involves putting the nano-TiO_2 in a water solution. You have to clean the glassware after the test. Washing the glassware sends the nano-TiO_2 down the drain. The waste water from your lab goes to a treatment facility, where it allows solids to filter out before the water goes to a nearby river. The river eventually reaches the ocean. Where is the nano-TiO_2 now? Should labs be severely restricted from such "dumping" of potentially hazardous materials? What if such restrictions become cost prohibitive to the research? What is your responsibility as a scientist or engineer in the safety of your research or the materials/products you develop? Who, if anyone, should be held liable if such releases result in environmental damage and/or human injuries?

Public Perception

2.11 As nanotechnology becomes better known, how do you think public opinion is likely to change? Will it become more positive, or could it become more negative? What factors will influence this trend?

2.12 How do you feel about nanotechnology and the quality of life? Will it improve our quality of life over the next 20 years, make the quality of life worse, or have no impact one way or another? In what areas might we see improvements? Where might we see new problems?

2.13 Think about the last time you confronted a new technology in your life—whether a new gadget you saw advertised or a new medicine your doctor prescribed. How did you decide whether it was likely to be more beneficial or more risky?

2.14 If you had control over several hundred millions of new government research dollars to invest, where would you invest it? Would it go into nanotechnology or would it go to other areas? What ones and why?

Future of Nanotechnology

2.15 Describe five cycles that apply to nanotechnology.
2.16 List five trends that you see affecting the development of nanoscience.
2.17 Identify two or three additional "attractors" that are related to nanotechnology and describe their probable path of development.
2.18 Select one of the nanotechnology attractors described earlier and discuss its potential impact on society.

Ancillary Problems

2.19 The space elevator is an elevator from Earth that goes beyond geosynchronous orbit (35,786 km above the equator). Make a list of pros and cons that are associated with such a concept.
2.20 Not in my backyard—NIMBY! Discuss the pros and cons related to the placement of a nanotechnology research institute in your neighborhood. What conditions must be agreed upon before such an undertaking gets off the ground? What are the benefits? The risks?
2.21 Bill Joy, the founder of Sun Microsystems, wrote an article titled, "Why the Future Doesn't Need Us" [35]. Read this article, go to a café, and discuss its content with classmates.

2.22 Eric Drexler (self-assembler technology) and Richard Smalley (traditional technology development) have had some heated discussions about nanotechnology. Let the Nanotech Wars begin! [36,37]. The public perception regarding the development of nanobots is a heated issue in terms of funding and in other ways. What do you think?
2.23 Geopolitical and economic forces are in effect in a big way today—forces that are able to change the way you and I live. Will nanotechnology have any impact on the course of these global forces (or vice versa)?
2.24 Public policy is a "system of laws, regulatory measures, courses of action and funding priorities concerning a given topic promulgated by a governmental entity or its representatives" [38]. Discuss how this topic relates to nanotechnology and rate its importance.
2.25 What do you foresee with regard to nanotechnology, education, and workforce development in the United States? In the world? Please answer with short sentences. List as many potential employment scenarios as you can that directly involve nanoscience or nanotechnology and that indirectly involve science and technology.
2.26 We mentioned earlier that nanotechnology requires that new partnerships between and among academia, business, and government be forged. Is this true? Why or why not?
2.27 Michael Crichton's novel, *Prey*, tells a horrific story about nanobots gone berserk [39]. What is your impression of the book? Is its premise feasible? Why or why not?

Section 2

Nanotools

CHARACTERIZATION METHODS

When you can measure what you are speaking about, and express it in numbers, you know something about it; but when you cannot measure it, when you cannot express it in numbers, your knowledge is of a meager and unsatisfactory kind: it may be the beginning of knowledge, but you have scarcely, in your thoughts, advanced to the stage of science.

WILLIAM THOMSON, LORD KELVIN,
Popular Lectures and Addresses (1891–1894)

THREADS

We are finished with the "Perspectives" division of the text. Hopefully, a generalized impression of nanoscience and nanotechnology has been acquired. Topics brought up in chapter 2, "Societal Implications of Nano," will be referred to throughout the text to reinforce concepts that are important to all that is outside the science. Before we journey on into nanoscience proper, it is timely, and without question obligatory, to familiarize ourselves with the tools of nanoscience. The "Nanotools" division of the text is comprised of two chapters: this chapter and chapter 4, "Fabrication Methods." Because reference is made throughout the text to images of nanomaterials, relevant spectra, and fabrication methods, placement of the two nanotool chapters early in the text is appropriate. These two chapters provide the language of nanoscience and prepare the student for what unfolds.

The purpose of this chapter is to serve as a general catalog for characterization techniques. Following chapters 3 and 4, we enter the physics division of the book, the properties and phenomena, and begin in earnest to delve into nanoscience with the surface of nanomaterials and its importance.

3.0   CHARACTERIZATION OF NANOMATERIALS

The need to sort, name, categorize, catalog, and detail the things, parts, and components of our world has accompanied humanity since the dawn of civilization. Characterization and reporting are fundamental to science. They are the means by which we communicate our scientific achievements. Measurement is characterization, the language of science. Measurement is accomplished with tools: the instruments, machines, equipment, and computer hardware and software. There are many types of characterization methods and most predate the advent of nanoscience and nanotechnology. Some methods have actually evolved alongside nanoscience and helped launch the *Nano Age* itself. The development of novel, integrated methods designed specifically to probe the nanoworld is an ongoing and evolutionary process.

In this chapter, we stay at an introductory level in presentation, but we hope that enough detail is provided to acquire the meaningful perspective. Although more space is allotted to electron and scanning probe methods, other techniques that provide *nano relevance* are also included in the discussion. Many methods are not mentioned beyond the attention given in **Tables 3.1** through **3.6**

TABLE 3.1	*Optical (Imaging) Probe Characterization Methods*	
Acronym	**Technique**	**Analytical value**
	Binocular microscopes	Imaging/gross morphology
	Compound microscopes	Imaging/fine morphology
CLSM	Confocal laser-scanning microscopy	Imaging/ultrafine morphology
SNOM	Scanning near-field optical microscopy	Rastered images
2PFM	Two-photon fluorescence microscopy	Fluorophores/biological samples
DLS	Dynamic light scattering	
PCS	Photon correlation spectroscopy	Particle sizing
QELS	Quasi-elastic light scattering	
BAM	Brewster angle microscopy	Gas–liquid interface imaging

TABLE 3.2	*Electron Probe Characterization Methods*	
Acronym	**Technique**	**Analytical value**
SEM	Scanning electron microscopy	Raster imaging/topology morphology
EPMA	Electron probe microanalysis	Particle size/local chemical analysis
TEM	Transmission electron microscopy	Imaging/particle size–shape
HRTEM	High-resolution transmission electron microscopy	Imaging structure chemical analysis
STEM	Scanning transmission electron microscopy	Biological samples
FEM	Field emission microscopy	Surface structure/molecular properties
	Electron diffraction	Crystal structure
RHEED	Reflection high-energy electron diffraction	Surface structure
LEED	Low-energy electron diffraction	Surface/adsorbate bonding
EBSD	Electron backscatter diffraction for SEM	Crystallographic information
SAED	Selected area electron diffraction	Local structure information
EELS	Electron energy loss spectroscopy	Inelastic electron interactions
AES	Auger electron spectroscopy	Chemical surface analysis
EBIC	Electron beam induced current	Transport properties in semiconductors

because we assume familiarity with most standard characterization techniques has been achieved by this time. Detailed discussion of characterization methods that measure engineering parameters such as elastic modulus, tensile, hardness and stiffness, etc. is discussed elsewhere. The same is true for biological and medical characterization methods.

3.0.1 Background

Scientists and thinkers over several millennia—Democritus, Hero, Boyle, Dalton, Newton, and many more—have speculated about the nature of matter and light and their interaction. We are now finally able to see atoms and molecules. However, long before the electron microscope could become a tangible tool, the electron had to be discovered. Long before the atomic force microscope became

TABLE 3.3	*Scanning Probe Characterization Methods*	
Acronym	**Technique**	**Analytical value**
AFM	Atomic force microscopy	Topology/imaging/surface structure
LFM	Lateral force microscopy	Surface/friction analysis
CFM	Chemical force microscopy	Chemical/surface analysis
MFM	Magnetic force microscopy	Magnetic materials analysis
STM	Scanning tunneling microscopy	Topology/surface structure/imaging
STS	Scanning tunneling spectroscopy	Electronic density of states
APM	Atomic probe microscopy	Three-dimensional imaging
FIM	Field ion microscopy	Chemical profiles/atomic spacing
IAP	Imaging atomic probe	Emitted ions for imaging surface
APT	Atomic probe tomography	Position-sensitive lateral location of atoms
POSAP	Position-sensitive atomic probe	Mass position resolution

TABLE 3.4	*Photon (Spectroscopic) Probe Characterization Methods*	
Acronym	**Technique**	**Analytical value**
UPS	Ultraviolet photoemission spectroscopy	Surface analysis
UVVS	UV–visible spectroscopy	Chemical analysis
AAS	Atomic absorption spectroscopy	
AES	Atomic emission spectroscopy	
ICP	Fluorescence spectroscopy	Elemental analysis
FS	Inductively coupled plasma spectroscopy	
SPR	Surface plasmon resonance spectroscopy	Surface/adsorbate analysis
LSPR	Localized surface plasmon resonance	Nanosized particle analysis
PLS	Photoluminescence spectroscopy	Elemental analysis
RS	Raman spectroscopy	Vibration analysis
SERS	Surface-enhanced Raman spectroscopy	Chemical analysis/bond structure
SERRS	Surface-enhanced resonant Raman spectroscopy	SERS coupled with electronic transition
smSERS	Single molecule detection SERS	Ability to probe single molecules
FT-IR	Fourier transform infrared spectroscopy	Asymmetrical vibration analysis
NIRS	Near-IR spectroscopy	Surface/IR analysis
DR-FTIR	Diffuse reflectance FTIR	Surface/adsorbate analysis
ATR	Attenuated total reflection	Adsorbate analysis
XRD	X-ray diffraction	Crystal structure
XRF	X-ray fluorescence	
EDX	Energy-dispersive x-ray spectroscopy	Elemental analysis
WDS	Wavelength dispersive x-ray spectroscopy	
XPS	X-ray photoelectron spectroscopy	Surface analysis/depth profiling
SAXS	Small angle x-ray scattering	Surface analysis/particle sizing (1–100 nm)
EXAFS	Extended x-ray absorption fine structure	Local surface atomic structure
CLS	Cathodoluminescence	Characteristic emission
NMR	Nuclear magnetic resonance spectroscopy	Analysis of odd no. nuclear species
EPR	Electron paramagnetic resonance	Analysis of paramagnetic species
STh-MRM	Scanning thermal microwave resonance microscopy	Thermal detection

a tangible tool, the record player had to be invented. We begin our background discussion with spectroscopy.

Spectroscopy (from the Latin *spectrum*, meaning "an appearance," from *spectare*, "to behold" + "scopy" from the Greek *skopein*, "to view") and other analytical

TABLE 3.5	*Ion-Particle Probe Characterization Methods*	
Acronym	**Technique**	**Analytical value**
MS	Mass spectrometry	Material composition
SIMS	Secondary ion mass spectrometry	Composition of solid surfaces
RBS	Rutherford back scattering	Quantitative–qualitative elemental analysis
NSS	Neutron scattering spectroscopy	Chemical structure adsorbate bonding
SANS	Small angle neutron scattering	Surface characterization
NRA	Nuclear reaction analysis	Depth profiling of solid thin films
FReS	Forward recoil elastic spectrometry	Hydrogen deuterium analysis
PIXE	Particle-induced x-ray emission	Nondestructive elemental analysis

TABLE 3.6	*Thermodynamic Characterization Methods*	
Acronym	**Technique**	**Analytical value**
TGA	Thermal gravimetric analysis	Mass loss versus temperature
DTA	Differential thermal analysis	Reaction heats heat capacity
DSC	Differential scanning calorimetry	Reaction heats phase changes
TMA	Thermal mechanical analysis	Thermal mechanical properties
PnDSC	Parallel nano DSC	Combinatorial analysis
NC	Nanocalorimetry	Latent heats of fusion
TPD	Temperature-programmed desorption	Surface adsorbate properties
TDS	Thermal desorption spectroscopy	Coupled with MS
SCAC	Single crystal absorption calorimetry	Absorption and adhesion energies
TDM	Linear–volume thermodilatometry	Dimensions as a function of °T
TL	Thermoluminescence	Surface states detrapping
BET	Brunauer–Emmett–Teller method	Surface area analysis
MP	Mercury porosimetry	Pore volume, pore size
BJH	Barnett–Joyner–Halenda method	Pore size distribution
Sears	Sears method	Colloid size, specific surface area

techniques possess a long and rich history [1]. It began about 2430 years ago, when Aristophanes demonstrated the first use of lenses. Aristotle speculated that in order to see color, light must exist, and Euclid made observations about the focusing properties of spherical mirrors. Ancient scientists such as Seneca, Cleomedes, and Ptolemy contributed to knowledge about scattering, reflection, and refraction. The rest of the history of spectroscopy is filled with too many milestones to mention here and can be found in any reliable reference. We will focus on the history of electrons, electron probe techniques, and scanning probe characterization methods.

G. J. Stoney in 1874 coined the term electron (from the Greek *electron,* meaning "amber," a substance when rubbed that causes sparks). An electron is a stable subatomic particle with a negative charge equivalent to 1.6022×10^{-19} C and with rest mass m_e equal to 9.1094×10^{-31} kg. In 1895, Jean Perrin determined that cathode rays carry a negative charge. During his study of cathode ray tubes in 1877, J. J. Thomson (**Fig. 3.1**) discovered that the electron was a subatomic particle, but he was not sure exactly what kind of particle: "I can see no escape from the conclusion that [cathode rays] are charges of negative electricity carried by particles of matter.... What are these particles? Are they atoms, or molecules, or matter in a still finer state of subdivision?" He also postulated that electrons were corpuscular (e.g., particulate) in nature. He came to this conclusion after applying a magnetic field to a cathode ray tube and found that the trajectory of an "electron" particle beam is deflected by the magnetic field. He then determined the mass-to-charge ratio (m/e) of the electron by measuring the degree of deflection. Thomson said upon his discovery, "We have in cathode rays matter in a new state." Since electron mass is difficult to measure, $-e/m_e$ is the quotient used by physicists where

$$-\frac{e}{m_e} = -1.758820150(44) \times 10^{11}\ \text{C} \cdot \text{kg}^{-1} \tag{3.1}$$

FIG. 3.1 *J. J. Thomson is known for his work with cathode ray tubes and many other notable scientific discoveries. He was awarded the Nobel Prize in physics in 1906.*

THOMSON, Joseph John
Nobel Laureate PHYSICS 1906
© Nobelstiftelsen

Source: Image courtesy of the Nobel Foundation. With permission.

In 1909, R. Millikan conducted the well-known oil-drop experiment and was able to measure the electron's charge by measuring to total charge q on an oil drop of predetermined mass m_{drop}.

$$q = \frac{m_{drop}\, g}{E} \qquad (3.2)$$

Where g is the gravitational constant and E is the strength of the applied electric field. By varying the strength of the ionizing x-rays, he found that q was always a multiple of 1.6×10^{-19} C. Louis de Broglie proposed that electrons possess wavelike qualities. In 1924, he theorized that the wavelength of a particle depends upon its momentum. De Broglie was awarded the Nobel Prize in physics in 1929 for his theory of the wave nature of electrons. From the contributions of

Thomson, de Broglie, and others, H. Busch developed the concept of the electromagnetic lens in 1926.

We introduced SEM, TEM, XRD, STM, and AFM methods in chapter 1 and discussed some of the highlights of their unique history. The versatility and power of these nanoscale tools have increased over the past few decades. The control, manipulation, and integration of individual atoms have at last been accomplished due to the advent of these methods. One of the greatest developments in the twentieth century was the application of computer capability to analytical tools. In addition to their value in crunching data, computers also assist researchers with positioning samples, focusing images, monitoring reactions, and data acquisition.

3.0.2   *Types of Characterization Methods*

Nanoscience and nanotechnology are interdisciplinary in nature. It is fair to say that an equally broad inventory of characterization and analytical techniques applies to nanostructured materials. There are approximately 700 single-signal characterization techniques and approximately 100 that are considered to be multisignal techniques [2]. These numbers are based on the types and combinations of three kinds of physical phenomena: (1) *primary* (1°) analytical probes (electrons, photons, neutrons, ions, etc.) and input stresses (heat, pressure, electric field, magnetic field, electricity, mechanical stress, etc.); (2) the types of measurable *secondary* effects (2°) (electrons, electromagnetic radiation, heat, pressure, volume change, mechanical deformations, etc.); and (3) the monitoring medium of choice (energy, temperature, mass, intensity, time, angle, or phase) [2]. The 1° probe, whether electron or photon, interacts with matter. The matter then responds to regain its equilibrium state. During this process, the 1° probe is modified. Excited electrons, phonons, excitons, or plasmons are some examples of how matter is altered. Alterations of the 1° probe result in the 2° effect—the signal that we measure [2].

Optical Probe Characterization Techniques.   The primary probe in optical methods (**Table 3.1**) is visible light of wavelength within the 400- to 800-nm range. The 2° signal from most 1° photon methods is usually another photon. The 2° photon signal arises from many possible physical phenomena, such as elastic scattering, inelastic scattering, or from emission (fluorescence).

Human eyes have great depth of field and depth of focus; however, they are not efficient at peering into nanostructured materials without assistance. Binocular scopes are useful for viewing gross morphology like a clump or large strands of carbon nanotubes. Compound microscopes offer more detail but resolution is limited to a few hundred microns. Confocal, near-field, and two-photon microscopes using visible and near-ultraviolet (UV) 1° photons have managed to push the resolution envelope of optical methods into the sub-100 nm range. Dynamic light scattering is an optical technique that is used to measure particle size, and Brewster angle microscopy (BAM) is used to analyze phase behavior at an air–water interface. Brewster's law is given by

$$\theta_{\mathrm{B}} = \arctan\left(\frac{m_2}{m_1}\right) \tag{3.3}$$

where m_2 and m_1 are the refractive indices of 2 media and is applicable to light that moves between those media. θ_B is the Brewster's angle, the point at which light of one polarization cannot be reflected. Selected optical microscope methods are listed in **Table 3.1.** Optical characterization methods are used primarily for imaging but also are useful in chemical analysis.

Electron Probe Characterization Methods. Electron probe characterization methods use high-energy electron beams. The major 2° effects (signals) of electron interactions with matter are discussed in more detail later in the chapter. Electron beams are used for imaging, chemical analysis, and determination of material structure. The most familiar electron probe methods are scanning electron microscopy (SEM) and transmission electron microscopy (TEM). In SEM, the electron beam is rastered across the analytical surface. A raster (from the German *raster*, meaning "screen," from the Latin *rastrum*, indicating "rake") is a rectangular pattern of parallel scanned lines from which an image is assembled. Electron guns in older televisions produced a picture by rastering electrons over a phosphor grid (the TV screen). Electron probe methods are listed in **Table 3.2.** SEM and TEM instruments are usually coupled to other types of analytical tools, primarily with energy-dispersive x-ray analysis (EDX or EDS).

Scanning Probe Characterization Techniques. All scanning probe microscopy (SPM) methods have two things in common: a finely sharpened probe tip and a system that enables the tip to scan (raster) over the surface of a sample. An image is acquired by moving the probe, or the sample, with a raster action—similar in concept to the SEM but utilizing completely different types of probes. Resolution on the order of atoms is possible with scanning probe techniques. Different types of scanning probe methods are listed in **Table 3.3.**

Spectroscopic Characterization Methods. There are many kinds of 1° photon probe techniques (**Table 3.4**). As in optical imaging, most photon probe methods involve a 1° photon in and a 2° photon out. UV–visible spectroscopy, based on absorption or emission (fluorescence) of photons, is the most common form of spectroscopy. Although the wavelength for UV–visible methods is within the nanoscale (e.g., <1 μm), the dimensions of most nanomaterials are much smaller. UV–visible techniques mainly involve electron transitions. Related parameters such as extinction, intensity, absorption, transmission, reflection, and wavelength are measured during UV–visible spectroscopy. Raman and Fourier transform infrared spectroscopy (FT-IR) are techniques that measure molecular and phonon vibrations of materials. In each case, 2° photon carrier waves are analyzed and information about vibrational states is acquired. Symmetrical vibrations (polarization) are exclusive to Raman spectroscopy, and asymmetrical vibrations (dipolar) reside in the domain of infrared analysis. In surface plasmon spectroscopy, the absorption of 1° photons causes surface electron plasmons to oscillate. The wavelength, λ_{max}, of the absorption is indicative of the surface state of the metal (e.g., adsorbed species).

X-rays form another important category of 1° photon excitation that is important to nanoscale analysis. The wavelength of x-rays ranges from 0.01 to 10 nm, corresponding quite well to the size of nanoparticles. The major 2° effects are photons or electrons depending on the type of analyzer installed in your instrument. The two most *nanorelevant* x-ray techniques are x-ray diffraction (XRD) and energy-dispersive x-ray (EDX), which reveal structural information and chemical analysis, respectively.

Nuclear magnetic resonance (NMR) and electron spin resonance (ESR) are techniques that rely on an applied magnetic field to induce signals. NMR is applicable to molecules that contain odd numbers of protons and neutrons (e.g., have a magnetic moment such as ^{1}H and ^{13}C). Electron paramagnetic resonance (EPR), commonly referred to as ESR, is a technique that is able to analyze materials that have unpaired electrons. There are numerous excellent sources that describe these techniques in more detail.

Microwave spectroscopy methods are applicable to nanomaterials. Local resolution for microwave spectroscopy by thermal near-field microscopy is less than 100 nm.

Ion-Particle Characterization Techniques. Mass spectrometry (MS) involves the breakup of a parent particle into ionized components. Daughter particles are produced by electron impact (1° electron) and by other techniques. MS used to detect the presence of sodium nanoclusters in particle beams showed that cluster abundance was dependent on size; the most stable clusters possessed a *magic number* of atoms. Smalley and his group used MS to prove the existence of C_{60}. Secondary ion MS (SIMS) is also a popular method. Neutrons and ions are able to modify matter by inducing thermal vibrations, rupturing bonds, and displacing atoms [2]. The 2° signals comprise secondary ions. In addition, 1° ion probes are capable of implantation and inducing ionization and de-excitation [2].

Thermodynamic Characterization Methods. Thermodynamic techniques are characterization methods that involve thermodynamic parameters. The primary probe is not a photon or an electron but rather a thermodynamic parameter such as temperature or pressure. In these cases, changes in temperature, pressure, phase, and volume are considered to be secondary effects. Bond energy, phase change (e.g., melting point), heat capacity, surface adsorption–desorption, volume change, surface tension, and vapor pressure are some effects measured by thermodynamic techniques.

Thermoluminescence is a technique in which a photoluminescence response is generated in a material from heating. Thermoluminescence provides information about surface states—in particular, the ability to detect trapped metastable electron–hole pairs in nanoparticles [3]. Heating the sample induces lattice vibrations that in turn provide energy for electron–hole recombination. Upon recombination, optical photons are emitted [3].

There are methods that have more "nano" relevance than others, but all apply with validity to nanomaterial characterization. The lists provided in **Tables 3.1** through **3.6** are by no means exhaustive. In general, two fundamental types of characterization exist: imaging by microscopy and analysis by spectroscopy [2]. Other conventional techniques, like thermal gravimetric analysis (TGA) and the Brunauer–Emmett–Teller (BET) surface area methods, do not employ electromagnetic radiation per se but can be, in the very broadest of senses, considered to be spectroscopic methods because x and y axes are involved.

Bulk Engineering Characterization Methods.   There are two general types of nanomaterials: (1) bulk materials that are composed of nanomaterials and (2) independent nanomaterials (e.g., a carbon nanotube resonator). Mechanical testing methods are used on both types of nanomaterials. Tensile properties exist at the nanoscale as does friction. Mechanical testing includes abrasion and scratch resistance, hardness, elastic modulus, tensile strength, fracture toughness, stiffness, and fatigue strength. Tensile stress in Pascal,

$$\sigma = \frac{P}{A} \tag{3.4}$$

where P is pressure in Neutons and A is area.

Stiffness K is defined as

$$K = \frac{P}{\delta} \tag{3.5}$$

where P is the applied force and δ is the deflected distance. Electrical testing methods include conductivity, capacitance, and resistivity. Material characterization tests include tensiometry (surface tension) and pycnometry (density). Energy dissipation tests include heat transfer, reflectivity, and thermal expansion. There are many more. All engineering parameters apply to nanomaterials.

3.0.3   *Optics and Resolution*

Due to the incredible contributions made by TEM and scanning tunneling microscopy (STM) techniques, we are now capable of imaging and detecting individual atoms and molecules! We have indeed achieved one of the holy grails of science by creating images of the "indivisible atoms" about which Democritus and others speculated over the millennia. However, just when we thought we could look no smaller, scientists at IBM Almaden Research Center in California on April 26, 2007, described a new microscope that is able to look *inside* an atom [4]. Some of the challenges associated with imaging smaller and smaller objects are discussed next.

The fundamental relationships that apply to imaging by optical wavelengths also apply to imaging with electrons. The main differences are the source of the 1° probe and the material of the lenses. Lenses in electron probe instruments consist of coordinated magnetic fields.

The relationship between focal length f (the distance from the lens center of a thin lens to the focal point) and the object and image distances is

$$\frac{1}{f} = \frac{1}{o} + \frac{1}{i} \tag{3.6}$$

where o and i are the object and image distances, respectively. The *focal point* is the place on the principal axis of the lens where parallel rays converge to a common point or focus. Graphical depictions of the focal point can be found in many references. Magnification, **M**, is simply the ratio of optical distances.

$$M = \frac{i}{o} \tag{3.7}$$

In the electron microscope technique, magnification is determined by measuring the size of the object displayed on a cathode ray tube (CRT) screen divided by the size of the scanned area. Electron microscopes automatically display magnification and a scale bar for reference. Magnification is simple to calculate in SEM. For example, if the CRT screen is 10×10 cm, a magnification of $10\times$ correlates with a specimen area equal to 1 cm^2 and $100\times \rightarrow$ 1 mm^2; $1000\times \rightarrow 100$ μm^2; $10,000\times \rightarrow 10$ μm^2; $100,000\times \rightarrow 1$ μm^2; and $1,000,000\times \rightarrow 100$ nm^2. Magnification is a one-dimensional measure and not the ratio of areas.

Resolution (from the Latin *resolutionem*, "the process of reducing things into simpler forms") is the ability to reduce or separate something into its components (e.g., the minimum distance between two distinguishable objects) or the finest detail that can be distinguished in an image from some set distance. Resolution is related to contrast. *Resolving power* is the ability to measure the angular separation of images that are close together. Counting pixels is another means of calculating resolution. Magnification and resolution, however, are not the same. Prove it to yourself by enlarging an image on your computer screen until you are able to see pixels.

Imaging of samples by optical or electronic means is dependent on the wavelength of the source or probe; the shorter the wavelength of the probe, the greater the resolving power of the instrument. The resolving ability of optical microscopes is ca. 0.35 μm. Resolution can be improved if an oil emulsion is applied to the objective lens to 0.18 μm (180 nm). As a general rule of thumb, the best achievable resolution of an optical microscope is equal to approximately half the wavelength of the illuminating beam. The wavelength of visible light lies roughly between 400 and 700 nm. On the other hand, the practical resolving ability of electron beam microscopes, depending on the applied accelerating voltage, is approximately 0.2 nm.

The quality of the hardware and the operational settings in a microscope also influence the resolving power. The *numerical aperture* of a lens factors significantly into the resolution equation. The numerical aperture (N.A.) of an optical system is a dimensionless unit that indicates the range of angles from which the lens system is able to accept or emit visible light and is defined by

$$\textit{Numerical aperture} = \text{N.A.} = n \sin \theta \tag{3.8}$$

where n is the refractive index of the medium between the lens and the sample and θ is the half-angle that is subtended by the rays entering the objective lens (circular aperture). The resolving power of the lens is proportional to wavelength divided by the numerical aperture (**Fig. 3.2**):

$$R \propto \frac{\lambda}{\text{N.A.}} \tag{3.9}$$

The resolution gets better (e.g., R gets smaller) when (1) the wavelength of the impinging radiation gets shorter (e.g., the λ of blue light is shorter than λ

Three different lens configurations highlighting the numerical aperture of electron micro-scopes are compared. The closer the aperture is to the sample the larger the angle enclosing the lens becomes. From the equation, N.A. is proportional to the sine of angle θ. For θ = 10° (a long focal length), N.A. = 0.17; for θ = 40°, N.A. = 0.64; and for θ = 60° (a shorter focal length), N.A. = 0.87. Small N.A.s translate into low resolution. High N.A. translates into high resolution.

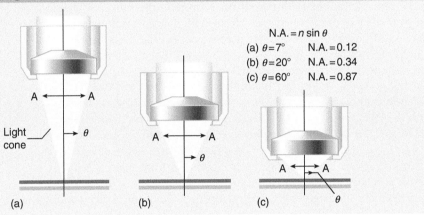

of red light), (2) the refractive index of the connecting medium gets larger (e.g., air with $n = 1.0$ vs. oil with $n \sim 1.5$), and (3) N.A. gets larger (e.g., the larger the half-angle θ is the bigger N.A. becomes). N.A. also increases as focal length f decreases.

Three phenomena are relevant to resolution. The Abbe diffraction barrier acts to limit resolution that in order to produce a true image, the N.A. must be large enough to transmit the entire diffraction pattern of a point source. The Rayleigh criterion is applied to judge the minimal resolvable detail of an object. An *Airy disk,* named after physicist George Airy, is the projection of light through a circular aperture (e.g., a pinhole) or is formed by focused light impinging upon a point source. The distance between point sources is an indicator of the resolution. The Airy projection appears to be made of concentric rings (orders) with decreasing thickness and increasing spacing between the rings progressing from the center (**Fig. 3.3**). A map of the intensity is shown adjacent to the projection. The diameter of the most intense (the central) Airy disk is inversely proportional to the diameter of the aperture [2]. Because each object in the image is capable of diffracting the impinging electron (or optical) beam, each object is able to form its own Airy disk. If the point images seem to be merged (e.g., unfocused, unresolved), a *disk of confusion* is formed.

The Rayleigh criterion states that two such images are considered to be *just resolved* when the central maximum of the first image falls into the first minimum of the diffraction pattern of the second image. For light transiting through a slit with width equal to d, the Rayleigh criterion is

$$d \sin \theta_R = \lambda; \quad \sin \theta_R = \frac{\lambda}{d}; \quad \theta_R = \frac{\lambda}{d} \tag{3.10}$$

FIG. 3.3

Three Airy discs are pictured. When light from a point source passes through an aperture, various points of the specimen appear to be a compilation of small patterns (small concentric circles) in the image. These are called Airy discs and are caused by the diffraction of light passing through the circular aperture of the objective. The Rayleigh criterion describes the limits of resolution at which an image can be resolved into separate entities. On the left is an unresolved image; in the middle, the condition of the Rayleigh criterion is depicted; and on the right, a resolvable image is shown.

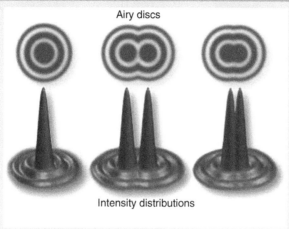

Airy discs

Intensity distributions

For small angles, the sine of an angle in radians is approximately equal to the angle itself: $\sin \theta_R \sim \theta_R$. Therefore, the Rayleigh criterion minimum angular resolution (in radians) for a circular aperture of diameter D is

$$Angular\ resolution = \theta_R = \frac{1.22\lambda}{2\text{N.A.}} = \frac{1.22\lambda}{D} = \frac{0.61\lambda}{\text{N.A.}} \qquad (3.11)$$

where D is the diameter of the circular aperture. Diffraction and resolution are different but related phenomena. It is the diffraction of light that restricts our capability to resolve features that are in close proximity. This factor is also important to optical lithography techniques.

For nanoscientists that use microscopes, the distance between two objects bears more relevance than the angular resolution. Angular resolution is converted into spatial resolution by multiplying by the focal length f. If the distance R is the arc subtended by θ_{min}, then R $= f\,\theta_R$ and

$$Spatial\ resolution = f\theta_R = \frac{(1.22\lambda)f}{D} \qquad (3.12)$$

Electron microscopes share similar characteristics with optical microscopes. Mathematical treatments of magnification, resolution, and diffraction for both are essentially the same.

The depth of focus (D_{Focus}) for an electron microscope is about 300 times that of an optical microscope. Why is this? Depth of focus is simply the distance

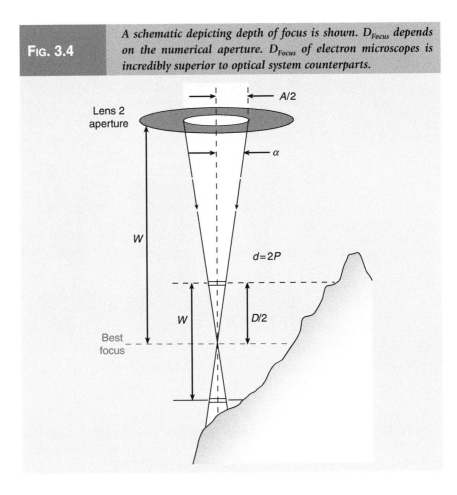

FIG. 3.4 *A schematic depicting depth of focus is shown. D_{Focus} depends on the numerical aperture. D_{Focus} of electron microscopes is incredibly superior to optical system counterparts.*

above and below the focal plane in which the object is clearly in focus. Aperture size and working distance (focal length of the objective lens) are two factors that exert influence over the depth of focus. The angle α is the angle formed between the aperture and the focal point on the sample. From **Figure 3.4**, notice that the D_{Focus} becomes larger as α gets smaller. Once again, because α is small we can safely assume that $\tan\alpha \approx \alpha$.

The angle α is the angle of divergence that is dependent on the ratio between the aperture radius and the working distance **W** [5]:

$$\tan\alpha = \frac{R_0}{\mathbf{W}} \sim \alpha \tag{3.13}$$

$$Depth\ of\ focus = D_{Focus} = \frac{0.2}{\alpha\mathbf{M}}\ \mu m \tag{3.14}$$

where **M** is the magnification as before. Another equation that is used to derive D_{Focus} relates the ratio of the wavelength to the square of the numerical aperture [6]:

$$Depth\ of\ focus = D_{\text{Focus}} \propto \frac{\lambda}{(\text{N.A.})^2} \qquad (3.15)$$

There are trade-offs, however. Making **W** larger yields a smaller α and thereby a larger D_{Focus}, but an overall decrease in R, the resolution of the image [5]. Increasing D_{Focus} by reducing the aperture diameter results in a decreased signal-to-noise ratio (e.g., a snowier picture). Depth of field (D_{Field}) is not the same as depth of focus. Depth of field is the lateral range within which images remain in focus and is described mathematically by [2]

$$Depth\ of\ field = D_{\text{Field}} = \frac{0.61\lambda}{n\sin\alpha\tan\alpha} \qquad (3.16)$$

where α is the semi-angle subtended by the aperture as before.

Marvin Minsky invented the confocal microscope in 1953 but it was not until the advent of lasers that the technique became useful. In a confocal microscope, images (optical sections) are illuminated through a pinhole aperture and a fluorescent signal is channeled via another pinhole near the detector. The image is then compiled from laser scans of the surface. In this way, the instrument is able to screen "out-of-focus" signals. In near-field microscopes, a subwavelength-sized light source (e.g., a glass fiber) hovering a few nanometers above the sample surface is able to acquire optical images on the order of 50-nm resolution. In two-photon microscopy (TPM), the optical section is acquired only from the focal plane where excitation occurs. The Osaka Bull depicted in **Figure 2.4** is an example of two-photon lithography.

The Abbe diffraction limit can be mitigated by application of the *near-field technique*. The basic concept of the procedure is to illuminate a sample through an aperture with subwavelength dimensions held close to the sample (e.g., the aperture-sample distance within half the diameter of the aperture). The resolution in this case depends on the aperture diameter and not the wavelength of the light. In order to acquire a whole image, the aperture is scanned (rastered) over the surface of the sample. Therefore, scanning near-field optical microscopy (SNOM) provided quite a breakthrough in optical imaging.

A *confocal microscope* is able to bypass some of the technical difficulties associated with wide-field optical microscopes. In particular, a confocal laser-scanning microscope (CLSM) utilizes a point-illumination system that allows only the light in the focal plane to be detected. The illuminated volume is on the order of $0.25 \times 0.25 \times 0.5$ μm. Just like with the near-field microscope, rastering is required in order to acquire a larger image. Confocal systems have the capability to block unwanted reflections or emissions. The apertures are designed to eliminate out-of-focus light from entering the detector. Confocal microscopes most commonly operate under the fluorescent mode but are also capable of assimilating reflected light. The resolution of confocal scopes is better than that of traditional optical microscopes (130 nm versus 350 nm, respectively).

The concept of resolution in atomic force microscopy (AFM) and other scanning probe methods is quite different from that of radiation-based microscopy. First of all, AFM imaging is a three-dimensional technique. Resolution limits of STM or AFM samples are quite impressive with resolution on the order of single atoms. The sharpness of the tip, the geometry of the sample surface, and the type of scanning mode (e.g., tapping versus contact) all play a role in determining the value of the resolution of the procedure.

Image Quality Factors.   Other factors contribute to the quality of the image. *Contrast* is the difference in intensity between an object and the background and other nearby objects. Because focal length for a specific lens system relies on wavelength, *chromatic aberrations* may occur if polychromatic radiation is used. Lens imperfections also contribute to chromatic aberration. *Spherical aberrations* arise due to imperfections in the lens by causing "rays" to fall away from the point of focus. "Stopping down" the lens (e.g., decreasing the size of the aperture) is a way to mitigate spherical aberration. *Astigmatism* is the result of rays originating from an off-axis to form images at alternative points. It is an aberration caused by unequal focus of vertical and horizontal lines at different points along the optical axis.

Optical Terms.   Because terms associated with optics will be used throughout the book, we will list a few of the most common. *Interference* is the superposition of electromagnetic waves. Depending on the structure of a material, electron microscopy (EM) (and particle) waves can add constructively, destructively, or anywhere in between. The colors seen in soap bubble thin films are based on interference phenomena. *Diffraction* is the behavior of a wave when it encounters an obstacle. Once again, depending on the material, interference patterns formed by the probing radiation yield information about its structure. Point objects are able to diffract waves to form Airy disks. *Reflection* of light occurs when an incoming wave is symmetrically bounced off a surface. *Refraction* occurs when impinging radiation is allowed to transit through a material, albeit in bent form. *Absorption* is a process by which electrons of a material interact with EM radiation— the excitation of electronic energy states. Relaxation back to the ground state results in emission that is characteristic of the transition. *Transmission* of radiation through a material is what is left after absorption and reflection have occurred. *Scattering* is a form of reflection and occurs off the surfaces of particles larger than 20–50 nm but smaller than visible bulk materials. Electrons and photons are differentially scattered off surfaces of materials depending on the type of material and the energy of the 1° probe.

3.0.4   The Nano Perspective

Without tools to characterize nanomaterials, we would be blind to the nanoscale. Methods to measure atomic scale phenomena essentially predate the formal designation of nanoscience as a new path in scientific exploration. Of course, techniques that measure macroscopic objects have been around since the beginnings of intellectual thought. One of the major distinctions to keep in mind is the following. Although we are more than able to measure the properties of a macroscopic quantity of nanomaterials—a clump of powder, for example—we are now able to measure the properties of a single atom, a single molecule, a single cluster, a single colloidal particle, and a single nanotube. We now challenge ourselves further to observe, measure, and probe materials that are not much larger than atoms and molecules and to develop integrated techniques that extract information simultaneously, quickly, efficiently, and accurately.

Characterization methods of the past have extracted molecular and atomic information by averaging the bulk response to some stimulation. This kind of methodology also applies to the analysis of nanomaterials. Raman spectroscopic analysis of carbon nanotubes relies on a relatively large amount of sample. However, the focus now is shifting to the characterization of individual nano-materials and molecules. With regard to this aspect of nanoscience, one must certainly agree that this change in the analytical paradigm is indeed a revolutionary one. Characterization methods have undergone dramatic changes over the past decade and there seems to be no limit of ingenious ways to probe the secrets of individual nanostructures.

3.1 ELECTRON PROBE METHODS

In this section we focus on characterization methods that rely on the electron as the primary probe. Once electrons interact with a specimen, secondary effects in the form of radiation (photons), electrons, and heat are detected and transformed into images or spectra. Electron probe characterization methods in particular are important to nanoscience due to their capability to analyze the structure of extremely small particles. Of these, the electron microscope is the most significant.

Electrons exhibit particle–wave duality. The de Broglie equation relates the momentum of an electron (a particle phenomenon) to wavelength:

$$p = \frac{h\nu}{c} = \frac{h}{\lambda} \tag{3.17}$$

where

p is momentum ($m\upsilon$)
ν is the frequency
c is the speed of light ($2.998 \times 10^8 \ \mathrm{m \cdot s^{-1}}$)
h is Planck's constant ($6.6262 \times 10^{-34} \ \mathrm{J \cdot s}$)
λ is the wavelength

Converting p into the conventional mass–velocity form, the wavelength of the electron is

$$\lambda = \frac{h}{p} = \frac{h}{m_e \upsilon} \tag{3.18}$$

Relativistic considerations may apply if the energy becomes very high (e.g., as electron velocity approaches the speed of light). The wavelength of the electron must be modified according to

$$\lambda = \frac{h}{p} = \frac{h}{\gamma m\upsilon} = \frac{h}{m\upsilon} \sqrt{1 - \frac{\upsilon^2}{c^2}} \tag{3.19}$$

where γ is the Lorentz factor for special relativity and τ the adjusted speed.

$$\gamma = \frac{dt}{d\tau} = \frac{1}{\sqrt{1-\beta^2}} = \frac{c}{\sqrt{c^2 - \upsilon^2}} = \frac{1}{\sqrt{1-\left(\dfrac{\upsilon^2}{c^2}\right)}} \tag{3.20}$$

β is the relative speed with respect to the speed of light c:

$$\beta = \frac{\upsilon}{c} \tag{3.21}$$

The Lorentz factor is always greater than 1. The relatavistic kinetic energy form is given by

$$KE = \left[\frac{1}{\sqrt{1-(\upsilon/c)^2}} - 1\right] mc^2 = (\gamma - 1) mc^2 \tag{3.22}$$

The classical form of the kinetic energy of the electron is given by

$$KE = \tfrac{1}{2} m_e \upsilon^2 \tag{3.23}$$

The wavelength of an electron can then be calculated by

$$\lambda = \frac{h}{\sqrt{2 m_e e V}} \tag{3.24}$$

where V (in $J \cdot C^{-1}$), the voltage, is converted to energy in the form of electron volts (eV).

In 1927, American scientists Davisson and Germer diffracted electrons from the surface of a nickel crystal and demonstrated the wavelike character of electrons [7–9]. This experiment validated the wave-particle nature of electrons. Bragg's law exclusively addresses wavelike-interference phenomena (more discussion is presented about Bragg's law in chapter 5):

$$n\lambda = 2d \sin\theta \tag{3.25}$$

Davisson and Germer's experiment showed that interference phenomena also apply to particle waves. Equations (3.26) and (3.27) illustrate the relationship between Bragg's law and wavelength:

$$\frac{1}{\lambda} = \frac{n}{2d\sin\theta} = \left(0.815\sqrt{V}\right) nm \tag{3.26}$$

$$\frac{1}{\lambda} = \frac{n}{2d\sin\theta} = \frac{p}{h} = \frac{\sqrt{2m_e E}}{h} = \frac{\sqrt{2m_e e V}}{h} \tag{3.27}$$

The discovery was accidental—not unlike the results of numerous other breakthrough experiments. The primary intent of their experiment was to measure the energies of electrons scattered from a metal surface. The *Faraday box* detector was rotated along an arc as a pattern emerged. Davisson and Germer also verified a periodic fluctuation in intensity of the scattered electron beam (e.g., wave behavior that conformed to Bragg's law).

3.1.1 Electron Interactions with Matter

Electrons are able to interact with matter in many different ways (**Fig. 3.5a** and **b**). In all cases, electrons serve as the primary probe (1°). An electron impinging on a solid surface may be scattered once, several times, or not at all. Multiple scattering occurs if the dimension of the material is greater than approximately twice the mean free path of the electron. The scattering may be elastic or inelastic and forward or backward. The probability of scattering is determined by the scattering cross-section and the mean free path of the electron—both dependent on the material properties (and size).

When a high-energy beam of electrons is incident upon a thick specimen, a pear-shaped region known as the *interaction volume* (volume of excitation) is formed. The interaction volume, which increases with increasing beam voltage (electron energy) and decreases with increasing specimen density (e.g., generally with increasing atomic number z), is normally between 1 and 5 µm. In general, electrons are able to penetrate solid materials to a depth proportional to $\sim E^2$ and laterally to $\sim E^{1.5}$ (or $\propto V^2$ and $V^{1.5}$, respectively) [2,10].

Types of electron–matter interactions in thick samples (SEM) are listed in **Table 3.7** and for thin samples (TEM) in **Table 3.8.** The important 2° effects that are exploited in the majority of investigations using SEM are secondary

Fig. 3.5	*Electron interactions with matter are shown in the image. On the left, the interaction volume corresponds to materials that have some meaningful thickness. Samples analyzed by SEM are generally thick and demonstrate interaction volume phenomena. On the right, interaction of electrons with a thin sample is shown. The TEM method is dependent on the capability of electrons to penetrate and pass through samples. For this reason, samples must be very thin.*

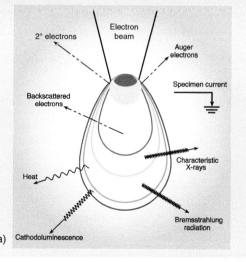

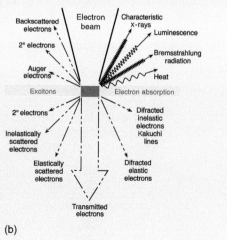

and backscattered electrons (for imaging) and characteristic x-rays (for simultaneous elemental analysis with EDX). The important 2° effects that are exploited in the majority of investigations using a TEM are transmitted electrons (for imaging), elastically scattered electrons (for simultaneous structural analysis by electron diffraction), and characteristic x-rays (for simultaneous elemental analysis with EDX).

TABLE 3.7	*Electron–Matter Interactions Caused by 1° Electron Probe: Thick Samples*		
2° Effects	**Physical description of effect**	**Value to analysis**	**Use**
2° Electrons	An electron transits near atom core and releases enough energy to excite a K-shell (2°) electron. A high population of 2° electrons is emitted per 1° electron (electron yield, δ). $$\delta = \frac{\text{Number of 2°}\,e}{\text{Number of 1°}\,e} \approx \frac{1}{\cos\beta} \quad (3.29)$$ where β is the incidence angle of the electron beam to the sample. δ values between 0.3 and 0.7 are typical. The energy of 2° electrons is considered to be low and ranges from ca. 5 to 50 eV. 2° Electrons are released from a depth of <1–5 nm. Others claim this depth is more on the order of 5–50 nm [10]. The emission of 2° electrons depends on the topography and to a lesser extent on atomic number. The greater the angle between the surface normal and the incident beam is, the greater is the path length of electrons and hence the larger the electron signal becomes (e.g., the contrast).	The SEM image is formed from low-energy 2° electrons after they exit the sample surface and are collected. Surface morphology, topography, and particle shape analysis are commonly performed by an SEM detecting and collecting 2° electrons. Resolution of 2° electrons depends on the diameter of the incident beam, the excitation volume, and the incident angle of the beam. 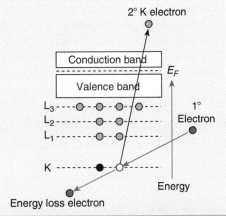 *Formation of 2° electrons. 1° Electrons penetrate a sample and have enough energy to excite a core electron. The 1° electron loses energy in the process.*	SEM EPMA
Backscattered electrons	1° Electrons collide with atoms normal to their path and scattered backward with large deflection (as high as 180°). The scattering is considered to be quasi-elastic. Backscattered electron yield, η, is low compared to 2° electrons and depends heavily on the local chemistry. $$\eta = \frac{\text{BSE current}}{\text{Incident } E - \text{Beam current}} \quad (3.30)$$	Backscattered electrons are highly directional and peak in the plane containing the surface normal and the incident beam. Topographical imaging used to differentiate areas of a sample (e.g., elements with higher atomic number) look brighter. Therefore, compositional analysis is accomplished based on brightness. One of the primary uses of backscattered electrons is in the analysis of mechanically alloyed materials. If the surface is flat, contrast can be obtained from regions differing in z by 1.	SEM EBSD

TABLE 3.7 (CONTD.)	*Electron–Matter Interactions Caused by 1° Electron Probe: Thick Samples*		
2° Effects	**Physical description of effect**	**Value to analysis**	**Use**
	Generation of backscattered electrons relies on specimen atomic number, interaction volume, and accelerating voltage. The larger the *z* of an element is, the larger is the angular deflection. The KE is similar to 1° electrons and sample depth of emission ranges from <10 to 100 nm. Depth on the order of 0.5 µm also is possible. The excitation volume for BSE is large and resolution thereby suffers. If only elastically scattered electrons are selected, resolution is on par with 2° electrons.	*Backscattered electrons. The trajectory of BSEs is back towards the beam after a slingshot action around the nucleus of the atom.*	
Surface plasmons	Surface plasmons are oscillating waves of electrons on metal. Since surface plasmons are activated by relatively low-energy sources, low-energy electrons (or photons) are responsible for generating surface plasmons on metals undergoing SEM analysis.	Surface plasmon spectroscopy is not performed in SEM analysis. *Surface plasmon oscillations of free electrons on the surface of metals are depicted. A dipolar resonance is established as the free electrons on the metal surface oscillate in response to EM stimulation.*	SPR
Auger electrons	Following release of an outer shell 2° electron, a higher energy electron falls into a vacancy created by the ejected inner shell (usually K-shell electron). Following ionization and relaxation, enough energy is released to emit low-energy outer (surface layer) electrons called Auger electrons. Simultaneous production of visible photons of energy	Auger electrons are used for surface chemical analysis—an electron energy fingerprint. They are characteristic of elements and are also capable of contributing information about the bonding state of that element. Analysis of quantity (intensity, % composition of low *z* elements) and kinetic energy of ejected electrons lead to semiquantitative and qualitative data.	AES SEM

(*continued*)

TABLE 3.7 (CONTD.)	*Electron–Matter Interactions Caused by 1° Electron Probe: Thick Samples*		
2° Effects	**Physical description of effect**	**Value to analysis**	**Use**
	$$\lambda = \frac{hc}{\Delta E} \qquad (3.31)$$ where ΔE is the difference in energy between the vacant shell and the higher energy level [2]. Although the yield is small, Auger electrons are characteristic of the element of origin. Auger electrons are of low energy: 100–1000 eV. Others say it is more like 2–50 keV. Auger electrons are emitted from or near the surface with sample depth <0.3–5 nm (1–3 nm).	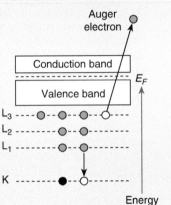 *Formation of Auger electrons. Valence band electrons are emitted following stimulation and relaxation (filling empty core shells). Auger electrons are characteristic of the element.*	
Characteristic x-rays	Following release of a 2° electron, usually K-shell, a vacancy is created. Electrons with higher energy drop down (cascade) to fill the void. In the process, x-rays are emitted that are characteristic of the energy transition and specific to the element. The characteristic x-ray peaks are superimposed on a continuous background spectrum consisting of Bremsstrahlung (see following effect). X-rays are emitted from a sample depth ranging from ~1 to 3 μm. Detection of x-rays is also semiquantitative because x-ray intensity can be converted into percentage composition. ZAF (atomic number effects) correction is applied with the assistance of a computer.	X-rays originating from K-, L-, M-shell x-rays are element specific. The emitted x-ray impinges upon a Li-drifted Si solid-state detector that is able to differentiate between x-rays of differing energy. The x-ray excites electrons in the detector into the conduction band. The greater the energy of the x-ray is, the higher is the number of electrons. Elemental analysis of micron diameter particles and larger materials is routinely accomplished. Silicon detectors are inefficient at low energies and therefore do not detect any element below Na. However, if a wavelength dispersive crystal spectrometer is coupled with the Si detector, elements down to B can be analyzed. 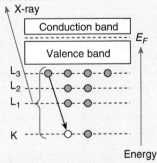 *Formation of characteristic x-rays from electron probe stimulation. Filling a shell vacancy by a higher energy electron results in release of x-rays that are characteristic of the element.*	EDX in SEM EDX in EPMA

TABLE 3.7 (CONTD.)	*Electron–Matter Interactions Caused by 1° Electron Probe: Thick Samples*		
2° Effects	**Physical description of effect**	**Value to analysis**	**Use**
Bremsstrahlung continuum radiation	When incident electrons undergo inelastic collisions, they slow down due to strong electrostatic forces applied by the nuclei. The lost energy appears in the form of EM radiation. This is the reason it is called "breaking radiation" and covers the entire range of the x-ray region of the spectrum. Bremsstrahlung is the major component of loss for high-energy electrons. Bremsstrahlung radiation is energy emitted by a moving charged particle and is proportional to m_e^{-2}, the rest mass of the particle.	Bremsstrahlung is considered to be background noise upon which the characteristic x-ray spectrum is superimposed.	EDX
Cathode ray luminescence (cathodoluminescence)	Visible light fluorescence following inelastic scattering of 1° electrons is released after ionization of valence shell electrons. In semiconductors, holes are produced after excitation. Recombination of electron–hole pairs results in emission of a photon. UV, visible light fluorescence, IR are produced depending on energy of 1° electron and penetration depth. The sample depth of electron penetration that results in cathodoluminescence ranges from <2 to 8 μm.	Crystal field influence Trace element abundance Semiconductors Geological samples Considered to be a "beam-injection" technique	CL-SEM EBIC
Induced current	Induced current can be generated in a semiconductor sample upon exposure to electrons. Excess carriers are generated with an electron beam and are assembled within an electric field inside the specimen. The electron-beam-induced current induces carriers recombine at defects or are collected at the contact, resulting in a current.	Induced currents typically are in the nano- to microampere range. They are used to locate and analyze p–n junctions in semiconductors, as well as to measure the diffusion length of minority carriers from Schottky or p–n junctions. Usually a thin evaporated Schottky contact made of aluminum is applied.	EBIC
Heating	Inelastically scattered electrons bounce back and forth within the interaction volume until energy is lost. Phonon excitation of lattice can result. Phonon vibration is then converted into heat. Low-energy continuum photons and low-energy electrons do not escape the lattice and energy is transformed into vibration of bonds and lattice (phonons).	The temperature rise is given by $$\Delta T = \frac{4.8E_o b_i}{C_\tau d_o} \qquad (3.32)$$ where E_o is the accelerating voltage in keV, b_i is the beam current, C_t is the thermal conductivity of the sample, and d_o is the beam diameter	SEM

TABLE 3.8	*Electron–Matter Interactions Caused by 1° Electron Probe: Thin Samples*		
2° Effects	**Physical description of effect**	**Value to analysis**	**Use**
Transmitted unscattered electrons	No interaction between electrons and sample as electrons transverse the specimen. Acceleration voltage in TEM is on the order of 100 keV or higher. Transmission capability is inversely proportional to thickness. Low atomic-molecular weight materials transmit electrons and appear to be light on the phosphor image screen. Heavy elements interact more strongly with electrons and appear as dark regions on the phosphor image screen.	Light areas equate to more transmission and dark areas to electron-dense areas. The result is a TEM image. Particle size can be measured in TEM. It is good practice to corroborate the measured size with those determined by other methods. TEM allows for direct observation of crystal structure. *Elastically scattered electrons in TEM are the bases for the image. Contrast is generated when dense materials absorb electrons.*	TEM HRTEM STEM
Elastically scattered electrons (ESEs)	Elastically scattered electrons are deflected with no loss of energy and transmitted obliquely though a sample. Elastically scattered electrons adhere to Bragg's law of diffraction.	Elastically scattered electrons are used in generating electron diffraction patterns. Information concerning crystal structure and orientation about the sample is acquired from the patterns.	Electron diffraction
Inelastically scattered electrons (ISEs)	Inelastically scattered electrons lose energy by interactions with the sample. Eventually, they too are transmitted through the thin sample. ISEs are transmitted obliquely through sample	Electron loss value typical of element Unique to bonding state (e.g., oxidation) Elemental analysis Kakuchi bands related to atomic spacing are formed. The bands consist of alternating dark and light lines. The inelastically scattered electrons that participate in the formation of Kakuchi lines are related to the atomic spacing of the specimen. The width of the Kakuchi line is inversely proportional to the atomic spacing and can track back to elastically scattered electron diffraction patterns.	EELS TEM HRTEM Electron diffraction
Characteristic x-rays	See **Table 3.7**	See **Table 3.7**	EDX (TEM)

3.1.2 Scanning Electron Microscopy and Electron Probe Microanalysis

The main components of the scanning electron microscope are shown in **Figure 3.6**. The electron gun at the top of the electron "optical" column produces the electron beam. The beam is focused to a diameter of ∼50 Å at the foot of the column by a series of magnetic lenses and is scanned in a square TV-type raster across the surface of the specimen. The various 2° effects emitted from the surface of the specimen (e.g., secondary electrons, backscattered electrons, x-rays) may be detected and the signals amplified and displayed.

Components.   Generally, two types of materials are used to produce electrons: filaments made of either tungsten (W) or lanthanum hexaboride (LaB_6). The tungsten cathode is a wire filament bent into the shape of a hairpin with a

FIG. 3.6   *A schematic representation of a scanning electron microscope (SEM) is shown. A high-powered electron beam originating from an electron gun near the top of the column is accelerated down a column through sets of collimators and magnetic focusing lenses. Detectors are placed near and around a sample. Many kinds of 2° signals are collected and assimilated. If an SEM has EDX capability, in situ elemental analysis is also conducted.*

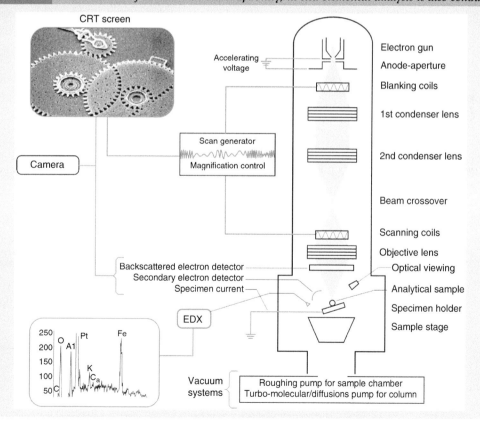

V-shaped tip that has a radius of 100 µm. The cathode is heated directly by a filament current, i_f. Electrons exit the filament with average energy:

$$E_e \sim kT \tag{3.28}$$

The energy necessary to facilitate the ejection of an electron from a material surface is governed by the work function of that material. Metals are the best materials for electron emission due to the presence of free electrons on their surface. The energy required to emit electrons is derived from the heat produced by the filament current. Field emission electron guns add extra energy by application of an electric field to the filament. The temperature of field emission electron guns (a.k.a. cold cathode emission) is much lower than that required for thermionic filaments. Richardson's law (Richardson–Dushman equation) relates the current density J_c obtained by thermionic emission of the filament:

$$J_c = \mathcal{A}_c T^2 \exp\left(\frac{-\Phi_W}{k_B T}\right) \tag{3.33}$$

where

 $\mathcal{A}_c$ is a constant that is characteristic of the material
 T is the absolute temperature
 Φ_W is the work function of the material
 k_B is Boltzmann's constant $(1.38 \times 10^{-23} \; J \cdot K^{-1})$

Use of a material with a lower work function Φ_W or a higher constant $\mathcal{A}_c$ results in an increase in the cathode current density (e.g., the brightness of the beam). The most important cathode material to be developed so far is sintered, or single crystal, LaB_6 for which $\mathcal{A}_c = 40 \; A \cdot cm^{-2} \cdot K^2$ and $\Phi_W = 2.4 \; eV$. A Wehnelt cap surrounds the filament and applies a negative potential that repels electrons and channels them through an aperture where a space charge is formed. The electrons are emitted as a point source called a space charge that is centered along the horizontal optical axis of the microscope. A positively charged anodic plate serves to accelerate electrons through the aperture.

Just like with optical imaging systems, electron microscopes are equipped with condenser and objective lenses; the main difference between the two systems is that the lenses in an electron microscope are not made of a solid material but rather controlled by a magnetic field. The electron beam exiting the anode plate of the electron gun assembly is divergent. A condenser lens is responsible for collimating the divergent beam, focusing it down the column and regulating the amount of current. Many SEMs have two sets of condenser lenses. The objective lens is located near the sample and is known as the "probe-forming" lens. Astigmation of the image is caused by many factors (e.g., imperfections in the magnetic lenses and contaminants in the column). Stigmators are special lenses that are able to compensate for the distortions and are housed in the objective lens.

Secondary electrons are collected and examined by a scintillator-photomultiplier *Everhart–Thornley* detector. The secondary electrons (SEs) are first collimated by a grid with an applied bias and then impacted upon the detector surface— a short-lifetime phosphor that is very efficient in converting electrons into ~400-nm ultraviolet photons. The number of electrons that reach the detector is a function of the surface topography. High points and surface features that face the detector produce more 2° electrons and hence a greater signal. Backscattered electrons (BSEs) are detected by a semiconductor array located at the bottom of the column. A current is produced when BSEs strike the semiconductor array. Tapping into the specimen current is another way to acquire an image. Most SEM samples are conducting, and all SEM specimens are fixed to a metal stage with conducting cement or tape. A specimen current is generated by inelastic scattering and absorption of impinging electrons after the SE, BSE, transmitted, and Auger electrons have left the sample. This absorbed current is a function of the atomic number of the elements in the specimen [10].

Image Generation.   Image generation occurs by rastering the electron beam (of spot size ca. 50 μm or less) across the sample surface. A reconstructed image is sent to a CRT and viewed. During the rastering process, there is a "dwell time," at which point the beam is paused. During the dwell time, the numerous types of secondary electron effects are expressed. The lifetime of the secondary effects is shorter than that of the dwell time. Before the beam moves to the next segment, the secondary effects have been detected, recorded, and dissipated. Magnification, brightness, and contrast all affect the image quality. Brightness is a function of the entire image and is controlled by the amplitude of the signal. Contrast is the variation of the signal from one rastered point to the next and is expressed by

$$C = \Delta S \cdot S_{\text{Average}}^{-1} \qquad (3.34)$$

where ΔS is the change in the strength of the signal between two points and S_{Average} is the average signal strength. Cathodoluminescence is an imaging technique used mostly in investigating luminescence from mineral specimens. A camera housed in the column system takes a photograph of the emitted light from the sample.

Operation.   The operation of an SEM is quite straightforward, although training is required to protect the high-maintenance instrument and its expensive components. Sample loading and mounting begin by fixing the specimen to a metallic stub with graphite cement or conducting tape. The stub is then attached to the metallic stage and placed inside the sample chamber. The chamber is evacuated with a roughing pump. The column is evacuated to a pressure less than 10^{-6} torr in order to begin the filament warm-up process. A special procedure must be followed to engage the filament (usually of the thermionic variety) and to align the beam. The acceleration potential is stepped up slowly. Once the beam current is detected and adjusted, the sample is introduced into the column and is ready for analysis.

Figure 3.7 depicts a JEOL JSM-7700F field emission SEM [11]. The resolution of this instrument is 0.6 nm at 5 kV and 1.0 nm at 1 kV. Magnification capability ranges from 25× to as high as 2 million×. Accelerating voltage ranges from 0.5 kV to approximately 30 kV. New lens systems are able to correct spherical and chromatic aberration, thereby improving the resolution of objects. This modern SEM is capable of performing high-resolution backscattered electron imaging as well as scanning transmission electron microscope imaging of thin samples [12]. As one can imagine, there is a trend to combine capabilities within one instrument.

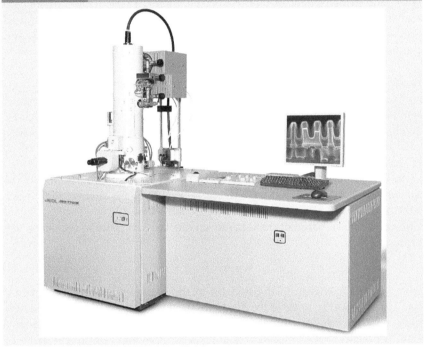

FIG. 3.7 *A JEOL JSM-7700F field emission SEM is shown. The resolution limit of this beautiful instrument is less than 1 nm. Magnification is on the order of 2× million. The resolution is 0.6 nm at 5 kV and 1.0 nm at 1 kV. Magnification capability ranges from 25× to as high as 2.0 million. Accelerating voltage ranges from 0.5 to approximately 30 kV.*

Source: Image reproduced with permission from JEOL, Ltd.

Not only does nano converge with regard to disciplines, nano-instrumentation also converges with regard to capability.

Electron Probe Microanalysis.   Henry Mosely in 1913 found that the wavelength of x-rays emitted after excitation was inversely proportional to the square of the atomic number of an element:

$$\lambda \propto \frac{1}{z^2} \tag{3.35}$$

This was the first time the composition of any material (brass in this case) was identified by characteristic x-rays. In 1922, Assar Hadding applied x-ray spectrometry to analyze minerals. The father of electron probe microanalysis (EPMA) is considered to be James Hillier, who patented the process in 1943.

EPMA is designed for nondestructive x-ray imaging and microanalysis of samples. Some special features of the EPMA include capability of ultrafine spot focusing (e.g., down to 1 μm), optical microscope imaging, and enhanced sample positioning [12]. The EPMA has spatial resolution capability of ca. 1 μm

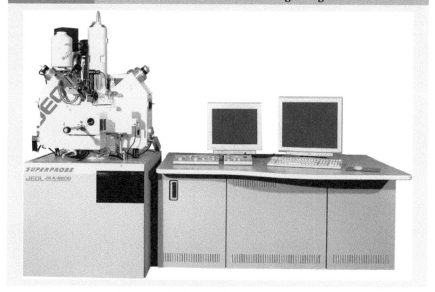

FIG. 3.8 *A JEOL JXA-8100/8200 EPMA is pictured. This instrument is capable of analyzing light elements like boron. Elemental characterization is accomplished with wavelength dispersive x-ray techniques (WDS). A comparison between EDX and WDS is given in section 3.3.3, "X-ray Methods." The backscattered electron image detection limit is 6 nm with a working distance (WD) of 11 mm and under 30 kV accelerating voltage conditions.*

Source: Image reproduced with permission from JEOL, Ltd.

(considered to be excellent), is highly sensitive (on the order of 0.5% for the major elements), and has a detection limit of ca. 100 ppm. Some EPMA have five crystal-focusing wavelength dispersive x-ray spectrometers on board. EDX detectors are also included for quick analysis. X-ray energy in the range of 0.1–15 keV is typically analyzed.

A JEOL JXA-8100/8200 electron probe microanalyzer is shown in **Figure 3.8** [12]. Image acquisition occurs by collection and assimilation of backscattered electrons. The detectable element range begins with very light elements such as $_5$B (with $_4$Be analysis capability also possible) to very heavy ones such as $_{92}$U. The detectable x-ray wavelength range (by wavelength dispersive spectroscopy, WDS) is from 9.3 to 0.087 nm. Large specimen size can be accommodated with dimensions of $100 \times 100 \times 50$ mm (height) and an analyzable area of 90 mm^2. The probing current ranges from 1 pA to 10 µA. The backscattered electron image detection limit is 6 nm with a working distance (WD) of 11 mm and under 30 kV accelerating voltage conditions [12].

3.1.3 *Transmission Electron Microscopy*

The TEM functions by the same principles as the SEM except that the detector is a phosphor plate that is able to capture images formed by transmitted electrons. Another major difference between TEM and SEM is that the accelerating voltage in TEM is usually far greater: 300 kV compared to ~50 kV. The wavelength of the electron beam is in the picometer range, 6.13–2.24 pm. The thermionic electron gun consists of a filament, usually made of tungsten or lithium hexaboride. The filament is heated under an applied potential until electrons are produced. Another difference from SEM is that TEMs have a projector lens system at the base of the column. The image that the TEM operator observes on the phosphor screen is the projected image of the sample.

Components. A schematic version of a generic TEM is shown in **Figure 3.9**. A thermionic or field emission gun is used to accelerate electrons between 100 and 400 kV. The electron beam is accelerated by the anode plate and then collimated via an aperture. The electrons pass through a double condenser lens system and focus upon a sample. In order to be transparent to the electron beam, specimens

FIG. 3.9

A schematic representation of a transmission electron microscope (TEM) is shown. TEMs are capable of significantly higher accelerating voltage; hence, the resolving power of such instruments is in the range of 0.05 nm. TEMs are also equipped with in situ EDX and electron diffraction capability. In many ways TEMs are considered to be the crown jewel of nanotools.

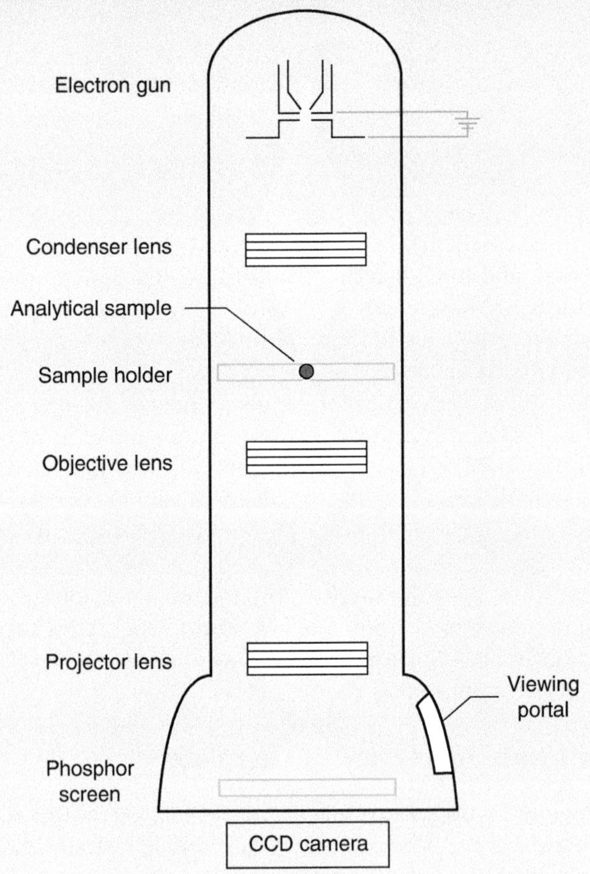

Electron gun

Condenser lens

Analytical sample

Sample holder

Objective lens

Projector lens

Viewing portal

Phosphor screen

CCD camera

typically are 1000 atoms or less in thickness (e.g., a few nanometers to >100 nm). Electrons that pass through the sample are then focused by an objective lens, channeled through another aperture, and then projected onto the phosphor fluorescent screen by the action of the projector lens for the purposes of viewing. The image is captured with a photographic Polaroid camera (outmoded now) or by a charge-coupled device (CCD) camera.

The CCD camera or photographic paper captures the "shadow image" of the specimen depending on the density of its various components in the sample. CCDs are integrated circuits that are composed of close-packed photodiodes. They function by converting photons into an image. In the TEM, the photons arise from the action of electrons impinging on a phosphorescent screen. When a specimen image is ready to be recorded, the viewing phosphor screen is tilted to open access to the CCD camera. Upon completion of image acquisition, the viewing screen is replaced to its original position.

In 1969, Willard Boyle and George Smith of Bell Labs invented the CCD [13]. They discovered that it was able to interact with electronic charges by way of the photoelectric effect. In 1974, a 100×100 pixel CCD prototype device was manufactured and the Sony Corporation began mass production of CCDs for camcorder use. The first CCD camera was used in a TEM in 1982 (the 100×100 pixel device directly exposed to 100-keV electrons) [14]. Direct detection, however, presented some serious issues. J. C. H. Spence and J. M. Zhou proposed an indirect detection strategy that employed a scintillation screen with an optical coupler [15]. CCD technology has developed significantly since 1974 with now 4000×4000 pixel arrays available for TEM imaging tasks. Thus, the Polaroid method of image acquisition gave way to digital imaging technology once again.

Resolution. The resolution of a good TEM is ca. 0.2 nm, on par with the distance between two atoms and the atomic radii of some heavy metals. Carl Zeiss SMT announced a major breakthrough in TEM resolution in 2005, achieving 0.07 nm with an experimental ultrahigh-resolution 200-kV field emission gun transmission electron microscope (FEG-UHRTEM). This level of resolution is near the theoretical limit of TEM analysis.

Image Generation. There are generally four types of images that are produced by TEM:

1. Transmitted electrons are responsible for generating a *bright field image* of the specimen. When the objective aperture is positioned just below the specimen, only the transmitted electron beam is allowed to pass down the column and form an image of the specimen on the phosphor screen. Any diffracted or inelastically scattered electrons are excluded from passage.
2. To acquire a *low-resolution dark field image*, the objective aperture must be repositioned (e.g., adjusted to one side or another of the main axis of the column) while a metal plate is inserted to block the major beam. Although spherical aberrations are formed in this configuration, a quick identification of specimens that produce diffraction patterns is accomplished.

3. A *high-resolution dark field image* can be formed. In this configuration, electron diffraction patterns are acquired for analysis.
4. If, on the other hand, the aperture is opened to allow many beams, including the direct electron beam, to pass, images are formed by interference of the direct beam and the diffracted beams. These are called high-resolution lattice images and are formed in a high-resolution (HR)-TEM. With HR-TEM, which is a combination of bright-field and lattice imaging, real and reciprocal space can be observed simultaneously.

Operation.   Samples are restricted to thickness generally less than 100 nm. The specimen is usually placed on a copper grid that is less than 100 μm thick and 3 mm in diameter. The specimen is prepared in several ways. Initially, for hard inorganic samples, sawing, grinding, and polishing are done to thin the sample. Such mechanical polishing proceeds until a thickness of several microns is achieved. The polished sample is then placed in an ion mill to undergo a finer level of thinning. Argon ions bombard the sample and remove material according to the accelerating voltage of the ion mill system. Milling continues until breakthrough is achieved. The circumferential area surrounding the zone of breakthrough can be extremely thin, less than a few nanometers. Chemical and electrochemical thinning and reactive ion etching are also techniques used to reduce the thickness of samples for TEM analysis.

Another procedure involves embedding the sample in a polymeric resin and sectioning with a microtome by the action of a diamond knife. This is usually accomplished for softer samples—biological samples in particular. The microtome instrument is capable of producing films with nanometer thickness. Following sectioning, the samples float on water and are collected on a copper grid, dried, and placed into the TEM sample holder in the middle of the column. The thickness of sectioned samples is estimated by observation of interference colors. Films that are thinner than a few hundred nanometers appear gray as they float on the water collector. Sections with thickness between ca. 200 nm and 1 μm display all the colors in the spectrum. The phenomenon of interference colors shows how bulk material properties change as size approaches smaller dimensions. Thicker sections look like the parent bulk embedding material. Staining samples, particularly those of biological origin, with heavy metal dyes that contain osmium or lead significantly enhances the structural detail during TEM analysis. The heavy atom sites appear darker because of better scattering or absorption of electrons. Direct deposition of sample materials sonicated in acetone or other volatile solvents onto carbon-coated copper grids without prior preparation is also common. Sonication of the sample beforehand breaks up and disperses the sample.

Limitations of the TEM procedure include the following:

1. Extensive sample preparation is time consuming and this limits the number of samples that can be imaged and analyzed.
2. In addition to the small physical size of samples, the field of view is relatively small as well. For example, the section under analysis may not be representative of the sample as a whole.

3. Sample structure and morphology have been known to change drastically under exposure to electrons with extremely high energy, and damage to biological samples in particular can occur.

4. TEM is a high-vacuum instrument that is costly to operate and to maintain, certainly playing a role in the high barrier entry required to conduct nanoscience research.

However, the advantages of the TEM technique are numerous. Any type of sample, whether electrically insulating, semiconducting, or conducting, is able to be imaged by TEM. The incredible resolution capability allows for atomic level inspection.

An image of a JEOL JEM-3200FS field emission transmission electron microscope is shown in **Figure 3.10** [11]. The TEM is equipped with a 300-kV FEG (a zirconium–tungsten Schottky FEG). It is also capable of conducting electron

FIG. 3.10 *A JEOL JEM-3200FS field emission TEM is pictured. The TEM is equipped with a 300 kV FEG (field emission gun). It is able to conduct EDX, EELS, and electron diffraction analysis. The element in the electron gun is ZrO/W and its point-image resolution capability is 0.17 nm. Beam spot size can be focused down to 0.4 nm in diameter. Magnification ranges from 100× to 1.5 million.*

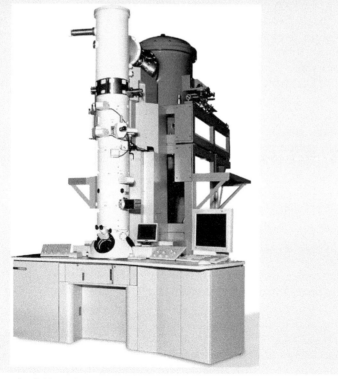

Source: Image reproduced with permission from JEOL, Ltd.

energy loss spectroscopy (EELS), EDS, and electron diffraction. At 300 kV, the TEM is capable of 0.17-nm point-image resolution and energy resolution of 0.9 *e*V. The spot size can be made as small as 0.4 nm. The magnification ranges from 100× to as high as 1.5 million× the size of the specimen structure.

3.1.4   *Other Important Electron Probe Methods*

Auger Electron Spectroscopy (AES).   Lise Meitner and Pierre Auger independently discovered Auger effect in 1923. Although her paper on the procedure predated Auger's, the procedure is known by the name of the latter. Lise Meitner should have also shared a piece of the Nobel Prize in chemistry for her contribution to the discovery of nuclear fission, but Otto Hahn was awarded the distinction in 1944.

Auger electrons, briefly described in **Table 3.7**, are produced from radiation (usually x-rays) or high-energy electron excitation of core electrons. With x-ray excitation, the kinetic energy of the ejected Auger electron is equivalent to

$$E_{\text{KE,Auger}} = h\nu - E_{\text{b}} - \Phi_W \tag{3.36}$$

where $h\nu$ is the energy of the x-ray; E_{b} is the binding energy of a K-, L-, or M-shell electron; and Φ_W is the work function of the material. Technically, the kinetic energy of an Auger electron is independent of the energy of the excitation source. Each element has a diagnostic Auger electron fingerprint. The kinetic energy of the electron is independent of the excitation energy. Chemical analysis of surfaces is attainable by Auger electron spectroscopy. AES consists of three steps: (1) ionization by removal of a core electron, (2) emission of Auger electrons or x-ray fluorescence, and (3) analysis of the Auger electrons (or x-rays).

AES is highly sensitive for all elements except H and He and is used to monitor surface cleanliness of samples. AES also is capable of semiquantitative analysis of surface composition. Auger depth profiling, in combination with argon ion sputter etching (500 *e*V–5 keV ions), provides semiquantitative compositional data as a function of sample depth. Scanning Auger microscopy (SAM) is capable of providing compositional information about surface heterogeneity. Auger analysis must be conducted under ultrahigh vacuum conditions. SEM provides better surface resolved images; AES is better equipped than XPS to produce depth profiles and its sampling depth is ca. 2 nm compared to EDX, which is capable of analyzing elements at depths of 1–2 μm.

An image of a JEOL FE-Auger JAMP-9500F field-emission Auger microscope is shown in **Figure 3.11** [11]. A Schottky field emission gun acts as the electron source that is capable of delivering 0.5–30 kV accelerating voltage. A Faraday cup serves as a detector for secondary electrons at a resolution of 3 nm (at 25 kV and 10 pA). The probe diameter for Auger analysis is 6 nm (at 25 kV and 1 nA). The range of analytical Auger electrons is 0–2500 *e*V.

Low-Energy Electron Diffraction Spectroscopy.   LEED is one of the primary methods of investigating surface structure. It can be applied qualitatively (size, symmetry of adsorbate unit cells) or quantitatively (information about atomic position). The principles of LEED were uncovered accidentally in

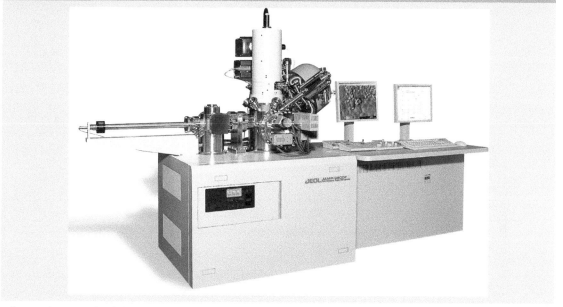

FIG. 3.11 *An image of a JEOL FE-Auger JAMP-9500F field-emission (Schottky) auger microscope is shown. Accelerating voltage capability ranges from 0.5 to 30 kV. Resolution is ca. 3 nm. The probe diameter for Auger analysis is 3 nm. Magnification ranges from 20 to 500,000×. This Auger instrument is equipped with an ion gun for depth analysis of samples and is capable of EDX and EBSD analyses.*

Source: Image reproduced with permission from JEOL, Ltd.

1924 by Davisson and Kunsman during a study of the 2° emission of electrons from a nickel surface. Low-energy electrons (10–500 *e*V) normal to the sample surface are elastically scattered by the electron density surrounding surface atoms [16]. This range of energy corresponds to distance between surface atoms or molecules (de Broglie λ between 0.39 and 0.055 nm) and therefore is able to generate diffraction patterns from the two-dimensional surface structure. The scattered electrons are collected by a retardation grid and analyzed. When analyzing two-dimensional surfaces, one of the Miller indices can be dropped; instead of *h*, *k*, and *l*, one only need be concerned about *h* and *k*. Miller indices are discussed more fully in chapter 5.

Electron Energy Loss Spectroscopy. Hillier and Baker described the EELS technique in 1944 [17]. It only became popular, however, after better microscopy techniques evolved by the 1990s and with the development of ultrahigh vacuum systems. Spatial resolution of 0.1 nm is obtainable with this method. The EELS technique is based on inelastic interactions of electrons with the specimen. The magnitude of energy loss is measured by an electron spectrometer and is manifested in the form of plasmons, inner shell ionization, inter- and intraband transitions, and phonons. EELS works well for low *z* elements and is therefore a complementary method to EDX but with better resolution.

3.2   SCANNING PROBE MICROSCOPY METHODS

Scanning probe microscopes are able to create detailed three-dimensional images of specimen surfaces with atomic resolution [18]. Two basic types of scanning probe microscopes are used in nanoscience. The first, developed in 1981 by Binnig et al., is called the scanning tunneling microscope (STM) [19]. In the STM technique, the magnitude of the tunneling current between the probe tip and the atoms of a substrate surface is monitored. STM samples must therefore be electrically conducting. AFM and its derivatives comprise a second class of SPMs. In AFM, the size of the force between a probe tip and the atoms of substrate surface is monitored. The AFM also developed by Binnig et al. was developed in 1985 specifically to address materials that were insulating. The AFM is able to measure forces on the order of 1 μN and less [20]. A third type of scanning probe method, called *scanning atom-probe microscopy* (APM), is based on a different approach and is discussed in section 3.2.4 in more detail.

There are many similarities between AFM and STM. Both are equipped with a probe tip fastened to a cantilever, a scanning (motion) mechanism, and a detector system. Four achievements contributed to the development of scanning probe methods:

1. Physical cushioning mechanisms (e.g., bungee cords hanging from a ceiling), motion-dampening eddy currents induced by magnets and copper plates, and nitrogen-gas-regulated suspension systems all serve to isolate the SPM from external vibration sources. Protective enclosures shield the microscope from room drafts. If extremely high-resolution work is required, cryogenic temperatures eliminate atomic and molecular movement.
2. Computerized feedback system control of piezoelectric devices with nanometer level precision guides the probe during its descent to the surface—less than a nanometer above where van der Waals attractions exert force on the probe tip (in AFM) or tunneling current is able to flow (in STM).
3. Computerized control of piezoelectric scanners moves the probe tip across the sample surface with nanometer precision.
4. The fabrication of sharper probes allows for better resolution of surface features. For example, carbon nanotube probes with sharpness of less than 1 nm have shown promise in the past few years.

Scanning Probe Tips.   The probe tip of a scanning probe microscope is important to the feature-resolving ability of atomic force and scanning tunneling microscopes. The probe is usually made of silicon or Si_3N_4 for AFM and tungsten for STM. In both cases, the probe is sharpened to a fine tip. Atomic probe methods rely on sharpened probes that are positioned as close to a sample surface as possible. Ultimately, a probe tip sharpened to one atom provides the best resolution (**Fig. 3.12**).

The probe tip has a rich history. The stylus profiler (profilometer) (from the Latin *pro-*, "forth" + *filare*, "to draw out," *filum*, "thread, spin" + "measure") is the progenitor of the atomic force microscope. A profilometer is an instrument designed to measure the topography of industrial material surfaces. Shmalz

| **FIG. 3.12** | *Two probes manufactured by Veeco Instruments are shown. Probe tips are made of many materials that range from silicon to harder materials like tungsten. The resolving power of AFM and STM depends on the sharpness of the tip (the curvature). Sharper tips result in more clarity of image.* |

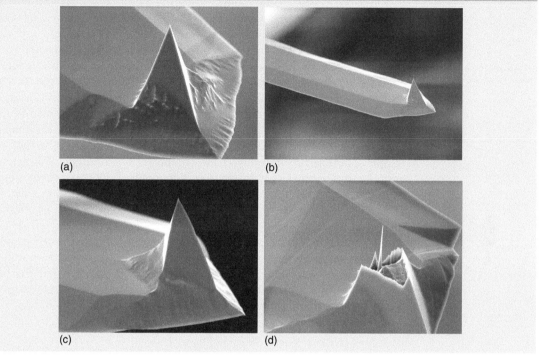

(a)   (b)

(c)   (d)

Source: Courtesy of Veeco Instruments, Inc.

invented the first profilometer in 1929 [21–23]. He used an optical lever arm to detect the motion of a sharpened probe attached to a shiny cantilever during a surface raster process. The magnification capability of this instrument was ca. 1000×. Early profilometers were equipped with a diamond-tipped stylus attached to a coil in an electric field. Raster of the profilometer tip across a surface induces a current that is proportional to the surface roughness [24].

A stylus (from the Greek *stylus,* meaning "stake, pillar" adapted to "instrument of writing") is a writing utensil. In ancient times, a small rod with a pointed end was used to scratch letters in wax tablets. Diamond or sapphire-tipped styluses were commonly used in phonographs to replay music. The SPM probe is a stylus that is able to read (and write?) at the atomic level.

Modern profilometers are capable of measuring surface roughness down to 1 nm [25]. In such mechanical probes, the tip is in direct physical contact with the surface of the sample. In 1971, Young developed a noncontact profiler by application of an electron field emission current between the probe and the surface [26]. The AFM evolved from the prosaic profilometer of the 1920s into a powerful and versatile tool of nanoscience. The major difference between the two is the magnitude of the applied force (e.g., the AFM utilizes much smaller forces).

Piezoelectric Materials.   Pierre Curie and his brother, Jacques, discovered piezoelectricity in 1883. They showed that electricity was produced when pressure was applied to selected crystallographic orientations. They also demonstrated the converse to be true. Piezoelectricity (from the Greek *piezien*, "to press, squeeze" + electricity) is the induction of electrical polarization in certain types of crystals due to mechanical stress [27]. Piezoelectric devices are made of dielectric components that are able to convert mechanical stresses (e.g., sound waves) into electrical signals and vice versa [27]. Common piezoelectric materials include quartz, various ceramics, and Rochelle salt. Transducers such as phonograph cartridges, microphones, radios, and strain gauges make use of piezoelectric materials.

Electromechanical piezoelectric ceramic transducers are responsible for creating the mechanical motion required to scan AFM and STM specimens. The motion, or change in geometric shape, depends on the type of crystal, its shape, and the strength of the applied electric field [23]. Typically, a standard piezoelectric crystal will expand about 1 nm per AC volt applied [23]. Therefore, to accommodate all the variables of motion required to map the surface of a sample, piezoelectric transducers are comprised of hundreds of layers. For example, if a piezoelectric material is composed of 1000 layers, motion of 1000 nm·V^{-1} is possible. With 50 V applied, motion of 0.05 mm is possible [23]. On the nanoscale, that is a lot of territory. The expansion of a stacked piezoelectric device is shown in **Figure 3.13**.

Depending on the manufacturer, two types of scanning motions are possible. In one, the sample is moved relative to the probe tip. In the other, the sample is kept stationary while the probe is rastered across the surface. For the most common configuration, the piezoelectric device controls the *x* and *y* displacement of the sample and the *z* motion of the probe tip.

3.2.1   *Atomic Force Microscopy*

One of the most powerful tools of nanoscience is a relatively simple mechanical device that is capable, however, of imaging atoms—a device that requires no special atmosphere, no expensive high-energy radiation or beam source, and can be operated under ambient conditions. Atomic force microscopy relies on the mechanical deflection of a cantilever to relay information about the contour of a sample surface. An atomically sharpened probe tip (50–20 nm or less) descends perpendicularly from the distal end of a cantilever. The tip-to-sample

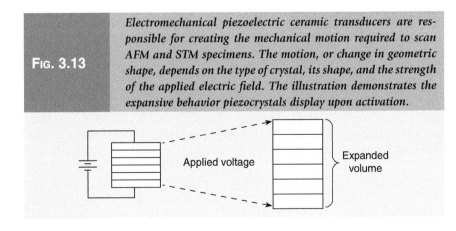

FIG. 3.13   *Electromechanical piezoelectric ceramic transducers are responsible for creating the mechanical motion required to scan AFM and STM specimens. The motion, or change in geometric shape, depends on the type of crystal, its shape, and the strength of the applied electric field. The illustration demonstrates the expansive behavior piezocrystals display upon activation.*

Applied voltage

Expanded volume

distance is fixed by means of a feedback mechanism that maintains constant force between them (i.e., constant height). A laser beam is focused on the top of the shiny cantilever and is reflected into a photodiode detector (**Fig. 3.14a**). Differences in the reflected beam are recorded by a split photodiode and recorded as changes in topography. The photodetector is able to discriminate motion with accuracy less than 1 nm. The surface is scanned grid-like by the probe to produce a three-dimensional image of the surface. A Veeco Instruments Dimension V Nanoscope multifunctional scanning probe microscope is shown in **Figure 3.14b.**

The AFM technique relies on a balance between attractive van der Waals and repulsive electrostatic forces between the probe tip and the surface. The net force is a function of the distance between the two (**Fig. 3.15**). The force is also a function of the dielectric constant ε of the medium. For example, the forces between the tip and the surface are much weaker if the probe is submerged in a liquid. Such versatility allows for analysis of biological samples by the AFM technique. Hand in hand with atomic-order resolution, the theoretical magnification potential of the AFM technique is on the order of 10^9 [18]. The versatility of the AFM is demonstrated further by its ability to image large objects that are tens of microns or more in size. The tobacco mosaic virus is 300 nm in length and the human hair is 50–100 μm in width. Both have been imaged by AFM. Several modes of AFM operation are listed and described in **Table 3.9.** There are three major modes of operation: contact, tapping, and noncontact.

Different cantilever–probe tip combinations are available for different types of AFM techniques. Cantilevers are fabricated from single crystal silicon. The length of AFM contact and tapping mode cantilevers is between 100 and 500 μm; the width is between 25 and 40 μm, and the thickness is between 1 and 10 μm. The resonant frequency of these cantilevers dwells between 10 and 300 kHz and the spring constant between 0.1 and 50 N·m⁻¹. The probe tip diameter ranges from a few nanometers for finer resolution to 20 nm for an average AFM to as large as 50 nm. Noncontact AFM mode cantilevers are made of highly doped single crystal silicon.

The Force in AFM. A sensor measures the force that is generated between the probe tip (cantilever) and the electron clouds of the sample. Hooke's law describes the relationship between the cantilever and the applied force:

$$F = -k\,z \tag{3.37}$$

where F is the force, k is the "spring constant" of the cantilever, and z is the distance of the deflection of the cantilever. Laser light is reflected off the shiny top surface of the cantilever into the photodiode detector. Therefore, the signal from the photodiode detector is proportional to the mechanical displacement of the cantilever [23].

Resolution. Image resolution in AFM is acquired in three dimensions: the x–y plane (or in-plane) typical of optical microscopes and in the z, or perpendicular, direction. The resolving power of an AFM is dependent on the radius of curvature and size of the tip. For example, a thin probe sharpened to a few atoms has better resolving capability than its larger, smoother counterpart. Broadening of features occurs when the tip radius of curvature is on par with the dimension of

FIG. 3.14

(a) A generic AFM system is shown. A focused laser beam is reflected off the back of the cantilever equipped with a sharpened probe and into a photodiode detector. Any deflection of the beam is translated into a topographical feature. Rastering of the tip over the surface (or vice versa) produces a topographical image. The photodiode is split into two compartments and is able to detect differences in beam position to the level of a nanometer. (b) Veeco Instruments' Dimension V Nanoscope multifunctional scanning probe microscope. The controller for this instrument is able to measure tip-sample/cantilever dynamics. Pixel density is 5120 × 5120. The scan range is 90 × 90 μm and can accommodate a sample size of 150-mm diameter. Modes: contact, tapping, lateral force (LFM), magnetic force (MFM), force–distance spectroscopy, electric force (EFM), scanning capacitance (SCM), scanning spreading resistance (SSRM), tunneling atomic force (TUNA), conductive atomic force (CAFM), scanning tunneling (STM), and torsional resonance mode microscopy (TRmode). Veeco probes include tapping (42 N·m⁻¹, 320 kHz, no coating), electrical (0.2 N·m⁻¹, 13 kHz, Pt-Ir coating), magnetic (2.8 N·m⁻¹, 75 kHz, –400 Oe), conductive (2.8 N·m⁻¹, 75 kHz, doped diamond coating), TUNA (2.8 N·m⁻¹, 75 kHz, pt-Ir coating), and SSRM (42 N·m⁻¹ < 320 kHz, doped diamond coating).

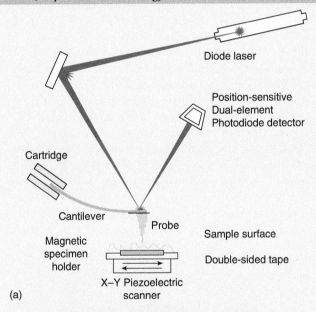

(a)

(b)

Source: Image courtesy of Veeco Instruments, Inc.

FIG. 3.15   *The form of this graph is similar to that of a Lennard–Jones pair interaction potential function and indicates a fundamental behavior characteristic of all physical interactions: There is always attraction and repulsion. Surface profilometers operate in hard contact mode regime of the graph.*

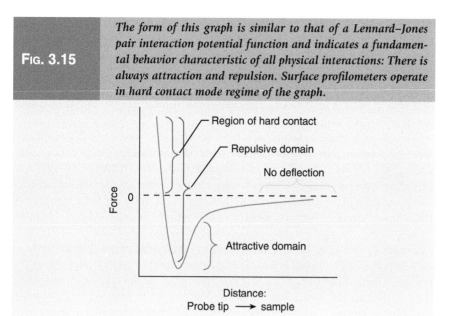

TABLE 3.9	*Atomic Force Microscopy Primary Modes*	
AFM mode	**Configuration**	**Capability**
Contact mode	The probe-cantilever assembly applies a constant force to the surface; the force constant (or spring constant) is <1 N·m⁻¹. The deflection of the cantilever is due to topographical changes characteristic of the surface. The mechanical deflection is translated into an optical signal by reflection of aligned laser light from the top surface of the cantilever into an aligned-calibrated dual photodiode collector. Due to intimate contact with the surface, large lateral forces can influence the probe. Surface contamination, electrostatic forces, and heterogeneous surfaces are able to impact the action of the probe. AFM is capable of imaging nanomaterials in air, vacuum, and liquids.	Samples with hard surfaces are appropriate for contact mode AFM analysis. AFM is capable of imaging insulator, semiconductor, and conductor surfaces. Qualitative information includes three-dimensional visualization and material sensing by phase contrast (chemical contrast of features with the same z-displacement). Quantitative information includes topographic mapping, particle and pore size, particle and pore morphology, surface roughness and texture, particle count, size distribution, surface area distribution, and volume–mass distribution [23].
Tapping mode	Strong repulsive regime Oscillating probe (>100 kHz) Intermittent contact with surface Lateral forces reduced significantly Amplitude and phase imaging	Three-dimensional topography Used for soft samples or those weekly bound to surface Biological materials DNA Carbon nanotubes
Noncontact mode	Weak-attractive regime Oscillating probe Can be applied with water layer	Soft surfaces

the object under analysis. The radius of curvature for typical contact mode AFM tips is quite large, on the order of 50 nm [2]. During a scan, the side of the probe makes contact before the tip. Probe tips sharpened to a single atom and close to the substrate surface provide the best configuration for image acquisition. Resolution decreases the farther a tip is from the surface. The lateral resolution of a typical AFM is ca. 1.5 nm. The vertical resolution of a typical AFM is even better at ca. 0.05 nm.

The use of carbon nanotube tips for AFM and STM has received a significant amount of attention during the past few years. Multiwalled (and even better single-walled) carbon nanotubes have shown promise as probe tip materials. For example, probe tips made of single-walled carbon nanotubes have shown excellent mechanical properties (e.g., stiffness), vibrational properties, and, for STM application, single-point electron discharge perfect materials for AFM and STM use.

Operation.   The sample material is fixed to metal disc with a spot adhesive or two-sided tape. The metal + sample is then snapped to position on top of the scanner assembly by the action of a magnet. Care must be taken not to damage the piezo material. A cantilever + probe (100–200 μm in length) is placed in the holder (cartridge), secured in the mount, and aligned with the laser-photodiode sensing apparatus. An optical microscope is used to position the probe over the sample surface. The image of the cantilever as it hovers over the sample is visible on a nearby CRT screen. The probe assembly is lowered to the sample surface by an automated feedback control system to a vertical distance of a micron or less. Obviously, we are not able to judge such a small distance and require the use of computerized electromechanical technology to guide the tip down the z-axis.

Parameters such as scan area, probe force, and scan rate are input into the system and the scan is put into motion. Successive line scans (rasters) are compiled to generate a three-dimensional image of the surface. There are two generic configurations for scanning. In the first, the sample is mobilized underneath a stationary probe. In the second, the reverse configuration is applied. Numerous computer programs are available for image manipulation.

In contact mode imaging, a soft, deflectable cantilever with a sharp probe tip is brought close to the surface (e.g., in the repulsive regime of **Fig. 3.15**). The cantilever is deflected according to Hooke's law with a spring constant that is typically in the range of 1×10^{-3} to $100 \ \text{N} \cdot \text{m}^{-1}$. The degree of deflection (usually along the z-axis) is recorded via an array of positional photodiodes that capture laser light reflected from the surface of the cantilever. Constant force or distance between the tip and the surface is maintained by a feedback loop. The magnitude of the forces applied to the tip range from 0.01 to 100 nN. Piezoelectric elements hold and position the sample in all three spatial domains—x, y, and z—and participate in the feedback loop process. A three-dimensional rendition of the topography of the surface is achieved by the contact mode of the AFM. Selected AFM images are displayed in **Figure 3.16.**

3.2.2 *Scanning Tunneling Microscopy*

"The wavelike properties of electrons permit them to 'tunnel' beyond the surface of a solid into a region of space forbidden to them under the rules of classical physics"—so stated Binnig and Rohrer in 1981. Electron (or quantum) tunneling is attained when a particle (an electron) with lower kinetic energy is able to exist on the other side of an energy barrier with higher potential energy, thus disobeying a fundamental law of classical mechanics. Tunneling is the penetration of an electron into a classically forbidden region [28]. Electrons exhibit wave behavior and their position is represented by a wave (probability) function. The wavefunction represents a finite probability of finding an electron on the other side of the potential barrier. Since the electron does not possess enough kinetic energy to overcome the potential barrier, the only way the electron can appear on the other side is by tunneling through the barrier. The principle of quantum tunneling was also used to explain exponential radioactive decay rates (half-lives). In 1928, the great physicist George Gamow showed that alpha decay occurs via a tunneling process. Considering only classical principles, it takes too much energy to pull a nucleus apart; however, in quantum mechanics, a finite probability is allocated to tunneling and hence nuclear decay is allowed. The inversion of the conformation of ammonia is an example of tunneling by a particle, in this case the hydrogen atom [28].

STM relies on an electronic signal to relay information about a sample—the strength of a tunneling current potential that exists between the probe tip and the substrate surface. Small changes in the distance between the probe tip and the substrate surface translate into large changes in the tunneling current. By this phenomenon, atomic scale resolution by STM is possible in the *x*, *y*, and *z* directions. The density of states (DoS) of solid-state materials can be also mapped by the technique called scanning tunneling spectroscopy (STS). Chemical reactions induced and oriented by the STM probe are also available by means of this technique.

There are two kinds of electron microscopes, and we do not mean TEM and SEM because we have already discussed them. The differences between SEM and STM are that the magnitude of current is quite diminutive in STM and that current originates from electron tunneling. Electron tunneling occurs between the conducting sample and the tip of the STM. The tip is very close to the substrate but not in actual physical contact (**Fig. 3.17**).

When an electron interfaces with a finite potential barrier with potential energy U_o that is greater than its own kinetic energy, the electron stays within the "box". Electrons are small enough and "quantum" enough to make tunneling happen. Electron tunneling current between two flat plates separated by a vacuum is given by

$$I \propto V e^{(-2kW)} \tag{3.38}$$

where *I* is the tunneling current, *W* is the distance between the surfaces (the width of the barrier), and *k* is a term that is related to the potential across the vacuum:

$$k = \frac{\sqrt{2m_e (U - E_e)}}{\hbar} \tag{3.39}$$

where m_e is the electron rest mass as before, *U* is the potential energy between the surfaces, E_e is the energy of the electron, and $\hbar$ is the Dirac constant ($\hbar = h/2\pi$). Typically, the potential difference, $U - E_e$, is equal to ~4 eV.

Fig. 3.16   *See caption on page 151.*

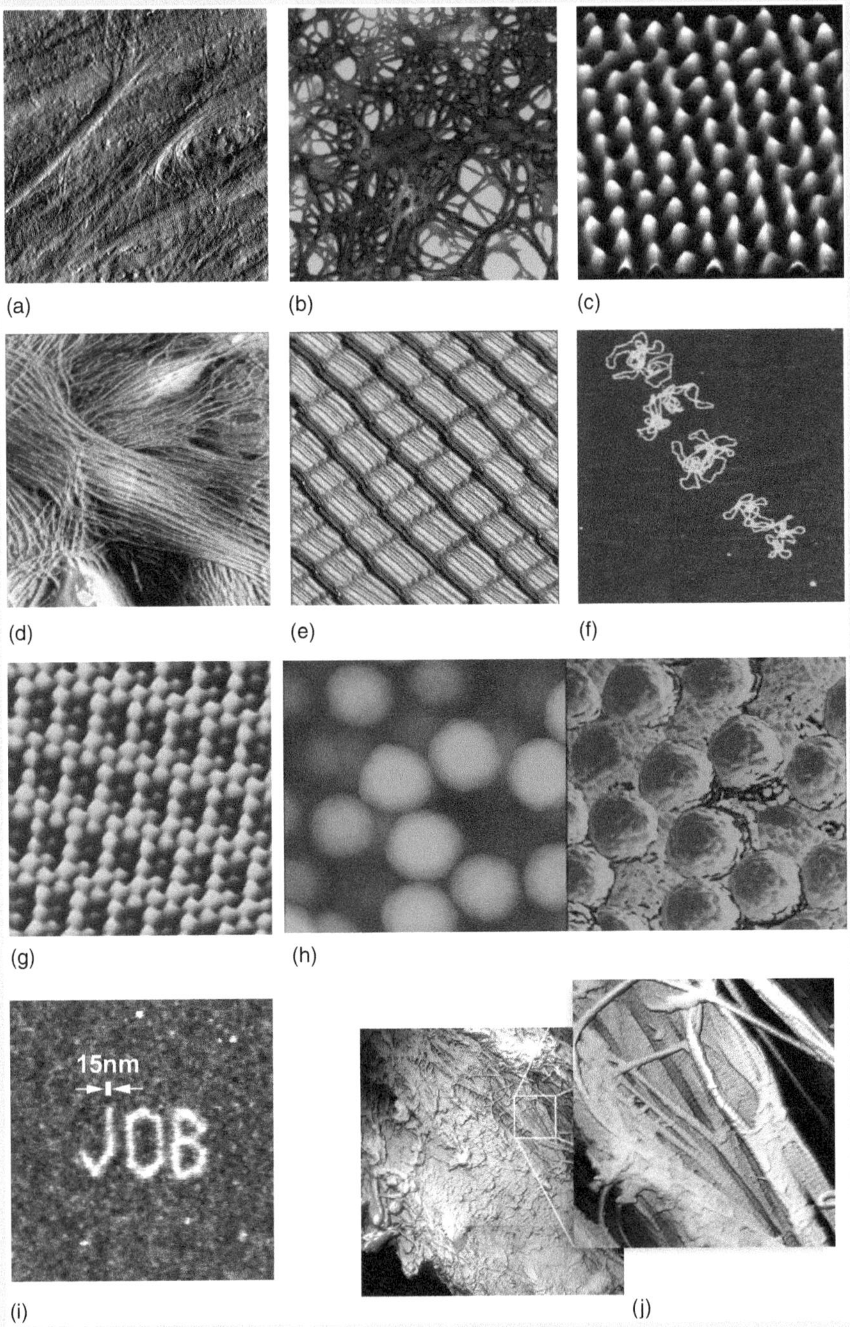

(a)

(b)

(c)

(d)

(e)

(f)

(g)

(h)

15nm

JOB

(i)

(j)

Source: Courtesy of Veeco Instruments, Inc.

Selected AFM images acquired by various Veeco instruments are displayed. (a) The image shows unmineralized collagen fibrils on the outer surface (periosteum) of trabecular bone. The 67-nm d-banding across the collagen fibrils is visible. The image was acquired by tapping mode with a MultiMode V AFM. Scan size is 8.6 μm. (b) A nanoscale scaffold made of nanofibers fabricated by a template method from a mixture of gelatin and alginate. The structure has potential application as a biosensor. The scaffold has features at both nano- and microscales that mimic the topography of natural extracellular matrices. Image was acquired with a Dimension 31000 AFM. The scan size is 20 μm. (c) A low-pass filtered image of mica surface atoms taken with a BioScope AFM system. The mica surface was imaged in contact mode. A 5-nm scan is shown. (d) Image of live MC3T3F osteobalst cells; networks of actin fibers of the cytoskeleton are portrayed. Scan size is 100 μm. The image was taken with a BioScope II AFM. (e) The periodic three-dimensional structure of the wing elements in Morpho peleides *butterfly. Image scan size is 10 μm. The structures impart color to the butterfly wing via nanophotonic properties due to the nanostructure of the wing. We have already seen TEM renditions of another* Morpho *species photonic structure on the front cover of the text. (f) DNA strands are imaged with tapping mode. Scan size is 700 nm. (g) STM image of oxygen atom lattice on rhodium single crystal. The enclosed image is the result of a 4-nm scan. (h) Images of poly(methylmethacrylate): AFM height on the left and phase on the right. (i) AFM point-and-click nanolithograpy with the low-noise Dimension CL SPM head (700-nm scan). (j) AFM image of a fibrous structure of naturally aged nineteenth century goat parchment. The uppermost layer was removed by a microtome. The characteristic periodicity of the collagen fibers is distinctly visible. The image was acquired with a MultiMode AFM. Scan size = left: 20 μm; right: 3 μm.*

FIG. 3.16

The current is defined by a modified version of equation (3.38) that takes into consideration the geometry and electronic structure of the probe tip:

$$I = C p_{tip} p_{sample} e^{-W\sqrt{k}} \qquad (3.40)$$

FIG. 3.17

A schematic rendition of a scanning tunneling microprobe is depicted. The tunneling current is a quantum phenomenon. Direct contact with the surface is avoided. Localized electronic structure (density of states) of a specimen is obtained by scanning tunneling spectroscopy (STS).

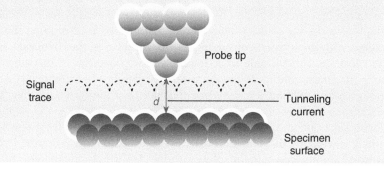

Signal trace

Probe tip

d

Tunneling current

Specimen surface

where C is a voltage-dependent constant and the p terms represent the electronic structures (density of states) of the tip and the sample, respectively [29].

The tip potential is biased with respect to the substrate. The direction of electron flow is determined by the direction of the bias. If the tip is negatively biased with respect to the substrate, then electrons will flow from the substrate to the probe. If the reverse configuration is applied, the electrons will flow from the probe to the surface.

Scanning Tunneling Spectroscopy.   A complementary method to STM is scanning tunneling spectroscopy. The purpose of STS is to generate a map of the localized electronic structure of a specimen—ideally with atomic level resolution, given that the probe tip conforms to the necessary criteria. The spectrum is a map of the electronic structure (e.g., the DoS). Typically, an STS spectrum is found by measuring the change in tunneling current as a function of voltage yielding an I–V curve of a selected section of the sample. The I–V curve acquired from a bulk metal, for example, is shaped in the form of a smooth continuum. For an atom, its density of states resembles a series of lines. Somewhere in between are the electronic states of nanomaterials, which are expressed as step functions.

The operation of the STM in this technique is straightforward. The tip of an STM is held in place over a region of interest while the bias voltage is manipulated. Other permutations of STS operation exist. In one, current is held constant during a scan while height is recorded. In another, the reverse configuration is applied [2]. Material conductivity (dI/dV) and work function (dI/dz) plotted against applied V are also measured by the STS technique [2].

3.2.3   *Other Important Scanning Probe Methods*

The development of scanning probe methods has proliferated since the advent of the *nano age*. Offshoots of AFM include *lateral force microscopy* (LFM), *force modulation microscopy* (FMM), *electrostatic force microscopy* (EFM), *magnetic force microscopy* (MFM), *scanning capacitance microscopy* (SCM), *scanning thermal microscopy* (SThM), and *chemical force microscopy* (CFM).

The following three techniques are operated in the contact mode. We understand how the AFM technique measures deflections primarily in the z-direction. On the other hand, LFM is able to detect lateral or torsional deflections of the cantilever. These deflection modes arise from forces that run parallel to the surface of the sample. LFM, therefore, is a useful tool in characterizing the frictional disposition of a sample surface. The CFM technique is able to probe the chemical nature of a selected region of a sample surface. Special probe molecules are adapted to the tip of the probe. During a scan, the fixed probing molecules interact with the substrate in specific ways depending on the chemical nature of the substrate. FMM is able to access the elastic properties of the substrate by measuring the amplitude following an applied cantilever oscillation [2]. Another contact mode method uses an *ambient atomic force microscope*. In this case the AFM probe is made of metal, negatively biased with respect to a p- or n-doped silicon substrate and surrounded by a meniscus of water. The AFM is able to write by oxidizing the silicon under the probe. Oxidation is accomplished by hydroxyl anions present in the solution induced by the electric charge (**Fig. 3.18**).

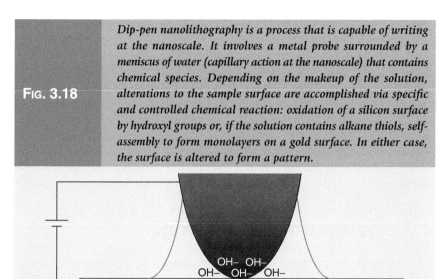

FIG. 3.18 *Dip-pen nanolithography is a process that is capable of writing at the nanoscale. It involves a metal probe surrounded by a meniscus of water (capillary action at the nanoscale) that contains chemical species. Depending on the makeup of the solution, alterations to the sample surface are accomplished via specific and controlled chemical reaction: oxidation of a silicon surface by hydroxyl groups or, if the solution contains alkane thiols, self-assembly to form monolayers on a gold surface. In either case, the surface is altered to form a pattern.*

A form of nanolithography, nano-dip-pen lithography, is discussed in more detail in chapter 4.

The following four techniques operate in noncontact mode and rely on changes in the resonant frequency of the cantilever to produce a variable signal. MFM, as its name implies, is a technique designed to measure localized magnetic domains on a material surface. In this technique, a tip coated with a ferromagnetic material is employed. EFM is a technique designed to measure localized charge domains on a material surface. SCM is a technique designed to measure localized capacitance on a material surface. The result is a map of the dielectric outlay of the specimen. The SThM technique uses a heated wire to determine the thermal conductivity of localized regions of a sample surface. Just like with its macroscopic counterpart, thermal conductivity is determined by measuring the change in current as a function of temperature for a specific mass.

3.2.4 Atom-Probe Methods

E. W. Müller, J. A. Panitz, and S. B. McLane invented the atomic probe microscope (APM) in 1967 by combining a time-of-flight mass spectrometer with a field ion microscope (FIM) [30]. APM, like FIM, consists of a movable sharpened tip but is operated in ultrahigh vacuum at cryogenic temperature. Ionization is induced at the surface, and the probe tip repels ions towards a detector aperture. A chemical profile of the surface is obtained in this way with resolution of one-atom spacing. J. A. Panitz went on to develop the imaging atomic probe (IAP) in 1974. The IAP does not require a moving tip. Atom-probe tomography (APT) also uses a detector that is sensitive to the position of the selected area [31]. Tomography (from the Greek *tomos,* "to slice, section") is a technique that is able to analyze cross-sections of a material. This development led to the position-sensitive atom probe (POSAP) invented by A. Cerezo, T. Godfrey, and G. D. Smith in 1988.

Modern atomic probes are capable of providing three-dimensional atomic resolution, compositional imaging, and chemical analysis (similar to SIMS) [32]. Individual atoms of a specimen can be removed and analyzed! Atoms are isolated by a combination of a high electric field with the application of either a laser or voltage pulse. Thin films deposited on the tip itself are readily analyzed. The LEAP 3000X Si, manufactured by Imago Scientific Instruments, has a 200-nm field of view (FOV) and a data collection rate of 5×10^6 ions per minute [32]. Lateral resolution of 0.1 nm and depth resolution of 0.5 nm are typical of this instrument.

3.3   SPECTROSCOPIC METHODS

The definition of spectroscopy (from the Latin *spectare*, meaning "behold," appearance + from the Greek *skopos*, meaning "watcher," examine) is as follows [33]: "Spectroscopy is a branch of science concerned with the investigation and measurement of spectra produced when matter interacts with or emits electromagnetic (EM) radiation." J. C. Engle, Jr., and S. R. Crouch provide an excellent definition of the spectrum [34]: "A spectrum is a display of the intensity of radiation emitted, absorbed, or scattered by a sample versus a quantity related to photon energy, such as wavelength or frequency."

Spectroscopy, then, in its most basic sense implies the study of phenomena created by the interaction of electromagnetic radiation with matter or, in its purest form, the study of the spectra itself. However, to some, spectroscopy may indicate a broader spectrum of analysis, including those that do not involve electromagnetic radiation. In the context of this chapter, spectroscopic methods are limited to those involving some sort of EM radiation, whether 1°, 2°, or higher order or interaction.

The interaction of radiation with matter assumes many forms. Molecular absorption of radiation is described by

$$E_{\text{Total}} = \sum E_{\text{Interactions}} = E_{\text{Electron}} + E_{\text{Rotation}} + E_{\text{Vibration}} \qquad (3.41)$$

For single atoms, these options are limited to electronic transitions. Electronic transitions between energy states are observed by measuring absorption, emission, fluorescence, or luminescence of radiation. Other types of mechanisms include scattering (elastic and inelastic), reflection, refraction, and diffraction. Nonradiative mechanisms such as vibration, rotation, ionization, and heat transfer also factor into many types of EM interactions.

Molecular-level electron transitions exist in several forms. We are quite familiar with the atomic and molecular ground-to-excited state electronic transitions such as $\sigma \to \sigma^*$, $\pi \to \pi^*$ and $n \to \sigma^*$, $n \to \pi^*$, where σ, π, and n are the sigma-bond, pi-bond, and the nonbond (electron pairs). Nanomaterials also show quantized transitions, especially in small clusters. Larger colloids are able to scatter light but metal quantum dot clusters have HOCOs and LUCOs (highest occupied and lowest unoccupied cluster orbitals, respectively). Nanoscale materials present a new domain of energy transitions: just between the discrete atomic-molecular quantum transitions and continuous transitions of classical bulk domains.

3.3.1 *UV-Visible Absorption and Emission Spectroscopy*

The ultraviolet–visible spectrum ranges from ca. 350 (ultraviolet) through 770 nm (reds) [34]. Many nanomaterials interact with UV–visible photons to produce a variety of effects: plasmon oscillations on the surface of nanometals (Ag and Au), electron transfer cascades (involving supramolecules like chlorophyll), exciton pairs in semiconductors (nano-TiO_2), particle size-dependent fluorescence (semiconductor quantum dots), diameter-dependent fluorescence (chiral single-walled carbon nanotubes), and interference colors (porous alumina or titania thin films). There are many more examples. We begin by reviewing some basic concepts.

Light can be transmitted, absorbed, or reflected:

$$I_o = I_{Ref} + I_{Abs} + I_{Trans} \tag{3.42}$$

where I_o is the intensity of the incident light and I is the intensity of light after interaction with the specimen. *Transmittance T*, another unitless parameter, is the ratio of radiant power I emerging from a material to the incident radiant power I_o.

$$T = \frac{I}{I_o} \tag{3.43}$$

The Beer–Lambert law is used to describe the phenomenon of *absorbance*, a unitless parameter:

$$A = -\log_{10} T = \varepsilon b c \tag{3.44}$$

where ε is the molar absorptivity ($L \cdot mol^{-1} \cdot cm^{-1}$), b is the path length of the sample in centimeters, and c is the concentration of the compound in the solution. The wavelength of maximum absorption is symbolized by λ_{max}. Emission of light (*luminescence* or *photoluminescence*) after excitation is called *fluorescence* if it happens quickly and *phosphorescence* if it is delayed [3]. The optical response of bulk and nanometals is shown in **Figure 3.19.**

Photoluminescence in Carbon Nanotubes. Bandgap photoluminescence of semiconducting single-walled carbon nanotubes (SWNTs) is due to excitons. Semiconducting SWNTs demonstrate specific electronic absorption, and a strong relationship between optical transitions and SWNT diameter has been demonstrated [35]. On the other hand, nanotubes that are metallic in character do not fluoresce [35]. The optical absorption by π-electrons in carbon nanotubes has been shown to be dependent on the diameter but also the chirality of the nanotubes [36]. Luminescence is related to a phenomenon known as a *van Hove singularity*, a subject that is discussed briefly in chapter 9.

Dipolar Plasmon Resonance in Nanometals. Dipolar plasmon, higher order plasmon excitations, and Mie scattering are phenomena responsible for the varied colors of gold and other metal nanoparticles and colloids. The optical response of gold nanoparticles formed by electroplating into porous alumina

FIG. 3.19

On the left, most of the visible light is reflected from a bulk metal surface. For example, the absorption–reflection of light is why bulk gold appears to have a yellow-reddish luster. The middle image depicts how a thin particulate metal film (beyond the percolation threshold) scatters and reflects light. Some light is transmitted, some absorbed, and some reflected. The color of a thin film of Au may appear green if it is microgranular. On the right, a large fraction of the light is transmitted and scattered through a transparent nanometal composite material. Absorption of light by nanogold yields many colors, depending on the size, shape, and orientation of the particle and the dielectric function of the surrounding medium.

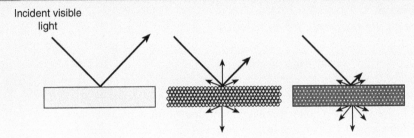

Incident visible light

templates shows a definitive size, shape (aspect ratio), and orientation dependence (**Fig. 3.20**) [37]. If the radius of the nanometal particle is approximately 1/100 of the wavelength ($r < 0.01\lambda$), dipolar plasmon resonance is responsible for the optical response. With regard to classical Drude optical theory, smaller particles demonstrate a pronounced surface effect. Specifically, electrons have a better

FIG. 3.20

Transmission colors produced by gold nanoparticles formed in porous alumina templates. Gold in the form of spherical to cylindrical nanoparticles was plated into the pore channels of anodically formed porous alumina membranes of varying diameters. Particle aspect ratio was controlled by the duration of electroplating. Particle size was controlled by the diameter of the pore channel. A red shift in λ_{max} occurs as particles become larger (increased diameter). In addition to the plasmon resonance, the onset of Mie scattering occurs with larger particles. The image is an approximate reconstruction from physical data [37]. In the figure, particle size increases from bottom to top and particle aspect ratio increases from left to right. The combination of the two independent parameters yields a color array—all consisting of gold nanoparticles.

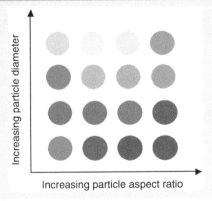

Increasing particle diameter

Increasing particle aspect ratio

chance of colliding with the surface boundary if the particle becomes smaller. This process increases the energy of the electrons and causes a blue shift in λ_{max}. On the other hand, as the particle becomes larger, scattering and higher order plasmon resonances are excited, resulting in a red shift in λ_{max}.

Composites consisting of a dielectric core coated with a metal shell also demonstrate tunable optical properties. Specific tuning of the optical response was also demonstrated by gold shell–silica–dielectric core composite nanoparticles [38]. Optical resonance of the surface plasmon from the visible to the infrared range was achieved by varying the thickness of the nanoshell and the dimensions of the dielectric silica core material [38].

Quantum Dots. A quantum dot is a nanoscale material within which the motion of electrons is confined in the *x, y,* and *z* directions. The definition can be expanded to include the confined motion of conduction band electrons and excitons. Technically, metallic nanoparticles are also considered to be quantum dots—especially upon losing their metallic character with smaller dimensions. The size-dependent emission of visible light from semiconductor quantum dots has fascinated scientists for the past few decades. Quantum dots are usually made of CdSe and other semiconductor materials such as InGaN, InGaAs, CdTe, and ZnSe. A progression in emission color is related to particle size in **Figure 3.21.**

3.3.2 Infrared and Raman Spectroscopy

Far-infrared, infrared, near-infrared, Raman spectroscopy, and surface-enhanced Raman spectroscopy (SERS) are extremely relevant methods used to characterize nanomaterials. These methods provide information about vibrational energy; IR methods rely on asymmetrical (dipolar) vibrations and Raman methods on symmetrical (polarizable) vibrations. In this way, the two methods complement each other.

Raman Spectroscopy. C. V. Raman and K. S. Krishnan in 1928 discovered that light scattered off certain molecules changed the polarization state of the molecules [39]. The Raman effect is a scattering phenomenon that links the vibrational frequencies of the molecule to the energy difference between the incident

| FIG. 3.21 | *The fluorescent emission from quantum dots of different size made of the same material is represented in this crude reconstruction. It is perhaps the quantum dot that embodies best the spirit of nanoscience—the exhibition of a physical property that is dependent on size and size alone. The figure does not do justice to an actual image of an array of fluorescing quantum dots (Fig. 1.20). The dots range in size from a few nanometers to greater than 5 nm. Excitation is by long-wave ultraviolet and emission ranges from less than 500 to greater than 600 nm.* |

and scattered light [34]. However, only molecules with symmetric vibrational (polarization) modes and transitions were amenable to Raman spectroscopic analysis. The SERS process depends on nanometal facets or particles that are able to enhance signal intensity by multiple orders of magnitude—an example of a spectroscopic method enabled by nanotechnology.

Two types of photon scattering exist—elastic scattering and inelastic scattering. Elastic scattering, in which there is no energy loss, is called Rayleigh (Rayleigh–Debye) or Mie scattering. Such scattering occurs when particle dimensions are on the order of or smaller than the wavelength of the incident radiation. Remarkably, then scattering is a true nanoscale phenomenon. If particle size does not conform to this limit (e.g., larger materials), then reflection and refraction also occur [34]. Rayleigh scattering (elastic) involves no change in the polarizability of the molecule [34]. However, a small portion of the incident photons is scattered at frequencies that are different from the incident light. This is called inelastic scattering and the result is exhibited in the form of vibrational, rotational, or electronic energy changes in the molecule. Another form of Raman spectroscopy called resonant Raman spectroscopy (RRS) occurs when there is resonance between the incident radiation and an electronic transition of the molecule [39].

The Raman effect, based on the interaction of a photon with the electronic polarization of the molecule, occurs at energies that are lower than electronic state energies and are known as "excitation to a virtual state." The Raman effect is the inelastic scattering of photons by the sample [40]. Photons that are scattered inelastically by molecules may gain or lose energy depending on the preexisting energetic state of the molecule [40]. Stokes lines are generated when the shifted frequencies are lower than the excitation frequency of the Raman laser. Anti-Stokes lines are formed when the opposite condition exists [40]. Please consult any number of excellent references for further clarification.

SERS is a technique that exhibits incredible sensitivity and has great importance to nanoscience. Some aspects of this method were discussed in chapter 1. Substrates in the SERS technique are noble metals like Au and Ag that happen to be in nanoparticulate form. The primary mechanism of SERS relies on the surface plasmon resonance of the nanometal particulates. An electromagnetic enhancement on the order of 10^6 of the vibrational signal from an adsorbed species is imparted to the spectrum. Analysis of SWNTs by SERS is diagnostic of their presence. Single-molecule detection is also possible with this method.

3.3.3 X-Ray Methods

X-ray methods involve sample excitation by x-rays (creating more x-rays) or by electrons (creating x-rays). X-rays can also be generated from a sample by bombardment by electrons or alpha particles. The energy of emitted x-rays is equal to the difference between the binding energies of the electrons involved in the transition [40]. X-ray fluorescence (XRF) is an elemental analysis procedure that measures 2° x-ray emission. The secondary emission is lower in energy (longer wavelength) than that of the stimulation radiation, and data acquired from XRF can be used in a semi-quantitative manner. X-ray scattering techniques include x-ray diffraction (XRD), small-angle x-ray scattering (SAXS), and x-ray absorption fine structure (XAFS).

X-ray Diffraction.   One of the most significant analytical methods to emerge in the late nineteenth and early twentieth centuries is XRD. With XRD, the structure of crystalline materials is revealed and analyzed. Just like with electrons, diffraction of x-rays is a result of scattering from atoms configured in regular arrays. The spacing between atoms and planes of atoms is on the order of the wavelength of the x-rays. Bragg's law forms the foundation for x-ray diffraction:

$$n\lambda = 2d \sin\theta \qquad (3.45)$$

The spacing between planes of atoms essentially functions as a three-dimensional diffraction grating [41]. Crystalline solids show long-range periodic structure, exhibited as Bragg diffraction peaks. The Bragg peaks are symmetrically distributed diffraction patterns of focused points [42]. The Cu–K_α (0.15416 nm) source is used for most XRD analyses.

Traditional x-ray diffraction is appropriately used to obtain structural characteristics of bulk crystals. For nanocrystals, traditional XRD is not always appropriate because the coherence length of the structure is limited [42]. Diffuse XRD patterns are formed from bulk materials that have short coherence range and from glasses—materials with no long-range order. Nanomaterials can exhibit a range of periodicity and structural coherence of short size [42]. This also results in diffuse peaks. Extra special techniques need to be applied. High-energy x-ray diffraction (HEXRD) and atomic pair distribution function (PDF) analysis have been shown to be extremely accurate in determining the fine structure of nanomaterials [42]. Nanomaterials in powder form are a popular means of conducting XRD analysis. Nanowire arrays of α-Fe_2O_3 have been analyzed by XRD [43].

Small Angle X-ray Scattering Analysis.   SAXS is based on the principle of scattering of x-rays by crystalline or amorphous but uniformly sized small particles. Particle size on the order of 1–100 nm can be measured by this technique. SAXS is applicable to powders in the dry state or suspended in a medium [44]. Particle size can also be estimated by analyzing the width of Bragg peaks in x-ray diffraction spectra [3,45]. Broadening in x-ray diffraction spectra is due to lattice imperfections (microstrain), instrument effects, and crystallite size.

3.4   NONRADIATIVE AND NONELECTRON CHARACTERIZATION METHODS

3.4.1   *Particle Spectroscopy*

Mass spectrometry is an instrumental method in which the mass-to-charge (m/e) ratio of fragmented atomic or molecular clusters is measured. The purpose of MS is to determine the composition of a sample material by analyzing the resultant mass spectrum of the fragmented components of the sample. MS analysis is capable of providing information about isotopic ratio, trace gas analysis structure, and qualitative analysis. E. Goldstein noticed in 1886 that the anode of a cathode ray tube attracted what he eventually named canal rays (Kanalstrahlen). W. Wien later found that the rays could be deflected by an applied electric field and J. J. Thomson in 1913 showed that neon consisted of two isotopes: ^{20}Ne and ^{22}Ne [40]. Thomson described in his book, *Rays of Positive Electricity and Their Application to Chemical Analysis,* how *kanalstrahlen* can be used to analyze chemicals. F. W. Aston constructed the first relatively modern MS in 1919. The first MS to apply electrical detectors was constructed by A. J. Dempster in 1918. The first commercial MS was made available to scientists in 1942 [40].

The technique of MS is straightforward. A sample is vaporized, ionized, and fragmented:

$$M + e^- \rightarrow M^+ + 2e^- \tag{3.46}$$

where M is a molecule of the sample material and M^+ is what is known as the *parent ion*. This reaction is the most common. Other reactions may produce multiple charged positive ions or even negative ions. The parent ions then undergo further fragmentation into lower mass ions. The potential energy of the fragments, electron volts, is converted into kinetic energy according to

$$eV = \tfrac{1}{2}mv^2 \tag{3.47}$$

Deflection of the fragment is caused by an applied magnetic field of strength H (in gauss). If r (in centimeters) is the radius of the curvature of the sector, then the centripetal force experienced by the ions (Hev) is balanced by the centrifugal force of the ions (mv^2/r):

$$\frac{mv^2}{\text{r}} = Hev \tag{3.48}$$

And, substituting $v = Her/m$ into equation (3.48), mass-to-charge ratio m/e is calculated from

$$\frac{m}{e} = \frac{H^2\text{r}^2}{2V} \tag{3.49}$$

The resultant ions are separated by their respective m/e and then detected accordingly. It is easier to vary the applied voltage, inversely proportional to m/e. The semicircular path that each ion undertakes in the analytical sector is a function of H, r, m, e, and the accelerating voltage, V. Each parabolic path is characteristic of some specific ion [40].

There are several types of MSs. The major type of mass spectrometer is the sector MS in which the ion path is deflected. The larger charged and lighter ions are deflected first by an applied magnetic or electric field. The time-of-flight (TOF) MS utilizes an electric field to accelerate ions through an equivalent potential. The time it takes for an ion to reach the detector is then measured. In this scheme, all ions have the same kinetic energy, but the velocity of each individual ion depends on its mass (e.g., ions with lighter mass will arrive at the detector before those with larger mass). The quadrupole MS applies an oscillating electric field that serves to stabilize or destabilize ions in transit [40].

Clusters of extremely small size—a few to several atoms—were first characterized by mass spectroscopic methods. Due to chemical and physical reactions rather than a change in the state of the energy of a molecule, ionization, both positive and negative, forms the basis of mass spectrometry [40]. Fragmented materials in the form of positive and negative species are produced by electron impact from an ion source [40,46].

Forward recoil elastic spectrometry (FReS) is a depth profiling method used to measure the concentration of hydrogen or deuterium in solids. An energetic beam of α-particles is directed 75° from the normal on a sample surface. PIXE is based on particle induced x-ray emission [47]. Inner shell ionization is caused by bombardment with high-energy protons (~MeV), the primary probe material. The protons are produced in an accelerator. The 2° effect of the protons is x-rays. PIXE is similar to EDX but offers much better sensitivity. Characteristic x-rays are produced after activation of the sample with high-energy ions.

3.4.2 Thermodynamic Methods

Several thermodynamic methods are used to evaluate nanomaterials. Temperature is the primary independent variable in such methods. Thermogravimetric analysis (TGA) measures the change in mass as a function of temperature under selected environments. TGA data plots, generally shown as percent weight loss versus temperature, reveal information concerning thermal stability, composition (purity), and reaction rates [40]. This procedure is often used to check the purity of single- and multiwalled carbon nanotubes. Differential thermal analysis (DTA) monitors the difference in temperature between a sample and a reference material as a function of temperature. Heats of reaction (endothermic and exothermic), phase transition, and reaction kinetic data are amenable to analysis by DTA [40]. Differential scanning calorimetry (DSC) measures the isothermal differential power between a sample and a reference. Data such as heats of reaction and heat capacity can be analyzed from DSC plots [40].

Recently, nanoscale thermal analysis was accomplished on energetic polycrystalline materials. These substances release stored chemical energy in the form of thermal and mechanical energy [48]. A heated AFM tip was used to initiate local melting, evaporation, and decomposition on materials as small as 100 nm. The temperature range in these experiments was controlled from 25°C–500°C.

3.4.3 Particle Size Determination

There are several ways to measure dimensions of nanomaterials. Transmission, scanning electron, atomic force, and scanning tunneling microscopy yield direct data on particle size. Spectroscopic methods based on light scattering phenomena are also used. Each method has its own set of issues when it comes to size determination e.g. artifacts and other systematic errors.

Light Scattering.   Although light scattering is not a thermodynamic method per se, the principles behind it, like diffusion, are based on thermodynamics. Dynamic light scattering (DLS), static light scattering (SLS), photon correlation spectroscopy (PCS), and quasi-elastic light scattering (QELS) are analogous methods that measure particle size. Particle size determination by the scattering of laser light has been around for quite some time. The sophistication of the method has increased dramatically over the years as the sizing of particles less than 1 nm is routinely accomplished for micelles, colloids, proteins, and other nanoparticles. The process is based on the principle of light scattering from the surfaces of small particles. If the scattered light is collected as a function of direction, the SLS technique is valid. If the correlation of light scattered intensity is recorded as a function of time from several directions, DLS is appropriate. DLS measures Brownian motion and correlates that information with the size of the particles. Dilute suspensions ranging from 0.0001 to 1% v/v prepared with suitable wetting agents in 40-μL flow cells are analyzed by a 35-mW laser at two scattering angles. The random motion of the particles is correlated with the scattered light intensity fluctuations. From this correlation, the particles' diffusion coefficient is obtained. The equivalent sphere particle size is then calculated from the Stokes–Einstein equation:

$$r = \frac{k_B T}{6\pi\eta D} \tag{3.50}$$

where

r is the van der Waals radius of the molecule in meters
k_B is the Boltzmann constant (1.380×10^{-23} J·K^{-1})
T is the absolute temperature in kelvins
η is the viscosity in pascals per second (or centipoises)
D is the self-diffusion coefficient in square meters per second

Although a spectroscopic method, it is clear from equation (3.50) that DLS has strong thermodynamic roots.

For example, if the self-diffusion constant of 9,10-diphenylanthracene in THF (tetrahydrofuran) at 300 K is equal to 1.04×10^{-9} m^2·s^{-1} and η is equal to 0.501 mPa·s^{-1}, then the radius of the molecule from equation (3.50) is calculated to be 0.42 nm (the actual van der Waals radius is 0.41 nm) [49].

3.4.4   *Surface Area and Porosity*

Surface area, pore volume, and pore size distribution and pore density measurements are required to characterize nanoparticles, whether porous or otherwise. Seemingly, any and every research paper that involves particles or porous materials includes a section on characterization of surface area and porosity. Several methods are used to acquire information about these parameters. We will present only a few, although every student of nanoscience should become familiar with as many as possible. The most popular methods of surface area calculation are those based on gas adsorption. Other methods include SAXS, small angle neutron scattering (SANS), electron and atomic force microscopy, NMR methods, and mercury porosimetry.

BET Method.   Adsorption is the attachment of atoms or molecules to the surface of a solid. The reverse process of adsorption is called desorption. The *adsorbent* is a solid substrate of high surface area upon which the *adsorbate*, a liquid or gas, is adsorbed. Physisorption and chemisorption are two types of adsorption processes. Physisorption is governed by van der Waals forces and usually results in multilayers of adsorbed atoms or molecules. Physisorption occurs at low temperatures with low selectivity, and the heat of adsorption is usually small, $10 < \Delta H_{abs} < 40$ kJ·mol^{-1}. Chemisorption, on the other hand, occurs at higher temperatures with higher selectivity and involves stronger interactions between adsorbates and the surface with $\Delta H_{abs} > 40$ kJ mol^{-1}. The surfaces of nanomaterials are quite large compared to their volume and therefore readily adsorb substances (more discussion to follow in chapters 5 and 6). Surface coverage is defined as the proportion of surface sites on an adsorbent that are partially or completely covered by an adsorbate and is designated as Θ—the fraction of adsorption sites occupied by an adsorbate at equilibrium.

The Brunauer–Emmett–Teller (BET) method to measure surface area was proposed in 1938 [50]. The BET method, however, is based on some broad assumptions [51]. Nonetheless, it is said that this method "serves as a monument to the achievements of imperfection due to its heavy use in determination of surface area of materials."

Regardless of the underlying assumptions, the BET method is extremely valuable in determining relative surface area. The assumptions are that (1) the surface is energetically homogeneous (although most surfaces are rough in nature), (2) only vertical interactions between adsorbed molecules are considered (e.g., any lateral interactions are neglected), and (3) the molecules adsorbed on the substrate demonstrate the strongest energy of adsorption and that the heat of adsorption (ΔH_{ads}) of subsequent layers is the same as the latent heat of condensation (ΔH_{cond}) of the adsorbate gas (e.g., $\Delta H_{ads} > \Delta H_{cond}$). BET is an extremely popular (and useful) means of determining surface area, especially in terms of relative surface area, and we will proceed to discuss its merits.

Specific surface area (m^2·g^{-1}) is an important parameter in nanoscience research—especially in the study of catalysts, gas separation, and purification materials. Gas adsorption methods enable us to evaluate the surface area, pore size, and pore size distribution of a material. The BET formula is as follows:

$$\frac{P}{V(P_o - P)} \equiv \frac{P/P_o}{V(1 - P/P_o)} = \frac{1}{cV_m} + \frac{c-1}{cV_m}(P/P_o) \tag{3.51}$$

where

P is the equilibrium experimental pressure
P_o is the vapor pressure of the adsorbate gas at the experimental temperature
V (m^3·g^{-1}) is the standardized experimental volume of the adsorbed gas per gram of adsorbant
V_m (m^3·g^{-1}) is the volume of the adsorbate monolayer per gram of adsorbent
c is a constant that relates the heat of adsorption ΔH_{ads} (for the first physisorbed layer) with ΔH_{cond}, the latent heat of condensation (additional layers):

$$c = \exp\left[\frac{\Delta H_{ads} - \Delta H_{cond}}{RT}\right] \tag{3.52}$$

Once c is known, the calculation of ΔH_{ads} is straightforward. This relation works well except for cases of high relative pressure, $(P/P_o) > 0.35$, or very low relative pressure, $(P/P_o) < 0.05$. At relative pressure $0.05 < (P/P_o) < 0.35$, equation (3.51) is approximately proportional to (P/P_o) and transforms readily into the equation of a line, $y = mx + b$. A plot of $(P/P_o)/[V(1 - P/P_o)]$ versus (P/P_o) will yield a straight line with slope m equal to $[(c - 1)/(cV_m)]$ and intercept b equal to $(1/cV_m)$. **Figure 3.22c** shows a sample BET plot consisting of adsorption and desorption isotherms. Specific surface area is calculated by

$$S_s = a_m \left(\frac{P_o V_m}{RT} \right) \mathcal{N}_A = a_m \left(\frac{V_m}{V_{Gas}} \right) \mathcal{N}_A \qquad (3.53)$$

where

S_s is the specific surface area in $m^2 \cdot g^{-1}$

a_m is the area of solid surface for adsorption of one gas molecule (0.162 nm^2 for N_2)

V_{Gas} is the molar volume of the gas in its standard state (2.24×10^{-2} m^3)

$\mathcal{N}_A$ is Avogadro's number (6.022×10^{23} mol^{-1})

Figure 3.22a and **b** shows an Autosorb-I Series™ surface area analyzer manufactured by Quantachrome Instruments and its schematic of operation. **Figure 3.22c** and **d** shows examples of a BET isotherm (with some porosity) and its slope-form plot, respectively. **Figure 3.22e** and **f** shows plots obtained from BJH analysis (described below) that reveal information about pore volume distribution and size, respectively. Pressure capability ranges from less than 3×10^{-10} to 1000 torr; adsorbate gases, N_2, Ar, CO_2, butane, krypton, ammonia, and water vapor, among others, can be used in the BET procedure. **Figure 3.23** shows SEM images of sample materials that underwent BET analysis.

BJH Method.   Porosity is the ratio of void volume to the solid component of a material ($\varepsilon = V_S/V_T$). In addition to the specific surface area, nanoscientists also need to know the overall porosity of a material and the size of pores. Pore size distribution is determined by another thermodynamic method called the Barnett–Joyner–Halenda (BJH) method [52]. The BJH method is based on the Kelvin equation. This method exploits the phenomenon of capillary condensation in mesoporous systems and is valid at P/P_o greater than 0.35. Another mechanism involves capillary filling in micropores (e.g., like those of M-41S type zeolite), and is suitable for BJH at P/P_o between 0.1 and 0.5. Enhancements to BJH by microscopic methods based on statistical mechanics and molecular simulation are applicable for pore size analysis for micro- and mesosystems. Discussions of *density functional theory (DFT)* and *molecular dynamic simulation* methods are beyond the scope of this text.

According to the International Union of Pure and Applied Chemistry (IUPAC), pores are defined according to the following criteria: *Macropores* have pore diameter greater than 50 nm, *mesopores* have pore diameter between 2 and 50 nm, and *micropores* have pore diameter less than 2 nm. Pore channels and cavities also come in various shapes and conformations.

The Kelvin equation relates the equilibrium vapor pressure of a substance above a curved surface to the equilibrium vapor pressure of the same substance over a flat planar surface. With regard to the nanoscale, the Kelvin equation predicts the pressure of condensation or evaporation of an adsorbate in a cylindrical pore that already is coated with a multilayer. Pore size is calculated from the

FIG. 3.22

(a) An Autosorb-I Series™ surface area analyzer manufactured by Quantachrome Instruments and (b) its schematic of operation are shown in the figure. Pressure transducers in the instrument are capable of resolving a minimum of 0.00025 torr with a minimum resolvable relative pressure (P/P$_o$) of 3.2 × 10^{-7} torr of N$_2$ in the 0- to 1000-torr range; 2.5 × 10^{-6} torr and 3.2 × 10^{-9} torr in the 0- to 10-torr range; and from 0–1 torr, 2.5 × 10^{-7} and 3.2 × 10^{-10} of nitrogen respectively. In (c) and (d), a BET isotherm (with porosity) is shown with its extracted BET slope. The specific surface area is determined from the amount of adsorbate per monolayer. Figures (e) and (f) and derived from BJH and DFT analysis to yield information about pore volume and pore size.

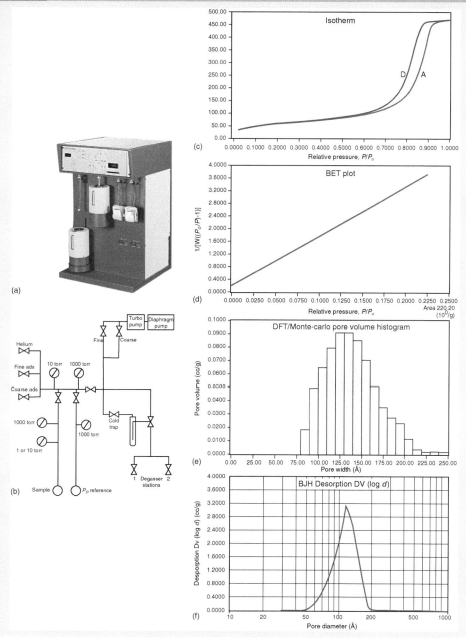

FIG. 3.23   *Carbon black (a), prickly gold (b), radiolarian (c), coal ash (d), and natural zeolite (e) are depicted.*

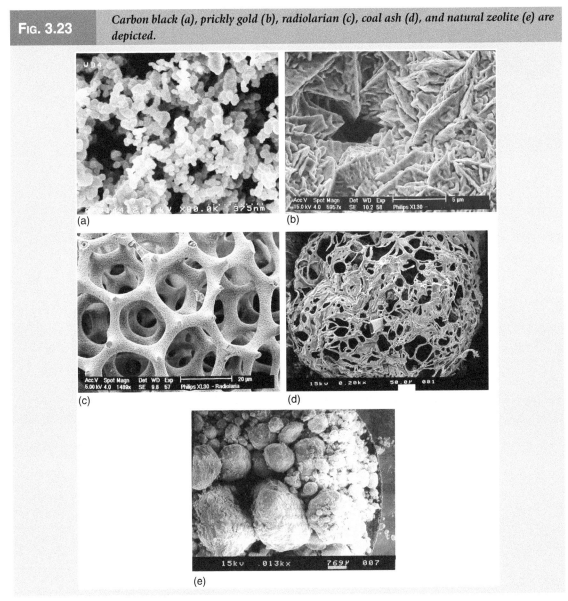

(a)

(b)

(c)

(d)

(e)

Source: Courtesy of Quantachrome Instruments.

Kelvin equation in conjunction with the statistical thickness derived from a *t-curve equation*:

$$\ln\left(\frac{P}{P_o}\right) = \frac{2\gamma V_m}{rRT} \tag{3.54}$$

where
   P is the experimental vapor pressure
   P_o is vapor pressure at saturation
   γ is the surface tension of the adsorbate

V_m is the molar volume as before
r is the radius of the droplet
RT has its usual meaning

Mercury Porosimetry.   Mercury porosimetry is another method by which porous materials are characterized. It is a pore filling method that is complementary to the BJH method because pore dimensions greater than 300 nm can be evaluated. BJH is valid down to ca. 0.3 nm. Pore size distribution is often measured by mercury porosimetry, although the method is not a geometric method per se. It is an indirect technique that works best if correlated to a model or other analytical techniques such as x-ray tomography or magnetic resonance imaging. Mercury porosimetry can be applied to acquire information about the porosity and pore interconnectivity parameters as well as pore size distribution. It is based on the transport and relaxation properties of liquid mercury. Mercury is used because it is nonwetting and has a high surface tension. A sample is evacuated at elevated temperatures and immersed into mercury. An external pressure is applied. As the pressure is increased, mercury fills the pore volume. The amount of mercury that finds its way into the pore volume is a function of the applied pressure—intrusion and extrusion of a nonwetting fluid in a porous solid.

Sears's Method.   In 1956, George Sears developed the Sears method to determine the specific surface area of colloidal silica particles by titration with sodium hydroxide (Sears). The surfaces of nonporous particles 5–100 nm in diameter are titrated in a 20% NaOH solution. At *pH* 9, 1.26 hydroxyl groups are adsorbed per square nanometer. The empirical relationship for the Sears method is

$$S_s = 26.4 \, (V_t - V_b) \tag{3.55}$$

where S_s is specific surface area as before, V_t is the milliliters of 0.1 *M* NaOH required for titration of 1.50 g SiO2, and V_b is the titration blank in absence of any silica. Particle size *d* in nm is determined by

$$d = \frac{6}{\rho S_s} = \frac{2.73}{S_s} \tag{3.56}$$

where the density of silica $\rho = 2.2$ g cm^{-3}.

NMR-Cryoporometry.   Porous materials can be characterized by determination of the melting behavior of fluids injected into pores [53,54]. Researchers have recently determined that the freezing point depression ΔT_f is a function of surface-to-volume ratio and that melting point shifts (ΔT_m) are influenced by the change in curvature of the pore surface [53,54]. The method used is called NMR-cryoporometry.

3.4.5   *Other Important Characterization Methods*

Quartz Crystal Microbalance (QCM).   A QCM has the capability to measure mass changes as small as a few nanograms per square centimeter, which is sensitive enough to detect monolayers of deposited materials. QCM is used to monitor the thickness of metal accumulating on a surface applied by evaporation or sputtering [55]. The QCM is used in biosensor designs due to this great sensitivity.

We have discussed the importance of vibrating piezomaterials in generating motion necessary for AFM and STM analysis. Another application of vibrating crystals is by the QCM technique. Change in mass can be measured by recording the change in frequency of the piezo crystal. The behavior of a QCM is defined by the Sauerbrey equation:

$$\Delta f = \frac{-2\Delta m f_o^2}{A\sqrt{\rho_q \mu_q}} = \frac{-2\Delta m f_o^2}{A\left(\rho_q \upsilon_q\right)} = -2.26 \times 10^{-6} f_o^2 \frac{\Delta m}{A} \qquad (3.57)$$

where f_o is the fundamental resonant frequency of the crystal (600 kHz–30 MHz for quartz and higher for overtones), A is the active area of the crystal between the electrodes, and ρ_q and μ_q are the density (2.648 g·cm^{-3}) and sheer modulus (2.947 × 10^{11} g·cm^{-1}·s^{-2}) of quartz, respectively. In another form of the equation, υ_q is the acoustic wave speed (3340 × 10^2 cm·s^{-1}).

Most QCMs employ AT-cut quartz (35°15′ across the growth direction of the α-quartz crystal) sandwiched between two electrodes, usually gold. The AT-cut yields a crystal with a shear perpendicular to the crystal face. The sensitivity of a QCM is ca. 1 ng·cm^{-2} compared to one of the best electronic microbalances that have sensitivity on the order of 0.1 μg. The QCM is therefore capable of measuring single layers of molecules or atoms.

A KSV Instruments, Ltd. QCM-Z500 (**Fig. 3.24**) has an acoustic sensor that is able to measure the change in mass of surface layers. It is an impedance-based instrument that is a versatile tool of nanoscience and able to monitor antibody (antibody–antigen) interactions, protein binding, lipid bilayer deposition, cell attachment–detachment, vesicle attachment, surfactant adsorption, polymer adsorption, nanoparticle adsorption, self-assembly, surface gels, metal thickness, and corrosion—to name a few applications. The frequency range of this particular instrument is 0.1–55 MHz with a resolution of 0.01 Hz. The mass resolution for a standard 5-MHz AT-cut crystal is 0.177 ng·cm^{-2}. The maximum mass load and thickness are ~0.1 mg·cm^{-2} and ~10 μm, respectively.

In one experiment successive layers of stearic acid ($C_{17}H_{34}O_2$) were deposited on the surface of a gold-coated QCM by the Langmuir–Blodgett technique [56]. Stearic acid is a long-chain aliphatic hydrocarbon with a carboxylic acid end group (**Fig. 3.25**). The frequency change was measured with a dissipative-QCM [55,56] following the deposition of each layer of stearic acid. A dissipative QCM was used because of the built-in capability to gauge another parameter important to monolayers: the softness of the stearic acid film. The correlation between the theoretical and experimental values with the change in frequency is shown in **Figure 3.26**.

Complying with the directive of converging evolution of instrumentation, a combination between QCM and AFM devices was applied to analyze electrodeposition processes [57]. The goal of the researchers was to monitor surface

FIG. 3.24	*The quartz crystal microbalance is a relatively unheralded workhorse for characterizing nanomaterials. Every evaporation instrument has one, but its versatility expands into many fields and applications. A QCM-Z500 quartz crystal microbalance manufactured by KSV Instruments, Ltd. is shown. It is capable of measuring changes in mass and viscoelastic properties of adsorbed or deposited layers. Mass change is proportional to the change in resonance frequency of the crystal and viscoelastic behavior is measured by the change in electrical resistance (impedance). The range of frequency is 0.1–55 MHz with resolution down to 0.01 Hz; mass resolution capability is 0.177 ng·cm^{-3}. The active sensor area is 20 mm^2. The instrument is capable of measuring film thickness as high as 10 μm and a load of 0.1 mg·cm^2.*

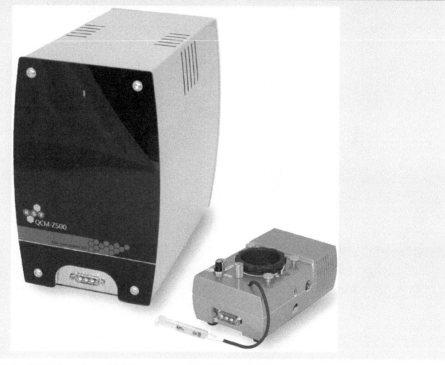

Source: Image and specifications are courtesy of KSV Instruments, Ltd. With permission.

characteristics at the micrometer and nanometer scales. In order to achieve this goal, they combined AFM with two kinds of acoustic wave devices: the QCM and SAW (surface acoustic wave). The QCM–AFM interaction consisted of longitudinal acoustic waves generated by the QCM that were reflected on the AFM cantilever. Sensitivity was calculated by

$$S \equiv \frac{\Delta f}{f_o} = \frac{A}{\Delta m} \qquad (3.58)$$

SAW has greater sensitivity (in the picograms per square centimeter range) than QCM. It uses an interdigitized electrode technology consisting of piezomaterials.

FIG. 3.25

Stearic acid layers are built on a gold-coated quartz crystal by the Langmuir–Blodgett technique. The frequency change induced following each cycle of deposition was measured with a dissipative QCM. The gold layer was deposited one layer at a time. The stearic acid monolayers were applied in successive deposition steps. Deposition of stearic acid on a gold thin film by the Langmuir–Blodgett process: The gold was first deposited on the surface of a quartz crystal microbalance. The procedure was repeated until 45 layers of stearic acid were deposited. The stearic acid monolayers were deposited with a KSV Minitrough 2 film balance from a 10^{-4} M $MnCl_2$ subphase with pH ~ 6 at a constant trough pressure of 30 mN·m^{-1} [55,56].

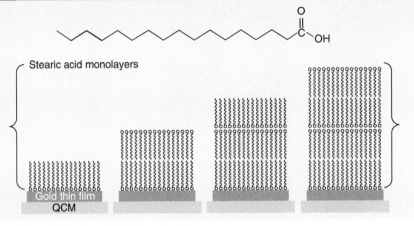

Source: Redrawn with permission from KSV Instruments, Ltd.

FIG. 3.26

The frequency change of the QCM signal is plotted as a function of the number of layers of stearic acid deposited on the gold surface. The absorbed mass can be calculated from the Sauerbery equation. The theoretical frequency change correlates well with the experimental values. The more material that was added to the monolayer assembly, the lower the resonance frequency became (e.g., the greater the change in frequency, Δf)[55,56].

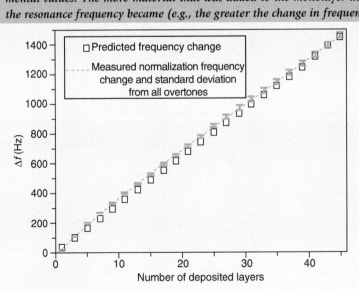

Source: Redrawn with permission from KSV Instruments, Ltd.

References

1. *History of spectroscopy*, College of Charleston, www.cofc.edu/~deavorj/521/History%20of%20Spectroscopy.htm (2007).
2. R. W. Kelsall, I. W. Hamley, and M. Geoghegan, *Nanoscale science and technology*, John Wiley & Sons, Ltd., London (2005).
3. C. P. Poole and F. J. Owens, *Introduction to nanotechnology*, Wiley-Interscience, John Wiley & Sons, Inc., New York (2003).
4. Nanoscience and Technology Institute (NSTI) Conference in Santa Clara, California, April 26 (2007).
5. J. I. Goldstein, D. E. Newbury, P. Echlin, D. C. Joy, C. Fiori, and E. Lifshin, *Scanning electron microscopy and x-ray microanalysis*, 2nd ed., Plenum Press, New York (1992).
6. M. V. Klein, *Optics*, Wiley, New York (1970).
7. C. J. Davisson, Are electrons waves? *Franklin Institute Journal*, 205, 597 (1928).
8. C. J. Davisson and L. H. Germer, Reflection and refraction of electrons by a crystal of nickel, *Proceedings of the National Academy of Science, USA*, 14, 619–627 (1928).
9. C. J. Davisson and L. H. Germer, Reflection of electrons by a crystal of nickel, *Proceedings of the National Academy of Science, USA*, 14, 317–322 (1928).
10. J. H. Wittke, Effects of electron bombardment, nau.edu/microanalysis/Microprobe/Interact-Effects.html, Northern Arizona University (2006).
11. Electron microscopes, JEOL, Ltd. http://www.jeol.com/HOME/tabid/36/Default.aspx (2007).
12. J. J. Donovan, Electron probe microanalysis, In *Scanning electron microscopy and x-ray microanalysis*, 2nd ed., J. I. Goldstein, D. E. Newbury, P. Echlin, D. C. Joy, C. Fiori, and E. Lifshin, eds., Plenum Press, New York (1992).
13. W. S. Boyle and G.E. Smith, Charge-coupled semiconductor devices, *Bell Systems Technical Journal*, 49, 587–593 (1970).
14. M. H. Ellisman, An historical perspective on digital imaging in transmission electron microscopy: Looking into the future, *Microscopy and Microanalysis*, 11, 604–605 (2005).
15. J. C. H. Spence and J. M. Zhou, Large dynamic range, parallel detection system for electron diffraction and imaging, *Review of Scientific Instruments*, 59, 2102 (1988).
16. C. J. Calbick, The discovery of electron diffraction by Davisson and Germer, *The Physics Teacher*, 91, 63–69 (1963).
17. I. Hillier and R. F. Baker, Microanalysis by means of electrons, *Journal of Applied Physics*, 15, 663–675 (1944).
18. B. Bhushan, Scanning probe microscopy—Principle of operation, instrumentation and probes. In *Springer handbook of nano-technology*, B. Bhushan, ed., Springer–Verlag, Berlin (2004).
19. G. Binnig, H. Rohrer, Ch. Gerber, and E. Weibel, Surface studies by scanning tunneling microscope, *Physics Review Letters*, 49, 57 (1982).
20. G. Binnig, C. F. Quate, and Ch. Gerber, Atomic force microscope, *Physics Review Letters*, 56, 930 (1986).
21. G. Shmalz, Uber Glatte und Ebenheit als Physikalisches und Physiologishes Problem, *Zeitschrift des Vereimes Deutscher Ingenieurte*, 1461–1467, Oct. 12 (1929).
22. http://www.pacificnano.com/afm-history_references.html
23. J. Scalf and P. West, Introduction to nanoparticle characterization with AFM, Pacific Nanotechnology, Inc., www.nanoparticles.pacificnanotech.com (2007).
24. D. D. Hsu, *Profilometer*, definition, www.chemicool.com/definition/profilometer.html (1996).
25. High Resolution Surface Profiler/Stylus-Type Profilometer, Ambios Technology, http://www.ambiostech.com/profiler.html (viewed 2007).

26. R. Young, J. Ward, and F. Scire, The Topografiner: An instrument for measuring surface microtopography, *Review of Scientific Instruments*, 43, 999 (1971).

27. Piezoelectricity, *Oxford American Online Dictionary and Thesaurus*, v. 1.0.0, Apple Computer, Inc. (2005).

28. I. A. Levine, *Quantum chemistry*, 4th ed., Prentice Hall, Upper Saddle River, NJ (1991).

29. G. Cao, *Nanostructures & nanomaterials: Synthesis, properties & applications*, Imperial College Press, London (2004).

30. E. W. Müller, J. A. Panitz, and S. B. McLane, The atom-probe field ion microscope, *Review of Scientific Instruments*, 39, 83–86 (1968).

31. M. K. Miller, *Atom probe tomography: Analysis at the atomic level*, Kluwer Academic/ Plenum, New York (2000).

32. Product specification, Imago LEAP 3000X-Si, Imago Scientific Instruments, Madison, WI (2007).

33. Spectroscopy, *Oxford American online dictionary and thesaurus*, v. 1.0.0, Apple Computer, Inc. (2005).

34. J. D. Ingle, Jr., and S. R. Crouch, *Spectrochemical analysis*, Prentice Hall, Upper Saddle River, NJ (1988).

35. S. M. Bachilo, M. S. Strano, C. Kittrell, R. H. Hauge, R. E. Smalley, and R. B. Weisman, Structure-assigned optical spectra of single-walled carbon nanotubes, *Science*, 298, 2361–2366 (2002).

36. A. Grünes, R. Saito, G. G. Samsonidze, M. A. Pimenta, A. Jorio, A. G. S. Filho, G. Dresselhaus, and M. S. Dresselhaus, Characterization of nanographite and carbon nanotubes by polarization dependent optical spectroscopy, *MRS Proceedings*, Symposium F, F3.47, Fall (2002).

37. G. L. Hornyak and C. R. Martin, Optical properties of a family of Au-nanoparticle-containing alumina membranes in which the nanoparticle shape is varied from needle-like (prolate), to spheroid, to pancake-like (oblate), *Thin Solid Films*, 303, 84 (1997).

38. S. J. Oldenburg, R. D. Averitt, S. L. Westcott, and N. J. Halas, Nanoengineering of optical resonances, *Chemical Physics Letters*, 288, 243–247 (1998).

39. A. M. Schwartzberg and J. Z. Zhang, Surface-enhanced Raman scattering (SERS). In *Encyclopedia of nanoscience and nanotechnology*, Marcel Dekker, Inc., New York (2006).

40. H. H. Bauer, G. D. Christian, and J. E. O'Reilly, *Instrumental analysis*, Allyn and Bacon, Inc., Boston (1979).

41. J. F. Shackelford, *Introduction to materials science for engineers*, 4th ed., Prentice Hall, Upper Saddle River, NJ (1996).

42. V. Petkov, T. Ohta, Y. Hou, and Y. Ren, Atomic-scale structure of nanocrystals by high-energy x-ray diffraction and atomic pair distribution function analysis: Study of Fe_xPd_{100-x} (x = 0, 26, 28, 48) nanoparticles, *Journal of Physical Chemistry C*, 111, 714–720 (2007).

43. X. Wen, S. Wang, Y. Ding, Z.-L. Wang, and S. Yang, Controlled growth of large-area, uniform, vertically aligned arrays of α-Fe_2O_3 nanobelts and nanowires, *Journal of Physical Chemistry B*, 109, 215–220 (2005).

44. A. Trunschke, *Modern methods in heterogeneous catalysis research*. Presentation, AC-FHI, Department of Inorganic Chemistry, w3.rz-berlin.mpg.de/~jentoft/lehre/trunschke_particlesize_031106.pdf (2006).

45. P. Lamparter, *Crystallite sizes and microstrains from x-ray diffraction and line profile analysis*. Power Point presentation, www.imprs-am.mpg.de/nanoschool2004/lectures-I/Lamparter.pdf (2004).

46. A. Markwitz, F. Lucas, J. Rusterucci, J. Kennedy, W. J. Trompetter, M. Rudolph, M. Ryan, V. White, and S. Johnson, A nuclear reaction analysis and optical microscopy study on controlled growth of large SiC nanocrystals on Si formed by low-energy ion

implantation and electron beam annealing, *Nuclear Instruments and Methods in Physics Research Section B: Beam Interactions with Materials and Atoms*, 249, 105–108 (2006).

47. S. A. E. Johansson, J. L. Campbell, and K. G. Malmqvist, *Particle induced x-ray emission spectrometry (PIXE)*, Wiley, New York (1995).
48. W. P. King, S. Saxena, B. A. Nelson, B. L. Weeks, and R. Pitchimani, Nanoscale thermal analysis of an energetic material, *Nano Letters*, 9, 2145–2149 (2006).
49. R. Hoffman, *Diffusion NMR*, Hebrew University of Jerusalem, chem.ch.huji.ac.il/nmr (2005).
50. S. Brunauer, P. H. Emmett, and E. Teller, Adsorption of gases in multimolecular layers, *Journal of the American Chemical Society*, 60, 309–319 (1938).
51. A. Seri-Levy and D. Avnir, The Brunauer–Emmett–Teller equation and the effects of lateral interactions. A simulation study, *Langmuir*, 9, 2523–2529 (1993).
52. E. P. Barrett, L. G. Joyner, and P. P. Halenda, The determination of pore volume and area distributions in pure substances, *Journal of the American Chemical Society*, 73, 373 (1951).
53. O. Petrov and I. Fúro, Curvature dependent metastability of the solid phase and the freezing—melting hysteresis in pores, *Physics Review E*, 73, 011608–011614 (2006).
54. O. Petrov and I. Fúro, Characterizing porous materials through the melting-freezing behavior of pore-filling fluids, 8th International Conference on Magnetic Resonance in Porous Media, September 10–14, Bologna, Italy (2006).
55. *Adsorption sensors: Impedance based QCM-Z500*, KSV, Ltd., Helsinki, Finland, 1–6 (2006).
56. *Langmuir–Blodgett deposition of stearic acid on gold in determining layer mass during repeated LB-film deposition by ex situ measurement of resonant frequency in air*, KSV, Ltd., Helsinki, Finland (2006).
57. J.-M. Friedt, L. Francis, K.-H. Choi, F. Frederix, and A. Campitelli, Combined atomic force microscope and acoustic wave devices: Application to electrodeposition, *Journal of Vacuum Science Technology A*, 21, 1500–1505 (2003).

Problems

3.1 Calculate the de Broglie wavelength of an electron under an acceleration voltage of 150 V.

3.2 Calculate the velocity of an electron moving at this accelerating voltage in problem 3.1. How does this value compare to the speed of light? The speed of light is 299.8×10^8 m·s^{-1}.

3.3 Calculate the wavelength of an electron beam in a TEM with accelerating voltage equal to 100,000 V. What is the velocity of the electron? Should we be concerned with relativistic effects? How would factoring in relativistic effect change your answer?

3.4 Compare the wavelength of a low-energy electron ($v = 0.01\ c$) to that of a gold atom. Use the de Broglie relation.

3.5 Name some important differences among emission, fluorescence, and luminescence.

3.6 1.8 eV of radiation is enough to produce an electron transition. To what wavelength does this correspond?

3.7 Explain the physical principles behind the mechanism of a scintillator. Do the same for a photomultiplier, a collimator, and a CCD camera.

3.8 Review the differences among secondary, backscattered, and transmitted electrons. Which ones are considered to be elastic?

3.9 Why is the wavelength of emitted x-rays following excitation inversely proportional to the square of the atomic number?

3.10 Explain in simple terms the phenomenon of Bremsstrahlung radiation.

3.11 What are Kakuchi lines? Explain this phenomenon.

3.12 What is the work function of a material? Which materials have the lowest work function? Explain the photoelectric effect.

3.13 Why is a phosphor screen used in TEM?

3.14 Explain the Auger equation in more detail. What kinds of analyses are best suited for Auger spectroscopy?

3.15 In Davisson and Germer's experiment, a crystalline Ni surface was exposed to an electron beam $\varphi = 50°$ relative to the normal of the crystal plane. A maximum in the intensity was observed at this angle and the acceleration voltage was equal to 54 V. What is an estimate of the d-spacing of the exposed crystal plane?

3.16 What is a Faraday box? Explain its mechanism of detection.

3.17 Interference from sectioned thin films (for TEM) on water: What is the thickness of a thin film of poly(methyl methacrylate) (PMMA) embedding material that looks blue ($\lambda_{incident} = 475$ nm)? The majority of the interference color is due to the PMMA embedding material with refractive index n equal to 1.489.

3.18 What is the factor of magnification if a 100-nm square field of view is projected to fill one fourth of a 17-in. diagonal square screen?

3.19 Convert an 850-nm wavelength into units of frequency, electron volts, wavenumbers, joules, and ergs.

3.20 (a) What is the d-spacing of a copper plane if it corresponds to x-ray wavelength from erbium-Lγ_3 x-rays (of wavelength 0.13146 nm) at diffraction angle of $2\theta = 50.5°$. Assume first-order diffraction. (b) What is the energy of these x-rays in kiloelectron volts?

3.21 Why would the sharpness of the tip (e.g., the curvature of the tip) play an important role in resolution of a scanned surface?

3.22 Describe the dual-detector system of an AFM in more detail.

3.23 Discuss the piezoelectric effect.

3.24 List the differences between atomic probe methods and atomic force methods.

3.25 The Lycurgus cup introduced in chapter 1 demonstrates dichroic optical behavior. Why? Why is King Lycurgus purple in the transmitted mode?

3.26 Why would the medium (or coating) of a metal affect the position plasmon resonance?

3.27 Before you read the upcoming chapters, why do you think quantum dots made of the same material but of different size demonstrate emission at different wavelengths?

3.28 C_{60} is a new form of carbon with 60 equivalent carbons. How many NMR peaks would you expect for the C_{60} NMR spectrum?

3.29 Describe the mechanism of operation of a time-of-flight MS.

3.30 TGA of a 1-g sample of a porous, anodically formed membrane formed in a 10% v/v sulfuric acid solution was conducted over a range of RT to 1200°C under a nitrogen atmosphere. The *mass versus T* curve is shown in the figure. Assume that the alumina contained water, alumina (as Al_2O_3), and H_2SO_4 only.

a. Using your chemical intuition, speculate upon the types of chemical/physical reactions that occur over this temperature range.

b. Calculate the percent mass loss of each volatile component.

c. What kind of thermal technique would be most suitable for analysis of phase changes?

3.31 DSC is a method that enables the inspection of phase changes. Describe how this is done. Also, discuss the theory behind DSC.

3.32 Calculate the BET surface area for the following generic nanopowder. The isotherm is of BET Type II (multilayer adsorption, no pores). All data are standardized to 1 atm and 273.15 K. Calculate P/P_o and $P/[V(1 - P/P_o)]$.

a. What is the specific surface area in terms of square meters per·gram for this material?

b. Assuming spherical shape, what is the diameter of the nanoparticles?

c. What is the heat of physisorption of N_2 gas on the nanoparticles?

Experimental (P/torr)	Experimental (V/[cm³ g⁻¹])
060	0.810
100	0.885
150	0.990
190	1.20
230	1.30
295	1.45
450	1.86
560	2.25
620	2.60

3.33 Discuss how BET, BJH, and mercury porosimetry are complementary methods.

3.34 How many gold atoms are contained in a nanogram? A picogram?

3.35 Do you expect the measured frequency to increase or decrease as more material is added to the surface of a QCM?

3.36 The most important methods to determine the size of nanoparticles are TEM and AFM. What are the advantages and disadvantages of the two methods? (Question courtesy of Günter Schmid, Uni-Essen.)

FABRICATION METHODS

Manufacturing takes place in very large facilities. If you want to build a computer chip, you need a giant semiconductor fabrication facility. But nature can grow complex molecular machines using nothing more than a plant.

RALPH MERKLE

Characterization methods have been presented, addressed, and discussed, albeit without providing significant detail. The catalog nature of chapter 3 is deliberately extended into this chapter, which is the last chapter in the "Nanotools" division of the text. Because reference is made continually to various kinds of fabrication techniques throughout the text, it is prudent to place introductory material concerning fabrication early in the book. In this way, the student should be able to establish a level of comfort with, perspective on, and understanding of fabrication methods when the subjects emerge time and time again later in the text. The physics division of the text—chapter 5 through chapter 8—engages the study of nanomaterial properties and phenomena.

Please take note that the fabrication methods listed in this chapter are but a few of the multitude that actually exist. We have tried to categorize in a generic sense the major forms and tried to illustrate the processes with commonly practiced fabrication techniques. Much can be learned about nanomaterials by understanding how they are made.

4.0   FABRICATION OF NANOMATERIALS

There is nothing more gratifying, arguably, than holding in one's hand the physical manifestation of an idea, concept, or theory. The link between the idea, concept, or theory and its physical form is the process of fabrication. The fabrication process begins in a laboratory with atomistic simulations, experiments, mock-ups, and prototypes. Eventually, after a battery of testing, the physical embodiment of the idea, concept, theory, simulation, mock-up, and prototype makes it way into a manufacturing facility. We have already listed several characterization methods. It is now time to discuss the fabrication of nanomaterials.

Nanomaterials are made by two generalized processes: *top down* (e.g., subtraction from bulk starting materials) or *bottom up* (e.g., addition of atomic or molecular starting materials). Each scheme has a unique set of advantages and disadvantages. We recommend that you make a checklist of the advantages, disadvantages, limitations, and issues confronting each method as we discuss them through the course of this chapter.

We also add a brief section on molecular modeling, which is a fabrication tool. It is part of the design process. Molecular modeling has become one of the most powerful tools in nanoscale research, development, and material design. There exists a perfect fit between simulation and nanomaterials since atoms and molecules in nanoscale materials are finite and countable, and computer capability in this day and age is still limited with regard to capacity. Depending on the quality of input parameters, molecular simulation is able to generate an accurate rendition of nanoscale material behavior. Low-energy states, structure, dynamical behavior, chemical reactions, fluxes and flows, and more have been modeled with some form of atomistic-molecular simulation.

4.0.1   *Background*

Like anything else that we present in this text, boundaries are drawn for the sake of convenience and clarity, although sharp ones are not always possible.

Boundaries defining fabrication methods are no different; however, we proceed unabated and present the first bifurcation in the road. Like the great baseball player Yogi Berra said, "When you come to a fork in the road, take it."

There are two generic strategies for nanomaterial fabrication: *top down* and *bottom up*. Top-down fabrication methods begin with bulk materials (top) that are subsequently reduced into nanoparticles (down) by way of physical, chemical, or mechanical processes. Bottom-up methods, on the other hand, begin with atoms and molecules (bottom). These atoms or molecules react under chemical or physical circumstances to form nanomaterials (up). Growth proceeds in zero, one, or two dimensions to form dots, wires, or thin films, respectively. There are two generic types of bottom-up procedures: In the first, nanomaterials retain some level of structural and functional independence; in the second, nanomaterials become identical components of a bulk material. An example of the former is an array of gold quantum dots in an electronic device. Examples of the latter case include bulk metals formed from nanocrystallites and the structure of bone tissue.

There is, of course, further blurring of boundaries. Two general kinds of overlap, and possibly a third kind, occur between the two types (bottom up and top down) of fabrication strategies. In one case, a technique may be designated as top down but its microscopic mechanism suggests otherwise. The best example of this is the formation of carbon nanotubes by laser ablation. The starting material is a target made of compacted graphite and catalyst particles—certainly considered to be a bulk material in a compacted form. However, carbon nanotubes form from atoms and molecules via a catalytic process—definitely from the bottom up. Bismuth metal, obtained in bulk form, is melted and subsequently evaporated into atoms that deposit on the surface of a template material. Evaporation of a melted metal source to produce atoms (and perhaps nanoclusters) is a top-down procedure, but the formation of the thin layer of bismuth from those evaporated atoms is certainly from the bottom up.

In the second case, a manufacturing process may consist of both top-down and bottom-up methods. During the course of the fabrication of a computer chip, application of a photoresist material by a process called spin-coating is top down (from a bulk liquid phase). Photolithography is top down; chemical etching of the photoresist or the silicon substrate to reveal features is top down, but chemical derivatization to form a monolayer comprising different materials is bottom up.

Hybrid fabrication technology is a combination of distinct top-down and bottom-up mechanisms that occur simultaneously. This category of fabrication is exclusive to the nanoscale, where top-down and bottom-up techniques converge at the 30-nm size scale [1]. At the 3-nm scale, even hybrid technologies will be challenged by supramolecular and molecular technology that in turn may give way to atomic and nuclear technologies at the subnanometer scale—the realm of the single atom, single electron, single spin, and single photon [1]. These developments will require that we redefine top-down, bottom-up, and hybrid fabrication technologies. In the final analysis, it matters not which designation is assigned to a specific procedure, but for the sake of pedagogical purposes, we will continue to explore many types of fabrication methods and label them as one or the other or both.

Nanofabrication methods, just like characterization methods, have a long history. Fabrication and synthesis processes are the descendents of well-developed chemical and physical techniques developed over millennia. Engineers tend to manufacture components from the top down and then assemble them to make a device. Chemists, on the other hand, have always made materials by reacting atoms and molecules to form chemicals in bulk quantities—from the bottom-up. Chemical synthesis is by definition a bottom-up process. With regard to the biological processes, all structures are formed from the bottom up. Are you able to think of any exceptions to this rule?

The convergent nature of nanotechnology is well represented by fabrication methods. Engineers, physicists, chemists, and biologists respectively bring top-down, top-down, bottom-up, and bottom-up methods to the same table. The future of fabrication will require more cooperation between and among the disciplines, and the design parameters of future *nanofabs* must include such forward thinking in order to accommodate diversity and to enhance interaction among all the participants.

It is not practical to build an automobile engine from the bottom up and, conversely, it is not practical to synthesize aspirin from the top down. However, in nanotechnology, similar structures can be built from either fabrication perspective [2]. Features on a silicon wafer can be produced by a standard top-bottom procedure called lithography (bulk wafer → application of a photoresist layer → mask-UV exposure → etch) or by a bottom-up procedure (bulk wafer → polymer or seed crystals → self-assembly) [2]. Once again, nanotechnology and nanoscience are changing the way we do things and fabrication methods are no exception.

4.0.2   Types of Top-Down Fabrication Methods

We begin our catalog of fabrication methods with top-down methods. Physical fabrication techniques are considered to be mostly from the top down. Top-down methods are extremely diverse. Nanomaterials are formed from the top down by mechanical-energy, high-energy, thermal, chemical, lithographic, and *natural* methods.

Top-Down Mechanical-Energy Fabrication Methods.   Cutting, rolling, beating, machining, compaction, milling, and atomization comprise a few examples of mechanical methods used to produce nanomaterials from the bulk. A mechanical method employs a physical process that does not involve chemical change—according to the traditional definition of chemical change (a reaction). Beating metals into a thin film is an ancient mechanical procedure used by the Egyptians and other pre-Hellenistic cultures to make swords, spear tips, and ornamental coatings. Mechanical energy methods such as ball milling operate on the principle of mechanical attrition. Kinetic energy, translated by hard, high-speed pellets, is imparted to samples by collision and friction. Samples are ground into fine powders by this method. An overview of mechanical top-down methods is shown in **Table 4.1.**

Top-Down Thermal Fabrication Methods.   In the purest sense, a thermal fabrication method employs a physical process (heating) that does not initiate a chemical change in the sample—according to the traditional definition of chemical change (a reaction). Once again, it has proven difficult to place

TABLE 4.1	*Top-Down Mechanical-Energy Fabrication Methods*
Method	**Comments**
Ball milling	Production of nanoparticles by mechanical attrition to produce grain size <5 nm [3] High-energy ball milling uses steel balls to transfer kinetic energy by impact to the sample. Highly polydisperse products and contamination are problems.
Rolling/beating	Traditional mechanical methods to minimize material thickness and refine structure. Gold can be beat into a 50-nm thick film [4].
Extrusion; drawing	High-pressure processes of forcing materials in a plastic phase through a die to form high-aspect ratio parts like wires. Bi metal forced through nanopore alumina is an analogous process at the nanoscale and can be considered a thermal–mechanical process.
Mechanical Machining, polishing, grinding, and ultramicrotome	Also known as conventional machining; resolution limit: 5 μm [5] Other techniques analogous to mechanical machining perform the same function with laser beams, focused ion beams, and plasmas. Mechanical grinders/cutters are used to thin TEM samples. These include dimple grinders, diamond saws, ultrasonic disc cutters, and ultramicrotomes (<100 nm sections).
Compaction; consolidation	Metal powder ball milled and compacted. Powders are considered to be bulk materials; therefore, compaction of powders to form bulk material is not considered to be a bottom-up method.
Atomization	Conversion of a liquid into aerosol particles by forcing through a nozzle at high pressure

specific thermal methods into this category. Some of the top-down mechanical methods also involve thermal exchange. During the ball mill process, heat is obviously generated and plays a role in the outcome of the nanomaterial structure. Heat may be deliberately added during ball milling. Several compaction methodologies involve heating of samples during processing. The methods listed in **Table 4.2,** although extremely diverse, involve direct and deliberate heating of the sample during the fabrication process; chemical change may happen or not.

The process of combustion occurs in the presence of oxygen and causes a chemical change. Thermolytic and pyrolytic methods imply a process called "lysis" (from the Greek *lysis*, "a loosening, setting free, releasing, dissolution," from *lysein*, "to unfasten, untie") and usually involve chemical changes to the starting materials. Pyrolysis is the conversion of one material into another material by the application of heat in the absence of oxygen. Thermolysis or thermal reaction is often used synonymously with pyrolysis. Nanomaterials are routinely formed during the combustion of bulk organic materials. In the absence of oxygen, polyaromatic hydrocarbons (PAHs) with nanometer dimensions are formed by chemical reaction in pyrolytic processes. Sublimation, on the other hand, is the process of a solid phase of a material becoming a gaseous phase without experiencing an intermediate liquid phase (top down).

Top-Down High-Energy and Particle Fabrication Methods. High-energy sources such as electric arcs, lasers, solar flux, electron beams, and plasmas are commonly used to produce nanomaterials from the top down. A by-product of high-energy methods is superheating: a desirable or undesirable outcome depending on the objective. Although heat is produced during operation, these are not labeled as thermal methods because the origin of the heat is not a conventional thermal source per se. **Table 4.3** lists several commonly used high-energy methods that

TABLE 4.2	*Top-Down Thermal Fabrication Methods*
Method	**Comments**
Annealing	There are two applications of annealing: (1) anneal of bulk materials to form nanocrystallites, and (2) transformation of nanomaterials into another physical phase [6]. Microphase separation to form nanoscopic structures occurs in copolymer bulk materials upon application of thermal anneal above the glass transition temperature.
Chill block melt spinning	Metal is melted with RF coil and forced through nozzle on rotating drum, where it solidifies; strips formed with nanostructure [7].
Electrohydrodynamic atomization (EHDA)	Production of monodisperse droplets; melt or liquid materials at nozzle with electric field between nozzle and surface: cone → thin jet → droplets EDHA + pyrolysis to produce 10-nm Pt nanoparticles [8]
Electrospinning	A high voltage is applied to a polymer melt solution to induce charging. Polymer solutions at room temperature are also used routinely. At an acquired threshold, an electrospun fluid jet emerges from a needle tip to form a Taylor cone. The substrate, held at a lower potential, is covered by the charged polymeric solution
Liquid dynamic compaction (LDC)	Molten stream of metal is atomized by high-velocity pulses of an inert gas and the semisolidified droplets are collected on a chilled, metallic substrate [9].
Gas atomization	Molten metal is subjected to high-velocity inert gas impact that forms metal droplets [7,10]. Kinetic energy is transferred to metal, resulting in small droplets upon solidification to form powders. Powders are then compacted to form high-strength bulk materials.
Evaporation	Evaporation of solid metal or other material samples to form thin films; usually performed under high vacuum (10^{-6} torr). Heat is produced by electrical resistance. If nanoclusters are formed during the evaporation process, it is top down. If atoms or molecules are formed during the evaporation process that recombine to form a thin layer without any chemical reaction, it is a crossover technique.
Electrospinning	The process of electrospinning utilizes electricity to form thin layers of filaments from bulk polymer, composite, or ceramic solutions; fibers with nanoscale diameter can be fabricated [11].
Extrusion	Nanowires by extrusion of bismuth melt by pressure injection into porous template material such as alumina [12]. Parallel Bi nanowires with diameter ~13 nm
Template synthesis + evaporation	Formation of single-crystal Bi nanowires by a vapor-phase technique into porous alumina template—7-nm Bi nanowires [13]; 400–500°C with N_2 trap. Only phase changes are involved in this process.
Sublimation	The physical process of sublimation involves a phase change from a solid into gaseous form without a liquid intermediate phase. If nanoclusters are formed by this process, it is considered to be a top-down process. If atoms or molecules are formed first and then agglomeration into nanoparticles occurs, it is considered to be a crossover technique in which both top-down and bottom-up processes occur nearly simultaneously. Sublimation does not involve a chemical change of the material.
Thermolysis; pyrolysis	Decomposition of bulk solids at high temperature (top-down). These terms are also applied to the decomposition of molecules—nanomaterials are formed after decomposition in a bottom-up way by agglomeration. Because of this crossover, it is hard to place pyrolysis/thermolysis into one category or the other. The most common sense of the terms implies that molecules are simply converted into other molecules. In this sense, pyrolysis and thermolysis are neither top-down nor bottom-up methods. In such reactions (like decomposition), chemical change does occur.
Combustion	Chemical combustion is a top-down process in which there is chemical conversion of bulk organic materials + impurities into molecules like CO_2, H_2O, and nanomaterials such as ash with micron to submicron dimensions. The process of combustion involves oxygen.
Carbonization of copolymers	Spun fibers from polymethylmethacrylate (PMMA)--polyacrylonitrile (PAN) microspheres in PMMA matrix (top down? or bottom up?). Temperature treatment at 900°C removes PMMA and converts PAN into MWNTs [14]. Carbonization is another example of the difficulty encountered in cataloging such processes.

result in nanomaterials. Evaporation, a thermal method based on resistive heating, is considered to be a crossover technique in that a bulk material is converted into small particles (molecules or clusters)—a top-down process—that are then deposited to form a nanomaterial (thin film)—a bottom-up process.

Top-Down Chemical Fabrication Methods. If chemical transformations occur during a fabrication process, we shall designate that process as a chemical fabrication method. Although fabrication (a.k.a. synthesis) methods that employ chemical procedures rightfully reside within the domain of the bottom up, there are several that can be considered to be top down. Combustion is an ambiguous

TABLE 4.3	*Top-Down High-Energy and Particle Fabrication Methods*
Method	**Comments**
Arc discharge	High-intensity electrical arc discharge directed on a graphite target (anode) + catalyst to produce single-walled carbon nanotubes that accumulate on the cathode Temperature ~4000 K [15,16]
Laser ablation	High-intensity laser beam directed on a graphite target + catalyst to produce single-walled carbon nanotubes; sample warmed to 1200–1500°C by furnace, laser Sample is collected on water-cooled copper collector [17]. This process can be considered to be a thermal and a high-energy method.
Solar energy vaporization	Solar energy focused on graphite target + catalyst to produce single-walled carbon nanotubes Temperature ~3000+K [18]
RF sputtering	Ion bombardment of metal, oxide, or other material targets to form thin film coatings Usually performed under moderate vacuum (10^{-3} torr). Atoms, molecules, and clusters are formed by this process.
Ion milling	Argon ion plasma is used to subtract material from a surface. The purpose is to clean surface or remove (thin) materials for TEM. No change in the chemical nature of the sample happens during this process.
Electron beam evaporation	This is similar to evaporation in **Table 4.2** but uses an electron beam source to heat material. Evaporated material condenses on target substrate. High vacuum is required. Thin-layer antireflection, scratch-resistant coatings are formed by this technique.
Reactive ion etching	Sensitive materials are etched by reactive chemical species in charged plasma. Chemical change of the etched material takes place during this process. The etching process is guided by maskant materials.
Pyrolysis	Pyrolysis can also be considered a high-energy method. Application of high-energy source like fire to bulk hydrocarbon materials (like a steak) in the absence of oxygen creates polyaromatic hydrocarbons (PAHs)—a top-down process (or if considering intermediates—for example, carbon atoms—it can be considered to be a bottom-up process). Pyrolysis of solid refractory nanoscale materials like Si–C–N substrate to form nanotubes at 1500–2200°C is a crossover technique [19]. Large-scale synthesis of multiwalled carbon nanotubes occurs in flame environments by burning carbon sources such as methane, ethylene, or benzene.
Combustion	Combustion can be considered to be a high-energy, thermal, or chemical fabrication method.
High-energy sonication	Ultrasonication uses high-energy sound waves to make nanomaterials from bulk materials. The technique is also used to disperse carbon nanotubes in a suitable solvent. The dispersion of bundles of nanotubes into individual tubes is top down. Probe tips are made of titanium, vanadium, and other metals and alloys. Micron- to nanosized residual tip metal is introduced into solutions during the sonication process.

chemical top-down method, depending on the starting material. The chemical structure of solid constituents is completely altered following a combustion process. Nanosized PAHs and fly ash are by-products of a top-down pyrolysis process, e.g., the burning of coal.

Chemical etching of solid substrates like a silicon wafer (masked or otherwise) is a top-down chemical method. Chemical etching processes, on the other hand, adhere to a slightly different classification criterion—specifically, that chemical alteration occurs only in the layers exposed to, and subsequently removed from, the solid substrate. In other words, although nanofacets or porous structures are formed on or within the solid substrate, the chemical structure of the solid substrate remains intact. Only the surface is altered (passivated, oxidized). The process of chemical alteration is only applicable to substrate material removed during the etching process, e.g., transformation of the solid into a water-soluble oxide.

Anodizing is a chemical etching process that involves electricity (e.g., electrochemical etching). This process is a crossover technique and consists of four parts.

1. Metal is electrochemically removed top down from the surface and released into solution in ionic form, Al^{3+}, during the anodizing of aluminum metal. The cationic products of anodizing are not nanomaterials; they are ions.
2. Hexagonally distributed, monodisperse scalloped structures (nanofacets) are formed on the surface of the aluminum anode during anodizing. The diameter and curvature of individual nanoscale scallops are dependent on the applied anodic voltage. This is a true example of top-down fabrication. The other two parts of the anodize equation are bottom-up procedures.
3. The reaction of metal cations with anions originating from the cathode reaction or with solution anions leads to the formation of nanoscale colloidal oxides that eventually form the porous layer (from the bottom up). Anionic species include oxides, hydroxides, and other negatively charged species (phosphates, sulfates, oxalates, or chromates).
4. The hexagonal porous anodic oxide layer is formed from the bottom up by the electrochemical reaction of Al^{3+} cations with various oxide anions. The scalloped top-down metal surface structures direct the size, orientation, and distribution of the bottom-up pore channels.

Overall, if we had to choose we should probably consider anodizing as a top-down fabrication process. Top-down chemical fabrication methods are listed in **Table 4.4.**

Top-Down Lithographic Fabrication Methods.   Many powerful top-down techniques involve some form of lithography. Lithographic techniques are what made the integrated circuit industry what it is today, and it continues to be the most viable method to form nanostructures that actually has widespread applications. The history of lithography was presented briefly in chapter 1. Traditionally, electromagnetic sources ranging from the visible wavelengths are still the most popular—especially in MEMS and circuit fabrication. Ultraviolet and x-ray sources are increasingly in demand as smaller features are required. Electron

TABLE 4.4	*Top-Down Chemical Fabrication Methods*
Method	**Comments**
Chemical etching	Standard acid or base solution etching of silicon and other materials, usually guided by maskant materials. Materials with nanometer pore channels are produced by this method. Etching of a metal surface without substantial oxide growth results in nanofacets.
Chemical–mechanical polishing (CMP)	CMP utilizes abrasives with or without a corrosive chemical slurry. Purpose is to thin and flatten samples. Surface roughness depends on size of abrasive. Mirror finishes with nanometer-scale roughness are produced by CMP methods.
Electropolishing	Electropolishing is an anodic method for brightening and smoothing the surface of a metal, primarily used for aluminum. The purpose of electropolishing is to reduce the surface roughness of a metal to nanometer scale. Conditions are extreme: concentrated acids (or bases), elevated temperature, and elevated current.
Anodizing	Anodizing is considered to be a crossover method in that nanofacets are formed on a bulk aluminum surface from the top down that in turn direct the formation of an anodic oxide from the bottom up. Anodizing operates under the same principle as electropolishing except that film growth is favored instead of film dissolution. Conditions are mild in comparison: dilute polyprotic acids, low temperatures (ca. 0°C), and low current.
Combustion	Combustion is an irreversible and dynamic chemical process that is catalyzed by high-temperature flames. Trees burning in a forest fire is a top-down way to form nanoaerosols.

beam sources are used primarily in the manufacture of the masks used in subsequent optical lithographic applications. Direct writing of lithographic patterns by electron and particle beams has proven to be an effective means of pattern generation without the use of a mask. Several low-cost, high-throughput nanolithographic techniques have emerged in the past decade or so (e.g., dip-pen nanolithography, nanosphere lithography, nano-imprinting, and other allied forms).

The new forms of lithography present us with another parsing dilemma. Most lithographic methods are top-down methods. However, if an STM current is involved in sensitizing a surface, should it be called a top-down method even though it is accomplished at the nanoscale from start to finish? If an AFM tip transfers atoms or molecules onto a surface, should this lithographic process be called a bottom-up method? Top-down lithographic techniques are listed and described in **Table 4.5**.

Top-Down Natural "Fabrication" Methods.   Both top-down and bottom-up fabrication methods abound in the natural world. Most of these natural processes are quite familiar, and there is no need to allocate any more time or space to them. **Table 4.6** lists selected top-down natural processes to provide another relevant fabrication perspective.

4.0.3   Types of Bottom-Up Fabrication Methods

We will not go into significant detail about bottom-up methods in this section because many of them will be discussed more fully in the text in later chapters.

TABLE 4.5	*Top-Down Lithographic Fabrication Methods*
Method	**Comments**
LIGA techniques	LIGA is a German acronym for "Lithographie Galvanoformung Abformung," a microlithographic method developed in the 1980s. It was one of the first major techniques to demonstrate the fabrication of high-aspect ratio structures. Beam sources include x-ray, ultraviolet, and reactive ion etching. MEMS devices are fabricated using LIGA techniques.
Photolithography	Light is used to transfer patterns onto light-sensitive photoresist substrates. Photolithography is primarily used in the manufacture of integrated circuits and MEMS devices. The wavelength range of optical lithography techniques ranges from the visible to the near ultraviolet—ca. 300 nm. The resolution of photolithography techniques is ~100 nm [20].
Immersion lithography	Just like with immersion optical microscopy, resolution can be enhanced by 30–40% with application of a liquid medium between the aperture and the sample with higher refractive index. The medium needs to conform to the following criteria: (1) refractive index $n > 1$, (2) low optical absorption at 193 nm λ, (3) immersion fluid compatible with the photoresist and the lens, and (4) be noncontaminating.
Deep ultraviolet lithography (DUV)	Resolution with deep ultraviolet with $\lambda = 248$–193 nm, resulting in features on the order of 50 nm
Extreme ultraviolet lithography (EUVL)	Short wavelength ultraviolet, $\lambda = 13.5$ nm. EUVL resolution: ~30 nm [20]. The major problem with EUVL is that all matter absorbs EUV and damage to substrates is very likely. High vacuum is required and mask must be made of Mo–Si.
X-ray lithography (XRL)	X-rays are produced by synchrotron sources. XRL is capable of producing features down to 10 nm. Problems include damage to substrate materials.
Electron beam lithography (EBL)	An electron beam source is used instead of light to generate patterns. Although e-beams can be generated below a few nanometers, the practical resolution is determined by the electron scattering of the photoresist material. Just like in SEM, electron interaction volumes are generated during exposure. Line width <20 nm and electron energy: 10–50 keV
Electron beam writing (EBW)	EBW is a direct-writing procedure and, therefore, no pattern masters are required. Direct-write e-beam resolution is ca. 20 nm with lateral dimensions <10 nm [20]. Operation of electron beam parameters and patterning are computer controlled.
Electron beam projection lithography (EPL)	This technique is similar to TEM in that electrons are focused through a lens and projected onto a surface. In this case, however, a pattern is placed near the aperture. EPL is a high-throughput technique. A diamond membrane is used as stencil mask material. The process is not limited by diffraction as are the photolithographic techniques [21].
Focused ion beam lithography (FIBL)	Utilizes a liquid metal ion source (LMIS) with beam size of 10+ nm. FIBL resolution is 30 nm [20]. There is less backscattering than EBL and FIBL resists are more sensitive. FIBL, however, is restricted by limitations in reliable ion sources, difficulty in focusing, shorter penetration depth, swelling of resist, and whimsical ion implantation episodes. FIBL is also more expensive and slower that optical methods.
Microcontact printing methods	The George Whitesides group at Harvard University invented the lithographic method of microcontact printing. A topographical master is created by standard lithographic techniques that employ electron, ion, or electromagnetic beams. A negative replica of the master is made by pouring an elastomeric polymer, usually polydimethylsiloxane (PDMS), over the master. Upon curing, the elastomer is removed and coated with a self-assembled monolayer such as hexadecanethiol. Application of gold then reproduces the master pattern. Sub-100 nm features are possible by this technique [20].
Nano-imprint lithography (NIL)	Nano-imprint lithography is used to fabricate nanometer-scale patterns. It is a straightforward economical process with high throughput and high resolution. Patterns are created by stamping a resist material with a prefabricated stamp. The stamp can be used over and over. There are two types: thermoplastic (TNIL) [22,23] and photo (PNIL) [22,23]

TABLE 4.5 (CONTD.)	*Top-Down Lithographic Fabrication Methods*
Method	**Comments**
Nanosphere lithography (NSL)	NSL is used to fabricate nanometer-scale patterns. It is a straightforward economical process with high throughput and high resolution. It is difficult to categorize this technique as top down or bottom up. Micron-scale latex spheres are often used as the template material. The interstices are nanoscale in size.
	NSL utilizes nanospherical materials in close-packed configuration as a mask to aid in the fabrication of periodic particle arrays (PPAs). Polymer nanospheres (diameter <300 nm) are in a single or double layer over insulator, semiconductor, metal, inorganic ion insulator, or organic π-electron semiconductor materials. Depending on the sphere diameter, nanoscale facets on the order of 22 nm are easily formed [24].
Scanning AFM nanostencil	An evaporated particle beam source is focused through a hole in an AFM cantilever.
	The procedure is good for metal deposition. This technique combines the ability to pattern a surface simultaneously with the ability to image the surface with the same cantilever. It is difficult to classify this technique as top down or bottom up (e.g., as it is for the thermal evaporation technique discussed before).
Scanning probe nanolithographies	There exist several forms of scanning probe nanolithographies. Some impart mechanical stress via the probe tip to a sensitized surface, followed by a chemical treatment; others apply an STM current to a substrate to create dangling bonds that react further to produce nanofeatures. These methods can be considered as top down in that nanofacets and features are produced from a solid bulk substrate.
2-Photon polymerization	Photopolymerization causes polymer to solidify to form three-dimensional image. Resolution of ~120 nm, although the laser λ is 780 nm [25]. This is, in the clearest sense of the term, a top-down process.

TABLE 4.6	*Top-Down Natural Fabrication Methods*
Method	**Comments**
Erosion	Conversion of macroscopic mineral-based materials into micro- and nanoparticles.
Etching	Etching of silicate rocks by carbonic acid from the environment contributes to erosion.
Hydrolysis	The decomposition of organic (and inorganic) matter by hydrolysis is a common way to make nanomaterials in the natural world.
Volcanic activity	Formation of fly ash and other materials by volcanic activity. The dispersion of volcanic byproducts is mostly airborne. Volcanic by-products contribute to the formation of clays like *montmorillonite* (a nanostructured material discussed in chapter 13).
Forest and brush fires	Formation of combustion gases, nanometer scale PAHs, amorphous carbons, and particulates
Solar activity	Radiation degradation of bulk synthetic, inorganic, and organic materials
Pressure and temperature	Formation of diamond crystallites from pressure and temperature processes applied to bulk materials; application to bulk carbon deposits (coal)
Biological decomposition	Decomposition is a process that begins at the bulk, micro-, or nanoscale level and terminates at the nano, molecular, or atomic level. Biological decomposition is mitigated by bacterial and other life forms in addition to inorganic natural processes.
Digestion	Reduction of bulk biological materials into nanometer and subnanometer scale components by the action of acids and hydrolysis; the formation of nitrogenous wastes is a bottom-up procedure, so to speak.

Bottom-up fabrication techniques are divided into four general categories: (1) gaseous phase methods, (2) liquid phase methods, (3) solid phase methods, and (4) biological methods.

Just as with top-down methods, it is difficult to pigeonhole a technique into a general category. Many bottom-up processes are characterized by tandem applications of liquid and gaseous techniques onto solid substrates. There are three generalized states of matter: gaseous, liquid, and solid. The distance d between molecules in a gas is proportional to

$$d = \left(\frac{V}{N} \right)^{\frac{1}{3}} \qquad (4.1)$$

where V is the volume and N is the number of molecules. For an ideal gas at standard temperature and pressure (STP), $V = 22.4$ L and $N = \mathcal{N}_A$, Avogadro's number, 6.022×10^{23}. The distance between atoms or molecules, center to center, in an ideal gas is equal to 3.34 nm.

A liquid is a state of matter that has volume but not shape. Although the atoms and molecules in a liquid are compressed as tightly as a solid, the molecules in a liquid are free to move randomly and unfettered. The distance between molecules or atoms in a solid is like that of a liquid, but random movement is severely restricted due to structural factors. Solids, of course, constitute the most condensed form of matter.

A technique is designated as gaseous, liquid, or solid if the process takes place in that appropriate medium or if the active constituent from which nanomaterials are formed is a gas, liquid, or solid. Once again, some difficulty in nomenclature is encountered when more than one phase is present during synthesis, but from a practical point of view, such classification is relatively straightforward.

Bottom-Up Gas-Phase Fabrication Methods.   Gases represent a highly dispersed phase of atoms and molecules. Some nanomaterials formed in the gas phase, like clusters, remain in the gas phase. More commonly, gas-phase precursors interact with a liquid- or a solid-phase material. If one of the precursors of a nanomaterial originates from the gas phase or if the reaction takes place in the gas phase, we shall call it a bottom-up gas-phase fabrication method (**Table 4.7**).

Nonbiological Bottom-Up Liquid-Phase Fabrication Methods.   Bottom-up liquid methods are numerous and diverse (**Table 4.8**). The choice of solvent is an extremely important parameter in any liquid-based bottom-up fabrication method. The liquid medium can be hydrophilic or hydrophobic, ionic or anionic, or heterogeneous (e.g., for the purpose of phase transfer of product between two immiscible liquids). The new field of supramolecular chemistry is conducted entirely in liquid media. All bottom-up biological fabrication processes occur in liquid media. The liquid phase is also where most chemists feel at home, and it is also going to be one of the prime drivers of nanotechnology. Scale-up of liquid-phase fabrication methods is a relatively straightforward process and it is at the scale-up stage where the chemists turn over the reins of a process to the chemical engineers.

Bottom-Up Lithographic Fabrication Methods.   We add a special category for lithography once again, but this time featuring bottom-up lithographic methods.

TABLE 4.7	*Bottom-Up Gas-Phase Fabrication Methods*
Method	**Comments**
Chemical vapor deposition (CVD)	CVD involves the formation of nanomaterials from the gas phase, usually at elevated temperatures, onto a solid substrate or catalyst. Carbon nanotubes are formed by catalytic decomposition of carbon feedstock gas in inert carrier gas at elevated temperature. Single-walled carbon nanotube production by CVD requires nanoscale Fe, Co, or Ni catalyst plus Mo activator on high surface area support (alumina) at >650°C. Methane gas serves as the carbon source [26].
Atomic layer deposition (ALD)	ALD is an incredibly precise sequential surface chemistry layer deposition method to form thin films on conductors, insulators, and ceramics. The layer formed by ALD conforms to surface topography. Precursor materials are kept separate until required. Atomic scale control pinhole-free layers are formed. Al_2O_3 layers are generated from hydroxylated Si substrate + $Al(CH_3)_{3(g)}$, then H_2O vapor is applied to remove methyl groups. The process is repeated until a target thickness is attained. Layer thickness: 1–500 nm
Combustion	The formation of Si nanoparticles from the combustion of SiH_4 (silane gas) and other silicon-containing gases like hexamethyldisiloxane under low-oxygen conditions produces Si nanoparticles as small as 2 nm. Al_2O_3 and TiO_2 can also be formed by combustion.
Thermolysis; pyrolysis	Solid Si nanoparticles can also be formed by the thermal decomposition of silane gas in the absence of oxygen. The bottom-up decomposition of ferrocene to form Fe nanoparticles is one of the best examples of a bottom-up gas-phase fabrication method.
Metal oxide (MOCVD) Organometallic vapor phase epitaxy (OMVPE)	Chemical characteristics of precursor materials utilize reactive gas-phase-organometallic compounds that decompose to form nanometer-scale thin films or nanoparticles. H_2 carrier gas, group III metal–organic compounds + group V hydrides 500–1500°C at 15- to 750-torr pressure are representative conditions under which MOCVD is performed.
Molecular beam epitaxy (MBE)	MBE is a thin film growth process conducted under high vacuum. A heated Knudsen cell or effusion cell is used to introduce reactants by molecular beams. MBE is able to deposit one atomic layer per application. Examples include alternate layers of GaAs and AlGaAs with each layer of 1.13 nm in thickness and InGaAs quantum dots [27]. The temperature used in MBE is commonly 750–1050°C in H_2 carrier gas.
Ion implantation	This is a tough method to categorize. Nanovoids, for example, can be created by ion implantation of Cu ions into silica and subsequent annealing [28]. It is bottom-up action performed on a bulk material. If the ions come from a bulk source, it has a bottom up component. Once the ions are formed, ion implantation is bottom up.
Gas phase condensation; thermolysis	Formation of Fe nanoparticles by decomposition of ferrocene at 200°C is an example of gas-phase process to form nanoscale Fe. Formation of lithium nanoclusters by decomposition of LiN_3 is another example [7]. Temperature at decomposition depends on the material.
Solid template synthesis	Provides a solid template substrate for gas-phase deposition of materials on the solid substrate. This is considered to be a mixed bottom-up system. Final nanomaterial size, shape, and orientation are predetermined by template parameters.

Bottom-up lithography methods are limited to a few kinds, based on template processes or direct writing (**Table 4.9**).

Bottom-Up Biological and Inorganic Fabrication Methods. Biological processes are overwhelmingly formed from the bottom up (**Table 4.10**). More detail is allotted to this topic in chapter 14.

TABLE 4.8	*Nonbiological Bottom-Up Liquid-Phase Fabrication Methods*
Method	**Comments**
Molecular self-assembly	This generic process is supported in liquid media. From some perspectives, supramolecular chemistry is a subset of molecular self-assembly. Almost all molecular self-assembly takes place in liquids. The liquid plays a major role in supporting intermolecular interactions and intermediate metastable species.
Supramolecular chemistry	Supramolecular chemistry, for reasons to be explained in chapter 11, is conducted in liquid media. Weak intermolecular forces are supported in liquids that allow many kinds of intermolecular interactions to take place. All significant biological metabolic processes occur in a liquid medium.
Nucleation and sol–gel processes	Precursor chemicals in a supersaturated state combine by self-assembly or chemical reaction to form seed particles. Thermodynamics drives a nucleation process that forms nanoparticles. The nucleation process depends on prevailing conditions of pH, temperature, ionic strength, and time [5]. Due to van der Waals attractions, colloids are formed. Sol–gel methods are irreversible chemical reactions of homogeneous solutions that result in a three-dimensional polymer. Sol–gel methods yield nanostructured materials of high purity and uniform nanostructures formed at low temperatures [5]. Negative replicas of colloidal hierarchical structures, upon drying, yield aerogels or xerogels. Such gels can be back-filled to produce nanocomposites or hybrid materials [5]. These are all pure bottom-up processes.
Reduction of metal salts	Noble metal clusters and colloids are formed by the reduction of metal salts like $HAuCl_4$ and H_2PtCl_6. Common reducing agents come in the form of organic salts like sodium citrate—$Na_3C_6H_5O_7$. By means of phase transfer reactions (consisting of an interface between two immiscible liquids), metal clusters and colloids are stabilized by the addition of organic ligands. For example, phosphine or thiols are adsorbed onto gold-55 to produce a stable cluster [29].
Single-crystal growth	Nucleation process to form single crystals in liquid media
Electrodeposition Electroplating	Electrodeposition is direct deposition of metals from metal salt solutions to form thin layers or films on a solid conducting substrate. Electrodeposition is an electrolytic process that forms thin metal films on the cathode of the cell. The process conforms to Faraday's law.
Electroless deposition	Electroless deposition is the autocatalytic deposition of metals without electrical assistance. It requires metal cation + catalytic (activated) surface + reducing agents like formaldehyde, alkali diboranes, alkali borohydrides, or hypophosphorous acid. Pt, Ni, Co, Au, and numerous other metals can be deposited on many kinds of substrates, including plastics. Electroless deposition has been used to create negative or positive replicas of porous nanostructures [30].
Anodizing	We have already characterized anodizing as a top-down process. We mentioned earlier that anodizing method contains a top-down component (formation of scalloped structure). Here, we focus on the bottom-up formation of the porous alumina. Aluminum metal is made the anode in an electrolytic cell consisting of a polyprotic acid (usually sulfuric, phosphoric, or oxalic). Pore diameter of <5 nm → >200 nm; with pore density: 20–80+% and film thickness: <1 μm → >100 μm. Anodized titanium several nanometers thick generates bright interference colors.
Electrolysis in molten salt solutions	Utilization of molten alkali halide salts with graphite electrodes with 3- to 5-A current [31] Erosion at the cathode to form tubes The product is transferred to toluene.
Solid template synthesis	Provides a solid template substrate for electrochemical, chemical, polymerization, and other liquid-phase reactions. Most methods are accomplished in a liquid medium. Final nanomaterial size, shape, and orientation predetermined by template parameters.
Liquid template synthesis	Liquid templates (micelles and reverse micelles) are commonly used to make quantum dots from the bottom up.
Supercritical fluid expansion	Solvent removal under hypercritical conditions forms aerogels and xerogels that contain nanometer-sized voids. Supercritical conditions imply that the medium is in neither liquid nor solid phase.

TABLE 4.9	*Bottom-Up Lithographic Fabrication Methods*
Method	**Comments**
Nanolithography: Dip-pen methods (DPNL)	Nanoprobe lithography in the form of dip-pen nanolithography was invented by Chad Mirkin's group at Northwestern University in Chicago [32]. DPNL is considered as an AFM-based soft-lithography technique. The operation of this method is quite simple. A water meniscus is formed between an AFM tip and a substrate. The AFM tip, in conjunction with the water meniscus conduit, is able to transfer molecules to the surface. The method has high spatial resolution (<10 nm), has high registration capability (probe can both read and write), and is able to deliver complex molecules such as DNA to a surface [20]. The major disadvantage, like that of STM writing, is low throughput.
Nanosphere template methods	Nanosphere lithography is a template method for fabrication of nanomaterials. Latex spheres are arranged on a substrate surface in various configurations: hexagonal close packed, or into a square array. The interstitial spaces between latex spheres serve as sites through which deposition can occur—a very straightforward, simple process. Although the distribution and placement of the spheres can be considered to be a top-down process, the deposition of material through the interstices definitely occurs from the bottom up.
Nanopore template methods (shadow mask evaporation)	Use of porous alumina membrane templates as templates to form arrays of nanoparticles. The size of the nanoparticles can be controlled from 5 nm to >200 nm. The space between nanoparticles can also be adjusted. Nanoparticle aspects are adjusted by the height of the mask, the pore size, and the direction of evaporation [33]. This technique is good for direct patterning without the need for additional steps such as etching or lift-off. The combinations of masks, materials, and substrates are enormous, and the process allows for straightforward upscale. Arrays have been used in the secondary fabrication of memory devices and carbon nanotubes.
Block copolymer lithography (BCPL)	BCPs applied by spin-coating (top down) self-assemble into an ordered array of nanoscopic domains on a surface. Selective removal of one component yields an etch mask. The substrate pattern is formed by plasma etching. In a specific example: a 35-nm thick polystyrene–PMMA copolymer layer is applied to a Si_3Ni_4-coated Si wafer. Removal of the PMMA leaves an ordered array of polystyrene nanodots. Reactive ion etching (REI) with CHF_3 transfers the pattern to the Si_3Ni_4 layer. The Si_3Ni_4-formed pattern is etched again by REI with HBr. The result is an ordered array of silicon pillars (wires) [34]. Block copolymer lithography was able to produce periodic arrays of 10^{11} holes per cm^{-2} [35]. One problem that faces this procedure is long-range order.
Local oxidation nanolithography	A scanning probe tip (a dynamic AFM tip) is placed a few nanometers above a substrate surface. The environment consists of saturated water vapor. A bias voltage is applied between the tip and the surface. Oxidation of the surface, if silicon, produces lines of silicon oxide. The breadth of the meniscus and the distribution of the electric field within determine the size of the feature [36]. Features as small as 7 nm were produced. One-nanometer projections were formed in the *z*-direction.
STM writing	The IBM logo pictured in chapter 1 was fabricated by a bottom-up method. Starting with xenon atoms, each atom was manipulated by the scanning probe tip into its final position. Other examples of this technique include the *quantum corral*—a circular array of Fe atoms placed on a Cu surface [37]. All scanning probe fabrication methods are hindered by low throughput.

4.0.4 *Nebulous Bottom-Up Fabrication Categories*

Fabrication of nanoscale materials (structures, domains) within solids is difficult to pinpoint. It is difficult to track the history of an atom or molecule throughout the course of a solid material. Solids contain a number of diverse

TABLE 4.10	*Bottom-Up Biological and Inorganic Fabrication Methods*
Method	**Comments**
Protein synthesis	Formation of proteins from precursor amino acids by elaborate process of protein synthesis Transfer RNA transports amino acids to ribosomal RNA and link with peptide bonds.
Nucleic acid synthesis	Synthesis of nucleic material (RNA, DNA) from sugars, phosphate, and nuclides (adenosine, guanine, cytosine, and thymine) from the bottom up The processes of mitosis and meiosis are template (replication) methods.
Membrane synthesis	Bottom-up agglomeration of lipids, phospholipids to form organized membrane structures that make life possible
Inorganic biological structures	Mother of pearl (nacre) 95% Inorganic aragonite (platelets 200–500 nm thick) + organic biopolymer Deformable nanograins [38]
Crystal formation methods	Nucleation depends on P, T, concentration, and composition. Flaws reduce surface energy by nucleation. Direction of growth depends on nanostructure.

defects that have nanoscale dimensions. Are these considered to be "nanomaterials" or nanofacets? Or are they merely nanodomains of the bulk type material? Voids formed by ion implantation do agglomerate to form nanovoids from the bottom up. We address this nebulosity in more detail later.

4.0.5 The Nano Perspective

There are many kinds of nanomaterials. When discussing fabrication methods, it is essential that the nature of the end product be understood. For example, some types of nanomaterials retain their nanoscale dimensions (e.g., quantum dots). Others form into components of more complex structures (e.g., one-dimensional, two-dimensional, or three-dimensional arrays of quantum dots). In these instances, the quantum dot retains its identity as a unique nanomaterial. In other cases, nanomaterials form the structure of an integrated bulk material. An example of a bulk material that is composed of nanostructured components is a Cu–Fe alloy in which nanodomains of one or the other metal exist within a bulk material. Steel made of nanosized grains has better mechanical properties than steel made of micron-sized grains.

Silk, collagen, elastin, and keratin tissue found in animals are composed of a hierarchy of increasingly larger structures [39]. The hierarchy begins with sub-nanometer materials and ends with a functional macroscopic material [39,40]. The relationship of nanostructure, muscle fibers, and connective tissue is shown in **Table 4.11**. A similar table can be created for bone tissue and other organ systems in animal bodies. From the purely structural point of view, it is clear that nature begins from the bottom up to build any kind of macroscopic functional material.

Fabrication of inorganic nanomaterials is bottom up, but some well-known methods such as erosion certainly operate from the top down. The construction of a snowflake is a nucleation process that emphasizes eccentricities in the *unit cell* of each snowflake, a bottom-up process. With regard to nanoscience and technology, materials are constructed from the top down, bottom up, or a

TABLE 4.11	*The Nanostructure of Tendons*	
Structural component	**Dimensions**	**Description/function**
Amino acids	<1-nm	The building blocks of proteins
Collagen	1.5-nm Diameter	Primary structure polypeptide (the protein of connective tissue)
Triple-helix coil (tropocollagen)	1.5-nm Diameter; 300 nm length	Three polypeptide strands form a cooperative quaternary structure.
Microfibrils	<4-nm Diameter	
Subfibril	10–20-nm Diameter	
Fibril	50–500-nm Diameter	Connective tissue called *endomysium* based on collagen subunits that surround muscle fibrils
Fascicle	50–300 μm	Bundle of muscle fibrils (~10) surrounded by *perimysium* (connective tissue) [40]. The *endo-*, *peri-*, and *epimysia* converge to form the tendon.
Tendon	10–50 cm	Attachment of muscle to bone support structure Provides flexibility and strength

combination of both. Although some fabrication categories are placed in one of the two major types, note that several cross borders. Lithographic techniques, when taking in the whole, are a combination of several techniques.

There are numerous challenges facing any kind of nanofabrication technique. According to George Whitesides of Harvard [41], "In almost all applications of nanostructures, fabrication represents the first and one of the most significant challenges to their realization." As with any process, there are advantages and disadvantages. Problems with top-down approaches include alteration of the surface structure during the process [42,43]. Lithography is a method that is capable of causing undesirable changes to the crystal structure and more damage from subsequent chemical etching steps. Reduction in conductivity and generation of excess heat due to surface imperfections in nanowires, for example, could be problematic [43]. Top-down methods can be extremely energy intensive because energy is required to create new surfaces (chapter 6). Nanoscale materials made from bottom-up methods may lack long-range order and structural integrity. There are more disadvantages.

We started our civilization with stone, bone, and wood and then metal and ceramics. These materials are classified as *hard matter*. Biological materials like leather and gut, materials based on biological soft matter sources, were also put to good use. With semiconductors, advanced alloys, and other hard materials, the tradition continued. The advent of plastics, polymers, pharmaceuticals, and other organically based materials ushered in a new era of chemistry: the chemistry of the covalent bond. We have entered the age of *soft matter*.

4.1 TOP-DOWN FABRICATION

Top-down approaches remove, reduce, subtract, or subdivide a bulk material to make nanomaterials. Top-down methods, therefore, are considered to be subtractive. Top-down fabrication methods logically reside within the realm of engineering and physics. Top-down fabrication dominates nanotechnology today, although significant ground has been gained by bottom-up methods [44].

Although tried and true, there are many challenges that confront top-down methods as miniaturization continues unabated towards the nanoscale. Contamination, machine cost and complexity, clean room cost and complexity, physical limits (photolithography), material damage, and heat dissipation are a few of the issues that confront top-down methods. There seems to be a strong link between the cost of a procedure and the size of the intended product. Specifically, it becomes more expensive to make smaller materials and devices. According to pundits, however, once the R&D phase is accomplished and the manufacturing line is in place, the cost of nanomaterial-enhanced products should go down.

A few selected top-down processes will be reviewed in the following sections. There are many we leave out. For the purposes of this course, a representative sample has been compiled that should provide enough insight and information into top-down fabrication methods.

4.1.1   *Mechanical Methods (Mechanosynthesis)*

Any procedure that involves the action of a bulk implement, tool, or machine on samples made of bulk materials is a top-down mechanical method. Mechanical methods base their action on kinetic energy: a hammer falling, a canister revolving, a roller thinning, a die extruding, a compacter compressing, etc. Beating and rolling methods to form thin metal films with nanometer dimensions and extrusion of soft materials in plastic phase to form wires are widespread industrial practices [5].

Ball Milling.   One of the most important mechanical top-down methods is ball milling (and shaker milling), a technique that is able to produce nanoscale materials by mechanical attrition. In ball milling, the kinetic energy of a grinding medium (e.g., stainless steel or tungsten carbide ball bearings) is transferred to coarse-grained metal, ceramic, or polymeric sample materials with the direct purpose of size reduction [3]. Rotation or rapid vibration of a drum or canister imparts kinetic energy to the grinding medium (under controlled atmospheric conditions to prevent oxidation) [5]. During the ball mill process, severe plastic deformation of the sample material initiates the formation of defects and dislocations. Any type of mechanical deformation subjected to high sheer and strain conditions leads to the formation of nanograined material [45]. **Figure 4.1** displays a rendition of a generic ball mill.

The result of the procedure, however, yields nanoparticulate materials peppered with defects with a wide distribution of size. On the upside, mechanical attrition is one of the least sophisticated technical processes and hence the least costly. Although the process has roots in ceramic processing and powder metallurgy for several decades, it is considered to be a rapidly evolving field [3]. Ball milling, first accomplished by J. Benjamin in 1966, produces mechanically alloyed materials. Alloys, metastable phases, quasi-crystalline phases, and amorphous alloys are formed by such mechanical attrition techniques [3].

The principle of mechanical attrition is relatively straightforward. A sample material is placed in a canister filled with ball bearings. The canister is activated and begins to rotate at increasingly higher revolutions per minute. The ball

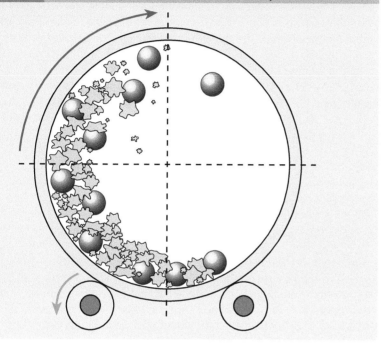

FIG. 4.1

There are two ends of the fabrication spectrum: At one end there is the high-priced lithographic equipment that requires a high-vacuum environment and expensive energy sources. At the other end there is the ball mill—a purely mechanical machine that fabricates nanomaterials by mechanical methods. Kinetic energy from a rotating or vibrating canister is imparted to hard spherical materials like ball bearings. The ball bearings in turn reduce bulk precursor materials into nanoparticles.

bearings impart significant kinetic energy to the samples, a much softer material. Several processes occur in the following order. The first event to happen is compaction and then rearrangement of particles. Secondly, elastic and plastic deformation and welding occur. Particle fracture and fragmentation further reduce the particle size. Griffith theory describes particle fracture in a mathematical sense:

$$\sigma_F \approx \sqrt{\frac{\gamma E}{c}} \tag{4.2}$$

where σ_F is the stress at which crack propagation leads to catastrophic failure, γ is the surface energy of the particle (joules per square meter), E is Young's modulus, and c is the length of the crack [3]. The tipping point is reached when the stress equals the strength of cohesion between atoms of an isotropic solid. As particles get smaller, due to enhanced surface energy, agglomeration forces (antifracture) predominate. A balance is struck among the stress, increased resistance to fracture, increased agglomeration, and maximum energy that is expended in milling.

There are several types of mechanical attrition devices. Shaker mills are the most popular form used by scientists and are able to produce particles <20 nm in diameter [3]. A back-and-forth high frequency (>1000 cycles·min^{-1}, ball velocity >5 m·s^{-1}) applied to a vial with milling balls ensures that samples pulverize properly. Planetary ball mills are commonly used in laboratories. In this form of mechanical attrition, rotational forces are the source of kinetic energy imparted to the grinding media and the sample.

Compaction and Consolidation.   Following a ball milling process (e.g., of a composition that consists of copper and iron metal constituents), materials are compacted with a tungsten-carbide dye under high pressure for extended periods of time [7]. After compaction, heat is applied, also under pressure, to the alloy. The result is a metal formulation that is characterized by an average grain size of 40 nm within a range of 15–75 nm. The whole point of this procedure is to produce a material with smaller grain size that demonstrates superior physical properties to that of a material with larger grain size. Nanograined alloys demonstrated fracture stress that was five times better than pure iron with larger grain size (50 nm–150 μm) [7,46].

Compaction of ceramic and superconducting nanomaterials by application of shock waves limits the grain growth [47]. Ceramic superconductor materials formed by such advanced techniques demonstrate higher current capacity, larger magnetic fields, and no energy loss through resistance.

4.1.2   Thermal Methods

A top-down method is considered to be thermal if an external source of heat is applied to the process. Melting a bulk material and converting the liquid into nanomaterials are considered to be a thermal top-down method. Many methods produce heat during operation, such as laser ablation and solar flux, but are considered to be high energy rather than thermal methods per se.

Chill Block Melt Spinning and Solidification.   This is a process that initially applies heat to bulk material with the intent of melting that material and performing an extrusion process. Quick solidification of the metal is induced to freeze the metal into a desired form. An RF (radio frequency) heating source is utilized to create a metal melt. The liquid metal is then forced through a nozzle in the form of a stream that is oriented over the surface of a rotating drum [7]. A bulk alloy material consisting of aluminum nanoparticles, 10–30 nm in size, made by this method demonstrated tensile strength in the gigapascal range.

Gas Atomization.   This is another top-down method that is suited for the manufacture of nanoparticulates. In this process, a high-energy stream of some inert gas is directed at a molten metal stream. Just like in the ball milling, kinetic energy is transferred to the sample—this time from the high-energy inert-gas beam. The impact initiates the formation of finely divided metal particles that upon solidification form into a finely divided powder. The nanopowder is then compacted to form a bulk metal with superior mechanical properties.

Electrohydrodynamic Atomization (EDHA).   Electrohydrodynamic atomization is an offshoot of electrostatic spray technology and is a subset of liquid disruption processes. The formation of a *Taylor cone* that terminates in a fine-stream jet forms the basic mechanism of EHDA. An electrostatic atomizer causes a net charge to develop on the surface of a droplet that causes dispersion due to coulombic repulsive forces. This process prevents agglomeration of droplets and hence particles are formed. The EHDA process is capable of producing particles as small as quantum dots.

The products of EDHA procedures depend on the flow rate of the liquid, the diameter of the needle orifice, the distance between the needle tip and grounded surface, and the strength of the applied AC field [48]. One of the primary goals of this procedure is to be able to synthesize nanoparticles rapidly and over large areas. The EHDA technique was used to atomize a solution of chloroplatinic acid $[H_2PtCl_6 \cdot (H_2O)_6]$ in ethanol. The purpose of the atomization procedure was to produce Pt metal particles. Droplets are sprayed on a Si–SiO$_2$ substrate and heated at 700°C for a short period of time. The dimensions of the Pt particles were on the order of 10 nm [8].

4.1.3 High-Energy Methods

Arc discharge, laser ablation, and solar vaporization are three high-energy top-down methods that are able to generate nanomaterials by the application of high energy electric currents, monochromatic radiation, or solar radiation to a solid substrate. Each method is capable of forming carbon nanotubes from graphite substrates that contain catalytic Fe, Mo, or Co particles. We consider any process that involves plasma to be a high-energy process. High-energy methods, with the possible exception of the solar version, are not practical to upscale due to the intense investment in energy that is required.

Arc Discharge Plasma Method. The first deliberate attempt to produce carbon nanotubes with an arc discharge method was accomplished with an arc plasma discharge method developed by Y. Ando in 1982 [15,16]. The formation of carbon nanotubes by arc discharge (plasma arcing) process is dependent on the pressure of He, the process temperature, and the applied current. Typical conditions utilize an applied voltage of 20 V, current ranging from 50 to 100 A, and He pressure of 50–760 torr. Two graphite rods are placed millimeters apart (**Fig. 4.2**). The sacrificial anode consists of graphite that is doped with metal catalyst particles. In this configuration, single-walled carbon nanotubes are fabricated. Multiwalled carbon nanotubes are formed if no metal catalyst is present in the graphite. At 100 A, carbon vaporizes in a hot plasma. Carbon cations are formed at the anode and the soot is collected at the cathode. The arc method, although relatively simple, produces an array of unwanted by-products. Samples originating from arc discharge methods often require extensive purification. Basing scientific conclusions on unpurified materials is not a recommended practice.

Laser Ablation of Solid Targets. In 1995, carbon nanotubes were synthesized by pulsed laser method. Graphite rods containing Co and Ni catalyst were heated to 1200°C and then exposed to laser pulses [17]. Heat is, therefore, generated by two means in this process—the furnace and the laser. The vaporized carbon is collected on a cooled finger downstream of the carbon targets. Continuous wave CO$_2$ (~2 kW) infrared, ultraviolet, or Nd:YAG lasers are the most common types of lasers used in the ablation method. A generic scheme is shown in **Figure 4.3**.

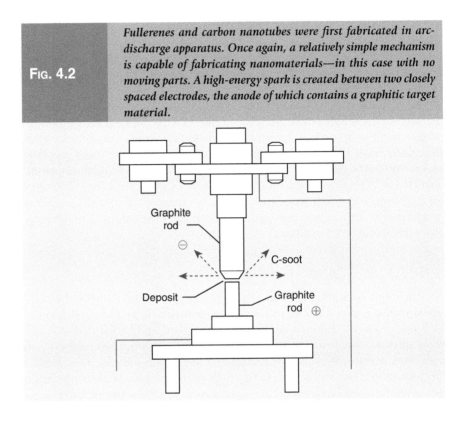

FIG. 4.2 *Fullerenes and carbon nanotubes were first fabricated in arc-discharge apparatus. Once again, a relatively simple mechanism is capable of fabricating nanomaterials—in this case with no moving parts. A high-energy spark is created between two closely spaced electrodes, the anode of which contains a graphitic target material.*

High-Flux Solar Furnace. Solar power has also been used to fabricate carbon nanotubes by a top-down procedure [18]. Since scale-up of the arc discharge and laser ablation methods is problematic, the goal is to increase the power of the solar furnace to a level of 500 kW [49]. At the National Renewable Energy Laboratory (NREL) in Golden, Colorado, researchers were able to produce fullerenes from a 10-mm diameter graphite pellet with a 10-kW high-flux solar

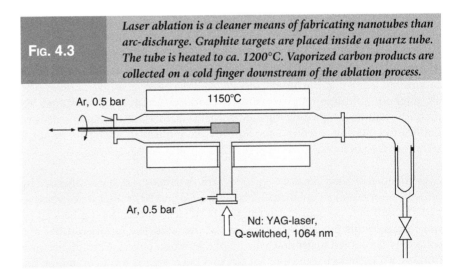

FIG. 4.3 *Laser ablation is a cleaner means of fabricating nanotubes than arc-discharge. Graphite targets are placed inside a quartz tube. The tube is heated to ca. 1200°C. Vaporized carbon products are collected on a cold finger downstream of the ablation process.*

furnace (HFSF) [18]. Temperatures in the range of 3000–4000 K were attained. NREL's high-flux furnace has 25 hexagonal mirrors to concentrate solar radiation that provide flux at 2500 suns or, with adjustments, 20,000 suns—quite impressive.

Plasma Methods.   We place ion milling, RF sputtering, plasma cleaning, and reactive ion etching into the category of high-energy methods. Plasma (from the Greek *plasma* or *plassein,* "to mold, to spread") is an ionized gas that is considered to be a distinct phase of matter. Plasmas contain ions and electrons and exist best in a vacuum environment for obvious reasons. Plasmas are electrically conductive and are strongly influenced by electric and magnetic fields. A simple reactive ion etching system is shown in **Figure 4.4**.

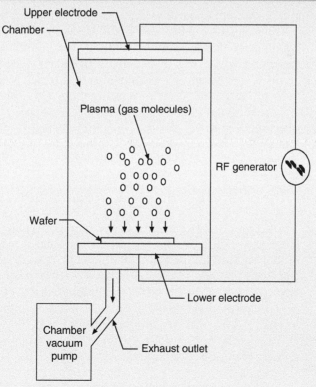

FIG. 4.4

Reactive ion etching (RIE) is an effective means of subtracting material from a substrate— hence, a top-down method. Molecules (usually oxygen, sulfur hexafluoride, fluorine, or other reactive species) are ionized to form chemically reactive plasmas by the action of an applied electromagnetic field (parallel plate configuration) under low-pressure conditions. The apparatus consists of a cylindrical chamber kept under a few millitorr vacuum conditions. Inductively coupled plasma (ICP) produced by RF magnetic fields is another mode of creating RIE plasmas. Combinations of parallel plate and ICP also exist. Since the trajectory of ions produced in RIE is mostly normal to the plane of a substrate, the process is capable of anisotropic etching—as opposed to chemical etching, which tends to act in an isotropic fashion.

RF sputtering is a physical (as opposed to chemical) vapor deposition (PVD) method. Atoms from a solid target source (hence the top-down designation) are ejected via the process of momentum exchange into the plasma by the action of high-energy ions, usually originating from argon. The ejecta are then deposited on a surface of a sample material to provide a coating. A radio frequency alternating current is commonly used to generate the plasma and a bias voltage is applied to the target to promote acceleration of ions.

Ion milling, another PVD process, is similar to RF sputtering except that no coating is formed. In actuality, the opposite is true. Material is removed to promote thinning or shaping of a sample material (e.g., formation of nanofacets). Thin films with dimensions on the order of a few nanometers for the purposes of TEM preparation are formed after exposure to ion mill plasmas. Reactive ion etching is a chemical process in which a reactive chemical species is added to the plasma mixture. Oxygen, fluorine derivatives, or etchant species that are known to react with targeted substrate materials are commonly used in reactive ion etching (RIE) procedures.

4.1.4 *Chemical Fabrication Methods*

Combustion of Bulk Materials.   Combustion is a top-down chemical method that is capable of producing nanomaterials. However, impurities in bulk carbon materials such as coal and oil contain contaminants that contribute to the formation of fly ash and acid aerosols. Polyaromatic hydrocarbon clusters (PAH) can be produced under incomplete combustion conditions. Pure hydrocarbons produce CO_2 and H_2O under efficient combustion conditions. Combustion also is a bottom-up method that is capable of producing nanomaterials.

Following the combustion of bulk Mg to MgO, a cluster-based nanoparticle bonding mechanism was the cause of agglomeration. This is apparently a common phenomenon that applies with equal validity to titania and alumina particles. For alumina, it was found that the primary Al_2O_3 aggregate was on the order of 1 μm in size, but that it was composed of clusters 10 nm in size [50].

Chemical Etching of Silicon.   Chemical etching is important in numerous industrial production procedures, lithography in particular. The anisotropic etching of silicon with KOH is a major industrial procedure. The reaction yields silicates [51]:

$$Si_{(s)} + 4KOH_{(aq)} \rightarrow [SiO_4]^{4-} + 4K^+_{(aq)} + 2H_{2(g)} \qquad (4.3)$$

The Si(110) surface undergoes the fastest etch rate of all the primary low-index planes surfaces. For example, the etch rates of Si in a 30% w–w solution of KOH at 70°C for the (110), (100), and (111) surfaces are equal to 1.5, 0.79, and 0.005 $\mu m \cdot min^{-1}$. A common isotropic etching solution used for silicon is HNA (HF + HNO_3 + CH_3CO_2H). Isotropic etchants operate independently of crystal direction. The trench profile following isotropic etching looks like an inverted "C" by cross-section; from anisotropic etching, the trench looks like a "V" with a flat bottom [52]. Etching with hydrofluoric acid is driven by the stability of the $[SiF_6]^{2-}$ complex:

$$Si_{(s)} + 6HF_{(aq)} \rightarrow [SiF_6]^{2-} + 2H^+_{(aq)} + 2H_{2(g)} \qquad (4.4)$$

As a result of lithographic procedures and subsequent top-down chemical etching, nano- to micron-sized features can be formed on the surface of silicon wafers.

Chemical–Mechanical Polishing. This method is a combination of a chemical etching and a mechanical attrition method. The process of polishing jade with a corundum-based abrasive has been traced back to Neolithic farmers in ancient China 6000 years ago [53]. Grinding is the planar removal of material from a target surface by a fixed abrasive. In polishing, the abrasive is allowed to roll. The surface roughness, determined by profilometers or AFM, is shown to be a few nanometers.

Chemical–mechanical polishing combines the mechanical grinding characteristics of abrasives with the chemical action of an etchant. Pressure is applied on the abrasive and hence on the surface through a conformal pad. This allows for free movement of the abrasive under the pad. The method is important to the lithography industry, where depth of focus (DoF) is ever shrinking with smaller wavelength sources and larger numerical apertures (N.A.). The smoother the surface of a Si wafer becomes, the better is the accommodation of shrinking DoF.

Anodizing and Electropolishing. These two techniques are integrally related and differ only with regard to purpose and conditions. Anodizing is a process that creates an insulating porous oxide layer on a conductive metal anode, usually aluminum, in an electrolytic solution, usually a dilute polyprotic acid. By providing hexagonally packed pore channels that are simple to fabricate and the ability to manipulate pore diameter and length during and after anodizing, the porous anodic film offers a perfect template for nanoscale material synthesis. Anodizing conditions consist of an electrolytic bath made of a polyprotic acid (H_2SO_4, H_3PO_4, $H_2C_2O_4$, or H_2CrO_4) at 0°C with applied voltage of 2–100 V dc. The formation of nanoscale pores with diameters ranging from a few nanometers to several hundred nanometers is the major product of anodizing. The chemical reactions in anodizing are

Anodic reaction $\qquad\qquad 2Al^{\circ}_{(s)} \rightarrow 2Al^{3+} + 6e^-$ $\qquad\qquad$ (4.5)

Oxide–electrolyte interface $\qquad 2Al^{3+} + 3H_2O \rightarrow Al_2O_{3(s)} + 6H^+$ $\qquad$ (4.6)

Cathodic reaction $\qquad\qquad 6H^+ + 6e^- \rightarrow 3H_{2(g)}$ $\qquad\qquad$ (4.7)

Overall oxide formation reaction: $2Al^{\circ}_{(s)} + 3H_2O \rightarrow Al_2O_{3(s)} + 3H_{2(g)}$ $\quad$ (4.8)

Anodizing, however, is a mixed fabrication method. Technically, it contains components that can be classified as top down or bottom up. The top-down component is the electrochemically assisted dissolution of bulk aluminum to form Al^{3+} cations. During this process, nanostructured scallops are formed in the surface of the aluminum metal. Pore diameter is directly proportional to the applied anodic dc potential ($d_{pore} \propto 1.4$ V) and is controlled by the diameter of the scallops on the metal surface. A schematic illustration of an anodic film is shown in **Figure 4.5.**

FIG. 4.5 *Porous alumina membranes formed by an anodic process can be considered to be the ultimate template material. They are insulating, optically transparent, chemically inert (near neutral pH), and thermally stable and possess reasonable mechanical properties. Once again, a relatively "low-tech" method is capable of fabricating nanomaterials (the pore channels).*

Electropolishing involves the removal of metal to form a smooth surface without forming a thick oxide layer. The conditions for electropolishing are rather severe compared to anodizing: elevated temperature (70–90°C), elevated level of current (10–20 A), and concentrated acid or base solutions. Electropolishing often precedes anodizing to prepare a smooth surface.

Hydrolysis Reactions.   These reactions can affect inorganic, organic, and biological materials. Hydrolysis occurs by the action of water to disrupt a bond. The bond can be as strong as a covalent bond, ionic bond, or any kind of intermolecular attraction. For example, dissolution of proteins from the top down by acid-catalyzed hydrolytic mechanisms is a common means to regenerate the constituent amino acids. The degree of hydrolysis determines the size of the final product.

4.1.5 Lithographic Methods

A brief history of lithography was presented in chapter 1. Lithographic methods are the most widely utilized industrial process in the high-technology sector. The computer industry, for example, depends heavily on lithography. Integrated circuits, microelectromechanical machines (MEMS), and numerous other applications require lithography during some phase of their manufacture. However, challenges facing lithography today are numerous as well. Fabrication of increasingly smaller features requires sources with smaller wavelength. With increasingly smaller wavelength (e.g., electron beams and x-rays), the resolving power of the procedure is enhanced but the substrate sustains more damage. Fabrication of increasingly smaller features also requires increasingly more expensive equipment. Wavelength-based lithographic techniques, although well established, are rather costly to operate.

Modern optical lithographic techniques utilize radiation sources with wavelength from a few to 300 or 400 nm. Nano-imprint and nanosphere lithography offer cost-effective facilitative alternatives to the high-vacuum, high-energy, high-maintenance processes. Once a few fundamental technical issues in these nanotechniques become better resolved, expect wavelength-based lithography fabrication to start giving way to these nanorevolutionary procedures. With the advent of nanosphere and nano-imprint lithography, both extremely simple methods capable of high resolution, the trend in operation costs may be reversed in the near future.

In general, the underlying operation of lithographic techniques has not changed much since the time of the inventor of the technique, Bavarian author Alois Senefelder, in 1796. Photolithography follows the general procedure of pattern transfer established by Senefelder but employs radiation or particle projection onto a resist material instead of writing on a limestone substrate (**Fig. 4.6**):

Deposition of thin layer on substrate → deposition of photoresist material → *exposure* via mask (the master) by energy source → *development* by etch (positive or negative replica) of excess material → *stripping* of all resist → chemical modification (additive or subtractive)

There are numerous energy sources employed in lithographic processes—visible to ultraviolet radiation and x-rays for photolithography. Electron and ion beams have also been applied in lithographic procedures. Top-down nanolithographic sources consist of photons (UV, DUV, EUV, and x-rays), particle beams (electrons and ions), physical contact printing (nano-imprint methods), and edge-based techniques (shadow evaporation). Bottom-up nanolithographic procedures like dip-pen lithography and self-assembly (surfactant systems and block copolymers) will be discussed in a later section.

There are three primary considerations for any lithographic process: resolution, registration, and throughput. Resolution, first discussed in chapter 3, is defined as the best attainable physical scale of a feature: the smaller the better. Registration

Lithography is the workhorse of the computer chip industry. It is the most common top-down manufacturing process and it is one nanomanufacturing technique that is widespread. A target material is first applied to the surface of a silicon substrate. Polymeric resist layer is then applied by spin coating. An energy beam, usually in the visible to ultraviolet wavelength range, is shined through a mask that contains a predetermined pattern. Regions exposed to the EM radiation are sensitized (positive resist) or protected (negative) to the subsequent etch step. Following etching, the resist is removed, transferring the pattern inscribed by the mask to the target material. Lithography is a rather expensive process that requires clean room conditions, high-vacuum conditions, and otherwise expensive equipment.

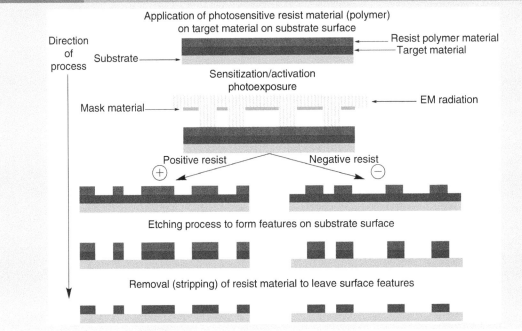

is the process of aligning one layer to another to form an integrated structure. Throughput is a gauge of the balance between cost effectiveness and the rate of production.

Optical Photolithography.   Optical lithography employs visible and ultraviolet radiation to transfer a pattern onto a receptive substrate. Ultraviolet radiation (deep ultraviolet lithography, DUV) is the most common kind in use today. Three general methods are used to expose wafers:

1. Contact printing, in which the mask lies on top of the resist (e.g., there is no wafer-mask gap), requires no magnification but resolution is limited (~500 nm). The mask degrades in this configuration resulting in loss of planarity.
2. Proximity printing places the resist in close proximity to the mask. There is no magnification with this configuration and resolution is even lower (~1 μm). Diffraction effects limit the accuracy of the pattern transfer process [54].

3. Projection printing is a widely adopted technique. An image is projected through a mask and reduced by a factor of four to ten times on the resist. Resolution is much better (~70 nm), but equipment is costly and accuracy is limited by diffraction [54].

A computer-generated pattern on a mask (from optical or electron beam generators) is transferred to a chromium surface (~100 nm thick) on fused silica [55]. The mask is then positioned over a substrate, usually silicon, silicon oxide, or a semiconductor material. The substrate is prepped beforehand with a thin layer of oxide, nitride, or other functional material, and then a photoresist material is applied by spin-coating—a photoresist material that is sensitive to the type of radiation used in the lithographic procedure. In optical lithography, the photoresist is illuminated through a mask and is rendered soluble (positive resist) or insoluble (negative resist) during the subsequent developer step. The exposed resist (positive) or the unexposed resist (negative) is removed by etching. For example, in a negative scheme, the exposed resist polymer becomes cross-linked after exposure to the radiation. Cross-linking implies that the resist material is more difficult to dissolve than areas that were unexposed. Following development, an additive process deposits material onto or into the etched areas. In subtractive processes, material may be removed by ion milling through the developed areas. Following these steps, the remnants of the resist are removed.

Resolution in projection lithography is diffraction limited but has improved over the years since the days of the first integrated circuits. Line widths of the late 1960s were on the order of 5 μm [56]. In 1997, this was reduced to 350 nm. Today, sub-100-nm line widths are commonly achieved. Some of the equations present below will look familiar.

For contact style printing, radiation interacts with the sample as a square wave with limited or no diffraction. The near field (or Fresnel diffraction limit), appropriate for proximity printing, and resolution are given by

$$W = k\sqrt{\lambda d_g} \qquad (4.9)$$

where d_g is the mask-to-wafer distance (gap), λ is the wavelength of the impinging radiation, and k is a constant that is close in value to 1 and depends on resist material and other technological parameters associated with the process. Fresnel diffraction occurs when

$$\lambda < d_g < \frac{W^2}{\lambda} \qquad (4.10)$$

and the minimal resolvable feature is

$$W_{min} \approx \sqrt{\lambda d_g} \qquad (4.11)$$

For the projection style of lithography (the most commonly applied form), the optical condition is called *far field* and the mask is called a *mask in the far field*. The optical description of far-field lithography is similar to other types of projection methods, whether optical or electronic. The minimal resolvable feature in a projection lithographic system is

$$W_{min} = k_1 \left(\frac{\lambda}{\text{N.A.}} \right) \tag{4.12}$$

where λ is the wavelength of the radiation used for exposure and N.A. is the numerical aperture of the optical lithographic instrument (usually equal to ~0.5). The factor k_1 is a constant for a specific lithographic procedure that depends on the index of refraction and thickness of the photoresist material (0.4–0.8, a quality descriptor). In general, the line width is approximately equal to the wavelength of the incident light.

The resolution limit for optical lithography is given by the following equations:

$$\text{N.A.} = n \sin \theta \tag{4.13}$$

where N.A. is the numerical aperture, n is the refractive index of the medium (if vacuum, $n = 1$), and θ is the half-angle of the cone of light that can enter or exit the lens. Does this look familiar? The numerical aperture is a function of the distance between the lens and the sample and is an indication of the resolving power of the system. The larger the numerical aperture is, the higher is the resolution capability of the instrument. The following equation should look familiar as well; it also applies to lithography:

$$R = \frac{1.22 f \lambda}{d} = \frac{1.22 f \lambda}{n(2f \sin\theta)} = \frac{0.61\lambda}{n\sin\theta} = \frac{0.61\lambda}{\text{N.A.}} \tag{4.14}$$

Depth of focus (DoF) (like *depth of field*) becomes a concern as resolution is increased in shorter wavelength tools. DoF is the distance from the objective lens that yields a focused image and it gets worse (smaller) as N.A. becomes larger:

$$DoF = k_2 \frac{\lambda}{(\text{N.A.})^2} \tag{4.15}$$

where k_2 is a constant associated with the photolithographic system and is traditionally equal to 1.

Contact, proximity, or projection modes are commonly used photolithography techniques. Contact type of photolithography (or shadow mode) is the case where the mask is right on top of the resists. Resolution is calculated from

$$2b_{min} = 3\sqrt{\frac{\lambda d}{2}} \tag{4.16}$$

where $2b$ is the grating period of a mask with equally spaced lines and d is the thickness of the resist material. In the proximity method, a gap exists between the mask and the photoresist and its resolution is found from

$$2b_{min} \approx 3\sqrt{\lambda d_g} \tag{4.17}$$

where d_g is the distance between the mask and the resist.

Particle Beam Lithography (IPL).   Because particle beams do not undergo diffraction and scattering is minimal, higher resolution can be achieved with IPL than optical, x-ray, or electron beam methods. Resist materials demonstrate greater sensitivity to ions than to electrons. Ion lithography is mostly used to repair masks in optical and x-ray lithographic procedures.

Extreme Ultraviolet Lithography (EUVL).   This technique applies radiation that is as short as 11–14 nm, significantly lower than those used in DUVL [57,58]. Features smaller than 50 nm have been achieved, but theoretically much smaller features are possible (e.g., <25 nm). EUVL is based on multilayer-coated optics that are able to reflect the intense UV radiation. Therefore, reflective coatings are applied on the optics and masks of an EUVL system. EUVL must also be performed in a high vacuum due to absorption of EUV by most forms of matter.

X-ray Lithography (XRL).   This method utilizes x-rays produced by a synchrotron source. Electrons are converted into x-rays in a synchrotron. XRL is an extremely expensive method, ca. $25 million for acquisition, setup and operation.

Advantages of XRL include: (1) the wavelength (4 nm) is well matched for nanoscale work, (2) scattering is limited by all materials in contact with x-rays, and (3) resist materials and wavelength can be matched to maximize absorption. XRL is a method that is scaled to the nanometer, minimizing scattering and maximizing resist absorption and image contrast (e.g., absorption without spurious scattering) [59,60]. Disadvantages of XRL include: (1) distortion of absorber (tungsten mask) material due to x-ray-induced internal stresses that result in warping, (2) difficulty in focusing x-rays by conventional lenses requires masks with ultrafine features—a difficult and time-consuming process, and (3) XRL is a very expensive process.

Electron Beam Lithography (EBL).   This technique enables patterns to be generated by an electron beam without the use of a mask [61]. Polymethylmethacrylate (PMMA) is sensitive to electron beam exposure. EBL is used to generate nanopatterns on PMMA film supported by a silica substrate. An electron beam at 10 kV with a beam current of 340 pA was used to construct lines on the surface of PMMA [61]. Line widths in the region of 50 nm were obtained [61].

Nano-imprint Lithography (NIL).   Nano-imprint lithography, considered to be a *soft lithographic* technique, is an economical method by which nanoscale resolution and high throughput (all desired by industry) are possible. NIL is a clearly defined top-down fabrication method. A stamp pressed into a soft film is responsible for creating a negative replica of the pattern. The film is hardened before the stamp is removed. Structures with resolvable features down to 5 nm have been produced in this way. UV-NIL uses a photopolymerizable thermoresin with a UV-transparent stamp. The liquid resin is easily stamped and then hardened with the application of UV-light. A generalized NIL scheme is depicted in **Figure 4.7**.

Stephen Chou of Princeton University is credited with discovery of NIL in 1996 [22,23]. The basic principle behind NIL is compression molding to create a thickness contrast pattern on a thin resist film on a metal substrate. Anisotropic etching is accomplished for pattern enhancement. The first patterns contained

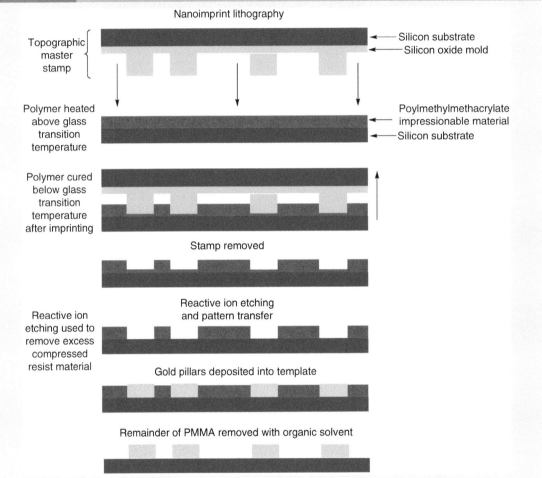

25-nm features spaced 70 nm apart. It is easy to understand why NIL is preferred over wavelength-dependent lithographic techniques: (1) NIL is able to achieve smaller features, (2) NIL takes less time, and (3) NIL is an inexpensive process that does not require ultrahigh vacuum and expensive radiation or electron beam equipment. The biggest problem with NIL is defectivity; although recent methods have driven the defect density to <0.1 cm^{-2}, a practical industrial level of $<0.01 \cdot$ cm^{-2} is desired.

Nanosphere Lithography. R. P. Van Duyne et al. of Northwestern University in 1994 developed the nanosphere lithographic (NSL) process [24,62]. NSL is a straightforward, versatile, high-throughput process that offers nanoscale resolution. Compared to other methods, NSL is quite fast and economical. The NSL process is able to create ordered arrays of differing configurations. In one case, materials are deposited through the open spaces between spheres to form an array (**Fig. 4.8**). In another application, the size of the spheres is

FIG. 4.8

Nanosphere lithography (NSL), like NIL, is another ingenious low-cost, high-throughput method to form nanomaterials and nanoparticle arrays. One simple method utilizes latex spheres that are close packed in a two-dimensional array. Deposition of metal between spheres, the interstitial spaces, forms star-shaped patterns of tetrahedrally formed nanostructured materials. RIE etching, depending on the type of active molecule, is able to reduce the size of the spheres (thereby creating wider gaps among the spherical matrix elements, or etch) in an anisotropic manner, the substrate under the spaces to form pore channels.

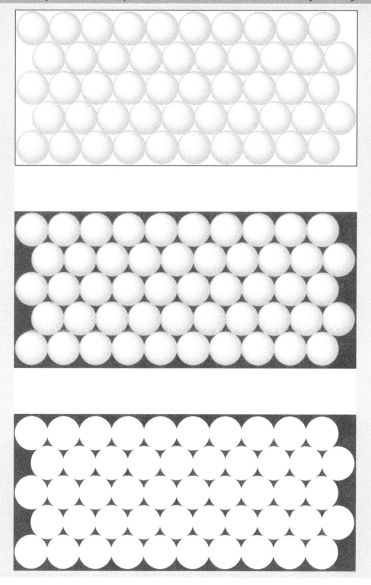

reduced by standard RIE with oxygen [63]. Columnar arrays several microns in height are then fabricated by application of a deep-RIE (Bosch) top-down process.

The NSL is a mixed bag. It is a bottom-up template method during the deposition of material at the base of the interstitial regions of the spheres, but it is a top-down method during the deep-RIE process.

Applications of NSL include its use in manufacturing size-tunable noble metal substrates in the range of 20–1000 nm. The optical response of NSL-formed Ag nanoparticles to their local environment was probed by localized surface plasmon resonance spectroscopy (LSPR) [62]. Results showed that zeptomole-level detection of adsorbed analytes was possible by LSPR spectroscopy [62]. Large-scale fabrication of protein nanoarrays based on NSL was demonstrated by Y. Cai et al. in 2005 [64]. Based on nanospheres with 300-nm diameter, protein islands were formed with ring shapes of 50-nm width and 118-nm diameter [64].

4.2   BOTTOM-UP FABRICATION

Bottom-up fabrication approaches selectively combine atoms or molecules to form nanomaterials. Bottom-up fabrication methods, therefore, are considered to be additive. Bottom-up fabrication methods reside within the realm of chemistry and biology. Nature, of course, has perfected bottom-up fabrication of nanomaterials.

Advantages of bottom-up methods are numerous. Self-assembly processes, for example, occur under thermodynamic control conditions. Because such processes exploit much weaker intermolecular interactions, as opposed to strong covalent bonds, nanomaterials are fabricated under milder conditions of temperature, pressure, and pH. The upscale potential of bottom-up methods is enormous. As with any other chemical process, it is "relatively" straightforward to scale up a process that takes place in a beaker on a lab bench (e.g., the domain of the chemist) to a batch production process in a manufacturing line (e.g., the domain of the chemical engineer). However, there exist significant challenges facing bottom-up methods. Overall robustness, long-range order (related to complicated patterns), and directed growth all leave something to be desired. In order for bottom-up fabrication of nanomaterials to become the dominant fabrication mode of industry, all of these concerns need to be overcome.

We divide bottom-up methods according to the phase within which the process occurs. We also add a special section discussing the solid state.

4.2.1   *Gaseous-Phase Methods*

Vapor phase reactions can be homogeneous (all reactants, products, and catalysts exist as a vapor) or heterogeneous (vapor–liquid or vapor–solid phases exist within the same sphere of reaction). If there exists a vapor (or any highly dispersed phase, e.g., a particle beam) in a process, we shall consider that process to be a gaseous-phase fabrication method.

Chemical Vapor Deposition.   Chemical vapor deposition (CVD) is one of the most effective procedures used to produce advanced materials. CVD is the best way to form carbon nanotubes because it is less energy intensive and more control is exerted over products. SiO_2, SiC, Si_3N_4, W, and other materials are routinely deposited on surfaces via CVD methods. In semiconductor industry practice, wafers are exposed to volatile precursor materials that react or decompose on the surface to form thin films. There are many kinds of CVD.

Chemical CVD (CCVD) is used in the fabrication of carbon nanotubes—single-walled and multi-walled operating temperatures range from the low 400°C to produce carbon fibers and multiwalled carbon nanotubes to temperatures >1000°C. The decomposition of methane, ethane, ethylene, propane, propylene, or acetylene or the disproportionation of carbon monoxide—all over catalysts—is an example of some of the carbon source materials (usually in gas form) used in CVD techniques. The decomposition of methane in the presence of catalysts (usually Fe, Ni, or Co) at temperatures of 700°C at atmospheric pressure yields SWNTs.

$$CH_{4(g)} \rightarrow SWNT + H_{2(g)} \tag{4.18}$$

Polysilicon thin films are formed by the decomposition of silanes in a low-pressure CVD (or liquid-phase CVD, LPCVD) chamber at ca. 650°C. If other gases, such as phosphine or arsine, are present in the stream, the silicon can be doped in situ. Silicon dioxide layers are formed by the gas-phase decomposition of tetraethylorthosilicate (TEOS). Since TEOS boils at ca. 168°C, the CVD process is conducted between the boiling point of TEOS and 750°C. TEOS breaks down into solid silica and gaseous diethylether:

$$Si(OC_2H_5)_{4(g)} \rightarrow SiO_{2(s)} + O(C_2H_5)_{2(g)} \tag{4.19}$$

Plasma-enhanced CVD (PECVD) is another bottom-up CVD fabrication method to produce thin films. The plasma is created by radio frequency or direct current discharge between electrodes [65]. Silicon dioxide, from silanes + O_2 or TEOS + O_2, can be formed with the PECVD technique at reasonably low pressure (~100 mtorr). Silicon nitride thin films are also deposited with plasma assistance. An example of a CVD apparatus is shown in **Figure 4.9.**

Metal oxide CVD (MOCVD) utilizes H_2 as a carrier gas, Group-III metal–organic compounds, and Group-V hydrides to make nanometer scale thin films or nanoparticles. Temperatures ranging from 500 to 1500°C at 15–750 torr pressure are representative conditions under which MOCVD is performed.

Atomic Layer Deposition.   Atomic layer deposition (ALD; a.k.a. atomic layer epitaxy, ALE) was introduced in 1974 by Tuomo Suntola of Finland with the intent of improving the quality of ZnS films used in electroluminescence displays. After a decade of development, high-quality phosphor layers and dielectric layers were produced, and the process has since acquired major importance to industrial manufacturing. ALD is the process of fabricating uniform conformal films through the cyclic deposition of self-terminating surface half-reactions that allows for thickness control at the level of the atomic layer [66]. ALD is a derivative of chemical vapor deposition (CVD), but one that differs from CVD in several notable ways [67]. The comparison is shown in **Table 4.12**.

ALD is a straightforward synthesis method that exploits specific chemical reactions with the intent of adding one molecular monolayer at a time. The process is

FIG. 4.9 *Chemical vapor deposition, especially in the case of carbon nanotubes, is yet another low-cost, "low-tech" method to form nanomaterials. A carbon source gas (usually methane, CO, acetylene, propylene, or ethylene) is introduced into a chamber (the quartz tube pictured) under reducing conditions. Upon contact with Co, Fe, or Ni catalyst particles, the gases decompose into C and H atoms. Nanotubes nucleate on the catalyst particle and grows out from the particle by either the tip-growth or base-growth mechanism. Typical CVD conditions use 10% methane, 5% hydrogen, 85% argon carrier gas at 700°C, and atmospheric pressure.*

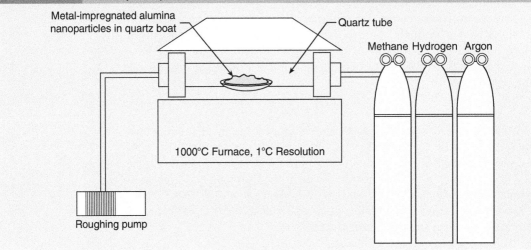

characterized by a binary reaction that is split into two half-reactions applied sequentially. ALD is characterized by the systematic use of self-terminating gas–solid reactions [68]. A self-terminating reaction depends on saturation of available surface sites and that precursors do not react with each other. The ALD process offers a powerful arsenal of properties that are specifically tailored for nanofabrication of thin films. First of all, conformal coatings can be applied to particulates or flat and curved surfaces of bulk materials. Secondly, atomic-scale control of thickness is possible by adding layers with stoichiometric scaling based on a chemisorption–saturation process. The process is broken down into the following general steps:

Surface activation → injection of A → purge → injection of B → purge → injection of A → purge → injection of B → purge → ⋯ → termination

Ultimately a film composed of a structure ABABABA … is formed. The thickness of the film can be estimated instantaneously by counting the steps in the process.

The process is depicted in **Figure 4.10**. In essence, the ADL sequential process alternates between chemisorption and saturation steps. Purging of the process follows each saturation step in the cycle. The ALD film growth process is referred to as *self-limiting* in that a stoichiometric process essentially terminates the reaction upon saturation. Excess reactants and products are purged from the chamber following each step.

TABLE 4.12	ALD–CVD Comparison	
Parameter	**Atomic layer deposition**	**Chemical vapor deposition**
Precursor reactivity	Highly reactive Self-limiting at saturation	Less reactive Can be autocatalytic
Potential materials	Metals, semiconductors, insulators Wide range of materials	Metal oxides, semiconductors, and carbon compounds
Selectivity	Highly selective	Low selectivity
Surfaces	Layers conform according to surface topography of substrate Surfaces capable of activation (limitation of application)	Layers conform according to surface topography of substrate.
Decomposition at reaction temperature	Reactants and products do not decompose.	Reactants can decompose.
Time of process	Few seconds per cycle	Variable
Uniformity	Saturation mechanism ensures uniformity over a broad area.	Uniformity control by process parameters (partial pressure of reactants, flow, pressure, temperature)—more difficult to execute
Thickness	Controlled explicitly by number of reaction cycles Deposition rate: ~6 nm·min^{-1}	Thickness control by process parameters— more difficult to execute
Conditions	Requires vacuum or inert atmosphere Lower temperatures (100–400°C but varies according to application) P, T, concentration, and gas flow distribution have little effect on the process.	Requires inert atmosphere and higher temperatures (>600°C) P, T, concentration, and gas flow distribution have significant effect on the process.
Upscale potential	Excellent	Good

The formation of alumina layers on a silicon surface will serve as an example of the ALD process. The first step in the process is the activation of a hydrogen-terminated silicon surface by exposure to water vapor:

$$:SiH + H_2O \rightarrow :SiOH + H_{2(g)} \qquad (4.20)$$

After evacuation of the chamber, trimethylaluminum (TMA) is added. The reaction occurs between the lone pairs of the oxygen atom and the empty p-orbital of the aluminum atom [66]:

$$Al(CH_3)_{3(g)} + :SiOH_{(s)} \rightarrow :Si-O-Al(CH_3)_{2(s)} + CH_{4(g)} \qquad (4.21)$$

Following purge of the reactants and products, water vapor is pulsed into the chamber. Water reacts vigorously and completely with the remaining methyl groups attached to the aluminum, replacing them with hydroxyl groups and releasing more methane. New *surface* hydroxyls are formed and aluminum forms bridge-oxygen bonds with its nearest neighbors:

$$:Si-O-Al(CH_3)_{2(s)} + 2H_2O_{(g)} \rightarrow :Si-O-Al(OH)_{2(s)} + 2CH_{4(g)} \qquad (4.22)$$

Oxygen bridges are formed between aluminum elements of the two-dimensional structure. The overall chemical reaction to form layers consisting of aluminum oxide is

$$2Al(CH_3)_{3(g)} + 3H_2O_{(g)} \rightarrow Al_2O_{3(s)} + 6CH_{4(g)} \qquad (4.23)$$

FIG. 4.10

Atomic layer deposition (ALD) technique is a straightforward method to manufacture, one monolayer at a time, a two-dimensional functional surface on a substrate. By the use of self-terminating gas–solid chemical reactions, high specificity is achieved. A self-terminating reaction depends on saturation of available surface sites and that precursors do not react with each other. Once saturation occurs over the surface layer, excess reactants are purged in preparation for the next step in the process. In the first step, a silicon surface is activated to generate hydroxyl groups. A highly reactive species, $Al(CH_3)_{3(g)}$, is introduced and one methyl group is readily displaced by a surface hydroxyl to form linkage with the surface. After purging, water vapor is added to initiate cross-linking and activation of the Al for the next round of $Al(CH_3)_{3(g)}$ application.

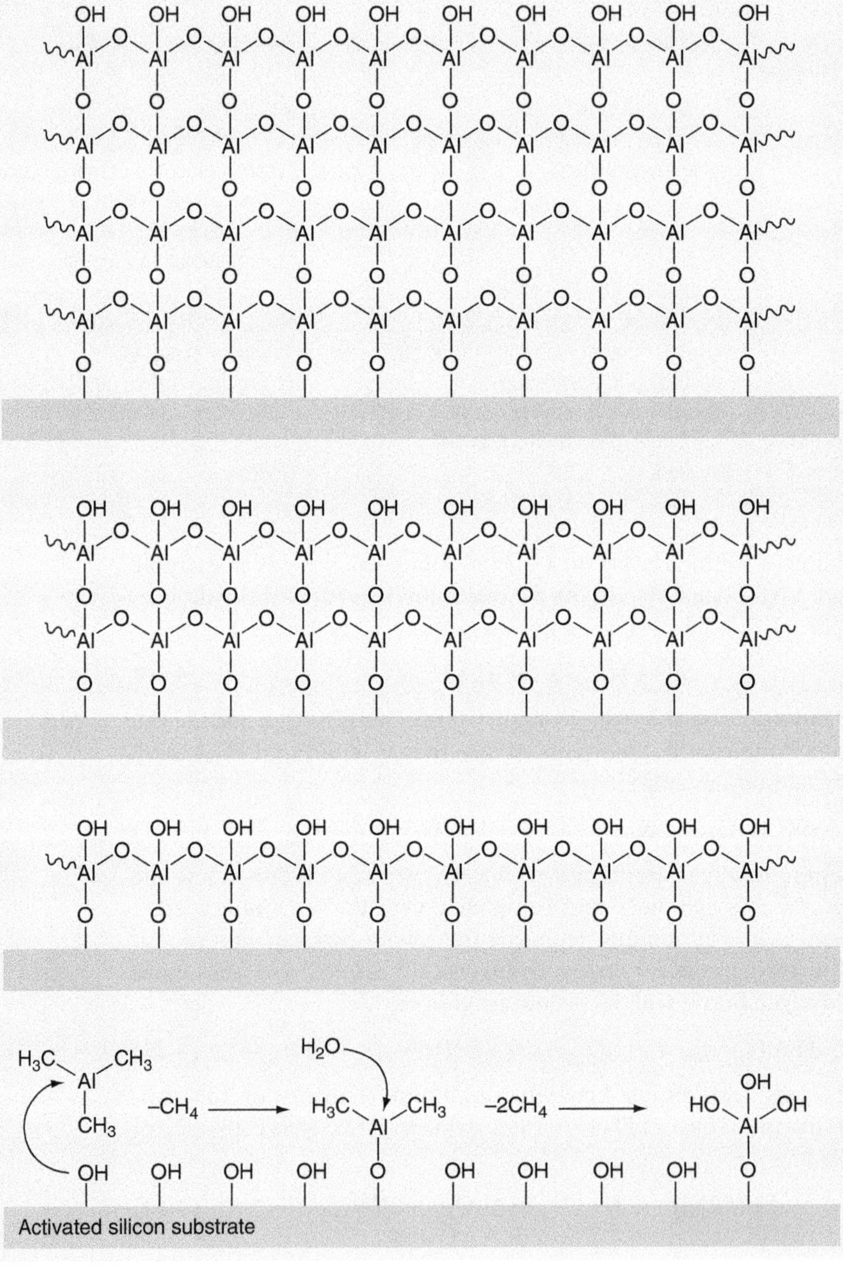

Activated silicon substrate

The mechanism of the chemisorption half-reactions was determined by cluster calculation hybrid density functional theory using TMA and water precursors [66]. Both the aluminum and water depositions were determined to be thermodynamically favorable (exothermic) and kinetically uninhibited [66]. STM and ab initio modeling studies of the three-dimensional structure of ultrathin aluminum oxide on NiAl(100) infer that Al is pyramidally and tetrahedrally coordinated [69].

Epitaxy.   Epitaxy (from the Greek prefix *epi-,* meaning "upon, placed or rested upon" + *taxis,* based on *taktos,* indicating "to arrange") is the directed growth of a crystalline substance on the crystalline face of a substrate. In this way, the crystal orientation of the deposited material matches that of the substrate. The growth is accompanied by binding of the atoms or molecules to one another to form a two-dimensional crystal. There are two kinds of epitaxial growth: *homeoepitaxy,* in which the layer and the substrate are the same, and *heteroepitaxy,* in which the two materials are different. There are three classes of epitaxial processes: vapor phase, liquid phase, and molecular beam (MBE). We will discuss the MBE method only.

MBE was introduced in the 1970s to make films of high quality—films made of semiconductors, metals, or insulators. The operation principle of MBE is straightforward. Atoms, molecules, or clusters of extremely pure form are produced top down by heating an extremely pure source material (evaporation) in an effusion cell (**Fig. 4.11**). Six to ten effusion cells, each containing a different

FIG. 4.11

Molecular beam epitaxy (MBE) is a single-crystal film growth technique [4]. MBE is the evaporation of one or more elemental or molecular species onto a heated target substrate material under ultrahigh vacuum [4]. In the figure, five atomic or molecular species are heated in effusion cells and directed towards a substrate. Shutters regulate the timing of release and the level of exposure of the MBE evaporated materials. A built-in RHEED system is able to monitor the development of the epitaxial film. An in situ mass spectrometer monitors vacuum conditions as well as the level of the evaporated species.

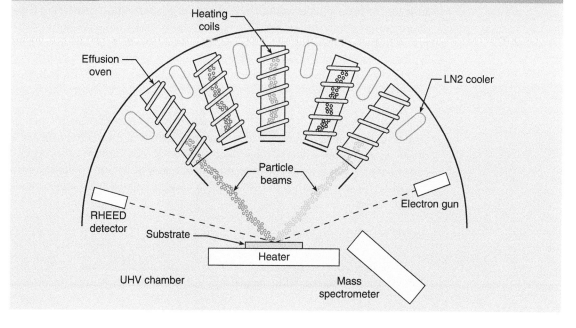

component, are focused onto the sample wafer. The crucibles containing source materials are made of pyrolytic boron nitride that can withstand temperatures of 1400°C; the shutters (that control the flux in the cell) are made of molybdenum or tantalum [70]. The chamber is usually baked at 200°C to remove contaminants for 24 h before use. Cryogenic screens surround the effusion cells to minimize spurious fluxes of atoms or molecules from the walls of the chamber. A RHEED in situ detector is used to monitor the status of the forming film.

The MBE process takes place in an ultrahigh vacuum (UHV), on the order of 10^{-10} torr, to ensure a contamination-free environment. The atomic flux (i.e., the "molecular beam") is directed towards the wafer surface where deposition occurs. The wafer is also heated to enhance diffusion along the surface. Films grow laterally on the surface (e.g., epitaxially). In order to create a film with minimal defects, the MBE process occurs at a slower rate than other deposition techniques like MOCVD. Typical MBE growth rate is on the order of $1\ \mu m \cdot h^{-1}$ [4].

One criterion for MBE operation is that the mean free path λ of atoms, molecules, or clusters be larger than the geometric dimension of the chamber. This prerequisite is fulfilled if a vacuum exists in the chamber. For example, at 1 torr, λ of N_2 is about 10^{-2} cm. At 10^{-11} torr, $\lambda > 10^8$ cm, thus exceeding the dimensions of any size chamber [70]. Another criterion for high-quality film growth is that the time it takes for surface-diffusion incorporation be less than the time required to deposit a monolayer [4]. If not, defects in the layer are bound to accumulate due to unincorporated atoms that become buried in the monolayer [4]. Doping of semiconductor thin films occurs with the timely opening of a shutter.

Ion Implantation.　　Ion implantation is simply a process by which ions of one material are inserted into a solid substrate. Applications of ion implantation are for surface finishing, semiconductor doping, silicon-on-insulator (SOI), or mesotaxy (crystal matching within the bulk). In SOI, an oxide layer is created within a silicon substrate. Oxygen is introduced by ion implantation and then annealed to form silicon oxide. The ion implantation method consists of three major components: an ion source, the accelerator, and a target. At the conclusion of acceleration, ion energies ranging from a few to 500 keV can be achieved. Lower energy ions are able to penetrate a few nanometers into the surface. A certain degree of crystalline damage occurs during ion implantation processes.

Combustion Processes.　　We revisit combustion processes once again but this time from the bottom-up perspective. The combustion of molecules, unlike the combustion of bulk materials described previously, is a bottom-up process if nanomaterials are formed. The formation of Si nanoparticles from the combustion of SiH_4 (silane gas) under low-oxygen conditions produces silicon oxide nanoparticles:

$$2SiH_{4(g)} + 2O_{2(g)} \rightarrow 2SiO_{2(s)} + 4H_2O_{(l)} \qquad (4.24)$$

Other precursor combustibles such as hexamethyldisiloxane ($C_6H_{18}OSi_2$; HMDSO) and hexamethyldisilazine ($C_6H_{19}NSi_2$; HMDSA) burned in air and propane fuel are also capable of forming SiO_2. The SiO_2 nanoparticles are quenched and collected on an aluminum plate [71,72]. Particles ranging from 2.5 to 25 nm were produced [71,72]. Combustion is the dominant pathway for the production of particulate matter into the environment—in particular, carbonaceous nanoparticles that are derivatives of benzene hierarchical structures. Al_2O_3 and TiO_2 nanoparticles can also be formed by combustion of the appropriate metal precursor compounds.

Single- and multiwalled carbon nanotubes can be produced in flames under the proper conditions. Catalysts are entrained into a flame stream by thermal evaporation simultaneously with the introduction of a carbon source gas like CO or ethane. Processes like these have evolved into a major industrial market for MWNTs in nanocomposite materials.

Thermal Decomposition. Solid Si nanoparticles can also be formed by the thermal decomposition of silane gas in the absence of oxygen:

$$SiH_{4(g)} \rightarrow Si_{(s)} + 2H_{2(g)} \tag{4.25}$$

The formation of Fe nanoparticle catalysts in the gas phase by the decomposition of ferrocene is a fundamental ingredient of the HiPCO™ process to manufacture carbon nanotubes—two bottom-up methods in tandem. At temperatures <500°C and at higher temperatures, the cyclopentadienyl decomposes further:

$$Fe(C_5H_5)_2 \rightarrow Fe + 2C_5H_5 \rightarrow H_2 + CH_4 + C_5H_6 + \cdots \tag{4.26}$$

The ferrocene decomposition process is also exploited in a lithographic writing technique utilizing an STM probe tip.

4.2.2 Liquid-Phase Methods

Much of chemistry takes place in solvents (e.g., liquid-phase media). Liquids provide an environment that is able to support molecules, intermediates, metastable materials, and other forms of materials that would otherwise be unable to react, agglomerate, or simply to exist. Biological phenomena, for the most part, are based on a very important liquid: water, which is one of the most important, versatile, and ubiquitous solvents known to science and to the life process itself. There are many kinds of liquid-phase fabrication methods listed in **Table 4.8** and we will discuss selected examples that highlight this diversity.

Molecular Self-Assembly. Molecular self-assembly is one of the most powerful methods to form nanomaterials en masse. Molecular self-assembly is considered to be a *soft-matter* phenomenon. Soft matter abounds all around us. Biological matter is mostly soft, except for bone, shells, teeth, and other hard structures. Organic phenomena like soap bubbles, micelles, polymers, colloids, amphiphiles, and liquid crystals are considered to be soft matter. The chemistry of soft matter is based on a fundamental understanding of intermolecular interactions.

Molecular-based self-assembly is a bottom-up fabrication process in the absolute and most pure sense of the term. The concept of self-assembly, however, does not stop at the molecular level. If a scientist is clever enough to place the proper molecular moieties on the surface of a nanoparticle, it is very likely that the nanoparticles themselves become capable of further self-assembly. What materials can you think of that exhibit this kind of behavior? Proteins involved in the immune system certainly qualify as a suitable answer to this question. Nature operates on the principle of self-assembly, and we are embarking on a similar path; nano is truly revolutionary in that regard.

Intermolecular interactions form the basis of molecular self-assembly, and they are discussed in some detail in chapter 10. Thermodynamic control versus kinetic control is another concept that is introduced in chapters 10 and 11. Examples of molecular self-assembly are all around us. In the basest sense, water achieves its low energy state as a liquid by forming hydrogen bonds. Although a liquid, however, water has structure. When liquid oil and liquid water are mixed, a self-assembly process is automatically initiated without the need for a trigger; it happens spontaneously at mild temperature, pressure, and *p*H. Micelles are another example. In this case, molecules with a hydrocarbon *hydrophobic* end and an inorganic *hydrophilic* end automatically self-assemble into micelles when placed in water (**Fig. 4.12**). This is also an example of self-assembly.

Supramolecular Chemistry.   Supramolecular chemistry is a relatively new field of chemistry. It is the chemistry of soft matter and is a bona fide bottom-up method to fabricate nanomaterials. Supramolecular chemistry is based on the *lock and key* hypothesis of Emil Fischer (**Fig. 4.13**).

Fabrication of supramolecular structures requires significant forethought and planning. Just to reiterate, these factors will be discussed more fully later, but we will mention a few salient aspects of supramolecular design and fabrication in this paragraph. To begin with, there exists a vast panoply of intermolecular interactions to choose from: van der Waals forces, hydrogen bonding, hydrophobic interactions, and various species of dipolar interactions, to name a few. Conventional "hard" bonding forms are also important to consider: covalent, ionic, metallic, coordination, etc. and cannot be neglected in molecular design protocol. Complementarity and "host–guest chemistry" are terms often applied in supramolecular chemistry jargon. They are essentially the supramolecular equivalents of the lock-and-key hypothesis, one of the most important and fundamental concepts of supramolecular chemistry. Examples of complementarity can be found anywhere in nature.

Supramolecular design begins with the covalent bond. Synthesis of precursors often requires the making and breaking of strong covalent bonds. This is called molecular chemistry—another bottom-up fabrication method. Once the precursors are made, after due consideration of intermolecular forces, they are allowed to react to form host–guest relationships. The fabrication of DNA, one of the most remarkable of all nanomaterials, is accomplished in this way; the backbone and the nucleotides are covalent bonded structures but the double-stranded DNA is held together with relatively weak intermolecular forces such as π–π interactions, hydrophobicity, and hydrogen bonds.

Sol–Gel Synthesis.   Colloids are what make sols. Colloids, a dispersed phase of one substance, exist as discrete entities within a continuous phase, usually water. A *sol* is a colloidal suspension of solid particles within a liquid. A *gel*, on the

FIG. 4.12	*Self-assembly is a fast expanding technique that is expected to make significant contributions to bottom-up manufacturing of nanomaterials. Various types of micelles are shown in the figure. Depending on solution conditions, micelles (the spherical structures depicted) or bilayers are commonly formed. Micelles are made of molecules called amphiphiles: single molecules that have both a polar and a nonpolar chemical group. In order for self-assembly techniques to become dominant, problems with self-assembly methods, including lack of long-range order and structural integrity, need to be overcome.*

other hand, is a solid that contains liquid within its pore structure. For example, a colloid made from the bottom up begins with the nucleation of appropriate atoms or molecules in a supersaturated solution. When enough colloids are formed, condensation occurs to form three-dimensional structures. The physical conformation of the structure is dependent upon the size of the colloids and the chemical nature of their surface. Supercritical extraction of the liquid and subsequent baking form compounds called *aerogels*. Extraction of the liquid at nonsupercritical conditions forms compounds called *xerogels*. Both forms are built from the bottom up and are bulk materials with nanoscale constituents. There are two general kinds of sol–gel precursor substances: the inorganic salts and the metal alkoxides that are organic materials containing silicon or metal coordinating centers.

Supramolecular chemistry is all about molecular recognition. Molecular recognition is achieved through complementarity—the ability of one molecule to interact with another in a highly specific manner. The "lock-and-key" hypothesis of Emil Fischer in the 1900s forms the basis of supramolecular chemistry.

FIG. 4.13

Hydrolysis of precursor molecules occurs first from TEOS:

$$Si(OC_2H_5)_4 + H_2O \rightarrow Si(OC_2H_5)_3(OH) + (C_2H_5)OH \qquad (4.27)$$

Subsequent hydrolysis yields reduced forms of TEOS (an alkoxide precursor):

$$Si(OC_2H_5)_3OH + H_2O \rightarrow Si(OC_2H_5)_2(OH)_2 + (C_2H_5)OH \qquad (4.28)$$

The next step is a condensation step that involves the hydrolysis products. The insipient form of the silicon dioxide structure is established at this time:

$$2Si(OC_2H_5)_3(OH) \rightarrow (OC_2H_5)_3Si\text{-}O\text{-}Si(OC_2H_5)_3 + H_2O \qquad (4.29)$$

Numerous other processes, many of them without the presence of water, are used to make SiO_2 nanoscale materials. Sol–gel synthesis proceeds with growth of increasingly larger particles via Ostwald ripening and other attractive mechanisms. In Ostwald ripening, larger particles that are energetically favored (due to

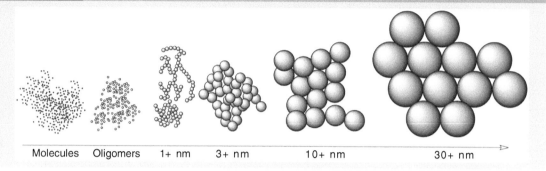

| FIG. 4.14 | *Sol–gel synthesis is an old technology that has incredible potential for nanomanufacturing. Starting from the absolute bottom with molecules and via the process of nucleation and Ostwald ripening, larger and larger particles are grown until the reaction is terminated. Following a sintering process, an array of close-packed spherical particles can be used to form aerogels or xerogels or act as a template to form other nanomaterials.* |

Molecules   Oligomers   1+ nm   3+ nm   10+ nm   30+ nm

curvature phenomena) grow at the expense of smaller, less stable particles. The process is schematically illustrated in **Figure 4.14.**

One means to obtain colloidal metal oxides is by the hydrolysis of the corresponding metal salt. TiO_2 colloids are formed by the base catalyzed hydrolysis of $TiCl_4$:

$$TiCl_4 + 2H_2O \rightarrow TiO_{2(s)} + 4H^+ + 4Cl^- \qquad (4.30)$$

The problem confronting procedures such as those depicted above is agglomeration. Small size is accompanied by high surface energy (chapters 5 and 6) and, therefore, nanosized particles are inherently unstable. Such coagulation can be prevented by addition of chemical agents or adjustment of the solution pH and other conditions.

More detail can be found in later chapters; however, the general class of sol–gel synthesis embraces a wide variety of nanoscale fabrication methods that deliver an equally wide variety of nanoscale materials. In addition to aerogels and xerogels, zeolitic structures and inorganic–organic hybrid materials have become increasingly useful.

Electrolytic and Electroless Deposition. These two methods are most appropriate for metal and semiconductor deposition. Thin films or nanoparticles can be fabricated from either method—both originating with atoms to form nanomaterials. Electrolytic deposition is based on Faraday's constant and Faraday's law. Faraday's constant is

$$\mathcal{F} = \mathcal{N}_A \cdot e \qquad (4.31)$$

where $\mathcal{F}$ is the Faraday constant ($96{,}485 \ C \cdot mol^{-1}$), $\mathcal{N}_A$ is the Avogadro number (6.022×10^{23} per mole), and e is the electron charge ($1.6022 \times 10^{-19} \ C \cdot electron^{-1}$). The modern form of Faraday's law of electrolysis is

$$m = \frac{Q}{e\upsilon}\left(\frac{M}{\mathcal{N}_A}\right) = \frac{1}{e\mathcal{N}_A}\left(\frac{QM}{\upsilon}\right) = \frac{1}{\mathcal{F}}\left(\frac{QM}{\upsilon}\right) \qquad (4.32)$$

where

m is the mass of metal plated on the cathode in grams

Q is the total number of coulombs accrued over the duration of the electro-plating process

v is the charge on the atom (the valence state)

M is the molar mass in grams per mole

The total charge per atom is equal to ev. Q is also found by integrating the measured current $I(t)$ over time t, where I is in terms of amperes per second:

$$Q = \int_{t=0}^{t=T} I(t)\, dt \tag{4.33}$$

or, if under constant current control conditions,

$$Q = I(\Delta t) \tag{4.34}$$

In its simplest form, electroplating (at the cathode) is represented by

$$M^{n+}_{(aq)} + ne_{(aq)} \rightarrow M_{(s)} \tag{4.35}$$

where M is a generic metal and n is an integer that indicates the charge on the metal atom ($n = v$).

Electroless deposition is, as its name indicates, a plating process for metals that does not require an outside electromotive force. It is a nongalvanic process. Electroless deposition is a purely chemical bottom-up process. The facility of electroless deposition is demonstrated by its ability to plate metal on nearly any substrate—metal, ceramic, plastic, or semiconductor. Electroless plating is an autocatalytic process that requires participation of a metal salt and a reducing agent. The formation of a nanometer scale film of Ag on a substrate is given by a multistep process:

$$AgNO_3 + KOH \rightarrow AgKOH + KNO_3 \tag{4.36}$$

$$AgOH + 2NH_3\ [Ag(NH_3)_2]^{1+} + OH^- \tag{4.37}$$

$$[Ag(NH_3)_2]^{1+} + OH^- + H_2COAg_{(s)} + 2NH_3 + H_2O \tag{4.38}$$

Electroless plating of gold is used extensively in the fabrication of semiconductor devices and metalized ceramic components. A formula for the autocatalytic plating of gold consists of 0.01 M gold chloride hydrochloride trihydrate, 0.014 M sodium potassium tartarate, 0.013 M dimethylamine borane, and 400 mg·L⁻¹. The pH of the solution (adjusted with NaOH) is 13.0 and the temperature is 60°C [73]. If you are a chemist, justify the components and conditions of this formula.

Biological Bottom-Up Fabrication Methods.   Last, but most certainly not least, we bring up nature's way to manufacture nanomaterials—yes, from the bottom up. Most natural fabrication processes occur in solution, mainly in aqueous media. Biological processes on the whole occur at mild temperatures, atmospheric pressure, and moderate pH. Now that this method has been introduced, please anticipate in-depth treatment to come in chapters 13 and 14.

4.2.3   *Solid-Phase Bottom-Up Fabrication?*

Solid-State Phenomena.   This category is difficult to define. Pure solid-phase bottom-up fabrication methods are "few and far between." By definition, a solid is a close-packed ensemble of atoms and molecules and, therefore, a condensed, bulk state. Any process that converts this bulk state into nanomaterials is rightly considered to be a top-down method. The formation of nanovoids from smaller cavities or vacancies could be labeled as a bottom-up, solid-state "fabrication" method. However, formation of defects, dislocations, twinning, and other material deformations in response to external stresses basically occur from the top down. Atoms do have the ability to diffuse through solids, assemble, and create nanodomains within the solid material. Would this process be considered as a bottom-up process? Defects occur in solids and they play important roles in the properties of bulk materials. It would be expected that such defects play a lesser role in nanomaterials. We will review a few general categories.

Zero-dimensional imperfections (or point defects) occur in homogeneous crystalline materials independent of chemical impurities [74]. One type of point defect is called a *vacancy*—an unoccupied atomic site. A vacancy can also be produced in response to a localized chemical impurity or a nonstoichiometric rendering of compounds within the solid. An *interstitial point defect* is the occupation of a site by an atom that is normally not occupied. A *Schottky* defect is made from a complementary pair of ionic vacancies, and a *Frenkel* defect is a combination of a vacancy and an interstitial point defect [74]. The dimensions of such defects are usually on the subnanometer scale. They are produced from thermal effects, chemical impurities, solid-state diffusion, and other sources of external stress—from the top down? They serve an important function because without such defects, diffusion through metals would be more difficult.

One-dimensional (linear) imperfections cover larger areas. A dislocation is a linear defect in a crystalline solid that is able to impart influence over a nanometer scale. There are *edge* dislocations and *screw* dislocations (a spiral stacking of planes around the dislocation line). Most linear defects consist of a mixture of the two [74]. Mechanical properties of materials are influenced greatly by such material imperfections. However, plastic deformation in metals, for example, is not likely without the presence of imperfections. One-dimensional imperfections are nanometer or greater sized structures. How are they "fabricated"?

Two-dimensional (planar) imperfections involve surface disruption of a crystal. To begin with, the surface is very different from the bulk. *Twin boundaries* (the plane that separates two identical mirror image crystal regions) are formed by deformation or by annealing (top down). The *grain boundary* is a two-dimensional imperfection. Grain size ranges from the nanometer scale to microns. The physical properties of grains in metals are dependent on top-down manufacturing conditions (e.g., quenching rate, temperature). Grain boundaries, like for the surface, represent enhanced areas of material diffusion. Grain boundary dislocations exist between grains with different orientations. Another two-dimensional imperfection, the surface step, has major importance in nanotechnology (discussed in more detail in chapter 6). The most important feature of engineering materials is the grain structure (e.g., grain boundary, size, and distribution) [74]. For example, the mechanical properties of materials containing nanograins outperform counterparts made of micron-sized grains [7].

Three-dimensional imperfections are characteristic of noncrystalline, or amorphous, solids [74]. An example of an amorphous solid is plate glass used in windows. There is no long-range order in amorphous solids. Quasi-crystals represent an intermediate structural state [74]. An offshoot of quasi-crystal research, the study of fractals, has given a new perspective to micro- (and now nano-) scale structures [74]. Some of the benefits of fractal research are the development of new kinds of thin films with tunable properties [74].

Self-Purification of Nanocrystals. Nanomaterials have the capability of self-purification. Because of this, the process of doping quantum dots becomes problematic. Self-purification occurs for several reasons. Because nanomaterials have small dimensions, the diffusion path length of atoms may exceed the physical dimensions of, for example, a quantum dot. Also, impurities can easily find their way to the surface and, once at the surface, impurities easily bind to the surface. The surface energy of any material, bulk or nano, is always higher than the cohesion energy of the volume. The surface energy of a nanomaterial is much higher than its planar bulk surface counterpart. The surface is always seeking ways to become more stable. Binding to complementary materials is one way to achieve stability. The bound atom may be an impurity (e.g., originating from a doping procedure) that has migrated from the interior.

Lastly, the binding energy of volume atoms in very small nanoparticles is lessened due to lowered coordination. The smaller the nanocrystal is, the less binding energy exists and, as result, the difficulty of doping increases. The argument presented here represents one of *antifabrication* rather than fabrication. In order to be able to dope quantum dots, the immediate environment surrounding the dots requires modification [75].

4.2.4   *Template Synthesis*

This process is perhaps one of the most facilitative in forming nanomaterials from the bottom up. A template (from the Latin *templum*, "plank, rafter," but also "a building for worship") is a material that acts as a gauge, pattern, or mold that is used to guide the manufacture of another piece. According to this definition, DNA serves as a template for the generation of another macromolecule—for example, RNA. This is an example of template synthesis. The mask used in lithography is a template. Porous aluminum oxide films with ordered arrays of pore channels are used as templates and masks. The electroplating (a bottom-up process) of Au into the pore channels of porous alumina results nanoparticles of gold. The aspect ratio and the size of the Au particles are determined by the pore diameter of the channel and the time of electroplating [76,77]. We dedicate a section in chapter 12 to this process.

In its simplest context, template synthesis involves either hard matter or soft matter. Porous alumina fits into the category of a hard matter template. Reverse micelles and DNA are soft matter templates. The basic process is illustrated in **Figure 4.15.**

FIG. 4.15 *Physical dimensions of nanomaterials formed in templates are directed by the template. Template synthesis makes use of a solid or liquid architecture that has the ability to isolate a chemical reaction within its physical boundaries. Templates can contain a cellular array, like the version depicted, or singular form such as that exhibited by micelles. Confinement of chemical reactions results in the formation of materials that would otherwise not exist. The process depicted demonstrates the facility of template synthesis. The pore channels of an anodically formed alumina membrane can act as host to a many kinds of materials.*

4.3   COMPUTATIONAL CHEMISTRY AND MOLECULAR MODELING

In today's laboratories, R&D firms, and places of manufacturing of nanomaterials and devices, the computer has proven to be an invaluable tool. Computational chemistry integrates theoretical chemical principles into computer programs. The results of computational chemical techniques are quantitative (exact) for hydrogen and assume more qualitative character as molecules become larger and systems become more complex. With increase in complexity, there is a concomitant increase in the level of input parameters, algorithms, and assumptions. Molecular modeling has become highly interdisciplinary as mathematicians, physicists, chemists, biologists, engineers, and computer scientists all work together to produce the desired outcome.

4.3.1 History

The content of this paragraph was obtained from excellent sources found on the Internet: Answers.com, Wikipedia, and the Sci-Tech Encyclopedia site of *Computational Chemistry*, to name just a few. The history of computational chemical methods begins in the early part of the twentieth century. Walter Heitler and German-American physicist Fritz London in 1927 are credited with performing the first theoretical chemistry calculations. The latter is also known for describing in 1930 weak intermolecular dispersion forces that bear his name. Several well-known physicists and chemists of the period contributed to the early development of computational quantum chemistry: Pauling and Eyring, to name a few. When computer technology emerged in the 1940s, solutions to wave equations became practical and, in the early 1950s, semi-empirical atomic orbital calculations were accomplished. The first ab initio Hartree–Fock computations that used a basis set of Slater orbitals were carried out in 1956 at MIT. Polyatomic calculations that employed Gaussian orbitals also were carried out during the 1950s and 1960s. Linear combination of atomic orbital methods (a.k.a Hückel methods) emerged in 1964. The discipline of scientific computational chemistry became a formal discipline around 1979.

The evolution of computational chemistry developed hand in hand with the development of the computer. Computational chemistry officially began in 1962 with the Quantum Chemistry Program Exchange (QCPE). This program helped scientists (chemists in particular) to develop, share, and apply quantum mechanical software [78]. With the advent of small computers, the ability to model real, but very simple, chemical systems began in earnest. In the 1980s, quantum mechanical methods and atomistic simulations were able to predict structures and behavior of small organic molecules and systems of >100 atoms.

Molecular modeling and nanomaterials combine to form a perfect match. There has never been a better, more compatible pair of complementary systems: Computers predict nanomaterial properties and behavior and advanced nanomaterials make computers better and more efficient! Both however are limited by the number of molecules in their structure (or program). Nonetheless, systems of hundreds to thousands of separate constituents (molecules, replicas, or ensembles) are routinely simulated. There are several approaches for computing nanothermodynamic phenomena. Molecular dynamics simulation is regarded as one of the best. In this approach all the atoms and molecules are assumed as "vibrating machines" that are programmed to interact in a predetermined fashion for a period of time under certain conditions to carry out virtual experiments. The results of the simulation are deconvoluted by various kinds of numerical methods.

Why is this important? Design will become one of the most important aspects of nanotechnology development [79]. Prognosticators state that while production costs will diminish significantly due to nanotechnology, the level of design will increase significantly due to the complexity involved [79]. According to Michael Riech at AIFT University of Karlsruhe, "Design will change radically under nanotechnology and for nano-engineers and nano-designers respectively, a broad knowledge will become even more important in the future.... As long as we are still far from the realization of complex nanotechnological applications, nano-engineering and nanodesign almost exclusively take place on computers."

4.3.2 General Types of Molecular Modeling Methods

There are four basic types of simulation methods: quantum mechanical ab initio, Monte Carlo, molecular dynamics, and molecular mechanics methods (**Table 4.13**). Finite element analysis has also become an increasingly popular and powerful tool for analyzing nanomaterial phenomena. Obviously, as stated earlier in this chapter, molecular modeling methods are not trivial in nature. We, therefore, will not offer great detail but rather a qualitative overview of the methods and relevance to nanotechnology.

Ab Initio Methods. Quantum mechanical ab initio (from the Latin *ab*, "from" + *initium*, "beginning") methods strive to simplify the solution of Schrödinger's equation (molecular Hamiltonian) for many particle systems and do not rely on a priori input in the form of empirical or semi-empirical parameters or experimental data. Common techniques include self-consistent field, LCAO (linear combination of atomic orbitals), and density functional methods. The Hartree–Fock version is the simplest in that average electron–electron repulsions are factored into the program. Density-functional theory is applied to calculate molecular electronic structure but empirical data are used to "grease the skids" for this technique. In this method, energy is expressed in the form of electron density and not in the form of a wave function.

The computational investment in ab initio methods is quite large and the technique is confined to the analysis of a few hundred atoms [79]. No dynamical or temporal aspects are considered in ab initio methods.

Molecular Mechanics and Dynamics Methods. Molecular mechanics (static) and dynamic methods, on the other hand, are grounded in classical force field theory [79]. The molecular mechanic method (electrons treated implicitly) is nonquantum mechanical (electrons treated explicitly). The objective of both ab initio and molecular mechanical methods is to develop the lowest energy configuration of the potential energy surface. Molecular mechanics simulations account for bonding energy a priori and the potential energy as the sum of all the forms of electrostatic interactions, stretching, bending, torsion, and van der Waals interactions. To date molecular dynamics (MD) simulations have been successfully applied to a narrow range of nanostructures that are energetically close to equilibrium [79]. Both methods lack information concerning quantum effects. Often, information gleaned from ab initio modeling is input a priori into molecular mechanics and dynamics programs.

Monte Carlo Methods. Monte Carlo methods (random walk, stochastic methods) rely on statistical ensembles based on Boltzmann style distributions. Monte

TABLE 4.13	Molecular Modeling Techniques	
Simulation method	**Description**	**Advantages and drawbacks**
Ab initio	Solution of Schrödinger's equation for many particle systems Types: Self-consistent field method, linear combination of atomic orbitals, and density-functional method Objective: describe the stable low-energy configuration—the local minima on potential energy surface Result is in the form of quantum effects.	Calculation time is a function of particle number; therefore, a great investment in computer space and time is required. Practical for ca. 10^2 atoms Many approximation algorithms exist to shorten calculation time. QM calculations lack a priori input (no interaction potential, chemical bond, or temperature input parameters). Dynamic aspects are not considered. No temperature data involved in the model. Relatively slow method
Monte Carlo	New particle configurations are created by a step-by-step random process. Systems configurations are sampled according to specific statistical ensemble. Method is grounded in Boltzmann distribution theory.	Calculations are valid for systems close to equilibrium. External input parameters such as temperature are required.
Molecular mechanics	Based on classical mechanics	Able to handle 10^5–10^7 atoms
Molecular dynamics (MD)	Based on Newton's laws of motion The potential energy of all components is calculated using force fields Harmonic, Morse, Gaussian, Lenard–Jones, and other potentials are used in the simulation. Each step is associated with a temporal element. Information: Particle position, velocity, momentum, kinetic and potential energies Utilizes mass points interacting through force fields derived from interaction potentials Describes electron interactions implicitly and considers molecules to be a collection of atoms held together by "sticky forces" Molecular mechanics derives a motionless structure. Molecular dynamics achieves a better model. MD is the choice of most nanotech modeling groups MD is the study of matter in motion. MD is an excellent program with which to study thermodynamics	Require detailed a priori input concerning bond energies; interaction potentials are chosen beforehand As a result, discovery of new molecular systems is not likely by this method. Fitting of potentials to experimental data is required to generate realistic scenarios. Such information is derived from ab initio calculations. No explicit quantum results are offered by these methods.
Finite element analysis	The method does not consider that materials have structure on the microscopic scale or below. Materials are treated like a continuum that has consistent properties when stretched, bent, made to conduct heat, or perturbed in other ways. Molecular level modeling is accomplished in a similar way with atoms and molecules acting as the points, nodes, mesh, etc.	Data depend on the quality and accuracy of applied input parameters, geometry. Properties like Young's modulus have been determined at the molecular scale.

Carlo (MC) methods are based on statistical mechanics rather than the molecular dynamic formulations discussed briefly earlier. These types of simulation techniques require input parameters (e.g., temperature) and, like MD methods, exist near the equilibrium point. Assignment of temporal markers to the evolving configurations is problematic in MC simulations. The five basic types of MC methods are

- classical, where samples are based on a probability distribution with the intent of finding minimum energy configurations and rate parameters;
- quantum, in which stochastic methods are used to predict quantum mechanical parameters based on the Schrödinger equation;
- volumetric, a method that generates numbers from random walk procedures to predict molecular volumes and molecular phase-space surfaces;
- path integral, based on quantum statistical mechanics and used to find thermodynamic properties by application of Feynman's path integral as the starting point; and
- simulation methods like kinetic and thermalization, which use stochastic algorithms to generate initial conditions.

These categories were described on the site of R. Q. Topper at <www.cooper.edu/engineering/chemechem/monte.html> (2002). Monte Carlo simulations make use of the Markov chain principle. The Markov chain is a computational process in which future states are conditionally independent of previous states.

Finite Element Analysis.   Finite element analysis (FEA) was first accomplished in 1943 by R. Courant, who conducted numerical analysis and minimization of variational calculus to seek solutions to vibrational systems. FEA methods were first applied to structural mechanics. In 1956, L. J. Topp, R. W. Clough, M. J. Turner, and C. Martin wrote a paper applying FEA to complex structures—in particular, stiffness and deflection relationships. Basically, with regard to engineering systems, FEA is a computer model that is able to analyze engineering parameters such as stress. It is often used in predesign work in the form of computer-aided design (CAD). In FEA, a mesh consists of system points called nodes. Inherent within the mesh are programs that contain structural properties, if engineering applications are the goal, that define how that structure will react to applied stresses. Mass, volume, temperature, strain energy, stress strain, force displacement, velocity, and acceleration are some parameters that are monitored or applied during the simulation [80].

The first step is to divide the structure (mechanical or atomic) into unique sectors called finite elements. Finite elements are joined to form nodes and form the mesh (or grid). Boundary conditions are often applied. Predetermined variables then act over each domain and the results of local equations are assimilated to give the system equation that describes the behavior of the whole system [80].

Examples of finite element analysis of nanomaterials are plentiful in the literature. In one case, the spring constants of AFM cantilevers made of sapphire and tipped with a diamond probe were modeled via nanomechanical FEA [81]. FEA was conducted to analyze quantum mechanical transport in strained quantum

FIG. 4.16

A *Morpho butterfly recreation by finite element analysis methods. The wing structure is made from carbon nanotubes and contains photonic crystals as in the natural form. FEA is a powerful tool in the nano domain; in many ways it can be considered to be a fabrication method. It is a valuable prefab tool involved in the design phase of nanosynthesis and assembly programs.*

Source: Courtesy of Halff Associates, Inc.

dots and wires [82]. It was found that strain is responsible for energy shifts of resonant current peaks and that strain causes additional fine structure in the current peaks. Finite element method (FEM) was used to model the heat generation and distribution and carbon migration in catalysts that form MWNT and SWNTs. From FEM, it was found that growth is mainly driven by a concentration gradient, as opposed to a thermal gradient, with the experimental temperature playing a major role in terms of activating the diffusion process [83].

Computer simulation methods are powerful ways to predict the behavior of nanomaterials. A system that is made of a countable number of atoms or molecules is a perfect candidate for computer-aided simulation techniques. Add computer scientists to the interdisciplinary crew—the continuum of qualified nanoengineers and scientists.

A finite element analysis rendition of the *Morpho* butterfly pictured on the cover of this text is shown in **Figure 4.16.**

References

1. B. K. Teoand and X. H. Sun, From top-down to bottom-up to hybrid nanotechnologies: Road to nanodevices, *Journal of Cluster Science,* 17, 529–540 (2006).
2. S. Price, Top-down and bottom-up processes. Presentation, EE 518, J. Ruzyllo, Pennsylvania State University (2006).
3. C. L. De Castro and B. S. Mitchell, Nanoparticles from mechanical attrition. In *Synthesis, functionalization and surface treatment of nanoparticles,* M.-L. Baraton, ed., American Scientific Publishers, Stevenson Ranch, CA (2002).
4. M. Ohring, *Materials science of thin films,* Academic Press, New York (2002).
5. R. W. Kelsall, I. W. Hamley, and M. Geoghegan, *Nanoscale science and technology,* John Wiley & Sons, Ltd., West Sussex (2005).
6. B. J. Y. Tan, C. H. Sow, T. S. Koh, K. C. Chin, A. T. S. Wee, and C. K. Ong, Fabrication of size-tunable gold nanoparticles array with nanosphere lithography, reactive ion etching, and thermal anneal, *Journal of Physical Chemistry B,* 109, 11100–11109 (2005).
7. C. P. Poole and F. J. Owens, *Introduction to nanotechnology,* Wiley-Interscience, New York (2003).
8. J. van Erven, R. Moerman, and J. C. M. Marijnissen, Platinum nanoparticle production by EDHA, *Aerosol Science Technology,* 39, 941–946 (2005).
9. S. M. Lee, *Dictionary of composite materials technology,* CRC Press, Boca Raton, FL (1989).
10. I. T. H. Chang, Rapid solidification processing of nanocrystalline metallic alloys. In *Handbook of nanostructured materials and nanotechnology,* Vol. 1, H. S. Nalwa, ed., Academic Press, San Diego (2000).
11. D. Li and Y. Xia, Electrospinning of nanofibers: Reinventing the wheel, *Advanced Materials,* 16, 1151–1170 (2004).
12. Z. Zhang, J. Y. Ying, and M. S. Dresselhaus, Bismuth quantum-wire arrays fabricated by a vacuum melting and pressure injection process, *Journal of Materials Research,* 13, 1745–1748 (1998).
13. J. Heremans, C. M. Thrush, Y.-M. Lin, S. Cronin, Z. Zhang, M. S. Dresselhaus, and J. F. Mansfield; Bismuth nanowire arrays: Synthesis and galvanomagnetic properties, *Physics Review B,* 61, 2921–2930 (2000).
14. D. Hulicova, K. Hosoi, S.-I. Kuroda, H. Abe, and A. Oya, Carbon nanotubes prepared by spinning and carbonizing core-shell polymer microspheres, *Advanced Materials,* 14, 452–455 (2002).
15. Y. Ando and M. Ohkohchi, Production of ultrafine powder of p-sic by arc discharge, *Journal of Crystal Growth,* 60, 147–149 (1982).
16. Y. Ando and S. Iijima, Preparation of carbon nanotubes by arc-discharge evaporation, *Japan Journal of Applied Physics,* 32, L107–L109 (1993).
17. T. Guo, P. Nikolaev, A. Thess, D. T. Colbert, and R. E. Smalley, Catalytic growth of single-walled nanotubes by laser evaporation, *Chemical Physics Letters,* 243, 49–54 (1995).
18. C. L. Fields, J. R. Pitts, M. J. Hale, C. Bingham, and A. Lewandowski, Formation of fullerenes in highly concentrated solar flux, *Journal of Physical Chemistry,* 97, 8701–8702 (1995).
19. Y. B. Li, Y. D. Yu, and Y. Liang, A novel method for synthesis of carbon nanotubes: Low temperature solid pyrolysis, *Journal of Materials Research,* 12, 1678–1680 (1997).
20. M. Di Ventra, S. Evoy, and J. R. Heflin, eds., *Introduction to nanoscale science and technology,* Kluwer Academic Publishers, New York (2004).
21. R. S. Dhaliwal, W. A. Enichen, S. D. Golladay, M. S. Gordon, R. A. Kendall, J. E. Lieberman, H. C. Pfeiffer, D. J. Pinckney, C. F. Robinson, J. D. Rockrohr, W. Stickel, and E. V. Tressler, PREVAIL—Electron projection technology approach for next-generation lithography, *IBM Journal of Research and Development,* 45, 615–638 (2001).

22. S. Y. Chou, P. R. Krauss, and P. J. Renstrom, Imprint lithography with 25-nanometer resolution, *Science*, 272, 85–87 (1996).

23. S. Y. Chou, P. R. Krauss, and P. J. Renstrom, Nano-imprint lithography, *Journal of Vacuum Science Technology B*, 14, 4129–4133 (1996).

24. J. C. Hulteen and R. P. Van Duyne, Nanosphere lithography: A materials general fabrication process for periodic particle array surfaces, *Journal of Vacuum Science Technology A*, 1553–1558 (1995).

25. K. Kawata, H. B. Sun, T. Tanaka, and K. Takada, Finer features for functional microdevices, *Nature*, 412, 697–698 (2001).

26. G. L. Hornyak, L. Grigorian, A. C. Dillon, P. A. Parilla, and M. J. Heben, A temperature window for chemical vapor deposition growth of single walled carbon nanotubes, *Journal of Physical Chemistry B*, 106, 2821–2825 (2002).

27. P. Petroff, A. Lorke, and A. Imamoglu, Epitaxially self-assembled quantum dots, *Physics Today*, 54, 46–52 (2001).

28. F. Ren, C. Z. Jlang, Y. H. Wang, Q. Q. Wang, and J. B. Wang, The problem of core-shell nanoclusters formation during ion implantation, *Nuclear Instruments and Methods in Physics Research B: Beam Interaction with Materials and Atoms*, 245, 427–430 (2006).

29. G. L. Hornyak, M. Kröll, R. Pugin, T. Sawitowski, G. Schmid, J.-O. Bovin, G. Karsson, H. Hofmeister, and S. Hopfe, Gold clusters and colloids in alumina nanotubes, *Chemistry, a European Journal*, 3, 1951–1956 (1997).

30. H. Masuda and K. Fukuda, Ordered metal nanohole arrays made by a two-step replication of honeycomb structures of anodic alumina, *Science*, 268, 1466–1468 (1995).

31. W. K. Hsu, M. Terrones, J. P. Hare, H. Terrones, H. W. Kroto, and D. R. M. Walton, Electrolytic formation of carbon nanotubes, *Chemical Physics Letters*, 262, 161–166 (1996).

32. X. Liu, L. Fu, S. Hong, V. P. Dravid, and C. A. Mirkin, Arrays of magnetic nanoparticles patterned via "dip-pen" nanolithography, *Advanced Materials*, 14, 231–234 (2002).

33. Y. Lei, L. W. Teo, K. S. Yeong, Y. H. See, W. K. Chim, W. K. Cjoi, and J. T. L. Thong, Fabrication of highly ordered nanoparticle arrays using thin porous alumina masks, http://dspace.mit.edu/bitstream/1721.1/3662/2/AMMNS009.pdf., ca. 2002, 1–6 (viewed 2007).

34. D. Zschech, D. H. Kim, A. P. Milenin, R. Scholz, R. Hillebrand, C. J. Hawker, T. P. Russell, M. Steinhart, and U. Gösele, Ordered arrays of <100>-oriented silicon nanorods by CMOS-compatible block copolymer lithography, *Nano Letters*, 7, 1516–1520 (2007).

35. M. Park, C. Harrison, P. M. Chaikin, R. A. Register, and D. H. Adamson, Block copolymer lithography: Periodic arrays of ~10^{11} holes in 1 square centimeter, *Science*, 276, 1401–1401 (1997).

36. R. Garcia, *Bridging nano and macro worlds with water Meniscii: Attomole chemistry and nanofabrication by local oxidation nanolithography*, keynote address, LITHO 2004, Agelonde-France (2004).

37. M. F. Crommie, C. P. Lutz, and D. M. Eigler, Confinement of electrons to quantum corrals on a metal surface, *Science*, 262, 218–220 (1993).

38. M. Berger, *Nature's bottom-up nanofabrication of armor*, Nanowerk, LLC (2006).

39. T. J. Deming, V. P. Conticello, and D. A. Tirrell, Biocatalytic synthesis of polymers of precisely defined structures. In *Nanotechnology*, G. Timp, ed., Springer-Verlag New York, Inc., New York (1999).

40. D. D. Chiras, *Biology: The web of life*, West Publishing Company, New York (1993).

41. Y. Xia, J. A. Rogers, K. E. Paul, and G. Whitesides, Unconventional methods for fabricating and patterning nanostructures, *Chemistry Review*, 99, 1823–1848 (1999).

42. B. Das, S. Subramanium, and M. R. Melloch, Effects of electron-beam induced damage in back-gated GaAs/AlGaAs devices, *Semiconductor Science & Technology*, 8, 1347–1351 (1993).

43. G. Cao, *Nanostructures & nanomaterials: Synthesis, properties & applications*, Imperial College Press, London (2004).

44. O. Saxl, Opportunities for industry in the application of nanotechnology, The Institute of Nanotechnology, London, nano.org.uk (2000).

45. C. C. Koch, Top-down synthesis of nanostructured materials: Mechanical and thermal processing methods, *Review of Advanced Materials Science*, 5, 91–99 (2003).

46. L. He and E. Ma, Nanophase Fe alloys consolidated to full density from mechanically milled powders, *Journal of Materials Research*, 15, 904–912 (2000).

47. A. G. Mamalis, Advanced manufacturing of advanced materials. In *Innovative superhard materials and sustainable coatings for advanced manufacturing*, J. Lee, N. Novikov, and V. Turkevich, eds., NATO Science Series, II. Mathematics, Physics and Chemistry, 200, 63–80 (2006).

48. K. Sung and C. S. Lee, Factors influencing liquid breakup in electrohydrodynamic atomization, *Journal of Applied Physics*, 96, 3956–3961 (2004).

49. T. Guillard, G. Flamant, D. Laplaze, J.-F. Robert, B. Rivoire, and J. Giral, Towards the large scale production of fullerenes and nanotubes by solar energy, *Proceedings of the Solar Forum 2001: Solar Energy the Power to Choose*, Washington, D.C. (2001).

50. V. V. Karasev, O. G. Glotov, A. M. Baklanov, A. A. Onischuk, and V. E. Zarko, Alumina nanoparticle formation under combustion of solid propellant, www.kinetics.nsc.ru/private_page/en/papers/on2.pdf, Institute of Chemistry Kinetics & Combustion, Russian Academy of Science (2003).

51. Wet-chemical etching and cleaning of silicon, Virginia Semiconductor, Inc., www.virginiasemi.com (2003).

52. W. C. Crone, A brief introduction to MEMS and NEMS. In *Springer handbook of experimental solid mechanics*, W. N. Sharpe, ed., Springer-Verlag, New York (2008).

53. J. Rawson, *Chinese jade from Neolithic to the Qing*, British Museum, London (1995).

54. A. Doolittle, Lithography and pattern transfer, Georgia Tech University, users.ece.gatech.edu/~alan/ECE6450/Lectures/ECE6450L7b-Lithography%20Steps%20for%20a%20CMOS%20Inverter.pdf

55. B. Bushan, ed., Introduction to micro/nanofabrication. In *Springer handbook of nanotechnology*, Springer-Verlag, Berlin (2004).

56. G. L.-T. Chiu and J. M. Shaw, eds., Optical lithography: Introduction, *IBM Journal of Research & Development*, 41, 3–6 (1997).

57. D. Atwood, *Extreme ultraviolet (EUV) lithography based on multilayer coated optics.* Presentation, Virtual National Laboratory, UC Berkeley (2004).

58. V. Banine and R. Moors, Plasma sources for EUV lithography exposure tools, *Journal of Physics D: Applied Physics (London)*, 37, 3207 (2004).

59. B. Braun, *Producing integrated circuits with x-ray lithography*, University of Wisconsin at Madison, http://tc.engr.wisc.edu/UER/uer97/author7/index.html (1997).

60. H. Smith and F. Cerrina, X-ray lithography in ULSI manufacturing, *Microlithography World*, Winter, 10–14 (1997).

61. M. K. Mundra, S. K. Donthu, V. P. Dravid, and J. M. Torkelson, Effect of spatial confinement on the glass-transition temperature of patterned polymer nanostructures, *Nano Letters*, 7, 713–718 (2007).

62. C. L. Haynes and R. P. Van Duyne, Nanosphere lithography: A versatile nanofabrication tool for studies of size-dependent nanoparticle optics, *Journal of Physical Chemistry B*, 105, 5599–5611 (2001).

63. C. L. Cheung, R. J. Nikolic, C. E. Reinhardt, and T. F. Wang, Fabrication of nanopillars by nanosphere lithography, *Nanotechnology*, 17, 1339–1343 (2006).

64. Y. Cai and B. M. Ocko, The large-scale fabrication of protein nanoarrays based on nanosphere lithography, *Technical Proceedings of the 2005 NSTI Nanotechnology Conference and Trade Show*, 1, Chapter 7: DNA, Protein, Cells and Tissue Arrays (2005).

65. Chemical vapor deposition, Wikipedia (viewed 2007).

66. M. Halls and K. Raghavachari, Atomic layer deposition growth reactions of Al_2O_3 on Si(100)-2x1, *Journal of Physical Chemistry B*, 108, 4058–4062 (2004).

67. *Atomic layer deposition tutorial*, Cambridge NanoTech, Inc., www.cambridgenanotech.com (viewed 2007).

68. R. L. Puurunen, Understanding the surface chemistry of atomic layer deposition: A case study for the trimethylaluminum/water process, *Journal of Applied Physics*, 97, 121301–121352 (2005).

69. G. Kresse, M. Schmid, E. Napetsching, M. Shishkin, L. Köhler, and P. Varga, Structure of ultrathin aluminum oxide on NiAl(110), *Science*, 308, 1440–1442 (2005).

70. F. Rinaldi, Basics of molecular beam epitaxy (MBE), Annual Report, Optoelectronics Department, University of Ulm, 1–8 (2002).

71. C. L. Yeh and E. Zhao, Combustion synthesis of SiO_2 on the aluminum plate, *Journal of Thermal Science*, 10, 92–96 (2001).

72. H. K. Ma and C. L. Yeh, The formation of nano-size SiO_2 thin film on an aluminum plate with hexamethyldisilazane (HMDSA) and hexamethyldisiloxane (HMDSO), *Journal of Thermal Science*, 12, 89–96 (2003).

73. J. Henry, Electroless (autocatalytic, chemical) plating. In *Metal Finishing: The Industry's Recognized International Technical Authority since 1903, 54th Guidebook Directory*, 84, 190–191 (1986).

74. J. F. Shackelford, *Introduction to materials science for engineers*, 4th ed., Prentice Hall, Inc., Upper Saddle River, NJ (1996).

75. G. M. Dalpian and J. R. Chelikowsky, Self-purification in semiconductor nanocrystals, *Physics Review Letters*, 96, 226802 (2006).

76. G. L. Hornyak and C. R. Martin, Optical properties of a family of Au-nanoparticle-containing alumina membranes in which the nanoparticle shape is varied from needle-like (prolate), to spheroid, to pancake-like (oblate), *Thin Solid Films*, 303, 84 (1997).

77. G. L. Hornyak, C. J. Patrissi, and C. R. Martin, Optical properties of gold-porous alumina composite membranes: The Maxwell-Garnett limit, *Journal of Physical Chemistry B*, 101, 1548 (1997).

78. *Materials Modeling and Simulation History*, Accelrys Software, Inc., http://www.accelrys.com/technologies/modeling/materials/history.html

79. M. Rieth, *Nano-engineering in science and technology—An introduction to the world of nano-design*, World Scientific Publishing, Hackensack, NJ (2003).

80. P. Widas, *Finite element analysis*, Virginia Tech University, www.sv.vt.edu/classes/MSE2094_NoteBook/97ClassProj/num/widas/history.html (1997).

81. T. Gang and F. Sansoz, *Determination of atomic force microscope cantilever spring constants via finite element modeling for nanomechanical analysis*, University of Vermont, Burlington, www.vtspacegrant.org/ESMD/Travis%20Gang%20URECA_Poster2.ppt (viewed 2007).

82. H. T. Johnson, L. B. Freund, A. Zaslavsky, and C. D. Akyüz, Finite element analysis of quantum mechanical transport in strained quantum dots and wires, *Journal of Applied Physics*, 84, 3714–3725 (1998).

83. C. Klinke, J.-M. Bonard, and K. Kern, Thermodynamic calculations on the catalytic growth of multiwall carbon nanotubes, *Physics Review B*, 71, 035403: 1–7 (2005).

Problems

4.1 Define top-down and bottom-up fabrication. Are there nebulous regions between these extremes?

4.2 Classify the following techniques as top down or bottom up:

 a. Plasma etching
 b. Epitaxial growth
 c. Formation of a hierarchical structure with carbon nanotubes
 d. Transformation of amorphous carbon into carbon nanotubes by thermal methods
 e. Transformation of a 1-kg diamond into a 1-kg block of graphite
 f. Formation of carbon nanotubes in an arc process with a target that contains no catalyst
 g. Aspiration of a liquid to form nanoparticulate aerosols that condense to form a bulk material
 h. Formation of a very thin layer of paint that contains latex (paint) particulates with 2 µm diameter
 i. "Big Bang" formation of the universe and the formation of a black hole
 j. Compaction of micron-sized particles
 k. Using clay to make a pot

 Can you name any other methods employed by nature that are top down?

4.3 What techniques include a combination of distinct top-down and bottom-up methods?

4.4 What is the resolution limit for a projection type of photolithographic system if the incident λ is 365 nm (the i-line of Hg), N.A. = 0.63, and k_1 = 0.6. What is the DoF needed for the best resolution?

4.5 Would etch rate of silicon depend on the type of surface exposed to the etchant? Why or why not?

4.6 Is energy required to reduce a bulk material into nanoparticles? Qualitatively (and relatively speaking), how much energy do you think is required?

4.7 What is the limit on feature size if 600-nm radiation is used in a photolithographic procedure?

4.8 Are there any solid-state "bottom-up" fabrication processes you can think of that result in the fabrication of nanomaterials?

4.9 What is the minimum feature size, W_{min}, possible in a proximity photolithography system with the following criteria:

 λ = 320 nm; d_g = 10 µm, 5 µm, 1 µm, or 0.5 µm; and k = 1?

 Is there any way to improve the resolution?

4.10 How can you improve the resolution of a projection photolithographic system?

4.11 Trace the bottom-up hierarchy of a biological structure in the human body (akin to the tendon example) starting with any of the three primary precursor molecules (amino acids, lipids, or sugars).

4.12 Can you think of any structures in biology that do not arise from nanomaterials?

4.13 Can thermal top-down methods be considered a unique category or are thermal methods in general components of other top-down methods?

4.14 What is the basic mechanism of RF sputtering? Is it top down or bottom up?

4.15 Anodizing efficiency is defined as the degree of conversion of aluminum into aluminum oxide (Al_2O_3). What factors are able to reduce anodizing efficiency?

4.16 Anodizing is a means of forming nanoporous structures. Is this a top-down method or a bottom-up method?

4.17 What would be the breadth of the features formed by nanosphere lithography (diameter = 1 µm) if the particles' transit is normal to the plane of the spheres? How could you improve the system?

4.18 Self-assembly is a bottom-up method. Do you believe that nanotechnology of the future would be based on self-assembly methods? Why do you think biological structures are based on self-assembly? Are there examples of biochemical processes that are not based solely on self-assembly?

Section 3

Physics: Properties and Phenomena

MATERIALS, STRUCTURE, AND THE NANOSURFACE

Everything you've learned in school as "obvious" becomes less and less obvious as you begin to study the universe. For example, there are no solids in the universe. There's not even a suggestion of a solid. There are no absolute continuums. There are no surfaces. There are no straight lines.

R. BUCKMINSTER FULLER (1895–1983)

Chapter 5

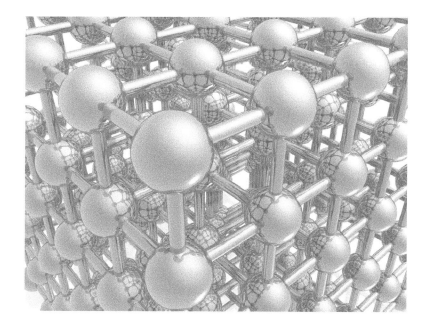

The nano perspective, an overview of societal implications, a brief foray into the nanoscale world, characterization tools, and fabrication strategies have been inventoried and discussed in a cursory manner. All of these topics dwell within the domain of well-established fields that significantly predate the time nanoscience and nanotechnology were officially anointed the "next big thing." It is time now to get technical and get into the essence of nanoscience. In other words, we are ready now to study the "physics of small" and familiarize ourselves with the properties and phenomena associated with nanomaterials.

Three important aspects of the nanoscale permeate all of nanoscience: the pronounced effects of the surface, the importance of the quantum-bulk boundary, and the inapplicability of bulk scaling laws—three different ways to state a similar message. In chapter 5, we focus on some materials science and the surface and get a feel for the importance of singular and collective surface area to nanomaterial behavior as particles get smaller. In chapter 6, the energetics of nanomaterials is discussed. Scaling laws (and their breakdown) and some basic quantum mechanics are discussed in chapter 7. Chapter 7 also describes zero-, one-, two-, and hierarchical nanostructured materials. Chapter 8 is about nanothermodynamics and potential violations of the well-established thermodynamic laws and with that chapter, the "Physics" division of the text comes to an end, at least in the formal sense of the term.

5.0   IMPORTANCE OF THE SURFACE

Everything has a surface (or interface) and both physical and chemical properties depend on the nature (and breadth) of that surface, regardless of whether it is a bulk material or a nanoscale material. Catalytic activity, electrical resistivity, adhesion, protective coatings, and gas storage are just a few phenomena that depend on the nature of the surface. The surface is important regardless how large or how small the material is. The icy surface of a comet, for example, made of amorphous solid water and known to be efficient at storing and releasing large amounts of gas, has been found to be nanoporous and of high surface area [1]. In upcoming sections, we look into surfaces from a geometric viewpoint and we also look at nanomaterials from a structural viewpoint, once again; no chemical or energetic topics are presented for discussion. Engineering materials are briefly reviewed to provide a perspective for what is to unfold.

5.0.1   Background

Surface science is a discipline that is devoted to the investigation of two-dimensional structures that serve as the interface between physical phases. Surface science involves experimental and theoretical study of the physics, chemistry, and biology of surfaces. The surface of a material is the outermost boundary of any object or substance. The etymology of the word surface (from Old French *sur*, "above" + *face*, from the Latin *facia*, "appearance, form, figure") indicates as much. Irving Langmuir was one of the founders of the discipline of surface science. The advent of surface analysis techniques such as AFM, STM, XPS, Auger electron spectroscopy, LEED, EELS, and SIMS helped establish this field as one of great importance to all of science and technology. Surfaces perform numerous vital functions. They keep things in; they keep things out; or they allow the flow of material and energy across an interfacial structure. Surfaces are also capable of initiating or terminating chemical reactions, as in the case of catalysts.

5.0.2 Natural Perspective

When discussing nanoscience, references made to nature always yield important information and guidelines. Nature, as we have hinted so far, is the ultimate nano-technologist, at the nano-, the micro-, and the macroscales. At the microscale, single-celled organisms or cells existing within a larger body depend on passive diffusion to obtain nutrients and remove wastes. If cell volume becomes too large, simple diffusion is hindered. Cells developed mechanisms such as pumps to circumvent diffusion obstacles brought about by increases in volume. In the context of evolution, was the size of the cell limited by diffusion kinetics?

On the macroscopic scale, organisms have developed numerous means of increasing surface area to meet metabolic demand. Examples include leaves and the fractal nature of root networks of plants; hair on mammals; microvilli in the small intestine of animals; capillaries in the circulatory system, the liver, and the kidneys; the nervous system; alveoli in the lungs; and so forth. Red blood cells, for example, possess concave surfaces, an adaptation that enhances surface area. At the nanolevel, the level at which all biochemistry occurs, increased surface allows for a greater number of substrates to participate in reactions that are mediated by nanoscale catalytic proteins known as enzymes.

What are the benefits of increased surface area? Surface area in biological terms is a measure of the scale of exposure that a biological entity, whether cell or organism, has with its environment. At the macroscale, small animals have a higher surface-to-volume ratio than larger animals and therefore require a higher metabolic rate to balance heat loss to the environment. Once again, in the context of evolution, did surface-to-volume ratio play a role in limiting the size that a cell can achieve? Is the process of mitosis itself a manifestation of the need for more surface area coupled with the survival advantage that larger size offers? Is it any wonder that most of the cells in an organism are micron sized even though, overall, an organism can be quite large (e.g., the blue whale). The metabolic machinery of nature runs on nano power, function, and structure. The combined surface area of all enzymes must truly be an enormous number. We then ask a fundamental question. Were nanoscale phenomena the gatekeepers at the single- to multicellular threshold?

The surface is just a part of the overall equation. A nanoparticle, like its bulk counterpart, has substance and shape and exists, for the most part, within a specified set of physical dimensions. We begin the chapter with the surface, devoid of references to chemical structure or reactivity and physical properties. The next section introduces (or reintroduces) some basic materials science. Surface-to-volume ratio, specific surface area (surface-to-mass ratio), structure, and orientation follow suit to round out this chapter.

5.0.3 Inorganic Perspective

Surface area is important to inorganic materials, whether they are the size of stars and planets or as miniscule as nanosized catalysts. The surface plays a major role in any science and technology that we investigate or devise.

| EXAMPLE 5.1 | *Bulk Physical Example of a Surface Effect* |

Sublimation is the conversion of a solid directly to a gas without an intermediate liquid phase. For the purposes of this problem, ice sublimates at a rate of 1 cm·h^{-1} (temperature <0°C and low-pressure conditions). (a) How long would it take a cube of ice with volume equal to 1 m^3 to disappear? (b) for eight cubes derived from the original cube with the same total volume? (c) for a sphere with the equivalent volume?

Solution:
It will take 50 hours to "melt" the large cube. Within 1 m^3 of ice, there exist 50 "boxes" of ice of 1-cm wall thickness, similar to Russian Babuska dolls (2- × 2- × 2-cm cube in the center = 8-cm^3 volume plus 49 additional 1-cm-wall boxes that increase incrementally by 1 cm in the *x*, *y*, and *z* directions).
For the eight smaller cubes, each volume is equal to 0.125 m^3. Each cube has at its center the 8-cm^3 core plus 24 additional 1-cm-walled boxes. The total time to melt each small cube is 25 hours.
For the sphere of 1 m^3 equivalent volume, the radius is 0.620 m. Considering the innermost sphere of radius 1 cm, there are an additional 61 shells with thickness of 1 cm. It would take 62 hours to melt the sphere, a factor of ×1.24 over that of the cube of equivalent volume.

Why do ice cubes assume spherical shape when melting?

Inorganic catalysts are important to industry. The major objectives of catalytic processes are high yield, efficiency, and selectivity. Most inorganic catalysts are nanomaterials. Catalysts reduce the activation energy of a reaction without being consumed. The catalyst is able to impart free energy to a reaction that otherwise would not occur under milder conditions. Since we cannot create energy out of thin air, where does this excess energy come from? We find that the "excess energy" originates from the surface. Heterogeneous catalysis, a state in which the catalyst is a solid phase and the reactant is in the liquid or gaseous phase, is the type of catalysis that is most relevant to nanoscience.

Heterogeneous catalysts have high collective surface area, much like their biological enzymatic counterparts. Collective surface area is the sum of individual surface area of all the particles. Metal nanoparticles, for example, are supported on highly porous oxide materials, usually alumina or silica based. The alumina support also possesses high surface area. High surface area is required to drive the efficiency of catalytic processes, especially if industrial production rates are the goal. The rates of reaction depend intrinsically on the available surface area of the catalyst particles. For some reactions, catalyst size is tailor fitted to a particular reaction (e.g., nanocatalyst size is optimized), another feature of nanotechnology.

Bulk gold, as we all know, is a noble metal that is relatively inert to oxidation and other reactions. Nanoscale gold, on the other hand, is capable of catalyzing chemical reactions—a capability not characteristic of bulk gold or even gold with micron dimensions [2]. Nanosized gold particles on oxide supports and two-monolayer thick gold islands supported on titania exhibited significant catalytic activity towards the low temperature oxidation reaction of CO [2]. Another study showed that gold nanoparticles with low coordination number interacted synergistically with the support material to increase the rate of oxidation of CO at 273 K [3].

Not all implications of small size, however, are beneficial. For example, the inflammatory response in rats upon inhaling titanium dioxide was linked to increased particle surface area—smaller size [4]. The inflammatory response was linear over a range of particle size; 250-nm diameter particles showed 10% response but an over 30% inflammatory response was associated with 25-nm diameter particles. Other work has recently shown a similar dependence of inflammatory response to particle surface area [5]. The inflammatory response also depended on the type of material that was inhaled. A high-activity material such as SiO_2 showed nearly astronomical increases in inflammatory response as a function of increasing surface area.

5.0.4 The Nano Perspective

We are able to determine thermodynamic properties of materials by measuring macroscopic parameters such as pressure, temperature, and volume. The properties associated with any material are macroscopic manifestations of its composition. In addition to molecular makeup and the amount of substance in a material, the properties of nanoscale materials are influenced by size, shape, structure, and orientation. Optical properties in particular are influenced by orientation. Mechanical properties of composite materials depend on the orientation of materials within its matrix. Thermal conduction away from a source into a heat sink is accomplished quite efficiently by single-walled carbon nanotubes-anisotropic conductors of heat.

The past decade has witnessed the integration of biological nanomaterials with all standard classes of engineering materials: metals, semiconductors, ceramics, polymers, and composites. A brief review of these materials follows. All of these engineering materials can be expressed in nanoscale form. "We rebuild it. We make it faster. But we make it smaller."

5.1 ENGINEERING MATERIALS

Material structure and composition are highly integrated. One relies on the other. The structure of any material is based on its atomic and molecular makeup. Metals and alloys; elemental and compound semiconductors; ceramics, powders and glasses; and polymers and composites comprise the traditional types of engineering material categories. Any materials science course instructor will tell you as much, but because this is a course on nanoscience, biological and supramolecular materials must be added to the mix. Special materials based on carbon, like graphite, diamond, fullerenes, and nanotubes—all materials that possess strong directional covalent bonds—also play a major role in the development of advanced materials. A brief overview of the major categories of engineering materials is found in **Table 5.1.**

Delocalized metallic bonds, which involve electron sharing but are nondirectional, predominate in metals. Metals crystallize into three major forms: face-centered cubic (*fcc*), body-centered cubic (*bcc*), or hexagonal close packed (*hcp*). Ceramic materials, the silicates in particular, are held together by strong ionic bonds. Ceramics are more complex than metals and crystallize in many forms. Pure semiconductor materials like silicon crystallize in a diamond-cubic configuration

TABLE 5.1	*Common Engineering Materials*		
Material type	**Bonding structure**	**Physical properties**	**Examples**
Metals Alloys	Metallic bonds Delocalized (nondirectional) Close-packed structures at minimal energy	Group IB–VIII and heavy elements Crystal structure is *fcc, bcc,* or *hcp* CN: 6–12 Long-range order	Conducting, ductile, readily formable High melting with large thermal expansion Relatively reactive and high strength. Characterized by metallic luster	Cr, Mn, Fe, Co, Ni, Cu, Mo, Ru, Pd, Ag, Cd, W, Os, Ir, Pt, and more
Elemental semiconduc- tors *Intrinsic* *Extrinsic*	Covalent bonds (directional)	Group IVA Diamond cubic *fcc*, tetragonal structure CN: 4	Semiconducting	*Intrinsic:* Si, Ge, Sn (also B, Te) *Extrinsic* (dopants): *n*-type: P, As, Sb, Bi, Se, Te *p*-type: B, Al, Ga, In
Compound semiconduc- tors Oxides	Covalent/ionic	Ceramic-like materials Zinc-blende Rutile, anatase	Semiconducting High melting Some are semiconducting and transparent	AlSb, GaP, GaAs, GaSb, InP, InAs, InSb, ZnSe, ZnTe, CdS, CdTe, HgTe TiO_2 Indium tin oxide (ITO)
Ceramics Ionic solids	Ionic bonds (non-directional) Covalent (directional)	Less densely packed Crystalline Metal + C, N, O, P, or S	Insulating High melting Chemically stable Brittle	Al_2O_3, MgO, SiO_2, silicates, alumino- silicates, ZrO_2, SiC, WC, NaCl, $CaCl_2$, and more
Glasses		Noncrystalline Short-range order	Low melting, soft	Silicates, Na_2O
Carbon-based materials	Covalent bonds (directional) characterized by electron sharing	Group IVA Open structures Crystalline Amorphous	Insulating/conducting Semiconducting High melting	Graphite, SWNT, MWNT, Fullerenes, diamond, fibers, carbon black
Supramolecu- lar structures	Covalent bond backbone precursors, coordination bonds, hydrogen bonds Compounds held together by intermolecular forces	Group IVA–VIIA Light elements C, H, O, N, S, P + minerals Self-assembly capability	Known as "soft matter" Many forms decompose under less than mild conditions. Formed under thermodynamic control conditions rather than kinetic conditions	Biological chemicals like DNA, proteins, lipids. Micelles, self-assembled monolayers Precursors are organic chemical
Polymers	Covalent backbones held together by weak inter- molecular forces	Group IVA–VIIA Light elements C, H, O, N, S, F	Extremely wide range of physical properties. Can be insulating or conducting Polymers are low-melting; some possess high chemical reactivity; others inert. Poor packing	Polyethylene (PE) Polyaniline (PAN) Polyvinyl chloride (PVC) Polymethyl-methacrylate (PMMA) Teflon

TABLE 5.1 (CONTD.)	*Common Engineering Materials*			
Material type	**Bonding structure**		**Physical properties**	**Examples**
Composites	All classes of bonding are represented in composite materials	All elements are represented in composites. The matrix is usually a polymer resin containing metal/ceramic/carbon fillers	Properties are variable and encompass the entire range of physical properties	Graphite epoxy resins Kevlar Fiberglass Wood Concrete
Biological materials	All classes of bonding are represented in biological materials	Group IVA–VIIA Light elements C, H, O, N, S, P+ minerals	Biomaterials are capable of directed self-assembly Denature easily under nonideal conditions Physical properties span a large range and depend on structure	Proteins Lipids Carbohydrates Nucleic acids Shell, bone, teeth Wood

with covalent bonding, similar to carbon-based systems. Multicomponent semi-conductors are similar in structure to ceramic materials. Polymers possess complex structure that may contain a wide range of noncrystalline material and are held together by covalent and weaker secondary forces (intermolecular bonds) such as van der Waals forces. Composite materials, another major category of engineering materials, are more complex due to the contribution of at least two, but usually more, categories of the major engineering materials.

The engineering content of this section was guided by an excellent materials text written by J. F. Shackleford: *Introduction to Materials Science for Engineers* (Prentice Hall, 1996). Although written before the *Nano Age* exerted its impact, it served as an excellent source for the development of this section.

5.1.1 Metals and Alloys

The most dominant metal used in engineering today is iron, mostly in the form of steel. The Earth's core consists mostly of iron. Iron also plays an important role in the fabrication of nanoscale materials. Carbon nanotubes, for example, can be formed from iron catalyst particles. Iron plays a vital role in living things as well in the form of cations. The iron cation is the central metal of hemoglobin, upon which aerobic organisms depend. Therefore, from the atomic level to the nanoscale level to the bulk engineering level and at the geophysical level, iron is an extremely fundamental and important material. Aluminum metal also has widespread use. At the nanoscale, aluminum metal provides the substrate for the fabrication of porous alumina templates. In living things, aluminum is required at extremely low levels but can be quite toxic at higher levels.

Nanometals come in many forms and perform diverse functions. Gold clusters comprising handfuls to thousands of atoms (a.k.a. quantum dots) can be placed in electronic arrays or act in the capacity of catalysts. Nanoshells of Au or

Ag formed on polymer or silica nanoparticles have demonstrated tunable optical response [6–8]. Recently, gold nanoshell materials have proved to be a valuable asset in treating cancer [9]. Gold substrates have shown extreme utility and adaptation in self-assembly processes that include alkanethiols. Nanometals have found their way into tennis racquets, other nanocomposite materials, and thin films. Metals and alloys containing nanograins, for example, exhibit superior physical properties. Fuel cell electrocatalysts made of nanometals have demonstrated efficiency and good performance.

Although metals are characterized by good electrical and heat conductivity, ductility, malleability, reflectance, and strength, many of these traits break down at the nanoscale. If made small enough, the electronic behavior of a metal starts to resemble that of a semiconductor. If made small enough, gold starts to act like a catalyst. The color of nanogold is strikingly different from the color of its bulk form.

5.1.2 *Semiconductors*

Semiconductors, unlike metals, have a band gap. The band gap existing between the valence and conduction bands, has no electronic states. The electrical conductivity of a semiconductor material lies between that of an insulator (e.g., a large band gap material) and a metal (e.g., a material with no band gap). Semiconductor materials without impurities or defects are called *intrinsic* semiconductors and are found within the metal-insulator boundary on the periodic table. Examples include silicon, germanium, tin, boron, and tellurium. Semiconductor properties, the band gap in particular, are manipulated by addition of dopants: impurities able to donate charge carriers in the form of electrons (*n-type*) or holes (*p-type*). These are known as *extrinsic* semiconductors.

Many semiconductor materials fall into the class of covalent-ionic solids. Examples include titanium dioxide (TiO_2), cadmium sulfide (CdS), and zinc oxide (ZnO). More complex materials such as indium tin oxide (ITO) are "semiconducting materials" that are both conducting and optically transparent. ITO is a mixture of indium(III)–oxide (In_2O_3) and tin(IV)–oxide (SnO_2). When indium is doped with tin, the additional electron becomes "free" and ITO becomes conducting.

Titanium dioxide (titania or titanium-IV oxide, TiO_2) is of great importance to nanoscience, nanotechnology, and industry. Highly porous nanocrystalline TiO_2 sensitized by transition metal or organic dyes is a major component of the Grätzel cell, a photocell capable of ca. 7% efficiency. Thin films ranging from 0.2 to 5 μm in thickness consist of nanosized particles 3–10 nm in size with overall specific surface of 50–150 $m^2 \cdot g^{-1}$ are formed by dip-coating methods. TiO_2 formed by sol–gel methods and sintered under 800°C yields 10- to 50-nm diameter particles with surface ca. 50 $m^2 \cdot g^{-1}$. The photocatalytic activity of TiO_2 is affected by surface area, degree of crystallinity, size, and structure (Li).

Miniaturization of bulk semiconductor materials implies, as for metals, that physical properties undergo change. Confinement, perhaps, is one of the most celebrated examples of the visible change observed in semiconductor quantum dots as particle size is diminished. The size-dependent emission of light by CdSe quantum dots provides a vivid example of this process.

5.1.3 Ceramic and Glassy Materials

Ceramic materials by definition are ionically bound, hard materials that are electrically and thermally insulating. Examples of important ceramic materials include Al_2O_3, Si_3N_4, MgO, SiO_2, Na_2O, CaO, and ZrO_2. Ceramic materials are chemically stable under a wide variety of extreme conditions: chemical, thermal, electrical, and physical. They generally possess high melting points, a characteristic that makes them suitable as refractory materials. Bulk ceramic materials, however, are quite brittle. Thermal expansion coefficient matching between a metal and its oxide coating can be problematic. The strength and fracture toughness of ceramics is enhanced with the incorporation of nanomaterial substituents.

Ceramics exist in crystalline form or noncrystalline form. Zeolites are crystalline ceramics that contain silica, alumina, and alkali metal constituents. Noncrystalline ceramics include the broad family of glasses. The interaction of a host medium, usually a glass, with nanomaterials, usually a metal, gives rise to the dichroism phenomenon of the Lycurgus cup. Ceramics also exist in the nanoform. Silica beads; the cavities in zeolites, aerogels, and xerogels; and the pore channels in anodically formed aluminum oxide are examples of nanomaterials that are made of ceramic materials.

5.1.4 Carbon-Based Materials

Special materials based on carbon form a very important class of engineering materials; as a matter of fact, carbon-based materials are integral components of the materials described in the next three sections. The primary allotropes of carbon are shown in **Figure 5.1.** Most people are familiar with the general forms of carbon, like diamond, graphite, and amorphous carbon. However, an exciting new class of carbon materials has been discovered in the past few decades: the fullerenes and their derivatives. A special section is devoted to these materials as well as to diamondoids and diamond-like materials that have gained significant attention over the past few years.

5.1.5 Polymers

Metals, semiconductors, ceramics, and glasses are inorganic materials. Polymers are made of organic materials and are usually thrown into the category of soft matter. Polymer chemistry is rightly called a spin-off of organic chemistry. As a consequence, most polymeric materials consist of the same elements that define organic chemistry: C, H, O, N, F, Cl, S, P, and Si. The age of modern engineering materials came about because of polymers. Plastics, teflon, lotions, membranes, tires, and thousands of other raw materials and products are based on polymer materials. Thermoplastics (materials that become less rigid upon heating) and thermosetting polymers (materials that become stiffer with heating) form two major classes of polymers.

Examples of nanoscale polymers are numerous. A block polymer is made of one kind of constituent macromolecule (or monomer) that is connected directly by or through a constitutional unit that is not part of the block (e.g., an interlink) [10]. When blocks are made of a different species of polymers, the term *block co-polymer*

FIG. 5.1

Allomorphs of carbon are depicted. (a) Diamond forms in a tatrahedral (sp³ bonding) arrange-ment where each carbon is bonded to four others. (b) Graphite is the most stable form of carbon. It forms in planes that consist of hexagonal arrangement of carbon atoms. The bond-ing in graphite is sp². Van der Waals forces hold the planes together. (c) Another form of dia-mond called lonsdaleite has more hexagonal character to its structure. It is believed to form after a meteor impact that transforms graphite into lonsdaleite while creating the hexagonal form of graphite. (d)–(f) Fullerenes were discovered in 1985 by Richard Smalley and his group. Fullerenes are carbon's representative in the zero-dimensional class of nanomaterials (chapters 7 and 9). (g) Amorphous carbon does not have any crystalline structure (no long-range order). Much of amorphous carbon contains small domains of crystalline graphite (also called turbostratic graphite) with some level of amorphous carbon that holds them together. Coal and soot are amporphous carbons. (h) Carbon nanotubes are called fullerene pipes and comprise (arguably) one of the most remarkable of the carbon allotropic class of materials—or of all materials, for that matter. More will be said about carbon nanotubes in chapter 9. Glassy carbon is a nongraphitizing carbon that is used in electrodes and for high-temperature cruci-bles. Glassy carbon is a purely isotropic form of graphite. Carbon nanofoams are considered to be an allomorph of carbon that consists of a three-dimensional structure of carbon clusters. Nanofoams were discovered in 1997 by A. V. Rode et al.

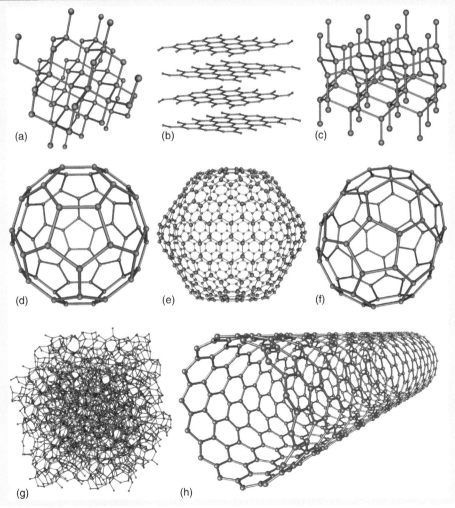

(a)   (b)   (c)

(d)   (e)   (f)

(g)   (h)

is used to describe the system. An application of block co-polymers—BCPL, is a type of lithography—was introduced briefly in chapter 4. Latex beads made of polystyrene are used in the manufacture of photonic crystals and in nanosphere lithography. Polymeric nanoparticles are used as a template to fabricate gold nanoshells. Membranes made of polycarbonate with track-etched pore channel matrices serve as templates to fabricate devices such as nanoelectrodes [11]. Lastly, polyvinyl alcohol (PVA), polyvinyl chloride (PVC), polystyrene (PS), polymethyl methacrylate (PMMA), polyanaline (PAN), polyethylene (PE), and so many more serve as the polymeric matrix for advanced nanocomposite materials, and all can be fabricated in nanoscale dimensions.

5.1.6 Biological Materials

Nanoscale materials transcend the entire spectrum of biological substances. Macroscale materials like wood and geologically altered materials like coal have been exploited by civilizations for millennia. This is also true for microscale materials like diatoms. DNA and various proteins are relatively new additions to the domain of engineering materials.

All biological materials are fabricated from the bottom up, and it is safe to assume that all macroscopic biological materials are comprised of nanocomponents. Biomimetics is the practice of copying nature. There are many examples of natural phenomena and products that we strive to emulate: polymers (in the substructure of nacre, e.g., the abalone shell); structural elements (wood, ligaments, and bone), electrical conduction (in eels and nervous systems), photoemission (by deep sea fish and glow worms); photonic crystals (butterfly and bird wings); hydrophobic surfaces (the lotus leaf and human skin); chemical sensors (the silkworm moth pheromone detection capability); adhesives (geckos' feet); high tensile strength fibers (spider silk); and artificial intelligence and computing ability (the human brain); the list goes on and on. The most remarkable contributions to engineering materials originate directly from biological nanoscale materials. Biological nanomaterials will be discussed in more detail in chapters 13 and 14.

5.1.7 Composites

Composites are materials that contain selected combinations of two or more of the other major classes of materials previously mentioned in this section: metals, semiconductors, ceramics, polymers, and, now, biological materials. A composite (from the Latin *compositus, componere*, "to put together" + "to place") is a material made of various parts or elements or other recognizable constituents—in essence, all the elements of the periodic table. There are many composites with which we are familiar. For example, fiberglass contains glass fibers embedded in a polymer matrix and collectively exhibits better properties than that of its components taken separately. The glass fibers offer high strength and the polymer material offers ductility. Together, they are considered to be a structural material that makes an efficient (and affordable) insulator. Concrete is an aggregate composite within which rock and sand interact with a silica cement glue matrix to form our bridges, buildings, and roads. Wood is a composite material that exhibits fiber reinforcement of its structure. Composite materials called graphite–epoxy resins are used in aircraft and spacecraft wings, bodies, and components. In addition to high strength, graphite composite materials offer enhanced thermal conduction, radiation ablation resistance, stealth technology, and electrical conduction. The bullet-stopping ability of Kevlar is well known.

An entirely new industry is on the horizon. The industry involves the research, development, application, and marketing of nanocomposite materials. In actuality, nanocomposites have already made an entrance into the global composite market. Multiwalled carbon nanotubes (MWNTs) are used routinely as fillers in nanocomposites and carbon nanotubes in tennis racquets.

5.2 PARTICLE SHAPE AND THE SURFACE

The surface of a material depends on its size and geometric shape. There is no attempt in this section to delve into atomic–molecular details or the chemical or physical nature of nanomaterials. Three factors that relate to surface area will be explored: (1) the effect of subdivision of a parent material on the overall surface area, (2) the effect of particle shape on surface area, and (3) the concomitant increase in surface-to-volume ratio as individual particles assume smaller dimensions.

Nanomaterials have a significant portion of atoms existing at surfaces, of which there are two general types: the exterior surface and the interior surface. Exterior surfaces comprise atoms that exist on the outside surface and edges. The external surface area of a powder, for example, takes into account the roughness of the constituent particulate surfaces. Generally speaking, cavities that are wider than they are deep are considered to contribute to the exterior surface roughness, as opposed to the inverse case in which they are considered to be pores and have internal surface area. Nanomaterials of importance that belong to the category of the exterior nanosurface type include metal and semiconductor clusters and colloids. Interior nanosurfaces refer to the surfaces of pore channels and cavities and the surfaces between grain boundaries of a crystalline or polycrystalline solid. Interior surfaces are represented by inorganic porous oxides found in anodically formed alumina, natural and synthetic zeolites, amorphous carbons, and silicas.

Collective surface area and surface-to-volume ratio are inversely proportional to the size of the particle. We define collective surface area as the sum total of surface area of individual nanomaterials—especially if subdivided from a bulk parent material. Collective surface area is an extensive (additive) property. Surface-to-volume (S/V) ratio is defined as the ratio of the surface area (in square meters) to the volume of the same particle (in cubic meters). As opposed to the collective sense, S/V ratio is relevant to a single entity—e.g., a particle or pore channel. The S/V ratio therefore is an intensive property. Although "surface-to-volume" is technically described in units of reciprocal meters (m^{-1}), S/V (or N_s/N_v) ratios are usually listed as a dimensionless number.

The volume of a sphere increases as the cube of the radius, r, while the surface area increases as the square of the radius. As r is decreased, the proportion of surface units (area or atoms) increases with respect to its volume. With regard to the collective sense, as in the case of a single bulk material divided into multiple clusters, surface area increases geometrically as more new surface is created,

FIG. 5.2 *(a) The number of cubes cut from a parent cube required to produce a specific area in square meters is plotted. It is clear from the figure that surface area increases quite quickly as cubes become smaller and smaller. (b) The proportion of surface atoms as a function of cube dimensions is shown graphically. Once again and without any chemical knowledge, an exponential relationship illustrates just why nanomaterials may really be unique.*

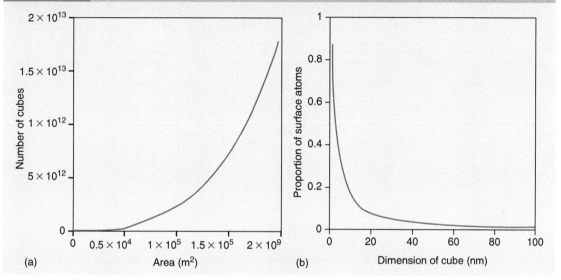

leading to incredibly large values (**Fig. 5.2a**). The trend in surface-to-volume ratio is depicted in **Figure 5.2b.**

5.2.1   Exterior Surface and Particle Shape

Without regard to the details involved in packing atoms or molecules, a cube of some material that is 1 m on each side has surface area equal to 6 m². If this cube were broken perfectly into smaller cubes that are 0.1 m on each side, resulting in 1000 additional cubes, the collective surface area becomes 60 m². For cubes with 1-cm edges, a million such cubes can fit into a cubic meter, and the collective surface area increases to 600 m². For cubes with 1-mm edges, the surface area expands to 6000 m². The trend is illustrated in **Figure 5.3.**

Surface area depends on particle shape in addition to size. Cubes, rectangular solids, spheres, spheroids, cylinders, pyramids, and discs are common shapes that nanomaterials assume. Tetrahedrons, icosahedrons, dodecahedrons, cuboctahedrons, tubes, hollow spherical structures, astral structures, and dendritic arrays are more examples of the incredible diversity found in synthetic and naturally occurring nanomaterials (**Fig. 5.4**).

A cube with the same volume as a sphere has a higher surface area. Cylindrical solids, with height h and diameter d, cover a wide range of surface area. The disk ($h \ll d$) and the wire ($h \gg d$) potentially have the highest surface area among the geometric series listed previously (**Fig. 5.5**). The sphere is the most remarkable geometric solid. It has an infinite number of symmetry axes. As result, a sphere

FIG. 5.3

A visual image of the concept of collective surface area is always helpful in acquiring the proper perspective. As one can readily see, it does not take too many divisions of the parent cube to reach a level of enormous surface area. Simple geometric progression from successive division of a parent cube illustrates the concept of collective surface area. If cubes are divided into eight equivalent cubes with each step, the surface area doubles after each division. For n equal to the number of cubes, surface area proceeds from 6 m² (for n = 1) to 12 (n = 8), 24 (n = 64), 48 (n = 512), and 96 m² (n = 4096). The length of a side of the n = 4096 cube is only 6.25 cm. One can only imagine the total surface area if these cubes are made to be 1 nm on a side.

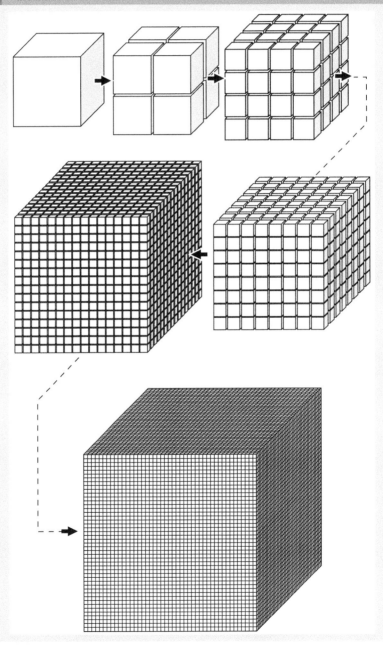

EXAMPLE 5.2	*Determination of the Collective Surface Area of Nanocubes*

How many cubes 1 nm on each side can be carved out of a cubic parent 1 m on each side? Find the collective surface area of the nanometer-sized cubes.

Solution:
This calculation is straightforward. We need to calculate the number of nanometer-sized cubes that are possible from a 1-m sized cube, find the area of one of the smaller cubes, and then sum for the total surface area. The surface area of a 1-m cube is 6 m².

Volume of nanocube: $\left(1\times10^{-9}\,\text{m}\right)^3 = \left(1\times10^{-27}\,\text{m}^3\right)$

Surface area of nanocube: $\left(1\times10^{-9}\,\text{m}\right)^2\times 6 = \left(6\times10^{-18}\,\text{m}^2\right)$

Number of nanocubes per cubic meter: $\dfrac{1\,\text{m}^3}{\left(1\times10^{-27}\,\text{m}^3\right)} = 1\times10^{27}$ nanocubes

Total surface area:

$$\left(1\times10^{27}\,\text{nanocubes}\right)\left(\frac{6\times10^{-18}\,\text{m}^2}{\text{Nanocube}}\right) = 6\times10^{9}\,\text{m}^2 = 6,000\,\text{km}^2$$

The total surface area of the nanocubes is a billion times that of the 1-m cube. This is equivalent to more than 2000 square miles—an area bigger than the state of Delaware (1954 sq. mi.). The power of nano!

made of a real material is characterized by an infinite number of degenerate energy states (e.g., all energy levels are equivalent). The sphere represents the lowest energy configuration of solid materials. The sphere also happens to represent the surface minimum in the continuum of geometric solids—a coincidence?

FIG. 5.4	*Nanomaterials can assume any shape imaginable—whether of the two-dimensional variety depicted in the first row, polyhedral shapes depicted in the middle row, or ellipsoidal forms depicted in the bottom row. Spheres, discs, wires, dots, and two-dimensional planar materials abound at the nanoscale.*

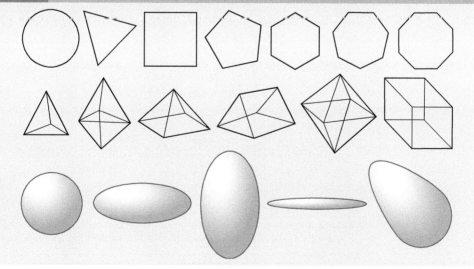

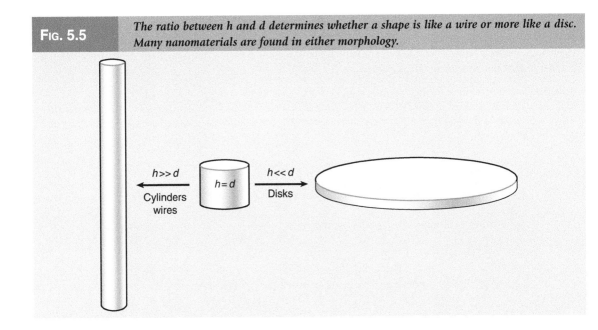

FIG. 5.5   *The ratio between h and d determines whether a shape is like a wire or more like a disc. Many nanomaterials are found in either morphology.*

Nanomaterials are found in all of the seven basic crystal systems and consequently in all the Bravais lattices. Crystalline forms derived from Bravais solids are also well represented in nanomaterials. Crystalline nanomaterials can assume the basic shapes of any of those based on Bravais lattices. Bravais lattices are discussed in more detail later.

EXAMPLE 5.3   *Shape Dependence of Surface Area*

Determine the relative surface areas of a cube (a^3) with the same volume as a sphere. Derive an expression that relates the surface area of a sphere to that of a cube of equal volume.

Solution:
Set the volume of the sphere equal to the volume of a cube (a^3); solve for volume in terms of a, and then calculate the surface area for each shape.

Volume of cube: $V_c = a^3$
Surface area of a cube: $S_c = 6a^2$
Volume of a sphere: $V_s = (4\pi/3)r^3$
Surface area of a sphere: $S_s = 4\pi r^2$
$V_{sph} = V_{cube} \rightarrow (4\pi/3)r^3 = a^3 \rightarrow r = 0.62a$
$S_{sph}(a) = 4\pi(0.62a)^2 = 4.83a^2$
Ratio of S_c to S_s: $6a^2/4.83a^2 = 1.24 \rightarrow S_c = 1.24\, S_s$

When comparing a sphere to a cube of the same volume, the sphere will have less surface area. Although a simple geometric phenomenon, this relationship has important consequences in nanoscience.

| **EXAMPLE 5.4** | *Surface Area of Right-Cylindrical Shapes of Constant Volume* |

Determine the surface area of right cylinders of unit volume where $V_{cyl} = 1$. Derive an expression that relates the surface area of the following cylinders normalized to that of a sphere with volume equal to 1:

    (a) $d = h$
    (b) $d \ll h$ (assume $h = 100$)
    (c) $d \gg h$ (assume $h = 0.01$)

Solution:
When the diameter d of a right cylindrical material is much greater than the height h of the cylinder, the shape is a disc. Conversely, if h is significantly greater than d, the shape is equivalent to that of a wire. Both types of shapes are exhibited by important nanomaterials. Use $V_{Cyl} = \pi r^2 h = 1$ and $S_{Cyl} = 2\pi r^2 + 2\pi r h$ and that $d = 2r$.

    $V_{cube} = 1.000$, $S_{cube} = 6.000$
    $V_{sphere} = 1.000$, $S_{sphere} = 4.836$
    $V_{Cylinders} = 1.000$, $S_{Cyl} = ??$

    (a) $h = d$, $r = 0.5419$, $S_{cyl} = 5.535$, normalized to the sphere = 1.14. When $d = h$, this cylinder has the minimum surface area within the family of cylinders with $V_{cyl} = 1$.
    (b) $h = 100.0$, $r = 0.5642$, $S_{cyl} = 35.47$, normalized to the sphere = 7.33. In the case of a wire, the second term in the surface area equation $S_{cyl} = 2\pi r^2 + 2\pi r h$ becomes more important as the height (length) of the wire increases.
    (c) $h = 0.01000$, $r = 5.642$, $S_{cyl} = 200.4$, normalized to the sphere = 41.4. In the case of a disk, the first term in the surface area equation $S_{cyl} = 2\pi r^2 + 2\pi r h$ becomes more important as the height, h, of the cylinder gets smaller.

5.2.2 Interior Nanoscale Surface Area

According to the International Union of Pure and Applied Chemistry (IUPAC), the definitions of pore size are as follows [12]: *Micropores* have widths smaller than 2 nm, *mesopores* have widths between 2 and 50 nm, and *macropores* have widths larger than 50 nm. We introduced this nomenclature earlier.

Obviously, all of the IUPAC designations could be considered as nanoscopic pores, depending on your point of view. The classification is somewhat confusing because the term "macropores" implies something quite larger in scale. Examples of microscopic pores are exhibited by the zeolites, which have 2-nm sized or less pores; mesoscopic pores are represented by the pore channels in anodically formed alumina that range from a few to a few hundred nanometers. Pores, cavities, and channels are found in crystalline structures (zeolites), assemblies of aggregated materials (ceramics), or in polycrystalline materials like anodically formed alumina.

Some degree of pore structure, whether in the form of cavities, channels, interstices, or grain boundaries, is inherent in most solids. Defects in solids such as voids and vacancies are cavities—albeit very small ones. A porous solid is defined as a solid with pores that are deeper than they are wide. In materials designed overtly to be porous, characteristics such as void volume (or porosity) are important considerations. These include polymers, ceramics, metals, and semiconductors. Pores can be closed or open. Closed pores influence properties such as density, electrical conductivity, and thermal conductivity. Open pores influence transport properties and are able to selectively pass materials based on size, whether gaseous or liquid, through a membrane. Void volume (porosity) ε of a porous material is defined as

$$\varepsilon = \frac{V_{\text{Pores}}}{V_{\text{Material}}} \tag{5.1}$$

Just like with exterior surface area, many chemical reactions and phenomena take place inside pores. Pores are able to confine substances within their boundaries and thereby influence kinetic control over the course of a chemical reaction. Recall that a substrate is confined within the active pocket of an enzyme—a very powerful means of initiating reactions under mild conditions. An entire industry is

---

EXAMPLE 5.5   *Surface Area of a Common Porous Material*

Determine the surface area in terms of cm^2 of an alumina membrane (10.0 cm $\times$ 10.0 cm $\times$ 40.0 μm thickness) with (a) no pores; (b) pore channels with diameter, $d_1 = 250$ nm, 55% porosity of membrane; and (c) diameter, $d_2 = 2.50$ nm, 55% porosity of membrane. Consider the pore channels to be right cylinders that are all oriented normal to the surface of the membrane.

Solution:
This example is solved primarily by meticulous bookkeeping.

(a) Surface area of membrane with no pores
    Apparent volume of membrane material:
    $V_{\text{Material}} = (10.0 \text{ cm})(10.0 \text{ cm})(40.0 \times 10^{-4} \text{ cm}) = 0.400 \text{ cm}^3$
    Surface area of the membrane: $S_{\text{Material}} = (10.0 \text{ cm})(10.0 \text{ cm})(2) + (40.0 \times 10^{-4} \text{ cm})(10.0 \text{ cm})(4) = 200 \text{ cm}^2$
(b) Total surface area of membrane with 250-nm diameter pores, $\varepsilon = 0.55$
    Total porous volume: $V_{\text{Pores}} = \varepsilon V_{\text{Material}} = (0.55)(0.40 \text{ cm}^3) = 0.22 \text{ cm}^3$
    Single pore channel volume: $V_{pc} = \pi r_1^2 T = \pi (125 \times 10^{-7} \text{ cm})^2(40.0 \times 10^{-4}\text{cm}) = 1.96 \times 10^{-12} \text{ cm}^2$
    Total number of pore channels: $N = V_p/V_{pc} = (0.22 \text{ cm}^3)/(1.96 \times 10^{-12}\text{cm}^3) = 1.12 \times 10^{11}$ pore channels
    Single pore channel surface area: $S_{pc} = \pi dT = \pi (250 \times 10^{-7} \text{ cm})(40 \times 10^{-4} \text{ cm}) = 3.14 \times 10^{-7} \text{ cm}^2$
    Total surface area of pore channels: $S_p = S_{pc}N = (1.12 \times 10^{11})(3.14 \times 10^{-7} \text{ cm}^2) = 3.52 \times 10^4 \text{ cm}^2$
    Total surface area of membrane not counting contribution from the pores:
    $S_m = (1 - V_p)(2L \times 2) + (T \times L \times 2) + (T \times L \times 2) = 0.45(200) + 4L(40.0 \times 10^{-4} \text{ cm}) = 90.2 \text{ cm}^2$
    Total surface area of material: $S_T = S_p + S_m = 3.53 \times 10^4 \text{ cm}^2$
    Obviously, the contribution to the total surface area from the membrane is negligible
(c) Total surface area of membrane with 2.5-nm diameter pores, $\varepsilon = 0.55$
    Total porous volume: $V_p = \varepsilon V_{\text{app}} = (0.55)(0.40 \text{ cm}^3) = 0.22 \text{ cm}^3$
    Single pore channel volume: $V_{pc} = \pi r_1^2 T = \pi (1.25 \times 10^{-7} \text{ cm})^2(40.0 \times 10^{-4} \text{ cm}) = 1.96 \times 10^{-16} \text{ cm}^2$
    Total number of pore channels: $N = V_p/V_{pc} = (0.22 \text{ cm}^3)/(1.96 \times 10^{-16} \text{ cm}^3) = 1.12 \times 10^{15}$ pore channels
    Single pore channel surface area: $S_{pc} = \pi dT = \pi (2.50 \times 10^{-7} \text{ cm})(40 \times 10^{-4} \text{ cm}) = 3.14 \times 10^{-9} \text{ cm}^2$
    Total surface area of pore channels: $S_p = (1.12 \times 10^{15})(3.14 \times 10^{-9} \text{ cm}^2) = 3.52 \times 10^6 \text{ cm}^2$
    Total surface area of membrane not counting the pores:
    $S_m = (1 - V_p)(L \times W \times 2) + (T \times L \times 2) + (T \times W \times 2) = 0.45(200) + 2(40.0 \times 10^{-4} \text{ cm})(L + W) = 90.2 \text{ cm}^2$
    Total surface area of the material: $S_T = S_p + S_m = 3.52 \times 10^6 \text{ cm}^2$
    The surface area of the membrane becomes even less important

What are the respective specific surface areas in terms of square meters per gram if the density of the alumina is 2.0 g·cm^{-3}?

based on the exploitation of the internal surface area of micro- and nanomaterials. Critical biological functions rely on cavities, pores, channels, and active pockets.

5.3 SURFACE AND VOLUME

By now you, the student, are certainly becoming aware of the power of the nanoscale, at least within the geometric frame of reference presented so far. However, the geometric development has not yielded any clues concerning physical properties and we have advertised at length the remarkable properties that nanomaterials possess based on smaller size and increased surface area. The following surface-to-volume ratio discussion will help pry open that door, if ever slightly. It is actually all about the surface atoms, which are the *first responders* in a reaction and form the interface with the exterior environment. The surface atoms are responsible for chemical behavior of a material. This is true for individual particles that assume new properties as size is reduced or for bulk materials that comprise nanosized grains. An example of the former includes depressed melting points for metal nanoparticles and of the latter enhanced tensile strength in steel. **Figure** 5.6 depicts the surface atom fraction of pseudospherical iron particles as size is reduced, ranging from 35 nm down. Particles over 35 nm have a significantly lower percentage of surface atoms.

FIG. 5.6 *Along with the same theme as before, the surface atom percentage (compared to the total volume) of pseudospherical metal particles is shown graphically. Pseudospherical implies that the metal particle is not in the form of a perfect sphere, but rather in the form of a cuboctahedron or other pseudospherical shape. It is apparent from the graph that the ratio of surface atoms increases dramatically as particles become smaller. In many nanomaterials, some actually can be considered to be all surface.*

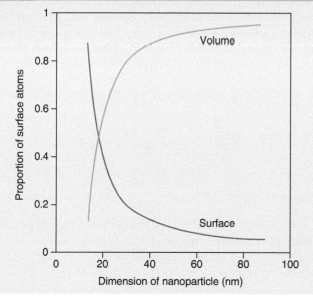

FIG. 5.7 *Gold atoms are arranged in a simple cubic crystal form, where the atoms in each layer are right on top of the atoms in the previous layer (called in register). Brute force implies direct measurement of edges, etc. to derive the ratios we seek with a minimal of sophistication.*

Source: Courtesy of Professor Anil K. Rao, Department of Biology, the Metropolitan State College of Denver.

The drumbeat delivering the message of surface area and its importance in nanoscience continues in this section. In the bulk world, properties such as electrical conductivity and optical response are averaged over a large area or mass. In the world of nanoparticles these properties are subject to change in ways that are quite remarkable (e.g., the Lycurgus cup). Specific surface area (square meters per gram) is an important parameter in the field of catalysis. The effects of particle size on the rate of catalyzed chemical reactions are well documented. The ratio of the number of atoms or molecules on the surface of a material to the number of atoms or molecules that exist in the volume of that material is of keen interest to us. At what point does this ratio become significant? In example 5.6, we count the number of surface atoms in a cube of a material called hypothetical gold—a structure formed in a primitive cubic lattice (**Fig. 5.7**).

5.3.1   *Geometric Surface-to-Volume Ratio*

Although we have touched upon the relationship the surface has with its volume for several standard geometric shapes, we now explore the finer details of that relationship.

5.3.2   *Specific Surface Area*

According to the *IUPAC Compendium of Chemical Technology* (2nd edition, 1997), specific surface area (usually symbolized by a, S, or A_s) is defined as the surface area divided by the mass of the relevant phase when the area of the interface between two phases is proportional to the mass of one of the phases (e.g., for a solid adsorbent, an emulsion, or an aerosol). The units of specific surface area

EXAMPLE 5.6	*Determination of the Percent of Surface by the Method of Brute Force*

This calculation features a material called "hypothetical gold." Calculate the number of surface atoms, N_s, edge atoms, N_e, and volume atoms, N_v, and the surface-to-volume atomic ratio N_s/N_v as size is decreased by order of magnitude increments, starting with a 1-cm^3 chunk of hypothetical gold (AAA...with in register packing throughout—a simple cubic structure—depicted in **Fig. 5.7**). Au: covalent radius = 0.144 nm (Note: Although gold exists in cubic-close-packed (ccp)/face-centered-cubic (fcc) structure, for this example we assume that the covalent radius is the same for both the real and hypothetical gold); atomic mass = 196.967 g·mol^{-1}, density ρ = 19.31 g·cm^3.

Solution:
First calculate the number of gold atoms per meter (and per centimeter, millimeter, micrometer, nanometer, etc.). Take the cube of your answer to find the volume in terms of number of atoms and the square × 6 to find the surface area in terms of the number of atoms. The calculation of N_s/N_v is straightforward.

A sample calculation is given and the remaining values and ratios are summarized in **Table 5.2**.

$$\text{Number of Au atoms in 1 m: } \frac{1\,\text{m}}{0.288\times10^{-9}\,\text{m}} = 3.47\times10^9 \text{ Au atoms}$$

$$\text{Volume of Au atoms: } (3.47\times10^9)^3 = 4.19\times10^{28} \text{ Au atoms}$$

$$\text{Surface area Au atoms: } (3.47\times10^9)^2\times6 = 7.22\times10^{19} \text{ Au atoms}$$

$$\text{Surface – volume ratio: } \frac{7.22\times10^{19}}{4.19\times10^{28}} = 1.72\times10^{-9}$$

are square meters per gram of material (m$^2\cdot$g^{-1}). Specific surface area is simply the ratio of S/V to the density of a material:

$$S_S = \frac{S_m}{\rho V_m} \qquad (5.2)$$

TABLE 5.2	*Surface–Volume Parameters for Hypothetical Gold of Diminishing Volume*

Cube dimension	Edge atoms (N_e)	Volume atoms (N_v)	Surface atoms (N_s)	Surface–volume ratio (N_s/N_v)
Meter	3.47×10^9	4.19×10^{28}	7.22×10^{19}	1.72×10^{-9}
Centimeter	3.47×10^7	4.19×10^{22}	7.22×10^{15}	1.72×10^{-7}
Millimeter	3.47×10^6	4.19×10^{19}	7.22×10^{13}	1.72×10^{-6}
Micrometer	3.47×10^3	4.19×10^{10}	7.22×10^7	1.72×10^{-3}
500 nm	1.74×10^3	5.23×10^9	1.82×10^7	3.47×10^{-3}
100 nm	3.47×10^2	4.19×10^7	7.22×10^5	0.0172
50 nm	1.74×10^2	5.23×10^6	1.82×10^5	0.0347
25 nm	86.0	6.54×10^5	4.44×10^4	0.0678
10 nm	34.7	4.19×10^4	7.22×10^3	0.172
4.90 nm	17	5.23×10^3	1.81×10^3	0.346
2.88 nm	10	1000	488	0.488
2.02 nm	7	343	218	0.635
1.15 nm	4	64	56	0.875

Note: The ratio of surface atoms to all the atoms of the volume increases from approximately one-billionth to 88%.

EXAMPLE 5.7 *Determination of Surface-to-Volume Ratio Relations of Selected Geometric Solids [13]*

Derive some generalized formulae of surface-to-volume ratios for the following shapes: cube, sphere, and cylinder with diameter d and height h where $d = h$, $d \ll h$, and $d \gg h$.

Solution:
The units of surface-to-volume ratio are in terms of reciprocal meters (m^{-1}), centimeters (cm^{-1}), or nanometers (nm^{-1}).

Cube with sides equal to a:
$$S_{cube} = 6a^2$$
$$V_{cube} = a^3$$
$$S_{cube}/V_{cube} = 6/a$$
Sphere:
$$S_{sph} = 4\pi r^2$$
$$V_{sph} = (4/3)\pi r^3$$
$$S_{sph}/V_{sph} = 3/r, \text{ with } 2r = d, S_{sph}/V_{sph} = 6/d$$
Cylinder with $h \gg d$ (the wire):
$$S_{cyl} = 2\pi(d/2)^2 + 2\pi(d/2)h \approx 2\pi(d/2)h$$
$$V_{cyl} = \pi r^2 h = \pi(d/2)^2 h$$
$$S_{cyl}/V_{cyl} = 2\pi(d/2)h / \pi(d/2)^2 h, S_{cyl}/V_{cyl} = 4/d$$
Cylinder with $d = h$:
$$S_{cyl} = 2\pi r^2 + 2\pi rh \rightarrow 2\pi(d/2)^2 + 2\pi(d/2)d \rightarrow 2\pi[(d^2/2)(1/2) + (d^2/2)]$$
$$V_{cyl} = \pi r^2 h = \pi(d/2)^2 d$$
$$S_{cyl}/V_{cyl} = 2[(1/2) + 1]/d, S_{cyl}/V_{cyl} = 3/d$$
Cylinder with $d \gg h$ (the disc):
$$S_{cyl} = 2\pi(d/2)^2 + 2\pi(d/2)h \approx 2\pi(d/2)^2$$
$$V_{cyl} = \pi r^2 h = \pi(d/2)^2 h$$
$$S_{cyl}/V_{cyl} = 2\pi(d/2)^2/[\pi(d/2)^2 h], S_{cyl}/V_{cyl} = 2/h$$

where S_s is the specific surface area in square meters per gram, S_m is the measured area in square meters, ρ is the density of the material (grams per cubic · meter), and V_m is the measured volume of the material in cubic meters.

Specific surface area is an important parameter to gauge the effectiveness of catalysts, filtration, and gas chromatographic efficiency, among many other applications. For example, in fish tanks, the specific surface area of the filter (as SSA) indicates how much fish waste can be metabolized by nitrifying bacteria in a 24-h period. In gas chromatography, specific surface area of carbon supports a range from 5 $m^2 \cdot g^{-1}$ for graphitic materials and to 1000 $m^2 \cdot g^{-1}$ for activated carbons. Diatomaceous earth, fire brick (calcined celite), teflon chips, and polymer beads are examples of high specific surface area supports used in chromatographic columns. Catalysis depends on specific surface area. A silicon carbide catalyst, β-SiC (150 $m^2 \cdot g^{-1}$), with 1% Pt doping showed 100% conversion of CO_2 at 175°C [14].

5.3.3 Spherical Cluster Approximation

The spherical cluster approximation (SCA) is based on several fundamental assumptions. First, cluster radius, surface area, and volume are calculated without consideration of packing fraction, coordination number, lattice constants, and other factors associated with structure and the real chemistry involved in its packing [15]. Obviously, the larger the number of atoms N becomes, the more accurate the SCA model becomes. Secondly, the SCA considers, as it name implies,

FIG. 5.8

The fraction of surface atoms is plotted against $N^{-1/3}$ for icosohedral geometric shell clusters. The dashed line represents the prediction of the spherical cluster approximation. The graph shows that the fraction of surface atoms calculated by the spherical cluster approximation approaches the limit $4 N^{-1/3}$ as cluster size increases. (R. L. Johnston, Atomic and Molecular Clusters, Taylor & Francis, London, 2002.)

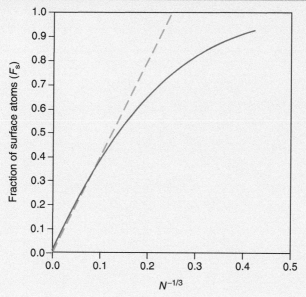

only hard spherical shapes [9]. Finally, in our case, we will designate that the clusters are made of one element: a metal. We begin with a cluster of N_a atoms with individual atomic volume equal to V_a. Our goal is to relate the number of surface atoms to the total number of atoms that comprise the cluster. Please refer to example 5.8.

The fraction of surface atoms is plotted versus $N^{-1/3}$ in **Figure 5.8**. The solid line represents the theoretical limit of the spherical cluster approximation [15]. The limiting value is given by equation (5.8) in example 5.8, $4N_a^{-1/3}$.

5.4 ATOMIC STRUCTURE

Fundamental knowledge of material structure is necessary in order to understand the behavior of nanomaterials. Therefore, a brief overview is presented in this section. Why go through this exercise? All nanomaterials seek a structural configuration that corresponds to minimal energy—highest possible stability. During this process, certain rearrangements within its structure are expected to occur. It is good to get acquainted with the generalized families of crystals to understand the driving forces behind such rearrangement.

5.4.1 Crystal Systems and the Unit Cell

Seven geometric crystal (axial) systems and 14 unique Bravais lattices, named after the French physicist Auguste Bravais in 1845, describe the position, direction, and

EXAMPLE 5.8 *Spherical Cluster Approximation*

Calculate the number of volume atoms, the number of surface atoms, and the percentage of surface atoms N_s to volume atoms N_v for a spherical cluster of N_a atoms. Relate the cluster radius (R_c), surface area (S_c), and volume (V_c) to the radius (R_a), surface area (S_a), and volume (V_a) of an individual atom in the cluster [15].

Solution:
If the volume of an individual atom is V_a and the number of atoms in a cluster is N_a, the volume of the cluster V_c is equal to

$$V_c = N_a V_a \tag{5.3}$$

Assuming that the cluster has a spherical shape with radius R_c, the volume of the cluster equals the volume of the constituent atoms, each with radius equal to r_a:

$$V_c = \frac{4}{3}\pi R_c^3 = N_a \left(\frac{4}{3}\pi r_a^3 \right) \tag{5.4}$$

The radius of the cluster then equals

$$R_c = N_a^{1/3} R_a \tag{5.5}$$

and

$$N_a = \sqrt[3]{\frac{R_c}{r_a}} \tag{5.6}$$

The surface area of the cluster, S_c, as it is related to the radius of the constituent atoms, then becomes

$$S_c = 4\pi R_c^2 = 4\pi \left(N_a^{2/3} R_a^2 \right) = N_a^{2/3} S_a \tag{5.7}$$

The number of cluster surface atoms is derived from the cross-sectional area of an atom A_a:

$$N_s = \frac{4\pi\, N_a^{2/3} R_a^2}{\pi R_a^2} = 4 N_a^{2/3} \tag{5.8}$$

This is the limiting expression for surface atoms as cluster size approaches large dimensions. The calculation of the fraction of atoms, F_s, is straightforward:

$$F_s = \frac{N_s}{N_a} = 4 N_a^{-1/3} \tag{5.9}$$

This approximation becomes better as cluster size approaches bulk dimensions [9].

structural planes of most crystalline materials (**Fig. 5.9**). Crystal structure is based on regularly repeating elements that form a pattern in three dimensions (e.g., the unit cell). The unit cell is described by a set of lattice parameters and is the irreducible representation of the crystal structure. For example, knowledge of lattice constants such as unit cell edge length (a,b,c) and crystallographic axis angle (α,β,γ) allows for facilitative quantification of material structure. In other words, translation of structurally equivalent positions of the unit cell over and over again results in a material with long-range order—ultimately, a crystal. In contrast, such long-range order is absent in amorphous materials such as glass, liquids, and gases.

FIG. 5.9	The seven basic geometric crystal systems are shown with corresponding Bravais lattices. We will just discuss the three cubic systems in this text.

Crystal system	Definitions	Generic rendition	Bravais lattice			
Cubic	$a=b=c$ $\alpha=\beta=\gamma=90°$	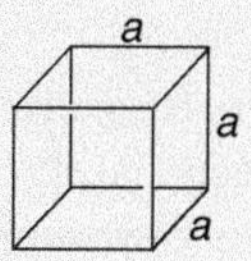	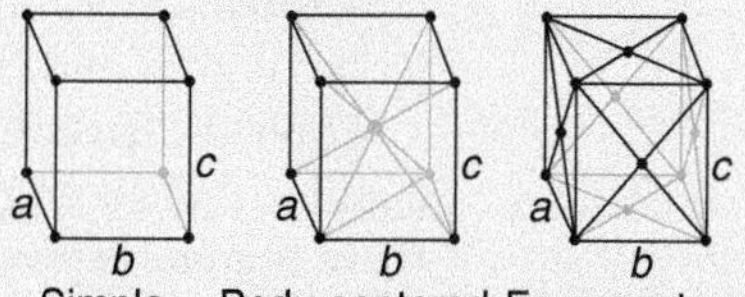 Simple	Body-centered	Face-centered	
Tetragonal	$a=b\neq c$ $\alpha=\beta=\gamma=90°$	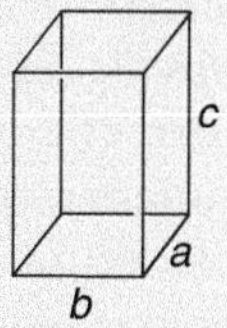	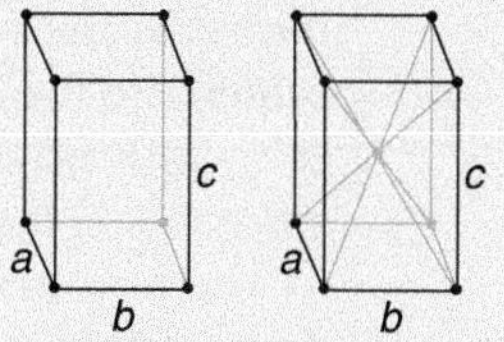 Simple	Body-centered		
Orthorhombic	$a\neq b\neq c$ $\alpha=\beta=\gamma=90°$	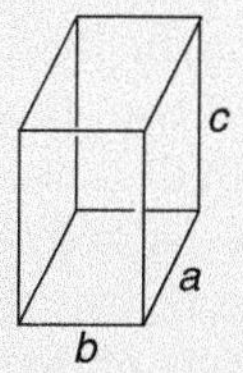	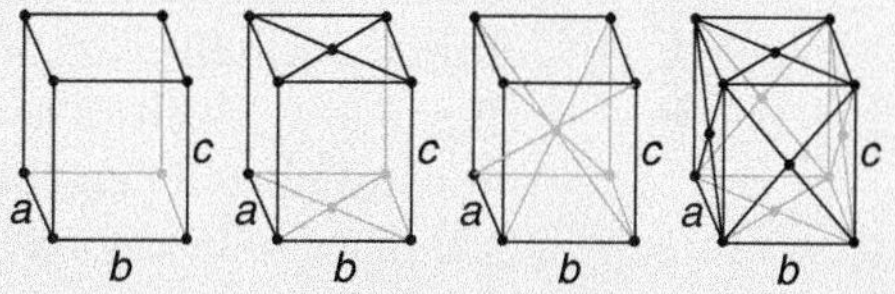 Simple	Base-centered	Body-centered	Face-centered
Rhombohedral	$a=b=c$ $\alpha=\beta=\gamma\neq90°$	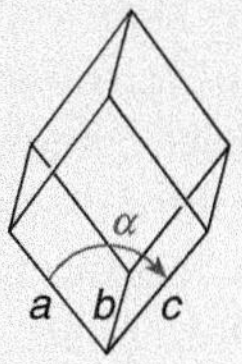	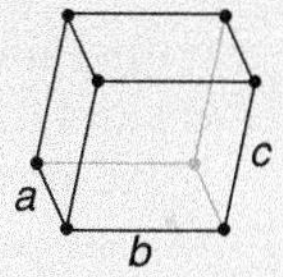 Rhombohedral			
Hexagonal	$a=b\neq c$ $\alpha=\beta=90°$ $\gamma=120°$	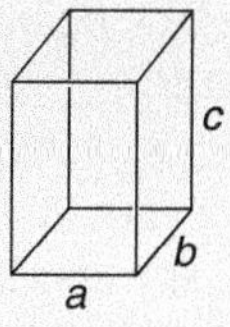	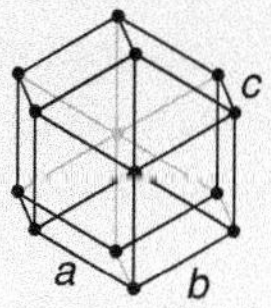 Hexagonal			
Monoclinic	$a\neq b\neq c$ $\alpha=\gamma=90°\neq\beta$	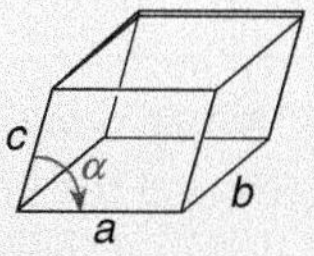	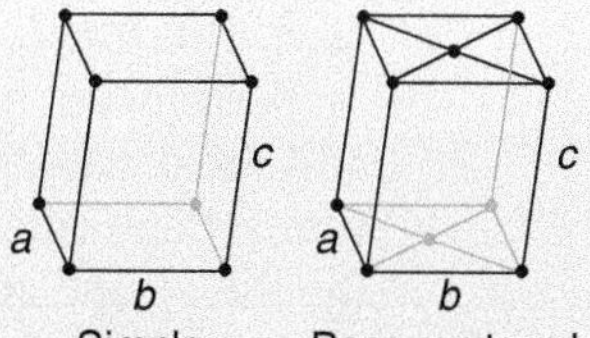 Simple	Base-centered		
Triclinic	$a\neq b\neq c$ $\alpha\neq\beta\neq\gamma\neq90°$	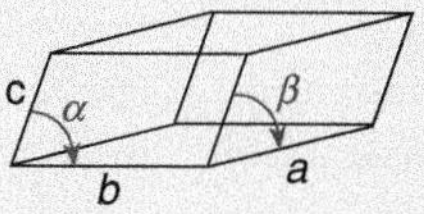	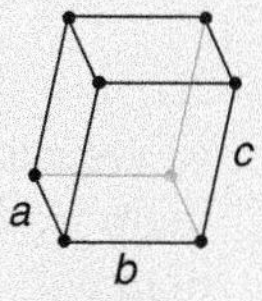 Triclinic			

5.4.2 Cubic and Hexagonal Systems

There are three Bravais lattices in the cubic (isometric) system: primitive (*cubic-P*), body-centered cubic (*bcc*), and face-centered cubic (*fcc*). Three cubic systems and the hexagonal close-packed system are depicted in **Figure 5.10a–c**. Metals crystallizing as *fcc* and *bcc* will serve as the models of choice because they are the least complex and easiest to visualize.

Three-dimensional structures have three lattice constants, *a, b,* and *c*. Between *a* and *b*, there is angle γ; between *a* and *c*, there is angle β; and between *b* and *c*, there is angle α. In cubic systems, systems that possess the highest symmetry, it follows that $a = b = c$ and $\alpha = \beta = \gamma$. The primitive cube has atoms at each apex of the cube; the *bcc* is the same except that an additional atom is placed in the center, and the *fcc* form has atoms placed in the center of each face of a primitive cube with no atom in the center. The number of nearest neighbors for each atom (a.k.a. the coordination number Z) is 6, 8, and 12, respectively.

There are several ways to pack atoms in a cubic system. As we know, atoms desire the greatest number of nearest neighbors to become stable. For example, in a two-dimensional plane, six atoms are able to completely surround a central atom to form the core of a hexagonally close-packed two-dimensional structure (**Fig. 5.11a**). Such close-packed two-dimensional planar structure has a maximum number of neighbors with the lowest possible void volume. Another identical plane of hexagonally close-packed atoms is placed on top of the first plane to fit into the holes or void volume (gaps) formed by the touching spheres in the first plane. This creates a structure of the AB type (**Fig. 5.11b**). There is another way to place the second plane: right on top of the first plane (on top of the atoms) and not in the gaps. This forms an AA structure (e.g., one that is *in register*) and it comprises the primitive cubic system (*cubic-P*) (**Fig. 5.7**).

We are now ready to place a third plane. Obviously if we stay in register in the AA-base system to create AAA, AAAA, and AAAAA, there are no options and we are finished with its structure (*cubic-P*). However, for the AB-base system, there are two ways to place the third plane, depending on which set of gaps gets covered in the third step. Two stacking arrangements are possible that yield different crystalline arrays: the ABABAB... or ABCABC... type, depending on which gaps get covered. An *hcp* structure ABA is formed if the third layer is placed directly on top of the first layer (**Fig. 5.11c**). The second is an *fcc* (cubic close-packed) structure (**Fig. 5.11d**). In this case, the third layer is placed above the uncovered voids (gaps) of the first layer—the gaps not covered by the original B layer. This forms an ABC structure. The two types are called *polytypes*.

Two types of holes are formed in close-packed structures: the tetrahedral hole and the octahedral hole (**Fig. 5.12**). A tetrahedral hole is a space enclosed by four spheres (three in one plane and the last one on top of the hole formed by the three), and an octahedral hole is a space enclosed by six spheres. The radius of an octahedral site is

$$r_{\text{oct}} = (1/4)(2 - \sqrt{2})a = (\sqrt{2} - 1)r = 0.4142r \qquad (5.10)$$

where *a* is the lattice constant and *r* is the radius of the sphere. From geometry, then, we see that an octahedral hole can accommodate another atom no larger than $0.4142r$. The number of octahedral holes equals the number of atoms (N_a) in a crystal. Tetrahedral holes are described by

Fig. 5.10

The crystal structures of the basic isometric cubic systems are displayed in point, space-filling, and hard sphere packed forms. (a) Simple cubic (or cubic-P, primitive) is shown on the top. The atoms at the corners are each shared with eight other cubes. Therefore, there is one atom per unit cell $(8)(\frac{1}{8}) = 1$. Each atom has six nearest-neighbors and therefore the coordination number Z is equal to 6. (b) Body-centered cubic (bcc) has two atoms per unit cell: one in the center and the same eight found in the simple cube at each corner of the cube: $1 + (8)(\frac{1}{8}) = 2$. The coordination number of atoms in bcc crystals is Z = 8. (c) Face-centered cubic (fcc) has one atom on each face (for a total of 6) that is shared by an adjacent cube and the same eight found in the simple cubic and fcc forms. The number of atoms per unit cell then is four per unit cell, one in the center and the same eight found in the simple cube at each point of the cube: $(6)(\frac{1}{2}) + (8)(\frac{1}{8}) = 4$. The coordination number of atoms in fcc crystals is Z = 12.

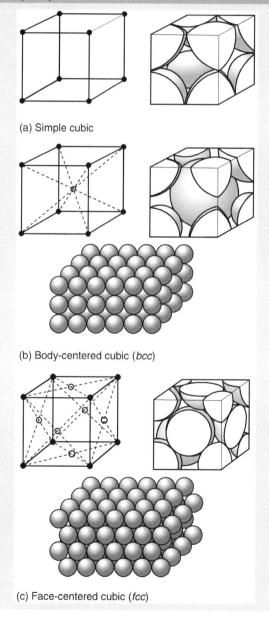

(a) Simple cubic

(b) Body-centered cubic (*bcc*)

(c) Face-centered cubic (*fcc*)

FIG. 5.11

Planar packing schemes for close-packed structures are shown. In (a), hard spheres are closely packed into one layer to form a six-coordinate structure (each sphere is surrounded closely by six others). (b) An identical layer is placed in the gaps among atoms of the first layer. Try this at home; can you find two (or more) possible identical configurations (degeneracies) in the placement of the second layer? (c) and (d) show two possible ways of placing the third identical layer. If layer three is placed directly over layer one, then you form the ABAB ... polytype characteristic of a hexagonal close-packed (hcp) crystal. Notice that the "holes" are the same holes shown in (b). If the third layer is not placed directly above the first layer, then an ABC ... polytype is formed. This is the fcc crystal structure. Notice that in this structure no holes (open gaps or interstitial spaces) are visible.

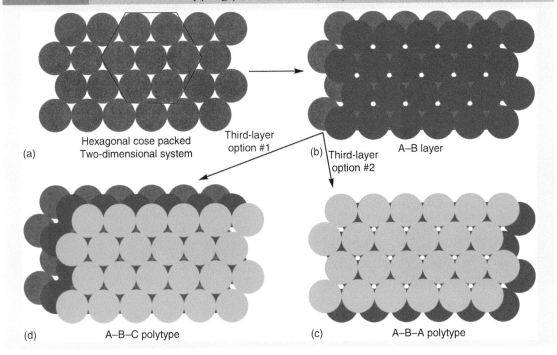

(a) Hexagonal cose packed Two-dimensional system

Third-layer option #1

(b) Third-layer option #2   A–B layer

(d) A–B–C polytype

(c) A–B–A polytype

FIG. 5.12

Two types of holes are displayed. On the left, the tetrahedral hole is illustrated. On the right, the octahedral hole is illustrated. In ABAB ... structures, octahedral holes are stacked on top of each other.

$$r_{tet} = (1/4)(\sqrt{3} - \sqrt{2})a = \left[\sqrt{(3/2)} - 1\right]r = 0.225r \qquad (5.11)$$

and there are twice the amount of tetrahedral holes in the close-packed solid than there are atoms ($2N_a$). Note that only smaller atoms are able to fit in the tetrahedral hole.

5.4.3 Packing Fraction and Density

During this brief discussion of packing fraction, we conveniently revert to the application of hard spheres as before. The spherical cluster approximation does not take packing fraction into account. Atomic packing factor or packing fraction, f_p, is a dimensionless quantity always less than one that relates the volume occupied by atoms to the total volume of the material:

$$f_p = \frac{N_a V_a}{V_c} \qquad (5.12)$$

where N_a is the number of atoms, V_a the volume of an individual atom, and V_c the volume of the cluster. Void volume, the spaces (gaps) between and among close-packed atoms, constitutes the remainder of the space in a crystal. The first step to determine packing fraction is to calculate the number of atoms in a unit

---

EXAMPLE 5.9 | *Packing Fraction for a Single Element Body-Centered Cubic Structure*

Calculate the f_p for ferrite, the α-iron allomorph that exists as a *bcc* crystal. What is the coordination number? What is the value of the lattice constant a?

Solution:
The body-centered cube is depicted in **Figure 5.11b**. *There are eight atoms at the corners of the cube and one atom in the center. Each corner atom shares bonds with the central atom and seven additional atoms within the unit cell and in adjacent unit cells. The coordination number, CN or Z, is the number of nearest neighbors for each atom. In a bcc crystal, each atom has eight nearest neighbors, CN = 8. This holds true for the central atom as well as the corner atoms. The bcc unit cell, as a result, contains just one eighth of the iron atom at each corner and one atom in the center. The total number of atoms in a bcc unit cell is then:*

N_a = *one central atom + (eight corner atoms × one eighth of each corner atom) = two atoms*

α-*Iron is therefore not a close-packed cubic structure. The lattice constant, a, is the length of the cell edges. Calculation of the value is accomplished by applying simple geometric relations.*

Because the center atom touches both corner atoms along any diagonal, the length of diagonal, d, of a cube in the bcc configuration is equal to d = 4r.

The length of the diagonal in terms of a is $d^2 = 3a^2$ to give $d = a(3)^{1/2}$; therefore,

$$a = 4r/\sqrt{3} \qquad (5.13)$$

The packing fraction is calculated by dividing the volume of the atoms by the volume of the cube:

$$f_p = 2 \text{ atoms } [(4/3)\pi r^3]/a^3 = 2[(4/3)\pi r^3]/(4r/\sqrt{3})^3, \ f_p = 0.68 \qquad (5.14)$$

The covalent radius of iron is 0.117 nm.

$$\text{The lattice constant } a = 4(0.117)/\sqrt{3}; \ a = 0.270 \text{ nm} \qquad (5.15)$$

cell. In the primitive cubic system, each corner atom is shared by eight adjacent cells. Thus, there is a total of one atom per unit cell (one eighth of each corner atom × eight corner atoms = one atom).

Calculation of theoretical densities can be found in a straightforward manner by applying

$$\rho = \frac{N_a M}{\mathcal{N}_A V_{uc}} \tag{5.16}$$

where N_a is the number of atoms per unit cell, M is the atomic mass in $g \cdot mol^{-1}$, V_{uc} is the volume of the unit cell in cm^3, and $\mathcal{N}_A$ is Avogadro's number [16].

5.4.4   Structural Magic Numbers

If small materials had the same physical properties as bulk materials, just on a smaller scale, our story would end right now. However, there is a break in the smooth curve of the continuum. Scaling laws break down at dimensions under 10 nm. In actuality, even particles much larger than 10 nm (e.g., colloids) start to display properties that meander away from bulk behavior (the color of gold colloids, for example) but for different reasons. The structure of a bulk material, for all practical purposes, is uniform down to micron-sized particles. When particles get smaller, clusters of similar size may have different structures. One or two atoms may make the difference between a stable cluster and one that is metastable or unstable.

Chemical reactivity of alkali and aluminum clusters is lower relative to open shell counterparts. Small alkali metal clusters are controlled by electronic structure. The reactivity of transition metal clusters, on the other hand, correlates strongly with geometric rather than electronic structure. Alkali and noble metals with an even number of electrons form clusters that are diamagnetic, while those with odd number of electrons form paramagnetic clusters. In nonmetallic clusters such as those composed of the rare gases, magic numbers are due exclusively to structural geometry and not electron considerations.

EXAMPLE 5.10   *Calculation of the Theoretical Density of Copper*

Calculate the theoretical density of copper from equation (5.16). Copper data: covalent radius = 0.128 nm; $M = 63.55 \ g \cdot mol^{-1}$; structure = *fcc*.

Solution:

Cu, fcc: $Z = 12$ and there are four atoms per unit cell, $a = 4r/\sqrt{2}$

Use $\rho = N_a M / V_{uc} \mathcal{N}_A$

Volume of the unit cell, $V_{uc} = a^3 = (4r/\sqrt{2})^3 = 4.74 \times 10^{-2} \ nm^3$; $V_{uc} = 4.74 \times 10^{-23} \ cm^3$

$\rho_{theoretical} = [(4 \times 63.55 \ g \cdot mol^{-1})]/[(4.74 \times 10^{-23} \ cm^3)(6.022 \times 10^{23})]$

$\rho_{theoretical} = 8.90 \ g \cdot cm^{-3}$

The actual value of the density of copper is $8.94 \ g \cdot cm^{-3}$.

Clusters consist of a few to several thousand atoms. A cluster is considered to be a quantum dot. Some clusters exhibit remarkable stability. These stable forms have what is called a *magic number* of atoms. In other words, completeness with regard to physical and electronic structure is achieved by clusters with a magic number of atoms. The concept of magic numbers is not new. Physicist Maria Goeppert-Mayer in 1948 explained why the nuclei of some atoms were more stable than others. Linus Pauling in 1965 added that nuclei with 2, 8, 20, 28, 50, 82, or 126 neutrons or protons were stable and were characterized by extremely long lifetimes. It turned out that the rationalization for nuclear cluster stability was based on a spherical quantum mechanical model called the *jellium model* that was especially suited for describing the stability of alkali metal clusters. Because the jellium model is able to explain both nuclear and electronic phenomena, it is truly one of the fundamental unifying concepts of physics [15].

If atoms are packed in such a way that all possible nearest-neighbor sites are filled to form a complete shell surrounding the central atom or preexisting shell, that cluster is considered to have a *magic number* of atoms. Based on mass spectrometry experiments in the 1980s, a periodic pattern of intense peaks was found in the mass spectra of sodium clusters [17]. The sodium clusters showed intense peaks at 2, 8, 20, 40, 58, and other sums of atoms. Adjacent peaks corresponding to smaller or larger sized clusters on either side of the magic number peak were far less intense [17]. Since MS intensity is proportional to the abundance of daughters of specific size and charge, we can safely conclude that magic number clusters are more stable than others with similar mass. Clusters that have a magic number of atoms have higher abundance, ionization potential, and binding energy. Neutral clusters with closed shells have low electron affinity while those with open shell structures have high electron affinity.

The structure of *fcc* close-packed structures like gold is a good place to begin. Starting with one atom, a maximum of 12 atoms (coordination number $Z = 12$) can be placed around the central atom to form a close-packed structure. The resulting structure is a geometric solid called a *cuboctahedron*—a 14-sided geometric solid with 12 vertices, 8 triangular faces, and 6 square faces (**Fig. 5.13**).

Thus, the first magic number—$\mathcal{M}^*$, the number 13—is achieved. There are 12 shell atoms and one core atom. There is no more available space to add more atoms, and if an atom is removed, a vacancy is created. Similar to the octet rule for electrons and nuclear closed shells, the magic number is a closed shell geometric configuration for atoms in a cluster and the same result is achieved in all cases: the stable, lowest energy configuration.

The next layer requires that an additional 42 atoms be packed around the first shell. The magic number for the cluster is now 55. The quantum dot cluster Au-55 is a cuboctahedral magic number cluster with 55 gold atoms. For the series of clusters that exist in icosohedron or cuboctahedron form, the appropriate magic number formula for the series is

$$\mathcal{M}^*(K) = 1, 13, 55, 147, 309, 561, 923, 1415, 2057, 2869, \text{ etc.} \quad (5.17)$$

Each shell K of the cluster, a pseudoconcentric shell, is described mathematically by

FIG. 5.13

Cuboctahedral gold clusters are shown. The progression represents a geometric series based on magic numbers: 13, 55, 147, 309, and 561. Notice how the faces retain their integrity regardless of the size of the cluster. As the cluster becomes larger, this figure also illustrates the concept of surface-to-volume ratio: The larger the cluster is, the more volume atoms seem to overwhelm the number of surface atoms.

Source: Courtesy of Professor Anil K. Rao, Department of Biology, the Metropolitan State College of Denver.

$$\mathcal{M}^*(K) = (1/3)(10K^3 + 15K^2 + 11K + 3) \qquad (5.18)$$

where K is the shell index number. The expression is based on the expansion of

$$\mathcal{M}^*(K) = 1 + \Sigma\,[k = 1 \rightarrow k = K]\,(10K^2 + 2) \qquad (5.19a)$$

In other words, every shell consists of $N_s = (10K^2 + 2)$ atoms. If that shell is the last shell to be added, then that shell contains the surface atoms, N_s. Another popular formula for magic numbers is one that includes the central atom as the first shell designated as $K = 1$ [13]:

$$\mathcal{M}^*(K) = (1/3)(10K^3 - 15K^2 + 11K - 3) \qquad (5.19b)$$

An expression that includes the central atom as a shell for the number of surface atoms is derived from equation (5.19b):

$$N_s = (10K^2 - 20K + 12) \qquad (5.20)$$

We now have another tool with which to calculate surface-to-volume ratio (**Table 5.3**). There are two ways to view surface-to-volume ratios. In one, the surface atoms N_s are related to the whole volume of the cluster (given as $N_s/\mathcal{M}^*$). In the other way, surface atoms are related to only the atoms that exist below the

TABLE 5.3	**Percent Surface Atoms of Cuboctahedral Magic Clusters**		
K	**$\mathcal{M}^*$**	**N_s**	**$N_s/\mathcal{M}^*$**
1	1	1	1.00
2	13	12	0.92
3	55	42	0.76
4	147	92	0.62
5	309	162	0.52
6	561	252	0.45
7	923	362	0.39
8	1415	492	0.35
9	2057	642	0.31
10	2869	812	0.28
50	4.04×10^5	2.40×10^4	0.06
100	3.28×10^6	9.80×10^4	0.03
500	4.15×10^8	2.49×10^6	0.006

surface—the true volume atoms (e.g., $\mathcal{M}^* - N_s$). For example, in the Au-55 cluster, only one atom is a true volume atom. We will only be concerned with the former expression.

The diameter of the cluster $2R_c$ is found from [13]

$$2R_c = d[2K - 1] \tag{5.21}$$

where $d = 2r$, the covalent diameter between two atoms for an *fcc* unit cell and is related to the lattice a constant by $d = a/\sqrt{2}$.

Body-centered cubic (*bcc*) clusters are formed as 14-vertex rhombic dodecahedrons. The magic number formula for *bcc* clusters is

$$\mathcal{M}^*(K) = (4K^3 + 6K^2 + 4K + 1) \tag{5.22}$$

Magic number series for *bcc* systems differ from the cubic close-packed *fcc* set (e.g., $\mathcal{M}^*(K_{bcc})$ = 1, 15, 65, 175, 369, 1105, 1695, 2465, etc.) [15]. For the *hcp* system, another set of structural magic numbers is generated: $\mathcal{M}^*(K_{hcp})$ = 1, 13, 57, 153, 321, 581, etc. [13]. All magic number clusters discussed here are of minimum volume, maximum density, and approximately spherical in shape—e.g., pseudospherical [13].

5.4.5 Miller Indices and X-Ray Diffraction

In 1839 British crystallographer W. H. Miller devised a systematic approach that is related to the geometry of the unit cell. The *Miller indices*, as they came to be known, are the fundamental language of crystallographers and engineers and extremely useful tools for predicting structural properties of materials. Miller developed a system of describing the orientation of crystallographic planes inside a crystal lattice by a set of integers. The integers are symbolized by (*hkl*). A similar notation, developed by Bravais and Miller, was invented to describe hexagonally packed systems: (*hkil*).

The index numbers are the inverses of the axial intercepts of the unit cell, e.g., the intercepts that a selected crystal plane makes with the crystallographic axes. Miller indices enclosed in parentheses, like (100), indicate the orientation of a specific plane with respect to a crystallographic axis. Braces denote planes that are related by symmetry. For example, the {100} family contains the (100), (010), and (001) planes and their negative counterparts. Brackets (e.g., [111]), indicate a specific lattice direction—specifically, the direction normal to the (111) plane. This would include the (222), (333), or (*nnn*) planes. The indices [*uvw*] in brackets are used to represent directions.

The Miller index algorithm is created by establishing a Cartesian coordinate system, *x*, *y*, and *z* or *a*, *b*, and *c*, and then placing the unit cell at the origin (0,0,0). The method for obtaining Miller indices is straightforward: (1) Set the origin of the unit sell at 000, (2) determine the value of intercepts of the crystal plane in terms of lattice constants, (3) take the reciprocal of each index, and (4) reduce them to the lowest terms. The use of reciprocals helps alleviate the awkward task of dealing directly with infinite intercepts, the case when a plane is parallel to an axis. **Figure 5.14** displays the fundamental low-index crystal planes of a simple cubic system. The orientation is the same for all cubic crystals. The only difference is the number of atoms and spacing within the low-index planes.

Crystallographic directions (e.g., [*uvw*]) are calculated by repositioning the directional vector through the origin and translating the projections into terms of the lattice constants *a*, *b*, and *c*. The values are then converted into the smallest integer values. Prove to yourself that no reciprocal relations are developed in this calculation. Linear and planar atomic densities are calculated in a manner similar to atom packing fraction. We find the number of whole atoms along a direction or in a plane and then divide that number by the appropriate dimension. Linear density is defined as the number of atoms per unit length centered along a given direction in a crystal. Planar density is defined as the number of atoms centered in a plane per crystallographic plane area.

EXAMPLE 5.11   *Miller Index Calculation*

Determine the Miller indices for the three simplest planes (low index) for any cubic lattice. What does a (111) plane look like? Do not be concerned with the atomic structure, only with the geometric structure of a cube.

Solution:
Consult **Figure 5.15** *for assistance with orientation.*

The lattice constants for a cubic crystal are the orthogonal series consisting of $a = b = c$ ($\alpha = \beta = \gamma = 90°$) and the coordinate system is set up accordingly.

The three simplest (low index) planes are those that intersect each major axis of the crystal (i.e., they intersect the designated axis but are parallel to the other two).

The plane (abc) that intersects the a axis at **a** *is the (a, ∞, ∞) plane. The symbol ∞ represents the parallel condition. The reciprocals are (1/a, 1/∞, 1/∞). With* **a** *= 1, the Miller indices (hkl) are designated as (100).*

The same logic is applied to the other two orthogonal planes that intersect the b and c axes at the points **b** *and* **c**, *respectively, yielding the planes (010) and (001), respectively.*

If the (111) plane intersects each major axis at the midway point, the plane is described by (1/2, 1/2, 1/2). Inverting and reducing to lowest terms, (hkl) = (111).

EXAMPLE 5.12	*Calculation of Linear Density*

Determine the linear density of atoms in the [110] direction for *fcc* gold (R_{mr} = 0.144 nm, a = 0.408 nm).

Solution:
The [110] vector runs along the bottom-face plane diagonal from coordinate points 0,0,0 to 1,1,0 of the primitive cubic fcc cell (e.g., the diagonal running under the fcc cubic unit cell). The length of the diagonal is equal to four times the metallic radius: $4R_{mr}$. Draw an fcc cubic unit cell. From inspection, there are two whole atoms centered in the line contained by the [111] direction. The linear density is then:

D_l = 2 atoms/$4R_{mr}$ = 1/(2R) or, in terms of the lattice constant a, D_l = 2/(a√2), D_l = 3.47 atoms·nm^{-1}

In 1895, W. C. Röntgen was the first to detect the arrangement of atoms in a crystal by means of x-ray analysis because the wavelength of the x-rays happened to coincide with the space between atomic planes—e.g., the *d*-spacing. X-ray diffraction (XRD) is important in studying the structure of nanomaterials, because of the need to measure altered *d*-spacing in nanoparticles. XRD depends on the Bragg equation given earlier in chapter 3 ($n\lambda = 2d \sin \theta$).

The *d*-spacing between planes for cubic systems is found from

$$d = \frac{a}{\sqrt{h^2 + k^2 + l^2}} \tag{5.23}$$

where *d* is the space between adjacent planes and *a* is a lattice constant.

FIG. 5.14	*The three low-index Miller hkl planes are shown in the figure for a simple-cubic system crystal. Higher index planes of the family of {100}, {110}, and {111} such as (200), (220), and (222), respectively, are not emphasized in this text. There are many hkl planes for every crystal. The (100) plane in the simple cubic system is framed by four atoms; the (110) is also framed by four, but notice how its orientation has been altered. The (111) plane and the (100) plane are displayed gloriously by the cuboctahedrons in Figure 5.13. Can you find the (100) planes in that figure?*

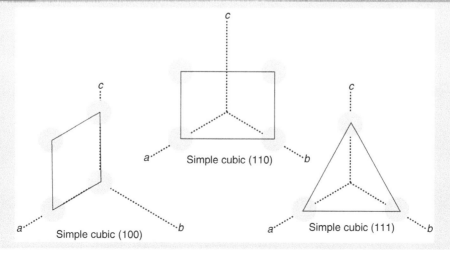

5.5   PARTICLE ORIENTATION

Nanoparticles, particularly those made of noble metals or semiconductors, are routinely placed within dielectric materials such as polymers or porous alumina membranes to form nanocomposite materials. Michael Faraday's experiments investigated solutions containing spherical nanogold particles. Beautiful colors were elicited from the constituent nanoparticles immersed in the surrounding water medium. Once embedded in the dielectric material, the orientation of the nanoparticles is fixed with regard to the plane of the composite. Exposure of the plane of the nanocomposite to impinging radiation (UV–visible light) results in an optical response by the combined effective medium. In other words, the optical response of the nanocomposite is due to a mixture of the dielectric constants of the metal, the surrounding medium, particle orientation, and their relative proportions. An example of such a composite is shown in **Figure 5.15.** The pore channels in a porous anodic alumina membrane (a transparent and insulating host material) were used to house gold nanoparticles of variable aspect ratio and size. The impinging UV-visible radiation was fixed normal to the plane of the membrane during the wavelength scan [18–20]; for example, the oscillating electric field of the EM radiation was oriented parallel to the plane of the membrane and perpendicular to the pore channels (**Fig. 5.15**).

FIG. 5.15   *Orientation of gold nanocylinders is shown with respect to the propagation vector of visible light. The electric field (and the orthogonal magnetic field) in the image is drawn perpendicular to the longitudinal axis of the ellipsoidal nanocylinders. In other words, electrons along the short axis are excited when exposed to the radiation producing plasmon resonances. Porous alumina is a perfect template for fabricating gold nanoparticles. When gold nanocylinders such as these are liberated from the confines of the alumina template and placed in a solution—for example, water—two resonance peaks are observed: one corresponding to the transverse axis (the short axis) and the other to the longitudinal axis. Spectra obtained have two absorption peaks. For ellipsoidal particles, if the particle are spheres, only one resonance is detected.*

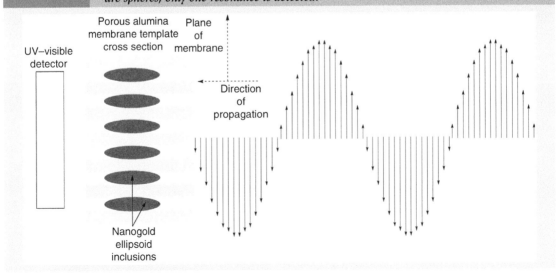

FIG. 5.16 *The family of ellipsoidal shapes originating from the parent sphere is shown. On the right, prolate ellipsoids are shown. For a sphere, a = b = c. Prolate ellipsoids are needle-like in shape with one dimension, say a (the major axis), much greater than the other two (the minor axes): a >> b = c. For oblate shapes, one dimension, say a once again, is much less than the other two axes: a << b = c.*

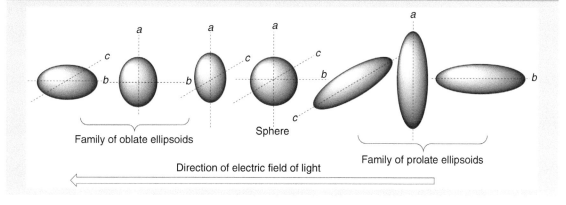

Family of oblate ellipsoids

Sphere

Direction of electric field of light

Family of prolate ellipsoids

This discussion is confined to ellipsoids of revolution. There exist three types of three-dimensional ellipsoids of revolution: *spherical* $(a = b = c)$, *prolate* $(a > b = c; b > a = c;$ or $c > a = b)$, and *oblate* $(a < b = c; b < a = c;$ and $c < a = b)$, where the orthogonal axes a, b, and c comprise a Cartesian coordinate system. Please consult **Figure 5.16** for reference. The prolate forms are similar to wires; the oblate forms are similar to discs. Ellipsoids of revolution are depicted in **Figure 5.16**.

Ellipsoids of revolution were chosen because the electric field that exists within such shapes is uniform throughout and therefore calculable. In this section, atom packing and structural factors are not considered—only the geometric shape is of importance. We are concerned with the maximum absorption λ_{max} of noble metal nanoparticles in the presence of an applied electromagnetic field, specifically in the visible region of the spectrum.

5.5.1 Surface Polarization in Metals

A plasmon (from the Greek *plasma*, "something molded or shaped" [like plastic] + the Latin *mons*, "mountain, body") is a *quasi-particle*—a particle that can be described by a collection of interacting particles—in this case, electrons. Plasmons occur on the surface of a metal, are quantized, and consist of collective longitudinal (or transverse) oscillations of the free electron gas (the conduction electrons). Plasmon oscillations are induced when excitation of the metal surface electrons by visible light or other sources occurs. The location (λ_{max}) and strength (absorption) of the oscillation depend on the wavelength of the incident radiation, particle composition (dielectric constants), size, shape, and orientation.

The plasmon energy is described classically as

$$\omega_p^2 = \frac{N_e e^2}{\varepsilon_o m} \qquad (5.24)$$

where
  ω_p is the plasmon excitation frequency
  e is the electron charge $(1.6022 \times 10^{-19}$ C)
  N_e is the electron density in reciprocal cubic meters
  m is the mass of the electron $(9.1096 \times 10^{-31}$ kg)
  ε_o is the vacuum permittivity of free space $(8.8554 \times 10^{-12}$ F·m$^{-1})$

What equation (5.24) implies is that metals with a higher density of electrons demonstrate λ_{max} that are blue shifted relative to metals with a lower density of electrons. It also implies indirectly that as particle size is diminished, the apparent electron density is increased because the frequency of collisions of electrons with the boundary increases—a higher energy state. This condition also results in a blue shift of λ_{max}.

Localized surface plasmon resonance occurs strongly on noble metal nanoparticle surfaces, particularly those of gold, silver, copper, and the alkali metals. Plasmon resonance and scattering are related phenomena, and we will visit with those phenomena shortly, but first a discussion concerning the effects of orientation on the plasmon resonance is in order.

An illustration of two ellipsoids of revolution exposed to an electric field is shown in **Figure 5.17**. The applied electric field is designated as E_o. The induced

FIG. 5.17

The polarization phenomenon of the surface electrons on a metal surface is shown. Upon exposure to the electric field of the light, the surface electrons generate an opposing field that is in turn opposed by an internal depolarization field. The strength of the oscillation depends on the shape and orientation of the nanoparticle (usually a noble metal or copper). The more pointed the nanoparticle is, the stronger the opposing field becomes. This is why molecules situated on the tips of high aspect ratio nanocylinders have significantly enhanced SERS enhancement compared to those placed on spherical particles—a nano-lightning rod effect of sorts.

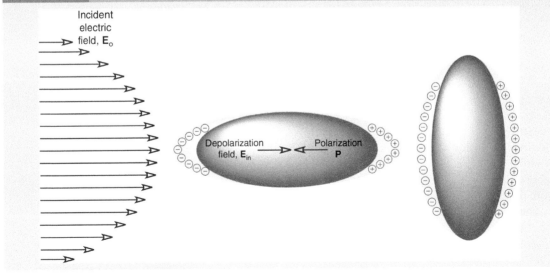

electric field, $\mathbf{E}_{in}$, is opposite in direction to the polarization, $\mathbf{P}$, parallel to the applied field. The strength of the depolarization field is dependent on the shape of the nanoparticle.

For spheres, the dipolar resonance, the most fundamental resonance, is described by the expression

$$\varepsilon_o = -2\varepsilon_m \tag{5.25}$$

where ε_m is the real component of the complex dielectric function of the metal and ε_o is the real component (refractive index) of the complex dielectric function of a surrounding insulator material. The factor 2 is typical for spheres—more about that later. The imaginary part of the two materials can be ignored at the wavelength range of interest. This relation is relevant only for particles smaller than 20 nm. For larger particles, higher order resonances and scattering need to be considered—topics beyond the scope of this text. Therefore, this is a case in which the size of a particle causes a change in the manifestation of a physical property.

There are two types of cluster size effects that affect the nature of the surface plasmon. Classical electrodynamic theory using bulk optical constants can be applied safely to describe the optical response of large clusters [21]. The atomic structure of a large cluster is similar to that of its bulk counterpart. This is called an *extrinsic* effect. Therefore, this is not to say that the optical response of large clusters is identical to that of bulk materials because it is not. With smaller dimensions comes the addition of restrictive boundary conditions. Collective electronic excitations known as Mie resonances are important in particles that are the size of colloids or larger. The color of a solution containing large metallic colloids is not the same as the color of the same material in bulk form; colloidal gold sols are purplish, for example.

The optical functions, however, for very small clusters are size dependent. This trend is due to *intrinsic* effects: Changes in the internal structure of the cluster relative to the bulk accompany smaller size. The optical functions (constants like the dielectric constant) do not apply to very small nanomaterials. These changes in electronic and structural properties as size is reduced occur at the threshold between the bulk and quantum domains. Intrinsic effects are not confined to optical properties. Many physical properties undergo radical alteration as size is reduced. They include ionization potential, binding energy, chemical reactivity, crystallographic structure, melting temperature, and optical properties (i.e., the plasmon).

The classification of cluster types shown in the following table for sodium clusters [21].

	Small	Medium	Large
Atoms, N_a	2–20	20–500	500–10^7
$2R_{Na}$ in nanometers	≤1.1	1.1–3.3	3.3–100
N_s/N_v	N_s/N_v inseparable	0.9–0.5	≤0.5

Other classifications exist as well [15] and, as you can see, it becomes a qualitative issue to some extent. In both tables, sodium atoms serve as the model.

	Small	Medium	Large
Atoms, N_a	$\leq 10^2$	10^2–10^4	$>10^4$
$2R_{Na}$ in nanometers	≤ 1.1	1.9–8.6	>8.6
N_s/N_v	>0.86	0.86–0.19	≤ 0.19

Surface enhanced Raman spectroscopy (SERS) is a method that exploits the surface plasmon resonance properties of noble metal nanoparticles. The resonance condition is achieved when equation (5.25) becomes very large. Giant electromagnetic enhancements for spheres:

$$G = \chi^{12} \left(\frac{\varepsilon_m - \varepsilon_o}{\varepsilon_m + 2\varepsilon_o} \right)_\lambda^2 \left(\frac{\varepsilon_m - \varepsilon_o}{\varepsilon_m + 2\varepsilon_o} \right)_s^2 \tag{5.26}$$

where $\chi = r/(r + d)$ with r equal to the radius of the spherical metal SERS substrate nanoparticles and d the distance of the analyte molecule from the surface of the metal particle. The λ and s represent the laser and Stokes fields, respectively. Obviously the SERS signal is enhanced significantly as the two denominators approach zero [22].

5.5.2   *Particle Depolarization Factor and Screening Parameters*

The sphere is a shape that possesses the highest level of symmetry of all geometric shapes. Although electron polarization occurs on spherical metal nanoparticles, the orientation of the spherical particle in an electric field is irrelevant as polarization is infinitely degenerate; that is, the sphere will have the same visible absorption maximum (λ_{max}) regardless of its three-dimensional placement in an electric field. That is why solutions containing spherical gold colloids have one absorption peak. Prolate particles in solution, on the other hand, have two absorption maxima that correspond to the longitudinal and transverse plasmon resonances [23].

We introduce three related terms in this section that help us understand how orientation of nanoparticles affects the optical response of metal nanoparticles. They are, in order of increasing complexity, the eccentricity ζ or f of the ellipsoid, the depolarization factor q, and the screening parameter κ. Eccentricity (from the Greek *ekkentros*, "out of the center," from *ek*, "out" + *kentron*, "center") is the measure of how much an ellipsoid deviates from the shape of a sphere. The eccentricity for prolate ellipsoids is defined as

$$\zeta^2 = \frac{a^2 - b^2}{a^2} = 1 - \frac{b^2}{a^2} \tag{5.27}$$

where (b/a) is the inverse of the aspect ratio of the particle (with b = semiminor axis, a = semimajor axis). The eccentricity of a sphere is $\zeta = 0$; for a prolate particle, $\zeta \to 1$ as the aspect ratio increases.

The concept of the depolarization factor q (the shape factor) emerged from the concept of the demagnetization factor first developed by J. C. Maxwell and

G. Mie and later embellished by R. Gans in 1915. The magnitude of the depolarization field depends on the shape of the particle and hence on the value of the depolarization factor q. The value of q ranges from $0 \rightarrow 1$ and the depolarization factors along the three coordinate axes of any particle conform to a sum rule

$$\Sigma q_i = 1 \tag{5.28}$$

where i indicates Cartesian coordinates a, b, c. With the electric field along the a-axis, the depolarization factor q_a is found from the eccentricity by

$$q_a = \frac{1-\zeta^2}{\zeta^2}\left[\frac{1}{2\zeta}\ln\left(\frac{1+\zeta}{1-\zeta}\right)-1\right] \tag{5.29}$$

For oblate particles with the electric field along the a-axis, the eccentricity f is equal to

$$f^2 = \frac{b^2-a^2}{a^2} = \frac{b^2}{a^2}-1 \tag{5.30}$$

and the depolarization factor q_a then is equal to

$$q_a = \frac{1+f^2}{f^2}\left[1-\frac{1}{f}(\arctan f)\right] \tag{5.31}$$

For prolate particles, q_b is found from the sum rule given earlier in equation (5.28). The depolarization factor is conveniently approximated by the following simpler relationship. We take a, b, and c to be the physical dimensions of the ellipsoid along the a, b, and c axes:

$$q_a = \frac{1/a}{1/a+1/b+1/c}; \quad \text{and if } b=c, \quad q_a = \frac{1/a}{1/a+2/b} \tag{5.32}$$

If the electric field is along the b-axis and from $q_a + 2q_b = 1$, then

$$q_b = \frac{1-q_a}{2} \quad \text{and} \quad \frac{1/b}{1/a+2/b} \tag{5.33}$$

The screening parameter κ indicates the amount of "screening" that takes place in a composite and is a function of the particle shape, size, and orientation of the material. The screening parameter is an indication of transparency. The calculation of κ depends on the depolarization factor q as follows:

$$\kappa = \frac{1-q_i}{q_i} \tag{5.34}$$

The screening parameter assumes any positive value from zero to infinity $(0 \le \kappa \le \infty)$ and is highly dependent on particle shape and orientation but not on the composition or distribution of particles. Composition is factored in by use of the complex dielectric constants ($\tilde{\varepsilon}$) of a material, and distribution is accounted for by the fraction of metal inclusion. All are displayed in the Maxwell-Garnett equation of effective medium theory [24,25]:

$$f_m \left(\frac{\tilde{\varepsilon}_m - \tilde{\varepsilon}_o}{\tilde{\varepsilon}_m + \kappa \tilde{\varepsilon}_o} \right) = \left(\frac{\tilde{\varepsilon}_c - \tilde{\varepsilon}_o}{\tilde{\varepsilon}_c + \kappa \tilde{\varepsilon}_o} \right) \tag{5.35}$$

where f_m is the volume fraction of the metal inclusion, the subscripts m and o are attached to the dielectric functions of the metal and the oxide, respectively, and subscript c is attached to the dielectric function of the composite (e.g., the effective medium). If the nanostructure of a composite consists of metal particles that are perpendicular to the electric field (e.g., perpendicular to the plane of the membrane) and the impinging radiation is also perpendicular to the membrane (e.g., the electric field of the light is in the plane of the membrane), as shown in **Figure 5.15,** then a screening charge is developed. This charge is able to exclude the photons from the metal and reflect the light back into the transparent medium. It is the screening parameter that we see in equations (5.25), (5.26), and (5.35). $\kappa = 2$ for spheres, and $\kappa = 1$ for wire-like (cylindrical) structures that are perpendicular to the electric field of the light (e.g., parallel to the impinging radiation); $\kappa = 0$, the condition of maximum screening (lowest absorption) in which parallel plates are perpendicular to the electric field (e.g., oblate particles along the c-axis), and $\kappa \rightarrow \infty$, the condition of minimum screening (maximum absorption) with plates parallel to the electric field (e.g., oblate particles along the *a* or *b* axes). Depolarization and screening values as a function of particle shape and orientation are summarized in **Figure 5.18.**

UV–visible spectra of gold nanoparticles of various aspect ratio and size are shown in **Figure 5.19.** The colors produced by gold–alumina composite membranes were illustrated in **Figure 3.32** in chapter 3.

The purpose of this exercise is to demonstrate how particle shape and orientation affect optical properties of nanoparticles. Unfortunately, we had to get at all of this in a roundabout, semi-ellipsoidal manner. We presented an expression for eccentricity and then used that to define the depolarization factor *q*. From *q*, the screening parameter κ was found—the term that imparts the effects of shape and orientation into the electrostatic expressions that describe the optical response of nanometals. Because we do not have to address electrodynamic factors (e.g., due to the quasi-static limit), all the working equations boil down to simple algebra. For detailed analysis, the imaginary components of the dielectric functions need also to be factored into the analysis.

5.5.3 *Quasi-Static Limit*

The quasi-static limit is achieved when the size of a nanoparticle is significantly smaller than the wavelength of the impinging EM radiation. A general rule of thumb is that the radius *r* of the particle should be on the order of 1/100 of the wavelength λ in order to qualify:

$$r \leq 0.01\lambda \tag{5.36}$$

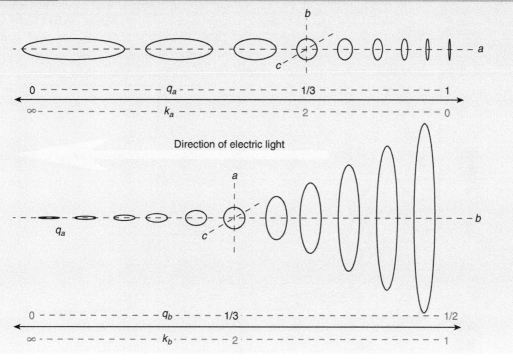

The progression of the depolarization factor and screening factor as a function of particle shape and orientation. The condition of the sphere is called three-dimensional isotropic screening (shape and orientation are uniform for a sphere and $\kappa = 2$). With $\kappa = 1$, the condition of two-dimensional isotropic screening is attained. This condition corresponds to the configuration depicted in Figure 5.15. The value κ equal to 0 corresponds to the condition of maximum screening (maximum transparency—e.g., the light is screened into the dielectric medium); κ equal to ∞ corresponds to the condition of minimum screening (maximum absorption).

Under these conditions, the oscillations of the time-dependent external field occur slowly relative to the motions of the free electrons comprising the plasmon. As a result, only the dipolar resonance of the plasmon contributes to the optical response. There are no phase shifts (retardation effects), multipole phenomena (e.g., quadrupole scattering and extinction), radiation damping, or dynamic depolarization. This condition lends itself to treatment by *electrostatic* principles, a far simpler method to predict absorption than if an electrodynamic treatment is required. The physical particle size regime that qualifies for quasi-static domain is $r < 10$ nm or so. Larger particles require Mie or Gans models to describe the optical response.

5.5.4 Orientation of Nanometals in Transparent Media

Composite membranes containing cylindrical gold nanoparticles ($a/b \approx 5$) embedded in polyethylene were analyzed with polarized light. In this way,

Experimental adsorption spectra (left) of gold–alumina composites are compared to simulated spectra (right) derived from Maxwell-Garnett effective medium theory relations. The diameters of the nanoparticles (b = c, the semi-minor axes) were equal to 16 nm in all experiments. The semi-major axis, a, was varied in the experiment and simulations, and the direction of the electric field of the light was perpendicular to the semi-major axis of the nanoparticles (e.g., the electric field of the light was parallel to the plane of the membrane), as shown in Fig. 5.15. The lengths a (in nm) correspond to spectra from top-to-bottom in the figures as follows (aspect ratios are given in parentheses): 165 (10.3), 104 (6.5), 69 (4.3), 39 (2.4), 24 (1.5), and 16 (1.0). The experimental positions of maximum absorption λ_{max} of spectra (in nm), from top-to-bottom in the figures, are as follows (theoretical values are in parentheses): 508 (508), 512 (509), 513 (516), 526 (520), and 539 (533). As aspect ratio was increased, the position of maximum absorption underwent a blue-shift (higher energy, shorter wavelength).

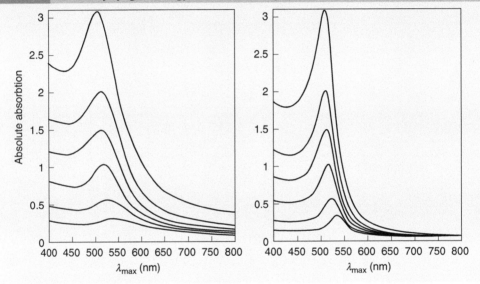

explicit investigation of particle orientation was possible. It is well known that ellipsoids with two perpendicular main axes are capable of generating two distinct λ_{max} [23]. In the case of the sphere, degeneracy exists with regard to plasmon oscillations and only isotropic optical response is expected. The prolate nanogold particles were oriented uniformly in a thin polyethylene film with the semimajor axis parallel to the plane of the film (e.g., the b-axis) (**Fig. 5.20a**) [26]. Spectral acquisition was conducted perpendicular to the plane of the nanocomposite film as the sample was rotated in-plane from 0 to 90°. The results are shown in **Figure 5.20b.**

When the incident electric field of the polarized light was held perpendicular to the major axis of the particles ($\theta = 90°$), λ_{max} occurred at 550 nm, yielding a reddish-purple appearance to the composite ($q \approx 1/2$, $\kappa \approx 1$). At the other extreme, with $\theta = 0°$, the case in which the major axis was rotated parallel to the electric field, λ_{max} shifted to an extremely broad peak at ca. 800 nm, nearly a 250-nm red shift, and the composite appeared greenish ($q \approx 1/10$, $\kappa \approx 10$). The condition of dichroism in this case is very different from that observed in the

FIG. 5.20

(a) Synthesis process of a nanogold–polyethylene composite material. Gold nanocylinders were first formed by electrodeposition inside the pore channels of a porous alumina membrane formed by the anodic process. The alumina was dissolved with NaOH, leaving gold nanocylinders on the surface of the polyethylene. The cylinders were oriented by applying light pressure with the tip of a pencil's eraser. The membrane was placed under a polarizer and rotated 90° (red film) to 0° (green film). The red indicated that the long axis of the cylinders was perpendicular to the electric field of the light and the green color indicated that the cylinders were parallel to the electric field of the light. (b) The corresponding spectra are displayed. The top curve (red, 550 nm) is derived from particles that are perpendicular to the electric field of the light. As the plane was rotated from 90 to 0°, incremental transition of the membrane color from red to green (800 nm) was observed.

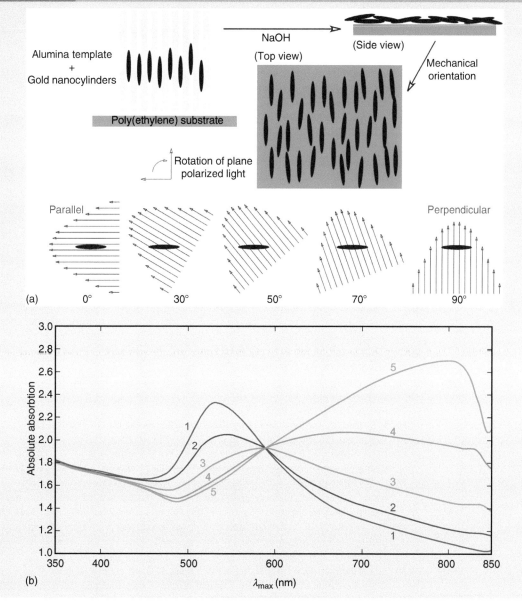

Lycurgus cup. Dichroism in this experiment was due to the orientation of identically aligned particles with respect to polarized light. In the Lycurgus cup, the apparent dichroism was due to absorption, transmission, and reflection phenomena (dependent on where the light source was positioned).

References

1. G. Koller, Study of ice leads to cool new research, Pacific Northwest National Laboratory, *U.S. Department of Energy Research News* (2001).
2. A. Sanchez, S. Abbet, U. Heiz, W.-D. Schneider, H. Häkkinen, R. N. Barnett, and U. Landman, When gold is not noble: Nanoscale gold catalysis, *Journal of Physical Chemistry A*, 103, 9573–9578 (1993).
3. N. Lopez, T. V. W. Janssens, B. S. Clausen, Y. Xu, M. Mavrikakis, T. Bligaard, and J. K. Nørskov, On the origin of the catalytic activity of gold nanoparticles for low-temperature CO oxidation, *Journal of Catalysis*, 223, 232–235 (2004).
4. G. Oberdörster, Toxicology of ultrafine particles in vivo, *Philosophical Transactions of the Royal Society of London Series A*, 358, 2719–2740 (2000).
5. A. D. Maynard, *Nanotechnology: Overview and relevance to occupational health*. Project on Emerging Nanotechnologies for the Woodrow Wilson International Center for Scholars (2005).
6. R. D. Averitt, S. L. Westcott, and N. J. Halas, The linear optical properties of gold nanoshells, *Journal of the Optical Society of America*, 16, 1824 (1999).
7. J. B. Jackson and N. J. Halas, Silver nanoshells: Variations in morphologies and optical properties, *Journal of Physical Chemistry B*, 105, 2743 (2001).
8. S. R. Sershen, S. L. Westcott, J. L. West, and N. J. Halas, An optomechanical nanoshell–polymer composite, *Applied Physics B*, 73, 379 (2001).
9. C. Loo, A. Lowery, N. J. Halas, J. West, and R. Drezek, Immunotargeted nanoshells for integrated cancer imaging and therapy, *Nano Letters*, 5, 709–711 (2005).
10. F. B. Calleja, and Z. Rosianiec, *Block copolymers*, CRC Press, Boca Raton (2000).
11. V. P. Menon and C. R. Martin, Fabrication and evaluation of nanoelectrode ensembles, *Analytical Chemistry*, 67, 1920–1928 (1995).
12. J. Rouquerol, D. Avnir, C. W. Fairbridge, D. H. Everett, J. H. Haynes, N. Pernicone, J. D. F. Ramsay, K. S. W. Sing, and K. K. Unger, Recommendations for the characterization of porous solids, *Pure & Applied Chemistry*, 66, 1739–1758 (1994).
13. C. P. Poole and F. J. Owens, *Introduction to nanotechnology*, Wiley-Interscience, John Wiley & Sons, Inc., New York (2003).
14. S. K. Singh, K. M. Parida, B. C. Mohanty, and S. B. Rao, High surface area silicon carbide from rice husk: A support material for catalysis, *Reaction Kinetics & Catalysis Letters*, 54, 29–34 (1995).
15. R. L. Johnston, *Atomic and molecular clusters*, Taylor & Francis, London (2002).
16. E. A. Irene, *Electronic materials science*, Wiley-Interscience, Hoboken, NJ (2005).
17. W. D. Knight, K. Clemenger, W. A. de Heer, W. A. Saunders, M. Y. Chou, and M. L. Cohen, Spectroscopy of metal clusters, *Physics Review Letters*, 52, 2141–2144 (1984).
18. G. L. Hornyak and C. R. Martin, Optical properties of a family of Au-nanoparticle-containing alumina membranes in which the nanoparticle shape is varied from needle-like (prolate), to spheroid, to pancake-like (oblate), *Thin Solid Films*, 303, 84 (1997).
19. G. L. Hornyak, C. J. Patrissi, and C. R. Martin, Optical properties of gold–porous alumina composite membranes: The Maxwell-Garnett limit, *Journal of Physical Chemistry B*, 101, 1548 (1997).
20. G. L. Hornyak, C. J. Patrissi, E. B. Oberhauser, J.-C. Valmalette, L. Lamaire, J. Dutta, H. Hofmann, and C. R. Martin, Optical properties of Au–Ag nanoparticle alumina composites, *Nanostructured Materials*, 9, 571 (1997).

21. U. Kreibig and M. Vollmer, *Optical properties of metal clusters*, Springer-Verlag, Berlin (1995).
22. A. Campion, Raman spectroscopy of molecules adsorbed on solid surfaces, *Annual Review of Physics*, 36, 549–572 (1985).
23. S. Link and M. El-Sayed, Shape and size dependence of radiative, nonradiative and photothermal properties of gold nanocrystals, *International Review of Physical Chemistry*, 19, 409–453 (2000).
24. J. C. Maxwell-Garnett, Colours in metal glasses and metal films, *Philosophical Transactions of the Royal Society of London A*, 203, 385–420 (1904).
25. J. C. Maxwell-Garnett, Colours in metal glasses, in metallic films and in metallic solutions—II, *Proceedings of the Royal Society of London*, 76, 370–373 (1905).
26. C. A. Foss, G. L. Hornyak, J. A. Stockert, and C. R. Martin, Template synthesis and optical properties of small metal particle composite materials: Effects of particle shape and orientation on plasmon resonance maxima, *MRS Symposium Proceedings*, 286, 431–436, Fall (1992).

Problems

5.1 List 20 examples of systems that demonstrate enhanced surface area, whether at the nanoscale or at the macroscale.

5.2 In example 5.2, we determined the collective surface area of nanocubes derived from a parent cube 1 m on each side. What is the total surface are of spheres inscribed within each cube (diameter = 1 nm)?

5.3 If the parent cube consisted of an *fcc* structure with spheres 1 nm in diameter, what would be the collective surface area of all the individual spheres? How does this compare to your answer in problem 5.2?

5.4 In example 5.6, we introduced hypothetical gold—a material packed in a simple cubic structure (AAA...). What is the density of hypothetical gold (atomic mass $= 3.27 \times 10^{-22}$ g)? How does it compare to the density of real gold, $\rho = 19.31$ g·cm³ (an *fcc* structure)? Assume atomic radii are the same.

5.5 Derive a generalized expression of surface area of right circular cylinders of constant volume normalized to the volume of a sphere. What does the graph look like? Hint: derive an expression of S_{cyl} as a function of (h/d).

5.6 Calculate the interior nanosurface area of a hypothetical ceramic material that is configured in an orthogonal network of interconnected tubes 20 nm in diameter. The tubes are spaced 100 nm apart center to center along x–y–z directions. What is the porosity of the material? The pore volume?

5.7 Some quantum dots exist in the shape of a pyramid. Derive a generic expression for the surface-to-volume ratio for such a pyramid. How does the S/V compare to that of a sphere with the same volume? Do the same for a cone.

5.8 Based on the spherical cluster approximation (SCA), calculate the surface fraction of sodium atoms of a 10-nm diameter cluster.

5.9 At what cluster diameter does the number of surface atoms drops below 10% of the total number of atoms in the cluster? (Adapted with permission from R. L. Johnston, *Atomic and Molecular Clusters*, Taylor & Francis, London, 2002.)

5.10 Considering packing fraction (not discussed in our presentation of the SCA), how is the number of surface atoms affected in a tungsten cluster that actually has a packing fraction? Assume packing fraction $f = 0.74$ for an *fcc* cluster. (Adapted with permission from R. L. Johnston, *Atomic and Molecular Clusters*, Taylor & Francis, London, 2002.) All other

SCA assumptions remain intact, and consider

$$N \cong \frac{V_c}{V_a^*} = \left(\frac{D}{2R_a}\right)^3 f$$

where
  V_a^* is the effective atomic volume
  N is the number of atoms in the cluster
  D is the diameter of the cluster
  V_c is the cluster volume
  R_a is the atomic radius (0.148 nm for W)
  f is the packing fraction (0.74 for *fcc* and 0.68 for *bcc*)

5.11 What is the diameter of a spherical *fcc* cluster that has 10^6 atoms with effective R_a = 0.25 nm?

5.12 Real clusters are not perfectly spherical. Why do you think this is the case? Would you expect them to become more spherical as the cluster size is increased?

5.13 Tabulate the specific surface areas (square meters per gram) of cubes, spheres, wires, cylinders, and discs made of gold (ρ = 19.31 g · cm^{-3}) using the surface-to-volume relations derived in example 5.7. Beginning with 1 nm and doubling the appropriate dimension until you reach 1052 nm, what trends can you make out from the table? What is unique about the relationship between the cube and the sphere? Is it different from the relationship we derived earlier in example 5.3? Why?

5.14 Using only the C–C bond distance in graphite and simple geometry, calculate (*by brute force!*) the diameter, the surface area, specific surface area, and density of a single-walled carbon nanotube pictured below that has 12 hexagons of carbon along the circumference with the long axis of the hexagon oriented along the long axis of the tube. Do not refer to chapter 9 for vector formulae. Assume that the C–C bond distance in single-walled carbon nanotubes is equal to 0.144 nm. The atomic weight of C is 12.01 g·mol^{-1}.

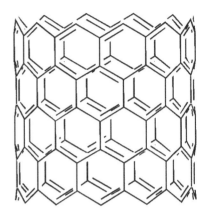

5.15 Calculate the planar density of low-index planes (100), (110), and (111) for an *fcc* crystal.

5.16 With regard to the [*uvw*] crystallographic direction, draw the [111] vector of a cubic system starting at the origin (*a,b,c* or *x,y,z*) = (0,0,0).

5.17 Define crystal density, planar density, atomic density, and linear density.

5.18 Calculate the crystal density of *fcc* gold. How does it compare to the experimentally determined value?

5.19 Calculate the linear density in terms of atom radius along the [110] crystallographic direction of an *fcc* crystal.

5.20 Calculate the planar density in terms of atom radius of the (110) plane in an *fcc* crystal.

5.21 Calculate the *d*-spacing for (111) planes in gold. Correlate these results to experimental data acquired by XRD (do some research).

5.22 Do some research and find out the value of the capacity of the average human lung. What would the size of the lungs be if there were no alveoli in them?

5.23 Concerning the resonance condition discussed in section 5.0, how does the magnitude of the screening parameter κ influence the wavelength of maximum absorption?

5.24 Discuss briefly how eccentricity, the depolarization factor, and the screening parameter are related.

5.25 What is a surface plasmon? Why do you think it is affected by the size of the material? What other factors are able to influence the surface plasmon?

5.26 What kind(s) of hole(s) is (are) found in a NaCl crystal? What is (are) its (their) size?

5.27 Calculate the linear and planar densities of *bcc* α-iron along the [100] direction and the (100) plane, respectively.

5.28 Starting with one cube (the first magic number) and then surrounding it with 26 cubes to generate the second magic number (27), derive a generic equation for magic numbers for such a cubic system. What is the formula for each additional layer?

5.29 Why are smaller nanoparticles usually more active in catalysis than larger ones? (Question courtesy of Günter Schmid, Uni-Essen.)

5.30 Solutions containing nanorods made of gold show 2 absorption maxima. Why is this? If the solution contained only spherical colloids, how many absorption maxima would be expected?

ENERGY AT THE NANOSCALE

When the attractive power of the solid is greater or less than half that of the liquid, the surface of the liquid must, at its origin, be in the same direction with that of the solid, instead of forming an angle with it, as it often does in reality....In this manner it may be shown that if the attractive power of the solid be equal to that of the liquid, or still greater, it will be wetted by the liquid, which will rise until its surface acquires the same direction with that of the solid; and that, in other cases, the angle of contact will be greater, in proportion as the solid is less attractive.

THOMAS YOUNG, M.D., *A COURSE OF LECTURES ON NATURAL PHILOSOPHY AND THE MECHANICAL ARTS* (PRINTED BY TAYLOR AND WALTON, LONDON, 1845)

*C*hapter 6

THREADS

Surface, geometric, and structural aspects of nano-materials, particularly of metals, have been introduced and briefly discussed in chapter 5. For the most part, not much discussion about energy or chemistry was presented in the previous chapter. It is time now to investigate the energetic character of nanoscale materials, particularly in the way it correlates with the nano surface. In this chapter, we begin to connect nanostructure with surface energy, and we wrap up the "What are nanomaterials?" and "Why are they interesting?" parts of the text. In chapter 7, discussions about why they have remarkable properties and phenomena begin in earnest. Chapter 7 proceeds to investigate the breakdown of bulk scaling laws at the nanoscale as numerous physical properties and phenomena are individually addressed. In chapter 8, a special section on nanothermo-dynamics is brought to the forefront—a new way to look at thermodynamics.

6.0   SURFACE ENERGY

The structure of nanomaterial surfaces contributes significantly to the energetic character of nanomaterials. Conversely, one can say with equal validity that the structure of nano surfaces is certainly due to energetic considerations. These statements correlating surface structure to energy also apply to bulk material surfaces. However, the net energetic impact of the bulk material is significantly reduced when compared to an equal volume of nanoparticles. This is not to imply that the surface is not important in bulk materials. It most certainly is important. Whether in nanoscale or bulk form, atoms and molecules that exist at the surface or at an interface are different from the very same atoms or molecules that exist in the interior of a material. The difference is manifested in altered structure, enhanced reactivity, and a greater tendency to agglomerate. The surface is the place where periodicity and chemical bonding of a crystal are disrupted and represent a phase that wishes to become something else.

The surface layer (selvage layer) of a bulk solid consists of about three or fewer layers of atoms or molecules, a few angstroms at the most [1]. These surface atoms on average have fewer nearest neighbors. Having fewer neighbors raises the energy and hence lowers the stability of those atoms and, consequently, of the surface itself. We know from our earlier discussion of electron shells, nuclear stability, and magic numbers that closed structures are preferred due to their enhanced stability (e.g., the state of minimal energy). In a bulk metal *fcc* crystal, the optimal coordination number Z for each atom is 12. Anything less than this ideal imparts an inherent instability to the surface (e.g., by way of the *dangling bond*). Dangling bonds spread gloom over the surface—a state of high anxiety. Dangling bonds must find ways to compensate for the apparent deficiencies by whatever chemical or physical means available to them. Please take note that we are only referring to bulk surfaces in this last discussion. If surface atoms do exist in a state of higher energy than fully coordinated volume atoms, we can safely conclude that it must take an investment in energy to create a new surface.

If surfaces were suddenly to become the energetically favored state, there would be driving forces to create more surfaces. That would indeed be a peculiar universe, one that consists entirely of surfaces. In reality, driving forces exist in materials and systems to actually minimize surface area. The smaller the surface-to-volume ratio is, the lower is the energy state of the material. Why then do we have nanomaterials? Nanomaterials are characterized by increased surface area, as we found out in the previous chapter. However, now in the context of this chapter, we try to link these behaviors and phenomena to the surface and hence to the energetic state of the nanomaterial. We have nanomaterials because we have found ways to ensure their presence—as has nature.

We introduced the discipline of surface science in the last chapter and continue here with the same theme. Surface science focuses on the interface between physical phases, whether solid–solid, solid–liquid, liquid–gas, or solid–gas. Therefore, the outermost atomic or molecular layers are involved in surface science investigations. Surface energy is an extrinsic (or *extensive*—more about this term in chapter 8) property of all materials, whether flat or otherwise. In other words, surface energy is an additive quantity. The surface energy of 10 identical nanoparticles is equal to the sum of the surface energy of each individual particle. If these particles were to agglomerate, the overall surface energy is reduced. **Figure 6.1** shows a simple geometric example of agglomeration and how surface energy is reduced overall as larger particles are formed.

EXAMPLE 6.1   *Surface Energy by a Simple–Brute Force Geometric Model*

Devoid of any chemistry or physics, show how surface energy can be reduced by agglomeration of cubes of unit volume to form larger cubic or rectangular solid shapes. Calculate (by brute force) the surface energy of resulting geometric shapes formed by the following agglomeration schemes. Assume that the surface energy of one face of the original cube is equal to γ.

(a) Zero dimensions (a dot): What is the total surface energy of one cube (**Fig. 6.1a**)? What is the surface energy of 27 independent cubes?

(b) One dimension (a wire): Determine the total surface energy of nine cubes agglomerated in one row (**Fig. 6.1b**). What is the surface energy of three rows consisting of nine cubes each?

(c) Two dimensions (thin film): Starting with one cube, completely surround the central cube in two dimensions with additional cubes until you arrive at a square configuration that is 3×3 cubes (**Fig. 6.1c**). What is its surface energy? What is the surface energy of three such squares?

(d) Three dimensions: Starting with one cube, form a larger cube by completely surrounding the central cube to form a large cube that is $3 \times 3 \times 3$ cubes (**Fig. 6.1d**). What is the surface energy?

Solutions:

(a) *If we assign the value γ to the surface energy of one face of a cube:*
The total surface energy of one cube is $6 \times \gamma = 6\gamma$.
The combined surface energy of 27 cubes is $27 \times 6\gamma = 162\gamma$.

(b) *The surface energy of a line of nine cubes is $[(9 \times 6) - (8 \times 2)]\gamma = 38\gamma$. The combined surface area of three such lines is $3 \times 38\gamma = 114\gamma$.*

(c) *The surface energy of a square of nine cubes is equal to $[(2 \times 9) + (4 \times 3)]\gamma = 30\gamma$.*
The surface energy of three such squares is $3 \times 30\gamma = 90\gamma$.

(d) *The surface of a cube consisting of 27 cubes is equal to $(6 \times 9)\gamma = 54\gamma$.*
Therefore, at constant volume, the surface energy decreases as the level of agglomeration increases:
162γ (independent cubes) $\rightarrow$ 114γ (linear cubes) $\rightarrow$ 90γ (square cubes) $\rightarrow$ 54γ (cubic cubes).

FIG. 6.1   *The surface energy of one side of these cubes is equal to γ. The combined surface energy of the each cube is 6γ. When two cubes are combined, two surfaces are lost: $(2 \times 6)\gamma - 2\gamma = 10\gamma$. Due to the fusion, the energy of two surfaces has been reduced to cohesion energy equal to 2γ. Please refer to example 6.1 for more details.*

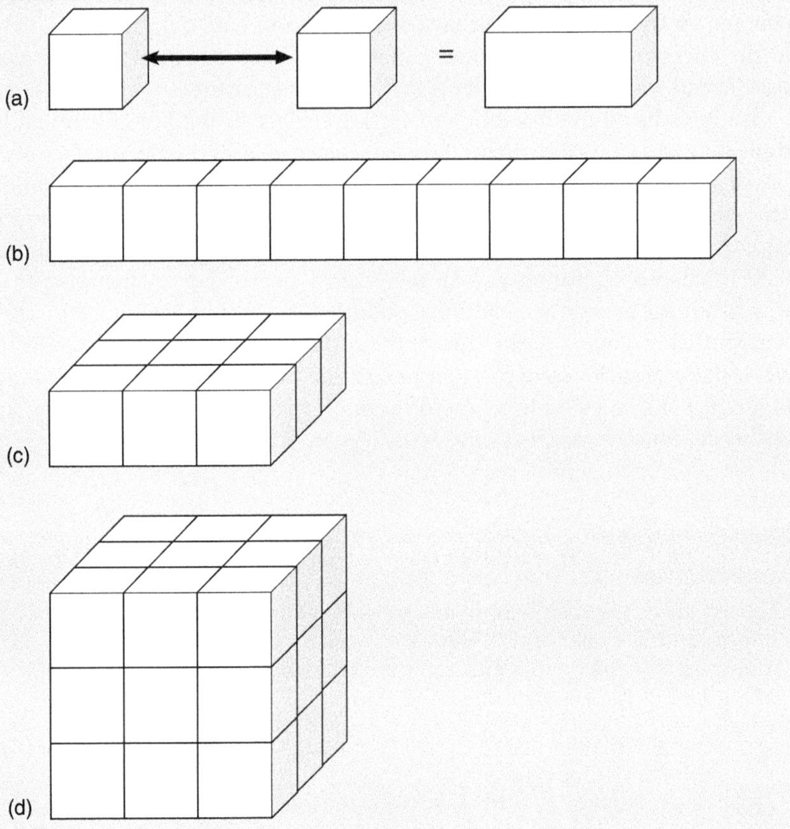

In addition to nanosized liquids and solids, surface phenomena govern much of the physical and chemical nature of colloids (ca. 10^{-6} to 10^{-9} m in size). The chemistry of colloids is really the chemistry of the surface. Surface phenomena are also vital to the functions of living things. Since the living cell is a complex of colloidal systems [2], there must be a significant number of surfaces available in cells to conduct daily business.

The surface is important at all size scales. At the nanoscale, surface phenomena become very interesting. Energy is a part of everything. The nanosurface is another boundary condition—one that actually interfaces with the macroscopic world. It is at this boundary that the currency of energy is exchanged.

6.0.1   *Background*

The phenomenon of capillarity has been known since the time of Leonardo da Vinci and probably earlier; however, the study of surface chemistry began

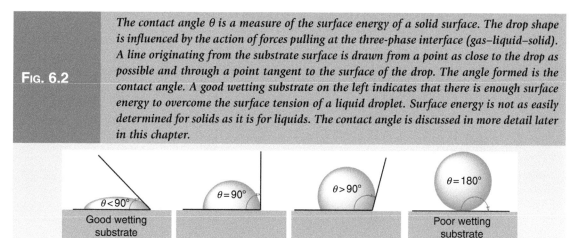

FIG. 6.2 *The contact angle θ is a measure of the surface energy of a solid surface. The drop shape is influenced by the action of forces pulling at the three-phase interface (gas–liquid–solid). A line originating from the substrate surface is drawn from a point as close to the drop as possible and through a point tangent to the surface of the drop. The angle formed is the contact angle. A good wetting substrate on the left indicates that there is enough surface energy to overcome the surface tension of a liquid droplet. Surface energy is not as easily determined for solids as it is for liquids. The contact angle is discussed in more detail later in this chapter.*

formally when Thomas Young in 1805 and Pierre Simon de Laplace in 1806 first investigated the surface tension (and curvature phenomena) of liquids. Young stated that the contact angle (of a drop of liquid on a flat solid substrate) that exists at the triple phase point of a solid–liquid–gas interface is a physical constant that depends on the materials involved (**Fig. 6.2**). Young is also noted for coining the word *energy* in the context of the "fundamental quantity created by heat which moved particles." Young was considered by some to be the last *polyhistor*, the "last man to know everything" [3].

Johann Carl Friedrich Gauss in 1830 introduced the concept of *surface energy*. He based his concept on the sum total potential of interacting pairs of molecules. In his *Principia Generalia Theoriae Figurae Fluidorum in statu Aequilibrii*, Gauss describes a mechanical theory of heat and claimed that his theory of potentials and methods of least squares connected nature with science. Josiah Willard Gibbs in 1877 developed the classical thermodynamic treatment of surface phenomena and the concept of the *Gibbs dividing surface*. G. Wulff in 1901 is credited with devising "Wulff plots," which depict the equilibrium surface of finite anisotropic solids—the minimum energy surfaces. Significant credit must once again be given to Irving Langmuir who, in the early twentieth century, developed an understanding of heterogeneous catalysis and adsorption kinetics and their associated surface phenomena. John Bardeen advanced the concept of the electric-double layer in 1936. The ability to fabricate a *clean* surface with the advent of ultrahigh vacuum systems enabled researchers to study reproducible crystal surfaces.

6.0.2 Nature

Just like in the discussion of the nanosurface and nanosize in chapter 5, we must always pay tribute to the remarkable examples that nature provides. Macroscopic systems exhibit surface energy phenomena that are based on nanoscale structure. The perfect (and often used) example to illustrate natural surface energy is the leaf of the lotus plant. The *lotus effect* (**Fig. 6.3**), a phenomena described by

FIG. 6.3

The lotus effect is depicted to demonstrate an extreme contact angle: ~180°. Solid surfaces that generate contact angles greater than 150° are considered to be superhydrophobic. For the lotus plant to achieve a contact angle of nearly 180°, hydrophobicity by itself is not enough. The drop is contacted by numerous structures that terminate in a point. In this way, the surface tension of the drop is never overcome and the surface behaves as if it were hydrophobic. Surface roughness of high-energy surfaces (hydrophilic) enhances good wetting. Surface roughness of low-energy surfaces (hydrophobic) enhances drying (e.g., the liquid must expend energy in order to conform to the topology of the surface). Synthetic surfaces called **Fakir** *carpets are roughened by facets 50 μm wide by 150 μm deep. Spacing between facets is 100 μm. On these surfaces, water droplets are removed without any wetting. If tilted, droplets roll right off.*

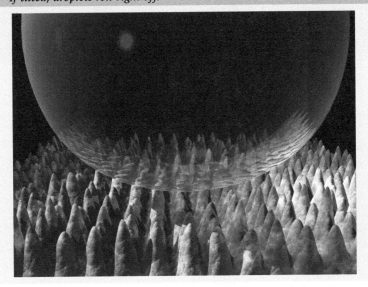

Source: Courtesy of Professor Anil K. Rao, Department of Biology, the Metropolitan State College of Denver.

W. Barthlott in 1990, is a property exhibited by the leaves of the lotus and other similar plants. The lotus leaf is a natural low-energy surface onto which water beads up into a nearly perfect sphere and rolls off the leaf, truly a benefit to a rain forest dweller [4]. Due to the intricate micro- and nanostructure of the leaf surface, enhanced drying and self-cleaning properties are also observed. A detailed presentation of the lotus effect is given in chapter 13, "Natural Nanomaterials." The water strider insect is an example of how another living organism takes advantage of the surface tension properties of water. The legs of the water strider are water repellant [5]. The macroscopic hairs possess nanofila-mentous structures that are able to support small air bubbles, thereby allowing the insect to stride over the surface of a pond or stream [5].

Microscale examples of surface phenomena in nature include the cell membrane and any other dividing surface found in eukaryotic cells and bacteria as well as the boundaries that enclose organelles and segregate vacuoles. The formation of hydrophobic or hydrophilic surfaces that are stabilized in complementary environments is a critical component in stabilizing all of life's surfaces. At the nanoscale, protein–surface interactions modulate cell adhesion, migration, pro-liferation, and differentiation [6]. For this reason, the adsorption of proteins at solid–liquid interfaces is the focus of much scientific endeavor. Depending on which amino acids extend to the surface, proteins can be soluble or insoluble in water.

6.0.3　Introduction to Surface Stabilization

At this introductory stage in the chapter, we assume that nanosurfaces are by nature higher in energy than surfaces of bulk material counterparts. In the next step, we assemble our collective knowledge of chemistry and state that "systems of high energy will strive to attain a state of lower energy, by whatever means possible." Fewer nearest neighbors for surface atoms and molecules result in atomic or molecular species with dangling bonds—bonds that require satisfaction. In **Tables 6.1** and **6.2**, the topic of surface stabilization is summarized along with examples of recourses available to systems to attain a minimal energy condition.

There are many chemical and physical mechanisms available to achieve compensation for bonding deficiencies: higher chemical reactivity, shorter bond length (accompanied by shrinking of interlayer distances), anisotropic structural distortion, coalescence and lowered melting points, and evaporation temperature,

TABLE 6.1	*Physical Modes to Stabilize High-Energy Surfaces*
Physical process	**Mechanism**
Surface relaxation	Inward forces reduce d-spacing. Consequences are reduced lattice constant and shorter bond length with surface atoms. Lateral forces create an asymmetric environment.
Surface restructuring	Formation of highly strained homogeneous bonds that impact crystal growth
Composition segregation	Diffusion of impurities to the surface Phase segregation
Physisorption	Formation of a monolayer (or more) of weakly bound atoms or molecules
van der Waals force	van der Waals forces are exerted at short distances. Does the gecko stabilize a surface by adhering to it?
Sintering	Formation of polycrystalline condensed materials at elevated temperatures. Grain boundary stabilization
Isotropic minimization	Formation of spherical particles to produce the lowest energy geometric solid
Anisotropic minimization	Equilibrium planes that form multifaceted structures. Concave versus convex surfaces
Maximization of packing efficiency	Low-index crystal planes enhance the number of nearest neighbors in three-dimensional solids; elimination of voids, defects, dislocations, etc. in nanomaterials.
Geometric agglomeration	Nonchemical agglomeration in the vapor and liquid phases. In liquid phase, this is called *Ostwald ripening*, where a larger particle grows at the expense of higher energy smaller particles.
Physical confinement	Confinement (kinetic) of metastable species in pore channels to preserve nanoparticle structure. This is a difficult category to precisely define because chemical stabilization may occur from the species in the wall of the cavity.
Artificial environments	Ultrahigh vacuum spaces keep metastable particles free of interaction with contaminants. The development of this technology allows R&D to proceed for high surface energy materials.

TABLE 6.2	*Chemical Modes to Stabilize High-Energy Surfaces*
Physical process	**Mechanism**
Chemical termination	Covalently bonded chemical species to surface atoms
Electrostatic stabilization	Formation of an electrical double layer to stabilize reactive species and particles in solution
DLVO theory	Stabilization by van der Waals attraction + electrostatic repulsion in solutions
Supramolecular stabilization	All intermolecular forces serve to stabilize supramolecular species in solution
Stabilization of biological molecules	Environmental controls (mild thermal conditions, protection from radiation); numerous intermolecular forces
Steric stabilization	Polymeric layered materials; ligand formation between sol and polymer chain. The process serves to isolate nanoparticles (e.g., stabilize).
Chemisorption	Chemisorption indicates a weaker chemical bond than a covalent bond. Chemisorbed molecules and atoms can be removed during TPD.
Surfactants	Surface stabilization of metastable phases by hydrophobic–hydrophilic interactions in solution
Kinetic stabilization	Metastable molecules that are thermodynamically unstable exist because there is no pathway available for chemical reaction.

just to name a few. Bulk surfaces, however, are not generally considered to be reactive. How do we reconcile what we just stated with reality? It is true that the surface achieves stability by distortions, passivation, and other processes, but overall, the surface of a bulk material contributes a miniscule amount to the overall stability of the material as a whole. A block of iron sitting in your garage is going nowhere, but it can rust. Once you remove the rusty layer, you basically have not lost much.

Surface energy is minimized in biological systems. One way to minimize the impact of the surface is to grow larger. For example, spherical particles are thermodynamically stable but biologically inefficient, unless they are one-celled organism. Some spherical algal species have evolved into extremely long cylindrical cells. Some of the largest cells in the world belong to algae species. Although extremely long (meters), the thickness of the cell is a micron or so—just about the size of a paramecium.

6.0.4 The Nano Perspective

Nanoscience is, to a great degree, a tribute to those who have been able to fabricate small materials as well as to those who devised ingenious methods that keep them small. These breakthroughs collectively form the basis for nanoscience and technology. Nanoscience is also about how macroscopic phenomena are transposed to the nanoscale and subsequently subjected to interpretation. All macroscopic phenomena retain relevance at the nanoscale; we just have to figure how and why its form has been altered and what it cost us. In this chapter, we discuss surface energy, the energy of adhesion, and the energy of cohesion, and the phenomena of capillarity, curvature, vapor pressure, and the Kelvin effect. These properties and phenomena were developed in the past century and even earlier in the nineteenth century. We now are able to measure and observe them at the nanoscale. Imagine if Thomas Young been able to peer through an electron microscope just once!

6.1 BASIC THERMODYNAMICS

Surface tension, γ, is defined as the force per unit length that opposes the expansion of surface area [7]. We will work in some basic concepts and derive surface energy expressions from thermodynamic relations—as always, a good place to start. To break up any condensed phase, whether liquid or solid, work has to be accomplished (energy has to be added) and only then can a new surface be created. Making a new surface involves taking atoms or molecules from the volume of a bulk material where they are stable and forcing them into a surface layer where they are inherently less stable but adapt to achieve stability. This relocation requires positive energy to accomplish. The work is measured in terms of surface energy (appropriate for solids) and surface tension (appropriate for liquids) and has units of joules per square meter or newtons per meter. Both forms are represented by the Greek letter γ:

Surface tension, $\gamma \rightarrow J \cdot m^{-2}$, $N \cdot m^{-1}$ (or $erg \cdot cm^{-2}$, $dyn \cdot cm^{-1}$) $\leftarrow$ surface energy, γ

The symbols σ and T are also used to represent surface tension. Liquid surfaces tend to contract to achieve a surface of minimum energy (the sphere) due to strong inward attractions. Take note: Once again the sphere makes an appearance to help us explain the nanoscale.

6.1.1 Derivation of Surface Tension, γ

Surface tension expressions can be derived entirely from mechanical thermodynamic principles. Application of the Helmholtz (A with V,T constant) and Gibbs (G with P,T constant) functions is required in order to understand the nature of surface tension. The incremental amount of work required to generate an incremental amount of new surface is

$$dW_{rev} = \gamma d\mathcal{A} \qquad (6.1)$$

where W is work in joules, $\mathcal{A}$ is the area in square meters, and γ is a proportionality constant that is always positive and equivalent to the surface energy (tension). The corresponding Helmholtz work function is then equal to [8]

$$dA = \gamma d\mathcal{A} \qquad (6.2)$$

The magnitude of γ is, of course, characteristic of the chemical nature of the surface material and $\gamma d\mathcal{A}$ is expressed in the form of energy.

To change the surface, clearly some form of pressure–volume work is required in addition to the usual definition of work, W—a term describing the incremental area of new surface ($+\gamma d\mathcal{A}$). With respect to a flat planar interface, the work required to create a new surface also involves changes in volume. The expression of reversible work required to add an increment of new surface is

$$dW_{rev} = -PdV + \gamma d\mathcal{A} \qquad (6.3)$$

where $(-PdV)$ is the work required to effect any volume change and P is the equilibrium pressure in the liquid and gas bulk phases. What exactly does surface tension in the form of newtons per meter (or dynes per·centimeter) indicate?

If a force (along the y-direction only) is applied by a piston (with area $\mathcal{A} = l_x l_z$) inside a rectangular container ($l_x \times l_y \times l_z$) and filled with one fluid, the resultant pressure of the system is equal to P, the equilibrium pressure [9]. The incremental amount of work required is simple pressure–volume work and is proportional to the incremental displacement of the piston in the y-direction, dl_y [9]:

$$dW_{rev} = F_{piston}\, dl_y = -P\,(l_x\, l_z\, dl_y) = -PdV \tag{6.4}$$

The expression for work then is adapted to accommodate the contribution of the interfacial layer:

$$dW_{rev} = F_{piston}\, dl_y = -P(l_x\, l_z\, dl_y) + \gamma l_x\, dl_y \tag{6.5}$$

$$F_{piston} = -P(l_x l_z) + \gamma l_x \tag{6.6}$$

$$P_{piston} = P - \gamma/l_z \tag{6.7}$$

In other words, a line of force is in place at the interface of the two phases—one that resists the force ($P\,\mathcal{A}$) of the piston. The surface tension is responsible for the resistance [9] and opposes the force of the piston.

Surface tension and surface energy are related to the differential form of the standard thermodynamic expression for free energy. Since energy is an extensive quantity, we are free to add terms as required. The surface energy requires us to add a surface energy term to the standard free energy expression:

$$dG = -S\,dT + V\,dP + \gamma\,d\mathcal{A} \tag{6.7a}$$

Differentiating with respect to temperature, we get the classical temperature dependence of the free energy (the surface free energy G_s) [17].

6.1.2 Surface Excess

Regardless which convention is used, energy is required to create a new surface and, on the flip side, energy in the form of heat is released when surfaces are merged. This happens when two drops of water merge to become one larger

---

EXAMPLE 6.2   *How Much Work Is Needed to "Uncontract" (Stretch) the Surface of Water?*

Neglecting the force of gravity, if you have a wire of length 2 cm, calculate the work required to lift a sheet of water held together by surface tension to a height of 5 mm. The surface tension of water is 72.8 dyn·cm⁻¹.

Solution:
This procedure has created a total surface equal to $2 \times l \times h = 2 \times 2$ cm $\times 0.5$ cm $= 4$ cm². The total work done to create this new surface is $(2\, l\, h) \times \gamma_{water} = 4$ cm² $\times 72.8$ dyn·cm⁻¹ $= 291$ dyn·cm⁻¹ $= 291$ ergs.

drop. This energy is called the energy of cohesion. Obviously, energy of cohesion of a homogeneous fluid is equal to twice the value of the surface tension. Surface energy, then, is also called *surface excess energy*. In other words, there is extra free energy available to drive chemical or physical processes on the surface.

For the reversible process at constant temperature, pressure (and amount of material), the surface energy (tension) is equal to

$$\gamma = \left(\frac{\partial G}{\partial A} \right)_{P,T} \tag{6.8}$$

where G is the Gibbs free energy in joules and A is the area in square meters. ΔG is negative, or spontaneous, when surface area is decreased and is positive, requiring energy input, when surface area is increased.

6.1.3 Kelvin Equation

Vapor pressure at the liquid–vapor interface is influenced by pressure. If pressure is applied (e.g., by an inert gas to the surface of a liquid in equilibrium with its vapor), the vapor pressure, p, of the liquid is increased. This is the Kelvin equation

$$p = p_o e^{V_m \Delta P / RT} \tag{6.9}$$

where p_o is the vapor pressure of the liquid under standard conditions and V_m is the molar volume of the liquid. The vapor pressure exerted by a curved liquid surface based on the *Kelvin equation* relates vapor pressure change to surface curvature:

$$\ln\left(\frac{p}{p_o} \right) = \frac{2\gamma V_m}{r_{critical}(RT)} \tag{6.10}$$

where p is the vapor pressure of the particle, p_o is the equilibrium vapor pressure of a planar bulk material, and V_m is the molar volume and $r_{critical}$ the radius of the particle. In other forms of the equations, atomic volume V_a and the Boltzmann constant are used:

$$\ln\left(\frac{p}{p_o} \right) = \frac{2\gamma V_a}{r_{critical}(k_B T)} \tag{6.11}$$

The Kelvin equation in essence gives us a free pass into the nanodomain. The key component is the radius term r in the denominator. As r gets smaller, the vapor pressure of the droplet becomes larger and larger. The Kelvin equation explains the growth of water droplets in the atmosphere and the condensation of cold gases in nanopores by surface area methods such as BET and BJH. The Gibbs–Thomson effect and the Kelvin equation are often used interchangeably. The Gibbs–Thomson effect states that the chemical potential μ and the vapor pressure p of a small particle are related inversely to the surface curvature of a particle, and hence the radius of the particle. Small droplets have high surface

EXAMPLE 6.3	*Work Required to Create More Surface Area*

A drop of water is forced through an aspirator at 20°C. Calculate the amount of work in millijoules required to increase the surface area of water from that of the 1 cm^3 drop to aerosol particles with 1 μm^3 volume. How many drops are formed from the parent 1 cm^3 drop? What is the new total surface area? What is the specific surface area in square meters · per gram (ρ_{water} = 1.00 g·cm^{-3})? Ignore any potential Joule–Thomson cooling effect. The surface tension of water is equal to 72.8 dyn·cm^{-1}.

Solution:

Calculate the new surface area, find the additional area, and apply equation (6.1), $W_{rev} = \gamma \Delta \mathcal{A}$.
Surface area, $S_{A,drop}$, of the 1 cm^3 drop = 4.84 cm^2
Surface area, $S_{A,aero}$, of the 1 μm^3 aerosol particle = 4.84 × 10^{-8} cm^2
Number of aerosol particles formed from parent drop: Aerosols ≡ 10^{12}
Total surface area of aerosol particles = 10^{12} (4.84 × 10^{-8}) = 4.84 × 10^4 cm^2
$\Delta \mathcal{A} = S_{A,aero} - S_{A,drop}$ = 4.84 × 10^4 cm^2 − 4.84 cm^2 ≈ 4.84 × 10^4 cm^2
Energy required to create new area = 72.9 dyn·cm^{-1} (4.84 × 10^4 cm^2) = 3.53 × 10^6 ergs
  = (3.53 × 10^6)(1 J/10^7 erg) = 353 mJ
Specific surface area, S_s = 4.84 m^2·g^{-1}

curvature and therefore, high vapor pressure. This is because the surface-to-volume ratio is larger than in a bulk material. This is also why depression in freezing point is caused by solutions that contain finely divided particles [10].

6.1.4 *Particle Curvature and the Young–Laplace Equation*

The Young–Laplace equation relates the pressure difference across the surface of a small particle:

$$\Delta p = \gamma \left(\frac{1}{R_x} + \frac{1}{R_y} \right) \qquad (6.12)$$

where R_x and R_y (or R_1 and R_2 or r_1 and r_2) are the radii of curvature along the x- and y-axes parallel to the surface. Because the value of the radius of curvature for a sphere is the value of the radius of the sphere, the Young–Laplace equation simplifies to

$$\Delta p = \frac{2\gamma}{R} \qquad (6.13)$$

Selected Δp values for water droplets at different radii in terms of atmospheres are as follows: 0.0014 (r = 1 mm), 0.0144 (r = 0.1 mm), 1.436 (r = 1 μm), and 143.6 (10 nm). A six-magnitude reduction in particle radius results in a five-magnitude increase in the internal pressure of the substance. The power of nano!

The pressure differential across a curved interface, in particular a drop of liquid or a cavity in a liquid with radius r, can be described as follows. Consider a cavity with radius r. The internal pressure, p, balances the tendency of the bubble to minimize its surface area due to contractive forces (e.g., the surface tension γ). The outward force associated with the internal pressure is $(4\pi r^2)$ [8]. The inward pressure bearing from outside the cavity is equal to $p_o(4\pi r^2)$. An incremental change in surface area $d\mathcal{A}$ as the radius changes by an incremental amount dr is

$$d\mathcal{A} = 4\pi \, (r + dr)^2 - 4\pi r^2 = 4\pi r^2 + 8\pi r dr + 4\pi(dr)^2 - 4\pi r^2 = 8\pi r dr \quad (6.14)$$

Therefore, the work needed to change the area by dr is equal to

$$dW = \gamma(8\pi r dr) \quad (6.15)$$

At equilibrium, all the internal and external forces are balanced and

$$p(4\pi r^2) = p_o(4\pi r^2) + \gamma(8\pi r), \quad (6.16)$$

thus simplifying to the Young–Laplace equation for spherical particles:

$$p - p_o = \Delta p = \frac{2\gamma}{r} \quad (6.17)$$

Once again we derive the Young–Laplace equation, albeit from another perspective—one that is purely mechanical. Because all forces sum to zero, $\Sigma F = 0$, this relation accounts for the difference between the internal and external pressure, equivalent to the surface tension times the distance that it is displaced. The Young–Laplace equation shows that the pressure inside a curved surface will always be larger than the pressure outside a curved surface [8]. As radius is increased infinitely, Δp approaches zero and the pressure differential across an infinite plane is achieved, which at equilibrium is equal to zero.

By substituting Δp derived previously into equation (6.9), we get the Kelvin equation for a droplet, equation (6.10). For a cavity surrounded by a liquid,

$$p = p_o e^{-V_m 2\gamma /r(RT)} \quad (6.18)$$

Because the pressure inside a cavity is less than the pressure outside, the only change in this expression from that of the Kelvin equation is the sign of the exponent [8].

Once again, as r increases to infinity, the vapor pressure of the liquid is expected to approach the equilibrium vapor pressure of a bulk planar surface. These relationships tell us that atoms or molecules on the surface of a small liquid droplet are held more loosely than the corresponding atoms or molecules that exist at a flat planar surface of the liquid. Reflecting back to chapter 5, we recall that atoms and molecules at the surface are deficient with regard to coordination number by fellow atoms or molecules. Higher curvature effectively reduces the number of nearest neighbors even further, resulting in higher vapor pressure over the curved surface.

Curvature. As a general rule, liquid materials in droplets with high positive curvature (convex) have a higher vapor pressure than the corresponding bulk material. Liquids in pores demonstrate negative curvature (concave) and have lower vapor pressure than their bulk counterparts. What exactly is curvature? As we progress to smaller dimensions, a parameter known as curvature assumes greater importance. A straight line, or flat bulk material, has no curvature. Curvature is a measure of the degree of deviation from a straight line. The curvature of a circle is the derivative of the angle ϕ (the angle made between some directional vector and the tangent line to the circle) with respect to arc length s (the change in the unit tangent vector per unit distance of the curve). The curvature is a parameter that is equal to the reciprocal of the radius [13]:

$$\kappa(s) = \frac{d\phi}{ds} = \frac{1}{r} \tag{6.19}$$

where κ is defined as the curvature and s is the arc length. Therefore, from equation (6.19), the smaller the radius becomes, the greater the extent of curvature is: $r \to \infty$, $\kappa \to 0$.

It is no surprise by now that physical properties are expected to change as particle size is decreased. In this case we relate how curvature influences physical properties. The pressure differential across the surface of a bubble, as shown previously, is a function of curvature. When a surface is curved, a pressure differential is instantly established across the air–liquid interface. From basic thermodynamics, we know that the vapor pressure of a liquid increases when the external pressure is increased. The Kelvin equation basically states that the vapor pressure at saturation is greater over a curved surface than it is over an infinite flat surface.

This kind of Gibbs surface is valid at the nanoscale if the effective surface free energy (surface tension) is a function of the radius of the nanoparticle [14]. In other words, in this case, classical thermodynamics can be extrapolated to the nanoscale.

6.1.5 Chemical Potential

The chemical potential of physical and chemical processes that involve small particles is dependent on the radius of curvature of the surface. We resort to another form of the Young–Laplace equation:

$$\mu - \mu_o = \Delta\mu = \gamma V_a \left(\frac{1}{r_1} + \frac{1}{r_2} \right) = \frac{2\gamma V_a}{r} \tag{6.20}$$

where V_a is the atomic volume and μ and μ_o are chemical potentials of the curved surface and bulk surface, respectively. Convex surfaces (like the surface of colloids) have a positive chemical potential. To form a convex surface from a flat bulk surface, an input of energy is required [19]. Concave surfaces have negative curvature. Correspondingly, concave surfaces represent low-energy surfaces. The surface of red blood cells is concave. Why do you think that is the case?

6.2 Liquid State

A liquid (from the Latin *liquidus*, "flowing, moist") is a state of matter that has no definite shape, is difficult to compress, and is able to flow. Liquids assume the shape of their container. These definitions can certainly be debated because at the nanoscale drops do assume a definite spherical shape. Liquids lack static long-range order, unlike solids. The atoms and molecules in a liquid are not bound tightly and have fewer neighbors with which to bind. This results in mobility of atoms and molecules that allows restoring forces to react readily in the event of any perturbation of the surface. As a result, liquids are able to respond quickly to any deformation and reassemble in a low-energy, minimum surface area state.

The surface layer is considered to be a few atoms or molecules thick in the absence of ions [9]. The surface of a liquid resembles an elastic skin that is always under tension. In the volume of liquids, balanced cohesive forces tug on atoms or molecules from all directions. In contrast, unbalanced cohesive forces at or near the surface tug on surface molecules and atoms laterally and inwardly (**Fig. 6.4**). In essence, the liquid contracts and compresses until it attains the lowest possible surface area per volume, the sphere. This configuration allows for elastic deformation of the surface. Surface atoms and molecules, therefore, are inherently less stable than their bulk counterparts. This instability is responsible for the enhanced vapor pressure of water droplets over their bulk counterpart.

Surface tension is best visualized as a stretched elastic membrane that tends toward contraction [11]. Drops, bubbles, cavities, and capillarity are all consequences of surface tension. A bubble is a region where gas is trapped by a condensed solid, usually a liquid. A bubble has two surfaces and therefore a factor

Fig. 6.4

Water molecules near the surface experience different forces. Notice in the figure that surface molecules have fewer nearest neighbors. Surface tension is the direct consequence of surface molecules trying to achieve the lowest energy state. In the bulk, each water molecule experiences the same cohesion forces that result in a balance (net force is equal to zero, $\Sigma F = 0$). Surface molecules are pulled inward by molecular attraction (there is no outward force). Resistance to compression by the liquid balances the inward force. Water is then squeezed until it achieves the lowest surface area possible; for a droplet, this is a sphere. The layer behaves as an elastic sheet that resists any change to its curvature (e.g., the action of a water strider).

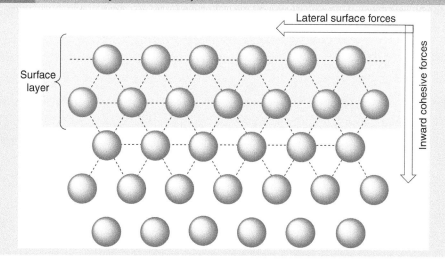

EXAMPLE 6.4   *Vapor Pressure of Nanosized Droplets*

Calculate the vapor pressure of $r = 90$ nm water droplets at 20°C. Assume the density of water at 20°C is 0.99823 $g \cdot cm^{-3}$; $\bar{V}_m = 18.047$ $cm^3 \cdot mol^{-1}$; $T = 293$ K; $R = 8.314$ $J \cdot K^{-1}$; $M_{H2O} = 18.015$ $g \cdot mol^{-1}$.

Solution:

Apply the Kelvin equation
$V_m = M/\rho = 18.047$ $cm^3 \cdot mol^{-1}$
$2\gamma V_m/rRT = [2(72.25$ $dyn \cdot cm^{-1})$ $(18.047$ $cm^3 \cdot mol^{-1})]/[(90 \times 10^{-6}$ $cm)$ $(8.314 \times 10^7$ $ergs \cdot mol^{-1} \cdot K^{-1})(293$ $K)] = 0.01068$;
     then $e^{0.01068} = 1.0107$
$p = (17.535$ $torr) \cdot exp[2\gamma V_m/rRT] = (17.535$ $torr) \times 1.0107 = 17.723$ $torr$.

of 2 has to be applied in surface energy calculations. A cavity, on the other hand, has just one surface. A drop or droplet is a liquid surrounded by a gas— only one surface. Liquids adopt shapes that tend to minimize surface area. This maximizes the number of atoms or molecules that exist within the volume of the drop.

The *energy of cohesion* is the energy that holds the liquid (or solid) together. It is also the energy that is released in the form of heat when two drops of the same liquid are joined. The energy of cohesion is equal to twice the surface tension energy because, in order to create a surface, the number of nearest-neighbor binding interactions must be halved (example 6.1):

$$E_{co} = 2\gamma \tag{6.21}$$

The cohesive energy density is defined as the ratio of the molar heat of evaporation (or molar heat of sublimation for solids) at standard pressure to the molar volume of the liquid: $\Delta H°_{vap}/V_m$. The molar heat of vaporization is the energy required to convert 1 mol of liquid into gas at its boiling point and is indicative of the forces that hold liquids together. **Table 6.3** lists cohesive energies of several common materials along with their respective surface tensions.

TABLE 6.3   *Cohesive Energies of Common Materials*

Material			Cohesive energy
Liquid	$\sim\Delta H°_{vap}$ (kJ · mol⁻¹)	**Boiling point (K)**	
Solid	$\sim\Delta H°_{sub}$ (kJ · mol⁻¹)	**Melting point (K)**	**Surface energy or tension (mJ · cm⁻²) of materials in liquid form**
NH_3 (l)	23.3	239.7	34.5 (at 20°C)
CH_3CH_2OH (l)	26.7	351.5	22.8 (at 20°C)
H_2O (l)	40.7	373.2	72.8 (at 20°C)
Hg (l)	60.3	356.6	485 (at 20°C)
Zn (s)	115.6	693	770 (at mp)
Au (s)	368	1338	1130 (at mp)

TABLE 6.4	Surface Tension and Interfacial Tension of Common Materials	
Bulk liquid	**Surface tension (γ_1)**	**Interfacial energy (γ_{12}) γ_2, water = 72.9 dyn · cm^{-1}**
Benzene	29	35
Chloroform	27	28
Carbon tetrachloride	27	45
Cyclohexanol	32	4
Diethyl ether	17	10.7
n-Octane	22	50.8
n-Octyl alcohol	27.5	8.5
Mercury	475	375

The *energy of adhesion*, on the other hand, is the surface tension of the interface between two nonsimilar liquids and this relationship has the following form:

$$\gamma_{12}\,\mathcal{A} = (1/2)W_{11} + (1/2)W_{22} - W_{12} = \gamma_1\,\mathcal{A} + \gamma_2\,\mathcal{A} - W_{12} \qquad (6.22)$$

where W_{11} and W_{22} are the cohesive energies, proportional to γ_1 and γ_2, respectively, of the pure components and W_{12} is the work of adhesion required at the interface. Just as before, the energy required to create an interfacial surface between two immiscible liquids is called the interfacial energy or interfacial tension. The shape of a material, nanomaterials in particular, depends heavily on the balance between cohesive and adhesive forces. Surface tension and interfacial tension of common materials are listed in **Table 6.4**. The term *interfacial tension* is also used if a surface separates two nongaseous phases.

6.2.1 Classical Surface Tension

Just like with water, the inner structural energy that holds solid materials together is called the *cohesion energy*, E_c. The behavior of bulk materials, considered to be relatively independent of the environment, is attributable to the cohesion energy [12]. We now come to understand that as surface-to-volume ratio is increased, surface energy plays an increasingly larger role in determining the behavior of the material. For the most part, we assume that the intrinsic cohesion energy (bond strength) remains at the bulk value even as size is diminished. Another assumption is that cohesion energy is the opposite of the sublimation energy (although this is true only at 0 K). The size dependence of both the inner cohesion energy and the surface energy is a subject that has received much attention recently [12].

Surface tension depends on long-range forces governed by dispersion, short-range forces such as hydrogen bonding in water, and entropic effects (e.g., the surface is more disordered than the bulk of the liquid) [8]. Surface tension, is a measure of the force required to break a liquid surface of 1 cm length. The surface energy of a solid is estimated by measuring a parameter called the *contact angle*, cos θ. The contact angle is the angle made by a liquid

TABLE 6.5	Contact Angles Formed by Selected Common Materials
Surface	**Contact angle**
Au	0°
Pyrex glass	ca. 0°
Pt	40°
Polymethyl methacrylate (PMMA)	60°
Teflon	110°
Lotus plant leaf[a]	~180°

[a] Water droplets on the surface of the lotus plant leaf are perfectly spherical.

drop at the three-phase boundary where the liquid, flat solid substrate, and gas phases meet (**Fig. 6.2**). This is considered to be a boundary condition. Surface tension relationships are described by Young's equation (often called the Young–Dupre equation):

$$\cos\theta = \frac{\gamma_{SV} - \gamma_{SL}}{\gamma_{LV}} \tag{6.23}$$

where γ_{SV}, γ_{SL}, and γ_{LV} are the surface energies at the solid–vapor, solid–liquid, and liquid–vapor phases, respectively. The contact angle is composed of the advancing contact angle θ_a, the receding contact angle θ_r, and the equilibrium contact angle θ_e. Angle hysteresis is said to occur when $\theta_a \neq \theta_r$. We will concern ourselves only with the equilibrium contact angle. The contact angle formed by a drop of water on some common substrates is shown in **Table 6.5**.

The contact angle is an important measure that correlates the surface energy of a solid to the surface tension of a liquid. The contact angle is indicative of a phenomenon known as surface wetting or *wettability*. Surface wetting is the ability of a solvent to spread on a solid substrate, a characteristic that is important, for example, in soldering. Water is a common liquid used in the determination of surface wettability. The magnitude of the spread of a droplet is an indicator of surface energy. The surface energy of a solid can be determined by contact angle measurements against a series of well-characterized wetting agents. If a drop spreads outward after application (e.g., on an unwaxed metal surface of an automobile), the interfacial area between the substrate and the drop is increased. In other words, the substrate surface has enough energy to overcome the surface tension of water and increase the interfacial area of the droplet.

From another perspective, the adhesive forces between the liquid and the solid are comparable in strength to the cohesive forces of the water. If the water forms up in a bead, it is obvious that the cohesive forces—the forces between like molecules—are stronger than the adhesive forces between unlike molecules. For example, if the drop remains spherical or does not spread to any great degree, there is not enough energy emanating from the weak attractive interactions that hold hydrocarbons together, as on a waxed surface, to overcome the surface energy of the water. Good wetting is said to happen if the contact angle lies between $0° < \cos\theta < 90°$. *Dewetting* is the opposite of wetting and is the case when a sheet of liquid is disrupted to form droplets (e.g., on a waxed metal surface of an automobile.

The critical temperature T_c is the temperature at which there is no more distinction between a liquid and its vapor. At the critical temperature, the value of γ becomes zero. The temperature dependence of surface tension of common liquids obeys the following empirical relationship of Guggenheim and Katayama (in E. A. Guggenheim, *Thermodynamics*, 5th ed., North-Holland, 1967):

$$\gamma = \gamma^\circ \left(1 - \frac{T}{T_c} \right)^n \tag{6.24}$$

where T_c is the critical temperature at which the liquid phase becomes indistinguishable from the vapor phase, n is equal to ca. $1.22 \approx 1$, and γ° is the surface tension of the liquid under standard conditions. Obviously, γ approaches zero as T approaches T_c. Surfactants are also able to lower the surface tension of water. For many compounds, the dependence of surface tension on temperature can be approximated by

$$\gamma = a - bT \tag{6.25}$$

where a and b are constants associated with a particular liquid. For example, for water, $a = 24.05$ (the actual value at $0°C$) and $b = 0.0832$.

Nanoscale Condensation. The equilibrium vapor pressure of a liquid and the curvature of the liquid–vapor interface are related by the Kelvin equation. The Kelvin equation predicts that unsaturated vapors will condense in nanochannels even into nanochannels. Meniscus radii below 4 nm have been validated for cyclohexane but not verified for water until 1981 when Fisher et al. demonstrated meniscus radii for water as low as 9 nm [15]. Fisher et al. stated in 1979 that "the application of the laws of thermodynamics, and the concept of a bulk surface tension, is valid in principle for such highly curved interfaces."

According to recent findings, the Kelvin effect and capillary rise are different aspects of the same phenomenon [16]. This is a profound demonstration of the validity of classical thermodynamics to nanoscale materials. In chapter 8, we find that classical thermodynamic theory is not enough to explain nanoscale phenomena.

6.2.2 Capillarity

Surface tension is responsible for capillary action. If a liquid has a tendency to adhere to the walls of the tube and spread out to cover more area (e.g., good wetting), energy is minimized. The height h that a liquid is drawn into a capillary tube is described in terms of surface tension:

$$h = \frac{2\gamma_{LV} \cos\theta}{\rho g r} \tag{6.26}$$

where

h is the height of the liquid
γ_{LV} is the liquid–air surface tension

r is the radius of the capillary
θ is the angle of contact
ρ is the density of the liquid
g is the gravitational acceleration constant

On glass, the material of most capillary tubes, water is able to wet the surface completely. In this case the contact angle θ is equal to zero and consequently $\cos\theta$ equals one and the expression falls out of the equation. This relationship is obviously most accurate when complete wetting of the tube material occurs.

If the cohesive forces that hold water together are weaker than the adhesive forces between the water and the glass wall of a graduated cylinder (e.g., there is wetting of the surface), the meniscus of the water forms a concave shape. This occurs when the meniscus contact angle is less than 90° (**Fig. 6.5**). On the other hand, if the liquid is not able to wet the glass surface, as is the case of elemental mercury (e.g., a liquid with extremely high surface tension equal to 485 dyn·cm⁻¹), there is limited wetting and a convex meniscus is formed. This occurs when the contact angle is greater than 90° (**Fig. 6.5**).

Capillary tube wetting can be described by the following adaptation of the Laplace form:

$$\Delta p = 2\gamma \cos\theta \left(\frac{1}{r_1} - \frac{1}{r_2} \right) \tag{6.27}$$

Capillary condensation is an important factor in determining the mechanical properties of micro- to nanoscopically sized matter, particularly of powders. Capillary condensation depends on the liquid surface tension γ_{LV}, the contact angle θ, and the radius of curvature of the solid surface R. The capillary force is then

$$F_{cap,max} = 4\pi R\gamma_{LV} \cos\theta \tag{6.28}$$

Capillary forces play an important role with microcontact AFM or STM tips. The principal mechanism forms the basis for several types of dip-pen nanolithography. Effectively, the probe tips act as the nucleation center for condensation. If the radius of curvature of the microcontact probe lies below the critical radius (~ the Kelvin radius), a meniscus is formed. The Kelvin radius is given by

$$r_K = \frac{2\gamma V_m}{RT} \ln\left(\frac{p_o}{p} \right) \tag{6.29}$$

where p_o is the saturation vapor pressure as before. The BJH method to determine porosity and pore diameter is based on the Kelvin equation.

6.2.3 *Surface Tension Measurements*

Methods to measure surface tension and energy are numerous. There are two principal methods to consider. One, based on a mechanical principle, is called tensiometry (from the Latin *tensio*, "stretching"). Tensiometry is a technique that is able to measure the interaction of a calibrated probe (usually made of a

FIG. 6.5

The meniscus of mercury (left) is compared to the meniscus of water (right). Mercury has one of the highest surface tensions of any liquid: 485 dyn·cm⁻¹. This is not surprising because metals have high cohesion energy. There is not enough surface energy available on the glass surface to overcome the high surface tension of mercury. The contact angle therefore is very large and the meniscus protrudes upward. For water, the silica glass is able to neutralize the whole of the surface energy of water. The contact angle is nearly 0° and the meniscus therefore is concave. The water is actually dragged upward. If the curvature is small enough, water will rise in the tube (e.g., by capillary action).

Mercury: High surface tension no wetting of glass surface

Water: Surface energy of glass enough to overcome surface tension of water—good wetting

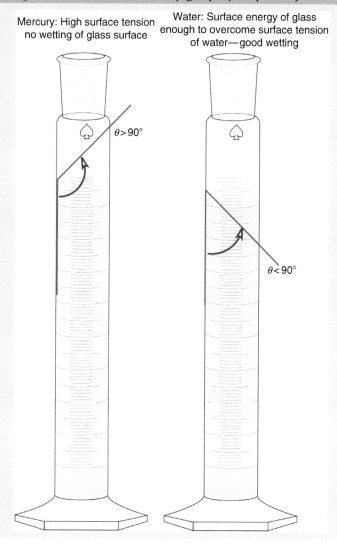

$\theta > 90°$

$\theta < 90°$

platinum–iridium alloy) of high surface energy with a liquid surface layer. The *Du Noüy ring* method uses a platinum ring that is submerged below the surface of a test liquid (**Fig. 6.6**). The probe is then lifted incrementally until all contact with the liquid is lost. The maximum force during this process is measured and is converted into the dimensions of surface tension. A KSV Sigma-700

The action of a Du Noüy ring is shown. The ring is made of a wettable metal (as most are)—in this case, platinum, a metal with a very high surface energy. The net force is measured as a function of time along the track of the ring. The graph shows the profile of the measurement step by step: (1)–(4): ring is pushed below the surface of the liquid and the system is allowed to equilibrate; (5)–(7): the ring is pulled upward; the force increases as the surface tension of the liquid exerts its influence. Moving the ring upward results in the creation of a new surface (requiring energy). At (8) the interface breaks as enough force has been exerted to overcome the surface tension.

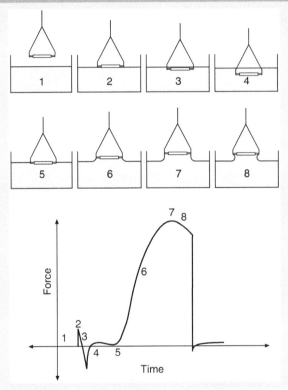

Source: Courtesy of KSV Instruments, Ltd.

tensiometer is shown in **Figure 6.7.** The 701 series is capable of measuring 0.01–1000 mN · m^{-1} with 0.001 mN · m^{-1} resolution. The maximum load for this instrument is 5 g with weighing resolution of 0.01 mg, force resolution of 0.01 µN, contact angle range of 0–180°, and contact angle resolution of 0.01°. The *Wilhelmy plate* technique, another method based on tensiometry, utilizes a calibrated platinum plate with a roughened surface. The surface of the probe is wettable by all liquids.

Another method is based on optical determination of the contact angle and is commonly referred to as a *contact angle goniometry* (based on the Greek *gonia*, indicating angle, + *metron*, "measure"), a process invented by W. Zisman of the U.S. Naval Research Laboratory in the 1950s. Goniometry is a complementary technique to tensiometry and is often used to independently verify results obtained by the latter.

Measurement of the surface energy of solids is not a simple task. The surface tensions against air of water, methanol, benzene, and mercury in air at 20°C are

Source: Courtesy of KSV Instruments, Ltd.

72.9, 22.5, and 485 dyn·cm⁻¹ (or mN·m⁻¹), respectively. In methanol, some hydrogen bonding is present and in mercury no hydrogen bonding is present but very strong metallic interactions predominate. The contact angles of a series of liquids with different surface tension values are placed on the solid. Contact angles are recorded and the surface energy of the solid is then calculated indirectly by this means.

6.3   SURFACE ENERGY (AND STRESS) OF SOLIDS

Use of the term "surface energy" rather than "surface tension" is appropriate for solids. We stick to homogeneous simple crystal forms in order to illustrate aspects of surface energy. The cubic *fcc* structure always serves as a good base. We also stay with (111), (100), and (110) low-index crystal planes in this section. Topics concerning *surface stress*, another common phenomenon associated with solid surfaces, will not be presented in this section. Liquids do not exhibit surface stress. It is a phenomenon associated strictly with solids only. Solids possess elastic stress up until the melting point [17]. To be complete, both surface energy and surface stress characterization are required to describe the state of the solid surface [17].

Nearest-Neighbor Models.   Nearest-neighbor atomistic models are 10–20% accurate but offer a simple physical interpretation of the equilibrium behavior of solids [17]. The N–N model does not consider next-nearest neighbors and higher order interactions. We shall also apply the atomistic perspective in describing the surface energy of solids. We begin our discourse by describing macroscopic condensed systems of bound atoms with pair interaction and their surface energy and then move on to nanoscale materials, *fcc* clusters in particular.

Descartes proposed that solid matter is held together by "hooks and other appendages." His mechanistic approach suggested that external pressure was responsible for solid cohesion. Newton speculated correctly that atoms interacted by attractive forces at long range and repulsive forces at short range. Based on his research of capillarity, Laplace speculated that atomic forces are short range in nature because the rise of a liquid in a capillary tube is independent of the thickness of the capillary wall.

Surface atoms and molecules of solids share similar properties with liquids except atoms and molecules remain relatively stationary. Solids cannot, for example, react spontaneously and change shape to configure a minimum surface. In its most simplistic form, the surface energy E_s of a crystalline solid is the number of surface atoms times the one-half cohesion energy per atom pair:

$$E_s = \tfrac{1}{2}\left(N_s \varepsilon_b\right) \tag{6.30}$$

where ε_b is the cohesion (binding) energy of atom pairs in the solid and N_s is the number of atoms that exist in the surface layer. The factor $1/2$ is included because making a new surface breaks bonds in two (in a perfect crystal). The assumption that surface atoms have the same binding energy as those within the solid is an approximation at best. The energy of a solid is determined by the strength of the chemical bonds within a crystal and its bonding configuration. The number of nearest-neighbors (and second-nearest, etc.) implicitly contributes to the bonding configuration and stability of the solid.

Phase changes at constant temperature and pressure generally involve a change in enthalpy. The enthalpy of sublimation at 0 K, ΔH_{sub} (or ε_{sub}) is the energy required to convert a solid directly into its vapor form without going through a liquid stage. The enthalpy of sublimation is a convenient approximation of the cohesive energy, or binding energy ε_b. The energy required to melt a solid is represented by the molar heat of fusion, ΔH_{fus}. Melting is an endothermic phase transition process in which energy is absorbed. The processes of heating and subsequent melting reduce the cohesive energy of a solid and, because all bonds are weakened, as a result, the surface energy is also reduced. With melting, the overall coordination of each atom in the bulk is also reduced. The magnitude of the intermolecular interactions in liquids is measured by the latent heat of vaporization, ΔH_{vap}—the energy at constant temperature required to convert liquid into vapor. Recall that at the boiling point of a liquid surface tension is nonexistent.

6.3.1 Interaction Pair Potentials

As any text in physical chemistry will show, there are basically two types of inter-actions between atoms: short-range repulsive interactions (nearest neighbors only) and long-range attraction interactions [18]. There are many models that illustrate the interatomic potential energy of solids. Hard-sphere potential models treat the atom as hard spheres with the same radius and are appropriate in modeling repulsive potentials. The potential energy between two hard spheres jumps to infinity as soon as the *collision diameter threshold* σ is attained [8]. The *Lennard–Jones (6–12) potential* is one such model that is useful in describing atomic behavior:

$$U(r) = \varepsilon_b \left[\left(\frac{\sigma}{r} \right)^{12} - 2 \left(\frac{\sigma}{r} \right)^6 \right] = \varepsilon_b \left[\left(\frac{r_e}{r} \right)^{12} - 2 \left(\frac{r_e}{r} \right)^6 \right] \qquad (6.31)$$

where ε_b is the binding energy between two atoms and σ represents the mini-mum separation. At equilibrium, the equilibrium radius $r_e = \sigma$ (approximately the sum of the average radii of the two atoms analogous to the lattice parameter a). The restoring force, $F = -dU/dr$, is equal to zero at equilibrium and $U(r_e) = -\varepsilon_b$, the cohesive (binding) energy for the pair interaction. The equilibrium value of the binding energy is equal to the depth of the energy well. When $r \geq 2\sigma$, the potential decreases by a factor of 32 or more and nearest-neighbor interactions can largely be ignored [17]. When $r < \sigma$, the r^{-12} factor dominates the r^{-6} term. When $r > \sigma$, the converse is true. Both attractive and repulsive forces are capable of reducing the potential energy of the system.

6.3.2 Surface Energy of Low-Index Crystals

The energetic state of surface atoms or molecules is one of incompleteness. Atoms with dangling bonds possess extra energy called excess surface energy—useful energy available to accomplish work. Resorting once again to a hard-sphere model, the surface energy of solids cleaved along low-index crystal planes can be approximated by applying the *nearest-neighbor–broken-bond model* introduced earlier [17]. The assumptions in our calculation are as before: that we use a hard-sphere model, consider only nearest-neighbor interactions, and the surface energy be equal to the average of the binding energy of the solid. Other assumptions include a rigid structure with no relaxation, surface strain, or any entropic or pressure–volume contributions [19]. Additional con-siderations include ignoring any potential surface strain or rearrangement asso-ciated with the new surface. Any such restructuring would, of course, contribute to lower surface energy [19]. We will apply the following formulae only to solids and also assume that the resulting surface energy of the newly formed surface is defined according to bulk parameters (e.g., the excess energy the surface atoms have over that of the interior atoms). The *fcc* metal will serve again as our model.

The total energy of a solid is estimated to be equal to the heat of sublimation at 0 K, ΔH_{sub}. The enthalpy and internal energy are considered to be equal. Enthalpy of vaporization (at 0 K), energy of formation, cohesive energy, dissociation energy, and enthalpy of fusion all are important factors with regard to the total binding of a solid. Cleavage of a crystalline solid is somewhat akin to radical sublimation—the abrupt removal of a plane of atoms from its neighbors in the adjacent plane. Calculation of the surface energy of a newly formed surface involves reference to ΔH_{sub} of the bulk material.

The *nearest-neighbor* model [17] estimates that the extensive overall binding (cohesion) energy of 1 mol of atoms E_b is

$$E_b = \tfrac{1}{2} Z \mathcal{N}_A \varepsilon_b \tag{6.32a}$$

where Z is the coordination number as before, $\mathcal{N}_A$ is Avogadro's number, and ε_b is the binding energy between two atoms in a solid. Rearranging and substituting ΔH_{sub} for E_b,

$$\varepsilon_b = \frac{2\Delta H_{sub}}{Z \mathcal{N}_A} \tag{6.32b}$$

For an *fcc* crystal with $Z = 12$, the cohesion energy of the crystal reduces to

$$\varepsilon_{b,fcc} = \frac{\Delta H_{sub}}{6 \mathcal{N}_A} \tag{6.33}$$

Crystals with faces that have the closest packing of atoms have the lowest surface energy [17].

The energy required to cleave an *fcc* crystal along the (111) crystal plane is calculated as follows [17]. One first counts the number of nearest neighbors in the (111) plane. Three atoms reside above the plane of cleavage and three atoms reside below the plane of cleavage. The result of the cleavage is two additional (111) planes. This means that three bonds are broken for every cleaved atom. The energy of every surface atom is then

$$\varepsilon_s = \frac{3}{2} \varepsilon_b \tag{6.34}$$

The energy of the (111) surface is

$$\varepsilon_s = \frac{3}{2} \varepsilon_b = \frac{3 \Delta H_{sub}}{12 \mathcal{N}_A} = \frac{\Delta H_{sub}}{4 \mathcal{N}_A} \tag{6.35}$$

Notice that this value is greater than the cohesion energy of the solid crystal by a factor of 3/2.

Another convenient way to calculate surface energy is to use

$$\gamma = \tfrac{1}{2} N_b \varepsilon_b \rho_a \tag{6.36}$$

where N_b is the number of atoms per unit area and ρ_a is the surface atom density [19]. This form of the calculation relates energy to the lattice constant a of the crystal [19]:

$$\gamma_{(100)} = \frac{1}{2}(4)(\varepsilon_b)\left(\frac{2}{a^2}\right) = \frac{4\varepsilon_b}{a^2} \tag{6.37}$$

The rest of the low-index plane surface energy values are listed in **Table 6.6**.

TABLE 6.6	*Relative Surface Energy of Low-Index fcc Planes*					
Cubic plane	Planar atomic density, σ_{hkl}	N–N broken bonds	Excess energy per atom	Surface energy[at] γ_{hkl} J·atom^{-1}·a^{-2}	Planar packing density	Surface energy[b] $\gamma_{hkl} \rightarrow \Delta H_{sub}$ J·atom^{-1}·a^{-2}
(100)	$2/a^2$	4	$4\,\varepsilon_b/2$	$4\,(\varepsilon_b/a^2)^\dagger$	0.785	$\sigma_{100}\,\Delta H_{sub}/3\mathcal{N}_A$
(110)	$\sqrt{2}/a^2$	5 (6[c])	$6\,\varepsilon_b/2$	$3\sqrt{2}\,(\varepsilon_b/a^2)$	0.555	$\sigma_{110}\,\Delta H_{sub}/2\mathcal{N}_A$
(111)	$4\sqrt{3}/a^2$	3	$3\,\varepsilon_b/2$	$2\sqrt{3}\,(\varepsilon_b/a^2)^\dagger$	0.901	$\sigma_{111}\,\Delta H_{sub}/4\mathcal{N}_A$

[a] G. Cao, *Nanostructures & Nanomaterials: Synthesis, Properties and Applications*, Imperial College Press, London (2004).
[b] J. M. Howe, *Interfaces in Materials: Atomic Structure, Thermodynamics and Kinetics of Solid–Vapor, Solid–Liquid and Solid–Solid Interfaces*, John Wiley & Sons, New York (1997).
[c] The sixth broken bond is difficult to find [17].

One would expect the (110) planes to have the highest surface energy because it has the lowest packing fraction of the three low-index crystal planes.

The value of the electronic work function depends on the plane of origin of the electron. **Table 6.7** lists the dependency of electronic work function of the crystal face [1]. Notice that the electron work function is highest for the electron that emanates from the (111) plane. In all three cases, the order of stability is (111) > (100) > (110), similar to the order of surface energy in **Table 6.6**. The close-packed (111) plane contains atoms with the most stability (lowest surface energy) and concentration of surface positive ion cores [1].

The correlation between surface energy γ of the solid–vapor interface and ΔH_{sub} is reasonable. The heat of sublimation, however, is not the best way to calculate the cohesion energy of a metal. A simple empirical means of calculating cohesion energy is to divide the metal melting point temperature by the constant -3.5 ± 0.3 [20]:

$$E_b = T_m/(-3.5 \pm 0.3) \qquad (6.38)$$

There is no way to obtain an absolute value of cohesion energy of metals, although better empirical relationships have been attained [20].

The equilibrium surface energy of a solid is the summation of the surface energies of individual facets. A cuboctahedron, for example, has 14 faces: 8 that

TABLE 6.7	*Electron Work Function Dependency on Surface Plane Face*	
Element	Surface plane	Work function, eV
Ag	(100)	4.64
	(110)	4.52
	(111)	4.74
Cu	(100)	4.59
	(110)	4.48
	(111)	4.98
Ni	(100)	5.22
	(110)	5.04
	(111)	5.35

are triangular (derived from {111} planes) and 6 that are square in shape (derived from {100} planes). Starting with a cube of *fcc* close-packed structure, if one were to incrementally chip off the corners (triangular pyramids), the following shape transition progression would be seen (consult **Fig. 6.19** later on):

Cube → cuboctahedron → regular truncated octahedron → octahedron

Without engaging in the complex details, the regular truncated octahedron will have the lowest energy surface of the four major categories of shapes [18]. Why do you think this is the case? Which one resembles a sphere more so than the others? Unless very large clusters are formed, perfect spherical shapes are not allowed in close-packed structures for obvious reasons.

6.3.3 *Surface Energy of Nanoparticles*

Although quantum confinement has not been formally introduced at this stage, for now let us be content with the following definition. Restriction of electrons in one or more dimension translates into quantum confinement. In other words, especially in the case of metals, a continuum of electronic states no longer exists. Quantum confinement in two-dimensional films occurs in the *z*-direction only.

Thin Films.   In bulk material surfaces, the relative surface energy is constant. In nanofilms, the surface energy was shown to fluctuate and that there was a preference of certain thicknesses over others; for example, layers were more stable if they possessed a specific thickness [21]. Layers of Pb were deposited on a Si(111) surface. The thickness of the Pb was varied from 6 to 18 monolayers [21]. Quantum confinement effects are expected to contribute to structural differences such as interlayer spacing. With the utilization of x-rays, information concerning the film thickness, atomic structure, and the interface with the Si surface were obtained. We have already discussed "magic numbers" as they apply to cluster stability. Are there any grounds for phenomena based on "magic thickness"? P. Czoschk, H. Hang et al. state:

> When confinement of the itinerant electrons in the metallic structure is taken into account, oscillations in the surface energy arise, leading to the preference of a certain thickness over others. It is seen that a smooth film represents a metastable state which, upon annealing, evolves through preferred or magic structures to a thermodynamically more stable, highly roughened state.

More Kelvin Effect.   The size-dependent evaporation of free Ag nanoparticles showed that the Kelvin effect is verifiable at the nanoscale [22]. Researchers predicted a surface energy value for Ag nanoparticles to be 7.2 $J \cdot m^{-2}$. A linear relationship exists between the onset of evaporation temperature and reciprocal particle size.

Dip-Pen Lithography.   There are many examples of surface energy phenomena involving nanoscale materials. Dip-pen nanolithography relies on capillary transport to deliver molecules from an AFM tip to a solid substrate. The capillary filling height and speed of silicon dioxide nanochannels are important in fabrication of wafers [23]. For example, capillary forces drive micropart self-assembly.

6.4 SURFACE ENERGY MINIMIZATION MECHANISMS

Surface tension in a liquid is reduced if the liquid assumes the shape of a sphere, the surface of minimum energy. Liquid surface tension is also reduced by contact with a solid of high surface energy. For example, water wets a gold surface completely as the contact angle of the water is reduced to nearly nothing. With liquids, water in particular, the addition of surface-active agents (surfactants) lowers the surface tension.

The mechanisms to reduce surface energy in solids are abundant. Nanomaterials in particular offer numerous solutions to attain low-energy surface states. Because surface area, both of the collective and singular variety, is an important aspect of nanomaterials, these energy minimization processes are what enable researchers to exert some modicum of control over nanomaterial stability. Part of the emergence of the *Nano Age* is due to this ability.

6.4.1 *Surface Tension Reduction in Liquids*

We focus our attention in this section on the mechanisms of surfactants and how they are able to change the magnitude of the surface tension, specifically of water. Water is a key ingredient of living things as well as the primary medium within which numerous industrial processes depend. There is a constant known as the Hamaker constant, A_H, that describes the dispersion energy of small particles.

A surfactant is a surface-active agent. Many types of chemicals and solid materials can act as surfactants. The seventh century ecclesiastical scholar Bede is known for saying, "Remember to throw into the sea the oil which I gave you, when straightaway winds will abate, and a clam and smiling sea will accompany you throughout your voyage."

In 1757, Benjamin Franklin also noticed that the wakes behind ships that had just dumped their greasy water were smoothed. And when he was at London's Clapham Common pond, Franklin is known for the following experiment:

> The Oil, tho' not more than a Teaspoonful produced an instant calm over a Space of several yards square, which spread, amazingly, and extended itself gradually, until it reached the Lee Side, making all of that Quarter of the Pond, perhaps, half an Acre as Smooth as a Looking Glass.

Surfactants. Surfactants (surface-active agents) are molecules that contain both hydrophobic (nonpolar, water insoluble) and hydrophilic (polar, water soluble) constituents. Long chain hydrocarbons or aromatic moieties contribute to the hydrophobic portion of a surfactant. The hydrophilic moiety can consist of cationic (e.g., amine terminated), anionic (e.g., sulfates, sulfonates, phosphates, or carboxylate terminated groups found in soaps), or non-ionic groups like ethers. When placed in water, the hydrophilic ends of the surfactant interact with the water molecules. On the other hand, the hydrophobic ends seek stabilization at the liquid–vapor interface (e.g., steric acid) or form bilayers in the form of lipid bilayers that are fundamental to biological processes. Packing character of the layer (bilayers) depends on the length of the hydrocarbon chain and the extent of unsaturation.

6.4.2 DLVO Theory

DLVO theory was developed in the 1940s by Boris **Derjaguin** and Lev **Landau** and independently by Evert **Verwey** and Theo **Overbeek** (hence DLVO). This theory explains the interactions between charged species in liquids. It is especially useful in describing the agglomeration, stability, and phase behavior of colloidal solutions. The stability of particles in solution depends on a total potential energy function, U_T. DLVO theory applies to colloids because most colloids carry an electrostatic charge. DLVO theory is based on a combination of attractive van der Waals forces (U_{vdw}), electrostatic forces (U_{es}), the concept of the electrical double layer, and the potential energy due to the solvent, $U_{solvent}$. The contribution of the solvent is minimal and takes place only within a few nanometers of the particles. An electric double layer (**Fig. 6.8**), initially devised by Helmholtz in 1879, is like a capacitor with the potential at the surface decreasing exponentially as counter ions are adsorbed. The double-layer model also has several assumptions: (1) ions are point charges, (2) only significant reactions are Coulombic, (3) electrical permittivity is constant through the double layer, and (4), the solvent is uniform at the atomic level. DLVO theory describes the interactions between two overlapping double layers.

The curve of a DLVO plot looks very much like one for a Lennard–Jones interaction potential except that the middle curve is positively inclined (**Fig. 6.9**):

$$U_T = U = U_A + U_R + U_{Solvent} \tag{6.39}$$

DLVO theory is another theory applied to the phenomenon of agglomeration. The attractive potential, U_A, is represented by

$$U_A = -\frac{A_H}{12\pi d^2} \tag{6.40}$$

where A_H is the Hamaker constant and d is the distance between colloidal particles. The Hamaker constant determines the effective strength of cumulative van der Waals interactions between colloids. Notice that this potential (attractive potential) is long range ($U \propto d^{-2}$). The repulsive potential, U_R, is given by

$$U_R = 2\pi\varepsilon a\xi^2 e^{-\kappa d} \tag{6.41}$$

where
   a is the colloid radius
   π is the solvent permeability
   ε is the dielectric constant of water
   κ is a function of the ionic composition
   ξ is the *zeta potential*

There are many assumptions in DLVO theory [19]: (1) flat surface, (2) uniform charge density, (3) constant surface electropotential, (4) static concentration

FIG. 6.8

In 1879, Helmholtz derived a mathematical treatment of a single layer of adsorbed ions on a charged surface. His treatment was based on a model of a capacitor. Gouy and Chapman in 1910–1913 derived a diffuse model in which they showed the potential decreases exponentially going away from the surface. The exponential decrease is due to adsorption of counter ions. Assumptions of this model include: that the ions are point charges, the significant reactions are all Coulombic, and the solvent is uniform with constant dielectric function. In most cases, the particle bears a negative charge. The figure shows a particle surrounded tightly by an inner layer (Stern layer) of cations in which the cations are bound strongly. A diffuse layer, made in this case of anions, is called the slipping plane, in which ions act in a uniform manner. The potential at this boundary is called the zeta potential ξ and decays exponentially away from the boundary. Electric double layers are also found on the surfaces of electrodes.

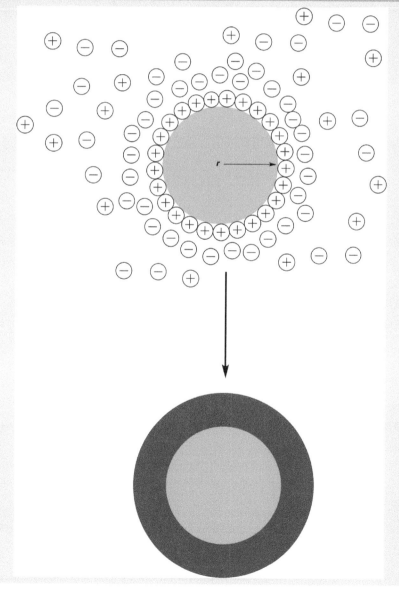

FIG. 6.9

A generic DLVO potential plot is depicted. The red curve on the left represents the balance between attractive and repulsive forces. DLVO theory integrates the attractive action of van der Waals forces and the repulsive action of electrostatic forces. In this way, it serves as a unifying theory. DLVO theory is actually a good way to describe the interaction between two approaching colloid particles. The maximum depicted in the figure represents the repulsive barrier of the system. If the barrier is greater than $10k_BT$, collisions caused by Brownian motion are not energetic enough to overcome the barrier and, as a result, agglomeration does not happen [19]. The primary minimum occurs after the double layer is totally overcome and agglomeration is the result. U_{max} is a function of the thickness of the double layer; the thicker the double layer is, the higher is the value of U_{max}. The secondary minimum is important in reactions that involve flocculation. The graph on the right represents the approach of two colloids after an increase in the ionic strength of the solution. Apparently, agglomeration becomes enhanced in such an environment.

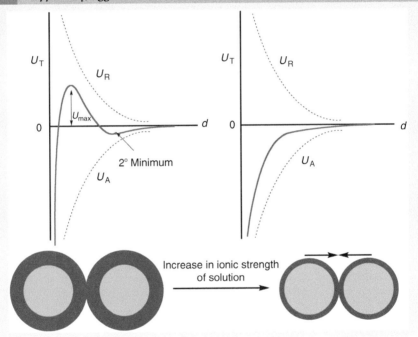

profiles, and (5) dielectric constant only solvent effect. Obviously, these assumptions are quite idealistic.

As the double layer between two colloidal particles starts to overlap, electrostatic repulsion exerts its effect. In such a solution, where colloidal approach is a consequence of Brownian motion, agglomeration occurs when colloidal particles collide with enough kinetic energy to overcome the barrier resulting from the repulsive forces. When this happens, the attractive forces predominate and the particles will agglomerate strongly and irreversibly. If the barrier is greater than $\sim 10k_BT$, then the collisions induced by Brownian motion will not provide enough energy to initiate agglomeration [19].

Cells tend to carry a negative charge. A layer of cations acts to stabilize the surface negative charge to form the electrical double layer. Adhesion of bacteria to cell surfaces can be explained by DLVO theory. Theoretically, bacteria should not be able to approach any closer than 10 nm to the surface of a cell depending on physiological conditions (e.g., approach is encouraged if the ionic strength of the medium is increased). Adhesion in cells is due to the combination of attractive and repulsive forces. At ca. 10 nm, there is some van der Waals attraction. At 2 nm, although the strength of electrostatic repulsion increases, van der Waals interactions start to dominate. At less than 1 nm, cell adhesion is nearly complete.

6.4.3 Polymeric (Steric) Stabilization

Steric repulsion keeps colloidal dispersions intact; that is, there is no agglomeration. The process serves to stabilize individual colloidal particles with additives that inhibit the coagulation of the suspended colloids. Materials used in this process include hydrophilic polymers and surfactants. It is simple to add polymer stabilizer to a solution. The polymer layer acts as a barrier to diffusion and therefore monodisperse solutions of nanoparticles can be synthesized. Polymers can bind tightly to the colloid by forming chemical bonds or form loosely adsorbed assemblies by weak physisorption mechanisms. Steric stabilization is compared to electrostatic stabilization in **Figure 6.10**.

6.4.4 Nucleation

The nucleation and precipitation process is an excellent, and classical, example of surface stabilization of nanoparticles. The process of nucleation is driven by thermodynamic parameters that are related to particle size. The free energy

Fig. 6.10 *Steric stabilization is also known as polymeric stabilization. The form of a steric stabilized particle that is independent of the electrical environment in a solution is compared to one surrounded by an electric double layer in the figure. There are scenarios in which there exist mixed electrostatic and steric stabilization. The importance of steric stabilization is that attached polymeric groups generate a diffusion barrier to further growth. As a consequence, nanoparticles with monodisperse size can be synthesized [19]. The dielectric constant of the solution is important. Obviously, a solvent in which the polymer is soluble is desired.*

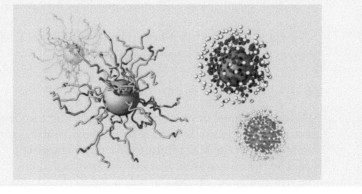

change for spherical particles (like a droplet of water in contact with its vapor) is given by [24]

$$\Delta G = \Delta \mu_\mathrm{v} + \mu_\mathrm{s} = \tfrac{4}{3}\pi r^3 \Delta G_\mathrm{V} + 4\pi r^2 \gamma \qquad (6.42)$$

where ΔG_V is the free energy per unit volume. The volume free energy $\Delta \mu_\mathrm{v}$ (the volume chemical potential) is the first term on the RHS of equation (6.42) and $\Delta \mu_\mathrm{s}$ is the chemical potential of the new surface. This is the classical equation for nucleation of a small particle in three dimensions. From this we can derive another form of the Gibbs–Thomson relation:

$$\Delta G_\mathrm{V} = \frac{k_B T}{V_\mathrm{a}} \ln\!\left(\frac{p_\mathrm{V}}{p_\mathrm{S}}\right) = \frac{k_B T}{V_\mathrm{a}} \ln(1+S) \qquad (6.43)$$

where S, the saturation, is defined as

$$S = \frac{p_\mathrm{V} - p_\mathrm{S}}{p_\mathrm{S}} \qquad (6.44)$$

If $p_\mathrm{V} > p_\mathrm{S}$, then the solution is supersaturated and there is nucleation and growth. If $p = p_\mathrm{S}$, an equilibrium state is attained. If $p_\mathrm{V} < p_\mathrm{S}$, the solution is not saturated and growth is not likely.

The calculation of the critical nucleus size is dependent on the surface tension—a barrier to nucleation—and relies on the $4\pi r^2 \gamma$ term in equation (6.42). The nucleation process is depicted graphically in **Figure 6.11.**

The critical nucleation size occurs when $\partial \Delta G / \partial r = 0$:

$$r_c = -\frac{2\gamma}{\Delta G_\mathrm{V}} = \frac{2\gamma V_\mathrm{a}}{k_B T \ln(1+S)} \qquad (6.45)$$

where V_a is the atomic volume (volume per atom) and S is the saturation as before. The energy barrier to nucleation is [19]

$$\Delta G_c^\ddagger = \frac{16\pi\gamma}{3\left(\Delta G_\mathrm{V}\right)^2} \qquad (6.46)$$

Therefore, the critical nucleus size depends on the degree of supersaturation; r_c is small for high supersaturation and large for low supersaturation [24]. In simple terms, particles that are larger than the critical radius will grow and particles that are smaller than the critical nucleus will shrink. In either case, elimination of the metastable intermediate "small particle" occurs.

6.4.5 Ostwald Ripening

Ostwald ripening, or coarsening, is the physical manifestation of the process described previously by which larger nanoparticles grow at the expense of smaller ones (**Fig. 6.12**). Larger crystals are energetically favored over smaller ones for the many reasons that we have discussed during the course of this chapter. The formation of small crystals is kinetically favored because they nucleate

The free energy profiles of a homogeneous nucleation and growth process as a function of particle size are depicted. The process starts with a supersaturated solution. There exist two thermodynamic contributions to the process: a volume term and a surface term. From the graph it is clear that there is a critical particle size threshold at which stability is achieved. The creation of larger particles following this threshold is a downhill process. The counterbalance in this process is the surface energy. The creation of a new phase and molecules coming together to form a particle with a surface results in increased surface energy. Anything to the left of the threshold then is also a downhill process to revert to the supersaturated state.

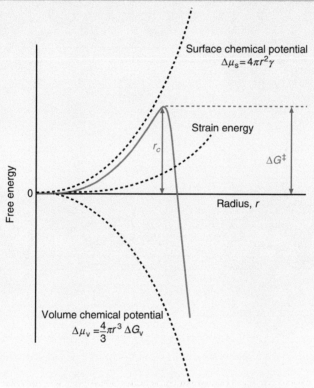

readily; nucleation has to start from the bottom up. Smaller crystals also have greater solubility and mobility in solution. Over a period of time, however, thermodynamic conditions favor the formation of larger crystals.

A dynamic equilibrium exists in solution between the rates of dissolution and precipitation of the dispersed phase to maintain the condition of saturation–solubility [25]. Smaller particles tend to dissolve and larger particles tend to grow. The process of Ostwald ripening is a function of particle curvature (as you might image). Equation (6.47) shows that the chemical potential of the process is dependent on the radius of the particles [19]:

$$\Delta \mu = 2\gamma V_a \left[\frac{1}{r_1} - \frac{1}{r_2} \right] \qquad (6.47)$$

FIG. 6.12

Ostwald ripening is the process of forming larger particles at the expense of smaller ones—driven by the reduction of surface energy. Larger particles are inherently more stable than smaller ones. Matter is lost and gained for particles of all sizes. However, the rate at which larger particles exchange matter is much lower than the rate at which smaller particles exchange matter. The net result is the extinction of the smaller particles. Ostwald ripening occurs in the solid, liquid, and gaseous phases and depends on the local environment. Homogenous structures result from this process.

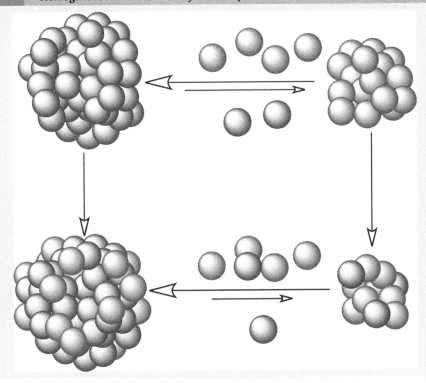

The mathematical formulation that relates solubility to surface curvature is

$$k_B T \ln\left(\frac{S}{S_o}\right) = \gamma V_a \left(\frac{1}{r_1} + \frac{1}{r_2}\right) \qquad (6.48)$$

where S represents the solubility of a particle with a curved surface and S_o is the solubility of the bulk form of the material [19]. Ostwald ripening is a form of sintering that occurs from the evaporation of atoms from one cluster and subsequent transfer to another. It so happens that the rate of "evaporation" is higher for smaller clusters than it is for larger ones—once again due to the numerous reasons we discussed in this chapter so far. The larger clusters then, over a period of time, get larger at the expense of smaller ones [26]. Coalescence implies the merging of similarly sized clusters without the dynamic interchange observed for Ostwald ripening (**Fig. 6.13**). Ostwald ripening occurs in the gas phase, in the liquid phase, and on a surface (surface-mediated Ostwald ripening sintering [SMORS]).

FIG. 6.13   *Coalescence implies the agglomeration of similarly sized particles. In this way, it is different from Ostwald ripening. The dynamic exchange between and among particles is not observed in coalescence and sintering.*

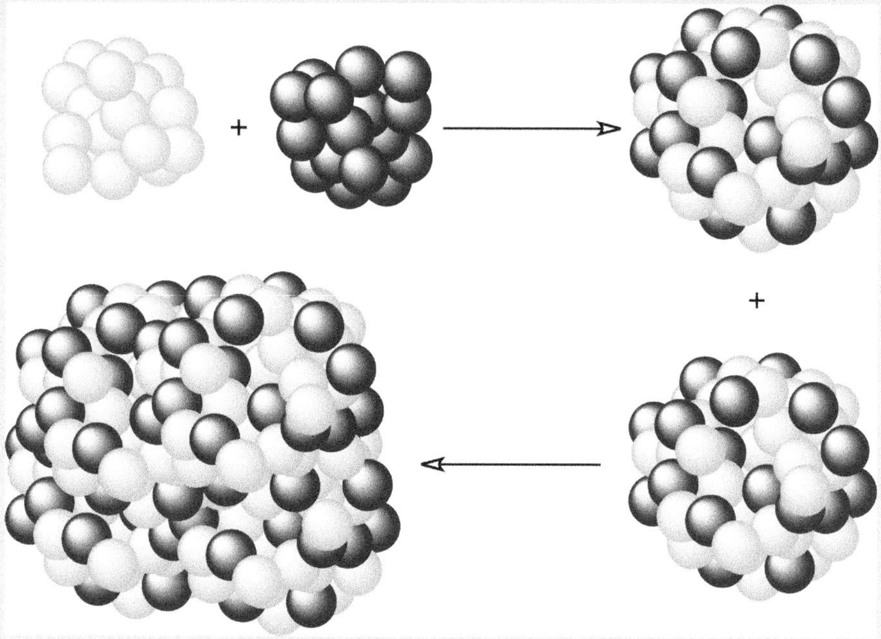

6.4.6 Sintering

The process of sintering is commonly associated with the solid phase, especially with regard to ceramic science. Sintering involves the agglomeration of similarly sized particles to form larger ones (**Fig. 6.13**), as opposed to Ostwald ripening (**Fig. 6.12**). Sintering involves fabrication of materials (usually ceramics, but also powder metallurgy) by heating powder precursors until agglomeration occurs. During sintering, there is a reduction of void volume as material flows into cavities (**Fig. 6.14**). This process results in a decrease in the overall volume of the material. In addition to reduction in porosity, sintering causes material transport from evaporation, condensation, and diffusion [27]. Sintering temperatures exceeding 1000°C are common.

Sintering is a complicated method of fabrication but one that is especially important at the nanoscale. In many cases, sintering is not desired. Sintering involves several stages during the process [19]: (1) diffusion (e.g., surface, volume, or grain boundary), (2) evaporation–condensation or dissolution–precipitation, (3) viscous flow, and (4) dislocation creep. Sintering is used to create dense materials. In sol–gel processes, drying is used to drive out all liquid phases, leaving a skeletal structure (e.g., xerogels). Application of sintering would produce a monolithic dense phase lacking a porous structure.

A generic version of sintering is shown in the figure. Following a compaction process that imparts some low-level order into the structure, the application of temperature converts individual particles into larger sized grains that adhere to one another very strongly. Sintering involves complex diffusion (surface, volume, grain), dissolution–precipitation flows, and evaporation–condensation processes [19]. Sintering is not necessarily desirable in nanomaterial fabrication.

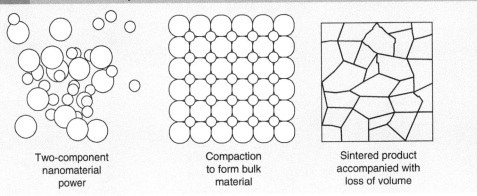

Two-component   Compaction   Sintered product
nanomaterial   to form bulk   accompanied with
power   material   loss of volume

6.4.7 *Structural Stabilization in Solids*

It is well known that the lattice parameters of metallic nanoparticles contract with decreasing particle size [28]. **Figure 6.15** illustrates rearrangement in a *primitive-C* material. Two mechanisms are shown: contraction of the surface layer into the bulk and lateral shift to increase the number of nearest neighbors. The lattice parameter *a* decreased in each case [19].

The lattice contraction of nanoparticles as a function of size can be predicted by [28]

$$\frac{\Delta a}{a_{bulk}} = -\frac{1}{1 + Kd} \tag{6.49}$$

where Δa is

$$\Delta a = a_{np} - a_{bulk}, \tag{6.50}$$

the lattice parameters of the nanoparticle (*np*) and the material in the bulk phase. The term *K* is related to the rigidity modulus *G* of the bulk material:

$$K = \frac{\pi G}{2\gamma_{bulk}} \tag{6.51}$$

γ_{bulk} is the surface energy of the bulk material and *d* is the diameter of the nanoparticle. *K* is in terms of reciprocal meters (or nanometers, angstroms). Plots of $\Delta a / a$ versus reciprocal diameter are linear with negative slope [28]. For palladium, G = 96.2 GPa, γ_{bulk} = 2.046 J·m^{-2}, and a_{bulk} = 3.8904 Å. From these values, *K* = 240.1 Å^{-1}. For a 2-nm particle, the theoretical percent contraction $\Delta a / a_{bulk} \times 100$ is expected to be

Fig. 6.15

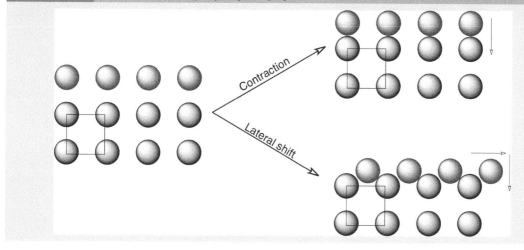

Restructuring of a simple cubic surface is shown. The coordination of a surface atom is reduced from 6 to 5 (one broken bond forming one dangling bond). Because of this, the surface atom is drawn inward in order to compensate for the lost bond. In this case, the overall structure remains relatively the same. There is only inward shift of the surface atom lattice. Lattice contraction is not just a surface phenomenon, as the whole of the nanoparticle may experience diminishing of its natural lattice constant. If more bonds are broken, as in the case of atoms with higher levels of coordination, the surface is actually restructured. In this case, formation of strained bonds compensates for the loss of coordination. For example, low-index {100} faces of silicon exist at a higher energy level than its low-index {111} sibling. However, restructured {100} faces have a lower surface energy than do natural {111} surfaces [19].

$$\left(\frac{\Delta a}{a_{bulk}}\right) = -\frac{1}{1+(240.1 \text{Å}^{-1})(20 \text{Å})} = -0.00021 = -0.02\% \; contraction \quad (6.52)$$

Reconstruction of outer layers also occurs by other mechanisms that involve loss of periodicity [29]. A surface layer that is perturbed is called a *selvedge* layer that perhaps consists of the upper two or so layers of the material. The surfaces of catalysts that accommodate surface carbon diffusion are also called selvedge layers. The selvedge layer may exist in a liquidus state during high-temperature catalytic activity. Atoms in a selvedge layer achieve stabilization by *outward relaxation* rather than contraction. The surface layers move outward and rearrange to preserve the configuration of the bulk atom packing symmetry. This layer is parallel to the layers of the bulk but not normal (original) (e.g., the layer has shifted). In another case, the surface atoms move outward but this time rearrange into a structure that has different symmetry from that found in the bulk of the material. This kind of rearrangement is called *reconstruction*. Following reconstruction, the surface properties of the material can be altered: electrical conductivity, chemical behavior, optical properties, and chemisorption–physisorption characteristics [29].

Surface stabilization is also achieved with the addition of adatoms, which are periodic arrays of an adsorbed monolayer of atoms that are directed by the preexisting surface lattice structure of the solid. Five "Bravais lattices" (instead of the 14 applied in volumetric crystals) characterize surface crystal structure (**Fig. 6.16**). We shall consider the simplest case—that of the simple primitive cubic surface (100) surface (**Fig. 6.17**). If $a = b$ in our simple square surface lattice, then the placement of the adatoms along x and y is related by vectors [28]:

FIG. 6.16	*Five two-dimensional "Bravais" lattices are depicted in (a). A fuller rendition that shows placement of atoms in the two-dimensional crystal is shown in (b).*

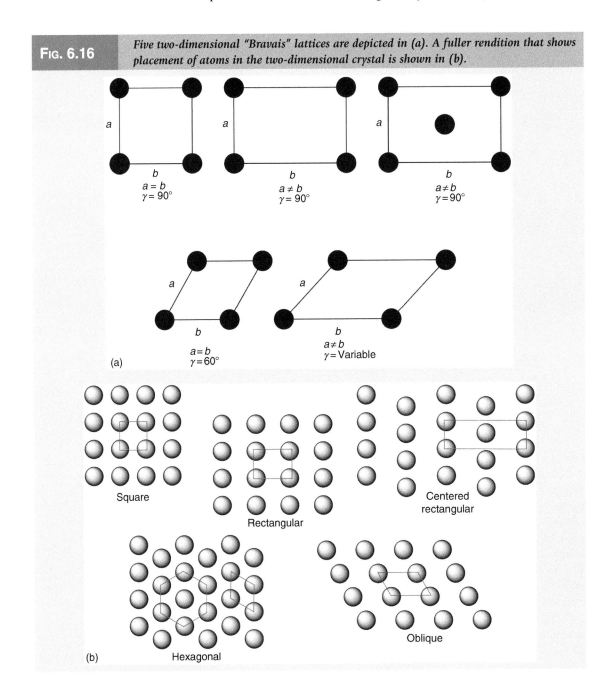

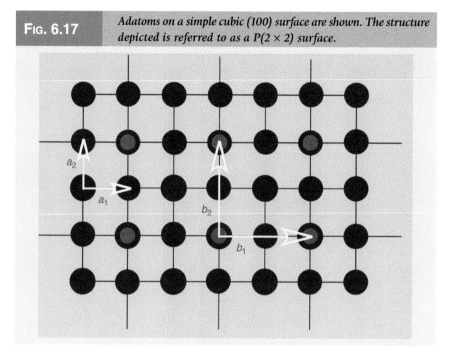

FIG. 6.17 *Adatoms on a simple cubic (100) surface are shown. The structure depicted is referred to as a P(2 × 2) surface.*

$$b_1 = M_{11}a_1 + M_{12}a_2 \tag{6.53}$$

$$b_2 = M_{21}a_1 + M_{22}a_2 \tag{6.54}$$

where, in the case of the primitive square, $b_1 = 2a_1$ and $b_2 = 2a_2$:

$$M = \begin{vmatrix} 2 & 0 \\ 0 & 2 \end{vmatrix} \tag{6.55}$$

The surface adatom array is designated as $P(2 \times 2)$ relative to the substrate crystal surface. The P stands for primitive.

Nanocrystalline solids are generally not isotropic. In other words, we do not expect nanocrystalline materials to be perfectly spherical in shape. The cuboctahedral structures shown earlier vividly demonstrate that anisotropic surface configurations exist in solid nanomaterials (**Fig. 6.18**). Recall that liquids tend to have isotropic surface energy (e.g., the surface energy is expected to be uniform at any point on the surface). In solids, different crystal planes (facets) on the surface have different surface energy. The individual facets on the surface contribute to the sum total equilibrium shape of the nanoparticle [19]. A general rule with regard to surface energy is that low-index planes—the $\{111\}$, $\{110\}$, and $\{100\}$ families of planes—have the lowest surface energy. We have already determined that the $\{111\}$ family of planes has the lowest surface energy of the group. When considering the packing structure of metals in particular, construction of a perfectly spherical shape requires more energy

FIG. 6.18	*The cuboctahedron structure is shown. The view of the structure through the upper-left vertex is along the [110] direction. The square faces represent {100} planes and the triangular faces represent {111} planes.*

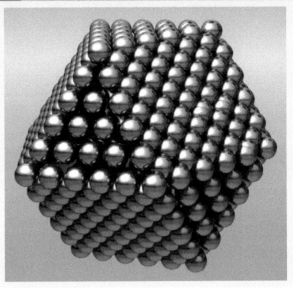

Source: Courtesy of Professor Anil K. Rao, Department of Biology, the Metropolitan State College of Denver.

than constructing one with a pseudospherical anisotropic crystal structure with facets. In this case, the spherical structure would possess higher surface energy.

Wulff plots represent the equilibrium shape of nanocrystalline materials. The general principle behind the Wulff plot is the relationship between the surface energy of a plane and its normal distance $(d_\perp)$ from a central point of reference—for example, that the equilibrium distance of the surface facet plane is proportional to the surface energy of that plane:

$$d_\perp = \frac{\gamma_{hkl}}{C} \qquad (6.56)$$

where C is a material constant. For example, a (111) plane, of lower surface energy, would be closer to the center of the crystal than would a (100) plane, of higher surface energy. The cuboctahedron is composed of 14 faces: 8 that are triangular (111) and 6 that are square (100). Oxidation of a NiO crystal that exists in the form of a cube with {100} faces goes through several morphological transformations during oxidation (**Fig. 6.19**): (1) cube, (2) truncated-cuboctahedron (contains {110} faces at vertices), (3) cuboctahedron at 75% oxidation with {100} and {111} faces, and finally (4) at complete oxidation into an octahedron consisting of {111} faces [30,31]. The changes in shape were simulated by using Wulff constructions of equilibrium morphologies [30,31].

Fig. 6.19

During oxidation of cubic NiO, the original cubic form undergoes several morphological transformations. Initially only (100) faces are visible on the surface, forming a perfect cube. After some oxidation of the crystal, the points of the cube are reacted (pyramids are removed) to form flat (110) surfaces and oblique (111) surfaces. As oxidation proceeds, a truncated cubic form and then a cuboctahedron are formed. The cuboctahedron is further oxided to form an octahedron. It also goes through a truncated state before achieving the final octahedron structure.

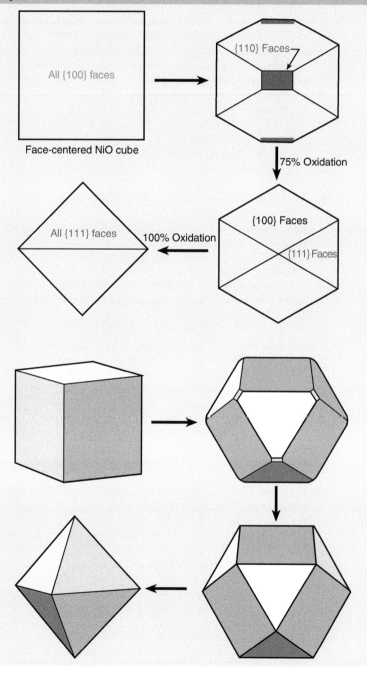

References

1. C. Kittel, *Introduction to solid state physics*, John Wiley & Sons, Inc., New York (1986).
2. W. J. Moore, *Physical chemistry*, 2nd ed., Prentice Hall, Inc., Englewood Cliffs, NJ (1955).
3. A. Robinson, Thomas Young: The man who knew everything, *History Today*, 56, 53–57 (2006).
4. W. Barthlott and C. Neinhuis, Purity of the sacred lotus, or escape from contamination in biological surfaces, *Planta*, 202, 1–8 (1997).
5. X. F. Gao and L. Jiang, Water-repellant legs of water striders, *Nature*, 432, 36 (2004).
6. M. Msksich, A surface chemistry approach to studying cell adhesion, *Chemical Society Review*, 29, 267–273 (2000).
7. J. A. Dean, ed., *Lange's handbook of chemistry*, 13th ed., McGraw–Hill Book Company, New York (1985).
8. P. W. Atkins, *Physical chemistry*, 4th ed., W. H. Freeman and Company, New York (1990).
9. I. A. Levine, *Physical chemistry*, 3rd ed., McGraw–Hill, New York (1988).
10. The Gibbs–Thomson effect, Wikipedia, en.wikipedia.org/wiki/Gibbs-Thomson_effect (updated February 2007).
11. *The American heritage dictionary of the English language*, 4th ed., Houghton Mifflin Company, Boston (2003).
12. M. S. Yaghmaee and B. Shokri, Effect of size on bulk and surface cohesion energy of metallic nanoparticles, *Smart Materials Structures*, 16, 349–354 (2007).
13. A. Shenk, *Calculus and analytic geometry*, Scott, Foresman & Co., Palo Alto, CA (1984).
14. V. M. Samsonov, A. N. Bazulev, and N. Yu. Sdobnyakov, On the applicability of Gibbs thermodynamics to nanoparticles, *Central European Journal of Physics*, 1, 474–484 (2003).
15. L. R. Fisher, R. A. Gamble, and J. Middlehurst, The Levin equation and the capillary condensation of water, *Nature*, 290, 575–576 (1981).
16. S. Siboni, Determination of the Kelvin equation in the presence of an arbitrary gravitational/inertial field, *American Journal of Physics*, 74, 565–568 (2006).
17. J. M. Howe, *Interfaces in materials: Atomic structure, thermodynamics and kinetics of solid–vapor, solid–liquid and solid–solid interfaces*, John Wiley & Sons, New York (1997).
18. B. M. Smirnov, *Clusters and small particles in gases and plasmas*, Springer-Verlag, New York (2000).
19. G. Cao, *Nanostructures & nanomaterials: Synthesis, properties and applications*, Imperial College Press, London (2004).
20. G. Kaptay, G. Csicsovszki, and M. S. Yaghmaee, An absolute scale for the cohesion energy of pure metals, *Materials Science Forum*, 414–415, 235–240 (2003).
21. P. Czoschke, H. Hong, L. Basile, and T.-C. Chiang, Quantum beating patterns in the energetics of thin film nanostructures, *Physics Review Letters*, 91, 226801 (2003).
22. K. K. Nanda, A. Maisels, F. Kruis, H. Fissan, and S. Stappert, Higher surface energy of free nanoparticles, *Physics Review Letters*, 91, 106102-1-4 (2003).
23. S. E. Jarlgaard, M. B. L. Mikkeslsen, P. Skafte-Pedersen, H. Bruus, and A. Kristensen, *Capillary filling speed in silicon dioxide nano-channels*, NSTI Nanotech, Boston (2006).
24. R. J. Hamers, *Materials growth*. Presentation, University of Wisconsin, hamers.chem.wisc.edu/chem630_fall2006/nucleation_and_growth/nucleation_and_growth2.ppt (2006).

25. D. J. Shaw, *Colloid & surface chemistry*, Butterworth–Heinemann, Oxford (1966).
26. M. Bowker, The going rate for catalysis, *Nature Materials*, 1, 205–206 (2002).
27. Sintering, Wikipedia, en.wikipedia.org/wiki/Sintering (2006).
28. W. H. Qi, M. P. Wang, and Y. C. Su, Size effects on the lattice parameters of nanoparticles, *Journal of Materials Science Letters*, 21, 877–878 (2002).
29. M. Ohring, *Materials science of thin films*, 2nd ed., Academic Press, San Diego (2002).
30. S. C. Parker, E. T. Kelsey, P. M. Oliver, and J. O. Titiloye, Computer modeling of inorganic solids and surfaces, *Faraday Discussions*, 95, 75–84 (1993).
31. P. M. Oliver, S. C. Parker, and W. C. Mackrodt, Computer simulation of the morphology of NiO, *Model Simulations in Materials Science and Engineering*, 1, 755–760 (1993).

Problems

6.1 Why are bubbles made of pure water not likely to occur?

6.2 For *fcc* gold, what is the surface energy of a freshly cleaved:

   a. (100) Surface?
   b. (001) Surface?
   c. (010) Surface?
   d. (111) Surface?

6.3 Calculate the work required to increase the surface area of mercury from 1 to 6 cm^2 at 20°C.

6.4 A bubble of air exists within a solution comprising acetone at 20°C and 1 atm. Calculate the pressure in a gas bubble if the bubble has a radius of 0.002 mm.

6.5 Find the cohesive energy (E_c), or binding energy, for graphite at 25°C from thermodynamic data (graphite, $\Delta H°_{f,298}$ = 0.00 kJ·mol^{-1}; gaseous carbon, $\Delta H°_{f,298}$ = 717 kJ·mol^{-1}). How does it compare to diamond under the same conditions ($\Delta H°_{f,298}$ = 1.897 kJ·mol^{-1})?

6.6 What is the cohesive energy of bulk water in the liquid state at 25°C?

6.7 Calculate the energy required to atomize one drop of water at 20°C in which the surface area is increased from 1 cm^3 to 1 m^3. Assume that no pressure–volume work is done. The surface tension of water is γ = 72.75 dyn cm^{-1} against air at 20°C. What is the radius of the droplets formed?

   Calculate the same problem for atomization of acetone (assume no Joule–Thompson cooling occurs during atomization) with surface tension γ = 23.70 dyn cm^{-1}.

6.8 For a simple cubic structure, what is the coordination number of each interior atom? Each surface atom? Each edge atom? Each corner atom? What amount of energy is required to split a cubic centimeter of the material in half along a (100) plane? Into eight equivalent cubes all along {100} planes?

6.9 For gold, an *fcc* metal, calculate the relative excess surface energy, ε_s in terms of joules per atom, of a freshly cleaved {111} plane. Z = 12 in the bulk. How many nearest-neighbor bonds are broken? What is this in terms of γ (in joules per square meter)?

6.10 Why are heats of vaporization of solids ~10% lower than the heats of sublimation of the same material?

6.11 Neglecting the force of gravity, if you have a wire of length 4 cm, calculate the work required to lift a sheet of acetone held together by surface tension to a height of 100 μm.

6.12 Is surface tension an extensive or an intensive property (consult your thermodynamics text)?

6.13 Would you agree that the Kelvin equation has built-in nano considerations before nanoscience became very popular? Explain your answer.

6.14 Why is the radius of curvature of a sphere equal to the radius of the sphere?

6.15 Young, Laplace, and others developed what is called a mechanical explanation for surface tension. What does this imply?

6.16 Calculate the vapor pressure of $r = 600$ nm water droplets at 20°C. Assume that the density of water at 20°C is 0.99823 g·cm^{-3}, $V_m = 18.047$ cm^3·mol^{-1}; $T = 293$ K, $R = 8.314$ J·K^{-1}; $M_{H2O} = 18.015$ g·mol^{-1}.

6.17 Define the concept of chemical potential. Give some examples of chemical potential.

6.18 What is the difference, qualitatively, between cohesive and adhesive energies?

6.19 Why is the contact angle of a water droplet on Teflon higher than that of the same sized droplet on gold surface?

6.20 What (intuitively) is the meaning of surface energy?

6.21 Make a list of all the factors that contribute to the higher surface energy of nanomaterials.

6.22 Calculate the height of a column of water in a nanocapillary tube (Pyrex™ glass) of diameter of 800 nm.

6.23 List some of the assumptions inherent in the *nearest-neighbor* model to calculate surface energy.

6.24 Why is the work function of a material dependent on the type of surface?

6.25 List as many ways to minimize surface energy as you can.

6.26 Explain DLVO theory.

6.27 Why does the term for the free energy of nucleation include a surface term?

6.28 Would you expect that "surface energy-to-volume energy" scales with surface-to-volume ratio as particles become smaller?

THE MATERIAL CONTINUUM

As in my conversations with my brother we always arrived at the conclusion that in the case of X-rays one had both waves and corpuscles, thus suddenly— … it was certain in the course of summer 1923—I got the idea that one had to extend this duality to material particles, especially to electrons.

LOUIS DE BROGLIE

*C*hapter 7

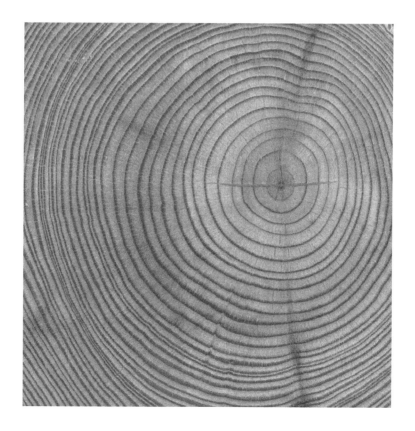

In chapter 5, we introduced topics that described the surfaces of nanomaterials and how size, shape, structure, orientation, and composition influence nanoscale material properties and phenomena. Although devoid of any real discussion of physics or chemistry, we gained a sense of how size and interfacial phenomena can potentially dominate behavior at the nanoscale. Chapter 6 addressed topics involved with the energetic character of nanoscale materials—clusters in particular—primarily from the viewpoint of the chemical bond and the mechanical–thermodynamic treatment of surface tension. In chapter 7, we now prepare to delve into the quantum domain and the effects of size confinement. We reaffirm that nanomaterial behavior does not scale in a linear fashion from bulk material behavior. We also find that nanomaterials, especially those on the scale of a few nanometers, do not necessarily obey the rules of quantum mechanics, although the fundamental relationships still hold true.

Nanotechnologists must be proficient in the basics of quantum mechanics, not only in its mathematics but also in its physical meaning. Indeed, a good intuitive understanding of quantum mechanics is a prerequisite for a thorough understanding of chemistry, physics, and materials science. Why? Because nanotechnology is all about a finite number of atoms or molecules that together are the basic building blocks of everything and so small that they are inherently quantum mechanical. For example, within an atom or molecule, negatively charged electrons are trapped in the neighborhood of one or more (positively charged) atomic nuclei because of the attraction the electrons have for the nuclei and repulsions for other electrons. These electrons only have certain energies that correspond to the 1s, 2p, etc. orbitals we learned in freshman chemistry. For some nanomaterials, electrons can also be trapped and exhibit only certain allowed energies.

Following this chapter, we present a special chapter about nanothermodynamics. We adhere to strict thermodynamic topics and hope to shed light on the energetic relationships of nanoscale materials: Are those relationships worthy of forming the foundation of a new discipline called nanothermodynamics? You decide!

7.0   MATERIAL CONTINUUM

The concept of "continuum" (from the Latin *continuus*, meaning "uninterrupted") is ingrained in everything surrounding us. The definition of a continuum is a continuous extent, or whole, no part of which can be distinguished from neighboring parts except by arbitrary division.

Many examples of common continua exist around us—those of language, mathematics, mechanics, time, radiation, and others. The matter–energy continuum implies that all matter and all energy are convertible through $E = mc^2$. The material continuum implies that there is a smooth transition from the bulk world into the quantum domain. This statement is completely true in the sense that all materials are made of atoms and molecules and that nanomaterials simply have progressively fewer of them. However, we have been preaching all along that nanomaterials have neither purely bulk nor strictly quantum properties. We then have to grapple with a dilemma: If there is a continuum of material and energy, then why all the fuss about nanomaterials? In order to come to terms with this apparent difference, it is our desire to understand how properties and phenomena *scale* with regard to diminishing size.

The continuum of interest in this section is that of materials, measured and scaled by size and size alone. The material continuum is

Extremely big things/trees → *macroscopic bulk materials/multicellular organisms* →
micron-sized colloids/crystals/single-celled organisms → **colloids/nanoparticles/**
nanocrystals/clusters/viruses/macromolecules → *molecules/atoms* → *subatomic*
particles → *subnuclear particles* → *much smaller things*

7.0.1 Material Properties and Phenomena

Table 7.1 lists properties and phenomena from two different perspectives: the macroscale and the nanoscale.

TABLE 7.1	*Material Properties and Phenomena*		
Property/ phenomenon	**Specific property**	**At the macroscale–microscale**	**At the nanoscale**
Structure and electronic configuration	Confinement	No confinement	Confinement in 0D, 1D, and 2D materials
	Surface area	Surface area of bulk materials, although important, is small compared to its volume	Exponentially enhanced Collective surface area can be enormous
	Surface-to-volume ratio	S/V is small; becomes insignificant as objects become larger	Approaches 1 when all atoms are surface atoms
	Lattice spacing	Characteristic of the bulk	Lattice spacing is altered. Spacing near surface contracts due to rearrangement. Ion vacancies larger
	Atom coordination	Coordination saturated except at surface where it is negligible	Coordination undersaturated at surface and in volume
	Electron orbitals	Continuous over the breadth of the material	Quantum: HOMO and LUMO Cluster: HOCO and LUCO Magic electronic numbers in alkali metal clusters
	Quantum mechanics	QM applies at the bulk level: bathtub waves	Nanomaterials exist at the quantum–classical interface
Electromagnetic properties	Radiation: absorption emission	Blackbody radiation Absorption–emission bands broad	Influenced by the Bohr radius
	Optical response	Metals reflect with partial absorption of light Micron-sized particles scatter light and conform to Mie theory analysis Higher order plasmon resonance and plasmon resonance is delocalized	Size-dependent absorption–emission. Environment dependent (effective medium theory) Quasi-static condition ($r \ll 0.01\ \lambda$). Dipolar plasmon resonance Localized surface plasmon resonance (i.e., transverse/longitudinal modes for prolate particles)
	Optical constants, ε, n, k	Bulk values apply and are valid in micron-sized particles.	Bulk optical constants no longer apply below 10–20 nm

(continued)

Table 7.1 (Contd.)	**Material Properties and Phenomena**		
Property/ phenomenon	**Specific property**	**At the macroscale–microscale**	**At the nanoscale**
	Bandgap	Traditional metal, semiconductor, insulator bandgaps. Bandgap independent of size	Bandgap is size dependent. Gold resembles a semiconductor in nanoparticles <2 nm in diameter
	Electrical conduction	Continuous and follows Ohm's law. Conductivity based on band structure Electron mean free path not significant with respect to surface Scattering by lattice defects and thermal phonons	Ohm's law does not apply (classically). Formation of discrete energy levels Coulomb staircase/blockade Tunneling currents are important Ballistic conduction (electron mean free path > dimensions)
	Magnetic memory coercive force	Size independent	Size-dependent magnetic properties. Gigantic magnetoresistance effects possible with stacked magnetic nanoparticle arrays
Thermo- dynamics	Discipline	Classical thermodynamics, statistical mechanics	Nanothermodynamics
	Equilibria	Macroscopic systems— thermodynamic equilibrium capability	Nanosystems subject to environmental fluctuations. Conditions of nonequilibrium/ steady state in living systems
	Number of atoms	Thermodynamic infinite limit: $N \to \infty$, $V \to \infty$, N/V constant	N is countable. Thermodynamic limits do not apply
	Intensive properties	Properties independent of amount of material Environment independent	Intensivity not always applicable as intensive properties can change with size. Serious environmental dependency
	Extensive properties	Properties dependent on amount of material	Altered definition of extensivity
	Entropy	Link between micro and macro domains	Violations of the second law. Is a fourth law of thermodynamics required?
	Melting point	Metals relatively high melting	MP drops precipitously below 20 nm Proportional to $1/r$
	Kelvin effect	Function of particle curvature Valid for bulk and nanomaterials	Particles <10 nm conform to Kelvin effect
	Surface tension/ energy	Classical surface tension and surface energy	Surface tension and surface energy a function of size
	Specific heat	Function of elemental makeup	Much higher than bulk counterpart
	Electron affinity	Function of elemental makeup	Electron affinity influenced by magic numbers in clusters
	Work function	Function of elemental makeup	Function of size
	Chemical bonds	The big three: ionic, covalent, and metallic predominate, although all others apply	Intermolecular forces are important: H-bonds, van der Waals, hydrophobic effect, dipole interactions, etc.
Chemical reactivity	Reaction scheme	Kinetic control at higher tempera- tures, harsh conditions; making and breaking of stronger bonds	Thermodynamic control occurs at lower temperatures, milder conditions; making and breaking of weaker bonds

TABLE 7.1 (CONTD.)	*Material Properties and Phenomena*		
Property/ phenomenon	**Specific property**	**At the macroscale–microscale**	**At the nanoscale**
	Adsorption	Surface adsorption by chemisorption/physisorption does not result in catalytic activity	Adsorption by small particles can result in catalytic activity
	Solubility	Large particles have limited solubility	Smaller particles have enhanced solubility. This is important in targeted drug delivery systems
	Sintering	Function of elemental makeup. Occurs at high temperatures	Occurs at lower temperatures
	Chemical activity	Function of elemental makeup. Reactivity takes place below the nanoscale	Reactivity of nanoparticles significantly enhanced due to excess surface energy
Mechanical properties	Tensile properties	Bulk materials made of micron-sized grains	Bulk materials made of nanograins have superior properties. Tensile properties can approach theoretical limit in carbon nanotubes
Scaling laws Classical continuum models	Density Power Frequency Efficiency Mechanical Dielectrics	Bulk scaling laws and continuum models apply to all materials down to micron-sized particles	Depends on the types of physical phenomena and the size of the particle. Bulk dielectric functions are valid to sub-100-nm particles

7.0.2 Background

Classical mechanics gives the perspective of trajectories of macroscopic particles while wave mechanics gives a view of the discrete energy levels associated with macroscopic particles without actually explaining the trajectories. Sometimes a complete explanation is difficult to obtain, but instead we can take advantage of each view and work on new windows of opportunity to explain different phenomena. Nanomaterials provide us with that new window.

Quantum mechanics assumes importance as material size is diminished. This is especially true in nanoscience; materials comprising a countable number of atoms and molecules are treated mathematically with wave functions. One way to look at the wave function is to consider particles as stationary waves (standing waves), and all that is needed to describe its behavior is this description of the standing wave (e.g., the wave function). We discussed the particle–wave duality in chapter 3, so it should come as no surprise that, at the nanoscale, the dual wave–particle nature of radiation becomes more important. The dual wave–particle nature was first propounded by Louis de Broglie (chapter 3) to address the apparent anomaly observed in black body radiation—specifically, the ultraviolet catastrophe.

In our universe, each particle (whether atom, molecule, or cluster) is characterized by a function called ψ. The only restriction placed on this function is the

Schrödinger equation—the fundamental equation of matter. From this equation, the average values of angular momentum, position, and energy are solvable by the application of mathematical operators (complex conjugates, integration) over all independent variables. Quantum mechanics is a branch of physics that was developed in the early part of the twentieth century. It is extremely effective in describing the behavior of atoms, molecules, nuclei, and, most recently, nanoparticles. Quantum mechanics is based on the concept that matter has wave properties. Quantum mechanics is used to calculate the energies of particles confined to a finite (and small) region of space. The uncertainty principle is a consequence of quantum mechanics, as is the phenomenon of electron tunneling. Quantum mechanics is about probability.

A particle located in a fixed point in space is localized in space with discrete physical properties such as mass. In contrast, a wave is inherently spread out over many wavelengths in space and could have amplitudes in a continuous range with waves superposing and passing through each other. On the other hand, particles are able to collide and bounce off each other. The dual wave–particle nature of radiation is all to do about tiny matter having the ability to behave as a wave as well as a particle (i.e., waves and particles have interchangeable properties). Mathematically, it can be explained by considering that if an object having momentum p has an associated wave whose wavelength is λ, then the momentum can be expressed as

$$p = \frac{h}{\lambda} = \hbar k \tag{7.1}$$

and

$$k = \frac{2\pi}{\lambda} \tag{7.2}$$

EXAMPLE 7.1 *A Convenient Form of the de Broglie Wavelength*

$$\lambda = \frac{hc}{pc} \tag{7.3}$$

where h is the Planck's constant, p the momentum, and c the velocity. Plugging in the Planck's constant value and the speed of light, $hc = 1239.84$ $eV \cdot nm$ and pc is expressed in electron volts.

This is particularly appropriate for comparison with photon wavelengths since for the photon, $pc = E$ (kinetic energy) and a 1-eV photon is seen immediately to have a wavelength of 1240 nm. For an electron with KE = 1 eV and rest mass energy 0.511 MeV, the associated de Broglie wavelength is 1.23 nm, about a thousand times smaller than a 1-eV photon. (This is why the limiting resolution of an electron microscope is much higher than that of an optical microscope.)

Some examples of physical phenomena that apply to the nanoscale:

Visible light (optical microscope) E ~ 4.1–2 eV): λ ~ 300–600 nm
X-rays (synchrotron, E ~ 1 keV): λ ~ 1 nm
Electrons (electron microscope, E ~ 100 keV): λ ~ 0.004 nm

Ions (He$^+$ ions, E ~ 100 keV): λ ~ 10^{-5} nm

where k is the wave vector (also called the *wavenumber*) and

$$\hbar = h/(2\pi) \tag{7.3}$$

the reduced form of Planck's constant. The wave vector is an energy vector of magnitude $2\pi/\lambda$ in the direction of the propagation of the wave.

7.0.3   Nano (Quantum) Perspective

The density of states (DoS) is a physical property of a material that indicates the structure and degree of packing of energy levels in a quantum mechanical system [1]. It is commonly symbolized by N and is a function of $g(E)$ or a function $g(k)$ of the wave vector k. Density of states is a phenomenon usually applied to the electronic levels of a solid. The densities of states for zero-dimensional (quantum dots), one-dimensional (quantum wires), and two-dimensional (quantum wells) materials compared to that of a bulk semiconductor material are shown in **Figure 7.1**.

The purpose of this chapter is to help illuminate the boundary between the quantum domain and that of the nanomaterial and, consequently, to make a link to the domain of the bulk material. The concept of the matter continuum is a valid one; all matter is described by the same equations: quantum, classical, or otherwise. It is simply a matter of complexity. Classical equations break down in many ways (e.g., the ultraviolet catastrophe, to name just one) but they are applied in modified form to describe nanomaterials. If there are enough quantum states, then we have a bulk material. If we have fewer quantum states, then

FIG. 7.1   *Top left: the DoS of bulk semiconductors continually increases with energy; top right: DoS of quantum wells is like a staircase—DoS changes abruptly for larger changes but is continuous for smaller changes; bottom left: DoS for quantum wires is nearly discrete with some continuity; and bottom right: DoS for quantum dots is discrete—very molecular-like in appearance. With increase in size of a nanostructure, DoS assumes continuity. Conversely, with diminishing size, DoS becomes discrete.*

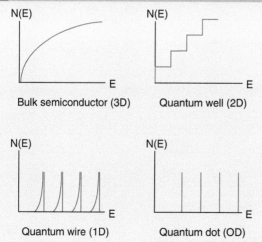

we have a nanomaterial. If we have even fewer quantum states, then we have molecules—a continuum for sure. This process continues to the atom and into the nucleus of the atom in which quantum states also exist. The matter continuum is in every sense of the term a true continuum.

7.1 BASIC QUANTUM MECHANICS AND THE SOLID STATE

Let us begin our description of the solid state by starting with a liquid, by considering a bathtub that is filled with water. If you swish your hand around in a bathtub, you will note that the only waves that persist for a long time are those that can "fit" inside the bathtub; other waves would simply spill water out of the tub. Actually, you will observe that many waves, measuring from crest to crest, fit into the bathtub! These waves are called "standing" waves as they just stand around instead of traveling forward like waves you would see at the beach. In other words, they have their crests and troughs always in the same places, and the nodes (areas midway between the crests and troughs where the water level is undisturbed) are always in the same places. The largest possible bathtub wave that has a crest at each end is shown in **Figure 7.2a.** Another wave with a crest at each end and in the middle is shown in **Figure 7.2b.**

The possible energies of standing waves in the bathtub are limited to certain specific (quantum!) values by the requirement that each wave needs to fit exactly into the bathtub. The human in the tub supplies the energy required to make waves. Now the energies of the bathtub waves can be likened to the wave functions that describe a particle in a box (like electrons trapped inside a quantum dot) that increases with the number of wave crests that are able to fit into the tub. If the water in the bathtub is swished around very gently and slowly, for example, then the only wave we will see would be the wave that has a crest exactly at each end of the tub with a node right in the middle of the tub. If we increase the agitation, we will see waves that have more and more crests in the middle, at least until all the water comes out onto the floor. The wave functions then would resemble the standing waves you observe in the bathtub. In fact, allowed energies are calculated easily by figuring out which waves "fit" into the

| FIG. 7.2 | *(a) A bathtub wave with a wave at either end. The node is in the middle. (b) A bathtub wave with two crests, one at each end, that occur simultaneously. Nodes exist at the walls of the tub.* |

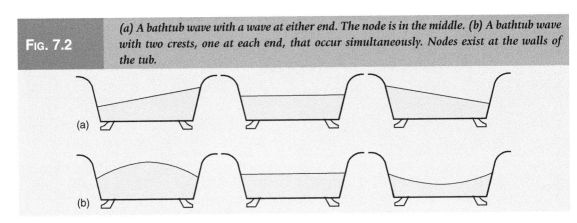

box and which ones do not! The meaning of the peaks and troughs of the wave function is somewhat akin to the waves in the bathtub, in the sense that something is going on at the peaks and troughs but not at the nodes.

7.1.1 Ubiquitous Particle in a Box

One of the simplest quantum mechanical problems is that of the ubiquitous "particle in a box." Solution of this problem is very useful for understanding of many properties of nanoparticles or quantum dots. In a one-dimensional particle-in-a-box problem, we consider a particle that is constrained to move in a single dimension, under the influence of a potential $V(x)$ [or $U(x)$] that is equal to zero for $0 < x < L$ and equal to ∞ for all other x. At the walls of the box, $V(x)$ is equal to infinity (hence, the infinite potential well) and the wave function equals zero (**Fig. 7.3a**). The square of the wave function at any given point is equal to the probability that the particle can be found at any point x within the boundaries. Hence, the particle is likely to be found at peaks or troughs but never at the nodes. Classically, if we put a particle in a box, and it is confined to move in only one direction, the particle will reflect off the walls and go backward and forward in the box (**Fig. 7.3b**).

At any time if we open the box we will find that the particle has equal probability of being somewhere in the box. In order to better understand the wave properties from the point of view of the wave equation, let us recall the harmonics of a string held at both ends (like in a guitar; see **Fig. 7.4**).

You already know that waves exist only when the wavelength is an integral multiple of the half wavelengths (**Fig. 7.5**):

$$L = n\frac{\lambda}{2} \quad \text{and therefore} \quad \lambda = \frac{2L}{n} \tag{7.4}$$

In this way, as you can see, the wavelengths of the standing wave of a string are *quantized*. As the particle is confined between two nonpenetrating walls and is confined to moving only parallel to the x-axis (our supposition), its linear momentum (mv, for a mass m, moving with velocity v) remains constant and

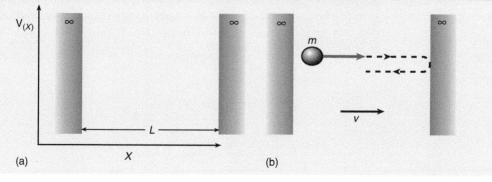

FIG. 7.3 *(a) Schematic representation of the wave function in an infinite well. The potential energy of the particle is infinite outside the box. (b) Classical representation of a particle in a box that reflects its motion in one dimension off the walls of the infinite well.*

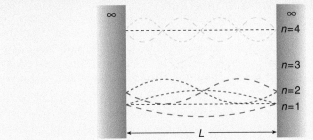

FIG. 7.4 *String harmonics: The strings are held at two ends. The quantum levels are represented by n = 1,2,3,...,∞. Waves only exist when there is an integral multiple of the half-wavelengths in the string.*

its kinetic energy (KE) is a constant as well. If we consider the wave to be described by a simple function for a standing wave:

$$y(x) = A\sin(kx) \tag{7.5}$$

where A is a constant that defines the amplitude. If we modify the value of k, the wavenumber is then

$$\text{If } k = \frac{2\pi}{\lambda} \text{ and } \lambda = \frac{2L}{n}, \quad \text{then } k = \frac{n\pi x}{L} \text{ and } y(x) = A\sin\left(\frac{n\pi x}{L}\right) \tag{7.6}$$

In 1926, Erwin Schrödinger proposed the wave equation that is used to describe how waves of matter (or the wave function) propagate in space and time. The wave function is similar to our identity card and contains all of the information that can be known about a particle. The wave function of a free particle moving along the x-axis is given by

$$y(x) = A\sin\left(\frac{2\pi x}{\lambda}\right) = A\sin(kx) \tag{7.7}$$

FIG. 7.5 *A vibrating string is shown. The waves exist only when the wavelength is an integral multiple of the half wavelength. At the nodes, the probability is zero. At the antinodes, the probability is maximized.*

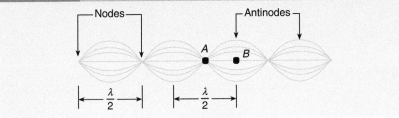

The waves, like those shown in **Figure 7.6,** represent a snapshot of a wave function at a given time. The probability of finding a particle in a position inside the box is given by

$$\left|\psi(x)\right|^2 = A\sin^2\left(\frac{2\pi x}{\lambda}\right) = A\sin^2(kx) \tag{7.8}$$

The probability is always positive and the sum total probability equals one. Because the de Broglie wavelength is quantized, the momentum is also quantized. Higher energy levels of the particle correspond to higher potential energy in the box as it climbs the walls of the infinite well. The rest of the energy of the particle is the kinetic energy, the energy due to its motion. The kinetic energy is also quantized:

$$E_n = \frac{1}{2}m\upsilon^2 = \frac{p^2}{2m} = \frac{(nh/2L)^2}{2m} = \left(\frac{n^2h^2}{8mL^2}\right); \quad n = 1, 2, 3, \dots \tag{7.9}$$

EXAMPLE 7.2   *Electron Confined within Atomic Dimensions*

Consider an electron in an infinite well of size 0.1 nm (typical of an atom). What is the ground state energy ($n = 1$) of the electron? What energy is required for the electron to jump to the third energy level ($n = 3$)?

Solution:
The electron is confined in an infinite potential well; therefore, its energy is given by

$$E_n = \left(\frac{h^2 n^2}{8mL^2}\right); \text{ for } n = 1 \text{ and } L = 0.1 \text{ nm} \tag{7.10}$$

$$E_1 = \frac{\left(6.626\times10^{-34}\,\text{J}\cdot\text{s}\right)^2 (1)^2}{8\left(9.110\times10^{-31}\,\text{kg}\right)\left(1\times10^{-10}\,\text{m}\right)^2} = 6.025\times10^{-18}\,\text{J} \text{ or } E_1 = 37.6 \text{ eV} \tag{7.11}$$

The frequency of the electron associated with this energy is

$$v = \frac{E}{\hbar} = \frac{6.025\times10^{-18}\,\text{J}}{1.055\times10^{-34}\,\text{J}\cdot\text{s}} = 5.71\times10^{16} \text{ radians}\cdot\text{s}^{-1} \text{ or } v = 9.092\times10^{15}\,\text{s}^{-1} \tag{7.12}$$

The third energy level (E_3 at $n = 3$) is

$$E_3 = E_1 n^2 = (37.6 \text{ eV})(3)^2; \ E_3 = 338.4 \text{ eV} \tag{7.13}$$

The energy required to take the electron from 37.6 to 338.4 eV is 300.8 eV.
This energy can be provided by a photon of exactly that energy—no more and no less. Since the photon energy is $E = h v = hc/\lambda$

$$\lambda = \frac{hc}{E} = \frac{\left(6.626\times10^{-34}\,\text{J}\cdot\text{s}\right)\left(2.998\times10^{8}\,\text{m}\cdot\text{s}^{-1}\right)}{(300.8 \text{ eV})\left(1.6022\times19^{-19}\,\text{C}\right)} = 4.12 \text{ nm} \tag{7.14}$$

which is the energy of an x-ray photon

EXAMPLE 7.3   *Particle-in-a-Box Application*

Consider a linear butadiene molecule. What is the length of the box? Calculate the energy of an electron confined in this molecular box? What is the expected absorption of butadiene $CH_2=C-C=CH_2$ [2]?

Also what is the energy of a dye molecule that is 0.8 nm in length with an electron $n_1 \rightarrow n_2$ transition?

Solution:
The C–C bond length is 1.54 Å and that of the C=C bond is 1.35 Å. The length of the box is then:
$L = 2 \times 1.35 + 1.54 = 5.78$ Å

$$E_n = \frac{n^2 h^2}{8mL^2}, \quad n = 1, 2, 3, \ldots \tag{7.15}$$

Since each state holds two electrons and there are four π-electrons, the $n = 1$ level is full and the $n = 2$ level is full. The promotion of the electron then must proceed as $n_2 \rightarrow n_3$.

FIG. 7.6   *The particle-in-a-box model is applied to the four π-electrons of the conjugated butadiene molecule. In reality, butadiene is not a linear molecule. The length of the box is equivalent to the length of the molecule that consists of two double carbon–carbon bonds and one single carbon–carbon bond. The absorption of butadiene is at 217 nm. The particle-in-a-box model derived value is 220 nm.*

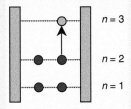

$$E = \frac{h^2}{8mL^2}\left(n_3^2 - n_2^2\right) = \frac{h^2}{8mL^2}\left(3^2 - 2^2\right) = \frac{\left(6.626 \times 10^{-34}\,\text{J·s}\right)^2}{8\left(9.110 \times 10^{-31}\,\text{kg}\right)\left(5.78 \times 10^{-10}\,\text{m}\right)^2}(5) = 9.02 \times 10^{-19}\,\text{J} \tag{7.16}$$

This energy corresponds to wavelength through

$$E = h\nu = \frac{hc}{\lambda}; \lambda = \frac{hc}{E} = \frac{\left(6.626 \times 10^{-34}\,\text{J·s}\right)\left(2.998 \times 10^8\,\text{m·s}^{-1}\right)}{9.02 \times 10^{-19}\,\text{J}}\left(\frac{10^9\,\text{nm}}{\text{m}}\right) = 220\,\text{nm} \tag{7.17}$$

The experimental absorption of butadiene is 217 nm—not a bad match from a particle-in-a-box calculation. A dye molecule with box length $L = 0.8$ nm has energy 2.8×10^{-19} J. This corresponds to $\lambda = 709$ nm.

The kinetic energy of the particle at any point inside the quantum box is equal to the relative magnitude of the curvature of the wave function (this is true for real waves as well—like in the sea wave, where the water moves fastest where the wave is most sharply curved at the very tip of the crest). One is able to observe that higher energy wave functions are sharply curved to give shorter distance between successive wave peaks. Hence, more wave peaks can fit inside

| EXAMPLE 7.4 | *Confinement of Proton and Electron in a Quantum Well* |

Consider a proton confined in a one-dimensional box of width 0.200 nm. (a) What would be the lowest energy of the proton? (b) What is the lowest energy of an electron confined in the same box?

Solution (a)

The energy of a particle of mass m in a 1D box of length L is

$$E_n = \frac{n^2\pi^2\hbar^2}{2mL^2} = \frac{n^2\pi^2(h/2\pi)^2}{2mL^2} = \frac{n^2h^2}{8mL^2} \qquad (7.18)$$

The lowest energy ($n = 1$) of a proton with $m_p = 1.67 \times 10^{-27}$ kg in a box with $L = 2 \times 10^{-10}$ m is

$$E_{1,p} = \frac{h^2}{8m_pL^2} = \frac{\left(6.626\times10^{-34}\,\text{J}\cdot\text{s}\right)^2}{8\left(1.67\times10^{-27}\,\text{kg}\right)\left(2\times10^{-10}\,\text{m}\right)^2} = 8.22\times10^{-22}\,\text{J}\left(\frac{\text{eV}}{1.6022\times10^{-19}\,\text{J}}\right) = 5.13\times10^{-3}\,\text{eV} \qquad (7.19)$$

Solution (b)

The lowest energy of an electron in a similar box is

$$E_{1,e} = \frac{h^2}{8m_eL^2} = \frac{\left(6.626\times10^{-34}\,\text{J}\cdot\text{s}\right)^2}{8\left(9.11\times10^{-31}\,\text{kg}\right)\left(2\times10^{-10}\,\text{m}\right)^2} = 1.506\times10^{-18}\,\text{J}\left(\frac{\text{eV}}{1.6022\times10^{-19}\,\text{J}}\right) = 9.40\times10^{-3}\,\text{eV}$$

Note that the electron energy is much larger than the proton energy because of the difference in the mass: $m_p \approx 2000\,m_e$.

the box. This reflects the physical condition that increasing the energy of the particle in the box increases its kinetic energy. The average kinetic energy of the particle for any of the allowed total average energy is just the difference between the horizontal line and the floor of the box.

The curvature of a function is just the *change in slope* of the function as one moves from left to right through any point. Since the slope of a function is expressed as its first derivative, the curvature must be the second derivative. That means if we represent the graph of the wave function by a function $\psi(x)$, then the curvature of the graph is given at any point x by

$$\frac{d^2}{dx^2}\psi(x) \qquad (7.20)$$

and the relative magnitude of the curvature (the magnitude of the curvature divided by the magnitude of the wave function) is

$$\frac{d^2}{dx^2}\psi(x) \bigg/ \psi(x) \qquad (7.21)$$

The expression gives the kinetic energy of the particle if it happens to be at the point x. To get the average kinetic energy of the particle we need to multiply by the probability that the particle actually is at the point that we are measuring (we already know that the probability is given by the square of the wave function).

	The quantized energy levels of a particle-in-a-box model are shown. The horizontal lines drawn across the box show the allowed energies of a particle in this box; the height of each line above the box floor corresponds to the value of the energy (a higher line means a higher energy).
FIG. 7.7	

Energy

n

4 ———————————————— $E_4 = 16E_1$

3 ———————————————— $E_3 = 9E_1$

2 ———————————————— $E_2 = 4E_1$

1 ———————————————— E_1

Hence, the average kinetic energy of the particle is given by multiplying the preceding expression by $\psi^2(x)$, like

$$KE(x) = \frac{1}{2}\psi(x)\frac{d^2}{dx^2}\psi(x) \tag{7.22}$$

7.1.2 *Two-Dimensional Quantum Systems*

The allowed energy levels inside the box are quantized and are shown in **Figure 7.7.** If the particle is electrically charged, it can emit a photon when it jumps from a higher state to a lower state. It can also jump from a lower to a higher state by absorbing a photon. The horizontal lines drawn across the box show the allowed energies of a particle in this box; the height of each line above the box floor corresponds to the value of the energy (a higher line means a higher energy). The lines on the figure show the wave functions that correspond to each allowed energy state.

The particle-in-a-box problem is extended from one dimension into higher spatial dimensions (e.g., two and three dimensions). Since the potential inside the box is always zero, the Schrödinger equation can be easily separated into two or three separate equations dependent on the spatial coordinates x, y, and z that have the exact same form as for the equation in the one-dimensional space. The solutions are the same for each individual dimension, and the total wave function is then a product of the individual wave functions, while the energy is a sum of the corresponding individual energies.

$$E_{n_1,n_2,n_3} = \frac{h^2}{8m}\left(\frac{n_1^2}{a_x^2} + \frac{n_2^2}{a_y^2} + \frac{n_3^2}{a_z^2}\right) \tag{7.23}$$

Because the walls are impenetrable for $x \leq 0$ and $x > L$, the wave function bouncing between the walls is $\psi(x) = 0$. The particle is never found outside the

EXAMPLE 7.5   *A Special Two-Dimensional Case*

Consider a rectangular infinite square well. The potential is zero inside the rectangle of dimension L_x by L_y. The potential is infinite outside the rectangle (**Fig. 7.8**). How do you go about deriving an expression for the energy of a particle in this well?

FIG. 7.8   *The classically allowed region exists inside this rectangular two-dimensional box. The same principles apply in this case except that now the particle is free to move in two dimensions.*

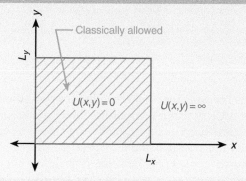

As before, we establish a boundary condition that y be zero in the infinitely disallowed region outside the rectangle. Inside the rectangle, where the potential is zero, we have the Schrödinger equation

$$-\frac{\hbar^2}{2m}\left(\frac{\partial^2}{\partial x^2}+\frac{\partial^2}{\partial y^2}\right)\psi(x,y)=E\psi(x,y) \tag{7.24}$$

Since the Hamiltonian is the sum of two terms with totally independent variables, we try a product wave function $\psi = X(x)Y(y)$ where

$$-\frac{\hbar^2}{2m}\left(\frac{\partial^2 X(x)}{\partial x^2}\right)=E_x X(x)\ \text{ and }\ -\frac{\hbar^2}{2m}\left(\frac{\partial^2 Y(y)}{\partial x^2}\right)=E_x Y(y)\ \text{ and }\ E_{Total}=E_x+E_y \tag{7.25}$$

The following differential equations have the usual solution:

$$X(x)=A_x\sin(k_x x)+B_x\cos(k_x x);\quad E_x=-\frac{\hbar^2 k_x^2}{2m}\ \text{ and }\ Y(y)=A_y\sin(k_y y)+B_y\cos(k_y y);\quad E_x=-\frac{\hbar^2 k_y^2}{2m} \tag{7.26}$$

Application of boundary conditions gives

$$\begin{aligned}
\psi(x=0)=0 &\Rightarrow B_x=0 & y(x=L_x)=0 &\Rightarrow k_x L_x=n_x\pi\\
\psi(y=0)=0 &\Rightarrow B_y=0 & y(y=L_y)=0 &\Rightarrow k_y L_y=n_y\pi
\end{aligned} \tag{7.27}$$

The resulting normalized wave functions are

$$\psi(x,y)=\frac{2}{\sqrt{L_x L_y}}\sin(k_x x)\sin(k_y y);\ k_x=\left(\frac{\pi n_x}{L_x}\right)\ \text{ and }\ k_y=\left(\frac{\pi n_y}{L_y}\right)\ \text{ and }\ E=\frac{\hbar^2\left(k_x^2+k_x^2\right)}{2m}=\frac{\hbar^2\pi^2}{2m}\left(\frac{n_x^2}{L_x^2}+\frac{n_y^2}{L_y^2}\right) \tag{7.28}$$

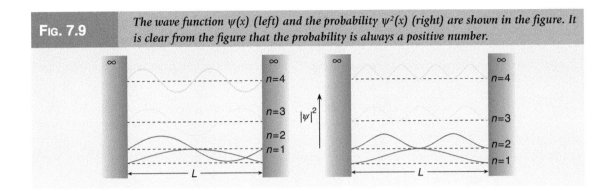

The wave function ψ(x) (left) and the probability ψ²(x) (right) are shown in the figure. It is clear from the figure that the probability is always a positive number.

box. The wave function is continuous everywhere—that is, for $\psi(x) = \psi(L) = 0$. The wave function that satisfies these conditions will be the only one that will be allowed inside the box and must have sinusoidal form. For $n = 1$, the probability function maximum of $\psi^2(x)$ is when $x = L/2$. For $n = 2$, the maximum occurs at $x = L/4$ and $3L/4$. The probability is zero for $x = L/2$. For the first three stationary states for the particle in a box ($n = 1,2,3, \ldots$), the wave function will be shown in **Figure 7.9** (left) while the probability for finding the particle in the box is shown in **Figure 7.9** (right).

7.1.3   Schrödinger Equation

The complete average energy comprises the sum of the average kinetic and average potential energies of the particle at some point x. If the potential energy is $V(x)$, then the average potential energy is $V(x)$ by the probability that the particle is at the point x:

$$V(x)\psi^2(x) \tag{7.29}$$

The total average energy including the kinetic energy of a particle at x is given by

$$\frac{1}{2}\psi(x)\frac{d^2}{dx^2}\psi(x) + V(x)\psi^2(x) \tag{7.30}$$

To find the all the energy at all x, we integrate over all x. E is

$$E = \int dx \left[\frac{1}{2}\psi(x)\frac{d^2}{dx^2}\psi(x) + V(x)\psi^2(x) \right] \tag{7.31}$$

This is a form of energy in terms of the wave function—the *variational principle*. It is also considered to be the second law of thermodynamics in disguise in that the wave function will adjust itself until the lowest possible energy minimum is achieved. This equation gives the relationship between the curvature of the wave function at any point x and the potential energy at that point. Another familiar form of the *time-independent* Schrödinger equation is

$$-\frac{\hbar^2}{2m}\frac{d^2}{dx^2}\psi(x)+V(x)\psi(x)=E\psi(x) \qquad (7.32)$$

$$H\psi_n=E_n\psi_n \text{ and that } H\,|\psi_n\rangle=E_n|\psi_n\rangle \qquad (7.33)$$

where H is a mathematical operator called the *Hamiltonian*, $|\psi_n\rangle$ are a set of *eigenstates*, and E_n is the *eigenvalue* of the Hamiltonian.

Boundary Conditions and Quantization. The solution of the Schrödinger equation is quite straightforward if we know where the value of the wave function is known. We know, for example, that the probability of a particle being inside the left wall of the box is equal to zero. Then we pick any slope for the wave function and use the Schrödinger equation to calculate the curvature. At this point, we find how the curvature behaves. We then calculate the slope and subsequently the wave function.

With Schrödinger's equation we can generate a wave function for any energy. In order to ensure that quantization of the energy occurs, the wave function must attain a certain value at the *boundary condition* that is forbidding of all but specific values of the energy. It is to these boundary conditions that the wave function owes its quantum nature, just as the walls of the bathtub enforce the quantization of waves within its porcelain boundaries (the open sea would not). In the case of the square well, our boundary condition is that the wave function will be zero as soon as it reaches the right-hand wall. If we require this, it turns out we can only choose certain specific values for the initial slope and curvature in slope of the wave function at the left-hand wall. This, in turn, means we can only use wave functions corresponding to certain energies.

Quantization of the Energy! Classically Forbidden Zones. When you raise the energy of the particle in a box high enough, the top energy level exceeds that of the allowed energy levels. Then a particle having this allowed total energy is forbidden in classical (nonquantum) mechanics from being inside the box since the potential energy at a point in this region is greater than the allowed total energy at this point. Hence, the particle, if it were here, would have to have a negative kinetic energy. This makes no sense classically because we cannot have kinetic energy less than that implied by zero motion (which is zero). But, from the quantum mechanical perspective, this makes good sense! Quantum mechanics only requires that the total average energy summed up over all possible positions of the particle be equal to the total average energy and hence it is a possibility to have a negative kinetic energy at some point as long as it is balanced by a positive contribution somewhere else. Thus, the quantum particle is allowed to visit *classically forbidden* regions of the box! The wave function in this region will, however, be greatly damped and more so if the level of the particle energy is raised further, indicating that the probability of the particle being in a classically forbidden region is greatly reduced.

This gives rise to the famous phenomena of *quantum tunneling*. Suppose that a particle is trapped in a well that has well-defined boundaries. If the classically forbidden region has finite width and there is a classically allowed region on the other side (as there is in this system, for example), then a particle trapped in the first allowed region can be found on the other side, having apparently traversed a region of space in which it was *not allowed*. This is called *tunneling*, so named because one imagines the particle tunneling through the barrier, instead of going over the top; it does not have enough energy to get over the top.

Erwin Schrödinger expressed the hypothesis of particle–wave nature in 1926. The continuous nonzero nature of its solution, the wave function which represents an electron or particle, implies an ability to penetrate classically forbidden regions and a probability of tunneling from one classically allowed region to another. In 1928, Fowler and Nordheim explained the main features of electron emission from cold metals by high external electric fields on the basis of tunneling through a triangular potential barrier. Conclusive experimental evidence for tunneling was, however, found in the 1960s by L. Esaki in 1957 and by I. Giaever in 1960. Esaki's tunnel diode had a large impact on the physics of semiconductors that led to important developments such as the tunneling spectroscopy and to increased understanding of tunneling phenomena in solids. They won a Nobel Prize along with B. Josephson in 1973.

With the ability to fabricate horizontal structures lithographically with dimensions of 50–100 nm, as well as the self-organization of different structures, it becomes possible to provide confinement in any or all three dimensions, leading to a variety of quantum plane, wire, and dot configurations. Because of the quantum effects, the number of electrons in the well of a quantum dot is quantized to an integer number of electrons, and even if the number is as large as tens to hundreds, single-electron changes can be observed. Combining quantum dots with very thin insulating layers, single-electron tunneling transistors have been implemented, and the possibility of single-electron logic is on the horizon. Research in this promising discipline has been active since its formulation in the mid-1980s. Even quantum molecules—that is, several closely spaced quantum dots interacting via quantum mechanical tunneling—have been explored.

7.1.4   Bohr Exciton Radius

A zero-dimensional material, called a quantum dot, is a material in which the crystallite size is negligible in all three spatial dimensions. The motion of electrons and holes, therefore, is also restricted in all three spatial dimensions. Thus, a quantum dot can be defined as a structure where all dimensions are comparable to the *Bohr exciton radius* (**Fig. 7.10**). The Bohr exciton radius is the distance between an electron in the conduction band and the hole it left behind in the valence band. As motion of electrons is restricted in all directions (like a particle in a box), it is called quantum confinement, and electrons exist exclusively in quantized levels of energy. A quantum dot is like a custom designed atom with a countable number of electrons. It is a semiconductor particle of usually 2–8 nm in diameter. The energy states of the electrons and holes in quantum dots are discrete and given by

$$E_n = \frac{3\hbar^2 n^2}{2m^* a^2}; \quad \text{where } n = 1, 2, 3, \ldots \tag{7.34}$$

where m^* is the effective mass of the electron (a is analogous to L for the linear box described previously).

FIG. 7.10

The Bohr exciton radius is shown in the figure. When an excited electron is promoted to a higher energy level in the conduction band, a hole is created in the valence band. In semiconductors, the bandgap is the region between the conduction and valence bands. The Fermi energy corresponds to a region inside the bandgap. The Fermi surface is the surface that separates the occupied level from the unoccupied level in metals. Technically, there are no states inside the bandgap of intrinsic semiconductors. Therefore, a semiconductor does not technically have a Fermi surface (or a Fermi energy). It is more accurate to refer to the "Fermi energy" of a semiconductor as its chemical potential.

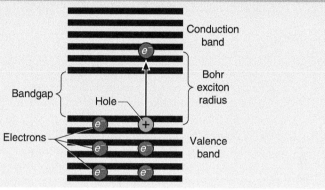

Source: N. W. Ashcroft and N. D. Mermin, *Solid State Physics,* Saunders College Publishing, Harcourt Brace College Publishers, New York, 1976. With permission.

The wave functions of the electrons and holes vanish on the particle surface, and the particle acts as a potential well with infinitely high walls. Each energy level is then associated with an atomic orbital. Two electrons can have the same energy in the orbital at the same time if they have different spins, but no more than two electrons (spin up and spin down) are able to occupy an energy level at the same time (*Pauli exclusion principle*).

Quantum confinement occurs as the semiconductor particle size, a, is reduced to that of the excitonic Bohr radius, which is the radius of the sphere defined by the three-dimensional separation of the electron–hole pair. The excitonic Bohr radius is given by

$$a_{exciton} = a_0 \left(\frac{\varepsilon}{m^*_{exciton}} \right) \tag{7.35}$$

where ε is the dielectric constant of the semiconductor, a_0 is the Bohr radius, and $m^*_{exciton}$ is the effective exciton mass given by

$$m^*_{exciton} = \frac{m^*_{electron} + m^*_{hole}}{m^*_{electron} m^*_{hole}} \tag{7.36}$$

The Bohr radius should not be confused with the Bohr exciton radius. The Bohr radius is the distance between an electron and its nucleus in the lowest energy level (smallest possible orbit) of the hydrogen atom and is given by

$$a_0 = \frac{4\pi \varepsilon_0 \hbar^2}{m_e e^2} = 5.29 \times 10^{-11} \, \text{m} = 0.53 \, \text{nm} \tag{7.37}$$

TABLE 7.2	*Bohr Exciton Radii of Selected Semiconductor Materials*						
Material	Si	Ge	GaAs	GaP	InP	CdSe	InSb
$a_{exciton}$ (nm)	4.9	17.7	14	1.7	9.5	3	69

where ε_o is the permittivity of free space, m_e is the electron rest mass, e is the elementary charge, and $\hbar$ is the reduced form of Planck's constant. The Bohr exciton radii of selected crystalline materials are given in **Table 7.2**.

7.1.5 Bandgaps

There are three general categories of materials based on their electrical properties: (1) conductors, (2) semiconductors, and (3) insulators (**Fig. 7.11**). In conducting materials like metals, the valence band (consisting of electrons in the outermost shell of the atom) and the conduction band overlap. This means that all valence electrons can participate in the conduction process. The valence band is analogous to the highest occupied molecular orbital (the HOMO) in a molecule, and the conduction band is analogous to the highest occupied molecular orbital (the LUMO) in a molecule, except, of course, that there are many more in the solid. The energy separation between the valence and conduction band is called E_g. The ability to fill the conduction band with electrons and the energy of the bandgap determine whether or not a material is a conductor, a semiconductor, or an insulator. In metals, thermal energy is enough to stimulate electrons to move into the conduction band.

The bandgap in semiconductors is a few electron volts. If an applied voltage exceeds the bandgap energy, electrons jump from the valence band to the conduction band, thereby forming electron–hole pairs called *excitons*. Insulators have large bandgaps that require an enormous amount of voltage to overcome the threshold. These materials therefore do not conduct electricity.

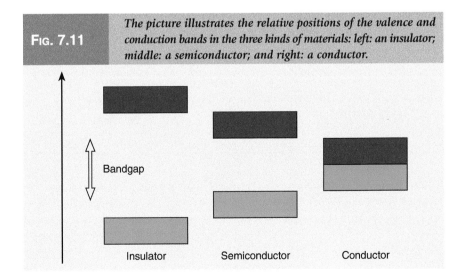

| FIG. 7.11 | *The picture illustrates the relative positions of the valence and conduction bands in the three kinds of materials: left: an insulator; middle: a semiconductor; and right: a conductor.* |

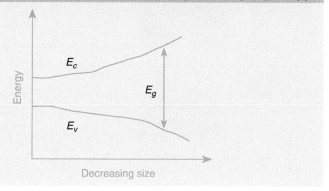

FIG. 7.12 *This figure provides a visual image of the change in the energy of the bandgap as particle size decreases. It is clear that the width of the bandgap increases with decreasing particle size. Because of this, smaller quantum dots are expected to emit light at higher energy (lower wavelength, a blue shift) and larger dots are expected to fluoresce (emit) at longer wavelengths (red shift).*

The increase in the effective bandgap energy is caused by quantum confinement effects; the photon energy is regarded as the sum of the bulk bandgap and the quantum confinement energy. At *very small dimensions* when energy levels are quantified, band overlap disappears in metals during transformation into a bandgap (**Fig. 7.12**). Normally, in the case of semiconductors, the bandgap increases with smaller dimensions due to quantization. Some states disappear from the bulk material entirely, including the highest levels of the valence band and the lowest of the conduction band. In the case of a metal, the widening of the gap actually causes the overlap to become a gap. This means that quantum dots can be made from materials that are either semiconductors or conductors as bulk materials.

The energy of the bandgap of a nanoparticle differs from the bulk by

$$E_{g,nanocrystal} = E_{g,bulk} + \Delta E_g = E_{g,bulk} + \frac{h^2 \pi^2}{2mr^2_{nanocrystal}} \tag{7.38}$$

As a particle becomes smaller, its bandgap increases. The relationship of shrinking size to the bandgap is shown graphically in **Figure 7.13** from semiconductor to a quantum dot to the atom.

Following absorption of energy (voltage or light source), excited electrons are able to jump from the valence band to the conduction band. If the energy of the light is equal to or higher than the bandgap, absorption occurs and an electron–hole pair is generated. When recombination of electrons and holes happens, energy is emitted. In some cases, energy is emitted in the form of phonons (vibrations in the material causing temperature to increase—also called nonradiative emission) or in the form of photons (electromagnetic waves—also called radiative emission). Some semiconducting materials have a direct bandgap (e.g., no states exist in the gap). Other materials have an indirect bandgap (e.g., one or more states exist in the bandgap).

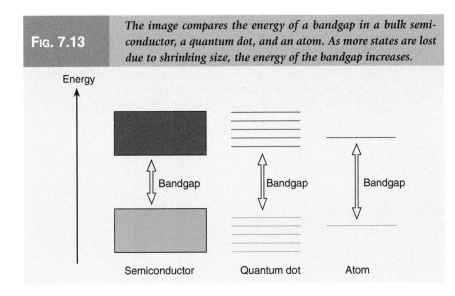

FIG. 7.13 The image compares the energy of a bandgap in a bulk semiconductor, a quantum dot, and an atom. As more states are lost due to shrinking size, the energy of the bandgap increases.

In order for an electron to be excited in an indirect bandgap material, the presence of both a photon and a phonon is essential because intermediate states, in addition to different energy from the valence band, also have different momentum. These materials can simultaneously emit both nonradiative and radiative emissions. One of the emissions is from the transition between the conduction band and the intermediate state and the other emission is from the transition between the intermediate state and the valence band. If light is absorbed and then electron–hole transition causes emission of another wavelength of light, the process is referred to as photoluminescence.

If the bandgap increases due to quantum confinement, the wavelength of light needs to be of higher energy (shorter wavelength) in order to be absorbed by the bandgap of the material. The wavelength of the fluorescent light emitted from the material also changes in the same manner (e.g., a blue shift in fluorescence is observed for smaller quantum dots). We therefore have a method of tuning the optical absorption and emission properties of a semiconductor over a range of wavelengths by control of the crystallite size. If the semiconductor in question is undoped and is a direct bandgap material, then both absorption and luminescence occur at the energy band edge. We can also introduce trap states within the bandgap to act as luminescent centers. A shallow trap can be described as a mobile charge orbiting the trap in a large $1s$ orbital, resulting in a susceptibility to quantum confinement effects. Deep traps have highly localized orbiting charges and are therefore relatively immune to confinement effects.

Semiconductor Quantum Dots.   The energy gap of a quantum dot depends on the size of the dot. The larger the size is, the lower is its absorption and fluorescence energy (red shift). The smaller the dot is, the higher is its absorption and fluorescence energy (blue shift). In addition to quantum size effects, doping nanoparticles with different materials is a means of controlling the dimensions of the bandgap of nanoparticle quantum dots. Doping introduces altered energy levels and defect sites within the bandgap of the semiconductor that can serve as electron "resting sites" during a transition. Therefore, semiconductors that normally absorb only UV light are now able to emit visible light. ZnS quantum dots, when doped with manganese, absorb in the UV range and emit orange light. In another study, Ge quantum dots on Si(100) showed narrow band luminescence as a function of dot size and proximity [3].

7.2   ZERO-DIMENSIONAL MATERIALS

A zero-dimensional materials exhibits quantum confinement in all three spatial dimensions—a quantum sphere as it were or, more correctly, a quantum dot (**Fig. 7.14**, left). In other words, no matter which direction one views the spherical nanoparticle, the electronic behavior along that dimension is quantized. In **Figure 7.14** (right), images acquired by TEM show gold nanoparticles.

7.2.1   Clusters

Bulk matter is made of atoms and molecules, and as such, they have been widely classified and their properties satisfactorily explained. An ensemble of atoms or molecules, the so-called *clusters*, is far from being well understood. Elemental clusters are held together by various forces depending on the nature of the constituent atoms: (1) inert gas clusters are weakly held together by van der Waals interactions, for example, $(He)_n$; (2) semiconductor clusters are held together by directional covalent bonds; and (3) metallic clusters are held together by fairly strong delocalized nondirectional bonds.

FIG. 7.14

Left: the electrons in a zero-dimensional material (a quantum dot) are confined in all three dimensions. In other words, quantum properties are expected to be observed in this material regardless of particle orientation. Right: a TEM image of gold nanoparticles on the order of 20-nm diameter. The optical absorption spectra of such particles are shown in Figure 7.16.

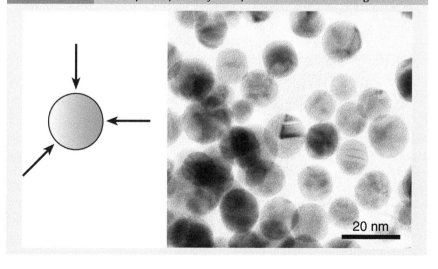

20 nm

TABLE 7.3	Cross-Section, Mass, N, and Surface Fraction—Revisited			
Size (nm)	Cross-section (10^{-18} m^2)	Mass (10^{-25} kg)	No. of molecules (N)	Fraction surface molecules (%)
0.5	0.2	0.65	1	
1	0.8	5.2	8	100
2	3.2	42	64	90
5	20	650	1,000	50
10	80	5,200	8,000	25
20	320	42,000	64,000	12

Elemental clusters, or a mixture of clusters of different elements, constitute the vast expanding field of materials sciences called *nanomaterials*. One has to be clear right away that clusters are not a fifth state of matter, as some may believe, but rather are simply intermediate between atoms on the one hand, and solid or liquid state of matter on the other, with widely varying physical and chemical properties. The number of atoms forming the cluster determines the percentage of atoms that are exposed on the surface of the cluster, with decreasing number of surface atoms with increasing size of the cluster, as we discovered in chapter 5. Some make a distinction among molecular clusters, nanoparticles, and nanomaterials. Physicists working with molecular beams have extensively studied agglomerations of a few atoms. Today, the mystery related to larger ensembles of atoms (in other words, *nanomaterials*) is fading due to active research being carried out across the world over the last few decades.

In **Table 7.3**, we see that smaller particles contain only a few atoms that are practically all at the surface. As the particle size increases from 1 to 10 nm, cross-section of the cluster increases by a factor of 100× and the mass of molecules by a factor of 1000×; concomitantly, the proportion of molecules at the surface falls from 100% to just 25%. For particles of 20-nm size, a little more than 10% of the atoms are on the surface. Of course, this is an idealized hypothetical case. If particles are formed by macromolecules (that are larger than the present examples), the number of molecules per particle will substantially decrease and their surface fraction increase accordingly. The electronic properties of such ensembles of atoms or molecules are the result of mutual interactions, so the overall chemical behavior is entirely different from individual atoms or molecules. The chemical behavior is also different from the macroscopic bulk state of the substance in question under the same conditions of temperature and pressure. We revisit the surface-to-volume relationships of chapter 5, but in this case comparing cross-section (which is proportional to r^2) to kilograms (proportional to volume).

The idea of tailoring properly designed atoms into agglomerates has brought in new fundamental work in the search for novel materials with uncharacteristic properties. Among various types of nanomaterials, cluster-assembled materials represent an original class of nanostructured solids with specific structures and properties based on structures between amorphous and crystalline materials. In fact, in such materials, the short-range order is controlled by the grain size, and no long-range order exists due to the random stacking of nanograins characteristic of cluster-assembled materials. In terms of properties, they are generally controlled by the intrinsic properties of the nanograins themselves and by the interactions between adjacent grains.

Cluster-assembled films are formed by the deposition of clusters onto a solid substrate and are generally highly porous with densities as low as about one half of the corresponding bulk materials densities. Both the characteristic nanostructured morphology and a possible memory effect of the original free cluster structures are at the origin of their specific properties. From recent developments in the cluster source technologies (thermal, laser vaporization, and sputtering) [4,5], it is now possible to produce intense cluster beams of any materials, even the most refractory or complex systems (bimetallic, oxides, and so on) for a wide range of size from a few atoms to a few thousands of atoms.

Cluster Particle in a Box. The de Broglie wavelength of an electron is found from de Broglie's relationship. The electron possesses a characteristic wavelength due to the particle duality exhibited by all matter. If a cluster's size is on the order of this wavelength, quantization of the energy of the electrons is likely. **Figure 7.15** illustrates the particle-in-a-box analogy for clusters. A ligand shell

FIG. 7.15 *The n = 1 and n = 2 levels are shown in a ligand-stabilized gold cluster. The cluster is the "box" and the ligands are the "walls." The de Broglie wavelength of the cluster is ca. 1.4 nm, corresponding to the diameter of the cluster. The complete particle is 2.1 nm in diameter [(2 × 0.35 nm = 0.7 nm) + 1.4 nm = 2.1 nm] [6,7]. If clusters are placed in a row to form a quantum wire, electrons are able to tunnel through the ligand barrier and conduct. In order for this to occur, the wavelengths need to match up constructively between clusters. The length of a two cluster wire is ca. 4.2 nm [6,7].*

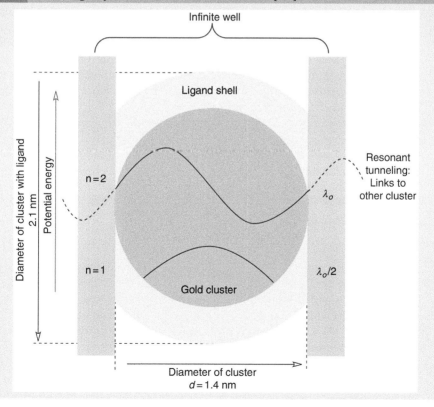

Infinite well

Ligand shell

Diameter of cluster with ligand 2.1 nm

Potential energy

n = 2

Resonant tunneling: Links to other cluster

λ_o

n = 1

$\lambda_o/2$

Gold cluster

Diameter of cluster
$d = 1.4$ nm

surrounds the cluster and in essence provides a barrier for the electronic wave functions of the cluster. Clusters in contact with one another (via the ligand shells) communicate electronic information by the phenomenon of tunneling (chapter 3).

7.2.2 Metal Clusters and the HOCO–LUCO

HOCO–LUCO. A gold nanoparticle that is 1 μm^3 in volume has 10^{10} atoms and is still considered to be a bulk material [6]. In bulk metals, energy levels are numerous and essentially continuous. If the diameter of the particle is reduced to some integral multiple of the de Brogile wavelength, λ_o, of an electron in the ground state of that material, quantum effects begin to emerge. If the size is reduced further to integral multiples of $\lambda_o/2$ (half the de Broglie wavelength), then electron energy levels in clusters become quantized (discrete). The ground state of the ligand-stabilized Au_{55} cluster has two electrons; the diameter of the cluster, $d_{cluster}$, is equal to $\lambda_o/2 = 1.4$ nm and the diameter of the cluster plus the ligand sphere is equal to ~2.1 nm (the walls of the infinite well, 0.35 nm thick) [6]. In the quantized electronic state, very small metal particles lose their ability to conduct electrons due to the existence of clearly defined highest occupied molecular orbitals (HOMOs) and the lowest unoccupied molecular orbitals (LUMOs) [7]. The HOMO in a cluster is aptly renamed the HOCO—the highest occupied cluster orbital. Concomitantly, the cluster LUMO is renamed the LUCO [7].

If naked clusters (on the order of $\lambda_o/2$) are in contact with one another, agglomeration occurs, particle size increases, and quantum properties are lost; all the HOCOs merge, as do all the LUCOs, to form a band structure. Quantization of electronic properties is maintained in cluster groups, however, if they have organic ligand shells and then are made to contact. The organic shells serve as the walls of infinite potential energy wells. Contact between clusters, therefore, occurs only by *resonant tunneling* through this barrier. This leads to electronic intercluster bands. A wire made of clusters therefore should be able to conduct electricity without resistance.

7.2.3 Optical Properties of Clusters

Several theoretical models predict the existence of so-called intrinsic size effects in the optical response of nanoparticles characterized by size that is significantly smaller than the electron mean free path [8]. The electron mean free path is defined by

$$\lambda = \upsilon \tau \tag{7.39}$$

At room temperature, λ is on the order of nanometers [9]. When particles are <40 nm (much smaller than the wavelength of light), the particles experience the quasi-static limit (chapter 5) and feel an electric field that is spatially constant with a time-dependent phase [9]. The metal nanoparticles absorb energy via the collective oscillations of free electrons (dipolar plasmons), interband transitions, or surface dispersion or scattering of the free electrons on a metals surface. Surface scattering occurs when the mean free path is comparable to the dimensions of the nanoparticle [9]. In metal nanoparticles, limitations of the electron mean free path as well as damping by the chemical interface increase the damping rate of the surface plasmon resonance, leading to shorter dephasing times. Due to these intrinsic size effects, the dielectric permittivity of a nanoparticle differs from that of the bulk metal because of the additional surface damping contributions. Experimental studies on ensembles of metallic nanoparticles revealed the existence of such effects. However, a quantitative description of these effects is difficult to acquire for ensembles, mostly because of inhomogeneous broadening in spectra.

The electronic structure of nanoparticles is unraveled by studying its interaction with electromagnetic radiation, specifically in the range of optical wavelengths. In absorption, the frequency of the incoming light wave is at or near the energy levels of the electrons. After absorption, the energy is (1) remitting by a photon or (2) retained and distributed by nonradiative mechanisms. Apart from absorption of light by objects, scattering of light also takes place. Gustav Mie developed a generalized theory of light scattering by a spherical particle at the turn of the nineteenth century. We discussed the quasi-static limit in chapter 5. The absorption scattering cross-section of a particle is proportional to r^3, while that of the scattering cross-section is proportional to r^6. If particle size is significantly less than the wavelength, then absorption is more important than scattering for very small particles ($2\pi r \ll \lambda$) [8]. Scattering becomes more important than absorption when the circumference of the particle is comparable to the wavelength of light ($2\pi r \approx \lambda$). Scattering does not occur for very small particles.

Dipolar Plasmon Resonance.   In metal nanoparticles, absorption by plasmon resonances arises from the large electron density. The collective particle plasmon mode strongly interacts with optical waves [8]. Plasmon resonance is observed in particles or planar surfaces that are metallic (e.g., silver or gold) nanoparticles or nanolayers that scatter optical light elastically with remarkable efficiency because of a collective resonance of the conduction electrons in the metal. The excitation of surface plasmons by light is denoted as a *surface plasmon resonance* (SPR) for planar surfaces or *localized surface plasmon resonance* (LSPR) for nanometer-sized metallic structures.

The position, width, and amplitude of the surface plasmon wave depend on the nature of the nanoparticles. The dipolar resonance was introduced in chapter 5. In local surface plasmon resonance, incident light strikes the surface of metal nanoparticles to stimulate the conduction electrons. This action creates a dipole that oscillates in phase with the electric field of the light and generates a restoring force in the metal nanoparticle as it tries to compensate for the change. The result is a unique resonance wavelength. The character of the dipolar plasmon is also influenced by the nature of the support material (or the surrounding dielectric medium). The position of the absorption, therefore, depends on particle size and shape, the type of metal, its orientation, and the surrounding medium. The plasmon is, not surprisingly, a quantized entity.

Noble metals of small size have a very strong visible absorption due to the resonant coherent oscillation of the free electrons in the conduction band. The brilliant ruby color in stained glass windows of cathedrals built in the seventeenth century is due to the strong absorption of gold nanoparticles. If the particle is not spherical (e.g. rod shaped), the optical (absorption and fluorescence) and physical properties of these gold nanorods become quite different from those of spherical ones.

In case of gold nanorods that contain nearly 10^5–10^6 atoms, there exists significant polarizability. In-phase excitation of the collection of these electrons with an incident light generates an electric field on the particle surface that is proportional to the dielectric of the metal and its volume. This enhanced field is a new property—one that is absent in the bulk gold.

Figure 7.14 showed gold nanoparticles of diameter of around 20 nm. Depending upon the size of the nanoparticles, the absorption of UV–visible light is different and, as such, gold nanoparticles show different colors. The optical spectra of those gold nanoparticles are shown in **Figure 7.16**.

7.2.4   Other Physical Properties and Phenomena

Surface Energy.   The trend in surface energy as a function of size is tabulated in **Table 7.4**. The surface energy of bulk calcite, $CaCO_3$, is 0.23 J·m^{-2} [10]. It is clear from the table that the materials we discussed in chapter 6 are relevant to nanomaterials. The energy of the edges exceeds the surface area.

Another stellar example of the concept of surface energy is given by a 1-g cube of NaCl. Starting with 1 g of material, the parent is subdivided into smaller cubes with increasingly larger surface area. The density of NaCl is 2.17 g·cm^{-3}. The average energy of surface atoms is 2×10^{-5} J·cm^{-2} and the average energy for edge molecules is 3×10^{-13} J·cm^{-1}. Notice that edge molecules have significantly higher energy than do surface atoms. From **Figure 7.17**, we see that the surface molecular density of NaCl per unit cell is equal to 2 molecules per a_o^2 or 6.27 molecules per square nanometer. The linear density is equivalent to 1 molecule per a_o or 1.77 molecules per nanometer. From this information, the average surface energy per molecule is calculated. One would expect that edge atoms (and corner atoms) have the highest surface energy (**Table 7.5**).

Thermal Properties.   One of the most popular illustrations of nanoparticle physical properties is that of the dramatic deviation of melting point from the bulk (**Fig. 7.18**). Recall our previous discussion concerning the coordination number of surface atoms—specifically, that surface atoms are undersaturated with regard to nearest neighbors. Although this physical condition is also true for bulk materials, the ratio of surface atoms is so small compared to the bulk volume that any discussion concerning surface effects, especially when it involves the melting of solids, is inconsequential.

The melting point change is inversely proportional to the radius r of the particle

$$\Delta\theta = \frac{2T_o\gamma}{\rho r(L)} \tag{7.40}$$

where
  $\Delta\theta$ is the change in melting point from the bulk temperature
  T_o is the melting point of the bulk material in kelvins
  γ is the surface tension of the solid–liquid interface in joules per square meter
  ρ is the particle density in kilograms per cubic meter
  L is the latent heat of fusion in joules per kilogram [13]

FIG. 7.16

(a) The optical spectra of increasingly larger spherical gold clusters (colloids) are shown. The transmission color of gold solutions is shown at the top left. The absorption spectra correspond to: clockwise from top right: <20 nm diameter, $\lambda_{max} \approx 518$ nm; $\lambda_{max} \approx 530$ nm; and $\lambda_{max} \approx 560$ nm. (b) The variation of size-dependent emission from quantum dots is graphically depicted in the figure. The larger the semiconductor dot is, the lower energy is the emission band. (c) The orange luminescence of manganese-doped zinc-sulfide is shown.

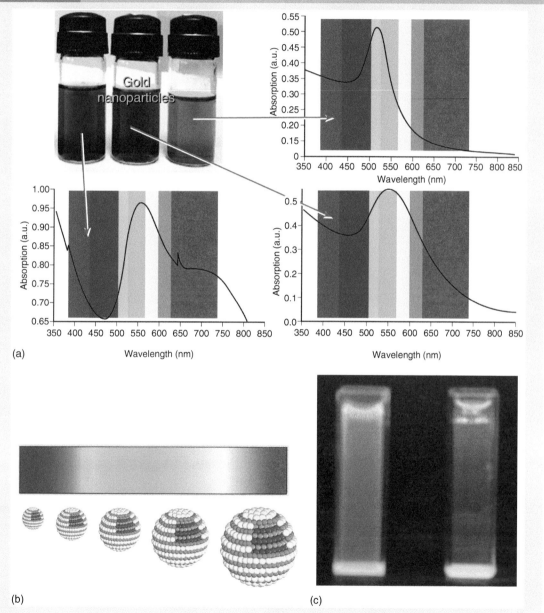

TABLE 7.4	Surface Energy of Calcite Particles	
Size (nm)	**Surface area (m·mol⁻¹)**	**Surface energy (J·m⁻²)**
1	1.11×10^9	2.55×10^4
2	5.07×10^8	1.17×10^4
5	2.21×10^8	5.09×10^3
10	1.11×10^8	2.55×10^3
20	5.07×10^7	1.17×10^3
100	1.11×10^7	2.55×10^2
1000	1.11×10^6	2.55×10

Source: K. Kamiya and S. Sakka, *Gypsum Lime*, 23, 163 (1979).

The smaller the particle is, the greater the range of melting differential can be. Nanomaterials are able to undergo sintering at lower temperatures.

Conductivity.   Electrical conductivity in bulk metals is continuous (e.g., the curve relating current to applied voltage is smooth and continuous). This is because there are an enormous number of electronic states in the conduction band of metals (**Fig. 7.19**, left). Electron mobility μ is described by [6]

Fig. 7.17   *The fcc structure of a sodium chloride crystal is depicted. The surface energy depends on the planar face of the molecule that is exposed. The (100) surface and the linear density in the [010] direction are shown. Edge atoms, due to their relative undersaturation of nearest neighbors compared to the surface atoms, should have higher energy. Many ligands attach to edges and corners (vertices) of metal clusters for this reason.*

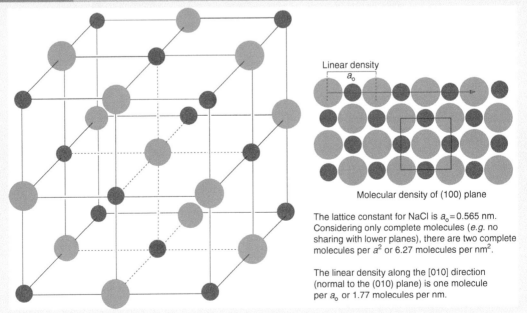

Linear density

a_o

Molecular density of (100) plane

The lattice constant for NaCl is $a_o = 0.565$ nm. Considering only complete molecules (*e.g.* no sharing with lower planes), there are two complete molecules per a^2 or 6.27 molecules per nm².

The linear density along the [010] direction (normal to the (010) plane) is one molecule per a_o or 1.77 molecules per nm.

TABLE 7.5	*Surface and Edge Energy*			
Size (nm)	Collective surface area (cm²)	Collective edge (cm)	Surface energy (J·g⁻¹)	Edge energy (J·g⁻¹)
1 nm	2.8×10^7	5.5×10^{14}	560	170
1 μm	28,000	5.5×10^8	0.56	1.7×10^{-4}
1 mm	2,800	5.5×10^6	0.056	1.7×10^{-6}
1 cm	280	55,000	0.0056	1.7×10^{-8}
1 dm	28	550	0.00056	1.7×10^{-10}
7.7 dm	3.6	9.3	0.000072	1.7×10^{-12}

Sources: G. Cao, *Nanostructures & Nanomaterials: Synthesis, Properties & Applications,* Imperial College Press, London (2004); A. W. Adamson and A. P. Gast, *Physical Chemistry of Surfaces,* 6th ed., John Wiley & Sons, New York (1997).

$$\mu_e = \frac{e\lambda}{4\pi\varepsilon_o m_e \upsilon_F} = \frac{e}{4\pi\varepsilon_o m_e}\tau \tag{7.41}$$

where

e is the electron charge in coulombs
λ is the mean free path of the electron
ε_o is the permittivity of space
m_e is the effective mass of the electron
υ_F is the Fermi velocity of the electron
τ is equal to λ/υ_F and is the time between collisions

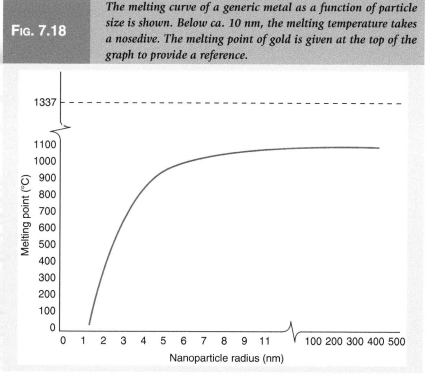

FIG. 7.18 *The melting curve of a generic metal as a function of particle size is shown. Below ca. 10 nm, the melting temperature takes a nosedive. The melting point of gold is given at the top of the graph to provide a reference.*

Generic (and idealistic) current–potential relationships are shown for a bulk metal (left) and a material capable of single-electron transport (right) [6]—like a gold cluster. The steps correspond to jumps in the current after a threshold level of voltage is applied. This image is also known as a Coulomb staircase or a Coulomb blockade.

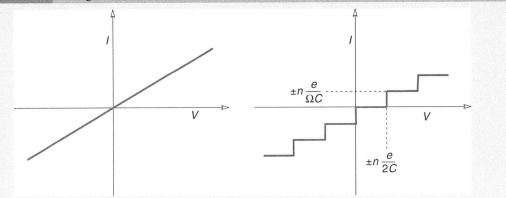

Scattering occurs in metal lattices due to electrons colliding with lattice defects (contaminants, vacancies, surfaces, grain boundaries, dislocations, etc.) and due to phonon vibrations [6]. For Cu, v is 1.6×10^6 m·s^{-1} and λ is 43 nm and t is 2.7×10^{-14} s [13].

Ohm's law states that electromotive potential is proportional to the product of current and resistance:

$$V = I\Omega \tag{7.42}$$

The magnitude of the coulombic energy E_C involved in electron transfer is

$$E_C = \frac{e^2}{2C} \tag{7.43}$$

where C is the capacitance of the particle in terms of charge per volt ($C \cdot V^{-1}$ or $C^2 \cdot J^{-1}$). In nanoparticles, charge transfer occurs via tunneling. Single-electron tunneling (SET) occurs if the thermal ambient energy is smaller than the coulombic energy [6]:

$$k_B T \ll \frac{e^2}{2C} \tag{7.44}$$

If a bias voltage is applied to the quantum dot ($V = e/C$), a tunneling current I_t is produced:

$$I_t = \frac{V}{\Omega_t} = \frac{e}{\Omega_t C} \tag{7.45}$$

where Ω_t is the tunnel current resistance. The threshold potential to electrical conduction (via tunneling current) is described by

$$V_{Threshold} = \pm \frac{e}{2C} \tag{7.46}$$

This relationship describes the phenomenon known as the *Coulomb blockade* (**Fig. 7.19**, right). The current-voltage transient shows no activity until this threshold is reached, at which point the electron is transferred. The staircase step height

$$\Delta I \propto e/\Omega C \qquad (7.47a)$$

and

$$\Delta V \propto e/2C \qquad (7.47b)$$

Catalytic Behavior of Nanoparticles. We have mentioned early on in the text how small gold nanoparticles demonstrate enhanced catalytic activity. Gold is not considered to be a catalyst, even in the domain of nanomaterials. Nanoparticles have catalytic properties due to the increased level of surface atoms—atoms that are eager to seek stability. Inherent high curvature is not a desired state and therefore the smaller the particle is, the more reactivity it tends to demonstrate. This does not always mean that smaller particles lead to increasingly faster reaction rates [14,15]. The dissociation rate of CO on Rh island aggregates peaks at 1000 atoms [14,15]. Smaller islands down to 100 atoms show decreased dissociation rates. The reaction rate of hydrogen gas with Fe nanoparticles was highest with 10 atom hydrogen-clusters; 17 atom clusters showed lower reaction rate, as did 7 atom clusters [15]. In another example, the turnover frequency of cyclohexene to cyclohexane was shown to increase with smaller Rh particle size [16].

The challenge involved in application of small catalyst particles is maintaining surface area and to keep agglomeration at a minimum. Ligand-stabilized clusters, although reducing overall surface area (that regulates the spacing of catalyst particles), are able to influence the selectivity in catalytic processes [6]. The semihydrogenation (and subsequent *trans-* vs. *cis-*directing capability) of alkynes by 3- to 4-nm Pd clusters serves as an example. The activity and selectivity of TiO_2-supported Pd clusters stabilized by two different kinds of ligands (1,10-phenanthroline vs. 2-*n*-butylphenanthroline) are strikingly different [6]. Application of the unsubstituted phenanthroline ligand yielded 95% of the *cis*-product after 75 min, but immediately transformed into the *trans*-product, which is not desirable. On the other hand, although the reaction took longer to complete (500 min), the *cis*-product comprised 100% of the product. Conversion to the *trans*-form occurred over a long period of time (ca. 2000 min) [6,17].

7.3 ONE-DIMENSIONAL MATERIALS

One-dimensional materials are those in which crystallite size is negligible in two dimensions but is not restricted in the third direction; that is, electron and hole motion are confined in two spatial directions while free propagation is allowed in the third direction. Perhaps one of the best examples of nanowires is the carbon nanotubes (**Fig. 7.20**). These are also called *quantum wires*. Thus, a quantum wire can be defined as a structure where two dimensions are comparable to the exciton Bohr radius (**Fig. 7.21**).

In other words, a quantum wire is an electrically conducting or semiconducting wire in which quantum effects control the transport properties in two dimensions. Due to the confinement of conduction electrons in the transverse direction of the wire, their transverse energy is quantized into a series of discrete values E_0 ("ground state" energy, with lower value), $E_1, ..., E_n$. One consequence

Fig. 7.20 *TEM images of single-walled carbon nanotube bundles and "wires" taken with a Philips CM-30 operated at 200 kV. Left: a global image of a tangled mass of SWNTs is shown. The dark regions of the image correspond to catalyst particles used in their synthesis. Right: A close-up image shows the bundling tendency of the SWNTs. SWNTs are actually known as "pseudoquantum wires."*

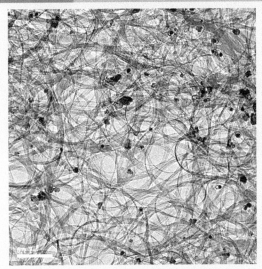 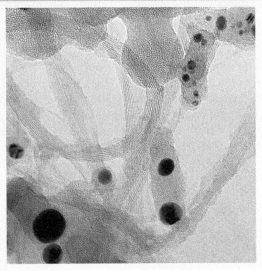

Source: Image printed with permission from Kim M. Jones, staff scientist in the Measurements and Characterization Division at the National Renewable Energy Laboratory (NREL) in Golden, Colorado.

of this quantization is that the resistance of a quantum wire cannot be deduced from the classical formula:

$$\Omega = \rho \frac{L}{A} \qquad (7.48)$$

where ρ is the resistivity, and L and A are length and cross-sectional areas of the wire, respectively. For such quantum wires, an exact calculation of the

Fig. 7.21 *A one-dimensional material is a high aspect ratio nanoparticle with electron confinement along its diameter (e.g., confinement in two dimensions). On the right, ZnO wires are shown. Note: The diameter of these wires is rather large—not exactly a one-dimensional material.*

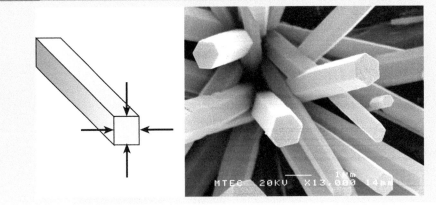

transverse energies of the confined electrons has to be accomplished to find the resistivity of the wire. Following from the quantization of electron energy, the resistance is, naturally, also quantized. The strength of the influence of quantization is inversely proportional to the diameter of the nanowire for any given material. For different materials, the quantized conductivity–resistivity is dependent on the electronic properties of the specific material—especially on the effective mass of the electrons. Stated simply, it will depend on how conduction electrons interact with the atoms within a given material. In practice, semiconductors show clear conductance quantization for wires with relatively large transverse dimensions ($\sim$100 nm). The energy of a simple square nanowire can be approximated by a two-dimensional potential well with wave function

$$\Psi_{n_x,n_y} = \sqrt{\frac{4}{L_x L_y}}\sin\left(\frac{n_x \pi x}{L_x}\right)\left(\frac{n_y \pi x}{L_y}\right) \tag{7.49}$$

in which the quantized energy is derived from

$$E_{n_x,n_y} = \frac{\hbar^2 \pi^2}{2m}\left[\left(\frac{n_x}{L_x}\right)^2 + \left(\frac{n_y}{L_y}\right)^2\right] \tag{7.50}$$

7.3.1 Types of Nanowires

The dimensions of nanowires made of metal, semiconductor, carbon, oxides, and organic materials range from a few nanometers to roughly 100 nm in diameter and a few to over 100 µm in length. Nanowires are made by single crystal growth, vapor–solid (VS), vapor–liquid–solid (VLS), vapor–solid–solid (VSS), laser ablation, chemical vapor deposition (CVD), metal oxide CVD (MOCVD), and molecular beam epitaxy (MBE) (chemical beam epitaxy [CBE]) procedures. Si nanowires surrounded by an oxide casing are formed from Si–Fe substrate at 1200°C. GaN nanowires (whiskers) can be grown epitaxially from sapphire surface by VLS of a nickel catalyst. The diameter of the catalyst dictates the dimensions of the nanowire and homogeneous nucleation events enable the control of wire length. Silicon nanowires were also grown by catalysts in the presence of silane and hydrogen gas. GaAs nanowires formed from Au catalyst with 50 nm diameter. Gold is not usually used in the role of a catalyst. Solution phase synthesis of ZnO vertical nanowire arrays using textured ZnO seeds generated wires 15–65 nm in diameter with 250–400 nm in length. Carbon nanotubes are discussed in more detail in chapter 9.

7.3.2 Physical Properties and Phenomena

Optical Properties. Nanowire optical properties are expected to be different from those of spherical clusters. Spherical particles are highly symmetric and their optical response is isotropic, regardless of the direction of the probing radiation. Orientation however is an important aspect of nanowire properties and allows them to be aligned in electric or magnetic fields. In 1912, Gans proposed that small ellipsoids abide by the dipole approximation—however that the surface plasmon mode splits into two distinct modes [18]. The curvature of the surface is responsible for this phenomenon due to its effect on the restoring force (or depolarization field) that affects the "confined" population of electrons [18]. The degree of the effect is a function of the aspect ratio of the particle. Spectra of solution containing gold nanorods reveal two absorption maxima: one for the longitudinal plasmon mode and the other for the transverse plasmon mode; both modes are independent of one another [19].

Surface modification of metal nanowires is able to alter the response. The optical response of metal particles is affected by the surrounding medium. Application of an organic material, essentially an insulating dielectric material, is expected to shift the wavelength of maximum absorption (λ_{max}) to lower energies. The dielectric material is able to diffuse the polarization and thereby cause the red shift in absorption.

Mechanical Properties.   There is much interest in the use of nanowires as connectors in electrical devices. Nanomechanical measurements using an AFM tip found that in Au nanowires, Young's modulus is independent of diameter and that the yield strength is the largest for the smallest tubes with strengths up to 100 times better than that demonstrated by bulk materials [20].

Ballistic Conduction.   Ballistic conduction occurs when the length of the conductor is smaller than the mean free path of the electron [11,21]. In other words, electrons are allowed to flow without collisions with phonons, impurities, etc. No energy is dissipated during the electrical conduction process [11]. The lack of impurities that cause elastic scattering in the conduit material is imperative for this to occur; loss of quantization results if this happens. Frank et al. first demonstrated ballistic conduction in carbon nanotubes [11,22].

7.4   TWO-DIMENSIONAL MATERIALS

Two-dimensional materials are those in which the crystallite size is negligible in one dimension and is not restricted in the other two directions; electron and hole motion is confined in only one spatial direction, whereas free propagation is allowed in the other two spatial directions. These are also called *quantum wells*. Thus, a quantum well can be defined as a structure where one dimension is comparable to the exciton Bohr radius.

Since particle motion is confined to two dimensions, the geometric correlation is to that of a plane with no thickness. When the quantum well thickness is comparable to the de Broglie wavelength of the carriers (i.e., electrons and holes), just like with zero- and one-dimensional materials, quantum confinement takes place in that material in just one dimension: that normal to the surface of the film (**Fig. 7.22**).

A simple approximation of the energy of square quantum well with thickness L that is comparable to the de Broglie wavelength of the carriers (generally electrons and holes) is

$$E_n = \frac{\hbar^2 \pi^2}{2mL^2} n^2$$

(7.51)

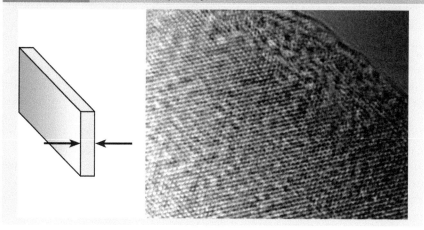

FIG. 7.22	A two-dimensional nanomaterial is a thin film that has electron confinement in just one direction. On the right, a thin film of MoS₂ is shown. The film is two to three layers thick—one that certainly qualifies as a two-dimensional nanomaterial.

Source: Image courtesy of Kostya Novoselov, School of Physics & Astronomy, University of Manchester. With permission.

7.4.1 Types of Thin Films

Two-dimensional materials are covered extensively in this text and therefore, we provide a brief overview of such materials. There are many ways to fabricate two-dimensional films: electrodeposition, evaporation, wet chemical methods, CVD, electroless deposition, and many, many more. More discussion of thin film types, fabrication, and modification ensues in later chapters.

7.4.2 Physical Properties

Optical Properties.   Optical properties depend on several parameters: (1) film composition, (2), film thickness, and (3) film structure (and orientation). These, in turn, affect properties such as interference, reflectivity, absorption, and the refractive index. All of these characteristics also depend on the kind of deposition mechanism. The films can be crystalline, polycrystalline, columnar, amorphous, or lacunar (fractal). If the film substructures are small enough, the film does not scatter light. Films can be hydrophilic or hydrophobic. Nanostructured thin films are anisotropic, usually in the direction normal to the surface. Films can be polarizing or neutral.

The optical behavior of *anatase* (TiO_2) thin films can be altered by addition of $ZnFe_2O_4$. Apparently, the zinc compound acts as a photosensitizer that enhances the photoresponse, and hence the photoactivity, of the TiO_2. The addition of $ZnFe_2O_4$ encouraged a strong photoluminescence emission due to enhanced localization of impurity- and defect-trapped excitons [23].

Mechanical Properties.   Mechanical properties of thin films differ from bulk material counterparts because of their unique nanostructure (and microstructure) and because they possess inherently large surface-to-volume ratio [24]. Recall that in chapter 4 we discovered that disc-shaped materials have the potential of largest surface-to-volume ratio. Thin film integrity is influenced strongly by its substrate. The substrate (thick and underlying) therefore plays a key role in determining the physical properties of the film that it supports.

7.5   HIERARCHICAL STRUCTURES

Hierarchical structures are structures that are synthesized (from the bottom up or top down—usually a combination) at different levels of complexity by altering growth conditions. Simply stated, hierarchical structures gradually grow from one parent structure into a more complex form (**Fig. 7.23**). Hierarchical nanostructures are three-dimensional materials. Biology abounds with examples of hierarchical structures. There are numerous examples of natural or man-made materials with hierarchical structures, such as bone or collagen networks. Bones in animal bodies are light and stable. That is because they are constructed optimally from the smallest to the largest levels. Their smallest elements are bound to fibrils that fold together to make lamellae. These, in turn, organize themselves into girders that form scaffolding—an inspiration, for sure, to structural engineers.

7.5.1   *Importance of Hierarchical Materials*

It is an exhilarating process to develop new hierarchical materials that will find applications that are important to humanity. Nanosized hierarchical structures

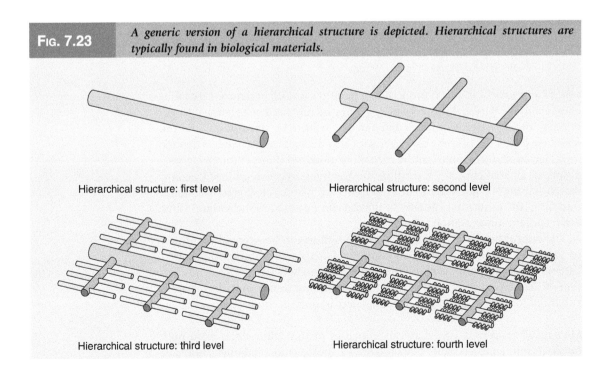

| **FIG. 7.23** | *A generic version of a hierarchical structure is depicted. Hierarchical structures are typically found in biological materials.* |

Hierarchical structure: first level

Hierarchical structure: second level

Hierarchical structure: third level

Hierarchical structure: fourth level

are good candidates for use in medical applications, environmental greening, and renewable sources of energy. Various materials like ZnO, TiO_2, C, Au, and Ag have been successfully applied in hierarchical structures. The morphology of ZnO nanostructures can be significantly altered by modifying the composition of the source materials. Nanowire ribbon arrays, for example, consist of a central axial nanowire surrounded by a radial distribution of nanobranches. The growth of the structure is in two stages: (1) rapid growth of the ZnO nanowire, and (2) nucleation and epitaxial growth of nanoribbons.

7.6 QUANTUM SIZE EFFECTS AND SCALING LAWS

Specific Heat. Specific heat is the amount of heat required to raise the temperature of a sample of mass m by ΔT [13]:

$$C = \frac{\Delta Q}{m\Delta T} \tag{7.52}$$

In 1912, Peter Debye proposed a quantum mechanical model for estimating the phonon contribution to the specific heat. A classical version was originally proposed by P. L. Dulong and A. T. Petit in 1819. They stated that the specific heat of a crystal is due to lattice vibrations equal to $3R \cdot M^{-1}$. (R is the gas constant and M is the molar mass in kilograms per mole). The model fails at low temperatures and is appropriate for solids with simple crystal structure [24]. The Debye model transforms the particle-in-a-box principle to phonons—"phonons in a box." The resonant modes of sonic phenomena of a cube of sides L have wavelength equal to

$$\lambda_n = \frac{2L}{n} \tag{7.53}$$

where n is an integer. The energy of a phonon is

$$E_n = h\nu_n = \frac{hc_s}{\lambda_n} = \frac{hc_s n}{2L} \tag{7.54}$$

where c_s is the speed of sound. This equation relates the energy of phonons to the size of the box.

The specific heat of crystalline nanoscale materials is higher than those of the bulk. C_v (the specific heat at constant volume) for bulk Pd is 25 J·mol^{-1}·K^{-1} (for a 6-nm nanocrystal, C_v = 37 J·mol^{-1}·K^{-1}). For 8-nm Cu, the C_v increases from 24 to 26 J·mol^{-1}·K^{-1}, and for 6-nm Ru nanoparticles, C_v increases from 23 to 28 J·mol^{-1}·K^{-1} [13].

Magnetism. Magnetic properties are size dependent in materials like Fe, Co, Ni, and Fe_3O_4. The coercive force behind magnetic memory required to reverse an internal magnetic field within a particle is size dependent. The strength of a particle's internal magnetic field is also size dependent.

Structural Fluctuations.   Independent clusters in particular have the ability to alter their structure rather quickly—a trait not characteristic of bulk materials. At high temperatures, these fluctuations cause breakdowns in symmetry that may result in the formation of a liquidus state in the particle [14].

7.6.1   Scaling Laws

Scaling laws offer us a means to place our universe into perspective. The following discussion is gleaned from an essay written by Chris Phoenix of the Center for Responsible Nanotechnology (CRN) and summarizes the importance of scaling to smaller sizes. Scaling laws tell us, for example, that small things are not as affected by gravity as are larger things [25]. Most importantly, scaling laws are important when it comes to the manufacture of small things. We are familiar with the scale-up process where a prototype is developed and placed into the hands of manufacturing engineers with the prospect of mass production in mind. This is the scale-up of a process from small to large. Scaling laws start from the large and predict properties of smaller things or vice versa.

Scaling laws with positive exponents become more important as things get larger (e.g., r^3). Scaling laws with negative exponents (e.g., r^{-1}) get more important as things get smaller. In nature, the muscles of elephants and fleas are very different. *Strength* is proportional to cross-sectional area ($\propto L^2$). The *weight* of the muscle is proportional to its volume ($\propto L^3$). Strength versus weight is proportional to L^2/L^3 or L^{-1}. This ratio gets 10 times larger as organisms get 10 times smaller—a scaling law. At the nanoscale, ca. 10^6 times smaller than a flea, there is no concern at all about weight [25].

Timescale considerations affect potential productivity. With regard to time, the speed of a bulk manufacturing process may be a few units per second, per hour, or per day depending on its size and complexity. An enzyme is able to transform substrates into products at a rate of thousands to millions times per second. The wingbeat frequency of dragonflies (~40 Hz) like Zygoptera (a larger species) is half as fast as that of Anisoptera (a smaller species) [26]. The smaller species is capable of quicker acceleration.

Power density is also important at any scale. Power density (force × speed) scales with L^2, just like strength. If a cubic engine 10 cm on one side is capable of producing 1000 W of power, then an engine of 1 cm on a side (10 times smaller along one dimension) should produce 1/100 of the power of the larger engine [25]. However, the volume of the smaller cube is 1000 times smaller than that of the larger cubic engine and 1000 of those smaller engines would end up producing 10,000 W of power—10 times as much power as the original 10-cm cubic engine. Power per volume scales with $1/L$. Therefore, scaling laws tell us that by manufacturing 1000 more parts that are 1000 times less in volume, 10 times the output of the original cubic engine is achievable with the same total mass [25]. Add to this that smaller things run at higher frequencies and, according to temporal scaling laws, the small cubic engines would be able to operate at a rate that is 10 times faster than the parent engine—the power of small!

Functional density is proportional to L^{-3}. This is simply the number of devices that can be crammed into a unit of space; the smaller the device is, the more of it one can pack into the usable space. The computer industry serves up the best example of the importance of this scaling law; each year, more and smaller transistors squeeze into the same universal-sized wafer. The result of such miniaturization is more memory and faster processing.

Forces operating on small things differ widely from those operating at the macroscopic level. Van der Waals forces originating from the setae of gecko feet are able to hold the creature to a ceiling. Electrostatic forces assume more importance at the nanoscale; at our scale, it is referred to as "static cling" of clothing materials [25]. As motors are miniaturized to the level of cells or molecules, power density increases, with nanoscale electric motors having a power density on the order of 10^6 W·cm^{-3} [25].

Friction at the nanoscale dwells in an altered reality. Friction scales with L^2 in the macroworld, indicating that frictional power is proportional to power: The more power that is exerted, the more wear due to frictional forces is expected to occur in large things. Wear life is proportional to L (thickness of a film or coating). Despite this drawback (e.g., thinner films are worse and therefore lead to lower operational lifetime), the intervention of *nonscaling mechanisms* saves the day. There should be no wear and tear as we know it at the nanoscale due to strong chemical bonds able to withstand localized heating forces.

7.6.2 *Classical Scaling Laws and the Nanoscale*

The content of this section was summarized from *Nanosystems*, chapter 2, "Classical Magnitudes and Scaling Laws," by Drexler [27]. The *magnitudes* of properties that characterize bulk materials, by definition, are not directly applicable to nanomaterials. Macroscale magnitudes, part of the classical continuum, are adjusted by the application of scaling laws, but they have a tendency to lose validity due to the effects of atomic structure, mean free path effects, and quantum mechanics [27]: "When used with caution, classical continuum models of nanoscale systems can be of substantial value in design and analysis. They represent the simplest level in a hierarchy of approximations of increasing accuracy, complexity and difficulty."

Scaling laws focus on some parameter that has an inherent characteristic of "scalability." This seems like a circuitous argument, but nonetheless it is an accurate one. For example, if we wish to predict the properties of an object that is larger than the one used in the model, we conduct the calculations based on increasingly larger values of some scale. If material strength and stress are considered to be constant, then several scaling laws relate to a dimension L [27]:

$$\text{Total strength} \propto \text{force} \propto \text{area} \propto L^2 \tag{7.55}$$

If one assumes a state of constant density, then

$$\text{Mass} \propto \text{volume} \propto L^3 \tag{7.56}$$

This is straightforward. The mass of a block of material with density $\rho = 3.5 \times 10^3$ kg·cm^{-3} has mass equal to 3.5×10^{-24} kg^{-1}, assuming that the density remains constant at that scale—a nonscaling parameter. For a more complicated property like acceleration,

$$\text{Acceleration} \propto \text{force/mass} \propto L^{-1} \tag{7.57}$$

A nanometer-sized object is capable of accelerating to a level of 3×10^{15} m·s⁻¹. Other relations are given: *power ∝ force · speed ∝ L^2; power density ∝ power/volume ∝ L^{-1}; frequency ∝ speed/length ∝ L^{-1}; speed ∝ acceleration · time = constant; time ∝ frequency⁻¹ ∝ L*, and *frequency ∝ acoustic speed/length ∝ L^{-1}*—all good scaling laws for bulk materials. According to Drexler [27],

> The accuracy of classical continuum models and scaling laws to nanoscale systems depends on the physical phenomena considered. It is low for electromagnetic systems with small calculated time constants, reasonably good for thermal systems and slowly varying electromagnetic systems, and often excellent for purely mechanical systems provided that the component dimensions substantially exceed atomic dimensions. Scaling principles indicate that mechanical components can operate at high frequencies, accelerations, and power densities.

7.6.3 Scaling Laws for Clusters

The following discussion is summarized from the book *Atomic and Molecular Clusters*, by Johnston [28]. Ionization energy, electron affinity, melting temperature, and cohesive energy vary with cluster size [27]. It is possible to model cluster size effects (CSEs) by application of the spherical cluster approximation (SCA) presented in chapter 5 in the case where an *N*-atom cluster is a sphere with radius *r*. Interpolation-based scaling law formulae that incorporate a scaling factor proportional to the power law of the cluster radius (like *L* before) are able to smoothly approximate the behavior of nanoclusters. A scaling law that incorporates the radius as a factor is

$$G(r) = G(\infty) + ar^{-\alpha} \tag{7.58}$$

where $G(\infty)$ is the value of the specific property *G* at the limit of the bulk material. Or, if it is by the number of atoms that potential scaling is focused (since properties at the nanoscale depend on the number of atoms and that surface fraction $F_s \propto N^{-1/3} \propto r^{-1}$),

$$G(N) = G(\infty) + BN^{-\beta} \tag{7.59}$$

The value of the exponent α in the previous equation is equal to 1; for β, the exponent is equal to 1/3.

Ionization energies of potassium clusters provide a good example to illustrate this process. The IE can be fitted with great accuracy to the following interpolation formula:

$$\frac{IP^K(N)}{eV} = 2.3 + 2.04\, N^{-1/3} \quad \text{or} \quad \frac{IP^K(r)}{eV} = 2.3 + 5.35\left(\frac{r}{\mathring{A}}\right)^{-1} \tag{7.60}$$

Figure 7.24 plots the result of the interpolation scaling of the ionization potential versus cluster size. The fit with experimental data is excellent [27].

This process also works well with melting temperature of clusters. The melting point of gold, $T_m{}^{Au}$, decreases with decreasing size (as we pointed out before) and follows the r^{-1} rule. A good fit between experiment and theory is given by the expression

FIG. 7.24 *Ionization energy of potassium is calculated from an empirically fitted scaling law (equation 7.62, left). Ionization energy is plotted against cluster size (by number of atoms). The experimental data fit better as the cluster gains in size, although the overall fit is certainly not bad. The dashed line indicates the bulk work function of the material [28].*

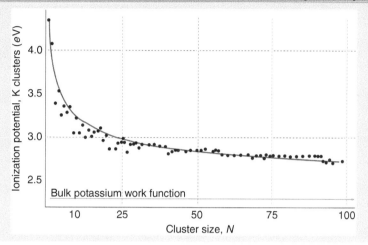

$$\frac{T_m^{\mathrm{Au}}(r)}{K} = 1336.15 - 5543.65 \left(\frac{r}{\overset{\circ}{A}} \right)^{-1} \tag{7.61}$$

Large clusters show good agreement with the scaling law but smaller ones deviated from the theoretical line (**Fig. 7.25**).

FIG. 7.25 *The melting temperature of gold is represented by an interpolation formula (equation 7.63) in the figure. Apparent oscillatory behavior in the melting point is observed as cluster size is decreased. The oscillations are due to quantum and surface (geometric) effects. The dashed line represents the melting temperature of the bulk material [28].*

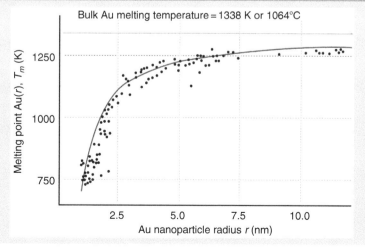

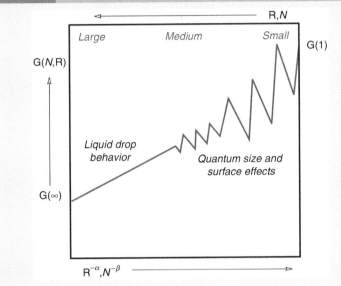

FIG. 7.26

The variation of a generic cluster property (G) with cluster size is graphically depicted in the figure. The oscillatory behavior of nanoparticles is exaggerated in the quantum domain of the material [28]. The smooth CSE "liquid drop" behavior is consistent with what we have been saying all along—that bulk materials exhibit smooth transitions in physical properties, due to the enormous population of numerous electronic states, and that small materials exhibit quantum behavior. Together they serve to illustrate the material continuum.

What are the reasons for these deviations? If one inspects the graph carefully, one sees that oscillatory behavior around the smooth CSE power-law curve is observed in the small cluster regime. These oscillations are typically observed for many kinds of phenomena [27]. These deviations are due to quantum size effects (QSEs). One example of a quantum size effect is the closing of electronic shells. Recall the stability of magic clusters that we discussed in chapter 6. A similar oscillation of energetic stability depended on the number of electrons in the cluster. The same is true here. In addition to the electrons, there is geometric shell closing that forms stable magic number clusters arising from surface effects (SEs). A schematic representation of cluster property (G) that exhibits the "smooth CSE liquid drop" behavior of large clusters is compared to the oscillatory behavior influenced by QSE and SE found in small clusters in **Figure 7.26.**

References

1. Density of states, en.wikipedia.org/wiki/Density_of_states (2007).
2. D. A. McQuarrie, *Quantum chemistry*, University Science Books, Mill Valley, CA (1983).

3. X. Ma et al., The size dependence of the optical and electrical properties of Ge quantum dots deposited by pulsed laser deposition, *Semiconductor Science Technology*, 21, 713–716 (2006).

4. P. Melinon, V. Paillard, V. Dupuis, A. Perez, P. Jensen, A. Hoareau, M. Broyer, J. -L. Vialle, M. Pellarin, B. Baguenard, and J. Lemme, From free clusters to cluster-assembled materials, *International Journal of Modern Physics B*, 9, 339–397 (1995).

5. H. Haberland, M. Moseler, Y. Qiang, O. Rattunde, T. Reiners, and Y. Thirmer, Energetic cluster impact (ECI): A new method for thin film formation, *Surface Review and Letters*, 3, 887–890 (1996).

6. G. Schmid, Metals. In *Nanoscale materials in chemistry*, K. J. Klabunde, ed., Wiley-Interscience, John Wiley & Sons, Inc., Hoboken, NJ, chap. 2 (2001).

7. G. L. Hornyak, S. Peschel, T. Sawitowski, and G. Schmid, TEM, STM and AFM as tools to study clusters and colloids, *Micron*, 29, 183–190 (1998).

8. U. Kreibig and M. Vollmer, *Optical properties of metal clusters*, Springer-Verlag, Berlin (1995).

9. C. Noquez, Surface plasmons on metal nanoparticles: The influence of shape and physical environment, *Journal of Physical Chemistry C*, 111, 3806–3819 (2007).

10. K. Kamiya and S. Sakka, Formation of calcium carbonate polymorphs, *Gypsum Lime*, 163, 243–253 (1979).

11. G. Cao, *Nanostructures & nanomaterials: Synthesis, properties & applications*, Imperial College Press, London (2004).

12. A. W. Adamson and A. P. Gast, *Physical chemistry of surfaces*, 6th ed., John Wiley & Sons, New York (1997).

13. M. Meyyappan, *Introduction to nanotechnology, II. Nanoscale properties*. Presentation, Center for Nanotechnology, NASA Ames research Center (viewed 2007).

14. C. P. Poole and F. J. Owens, *Introduction to nanotechnology*, Wiley-Interscience, John Wiley & Sons, Inc., Hoboken, NJ (2003).

15. R. L. Whetten, D. M. Cox, D. J. Trevor, and A. Kaldor, Correspondence between electron binding energy and chemisorption reactivity of iron clusters, *Physics Review Letters*, 54, 1494–1497 (1985).

16. G. W. Busser, J. G. van Ommen, and J. A. Lercher, Preparation and characterization of polymer-stabilized rhodium particles. In *Advanced catalysis and nanostructured materials*, Academic Press, San Diego (1996).

17. G. Schmid, V. Maihack, F. Lantermann, and S. Peschel, Ligand-stabilized metal clusters and colloids: Properties and applications, *Journal of Chemical Society, Dalton Transactions*, 589–595 (1996).

18. J. Pérez-Juste, I. Pastoriza-Santos, L. M. Liz-Marzán, and P. Mulvaney, Gold nano-rods: Characterization and applications, *Coordinated Chemistry Reviews*, 249, 1870–1901 (2005).

19. G. L. Hornyak, *Characterization and optical theory of nanometal/porous composite membranes*, Ph.D. dissertation, Colorado State University (1997).

20. B. Wu, A. Heidelberg, and J. J. Boland, Mechanical properties of ultrahigh-strength gold nanowires, *Nature Materials*, 4, 525–529 (2005).

21. S. Chappel and A. Zaban, Nanoporous SnO_2 electrodes for dye-sensitized solar cells: Improved cell performance by the synthesis of 18 nm SnO_2 colloids, *Solar Energy Materials & Solar Cells*, 71, 141–152 (2002).

22. S. Frank, P. Poncharal, Z. L. Wang, and W. A. der Heer, Carbon nanotube quantum resistors, *Science*, 280, 1744–1746 (1998).

23. Y. Jin, G. Li, Y. Zhang, Y. Zhang, and L. Zhang, Photoluminescence of anatase TiO_2 thin films achieved by the addition of $ZnFe_2O_4$, *Journal of Physics: Condensed Matter*, 13, L913–L918 (2001).

24. Dulong–Petit law, en.wikipedia.org/wiki/Dulong-Petit_law (2007).

25. C. Phoenix, Scaling laws—Back to basics, CRN Science & Technology Essays-2004, Center for Responsible Nanotechnology, crnano.org/essays04.htm#Scaling (2004).

26. G. Rüppell, Kinematic analysis of symmetrical flight maneuvers of Odonata, *Journal of Experimental Biology*, 144, 13–42 (1989).

27. E. K. Drexler, *Nanosystems*, John Wiley & Sons, New York (1998). Foresight Institute, www.foresight.org/Nanosystems/Ch2/chapter2_1.html

28. R. L. Johnston, *Atomic and molecular clusters*, Taylor & Francis, London (2002).

Problems

7.1 Why does classical blackbody radiation treatment break down at high energies?

7.2 Describe the photoelectric effect and provide the generalized equation. Why is it considered to be a quantum phenomenon?

7.3 Werner Heisenberg stated that two complementary properties of a system (like position and momentum) can never be simultaneously known. Is this principle applicable to nanoparticles? Use $\Delta p \Delta x \geq h$ as a frame of reference.

7.4 A nanoparticle with mass 5×10^{-27} g exists in a 1-nm, one-dimensional box. What is the wavelength of radiation that is emitted when the nanoparticle loses energy from the $n = 3$ level to the $n = 2$ level?

7.5 What does the term degeneracy imply with regard to energy levels?

7.6 Explain in the simplest of terms the phenomenon of quantum tunneling.

7.7 Define the terms wave, particle–wave duality, vibration, rotation, and momentum. Which possess quantized characteristics?

7.8 Why there are differences in bandgap energies between insulators, semiconductors, and metals?

7.9 Explain the presence of the zigzag lines in the material continuum trace shown in **Figure 7.26**.

7.10 Is there a material continuum or does the concept break down in the quantum domain?

7.11 Bulk gold is yellow, and gold nanoparticles are light red, purple, or blue. What is the reason for the change of color? (Question courtesy of Günter Schmid, Uni-Essen.)

7.12 What does the specific color depend on? (Question courtesy of Günter Schmid, Uni-Essen.)

7.13 Why is the plasmon resonance visible only for Cu, Ag, and Au as opposed to other metals? (Question courtesy of Günter Schmid, Uni-Essen.)

7.14 What is a "Coulomb blockade" and what do its properties depend on? (Question courtesy of Günter Schmid, Uni-Essen.)

7.15 How can Coulomb blockades be measured? (Question courtesy of Günter Schmid, Uni-Essen.)

7.16 When are quantum confinement effects observed in nanocrystallites?

7.17 What is meant by quantum confinement in a semiconductor quantum dot? What are the implications of this for the energy levels?

7.18 In how many dimensions free electron motion is restricted in quantum wires?

7.19 What do you understand by the wave–particle duality in nanotechnology?

7.20 A proton is confined to moving in a one-dimensional box of width 0.200 nm.

(a) Find the lowest possible energy of the proton.

(b) What is the lowest possible energy of an electron confined to the same box?

(c) How do you account for the large difference in your results for (a) and (b)?

7.21 An electron is confined to a 1 μm thin layer of silicon. Assuming that the semiconductor can be adequately described by a one-dimensional quantum well with infinite walls, calculate the lowest possible energy within the material in units of electron volt. If the energy is interpreted as the kinetic energy of the electron, what is the corresponding electron velocity? (The effective mass of electrons in silicon is 0.26 m_0, where m_0 = 9.11 × 10^{-31} kg is the free electron rest mass).

7.22 Estimate the minimum velocity of an apple of mass 100 g confined in a crate of size 1 m.

NANOTHERMODYNAMICS

I see evidence of the second law of thermodynamics all over my room. Are you trying to tell me that if I made my room much much smaller, it would clean itself?

UNKNOWN STUDENT, 2007

Chapter 8

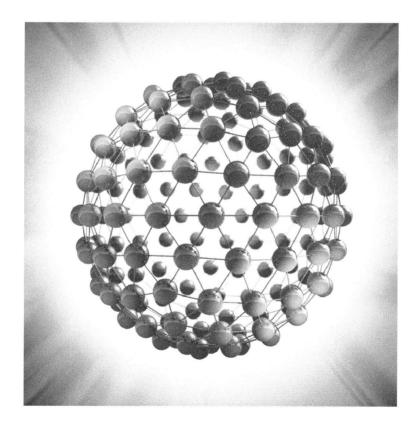

The physics division continues with a special chapter on nanothermodynamics. It is recommended to students with at least college level physical chemistry. For those who do not have such a background, we recommend that you at least read the chapter and discuss it with your instructor and classmates. Much of the chapter is nontechnical/nonmathematical and should acquaint you with the basic terminology and the concepts. In many ways the chapter will serve as a primer on thermodynamics, but familiarization with the subject matter, particularly as it pertains to living things, should provide for interesting reading.

So far we have investigated nanomaterial surface and structure and discussed the energetics of nanoscale materials in chapters 5 and 6, respectively. Chapter 7 showed us how scaling laws break down at the nanoscale as properties and phenomena acquire more quantum character with diminished size. We now present another viewpoint of the nanoscale—a view from a purely thermodynamic perspective. We begin by reviewing some obligatory fundamentals of classical thermodynamics and statistical mechanics and then discuss selected theories that apply specifically to nanomaterials; these theories essentially bring into question the validity of applying classical thermodynamics to small systems without considering them as small systems.

Special sections on fullerenes and carbon nanotubes are presented in chapter 9. With chapter 9 we begin the chemistry division of the book and we move on to intra- and intermolecular chemical bonding in chapter 10.

Although an in-depth discussion of nanothermodynamics is beyond the scope of this text, we feel strongly that the subject matter must be mentioned and introduced at this time. As a result, we limit our discussions to qualitative perspectives without much quantitative treatment. It is up to you the student to absorb, learn, and assimilate what you feel is interesting, necessary, or relevant and develop your own unique perspective on this complex subject.

8.0   THERMODYNAMICS AND NANOTHERMODYNAMICS

Nanothermodynamics, like atomistic simulation, is nontrivial, and understanding nanothermodynamics requires some tweaking of preexisting thermodynamic paradigms. We have referred to nanoscience and nanotechnology as disruptive to society. In this chapter, we assert that thermodynamic properties of nanomaterials need to be viewed in a different light as well.

Adhering to our historical theme, many of the concepts that support nanothermodynamics were developed well before the word "nanotechnology" became popular. Adhering to our continuum theme, we vigorously assert that nanothermodynamics is a part of disciplines belonging with the energy neighborhood. Lastly, adhering to our "nature is still the best teacher" theme, we once again place focus on nature's mastery of the thermodynamics of small things.

8.0.1   *Background*

Thermodynamics (from the Greek *thermos*, meaning "hot" and *therme*, indicating heat, + *dynamikos*, meaning "powerful") is a branch of science that tries to bring understanding and order to chaos—a somewhat paradoxical definition in that one of the basic drivers of thermodynamics is chaos. Thermodynamics is the study of energy and its interconversion. It is the study of heat and its relationship with physical parameters like temperature, pressure, density, and other macroscopic measurable properties of matter. Thermodynamics is a science that is grounded in experiment and theory and centered on the concept of equilibrium. Thermodynamics, along with statistical mechanics and the kinetic theory, provides tools that enable us to understand the relationships between a system and its surroundings.

Historically, thermodynamics developed from the need to increase the efficiency of early steam engines. The two Roberts, Boyle and Hooke, initiated an insipient form of thermodynamics when they created an air pump to measure the pressure–temperature–volume correlations and eventually to describe a relationship that we now know as Boyle's law. Nicolas Léonard Sadi Carnot in the early 1800s added significant contributions to early thermodynamic theory. Carnot, who is known as the father of modern thermodynamics, invented the Carnot engine and developed the working theory of the Carnot cycle. Concepts such as engine power and efficiency soon followed. James Joule coined the word *thermodynamics* in the mid-1800s. While working with steam engines, he strove to understand the relationship between heat, power, and mechanical work and later on with electricity.

Kelvin, Clausius, and Rankine took Boyle's law to the next level and the first and second laws of thermodynamics emerged by the mid-nineteenth century. Josiah Willard Gibbs was one of the most productive experimentalists and theoreticians of the mid- to late nineteenth century (**Fig. 8.1**). He published a fundamental

| **Fig. 8.1** | *(a) The U.S. Postal Service released this stamp in May 2005 as a tribute to Josiah Willard Gibbs. Gibbs was the first American to receive a doctorate of engineering in 1863. His contributions to the first and second laws of thermodynamics and the Gibbs equation are fundamental to thermodynamics. The background is made of thermal coordinates. (b) An interesting subtlety exists in the image. If one inspects the collar of his shirt, his famous equation will be found there: $dE = Tds - PdV$ (U.S. Postal Service and Zolti Spakovsky, Cambridge, Massachusetts, in Mechanical Engineering, 8 April 2006). The equation is reproduced exactly as it is found in Gibbs's paper "Graphical Methods in the Thermodynamics of Fluids," published in the Transactions of the Connecticut Academy in 1873.* |

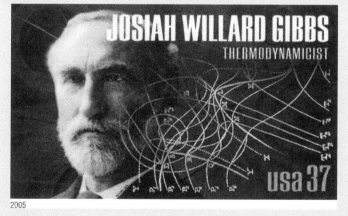

(a)   (b)

paper in 1873 detailing geometric representations of thermodynamic surfaces. By the age of 34, he changed the face of science. He is credited with formulating the concept of thermodynamic equilibrium of systems in terms of energy and entropy. His book, *Elementary Principles of Statistical Mechanics*, published in 1902, linked thermodynamics to statistical analysis. Gibbs was a fundamental force behind the development of nanothermodynamics, albeit without prior knowledge of the term. T. L. Hill, one of the pioneers of nanothermodynamics, gave credit to Gibbs for adding chemical potential terms to the basic energy–heat–work equations. It is from the chemical potential term, μ, that Hill's *subdivision potential term* to describe nanothermodynamic systems is derived.

The concepts of system, surroundings, internal energy, free energy, entropy, and enthalpy gradually matured in the early twentieth century and were fundamental not only to engine design but also to a wide range of various applications such as mechanical engineering, chemical engineering, aerospace engineering, materials science, and biology. The work accomplished by these early pioneers laid the foundations of macroscopic *classical thermodynamics*—an approach that lacks any obvious atomic (or nano) interpretation and assumes in its interpretation a state of equilibrium.

8.0.2 The Nano Perspective

Thermodynamics is defined within the framework of a continuum that is centered on the transfer of energy. Once again we have to draw lines to help us understand its meaning. We invented *classical thermodynamics* to understand how engines behave and then, from the top down, to understand how molecules behave, from the perspective of average properties. We invented *statistical mechanics* to illustrate how thermodynamics is constructed from the bottom up, thus adding a complementary component to the continuum. We invented *quantum mechanics* to describe the behavior of atoms and molecules knowing that all thermodynamics is based ultimately on quantum mechanics. These are "integrated into everything" within the greater space of thermodynamics, although each component may seem to be extremely segregated in its expression: classical thermodynamics, kinetic theory, statistical mechanics, quantum mechanics, pseudoequilibrium thermodynamics, quasi-equilibrium thermodynamics, nonequilibrium thermodynamics, nonergodic thermodynamics, and, now, nanothermodynamics.

Nature once again gives us an amazing number of paradoxes to ponder. We all understand that macroscopic behavior can be described by using classical thermodynamic principles. However, these principles do not seem to apply clearly to living systems. Although metabolic processes ultimately follow thermodynamic laws when considering the system as a whole, much of what goes on in the cell seems to be quite mysterious—quite nonthermodynamic—as if a *fourth law of thermodynamics* is needed to explain what actually transpires inside the living cell [1]:

> The sterile, mechanical universe of the nineteenth century theories was conceived as a closed, isolated system tending to equilibrium. As predicted by the second law of thermodynamics, it will inexorably expire as available energy is

spent doing work and converted to entropy … But a recent revision is underway by which life has become known as an open system infused and organized by a flow of energy and information …. An effort to articulate a "fourth law of thermodynamics" counter to the second law is now in progress.

We have forged strong links between nanotechnology and nature throughout the text. The preceding statement is another example of the revolutionary imperative of nanotechnology, once again grounded within the intricate machinations of nature.

Replication, for example, exists in biological and nonbiological systems. The process of replication is more a kinetic process than a thermodynamic one [2]. On the other hand, nonreplicating systems have more *thermodynamic* (thermostatic?) character and less kinetic character. Proponents of this train of thought believe that kinetic stability and kinetic selection are forces that drive replication processes in biology [2]. Unfortunately, this makes thermodynamic interpretation of such complex processes even more difficult.

Before we begin our quest, let us propose a few simple questions. When did biological nanomaterials start to display their special functions and what thermodynamic driving forces made this possible? How do they accomplish those functions and how do they fit into the complexity that characterizes all biological systems? And, why nanomaterials? What is it about the special relationship nanomaterials have to energy transfer that makes life work?

8.1 Classical Equilibrium Thermodynamics

We begin this section with a brief recapitulation of classical thermodynamics. We will strive to be as succinct as possible in our presentation.

8.1.1 Extensive and Intensive Properties and State Functions

Extensive and Intensive Properties. There are two basic categories of thermodynamic properties: *intensive* and *extensive*. Intensive and extensive properties and *intensivity* and *extensivity* are terms that describe physical parameters that are fundamental to thermodynamics and to nanothermodynamics, albeit in altered form. Properties and phenomena that do not rely on "how much" is involved are called *intensive properties* and are independent of the size of a system. Examples of intensive properties include density ρ (grams per cubic centimeter), pressure P (newtons per square meter), temperature T, and chemical potential μ. For example, the temperature of a system in equilibrium is the same throughout that system no matter how large or small. And, conversely, in the absence of a gradient, thermal equilibrium is achieved when all parts of a system, no matter how large or small, have the same temperature.

Extensive properties are those that have linear dependence on the amount of matter or size involved. Internal energy (U), enthalpy (H), Gibbs free energy (G), Helmholtz free energy (A), volume (V), area ($\mathcal{A}$), mole number or number (n or N), and entropy (S) are scaled by the amount of material on hand. Energy released during a forest fire depends on the number N_{Trees} that burned. The total energy of the fire is equal to the sum of the energy of each component (tree). The relationship between extensive and intensive variables is accomplished through the theory of homogeneous functions (in which all terms are of the same degree, e.g., polynomials with the same exponent).

From a mathematical perspective, thermodynamic functions are either homogeneous to the first degree ($\lambda = 1$) and are called *extensive* functions or homogeneous to the zeroth degree ($\lambda = 0$) and are called *intensive* functions. The straightforward mathematical representation is

$$f\left(x_1, x_2, \ldots, x_n\right) \Rightarrow f\left(hx_1, hx_2, \ldots, hx_n\right) = h^\lambda f\left(x_1, x_2, \ldots, x_n\right) \tag{8.1}$$

where $f(x)$ is a smooth continuous function with the following characteristics: (1) the arguments of the function are scaled by h, (2) the function f is scaled by the factor h^λ, and (3) the function is homogeneous to the degree λ [3,4]. Euler's theorem of homogeneous functions is expressed as a differential equation of the form

$$x_1\left(\frac{\partial f}{\partial x_1}\right) + x_2\left(\frac{\partial f}{\partial x_2}\right) + \cdots + x_n\left(\frac{\partial f}{\partial x_n}\right) = \sum_{i=1}^{n} x_i\left(\frac{\partial f}{\partial x_i}\right) = \lambda f\left(x_1, x_2, \ldots, x_n\right) \tag{8.2}$$

where λ is called the Euler exponent and $x_k\left(\frac{\partial}{\partial x_k}\right)$ is called the Euler operator. Stated in more complex terms, homogeneous functions are *eigenfunctions* of the Euler operator, with the degree of homogeneity as the *eigenvalue* [4].

The entropy, S, for example, is completely described by three extensive variables: internal energy, volume, and number: $S = S(U, V, N)$. If we state that the change in internal energy U is a function of extensive variables only, $U = U(S, V, N, X)$, a general case relating U and the extensive variables to the Euler exponent scaling factor λ is

$$U(S, V, N, X) \rightarrow U(\lambda S, \lambda V, \lambda N, \lambda X) \rightarrow \lambda U(S, V, N, X) \tag{8.3}$$

where X symbolizes all other types of extensive variables. Differentiating U with respect to λ yields

$$U = S\left(\frac{\partial U}{\partial S}\right)_{V,N,X} + V\left(\frac{\partial U}{\partial V}\right)_{S,N,X} + N\left(\frac{\partial U}{\partial N}\right)_{V,S,X} + X\left(\frac{\partial U}{\partial X}\right)_{S,V,N} \tag{8.4}$$

Breakdown of individual terms yields intensive terms T, P, μ, and xt, respectively, where x is a generalized intensive parameter that corresponds to the extensive variable X:

$$T = \left(\frac{\partial U}{\partial S}\right)_{V,N,X} ; \quad -P = \left(\frac{\partial U}{\partial V}\right)_{S,N,X} ; \quad \mu = \left(\frac{\partial U}{\partial N}\right)_{V,S,X} \quad \text{and} \quad x = \left(\frac{\partial U}{\partial X}\right)_{S,V,N} \tag{8.5}$$

The extensive variable X is planted at this time as a teaser to be explained later, and the form of the final term in equation (8.5) is explored further in section 8.4.3. The integrated form of U from equation (8.4) is given as

$$U = TS - PV + \mu N + xX \tag{8.6}$$

Notice the additive nature of extensive–intensive terms in classical thermodynamic formulations and that ratios of extensive terms yield intensive terms.

For the Gibbs free energy, $G = G(N, P, T)$, the extensive dependence of G is only on N (P and T are intensive properties). Therefore, $G(\lambda N, P, T) = \lambda G(N, P, T)$. G is a homogeneous function of degree 1 with its extensive argument as N. From Euler's theorem,

$$G(N, P, T) = N \frac{\partial G}{\partial N} = \mu N \tag{8.7}$$

Equation (8.7) is an important equation and serves as a starting point in our discussion of nanothermodynamics.

Although we will not go into any more detail, the description of a thermodynamic system requires at least one extensive function and at least one intensive function. It is also true that the ratio of two extensive quantities results in an intensive quantity. Entropy per mole number (S/N) is an intensive property based on the ratio of two extensive properties. Density, an intensive property, is defined by the ratio of mass to volume—both extensive quantities. The Gibbs–Duhem equation

$$S\,dT - V\,dP + N\,d\mu = 0 \tag{8.8}$$

is a combination of extensive and intensive variables. The intensive variables are not independent of one another. Knowledge of the values of two of them will yield the value of the third.

Conjugate variables consisting of an intensive property and an extensive property yield information about a thermodynamic system—for example, pressure-volume (mechanical parameters), temperature–entropy (thermal parameters), and chemical potential–particle number (material parameters). The terms *extrinsic* and *intrinsic* are used in physics and thermodynamics too but should not be confused with the extensive and intensive defined previously. In the general sense, the term extrinsic implies components, phenomena, and conditions outside a system while the term intrinsic implies the opposite.

State Functions.   The value of a property in a state of thermodynamic equilibrium is called a *state function* [5]. Properties that are independent of the path taken to reach a final state from an initial state are called state functions and are represented by ΔU for internal energy. Path-independent state functions contain exact differentials where i represents the initial state and f the final state (**Table 8.1**):

$$\Delta U = \int_i^f dU = U_f - U_i \tag{8.9}$$

The first law of thermodynamics (discussed more fully in the next section) is represented as $U = q + w$, where q and w are heat and work, respectively. The variables q and w are dependent on the path taken and therefore are not state functions. Therefore, the parameters dq and dw are not exact differentials (e.g., the integration of q or w does not result in a Δq or a Δw [6]).

TABLE 8.1	*State Functions—Summary of the Five Kinds of Energy*		
Name	**Euler form**	**Differential form**	**Environmental variables**
Internal energy	$U = q + w$ (8.10)	$dU = T\,dS - P\,dV + \mu\,dN$ (8.11)	S, V, N
Enthalpy	$H = U + PV$ (8.12)	$dH = T\,dS + V\,dP + \mu\,dN$ (8.13)	S, P, N
Helmholtz free energy	$A = U - TS$ (8.14)	$dA = -S\,dT - P\,dV + \mu\,dN$ (8.15)	T, V, N
Gibbs free energy	$G = H - TS$ (8.16)	$dG = -S\,dT + V\,dP + \mu\,dN$ (8.17)	T, P, N
Grand potential energy[a]	$\Omega = A - \mu N$ (8.18)	$d\Omega = -S\,dT - P\,dV - N\,d\mu$ (8.19)	T, V, μ

[a] Process at constant temperature and chemical potential. The grand potential energy parameter is a measure of nonchemical work (e.g., mechanical work).

8.1.2 The System, Its Surroundings, and Equilibrium

Thermodynamic equilibrium implies that only infinitesimal reversible changes of thermodynamic parameters are allowed. These infinitesimal changes occur in the forward or the reverse direction (a.k.a. a reversible process). A system is in thermodynamic equilibrium when temperature, pressure, and material are in equilibrium. For example, Gibbs free energy for a system is considered to be in thermodynamic equilibrium when its overall thermodynamic potential is minimized. Mathematically, the equilibrium state of Gibbs free energy at constant temperature and pressure (T, P) is represented by

$$\Delta G_{T,P} = 0 \tag{8.20}$$

or when all chemical potentials are equal (*diffusive* or material equilibrium)

$$\mu_1 = \mu_2 = \mu_3 = \cdots = \mu_n \tag{8.21}$$

where μ represents the chemical potential for a substance.

The macroscopic compartment in which a reaction takes place is called the system. Everything else belongs to the surroundings. If there is no exchange of mass, it is called a *closed system*. Conversely, if matter is exchanged, it is considered to be an *open system*. If there is no exchange of matter and energy with the surroundings, then it is called an *isolated system*. A *reversible process* is one that is in equilibrium and causes no change to the system or its surroundings. An *irreversible process* causes the system to permanently deviate from its original state—a characteristic of nonequilibrium systems. Factors that cause irreversibility are friction, mixing, convection, turbulence, and heat transfer.

If there is no exchange of heat, it is called an *adiabatic system*. An *isothermal system* implies constant temperature, and if the pressure remains constant, the system is called *isobaric*. If volume is held constant, it is called an *isochoric process*. A *steady-state process* is one in which the internal energy is kept constant. If the process is reversible and adiabatic, entropy is considered to be *isentropic*—a system at constant entropy.

Parameters that describe a thermodynamic state consist of combinations of intensive and extensive variables: $P, V, T, N, \mathcal{N}, \mu, \mathcal{E}, U, S, H, A,$ and G. The term $\mathcal{E}$, the subdivision potential, will be defined later in the chapter. Therefore, a thermodynamic state may be thought of as the manifestation of a system with a set number of variables held constant within its intrinsic and extrinsic environment.

8.1.3   Laws of Thermodynamics

The Zeroth Law of Thermodynamics.   If two thermodynamic systems are separately in thermal equilibrium with a third, they are also in thermal equilibrium with each other (**Fig. 8.2**). This law helps to compare the systems. The zeroth law is needed for the development of thermodynamics.

The First Law of Thermodynamics.   If it were not for Carnot's untimely death of cholera in 1832 at age 36, he would have certainly written both the first and second laws of thermodynamics. Unpublished notes reveal that he understood the concepts quite well: "Heat is simply motive power, or rather motion, which has changed its form … power is, in quantity, invariable in nature; it is … never either produced or destroyed."

The change in the internal energy of a closed thermodynamic system is equal to the sum of the amount of heat energy supplied to the system or created by the system and the work done on or by the system. It is about the conservation of energy:

$$E_{Total} = KE + V + U \tag{8.22}$$

where E_{Total} is the total energy, KE and V are the macroscopic kinetic and potential energies, respectively, and U is the internal energy. For a closed system, the first law is

$$\Delta U = q + w \tag{8.23}$$

and its differential form is

$$dU = dq + dw \tag{8.24}$$

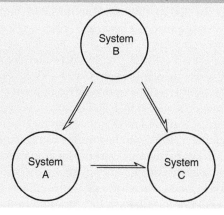

FIG. 8.2   *Illustration of the zeroth law of thermodynamics–similar to the transitive axiom of algebra. Two systems that are in thermal equilibrium with a third system are in thermal equilibrium with each other. The zeroth law is needed for the development of thermodynamics and helps to establish temperature as a state function.*

where q and w are heat and work, respectively. The following equation also presents the first law from another perspective:

$$\Delta E_{\text{Total-System}} + \Delta E_{\text{Total-Surroundings}} = 0 \qquad (8.25)$$

In most cases in nanoscience, K and V equal 0. Because U is a state function and there is no consideration of the path, the internal energy of a system in equilibrium is

$$\Delta U = 0 \qquad (8.26)$$

Specific heat (or specific heat capacity) is the amount of heat (energy) required to raise the temperature of 1 g of a substance by 1°C:

$$C = \frac{Q}{m\Delta T} \qquad (8.27)$$

where Q is the heat content, m is the mass of the substance, and ΔT is the change in temperature. Heat capacity is expressed in joules per kelvin, the same units as for entropy S. If it is measured at constant pressure, heat capacity is called *heat capacity at constant pressure* (C_p); if is measured at constant volume, then it is called the *heat capacity at constant volume* (C_v):

$$C_v = \left(\frac{dq}{dT}\right)_V \qquad (8.28)$$

and

$$C_P = \left(\frac{dq}{dT}\right)_P \qquad (8.29)$$

Due the higher proportion of surface atoms in nanoparticles, both volume and surface atoms make different contributions to the specific heat capacity [7].

The molecular nature of internal energy comes in many forms. Molecules have kinetic energy equal to

$$KE = \tfrac{1}{2}mv^2 \qquad (8.30)$$

At temperature K, $KE = {}^3\!/_2 RT$ where R is the gas constant ($R = 8.314 \text{ J} \cdot \text{mol}^{-1} \cdot \text{K}^{-1}$). At room temperature (ca. 300 K), this equals about 3.8 kJ · mol⁻¹. The level of this energy is important in nanomaterials because it represents the amount of energy required to overcome weak intermolecular attractions and send molecules flying off in different trajectories. If the energy of an intermolecular interaction is lower than this threshold value, then it is likely that the electrostatic attraction that holds two molecules together will be overcome by ambient thermal energy.

Intermolecular, rotational, vibrational, and electronic energy also contribute to the overall internal energy. Statistical mechanics shows that the rotational energy for a gas lies between RT and ${}^3\!/_2 RT$, depending on the configuration of the molecule. Atoms, for example, do not have a rotational or vibrational component. Vibration and electronic energy levels are described by quantum mechanics. Intermolecular forces are electrostatic in nature. Each hydrogen bond contributes a small amount of energy to the overall system. Chapter 10 is devoted to the presentation and discussion of intramolecular and intermolecular bonding. In summary [5],

$$U_{Total} = U_{Trans} + U_{Rot} + U_{Vib} + U_{Elec} + U_{Rest} \qquad (8.31)$$

where U_{Rest} is the *molar rest mass energy* of electrons and nuclei. Without exception, all of these forms of internal energy apply to the combined internal energy of nanomaterials.

The Second Law of Thermodynamics. William Thomson (Lord Kelvin) in the 1840s and Max Planck later are credited with the formulation of the second law. In the mid-1800s, Rudolf Clausius published his fundamental thesis on the second law, called the *Clausius statement*. J. Willard Gibbs stated that modern thermodynamics began with the Clausius statement. Clausius is also credited with coining the term entropy. Gibbs was the first scientist to apply the second law of thermodynamics to experiment and was awarded the Copley Medal of the Royal Society of London in 1901 for his work and accomplishment. Gibbs's award was for "the exhaustive discussion of the relation between chemical, electrical and thermal energy and the capacity for external work."

The essence of the second law is that, for an isolated system, conditions proceed to a state of maximum entropy. It also implies that 100% efficiency is impossible and that the existence of perpetual motion machines is equally impossible. The second law is about entropy and equilibrium. According to P. W. Atkins in his book, *Physical Chemistry*, the second law exploits the parameter of entropy to "identify the *spontaneous changes* among the *permissible changes*" (e.g., permissible with regard to the first law) [6]. In other words, irreversible changes increase the entropy of a system. Reversible reactions, by definition, do not generate changes in entropy. Can you think of any system in the real world that is ideal in this way?

Entropy is defined as a measure of order within a closed system. Entropy is also defined as the thermal energy unavailable to do useful work (please keep this statement in mind for later discussion). The total entropy of any isolated thermodynamic system tends to increase over time,

$$\frac{dS}{dt} \geq 0 \qquad (8.32)$$

where t is time, S is entropy (joules per kelvin) and approaches a maximum value. Equation (8.32) encapsulates the concept of the second law and signifies the irreversibility of most natural processes. Many experiments and simulations of nanomaterials focus on the validity of the second law—in particular, violations of the second law by small systems under short timescales [8].

The Third Law of Thermodynamics. The entropy of a perfect crystal approaches 0 K at absolute zero temperature. The third law quantifies entropy in the following form:

$$\Delta S \rightarrow 0 \text{ as } T \rightarrow 0 \text{ K} \qquad (8.33)$$

The third law was developed by Walther Nernst in the early part of the twentieth century. A modified version of the third law was proposed by M. Randall and G. N. Lewis. They added that not only does the change in entropy approach zero, but also the value of entropy itself becomes zero. According to the Boltzmann relation ($S = k \ln W$, where W is the number of ways energy can be distributed), when $W = 1$ (e.g., only one energy state), then $S = 0$. J. P Abriata and D. E. Laughlin in 2004 proposed that solids in equilibrium at 0 K exist in the form of pure elements or atomically ordered phases [9]. However, apparent violations of the third law have been exhibited by one-dimensional classical systems [10].

Application of the third law was the driving force behind imaging the physical structure of carbon nanotubes [11]. Obviously, colder samples provide for better scanning tunneling microscope (STM) images. At colder and colder temperatures, molecular vibrations diminish. Scanning tunneling microscope imagery of a bundle of single-walled carbon nanotubes was accomplished at 78 K and is shown in **Figure 8.3**.

The Fourth Law of Thermodynamics.   Although there is no clear declaration of the fourth law of thermodynamics, many researchers have tried to implement

FIG. 8.3

Three carbon nanotubes imaged by STM are shown in the figure. The image was acquired under the following conditions: constant current STM taken in UHV at 78 K, with a sample bias –1.5 V, tunneling current of 0.2 nA, and a scan size of 17-nm image of three carbon nanotubes on Au (111) surface. Recall that we discussed surface reconstruction in chapter 6. A herringbone pattern reconstruction on the Au (111) surface is also visible. The Au (111) reconstruction consists of partial dislocation ridges due to a uniaxial contraction (4.2%) along the [110] direction. There are 23 atoms for every 22 sites. The reconstruction results from atoms in bridge sites at elevated positions and other regions where atoms occur in the hollows of fcc ABCABC … and hcp ABABAB … crystal structures. The point of the image in this text, however, is to show how the third law of thermodynamics is demonstrated during the acquisition of this image. Quite impressive!

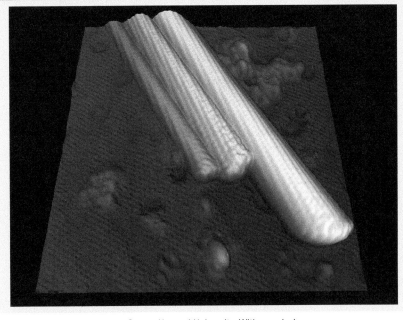

Source: Image is courtesy of Ping Xie, Lieber Group, Harvard University. With permission.

one. Keep yourself posted on the developments. Efforts to generate a fourth law of thermodynamics (and even fifth and sixth laws) have been ongoing since the late nineteenth century. Ludwig Boltzmann proposed that "the fundamental life-struggle in the evolution of the organic world is available energy" [12]. He was aware of a special thermodynamics that applied to living things.

The Onsager reciprocal relations, an attempt to apply the laws of thermodynamics to nonequilibrium systems, were developed by Nobel laureate Lars Onsager of Norway in 1931. His formulation was compelling to the extent that many scientists refer to it as the fourth law of thermodynamics [13]. Any detailed discussion of Onsager relations is way beyond the scope of this text. They will, however, be displayed later on in the chapter. The *maximum power principle* of A. Lotka in 1922 implied that various versions a fourth law exist [13,14]. These fourth laws were all based on the notion of reconciling thermodynamics with life processes.

8.1.4 Fundamental Equations of Thermodynamics

Pressure (P), volume (V), temperature (T), the number of moles or amount of material content (n or N), energy (E), and work accomplished (W) are familiar parameters. Terms like specific heat (C), chemical potential (μ), enthalpy (H), entropy (S), and free-energy (G) are also fundamental to thermodynamics.

Internal energy is the heat content of the system under constant volume and temperature. *Enthalpy* (H) is the heat content of the system under constant pressure and temperature. *Entropy* (S) contributes to the spontaneous change in the system and is the measure of disorder at the molecular level. The tendency towards greater disorder is a driving force in reactions. *Free energy* is defined as the amount of useful work that can be obtained from a system. There are two major types of free energy: the *Gibbs free energy*, G, and the *Helmholtz free energy*, A. The Gibbs form is valid within the frame of constant pressure and temperature; the Helmholtz form is valid at constant volume and temperature. Basic thermodynamic equations are listed in **Table 8.2** for reference without further ado and without further elaboration.

TABLE 8.2   *Thermodynamic Parameters*

Parameter—constant	Common form		Comments	
Ideal gas law	$PV = nRT$	(8.34)	$R = 8.314\ \text{J}\cdot\text{mol}^{-1}\cdot\text{K}^{-1}$	(8.35)
Work	$w = \int F_x\,dx + \int F_y\,dy + \int F_z\,dz$	(8.36)	$w = -\int P\,dV$	(8.37)
First law internal energy	$\Delta U = q + w$	(8.38)	$dU = T\,dS - P\,dV$	(8.39)
	$\Delta U = q - \int P\,dV$	(8.40)		
Enthalpy	$\Delta H = \Delta U + P\Delta V$	(8.41)	$dH = T\,dS + V\,dP$	(8.42)
Entropy	$\Delta S = \dfrac{1}{T}\int dq$	(8.43)		
Heat capacity	$C_P = \dfrac{dq_P}{dT} = \left(\dfrac{\partial H}{\partial T}\right)_P$	(8.44)	Constant P	

(continued)

TABLE 8.2 (CONTD.)	*Thermodynamic Parameters*			

Parameter—constant	Common form		Comments	
Heat capacity	$C_V = \dfrac{dq_V}{dT} = {}_P\left(\dfrac{\partial U}{\partial T}\right)_V$	(8.45)	Constant V	
Joule–Thomson effect	$\mu_{JT} = \left(\dfrac{\partial T}{\partial P}\right)_H$	(8.46)	Constant H	
Second law	$\Delta S \geq 0$	(8.47)	No such thing as 100% efficiency	
Gibbs free energy	$\Delta G = \Delta H - T\Delta S$	(8.48)	$dG = -SdT + VdP$ (8.49) Constant P, T	
Helmholtz free energy	$\Delta A = \Delta U - T\Delta S$	(8.50)	$dA = -SdT - PdV$ (8.51) Constant V, T	
Chemical potential	$\mu_i = \left(\dfrac{\partial G}{\partial n_i}\right)_{T,P,n_j}$	(8.52)	Constant T, P and n_j; intensive property Gives free energy after adding or removing material from an open system	
Third law	$\mathrm{Lim}_{T \to 0}\, \Delta S = 0$	(8.53)	by Walther Nernst	
Standard equilibrium constant	$\Delta G^\circ = -RT \ln K_P^\circ$	(8.54)	For chemical reactions	
Gibbs equation	$dG = -SdT + VdP + \sum_i \mu_i dn_i$	(8.55)	The fundamental equation of thermodynamics	
Clapeyron equation	$\dfrac{dP}{dT} = \dfrac{\Delta H}{T\Delta V}$	(8.56)	Phase changes for a one component system	
Clausius–Clapeyron equation	$\dfrac{d\ln P}{dT} = \dfrac{\Delta H}{RT^2}$	(8.57)	Phase equilibria	
Gibbs–Helmholtz equation	$\dfrac{d}{dT}\left(\dfrac{\Delta G^\circ}{T}\right) = \dfrac{-\Delta H^\circ}{T^2}$	(8.58)	Used to calculate the changes in Gibbs energy as a function of temperature	
van't Hoff equation	$\dfrac{d\ln K_P^\circ}{dT} = \dfrac{\Delta H^\circ}{RT^2}$	(8.59)	Good way to find ΔG°, ΔH°, and ΔS° if K_p, the equilibrium constant, is known from slope of $\ln K$ versus $1/T$	
Kelvin equation	$\ln \dfrac{P}{P_0} = \dfrac{2\gamma V_m}{rRT}$	(8.60)	Vapor pressure of liquid over curved surfaces	
Raoult's law	$P_i = x_i P_i^\circ$	(8.61)	Molecular view of vapor pressure depression— for liquids (usually volatile)	
Henry's law	$P_i = K_i x_i$	(8.62)	Partial vapor pressure of a solute in an ideally dilute solution	
Ideal solution chemical potential	$\mu_i = \mu_i^\circ + RT \ln x_i$	(8.63)	$x_i = \exp\left[\left(\mu_i^{id} - \mu_i^\circ\right)/RT\right]$ (8.64)	
Gibbs–Duhem equation	$\sum_i \mu_i \, dn_i = 0$	(8.65)	The increase of μ of one substance causes the decrease of μ for another Constant T, P	
Excess Gibbs functions	$G^{XS} = G - G^{id}$	(8.66)	Deviation from ideality	
Surface tension	$\gamma = \left(\dfrac{\partial G}{\partial A}\right)_{P,T}$	(8.67)	Units of $J \cdot m^2$	

8.1.5 Equilibrium Constant and Reaction Kinetics

The link between thermodynamics and kinetics for a generic reaction given here between species A and species B to produce species C,

$$A + B \underset{k_{-1}}{\overset{k_1}{\rightleftharpoons}} C, \tag{8.68}$$

is given by the following relation involving the equilibrium constant and the rate constants (k):

$$K = \frac{\Pi[\text{Products}]}{\Pi[\text{Reactants}]} = \frac{[C]}{[A][B]} = \frac{k_1}{k_{-1}} \tag{8.69}$$

The link between the equilibrium constant to Gibbs free energy under standard conditions is

$$\Delta G^{\circ} = -RT \ln K_P^{\circ} \tag{8.70}$$

The activation energy, E_a, of a generic chemical reaction can be found by the Arrhenius relationship:

$$\ln k = \ln A - \frac{E_a}{RT} \tag{8.71}$$

where k is a rate parameter of the reaction and A is the Arrhenius pre-exponential factor. A is approximated by classical collision theory:

$$A = \mathcal{N}_A \sigma \sqrt{\frac{8kT}{\pi\mu}} \tag{8.72}$$

where $\mathcal{N}_A$ is the Avogadro's number, σ is the cross-section, and μ is the reduced mass. Activation energy E_a is found from the slope of ($\ln k$) versus ($1/T$) equal to ($-E_a/R$).

Nanoparticle catalysts are used in many applications where the Arrhenius relationship is applied. Catalytic systems do not represent ideal classical thermodynamic systems due to the presence of steady-state and nonequilibrium factors. The catalytic synthesis of single-walled carbon nanotubes from nanosized catalysts is not adequately explained by invoking classical thermodynamic principles alone. We will discuss this in more detail shortly.

Reaction kinetics implies some degree of nonequilibrium character (e.g., steady-state reactions and irreversibility depending on the type of reaction). The steady-state approximation is derived from kinetic theory. One of the characteristics of steady-state processes is that the concentrations of intermediates remain constant throughout the reaction:

$$\frac{dC_i}{dt} = 0 \tag{8.73}$$

where C_i is the concentration of the intermediary component i.

Catalysts, a broad class of nanomaterials, are able to influence the rate of a chemical reaction. Ideally, catalysts are not consumed in the reaction and are able to cycle indefinitely. Catalytic reactions are often discussed from a kinetic point of view. Following activation and induction, catalytic processes endure a steady-state phase until the end of the run, when the feed material is stopped, the catalyst is deactivated, or the temperature is turned down. We are quite familiar with the catalytic function of enzymes in living things—at body temperature nonetheless! Once again, life is able to impart the kinetic, steady-state aspect to the nano perspective. What systems exhibited by living things fall in the domain of nonequilibrium processes, at least as defined in this section?

8.2   STATISTICAL MECHANICS

We are familiar by now with macroscopic properties such as heat capacity, internal energy, enthalpy, entropy, conductivity (heat, electrical), viscosity, surface energy, dielectric behavior, magnetism, and many more. On the other hand, molecular properties involve intramolecular and intermolecular forces and phenomena such as vibrations, rotations, geometry, and structure [5]. Statistical mechanics couples knowledge of atomic/molecular energy levels with the macroscopic expression of properties of bulk materials and serves as the link between quantum mechanics and classical thermodynamics. It is the bridge between classical bulk thermodynamics and molecular properties [6].

James Clerk Maxwell and Ludwig Boltzmann established the basis for statistical mechanics with their development of statistical distributions and the kinetic theory of gases in the mid- to late 1800s. The Maxwell–Boltzmann probability distribution function forms the foundation of the kinetic theory of gases. They proposed that the occupation (distribution) of energy states is proportional to $\exp(-\Delta E/k_B T)$. In 1889, Gibbs, applying classical thermodynamic principles, began to work in earnest on the subject of statistical mechanics. He had published much of that work in books in 1902, that provided a foundation for the later development of quantum theories. Albert Einstein also made contributions to statistical mechanics at the same time [5].

There is a field called *equilibrium statistical mechanics* and, not surprisingly, one called *nonequilibrium statistical mechanics*. Nonequilibrium statistical mechanics involves mathematical treatment of phenomena such as transport and chemical reaction rate. Due to its complexity, this field of nonequilibrium statistical mechanics is not as well developed [5]. We shall adhere to the former in this section. Our objective is to simply get a feel of the discipline without too much detail.

8.2.1   *Microstates and Macrostates*

The term "microstate" refers to the quantum state of a system. Ludwig Boltzmann, perhaps one of the most brilliant physicists ever, derived his famous equation, the *Boltzmann principle*, in the latter part of the nineteenth century. It links entropy (S) with probability and serves as the basis for statistical mechanics:

$$S = k \log \mathcal{W} \tag{8.74}$$

where $\mathcal{W}$, weight, represents the number of microstates (system energy levels) that have a significant probability of being occupied [5]. Another popular form of the equation uses the natural log rather than $\log_{10}$ ($S = k \ln \mathcal{W}$). $\mathcal{W}_i$ is often used to denote the degeneracy (distinct states with the same energy) of energy level E_i. Regardless which form is used, the equation states that the systems energy is proportional to the probability of population of energy states. As $\mathcal{W}$ increases, the system's entropy also increases. An inscription of the formula is carved into Boltzmann's gravestone in Vienna (**Fig. 8.4**).

The weight, $\mathcal{W}$, of a configuration comes from probability theory and is represented by

$$\mathcal{W} = \frac{N!}{n_0! \, n_1! \, n_2! \, \ldots} \tag{8.75}$$

where N is the total number of molecules and n represent individual microstates (specific sets of populations of molecules). Each molecule of population n_x is represented by an energy ε_x. Note that $n_0! = 1$ [6]. The ability to predict microscopic properties is the basic advantage of statistical mechanics over classical thermodynamics. Both theories are linked through the second law of thermodynamics by way of the concept of entropy. However, entropy in classical thermodynamics can only be known empirically, whereas, in statistical mechanics, it is a function of the distribution of the system on its microstates. The essence of statistical mechanics is known as the "equal a priori probability postulate": "Given an isolated system in equilibrium, it is found with equal probability in each of its accessible microstates."

This postulate is a fundamental assumption in statistical mechanics and indicates that any system in equilibrium will have an equal probability for any of its available microstates. In the thermodynamic limit of $N \rightarrow \infty$, the dominating configuration is overwhelmingly the most probable [6].

8.2.2 *Canonical Ensembles*

An *ensemble* is an infinite number of noninteracting (e.g., chemically noninteracting) systems of differing microstates that comprise the same macrostate [5]. According to some, an ensemble is composed of interacting (e.g., energetically) molecules that all have the same temperature [6]. A *canonical ensemble* (from the Latin *canon*, indicating "measuring line, rule, collection") is associated with constants N, V, and T (e.g., all systems have the same temperature); a *microcanonical ensemble* with N, V, and E (or U) constants (e.g., all systems have the same energy); and the *grand canonical ensemble* with constants μ, V, and T (e.g., it is an open system). We only address equilibrium statistical mechanics in this section.

For the canonical ensemble, one of the underlying principles of such a system is that, at constant T, V, and N, all quantum states with the same energy have an equal probability of occurring [5]. The canonical (molecular) partition function is

$$Z = \sum_j e^{-\beta E_j} = \sum_j e^{-E_j / kT} \tag{8.76}$$

FIG. 8.4   *Ludwig Boltzmann's grave in Vienna. His famous equation, S = k log W, is carved into the stone. Sadly, there was much controversy over his theories and perhaps that contributed to his suicide. Truly, Boltzmann was one of the most brilliant minds of all time—someone who understood the major driving force of the universe: entropy.*

Source: en.wikipedia.org/wiki/Image:Zentralfriedhof_Vienna_-_Boltzmann.JPG.

where Z is the sum over all states, β is equal to $1/k_BT$ (the thermodynamic beta), and E_j is the "quantum-mechanical energy" level of the macroscopic system [5]; $e^{-E_j/kT}$ of state j is proportional to the probability p_j of the state j:

$$p_j(E_j) = \frac{e^{-E_j/kT}}{Z} \tag{8.77}$$

This is the probability that a system of fixed volume, temperature, and composition is in quantum state j of energy E_j [5]. And for the system that has energy of E_i,

$$p_i(E_i) = W_i \frac{e^{-E_i/kT}}{Z} \tag{8.78}$$

where W_i is the number of degenerate quantum states with energy E_i. Statistical mechanics is absolutely grounded in probability theory.

The following statistical mechanical relations are used to express pressure, internal energy, entropy, the Helmholtz function, and chemical potential. Notice how the mathematical expressions are tied into classical thermodynamic forms but with the addition of a statistical component:

$$P = kT\left(\frac{\partial \ln Z}{\partial V}\right)_{T,N} \tag{8.79}$$

$$U = kT^2\left(\frac{\partial \ln Z}{\partial T}\right)_{V,N} \tag{8.80}$$

$$S = \frac{U}{T} + k \ln Z \tag{8.81}$$

$$A = -kT \ln Z \tag{8.82}$$

$$\mu = -RT\left(\frac{\partial \ln Z}{\partial N}\right)_{T,V,N} \tag{8.83}$$

According to I. N. Levine's *Physical Chemistry* [5], the steps needed to calculate macroscopic thermodynamic parameters are as follows: (1) solve the Schrödinger' equation to obtain quantum mechanical energies (E_j) for the system, (2), evaluate the canonical partition function Z, and (3) use $\ln Z$ to calculate the system's macroscopic thermodynamic properties [5]. Z is approximately equal to

$$Z = We^{-U/kT} \tag{8.84}$$

and

$$\ln Z = \ln W - \frac{U}{kT} \tag{8.85}$$

Thus, the Boltzmann principle (equation 8.74) can be derived by combining equations (8.81) and (8.85) [5]:

$$S = \frac{U}{T} + k \ln Z = \frac{U}{T} + k \ln W + k\left(-\frac{U}{kT}\right) = k \ln W \tag{8.86}$$

An interesting paradox to ponder is that entropy only has significance for a large number of atoms and molecules, yet it is derived from molecular properties [5]. How is this apparent paradox reconciled? We discover later that it is possible to estimate the entropy for a single molecule.

8.2.3 *Energy (Molecular) Partition Functions*

A small case z is used to represent molecular partition functions and the small case Greek epsilon is used to represent molecular energies. The molecular energy is the sum of all the forms of energy of a molecule presented in a manner similar to equation (8.31):

$$\varepsilon_{\text{Molecular Energy}} = \varepsilon_{\text{Trans}} + \varepsilon_{\text{Rot}} + \varepsilon_{\text{Vib}} + \varepsilon_{\text{Elec}} \tag{8.87}$$

and z, the molecular partition function, is the product of all the probabilities of the individual molecular partition functions:

$$z = (z_{\text{Trans}})(z_{\text{Rot}})(z_{\text{Vib}})(z_{\text{Elec}}) \tag{8.88}$$

or, given in the usable form,

$$\ln z = \ln z_{\text{Trans}} + \ln z_{\text{Rot}} + \ln z_{\text{Vib}} + \ln z_{\text{Elec}} \tag{8.89}$$

Relating Z to the molecular partition functions yields

$$\ln Z = N \ln z_{\text{Trans}} + N \ln z_{\text{Rot}} + N \ln z_{\text{Vib}} + N \ln z_{\text{Elec}} - N(\ln N - 1) \tag{8.90}$$

For example, internal energy, from equation (8.80) is then represented by

$$U = kT^2 \left(\frac{\partial \ln Z}{\partial T} \right)_{V,N} = NkT^2 \left[\left(\frac{\partial \ln z_{\text{Trans}}}{dT} \right)_V + \left(\frac{d \ln z_{\text{Rot}}}{dT} \right) + \left(\frac{d \ln z_{\text{Vib}}}{dT} \right) + \left(\frac{d \ln z_{\text{Elec}}}{dT} \right) \right] \tag{8.91}$$

There are many excellent sources available if you wish to pursue this fascinating field on your own. The purpose of the section on statistical mechanics is simply to show how thermodynamic properties can be derived from the bottom up using the tools of statistical mechanics.

8.3 OTHER KINDS OF THERMODYNAMICS

Near-equilibrium, nonlinear, nonequilibrium, steady-state, and pseudoequilibrium thermodynamics are terms you will encounter often—especially if you choose to study nanomaterials. There are also *quasi-equilibria*. Most nonequilibrium thermodynamic experiments are conducted at near-equilibrium conditions (conditions not too far off equilibrium). All types of thermodynamics are part of the *thermodynamic equilibrium continuum*. The discriminator among them is the level of complexity involved in describing their respective systems. Pseudoequilibrium (*pseudo-*, from the Greek meaning "false, feigned or erroneous") thermodynamics is a bit more difficult to define. The term implies that there is some kind of equilibrium apparent in the system but one that does not quite conform to the criteria required for a true equilibrium. Therefore, it is a false equilibrium! It may refer to a system that is kinetically confined but still demonstrates some degree of reversibility—perhaps called pseudoreversibility.

If one were to step back, contemplate, and pick one word to describe conventional thermodynamics, that word would probably be *static*. Mostly, classical thermodynamics is about macroscopic evaluation of a system in equilibrium by the use of extensive and intensive parameters. According to M. Agrawal of Stanford University in his 2005 article "Basics of Irreversible Thermodynamics" [68]:

> The dynamic part is a stretch indeed when the most dynamic happening is whether or not there is spontaneity between the initial and final state or if a change is so slow that the system is evolving through a series of quasi-equilibrium states…. The dynamics is certainly evident in non-equilibrium situations where demonstrable evolution is taking place.

Nonequilibrium thermodynamics implies irreversibility, fluctuations, dissipation, open systems, and "time independent (steady-state) thermodynamic systems" [14]. Nonequilibrium at the extreme implies convection, flow, and perhaps even turbulence. Steady-state thermodynamics is a configuration in which there is no accumulation of heat or matter within the system and extensive properties within the system are time independent. Steady-state thermodynamics and nonlinear thermodynamics are considered to be a subset of nonequilibrium thermodynamics. According to PBS.org, nonlinear, nonequilibrium thermodynamics is

> A branch of physics developed in the latter half of the twentieth century that deals with systems of particles far from the near equilibrium, conditions studied in classical thermodynamics and which are governed by complex, non-linear forces. Significant attempts have been made to extend this theory into the realm of living (self-replicating) organisms.

In the early 1990s, the development of micromanipulation methods allowed researchers to study the energy fluctuations of small systems [15]. In macroscopic thermodynamics, behavior is predictable and therefore fluctuations (e.g., deviations from ideal behavior) are averaged over the whole system and therefore rendered insignificant. In a nonequilibrium small system, fluctuations can dominate as materials acquire smaller dimensions [15].

8.3.1 The Onsager Relations

Onsager equations are in actuality fourth law interpretations of thermodynamic potentials, forces, and flows. These relations define systems out of equilibrium but suggest that there exist local equilibria. For example, evaluation of thermal conductivity by equilibrium methods is problematic and is based on $\Delta T \to 0$, although a real steady-state temperature gradient is in place. The best way to analyze this process is to quantify the rate of change of the entropy [16]. One fundamental expression in thermodynamics relates entropy to other extensive properties like internal energy and volume and shows how the change in entropy is a function of intensive properties such as temperature, pressure, and chemical potential:

$$dS = \frac{1}{T}dU + \frac{P}{T}dV - \sum_{i=1}^{n}\frac{u_i}{T}dn_i \tag{8.92}$$

The corresponding thermodynamic forces are the gradients $1/T$, P/T, and μ/T [16].

$$\frac{\partial S}{\partial t} = \sum_i \mathbf{J}_i \cdot \mathbf{X}_i \tag{8.93}$$

where $\mathbf{J}$ is a heat flux and $\mathbf{X}$ is a force. The force in this example is supplied by a temperature gradient, ∇T. The entropy is defined as a function of all extensive quantities where $\mathbf{I}_i$ represents a conjugate intensive variable involving the entropy (Ext_i) [17]:

$$\mathbf{I}_i = \left(\frac{\partial S}{\partial \mathrm{Ext}_i} \right) \tag{8.94}$$

$\mathbf{J}$ is defined by a linear matrix of coefficients that is expressed by an Onsager equation:

$$\mathbf{J}_i = \sum_j \mathbf{L}_{ij} \cdot \nabla \mathbf{I}_i \tag{8.95}$$

This expression is valid in the regime of small forces and slow variation. Quantification of entropy is a key factor in nanothermodynamics.

Onsager's PhD dissertation, titled *Onsager Reciprocal Relations*, was rejected by the faculty at the Norwegian Institute of Technology due to "insufficiency." His second dissertation, a solution of the *Mathieu equation*, was labeled as "incomprehensible." He was fired from his position at Johns Hopkins in 1928 for "incomprehensible teaching" and for the same reason at Brown University in 1933. When he was hired at Yale, a controversy erupted because he did not have a PhD He finally obtained a PhD in 1935 and then went on to win the Nobel Prize in chemistry in 1968.

8.3.2 *Nonequilibrium Thermodynamics*

We are all familiar with irreversible reactions. An example of an irreversible reaction of a material that is kinetically stable is the explosion of nitroglycerin. Nitroglycerin is thermodynamically unstable but kinetically stable unless you start shaking the liquid or drop it. The reaction is, for all practical purposes, irreversible. Fullerenes are thermodynamically less stable than graphite but are kinetically stable. It all depends on the environment. The thermodynamic behavior of nanomaterials is very much dependent on their immediate environment. Does this imply that nanomaterials possess inherently nonequilibrium characteristics?

A subtle example of kinetic stability is provided by the diamond–graphite equilibrium. Diamond is only stable at very high pressure. Graphite is stable at atmospheric pressure. Although diamond exists at STP conditions, the transformation of diamond into graphite is extremely slow (a kinetic process). A way to accelerate the process is to provide heat and more pressure.

The bond strain in C_{60} is about the highest known in carbon materials, yet the material does not spontaneously decompose at room temperature. Kinetic stability is related to the threshold energy required to start a reaction. The entropy produced in an irreversible reaction is overall positive:

$$\Delta S_{\text{Irreversible}} > 0 \qquad (8.96)$$

Kinetic stability is also demonstrated by colloids that are thermodynamically metastable or unstable with respect to their bulk counterparts because they do tend to exist for quite some time. Why is that? Obviously some form of environmental support must be in place to ensure that colloids persist in their nano- to microscale form.

An automobile engine is an example of a steady-state system. Material is added at a constant rate (unless you have a lead foot) and material is released in the form of gases—CO_2, H_2O, and others, also at a relatively constant rate. Heat is dissipated at a constant rate and the temperature of the system is held constant by transfer of heat to water (unless your car has bad hoses). Once the power cycle is terminated, the system goes to equilibrium with its surroundings and assumes the same temperature. In living systems, this would mean *cellular death*. Combustion itself is a highly nonequilibrium thermodynamic condition that is regulated by steady-state systems such as the automobile engine. The amount of energy released by the gasoline and the equilibration of the engine temperature with the environment are thermodynamic processes. The rate at which gasoline is burned and the scrubbing of exhaust gases by catalysts are kinetic parameters. The operation of the engine is an irreversible nonequilibrium steady-state process.

Use of the steady-state approximation (equation 8.73) simplifies rate equation calculations, as in the case of intermediates in catalytic processes. Rates in catalytic reactions are expressed as ($mol \cdot min^{-1} \cdot g^{-1}$) of catalyst or ($mol \cdot min^{-1} \cdot M^{-2}$) per surface area of catalyst. Steps in the catalytic reaction process, in order of occurrence, include diffusion of reactant to catalyst surface, adsorption of reactant on the catalyst surface, the targeted chemical reaction, desorption of products from the catalyst surface, and diffusion of the products away from the catalyst. Particle size and temperature affect the progress of the catalytic reaction. Although the change in energy between the starting material and the product is considered to be a state function and, as such, dwells in the domain of classical thermodynamics, it is obvious that catalytic processes are characterized by far more complexity. In other words, the path assumes a high level of importance in catalytic processes, especially if commercial factors are involved. Many steady-state processes exist in cellular processes.

Consider the nonequilibrium convection pattern mechanism of a Bénard cell, an inorganic system comprised of water and copper plates (**Fig. 8.5**). A thin layer of water is trapped between two petri-dish-sized Cu plates. As heat is added, molecular equilibrium is first established until a critical level is reached, at which point convection of water begins. A nonequilibrium self-organized structure, known as a *dissipative* structure, is formed. Upon elimination of the heat source, the self-organized patterns disappear and the equilibrium condition is restored. The capability to perform this kind of self-organized work is driven by the steady-state input of heat energy. Self-organization is an interesting consequence of nonequilibrium states. Biological entities are considered to be self-organizing, self-replicating, dissipative structures [18–20].

Fig. 8.5 A *Bénard cell is a flat, circular dish filled with a thin layer of water that is uniformly heated from below. If the water layer is of appropriate thickness, heat is dissipated rapidly through the self-assembly of hexagonal patterned convection cells (energy gradients) rather than turbulent boiling. The Bénard cell is a dissipative structure. Does the Bénard cell provide an adequate model for living systems that are also dissipative in nature? One criterion to consider is that of the enduring structure. Living systems are characterized by enduring structures that are capable of accomplishing work. Is a Bénard cell an enduring structure capable of accomplishing work beyond that of maintaining its cellular pattern? Living systems, for example, are able to create less entropy from ordered structures. "The 2nd Law cannot be violated, but it can be stalled"* (interesting discourse from J. Fournier, Evolution, Entropy and Work, *www.geoman.com/jim/entropy.html*).

Source: Courtesy of Professor Anil K. Rao, Department of Biology, the Metropolitan State College of Denver.

8.3.3   The Concept of Pseudoequilibrium

According to one definition, *pseudoequilibrium* is a state of a system in which the distribution of a component of interest exists between the solid and liquid phase (perhaps solid and gaseous phase) and does not change upon further equilibration [21]. The time of equilibration required to reach this state of pseudoequilibrium depends on experimental conditions. Pseudoequilibria phenomena are often encountered in sorption and catalytic studies that are eventually uncovered in subsequent kinetic analysis. In a similar vein, we can say that pseudoequilibrium implies a metastable state of a solid in which the content of free energy is higher than that characteristic of the equilibrium state. A pseudoequilibrium rate constant can be considered to be associated with a process that is constantly unbalancing the equilibrium.

Pseudoequilibrium is observed in the conditions of carbon plasmas for diamond deposition [22]. Pseudoequilibrium components are detected in the conversion of methanol to CO_2 and H_2 in the pores of a catalyst after analysis of the equilibrium constant in which significant deviations were found to occur. Kinetic confinement of the catalytic reaction and thermal gradients in the pores contributed to the pseudoequilibrium state [23]. The conversion of methane to single-walled carbon nanotubes under CVD conditions by Fe–Mo catalysts is reported to closely approximate pseudoequilibrium conditions [24]. The

Michaelis–Menten constant K_M in enzyme kinetics is considered to be a dynamic or pseudoequilibrium constant.

Protein adsorption phenomena are often associated with a pseudoequilibrium state. The protein system is considered to be a self-contained small thermodynamic system complete with associated water molecules. Such small systems are susceptible to thermal fluctuations in the immediate environment. Fluctuations in internal energy, volume, and conformational changes of the protein are therefore likely to occur when proteins are exposed to thermal fluctuations. Introduction of a solid substrate like silicon (e.g., a new environment) causes transient conformations to become stable at the solid–liquid interface during adsorption [25]. The protein is able to adsorb in many different configurations. The adsorption can be rapid and reversible but dependent on prevailing environmental and kinetic conditions.

One can safely conclude that true equilibrium most likely never exists in vivo in plasmas, aquatic environments, and other systems of great complexity. Ideal thermodynamics, in the natural world, is not the rule for dynamic systems but more like an exception. The closest a real system comes to the thermodynamic ideal is in the form of a dilute ideal gas or when two blocks of purely homogeneous materials are in thermal equilibrium.

8.3.4   *Cellular and Subcellular Systems*

We have already discussed terms such as irreversibility, nonequilibrium and kinetic stability, and pseudoequilibria and have yet to be introduced to concepts such as fluctuations, dissipation, nonextensivity, and nonintensivity. We have also indicated on several occasions how nanotechnology—nanomaterials in particular—is fundamental to life and its development. Living processes and structures cannot be explained by classical thermodynamics even though the specter of classical thermodynamics permeates all of biology. The fourth law of thermodynamics may be coming soon, but until then, let us take a brief look into nonequilibrium thermodynamics and apply some of its concepts to biological phenomena.

Stephanie E. Pierce of the Department of Biological Sciences at the University of Alberta stated in 2002 that the organization of information systems in living things follows the second law in that speciation occurs when these systems become complex (e.g., addition by mutation) and disorganized and are subjected to environmental forces [18]. She states that the "entropic drive to randomness underlies the phenomena of variation and speciation" [18]. How does this tie in with nonequilibrium thermodynamics? Stephanie Pierce goes on to say

They take in a[nd] give off energy from the environment in order to sustain life processes and in doing so function in a state of non-equilibrium. Although biological organisms maintain a state far from equilibrium, they are still controlled by the second law of thermodynamics. Like all physiochemical systems, biological systems are always increasing their entropy or complexity due to the overwhelming drive towards equilibrium. But unlike physiochemical systems, biological systems possess "information" that permits them to self-replicate and continuously amplify their complexity and organization through time.

Fritjof Capra in his book, *The Web of Life*, states [26]:

We emphasize that life is at its very center. This is an important point for science, because in the old paradigm, physics has been the model and source of metaphors for all other sciences. "All philosophy is like a tree," wrote Descartes. "The roots are metaphysics, the trunk is physics, and the branches are all the other sciences." Deep ecology has overcome this Cartesian metaphor...physics has now lost its role as the science providing the most fundamental description of reality. However, this is still not generally recognized today...Today the paradigm shift in science, at its deepest level, implies a shift from physics to the life sciences.

Kinesin. The biomolecular machine called kinesin serves as an excellent example of a nonequilibrium steady-state thermodynamic system. Most cellular processes function by random diffusion. Others require directed action. Molecular motors are able to direct cellular functions to produce desired outcomes. There are three types of molecular motors: myosins (muscle), dynesins (inward draggers), and kinesins (outward draggers). Members of the family of nonequilibrium steady-state systems include the steam and automobile engines, the motion of *E. coli* swimming in water, the smallest artificial motor, a 1-μm bead dragged in water, and the action of kinesin [15]. Because these systems require an input of energy and then dissipate energy continuously, they are considered to be in a state of nonequilibrium [15]. The function and operation of the molecular motor kinesin offers clues to unraveling some secrets of nonequilibrium thermodynamics at the nanoscale.

Kinesin is a microtubule-based ATPase motor that performs tasks like organelle transport (**Fig. 8.6**). Kinesin is involved in replication, transcription, translation, and repair of DNA/RNA and operates along microtubule tracks (protein filaments). It obtains energy from hydrolysis of ATP (adenosine triphosphate) at each step and is able to exploit energy from thermal fluctuations. Kinesin takes one 8-nm step every 10–15 ms while dragging its cellular load outward away from the nucleus [15].

The distinction between macroscopic nonequilibrium steady-state systems and nanoscale nonequilibrium steady-state systems is profound. Molecular machines like kinesin, unlike their big counterparts, are able to utilize thermal fluctuations to power motion. By a process known as rectification, kinesin is able to channel captured thermal fluctuation energy into motion only in the forward direction.

FIG. 8.6	*The action of kinesin is depicted in the figure. Kinesin is shown hauling a subcellular object away from the center of the cell. Its "feet" are able to extract energy from ATP as well as from thermal fluctuations in the surrounding cytoplasm.*

Source: Courtesy of Professor Anil K. Rao, Department of Biology, the Metropolitan State College of Denver.

EXAMPLE 8.1   *Efficiency and Heat Dissipation of Kinesin Action*

What is the efficiency of a kinesin motor and how much heat does kinesin dissipate in one second of work? Assume 20 $k_B T$ of energy in kilojoules per·mole is released during the hydrolysis of ATP and that the motor does about 12 $k_B T$ of work with each step [15]. The Boltzmann quantity is equal to 1.38×10^{-23} J·K^{-1} and body temperature is equal to 310.15 K.

Solution:
Efficiency ε is defined as

$$\varepsilon = \frac{W_{Max}}{\text{Energy Input as Heat}} = \frac{w}{q_h} = 1 - \frac{q_c}{q_h} = 1 - \frac{T_c}{T_h} \tag{8.97}$$

$$\varepsilon = \frac{12 k_B T}{20 k_B T} = 0.60 \, \text{Efficiency}$$

If 8 $k_B T$ are released as heat per step and the average time between steps is [(10 + 15) ÷ 2 = 12.5], then the energy dissipated per second is

$$\text{Energy Dissipated} = \left(\frac{1000\,\text{s}}{12.5\,\text{s} \cdot \text{step}^{-1}} \right) \frac{8 k_B T}{\text{step}} = 640\, k_B T \rightarrow 4.28 \times 10^{-21}\, \text{J} \rightarrow 2.57\, \text{kJ} \cdot \text{mol}^{-1} \tag{8.98}$$

The kinesin moves by extracting heat energy from the surroundings and uses the energy obtained from ATP to ensure that only forward fluctuations are utilized. This is called energy rectification of thermal fluctuations towards directed motion [Bustamante]. The kinesin-microtubule system is an example of a nonequilibrium steady-state system that requires a constant energy input to operate and is characterized by the constant dissipation of energy. Life itself is an assembly of nonequilibrium systems.

From the nano perspective, and again reflecting on the origins of life and the role nanotechnology played, this concept of nanothermodynamics is most certainly an intriguing one.

Notice that the calculations performed in this example were all based in classical thermodynamics on principles one would find in any generalized physical chemistry text. There is nothing exclusively nano about this treatment, although the acquisition of thermal fluctuations to power forward movement happens at the nanoscale. That form of energy acquisition is not addressed by classical methods.

Therefore, in addition to obtaining energy from ATP, kinesin is able to cultivate energy from thermal fluctuations—not a process explained adequately by classical thermodynamics. Rectifying behavior is behavior allowing heat (or electricity) to flow only in one direction.

According to C. Bustamante et al., RNA polymerase moves along a DNA strand to transcribe a complementary strand of RNA. During this process, the polymerase moves by extracting energy, like kinesin, from a thermal bath, and then uses the bond hydrolysis energy released by ATP to ensure that only forward fluctuations are captured (rectification). The motion is described as a nonequilibrium time independent steady-state system.

Thermal fluctuation forces are considered to be random, but in small systems, their effects can be significant and therefore "arbitrarily large" [15]. Local thermal fluctuations will cause fluctuations in ΔU, q, and w. For a macroscopic system, the ratio of the probabilities of absorbing heat and releasing heat in an equilibrium system is equal to one (e.g., there is no net thermal fluctuation).

$$\frac{P(+Q)}{P(-Q)} = 1 \qquad (8.99)$$

The average amount of heat $\langle Q \rangle$ is correlated with an average amount of entropy in a steady-state system:

$$\langle S \rangle = \frac{\langle Q \rangle}{T} \qquad (8.100)$$

The entropy production, σ, is synonymous with the rate of heat exchange with a surrounding bath ($\sigma = Q/Tt$). Reformulation of the probability distribution yields

$$\lim_{t \to \infty} \frac{k_B}{t} \ln \frac{P_t(+\sigma)}{P_t(-\sigma)} = \sigma \qquad (8.101)$$

This equation favors the probability of a steady-state system delivering heat ($+\sigma$) to the bath over that of the probability of the system absorbing heat ($-\sigma$). This is not surprising because nonequilibrium systems are dissipative. For large extensive systems, the probability of heat absorption by the system is insignificant [15,27]. For small systems, like molecular motors, the probability of absorption becomes significant—an apparent violation of the second law. Although molecular motors deliver heat to the "bath" (e.g., they too are dissipative), they also move by rectifying thermal fluctuations (e.g., absorption of heat from the bath) [15].

Times are changing and nanotechnology is a great of part of the driving force. The purpose of this section is to provoke thought—to alter your paradigm and perhaps instill a sense of curiosity. Nanomaterials are fundamental to living things. Nanothermodynamics, a topic discussed more fully in the next section, is not equivalent to our classical preconception. However, when evaluating the sum total of the energy transformations, as ΔS is always greater than zero, at the nanolevel, the self-organizational level, this postulate itself may deviate from the accepted norm—that ΔS on a local scale may be less than zero.

When one ponders the entropic aspects of living things, one realizes that order is made from disorder to make functioning living things. Although the overall entropy in the universe is constantly increasing and the energy from the sun is the external source that drives our steady-state engines, local entropy within living things is decreased or held in check. This also makes one wonder about the aging process. Is aging a process wherein entropy incrementally reasserts itself in living things due to the breakdown of regulatory "ordering" mechanisms that are genetically controlled? And why do organisms display such a wide range of lifetimes? What is the relationship of environment, information, and entropy?

8.4   NANOTHERMODYNAMICS

The thermodynamics of nanomaterials, existing somewhere between the bulk and the atomic, is a key component in the energy transfer continuum we refer to often in this text. It is also a key link between bulk thermodynamics and the apparent thermodynamics of living things, another theme of this text. Research over the past few years has revealed that nanomaterials do not follow the strict laws established by classical thermodynamics. As we stated before, classical thermodynamic approaches assume that a system is in a state of equilibrium. This assumption does not always apply to all nanomaterials because not all nanomaterials represent stable phases, but yet many persist. Some metastable phases are stable long enough for engineering applications. One reason for this is that a great amount of energy is spent on synthesizing nanomaterials; this is true especially for top-down fabrication methods.

Nanomaterials in metastable phases fabricated from bottom-up methods are done so under moderate temperature and pressure conditions, as opposed to the corresponding bulk version of the material where high temperature and pressure often exist or if top-down methods are employed. The theory of Wang and Wang suggests that size effects are responsible for the formation of metastable phases and that the capillary effect due to particle curvature is enough to drive metastable phases into new stable phases [28]. In other words, nanomaterials have higher free energy content than their bulk counterparts, as we learned in chapter 6.

Nanomaterials have higher levels of energy than their bulk counterparts, yet they can be stable. Since the goal of classical thermodynamics is to lower energy, how does this happen and why does it persist? Such a condition is possible if there exists a large proportion of disordered structures in crystals. Most of classical thermodynamic calculations do not take these into account. For example, the strain energy associated with a crystal lattice plays a crucial role in the thermodynamics of these materials. In fact many processes of synthesis of nanomaterials are based on introducing more defects into the structure of crystals like equal channel angular pressing (ECAP) and severe plastic deformation (SPD). Arresting a structure formed at a high temperature by suddenly lowering the temperature converts thermal energy into the crystals' defects. This technique is accomplished by CVD, PVD, and bulk metal glass (BMG) procedures.

Trapping chemical energy inside nanomaterials is also accomplished by wet chemical methods. The trapped energy is sequestered in crystal defects like twins (a signature of highly stressed crystals). Thus, the thermodynamic principles are modified from their classical appearance to accommodate factors such as strains and defects in order to explain the properties and phenomena of nanomaterials. Hence, there are several new approaches to compute the thermodynamic aspects of nanomaterials. The most common of them takes the grain size or the particle size and the defect concentrations into account. More sophisticated approaches employ higher orders of atomic interactions.

Nanosystems may not be large enough to assign extensive properties to them [29]. Particles can be considered to be polymorphs of the bulk material without statistical homogeneity, one of the tenets of the Euler theory [29]. In nanosystems, intensive properties can change in an instant, depending on the environment—a state of affairs not applicable to macroscopic equilibrium systems; the term equilibrium implies that a certain amount of built-in inertia is in effect. Therefore, transformation of one equilibrium state into another is an infinitely slow process that is time independent [30]. Transformations of nanosystems, on the other hand, are time dependent. The parameter λ, the Euler exponent, must deviate

TABLE 8.3	*Comparisons between Macro- and Nanothermodynamics*	
Parameter	**Classical thermodynamics**	**Nanothermodynamics**
Number of particles, N	N is infinite	N is countable
Number of independent variables (degrees of freedom)	Fewer degrees of freedom	More degrees of freedom
Surface-to-volume ratio	Miniscule	Significant
Thermodynamic limit	Yes	No
Heisenberg uncertainty	Not important	Important to set bin size for computation
Environment	Not important	Important
Size	Not important	Important
Geometry	Not important	Important
Fluctuations	Small	Large
Extensivity	Homogeneous to degree one ($\lambda = 1$) λ is the Euler exponent	Loses its significance—upon scaling, extensive parameters may change disproportionately
Intensivity	Homogeneous to degree zero ($\lambda = 0$)	Loses its significance due to large fluctuations—may not be retained
Materials of interest	Any bulk material	Droplets, bubbles, clusters, confined spaces, edges, defects, large molecules, kinesins, enzymes
Heat transfer	Heat dissipation	Heat dissipation and rectifying behavior

from unity in nanothermodynamics or even assume negative values. What all this means is that thermodynamic parameters like entropy for small systems depend on a small number of atoms and molecules, their environment, and temporal factors. Note that all of these characteristics deviate from the macroscopic factors that affect entropy (**Table 8.3**).

Another aspect to keep in mind while reading this chapter is that the second law of thermodynamics implies that it is impossible to extract heat from a reservoir and convert it to accomplish work [31]. We have noted in the last section that extraction of "unusable heat" is possible at the nanoscale, at least theoretically. Because of these fundamental distinctions, the classical thermodynamic approach may not be accurately transposable to nanomaterials. A new branch of thermodynamics, called *nonextensive thermodynamics*, offers some innovative theories that explain nanoenergy transfer [32]. We now are challenged to explain the entropic properties of small systems that depend on a countable number of atoms.

Fluctuations. What are fluctuations? Fluctuations are all about the second law of thermodynamics. In general, fluctuations are perturbations in a system—equilibrium or otherwise. Fluctuations are variations in extensive or intensive properties. With an infinitely sized ensemble, fluctuations are ignored. For microscopic systems, fluctuations cannot be ignored. Fluctuations have significant impact near phase transition boundaries where pseudoequilibrium conditions prevail. When a system is far removed from a phase transition, fluctuations are not important and it is possible to define the system with thermodynamics. Also, the larger the system becomes, the impact of phase boundaries (two-dimensional surfaces) diminishes. In small systems of less than a few thousand atoms or molecules, the boundaries become more important but are no longer clearly defined. Small metal clusters, for example, do not possess a clearly defined melting point because at the so-called "phase boundary" it is hard to distinguish whether the cluster is in liquid or solid form. This is the reason the parameter called the melting point is a macroscopic property.

Fluctuations within a large system that has an extremely large number of constituents tend to average out to zero. The energy and material in a subsystem may fluctuate by $\langle \Delta U \rangle$ and $\langle \Delta N \rangle$, respectively, but the statistically derived average values of fluctuations for a large system at the thermodynamic limit average out to zero.

$$\langle \Delta U \rangle = 0 \text{ for internal energy and } \langle \Delta N \rangle = 0 \text{ for material} \qquad (8.101b)$$

Statistically, one is able to grasp the idea that fluctuations in small systems assume greater weight. The best way to probe the internal structure of small systems is to apply entropic considerations. Expressions involving entropy are directly able to scale the probability of fluctuations involved.

Emergent Properties. The concept of *emergent properties*, applied universally to any system—not just ones based in thermodynamics—is quite intriguing. Insect groups like bees and ants, for example, rely on a few key chemical signals to trigger actions that are responsible for nearly all their behavior. An emergent property is based on simple properties that, when taken together and provided with the proper stimuli, result in a new, higher level of complexity. Such systems are characterized by a finite probability of pathways that may disobey the second law of thermodynamics that are capable of extracting order and information from the environment. The system may be open (material flow) and not isolated (energy flow) from the environment. Depending on the ambient conditions, atoms, molecules, and particles are able to self-organize to form phases. This forming of phases and boundaries is classified as an emergent process. The Bénard cell discussed earlier can be considered to be an emergent process, although no transfer of information involved.

If you are interested in acquiring a better understanding of nanothermodynamics, please refer to one of the best books on the subject matter, written by G. A. Mansoori and called *Principles of Nanotechnology: Molecular-Based Study of Condensed Matter in Small Systems*, World Scientific Publishing, 2005 [30].

8.4.1 Background

In the early 1960s, Terrell L. Hill wrote a book on thermodynamics of small systems related to his interest in colloidal systems, polymers, and macromolecules [33,34]. The notion of nanothermodynamics soon followed. His first book addressed the thermodynamics of metastable states in microenvironments [34]. A year later, Hill followed up with a book on the thermodynamic considerations of metastable nanostructured materials at microscopic levels [35]. R. S. Rowlinson at the 1983 Faraday lecture summarized that, for particles smaller than a few nanometers [36], "thermodynamics and statistical mechanics lose their meaning."

Although the field of nanothermodynamics, once again as with other things nano, has been around for a little while, it must be considered to have a spark of revolutionary element. T. L. Hill went on to introduce the term "nanothermodynamics" in 2001. Hill's abstract from the article follows [37]:

> Gibbs initiated his main contribution to thermodynamics by adding new chemical potential terms to the basic energy–heat–work equation. Nanothermodynamics is initiated, as a next step, by adding (at the ensemble level) a further subdivision potential term to the chemical potential terms of Gibbs. The basic equations of nanothermodynamics can then be deduced efficiently from this starting point.

Hill stated in 2002 that the "thermodynamics of a small system will usually be different in different environments" [38]. In other words, material properties may depend on the surrounding environment and not just on size [36]. Also in 2002, Hill, with R. V. Chamberlin, adapted classical thermodynamic principles to small particles by taking into account surface effects studied in the form of "small system ensembles" [38]. Modern nanothermodynamics began with Constantino Tsallis in 1988, who proposed the concept of Tsallis entropy and Tsallis statistics in his paper, "The Possible Generalization of Boltzmann–Gibbs Statistics" [39]. From this seminal work, the foundation for nonextensive thermodynamics was laid.

Latex Beads and Thermal Fluctuations.   In 2002, scientists at the Australia Northern University found that the second law can be broken for short periods of time in microscopic materials [8]. In some recent experiments, micron-sized latex beads were suspended in water and infrared lasers were used to track movement and actually drag the beads through the water for short distances; the IR laser beams acted as tweezers that were able to push the bead along a trajectory [40]. They found that the system showed negative entropy over short time frames (the latex beads absorbed energy from random movement of the water molecules, a.k.a. thermal fluctuations), but quickly returned to the normal positive entropy state when measured over longer time frames. Measurement of the motion of the latex beads and extrapolating the level of forces applied to it by the water demonstrated that the bead was "kicked" by the water molecules. This suggests that energy was transferred from the water (the thermal bath) to the bead and that energy in the form of heat was transferred from the water to the bead [40]. Denis J. Evans of the Research School of Chemistry at the Australian National University States [40]: the results imply that the fluctuation theorem has important ramifications for nanotechnology and indeed for how life itself functions where disorder can suddenly become order.

Microscopic systems are capable of becoming more ordered for short periods of time [31]. Because of this and related findings, researchers feel that limits could be placed on miniaturization because "nanoscale devices will not be a simple scaled-down version of bulk counterparts" in that they could possibly operate backwards [41]. Recall that macroscopic thermodynamics is a statistical rendition of the behavior of bulk materials. When the sample size is reduced to such a great degree, macroscopic statistics may not apply.

Nanoparticle Structural Transformation. In 2003, it was shown that water is able to drive the structural transformation of nanoparticles [42]. The thermodynamic behavior of small particles is different from the bulk by the additive quantity $\gamma \mathcal{A}$, where γ is the interfacial surface energy and $\mathcal{A}$ is the interfacial surface area. It was reported that a nanoparticle system is able to exhibit structural changes as a function of the surface environment rather than just particle size [42]. Earlier results have shown that polymorphs of the same material with different interfacial free energies can cause phase stability changes when particle size is reduced. In experiments, ZnS nanoparticles with average diameter equal to 3 nm exhibited reversible structural transformation (to a minimum energy configuration) upon removal of the synthesis solvent (methanol). Immersion into water revealed that the structure of the ZnS particles was substantially altered from the previous configuration. For example, the particles exhibited significant reduction in surface distortions. Zhang et al. concluded that structure and reactivity of nanoparticles depend on both particle size and the surrounding environment [42].

8.4.2 *Application of Classical Thermodynamics to Nanomaterials*

The machinery of classical thermodynamics can be extended successfully to nanomaterials, without consideration of small systems, and derive quantitative data. One major criterion is that the sample size be relatively large. Allotropes of carbon were shown in **Figure 5.1**.

Free energy of formation of selected carbon allotropes is compared to graphite and diamond in **Figure 8.7**. The energy data were acquired from calorimetric studies (a tried and true classical thermodynamic procedure), experimental CVD data, and atomistic simulations. Since fullerenes are "highly strained molecules," they are expected to be less stable (higher energy) than graphite. Therefore, it is expected that fullerenes acquire more stability as additional carbons are added to the structure. Hence, the ΔH°_{f} values of fullerene-like compounds approach the value of graphite in asymptotic fashion as more carbon is added [43,44]. Fullerenes and nanotubes exist in a kinetically stable *metastable phase* and are formed under *pseudoequilibrium* conditions. Single-walled carbon nanotubes (SWNTs), multiwalled carbon nanotubes (MWNTs), graphitic fibers and tubes, and fullerenes all exist at higher energy than graphite [44].

Single-Walled Carbon Nanotube Growth Mechanism. Classical thermodynamic principles can be extended to explain the energy associated with graphitic structures. A single-walled carbon nanotube is nothing more than a sheet of graphite (called graphene) rolled to make a cylinder. Multiwalled nanotubes are a series of concentric SWNT tubes of increasing diameter. However, an excess amount of energy is required to get this rolling accomplished. The trade-off in energy is a balance between forming an open structured planar sheet of graphite of limited size (more accurately, a graphene sheet that has numerous dangling bonds) and forming a closed structure with bond strain but with no dangling bonds. As we find out, the size of the catalyst particle has great influence over these two parameters.

FIG. 8.7

Relative enthalpies of formation of various carbon allotropes are graphically depicted. All values are relative to the enthalpy of formation of graphite (which is the most stable form of carbon, $\Delta H_f = 0.0$). "A" indicates a type of nanotube called the armchair nanotube (more detail provided in chapter 9). "Z" stands for zigzag carbon nanotube. "C" indicates a fullerene. The number in parentheses indicates the radius of the nanotube in nanometers. The asterisk (*) indicates that the enthalpy of formation of the fullerene was determined by calorimetry from references 43 and 44. ESWN is an experimental estimate of enthalpy of nanotubes. (Unpublished results, G. L. Hornyak et al., National Renewable Energy Laboratory, 2002.) All other results (all A, Z, and C data points) are from reference 48 and were estimated by quantum molecular dynamic (QMD) simulations.

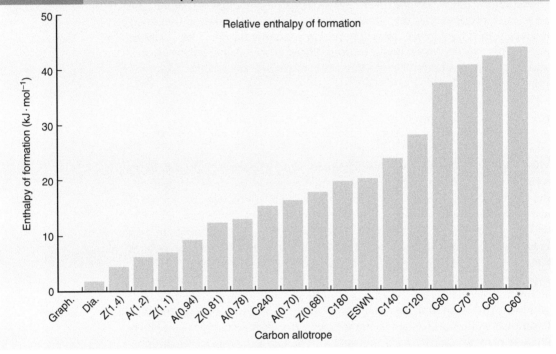

Richard Smalley et al. in 1999 proposed a growth mechanism of single-walled carbon nanotubes called the *yarmulke* mechanism [45]. Fe metal clusters with diameter of 0.7–1.4 nm (50–200 Fe atoms) are formed from the gas-phase decomposition of ferrocene. Upon the addition of a carbon source gas like CO, solid carbon is formed on the surface of the Fe catalyst from disproportion-ation of CO gas. Carbon nucleates on the surface of the Fe catalyst and then aggregates to form a hemifullerene cap. The cap, a structure that resembles a yarmulke (a skull cap), lifts off when additional carbon atoms are added, and a nanotube is formed. The diameter of the tube correlates with the diameter of the Fe catalyst particle. The 0.7-nm diameter tube is the smallest practical limit for SWNTs, although a tube of diameter of 0.4 nm was detected inside MWNT structures—constituting the theoretical and experimental limit [46]. The 0.7-nm value also happens to correlate with the diameter of a fullerene [45]. Larger catalyst particles formed *nanocages*, structures that encapsulate the metal catalyst. The process of SWNT propagation is by a base-growth mechanism (e.g., the catalyst remains anchored to a substrate, see chapter 9). From the kinetic perspective, SWNT growth depends on the number of active sites on the catalyst surface and the adsorption rate [47].

The growth dynamics of a single-walled carbon nanotube were monitored in situ with an ultrahigh-vacuum transmission electron microscope at 650°C [47]. M. Lin et al. found that SWNTs preferentially grow on catalyst particles less than 6 nm in diameter. They identified three distinct growth domains: (1) incubation (nucleation), (2) growth, and (3) passivation [47]. Larger particles generate *nanocages* that encapsulate (and therefore, inactivate) the catalyst. Following decomposition of a carbon source gas on the surface, two possible paths are available for carbon diffusion: along the surface (lower energy of activation due to lower coordination number) or diffusion through the bulk of the catalyst particle. Both thermodynamic and kinetic effects are responsible for the nucleation and growth process [47]. In smaller sized catalyst particles, however, the probability of diffusion through the catalyst increases due to three factors: (1) path length of diffusion on the order of the dimensions of the catalyst, (2) reduced coordination number within the volume of the catalyst, and (3) the phase of catalyst may be no longer solid but rather in the form of a liquidus—a factor that is highly related to the level of coordination.

Ab initio calculations have determined that growth of a graphene sheet is more likely (more stable) than an aggregate of isolated carbon atoms. In other words, a level of order is preferred over a loose agglomeration of carbon atoms that are seeking coordination. This result is not surprising in that the driving force to reduce the number of dangling bond sites is quite powerful. Therefore, the future fate of the graphene sheet relies on the size of the catalyst particle.

To review, there are two possible types of carbon products: a hemispherical graphene cap or a graphene sphere (carbon nanocage) that encapsulates the catalyst. The path that is selected is dependent on the catalyst size. The system seeks to find a balance between surface energy (a consequence of curvature and dangling bonds) and strain energy (a consequence of C–C bonds out of their normal orientation) [47]. Larger catalyst particles have less curvature, and the resulting graphene sheet that forms on its surface has lower torsional bond strain. Nanocage formation is favored in this configuration.

For smaller catalyst particles, bond torsional energy dominates the surface energy of the graphene layer and the hemifullerene cap is formed into a tube. As the tube is formed, the overall bond strain energy is relaxed as more carbon is added. The bonds in a tube are strained in just two dimensions rather than the three for a spherical structure of the same diameter.

Calorimetry. Calorimetry, a tried and true classical procedure, is the measure of the amount of heat released or absorbed in a chemical reaction, usually that of a combustion reaction. Heats of formation in particular are calculable by calorimetric methods. The heats of formation of fullerenes have been determined in an *isoperibolic aneroid microcalorimeter* [43,44]. They were found to be $\Delta H^\circ_f(C_{60}) = 42.5$ KJ·mol^{-1} and (of carbon atoms and not a mole of fullerenes) above that of graphite (**Fig. 8.8**) [44]. This was accomplished with no consideration of smallness—just the result of a standard calorimetric study. The goal

FIG. 8.8   *Chemical structure of polyhedral oligomeric silsesquioxane (POSS) and pentacene.*

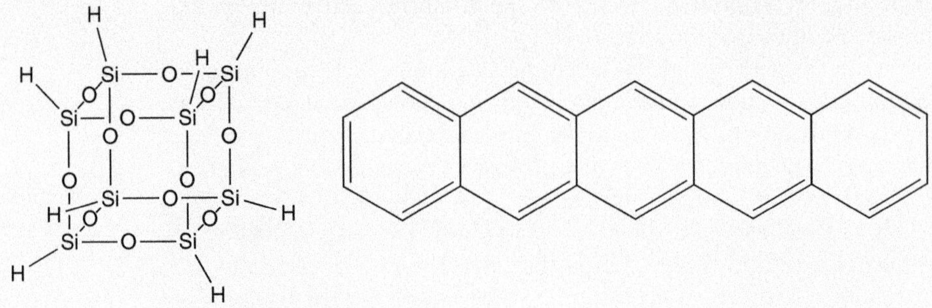

was to obtain the heat of formation of a macroscopic amount of fullerene a nanomaterial.

Molecular Modeling Method to Calculate ΔH_f^o. The equilibrium energies of fullerenes and carbon nanotubes were also estimated by first principles (ab initio) quantum molecular dynamics (QMD) calculations [48]. The single input parameter in the simulations was the *planarity* of the bonds in the fullerene. Planarity is defined by the π-bonding angle, ϕ_π. The interaction between parallel π-orbitals is directly proportional to cos ϕ_π. The definition of planarity was then refined to indicate the average value of cos ϕ_π. Therefore, fullerenes that exhibit more curvature naturally had more bond strain. The curvature is a function of particle diameter and the number of carbon atoms along a specified circumference. Data obtained for fullerenes and tubes by the QMD method are shown in **Figure 8.7** along with calorimetric, CVD, and other determinations.

Classical Thermodynamics and Carbon Nanotubes. Classical equilibrium thermodynamic analyses have been conducted on graphitic fibers formed over a nickel catalyst [49,50]. The studies were conducted to optimize conditions that reduced coking of catalysts. Ironically, the cause of the catalyst deactivation was growth of carbon fibers—an unwanted consequence of carbeurization reactions [49,50]. The fiber morphology was either multiwalled carbon nanotubes or solid fibers of graphite. Later it was discovered that the thermodynamic analyses of these early experiments were flawed. Specifically, the work of Rostrup-Nielsen and others measured the equilibrium conditions of an intermediate species (e.g., surface metal carbides) and not that of the carbon fibers [24]. Because carbon activity in the catalyst particle is not constant, the gas phase carbon activity is in *pseudoequilibrium* with the surface carbide rather than with the filaments [24]. The mechanism to form fibers and MWNTs occurs by a tip growth mechanism, in contrast to SWNTs that form by a base-growth mechanism. The question to pose while reading the remainder of this section is simply this: Are classical thermodynamics principles enough to explain all that is occurring during the SWNT growth process?

Recent studies adopted similar classical procedures. SWNTs are formed by a chemical vapor deposition (CVD) process. Methane–hydrogen mixtures at

temperatures from 600 to 1000°C were reacted with iron–molybdenum catalysts. Presence of SWNTs was verified by Raman spectroscopy. The endothermic
SWNT formation process based on the decomposition of methane is illustrated
in the following equations:

$$CH_4 \rightarrow C_G + 2H_2 + \Delta H_G(T) \tag{8.102}$$

$$C_G \rightarrow C_{SWNT} + \Delta H_C(T) \tag{8.103}$$

$$CH_4 \rightarrow C_{SWNT} + 2H_2 + \Delta H_{SWNT}(T) \tag{8.104}$$

$$\Delta H_C(T) = \Delta H_{SWNT}(T) - \Delta H_G(T) \tag{8.105}$$

$$K_{SWNT}(T) = \frac{(p_{H_2})^2}{(p_{CH_4})} \tag{8.106}$$

EXAMPLE 8.2   *Decomposition of Methane to Form SWNTs*

Calculate the enthalpy of reaction of SWNTs (ΔH_{SWNT}) from the decomposition of methane (Equations
8.102–106). Temperature and the equilibrium constant data are given below. Compare the enthalpy of
decomposition of SWNTs to that of the decomposition of methane to form graphite.

Data:

T/K	892	913	958	1005
K_P	0.713	1.281	2.766	4.053

The onset temperature of SWNT formation was chosen to be the thermodynamic threshold of interest. The
onset temperature, and hence the K_P of the decomposition–growth process, of SWNT formation depended
on the partial pressures of H_2 and CH_4.

A van't Hoff plot of $-\ln K$ versus $1/K \times 1000$ is required to determine the ΔH of the reaction.

Transformed data:

$1/K \times 1000$	1.121	1.095	1.044	0.995
$-\ln K_P$	0.3387	−0.2478	−1.0175	1.3995

Calculation of ΔH_{SWNT}:
*These data show how the equilibrium constant depends on the reaction temperature. The reaction enthalpy can be
extracted from the slope of the line plotting $-\ln K_P$ versus $(1/K) \times 1000$. The van't Hoff equation assumes that ΔH
does not change much over the temperature range in question.*

The equation of the line is

$$y = -15.107 + 13.655x \tag{8.107}$$

The slope of the relationship is proportional to $\Delta H/R$

$$\Delta H_{SWNT} = (+13.655 \times 10^3 \text{ K}) \times 8.314 \text{ J} \cdot mol^{-1} \cdot K^{-1} \div 1000 = +114 \text{ kJ} \cdot mol^{-1}$$

The decomposition of methane to form graphite is $\Delta H_G = 78.8 \text{ kJ} \cdot mol^{-1}$.

*The excess enthalpy required to form SWNTs is ca. 25 kJ · mol^{-1} more than required to form graphite, the most stable
allotrope of carbon. This value seems about 7 kJ · mol^{-1} greater than values predicted using QDM.*

What factors may contribute to deviations from predicted values?

ΔG and ΔS are also calculable from the preceding data using $\Delta G = -RT \ln K_P$ and $\Delta G = \Delta H - T\Delta S$.

where ΔH_C is the difference in enthalpy at constant pressure and temperature between graphite and SWNTs and p is the partial pressure of the gaseous reactants and products. Plots of $\ln K$ versus reciprocal temperature are linear (assuming that ΔH does not change significantly over the experimental temperature range—a.k.a. a van't Hoff plot—and that the system is an equilibrium system. If $(-\ln K_p)$ is plotted against $[(1/K) \times 1000]$, the slope of the line is proportional to $\Delta H/R - \Delta H = \text{slope} \times R \times 1000$. If there is a positive slope, the reaction is endothermic. A negative slope is indicative of an exothermic reaction.

The calculation of the Gibbs free energy also proceeds via classical methods.

$$\Delta G_{SWNT}(T) = -RT \ln K_{SWNT}(T) \qquad (8.108)$$

$$\Delta G_C(T) = -RT \ln \frac{K_{SWNT}(T)}{K_C(T)} \qquad (8.109)$$

Once ΔH is calculated, the values of ΔG and ΔS can be found from equation (8.108) and the equilibrium constant and $\Delta G = \Delta H - T\Delta S$, respectively. The decomposition of CH_4 over Fe–Mo catalyst is endothermic overall. The disproportionation of CO over nickel catalyst is exothermic. So far this treatment has been entirely from the classical perspective.

Gibbs Free Energy Estimate of SWNT Growth. L. M. Wagg et al. of the National Renewable Energy Laboratory found the Gibbs free energy of formation for SWNTs relative to the formation of graphite to be between 16.1 and 13.9 kJ · mol^{-1}. Gas feed composition (methane and hydrogen) and reaction temperature (700–1000°C) were varied to determine the thermodynamic threshold for nucleation and growth of SWNTs [24]. Experimental parameters for efficient growth of SWNTs by the decomposition of methane have been reported earlier [51].

Are explanations of the SWNT nucleation-growth process based only on classical thermodynamic principles valid? The answer to the question is an unequivocal "no." It is not appropriate to interpret the SWNT growth process by classical thermodynamic principles and methods alone. The authors claim that the concept of *pseudoequilibrium* needs to be invoked in order to explain the processes involved [24]. For one thing, the process resembles a steady-state process where there is a (material in $\Leftrightarrow$ material out) scenario with constant internal energy. There are potential nonequilibrium domains that exist below 700°C that, according to classical analyses, were detected at p_{H2} less than 10%. The purported "decomposition of methane" in these cases actually exhibited an exothermic response as opposed to the normal classical endothermic nature of methane decomposition at equilibrium conditions (higher temperature, higher p_{H2}) [52]. Extreme driving forces, nonequilibrium mixing of gases and thermal gradients may be responsible for exothermic character of the low hydrogen partial pressure domain [52].

Intermediates formed throughout the reaction are complex and rely on kinetic parameters as well as upon thermodynamic ones. As a case in point, iron carbides formed within the catalyst particles may be mistaken for the thermodynamic product (tubes) if weight gain–loss methods of tracking the reaction coordinate are employed [24]. Regardless of the ideality, this approach does provide a good starting point, but extreme caution must be practiced when interpreting experimental data based on purely classical principles.

8.4.3 Small System Thermodynamics (the Theory of T. L. Hill)

The work of J. M. Rubi et al. [53,54], H. J. F. Jansen [55], R. M. Baigi [56], G. A. Monsoori [3,30], A. K. Rajagopal [33], and J. -R. Roan [57] (and others) was extremely helpful in putting this section together by providing some of the best overviews of nanothermodynamics to be found in the literature. Two factors wrap themselves around this new thermodynamics: (1) modification of the terms *extensivity* and *intensivity* and (2) the dependence of nanomaterials on their immediate surroundings (e.g., their immediate environment).

A good place to start is with the Gibbs–Thomson effect and its description of the nucleation and growth of nanoparticles and nanodroplets:

$$\Delta G = \frac{4}{3}\pi r^3 \Delta G_V + 4\pi r^2 \gamma \qquad (8.110)$$

If this equation looks familiar it is because we covered it in chapter 6. It describes the free energy change of a small particle in which ΔG_V is the free energy change per volume, $4\pi r^3 \gamma$ is the free energy change per unit surface area, and γ is the surface tension (joules per square meter) as before. Notice that the surface tension (energy) term γ is assumed to be independent of size. In actuality, the surface energy increases dramatically once the radius dips below 3 nm [58]. What this relation indicates is that a smaller particle will have a higher relative energy than a larger one. It also illustrates the concept of additivity. The volume term alone is not able to describe the change in free energy for the nucleation process. However, the addition of a surface term provides the necessary correction to the energy. G, r, and V are extensive terms. G_V and γ are intensive terms. As with any classical thermodynamic parameter, the relation can be expressed in terms of entropy.

As experimental techniques were improved, researchers found that classical nucleation theory was incomplete and did not account for the temperature dependence of the rate of nucleation. Several theories were proposed to explain the variance: Hill theory (given next), Tsallis entropy, and a new approach proposed by Wang and Wang [28].

Hill Theory for Small Systems. Thermodynamic limits apply in classical thermodynamics and statistical mechanics: $N \rightarrow \infty$ and $V \rightarrow \infty$ where N/V is a constant. If volume V is doubled, then energy, entropy, and all other extensive properties are also doubled. Intensive properties remain the same. Rephrased, it means that thermodynamics is only valid if a system contains an infinite number of particles distributed in an infinite volume with a constant density [53]. This viewpoint guarantees existence of extensivity without fluctuations (independent of the environment). Statistical mechanics is a means by which the thermodynamic limit is approximated in an asymptotical manner. But what if there is no thermodynamic limit?

When particles assume small dimensions, as in the case of biological and colloidal systems, the concept of extensivity breaks down. The best example of this is provided by surface energy. We know from chapter 6 how the behavior of surface atoms differs significantly from that of the bulk. In the bulk, the proportion of surface atoms is small and the contribution of surface energy to the energy of the system as a whole is miniscule. However, in small materials, the surface assumes greater influence. A modification of macroscopic thermodynamics is in order. T. L. Hill presented this equation of the Gibbs energy for a small cluster [59]:

$$G = uN + aN^\beta \qquad (8.111)$$

where μ is the chemical potential, a is an intensive parameter constant, and β is less than 1. As $N \to \infty$, the second part of the equation becomes less significant, but for small particles like clusters, both terms are important. In the former scenario, $G \propto \mu N$ yields the energy per particle—an additive extensive parameter (e.g., G for the system is the sum of G of all the parts). However, if the second term in equation (8.111) (the exponential) becomes important, then the concept of extensivity for G is no longer valid [54]. A mathematical fix that restores extensivity is

$$G = \hat{\mu}N \text{ where } \hat{\mu} = \mu + aN^{\beta-1} \qquad (8.112)$$

$$\hat{\mu} - \mu = aN^{\beta-1} \qquad (8.113)$$

What this states is that for small systems, an ensemble of constituent equivalent replicas becomes a large system that has the usual thermodynamic infinite limit properties [54]. An ensemble such as this provides a macroscopic base from which to study the nanodomain. A consequence of this, according to Hill, is that thermodynamic parameters of different ensembles may not be equivalent [53,59].

The differential form of the internal energy and the other forms of the total system by combining the first and second laws of thermodynamics is

$$dU = TdS - PdV + \mu dN \qquad (8.114)$$

To start, let us visualize an ensemble of $\mathcal{N}$ equivalent components (identical, independent replicas of small systems) of a one material system. For a macroscopic system, $\mathcal{N} = 1$. The total energy, entropy and volume, etc. are a function of the number of replicas $\mathcal{N}$ in the system at large:

$$U_{\text{Total}} = \mathcal{N}\bar{U}; \quad V_{\text{Total}} = \mathcal{N}\bar{V}; \quad S_{\text{Total}} = \mathcal{N}S \quad \text{and} \quad N_{\text{Total}} = \mathcal{N}N \quad (8.115)$$

where the barred terms represent average values per replica [54]. Another term called the *subdivision* potential, $\mathcal{E}$, is a type of chemical potential that includes the additive contributions of small systems to the total energy. Therefore, for the set of small N systems for constant T, P, change in the energy of the ensemble is given by

$$dU_{\text{Total}} = TdS_{\text{Total}} - PdV_{\text{Total}} + \mu\,\mathcal{N}dN_{\text{Total}} + \mathcal{E}d\mathcal{N} \qquad (8.116)$$

Obviously, if we deal with a macroscopic system, $\mathcal{N} = 1$ and the $\mathcal{E}$ subdivision potential term would vanish. For this ensemble of small systems, $N \neq 1$ and $\mathcal{E} = 0$. Now, keeping T, P and N constant, integrating equation (8.116) yields

$$U_{\text{Total}} = TS_{\text{Total}} - PV_{\text{Total}} + \mathcal{E}\mathcal{N} \qquad (8.117)$$

Dividing equation (8.117) by $\mathcal{N}$ yields

$$\bar{U} = TS - P\bar{V} + \mathcal{E} \tag{8.118}$$

Identifying $\mathcal{E}$ with G gives [3]:

$$TS - P\bar{V} + \mathcal{E} \rightarrow TS - P\bar{V} + G \tag{8.119}$$

Recalling that $G = \hat{\mu}N$,

$$\bar{U} = TS - P\bar{V} + \hat{\mu}N \tag{8.120}$$

Rewriting, we get [3]

$$\bar{U} = TS - P\bar{V} + \mu N + (\hat{\mu} - \mu)N \tag{8.121}$$

At the thermodynamic limit, G equals μN and the last term of equation (8.119) disappears (as $\hat{\mu} \rightarrow \mu$) [54].

This is the *Gibbs–Hill* adaptation to small system thermodynamics. In other words, it is an adaptation of macroscopic thermodynamics with additive terms that address small system effects. This brand of thermodynamics does not wander too far off equilibrium. There are, of course, a handful of recent interpretations that differ from this approach.

EXAMPLE 8.3 *Evaluation of $\mathcal{E}$*

Derive a form of the subdivision potential $\mathcal{E}$ from the linear homogeneous Euler form of U_{Total} at T, P. U_{Total} for a small one-material system is

$$U_{Total} = TS_{Total} - PV_{Total} + \mu N_{Total} + \mathcal{E}\mathcal{N} \tag{8.122a}$$

Solution:

Division by $\mathcal{N}$ small systems in the ensemble yields the average quantities:

$$U = TS - PV + \mu N + \mathcal{E} \tag{8.122b}$$

Differentiating gives:

$$dU = d(TS) - d(PV) + d(\mu N) + d\mathcal{E}$$

$$dU = T\,dS + S\,dT - [P\,dV + V\,dP] + \mu\,dN + N\,d\mu + d\mathcal{E}$$

From the definition, $dU = T\,dS - P\,dV + \mu\,dN$

$$T\,dS - P\,dV + \mu\,dN = T\,dS + S\,dT - [P\,dV + V\,dP] + \mu\,dN + N\,d\mu + d\mathcal{E}$$

Therefore:

$$d\mathcal{E} = -S\,dT + V\,dP - N\,d\mu \tag{8.123}$$

Notice from equation that $(-S\,dT + V\,dP) = dG - \mu\,dN$

$$d\mathcal{E} = (dG - \mu\,dN) - N\,d\mu = dG - d(\mu N)$$

$\mathcal{E} = G - \mu N$ *and from equation 8.112, $G = \hat{\mu}N$, restoring extensivity to the expression*

EXAMPLE 8.3 (CONTD.)　　*Evaluation of ℰ*

$$\mathcal{E} = (\hat{\mu} - \mu)N \tag{8.124}$$

Deviations from large systems are due to $(\hat{\mu} - \mu)$, which becomes insignificant at the thermodynamic limit. In the form below, it is easy to see why ℰ vanishes as N gets infinitely large and $\hat{\mu} \to \mu$.

$$\hat{\mu} = \mu + \frac{\mathcal{E}}{\mathcal{N}} \tag{8.125}$$

The subdivision potential ℰ is a form of the chemical potential. Surface, edge, and other small system effects contribute to ℰ. In different environments, ℰ assumes different forms. Macroscopic thermodynamics is recovered as $N \to \infty$ (the thermodynamic limit) and $\mathcal{E}/N \to 0$. Therefore, ℰ is the limit for small system thermodynamics. Another form that you may find for ℰ is given by

$$\mathcal{E} = \left(\frac{\partial U_{Total}}{\partial \mathcal{N}} \right)_{S_{Total}, V_{Total}, \mathcal{N}_{Total}} \tag{8.126}$$

where ℰ is an "intensive parameter" in the form of a chemical (subdivision) potential that represents all other extensive variables that can be measured and specified to determine the thermodynamics of the system.

8.5　MODERN NANOTHERMODYNAMICS

In 1876, physicist Josef Loschmidt, in response to the prevailing thought that the universe will suffer "heat-death" as a result of the second law of thermodynamics, submitted the "reversibility paradox" to the scientific community: "If the motion of individual particles is considered to be reversible, why then is their collective behavior irreversible?" This critique encouraged Boltzmann to create his statistical concept of entropy shown in equation (8.74). Equations of thermodynamics are based on the random motion of many Avogadro's numbers of particles (in Germany, this value was referred to as Loschmidt's number). Heat always flows in one direction towards the cold pole and entropy always increases. Right? Boltzmann, upon reformulation of his classical version of the second law into the widely accepted statistical form, understood the importance of statistical fluctuations.

8.5.1　Nonextensivity and Nonintensivity

Tsallis Formulation and Entropy.　C. Tsallis proposed a new definition for entropy in 1989 [3,39]. His concept is summarized in equations (8.127) and (8.128). The idea that thermodynamics of small systems have nonextensive and nonintensive components is quite an out-of-the-box viewpoint. Its intent was not to replace Gibbs–Boltzmann statistics, but rather to offer explanations to anomalous systems characterized by nonergodicity or metastable states (fluctuations). We have seen, by way of classical thermodynamic precepts, that extensivity and intensivity explain thermodynamics and cooperate with differential and integral formulations. This is evidenced by corrections such as $G = \hat{\mu}N$ with the purpose of restoring extensivity into classical formulations.

The mathematical formalism of Tsallis's new definition of entropy is given by

$$S_q = k_B \frac{1 - \sum_{i=1}^{W} p_i^q}{q-1} \qquad (8.127)$$

where p_i is the probability of microstate i and where $0 \leq q \leq 1$ [3,39]. The entropy for two independent systems A and B with independent probability of occurring is

$$S_q(A+B) = S_q(A) + S_q(B) + \frac{1-q}{k} S_q(A) S_q(B) \qquad (8.128)$$

The entropy, as defined in the preceding equation, is nonextensive (nonadditive) for $q < 1$. As $q \to 1$ and when $q = 1$, the entropy according to this relation is reduced to the Boltzmann–Gibbs form and becomes extensive [3]. Others have shown that Tsallis and Hill thermodynamics can be reconciled—specifically, that the nonadditive property of the Tsallis entropy forms the basis of the subdivision entropic potential of Hill [60]. This last statement provides some level of support for the continuum theory of energy we espouse.

Non-Tsallis Formulations. However, Mohazzabi et al. in 2005 showed, by way of molecular dynamic simulations, that nonextensivity and nonintensivity can be explained within the framework of Boltzmann–Gibbs formalism without consideration of Tsallis's thermodynamics [3]. They demonstrated that internal energy and entropy are nonextensive and that temperature and pressure are nonintensive [3].

In the study of Mohazzabi et al. that we mentioned in section 8.4.3, systems of variable number of identical particles in three dimensions were simulated by a molecular dynamic program [3]. The parameters of the simulation were as follows: (1) argon-like particles of mass m (the unit mass) were made to interact according to a pair-wise Lenard–Jones (6–12) interatomic potential energy function potential: $u(r) = 4\varepsilon[(\sigma/r)^{12} - (\sigma/r)^6]$, where ε (the unit energy) is the depth of the well and σ (the unit length) is the "hard core radius", (2) the number of particles within each system varied from 8 to 1000: $n(2,3,4,\ldots,10) = n^3(8,27,64,\ldots,1000)$ that were distributed in a square lattice, (3) the particles were randomly assigned velocity v and equations of motion were solved, and (4) internal energy was calculated from kinetic and potential terms, temperature from ε/k_B, and pressure from the change in momentum of particles per unit time per unit area after collision with the container walls.

The unit of entropy was the Boltzmann constant k_B. The computation of entropy was not straightforward [3]. Entropy was not calculated from classical thermodynamic procedures that assume a finite number of microstates, that all microstates in an ensemble are equally likely, and that entropy is an extensive parameter where $S(N, V, U) = k \ln \mathcal{W}(N, V, U)$. Calculation of S from particle trajectories for continuous microstates (e.g., from positions and velocities) cannot be accomplished by classical methods. Therefore, entropy was calculated from particle trajectories by choosing a "bin size" for each coordinate of momentum (e.g., setting a limit on bin size) [3]. This example demonstrates the level of thought and expertise required to produce a molecular dynamic simulation. Obviously, the results depend on the number and quality of the input parameters.

The molecular dynamics (MD) simulation consisted of small systems interacting via Lenard–Jones type potential energy functions ranging in size from a few argon-like particles to 1000 interacting argon-like particles. The authors found deviations from macroscopic thermodynamics even in systems as large as 1000 particles [3]. Conclusions from this work based on MD simulations are [3]:

- Plots of internal energy versus N were not linear (to be linear, slope = U per particle), indicating a small nonextensivity. U was found to be slightly subextensive (more negative).
- With regard to internal energy, the nonextensive behavior of simulated particles is related to the surface-to-volume ratio. This implied that as surface-to-volume ratio increased, nonextensivity became more pronounced.
- $U_2 \neq 2U_1$. This means that U is nonextensive in that two particles did not result in twice the energy. Extra energy was obtained from particle interaction—converting external potential energy into internal energy.
- Entropy was shown to be nonextensive. Because it increased for smaller systems, entropy was classified as superextensive.
- Neither pressure nor temperature demonstrated intensivity and they are therefore nonintensive properties.
- All parameters approached extensive or intensive states as N was increased.

According to preceding discussion, extensivity and intensivity for small systems exist in an altered state from that of the standard macroscopic state. The smaller the system was, the more pronounced became the deviation [3].

Extension of the Young–Laplace Equation.   Wang and Wang proposed in 2005 that a universal quantitative thermodynamic nanoscale model based on the extension of the Young–Laplace equation is sufficient to explain deviations from classical nucleation theory. The theory is based on computer simulations of diamond and C–BN (amorphous carbon boron nitride) nanocrystal nucleation processes. Correlation of Young's equation with equilibrium phase diagrams was used to describe the thermodynamics of metastable phase nucleation [28]. According to the authors, no adjustable parameters in simulations were required to explain the phenomena and simple extrapolation of the phase equilibrium was successful in predicting nanothermodynamic behavior from macroscopic thermodynamic data [28]. Simulations were also accomplished for one-dimensional structures formed under vapor–liquid–solid mechanisms. From their method, Wang and Wang were able to predict thermodynamic and kinetic size limits of nanowires as well as nucleation thermodynamics and diffusion parameters of catalyst nanoparticles at the tip of the nanowire [28].

8.5.2 Nanothermodynamics of a Single Molecule

It is inevitable that we ask this question: Is thermodynamics valid at the single molecule level? We stated earlier that entropy only applies for a great number of atoms and molecules and is meaningless for a single one. J. M. Rubi et al. in 2006 studied this phenomenon [54]:

> We show how to construct non-equilibrium thermodynamics for systems too small to be considered thermodynamically in a traditional sense … We show in particular that the Gibbs equation, when formulated in terms of average values of the extensive quantities, is still valid, whereas the Gibbs–Duhem equation differs from the equation obtained for large systems due to the lack of a thermo-dynamic limit … The potentials of mean force and mean position correspond respectively to our Helmholtz and Gibbs energies. The results show that a ther-modynamic formalism can indeed be applied at the single-molecule level.

8.5.3 Modeling Nanomaterials

Examples of computer simulations of nanomaterials are abundant in the litera-ture. Selected examples of nanomaterial modeling gleaned from the literature are presented next.

Polyhedral Oligomeric Silsesquioxane. The bottom-up design of self-assem-bling nanoparticles consisting of polyhedral oligomeric silsesquioxane (POSS) (**Fig. 8.8**) was demonstrated by a "multiscale computer simulation method" that included four types of modeling systems: (1) ab initio quantum mechani-cal calculations, (2) molecular dynamics simulations with classical and reactive force fields, (3) Monte Carlo simulations, and (4) coarse-grained mesoscale simulations [61,62]. The purpose of the study was to optimize optical, thermal, and mechanical properties of nanocomposite materials that contain the func-tionalized POSS constituents. Thermal, vibrational, mechanical, and structural properties were first studied by ab initio and MD simulations. Results were then compared to preexisting experimental data. It was found that the POSS cages have a unique rhombohedral packing that implies parallel packing in the crystal form. The consequences of such packing are higher charge carrier capability when functionalized with acene type molecules (pentacene) (**Fig. 8.8**). Evaluation of the electronic properties of the composite organic–inorganic hybrid semiconductor simulations revealed similarities to acenes but predicted superior thermal and mechanical bulk characteristics [62].

Another study involved the modeling of POSS dissolved in hexadecane at 400–1000 K. Radial distribution functions, potentials of mean force, and self-diffusion coefficients were determined by molecular dynamic simulations [63]. Canonical (@ N, V, T) molecular dynamics simulations were used to compute the thermodynamic and transport properties of the POSS. Substitution parame-ters included temperature and atomic replacement or removal. The simulation showed that the mechanical properties of synthetic polymers are enhanced with the incorporation of nanoscale particulate materials.

Dodecanethiol SAMs. In another example, equilibrium structures and thermo-dynamic properties of dodecanethiol self-assembled monolayers were investigated by MD simulations [64]. It was found that compact passivating monolayers were formed on (111) and (100) Au surfaces. At lower temperatures, the passivating molecules organize into parallel bundles. When the temperature is raised, the bundles melt and cover the gold in a uniform way [64]. The "melting temperature" is much lower than that found in a conventional self-assembled monolayer of dodecanethiol on a flat gold surface. Phase transitions such as the one described in this example are commonly studied with MD simulations.

Nanovoids. Nucleation and melting of materials containing nanovoids were also investigated by MD simulations [65]. It was shown that the behavior of four melting stages in nanomaterials differs significantly from the bulk form. The melting in each stage depends on the interactions among thermodynamic mechanisms that arise from changes in the interfacial free energy, the curvature of the interface, and the elastic energy induced by the density change during melting [65]. Ultimately, melting was shown to depend on the internal defects of a material (e.g., the nanovoids). The parameters of the simulation were as follows: (1) zero external P and constant T, (2) standard Lenard–Jones potentials, where $\phi(r) = 4\varepsilon - [(\sigma/r)^{12} - (\sigma/r)^6]$, where $\varepsilon = 119.8\ k_B$ potential well (where $\varepsilon/k_B \approx 120$ K) and $\sigma = 3.405$ Å length parameter, (3) cutoff distance was 2.5σ, (4) *fcc* structure lattice, (5) three different size systems: $20 \times 20 \times 20$, $30 \times 30 \times 30$, or $40 \times 40 \times 40$ unit cells, (6) periodic boundary conditions applied in all directions, (7) spherical voids created by removing atoms from the center, (8) system relaxed to $T = 0.2\ \varepsilon/k_B$, (9) initial void radius ranges = 0.58–0.6.62 nm, and (10) void sample was heated and held isothermally for 200,000 time steps and observed nucleation and growth processes of melting from the nanovoid [65].

Lipid Bilayer Property Modeling. We also add a biological example. Classical MD simulations of atomistic models of interactions between combustion-formed carbon nanoparticles and lipid bilayers were accomplished to determine structural, dynamical, and thermodynamic effects on biological membranes [66]. The authors found that the membrane acts as a discriminator of particle shape and structure, resulting in altered solvation, mobility, adsorption, and permeation [66]. For example, from simulations, it was found that spherical particles tend to stay near the surface of the biological membrane, whereas higher aspect-ratio particles sequester themselves, although not trapped, within the hydrophobic tails of the hydrocarbon region of the membrane [66]. Particle size was also modeled and found to be a factor in the strength of adhesion to the membrane (e.g., larger particles adhered more strongly to the membrane). The dynamical aspects of this study were quite intriguing. Combustion-generated nanoparticles, when inside the membrane, perturbed the structure of the natural lipid and, when coupled with thermal fluctuations, induced pore formation in the membrane [66].

Monte Carlo and MD methods were employed for this study. All dangling bonds of the combustion-generated carbon moieties were terminated with hydrogen in simulations. Applied force fields were obtained for generic nonmetallic main group elements. A preprogrammed lipid bilayer model was used in the MD simulations; 64 lipids were agglomerated to form a membrane 34 Å in thickness and 200 water molecules were included. The longest carbon nanoparticle was 17 Å.

8.5.4   Modern Non-nanothermodynamics?

No theory, however, is embraced and accepted fully. S. Gheorghiu et al. state that application of Occam's razor is the best policy when it comes to the explanation of the *nonergodic* theories proposed earlier [67]. William of Ockham, a fourteenth century logician and Franciscan friar, stated (simply) that "the explanation of any phenomenon should include as few assumptions as possible, eliminating those that make no difference in the observable predictions of the explanatory hypothesis or theory," or, "all things being equal, the simplest solution tends to be the best one." This philosophy is also expressed as the "principle of parsimony."

Gheorghiu et al. stated emphatically that phenomena characterized by *power-law distributions* are plentiful in nature and in the applied sciences [67]. They go on to state that "complicated arguments based on long-range correlations or non-ergodicity are often incorrect or misleading in explaining many naturally observed power laws, in particular those described by the Student distribution." The Student's distribution (*t*- or Student's *t*-distribution) is a means of quantifying deviations from the mean without knowledge of the standard deviation. Its objective is to estimate the mean value of a normally distributed population when the sample size is small. The Student's *t*-test was derived by W. S. Gossett in 1908 while employed by the Guiness Brewery in Dublin.

These claims in effect challenge the assertions of Tsallis et al. and the theory of nonergodic systems in general—that nonextensivity exists for small systems. The authors claim that the Boltzmann factor itself has a Student form and that the *t*-distribution provides an accurate description of power-law statistics associated with polydisperse heterogeneous systems—for example, that the representation is completely thermodynamically classical without the need for spatial or temporal correlations [67].

Concluding Thoughts.   Please recall that this was a special chapter. Its purpose was to introduce concepts in thermodynamics that differ (perhaps drastically) from the well-established paradigms with which we are all so familiar. The purpose was to bring awareness and to provoke thought. Although we authors do not claim to be experts in this new, dynamic field, rest assured that due diligence was practiced during the acquisition of references, consultations, and the writing. Therefore, any feedback from students and professors would be greatly valued. In future editions, we hope to present a formal treatment of the subject matter complete with a challenging set of problems.

The message to take away with you after completing this chapter is a simple one: Thermodynamic behavior of materials at the nanoscale may not be the same as thermodynamic behavior of the same materials at the macroscopic scale. Nanomaterials are subjected to their environment, especially to thermal fluctuations, which may not be considered large on the macroscopic scale but are significant at the local level. Such behavior is expressed as nonextensive and/or nonintensive—definitely anathema to thermodynamics at the big level. Terms such as nonequilibrium, pseudoequilibrium, quasi-equilibrium, and metastable are associated with nanomaterials and a deeper level of understanding of those basic concepts is required in order to understand what transpires at the nanoscale.

Lastly, the link between thermodynamics and living things was explored from yet another frame of reference. Throughout the text we have emphasized the importance of nanomaterials to living things. We believe that this chapter, more than any other, has helped to frame that special and fundamental relationship in a proper perspective.

References

1. E. J. Chaisson, *Cosmic evolution—The rise of complexity in nature*, President and Fellows of Harvard College, Boston (2001).
2. A. Pross, Stability in chemistry and biology—Life as a kinetic state of matter, *Pure & Applied Chemistry*, 77, 1905–1921 (2005).
3. P. Mohazzabi and G. A. Mansoori, Nonextensivity and nonintensivity in nano-systems: A molecular dynamics simulation, *Journal of Computational Theoretical Nanoscience*, 2, 1–10 (2005).
4. Euler's theorem on homogeneous functions, PlanetMath.org, planetmath.org/encyclopedia/EulersTheoremOnHomogeneousFunctions.html (2007).
5. I. N. Levine, *Physical chemistry*, 3rd ed., McGraw–Hill, Inc., New York (1988).
6. P. W. Atkins, *Physical chemistry*, 4th ed., W. H. Freeman and Company, New York (1990).
7. B.-X. Wang, L.-P. Zhou, and X.-F. Peng, Surface and size effects on the specific heat capacity of nanoparticles, *International Journal of Thermophysics*, 27, 139–151 (2006).
8. G. M. Wang, E. M. Sevick, E. Mittag, D. J. Searles, and D. J. Evans, Experimental demonstration of violations of the second law of thermodynamics for small systems and short time scales, *Physics Review Letters*, 89, 050601(2002).
9. J. P. Abriata and D. E. Laughlin, The third law of thermodynamics and low temperature phase stability, *Progress in Materials Science*, 49, 367–387 (2004).
10. S. Lovesey, Condensed matter theory, The Rutherford Appleton Laboratory, http://www.isis.rl.ac.uk/ISIS97/theory.htm (1997).
11. P. Xie and C. M. Lieber, *Carbon nanotubes on gold (111) surface*, Lieber Group, Harvard University (2007).
12. Tentative fourth law principles. In laws of thermodynamics, Wikipedia, en.wikipedia.org/wiki/Laws_of_thermodynamics (2007).
13. Lars Onsager, Biography, Answer.com, www.answers.com/topic/lars-onsager?cat=technology
14. Nonequilibrium Thermodynamics, Wikipedia, en.wikipedia.org/wiki/Non-equilibrium_thermodynamics (2007).
15. C. Bustamante, J. Liphardt, and F. Ritort, The nonequilibrium thermodynamics of small systems, *Physics Today*, 58, 43–48, July (2005).
16. G. D. Mahan, *Many-particle physics*, 3rd ed., Physics of Solids and Liquids Series, Kluwer Academic/Plenum Publishers, New York, (2000).
17. Onsager reciprocal relations, Wikipedia, en.wikipedia.org/wiki/Onsager_reciprocal_relations (2007).
18. S. E. Pierce, Nonequilibrium thermodynamics: An alternate evolutionary hypothesis, *Crossing Boundaries*, 1, 49–59 (2002).
19. D. R. Brooks and E. O. Wiley, *Evolution as entropy: Towards a unified theory of biology*, University of Chicago Press, Chicago (1986).
20. D. R. Brooks, D. D. Cumming, and P. H. LeBlond, Dollo's law and the second law of thermodynamics: Analogy or extension? In *Entropy, information, and evolution*, Weber, Depew, and Smith, eds., The MIT Press, Cambridge, MA (1988).
21. M. Kosmulski, *Chemical properties of material surfaces*, Surfactant Science Series, 102, CRC Press, Boca Raton, FL (2001).
22. H. Zhu, S. F. Webb, and R. Blumenthal, The chemical composition of diamond plasmas, *Journal of Vacuum Science Technology A*, 14, 952–959 (1996).

23. T. A. Brabbs, Catalytic decomposition of methane for onboard hydrogen generation, NASA technical paper 1247, 1–48, June (1978).

24. L. M. Wagg, G. L. Hornyak, L. Grigorian, A. C. Dillon, K. M. Jones, J. Blackburn, P. A. Parilla, and M. J. Heben, Experimental Gibbs free energy considerations in the nucleation and growth of single-walled carbon nanotubes, *Journal of Physical Chemistry B*, 109, 10435–10440 (2005).

25. J. D. Andrade, V. L. Hlady, and R. A. Van Wagenen, Effects of plasma protein adsorption on protein conformation and activity, *Pure & Applied Chemistry*, 56, 1345–1350 (1984).

26. F. Capra, *The web of life: A new scientific understanding of living systems*, Anchor Books, Doubleday, New York (1996).

27. G. Gallavotti and E. G. D. Cohen, Dynamical ensembles in nonequilibrium statistical mechanics, *Physics Review Letters*, 74, 2694–2697 (1995).

28. C. X. Wang and G. W. Wang, Thermodynamics of metastable phase nucleation at the nanoscale, *Materials Science and Engineering: R Reports*, 49, 157–202 (2005).

29. D. R. Baer, J. E. Amonette, and P. G. Tratnyek, Small particle chemistry: Reasons for differences and related conceptual challenges, Interagency Grantees Meeting/ Workshop—*Nanotechnolgy and the environment: Applications and implications*, Pacific Northwest National Laboratory, September 15–16 (2003).

30. G. A. Mansoori, *Principles of nanotechnology—Molecular-based study of condensed matter in small systems*, World Scientific Publishing Co. Pte. Ltd., London (2005).

31. P. F. Schewe, B. Stein, and J. Riordon, Nanothermodynamics, physics news update, *The American institute of physics bulletin*, 598, July 17 (2002).

32. G. R. Vakili-Nezhaad, Euler's homogeneous functions can describe non-extensive thermodynamic systems, *Journal of Pure & Applied Mathematical Science*, 1, 7–8 (2004).

33. A. K. Rajagopal, C. S. Pande, and S. Abe, Nanothermodynamics—A generic approach to material properties at the nanoscale. Invited presentation, Indo-U.S. Workshop on *Nanoscale materials: From science to technology*, Puri, India, April 5–8 (2004).

34. T. L. Hill, *Thermodynamics of small systems, I*, W. A. Benjamin, New York (1963).

35. T. L. Hill, *Thermodynamics of small systems, II*, W. A. Benjamin, New York (1964).

36. R. S. Rowlinson, Molecular theory of small systems, Faraday Lecture, *Chemical Society Reviews*, 12, 251–265 (1983).

37. T. L. Hill, A different approach to nanothermodynamics, *Nano Letters*, 1, 273–275 (2001).

38. T. L. Hill and R.V. Chamberlin, Fluctuations in energy in completely open small systems, *Nano Letters*, 2, 609–613 (2002).

39. C. Tsallis, Possible generalization of Boltzman–Gibbs statistics, *Journal of Statistical Physics*, 52, 479–487 (1988).

40. P. Weiss, Law and disorder: Chance fluctuations can rule the nanorealm, *Science News*, 162, 51 (2002).

41. M. Chalmers, The second law of thermodynamics "broken," *New Scientist*, 09, 21, July 19 (2002).

42. H. Zhang, B. Gilbert, F. Huang, and J. F. Banfield, Water-driven structure transformation in nanoparticles at room temperature, *Nature*, 424, 1025–1029 (2003).

43. H.-D. Beckhaus, C. Rüchardt, M. Kao, F. Diedrich, and C. S. Foote, The stability of buckminsterfullerene (C_{60}): Experimental determination of the heat of formation, *Angewandte Chemie International Edition in English*, 31, 63–64 (1992).

44. H.-D. Beckhaus, S. Verevkin, C. Rüchardt, F. Diedrich, C. Thilgen, H.-U. ter Meer, H. Mohin, and W. Müller, C_{70} is more stable than C_{60}: Experimental determination of the heat of formation of C_{70}, *Angewandte Chemie International Edition in English*, 33, 996–998 (1994).

45. P. Nikolaev, M. J. Bronikowski, R. K. Bradley, F. Rohmund, D. T. Colbert, K. A. Smith, and R. E. Smalley, Gas-phase catalytic growth of single-walled carbon nanotubes from carbon monoxide, *Chemical Physics Letters*, 313, 91–97 (1999).

46. H. Y. Peng, N. Wang, Y. F. Zheng, Y. Lifshitz, J. Kulik, R. Q. Zhang, C. S. Lee, and S. T. Lee, Smallest diameter carbon nanotubes, *Applied Physics Letters*, 77, 2831–2833 (2000).

47. M. Lin, J. P. Y. Tan, C. Boothroyd, K. P. Loh, E. S. Tok, and Y.-L. Foo, Direct observation of single-walled carbon nanotube growth at the atomistic scale, *Nano Letters*, 6, 449–452 (2006).

48. G. B. Adams, O. F. Sankey, J. B. Page, M. O'Keefe, and D. A. Drabold, Energetics of large fullerenes: balls, tubes and capsules, *Science*, 256, 1792–1795 (1992).

49. I. Alstrup, A new model explaining carbon filament growth on nickel, iron and Ni–Cu alloy catalysts, *Journal of Catalysis*, 109, 241–251 (1988).

50. J. R. Rosrtup-Nielsen, Equilibria of decomposition reactions of carbon monoxide and methane over nickel catalysts, *Journal of Catalysis*, 27, 343–356 (1972).

51. G. L. Hornyak, L. Grigorian, A. C. Dillon, P. A. Parilla, and M. J. Heben, Temperature window for optimized single walled carbon nanotube synthesis by CVD, *Journal of Physical Chemistry B*, 106, 2821–2825 (2002).

52. G. L. Hornyak, Thermodynamic parameters of SWNT formation from the decomposition of methane over Fe–Mo catalysts, unpublished results (2002).

53. J. M. G. Vilar and J. M. Rubi, Thermodynamics "beyond" local equilibrium, *Proceedings of the National Academy of Science USA*, 98, 11081–11084 (2001).

54. J. M. Rubi, D. Bedeaux, and S. Kjelstrup, Thermodynamics for single-molecule stretching experiments, *Journal of Physical Chemistry B*, 110, 12733–12737 (2006).

55. H. J. F. Jansen, Thermodynamics, Dept. Physics, Oregon State University, www.physics.oregonstate.edu/~tgiebult/COURSES/ph642/Jansen.pdf, December 14 (2003).

56. R. M. Baigi, *Nanothermodynamics: A subdivision potential approach*. Presentation, Computational Physical Sciences Research Laboratory, Department of Nano-Science, IPM. Viewed (2007).

57. J.-R. Roan, Introduction to biophysics I. Presentation, Fall 2004, *Bioenergetics*, Chapter 3, Energy, 140.120.11.15/vedio/biophysics/Chinese/teachning/Fall%202003/Lecture_Notes.files/Lecture4_Bioenergetics.ppt (2004).

58. C. T. Campbell, S. C. Parker, and D. E. Starr, The effect of size-dependent nanoparticle energetics on catalyst sintering, *Science*, 298, 811–814 (2002).

59. T. L. Hill, *Thermodynamics of small systems*, Dover, New York (1994).

60. V. Garciá-Morales, J. Cervera, and J. Pellicer, Correct thermodynamic forces in Tsallis thermodynamics: Connection with Hill nanothermodynamics, *Physics Letters A*, 336, 82–88 (2005).

61. T. C. Ionescu, F. Qi, C. McCabe, A. Striolo, J. Kieffer, and P. T. Cummings, Evaluation of force field for molecular simulation of polyhedral oligomeric silsesquioxanes, *Journal of Physical Chemistry B*, 110, 2502–2510 (2006).

62. J. Kieffer, Simulation of self-assembly of functionalized silsesquioxane molecules, http://meetings.aps.org/link/BAPS, MAR.H14.3 (2005).

63. A. Striolo, C. McCabe, and P. T. Cummings, Effective interactions between polyhedral oligomeric silsesquioxanes dissolved in normal hexadecane from molecular simulation, *Macromolecules*, 38, 8950–8959 (2005).

64. W. D. Luedtke and U. Landman, Structure and thermodynamics of self-assembled monolayers on gold nanocrystallites, *Journal of Physical Chemistry B*, 102, 6566–6572 (1998).

65. X.-M. Bai and M. Li, Nucleation and melting from nanovoids, *Nano Letters*, 6, 2284–2289 (2006).

66. R. Chang and A. Violo, Insights into the effect of combustion-generated carbon nanoparticles on biological membranes: A computer simulation study, *Journal of Physical Chemistry B*, 110, 5073–5083 (2006).
67. S. Gheorghiu and M.-O. Coppens, Heterogeneity explains features of "anomalous" thermodynamics and statistics, *Proceedings of the National Academy of Science USA*, 101, 15852–15856 (2004).
68. www.stanford.edu/rmukul/tutorials/irreversible.pdf

Problems

8.1 Does the concept of continuum apply to thermodynamics?

8.2 List as many intensive properties as you can. Do the same for extensive properties.

8.3 Define a state function.

8.4 A great thermodynamic review is achieved by deriving a thermodynamic relation from first principles. Do this for the Clausius–Clapeyron equation.

8.5 Explain to a classmate the phenomena of reversibility and irreversibility.

8.6 Explain in more detail Boltzmann's famous equation: $S = k \ln \mathcal{W}$.

8.7 Construct a case for the existence of non-equilibrium thermodynamics in living things.

8.8 List some examples of the successful application of classical thermodynamics to nanosystems.

Chemistry: Synthesis and Modification

Carbon-Based Nanomaterials

Chemists are a strange class of mortals, impelled by an almost maniacal impulse to seek their pleasures amongst smoke and vapour, soot and flames, poisons and poverty, yet amongst all these evils I seem to live so sweetly that I would rather die than change places with the King of Persia.

JOHANN JOACHIM BECHER,
Physica Subterranea (1667)

It is disconcerting to reflect on the number of students we have flunked in chemistry for not knowing what we later found to be untrue.

ROBERT L. WEBER,
Science with a Smile (1992)

Chapter 9

Every attempt to employ mathematical methods in the study of chemical questions must be considered profoundly irrational and contrary to the spirit of chemistry...if mathematical analysis should ever hold a prominent place in chemistry—an aberration which is happily almost impossible—it would occasion a rapid and widespread degeneration of that science.

AUGUSTE COMTE,
Cours de Philosophie Positive (1830)

This special chapter is devoted to carbon-based nanomaterials. Carbon is the basic element of life. Carbon is the basic element of organic chemistry. Why should not carbon take the spotlight, albeit in a shared capacity, in nanoscience as well? Arguably, there are many equally fascinating nanomaterials and who is to say that quantum dots are not more fascinating? In answer, quantum dots are rather one dimensional (please pardon the pun), whereas carbon-based nanomaterials have a wide range of applications. We understand how size affects physical properties. That concept is rather clear by now—how a material is transformed into a semiconductor as small dimensions are achieved. We now ask you to believe the unbelievable: Materials made of one component (carbon), all of the same relative size with similar structure (e.g., a tube), can act as an insulator, a conductor, or a semiconductor. How can this be? How is this possible with an element that makes the strongest bonds and

one that has the ability to bond in so many ways? How can materials that consist entirely of carbon be so different?

With chapter 9, we begin the chemistry division of the text. This chapter represents the perfect transition between the physics and chemistry divisions of the book. Although there is plenty of material describing properties and phenomena (a.k.a. the physics of carbon nanotube structure and properties), the chapter includes plenty of material describing synthesis and chemical modification. The chapter then also serves to introduce intermolecular interactions without actually addressing them in detail; that is reserved for chapter 10. Chapters discussing the chemical bond (chapter 10) and supramolecular chemistry (chapter 11) follow; chapter 12 is dedicated to chemistry of nanomaterials in general. These four chapters comprise the chemistry division of the text. Biological topics are presented and discussed in chapters 13 and 14.

9.0   CARBON

The term carbon (from the French *charbone,* "glowing coal," based on the Latin *carbo*) was coined in the 1780s by French chemist Lavoisier. J. J. Berzelius of Sweden in 1824 also played a role by designating the letter "C" for carbon in the periodic table. Carbon is the most important element found in living things. Without carbon, there would be no life (at least as we know it). Carbon is special because of its ability to bond to many elements in many different ways. Its four-coordinate bonding capability allows for the synthesis of unique structures. Although silicon also is capable of four-coordinate bonding, its versatility is limited compared to carbon. Carbon is smaller than silicon. Ironically, silicon is the most important element in the semiconductor industry and a quite important player in polymer chemistry and sol–gel synthesis.

Potential applications of inorganic carbon materials are well known: lead pencils, graphite fibers, highly ordered pyrolytic graphite (HOPG), diamonds, amorphous carbon filters, chromatographic column packing materials and sieves, hardeners in steel, moderator material in nuclear reactors, plastics, paints, and lubricants. Organic chemistry has given us the polymer and its multitude of applications. Organic carbon, in the form of dinosaurs, Paleozoic and Mesozoic plants, and plankton, was transformed over the ages into oil that powers our civilization.

Inorganic carbon in the form of CO_2 is converted into high-energy sugars by the action of enzymes in plants. The sugars in turn are broken down during metabolism back into CO_2. Such inorganic carbon is extremely important on the global scale due to its strong ability to absorb infrared radiation. One can state with great confidence that all of supramolecular chemistry, an offshoot of biochemistry and organic chemistry, is a chemistry that relies heavily on the carbon atom. Biological materials are of course based on carbon; all life is carbon based. Starting with amino acids, lipids, proteins, nucleic acids, and a host of biological vitamins and cofactors all possess a carbon backbone. Because biological materials are nanomaterials—at least materials that are made of nanomaterials—carbon also is important to nanoscience. Many kinds of inorganic nanomaterials are based on carbon, which is the topic of this chapter.

The abundance of carbon in the universe is ca. 0.5 ppm; it is the sixth most abundant element in the universe. Its atomic mass is 12.011 a.m.u. Its electron configuration is $1s^2 2s^2 2p^2$. Carbon oxidation states are ± 4 and ± 2 with the valence state of 4 the most common. Its covalent radius is 0.77 Å, its ionic radius is 0.91 Å, and its atomic volume is 4.58 $cm^3 \cdot mol^{-1}$. The density of C is 2.62 $g \cdot cm^{-3}$, but of course this value varies significantly depending on the type of carbon.

Sir Harold W. Kroto of the University of Sussex and Richard E. Smalley of Rice University were awarded the Nobel Prize in chemistry, along with Robert Floyd Curl, Jr., also of Rice University, in 1996. In 1985, they discovered physical evidence for the existence of fullerenes [1]. Experimental mass spectrometry of molecular beam products showed discrete peaks that corresponded to 60 carbons. Higher order carbon peaks, corresponding to larger fullerenes, were also detected. Smalley was a molecular beam expert, specializing in the synthesis of laser vaporization techniques applied to semiconductor materials as well as carbon. Kroto was interested in stellar processes and how the vaporization of graphite may provide clues to their origin. The combination helped to kick start the *Nano Age*. The archetypal fullerene is C_{60}, a spherical structure that is composed of 60 carbon atoms configured as a soccer ball (**Fig. 9.1**).

Carbon nanotubes have been manufactured for centuries on Earth by natural pyrolytic processes, but they were formally discovered by Japanese electron microscopist Sumio Iijima of NEC Laboratories in Japan in 1991 [2]. Iijima and Toshinari Ichihashi (and, independently, Donald Bethune of the IBM Almaden Research Center) discovered single-walled nanotubes (SWNTs) in 1993 [3,4]. Multiwalled tubes have been around for decades, an undesirable by-product of catalytic processes.

9.0.1 *Types of Carbon Materials*

We are interested in four types of carbon materials in this chapter: diamond (specifically diamondoids and thin layer diamond-like films), graphite, fullerenes, and carbon nanotubes. Diamond is found in cubic and hexagonal forms. In cubic diamond, each carbon atom is linked to four others by sp^3 bonds to form a tetrahedral array. The bond length of each C–C in diamond is 1.544 Å, almost 10% longer than the C–C bond length in graphite. This, of course, makes sense because double or partially double carbon bonds are shorter than single bonds. Allotropes of carbon were displayed in **Figure 5.1**. There are many kinds of carbon materials: (1) diamond, Lonsdaleite, and diamondoid materials; (2) graphite (Bernal and others), turbostratic graphite, highly ordered pyrolytic graphite

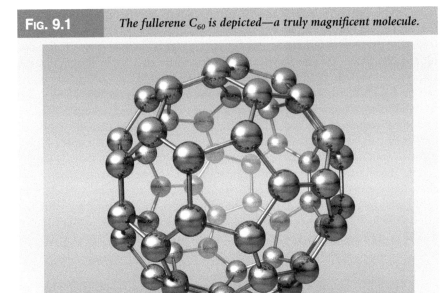

Source: Image courtesy of Professor Anil K. Rao, Department of Biology, the Metropolitan State College of Denver.

(HOPG), and lampblack; (3) fullerenes (spherical and elongated), nanobuds, nanohorns, and carbon nanotubes (MWNTs, SWNTs, and DWNTs); (4) glassy carbons; and (5) amorphous carbons. Carbon allotropes span a wide range of properties, in many cases representing the extremes of physical properties.

Graphite and Diamond.   No two materials made of the same element can be more different than graphite and diamond or more extreme in their differences. Graphite is one of the softest materials; diamond is one of the hardest. As a result, diamond is used as an abrasive and graphite is used as a lubricant (interestingly, graphite has recently been shown to be a poor lubricant in space—no intercalated moisture?). Diamond is also a good lubricant that is used in the space shuttle on bearings. Graphite is electrically conducting and diamond is electrically insulating; however, diamond can be made semiconducting by adding dopants (*p*-doping only). Diamond is optically transparent and graphite is optically dense. Diamond crystallites are members of the isometric (cubic) system; graphite crystals are members of the hexagonal system. Diamond has an isotropic structure, whereas graphite structure is anisotropic. Diamond conducts heat extremely well (one of the best in that capacity), but graphite, taken as a whole, does not (graphite$_{\parallel}$ conducts heat very well "in plane"; interplane conduction graphite$_{\perp}$ is poor). Interestingly, graphite is the stable form of the two. Diamond is constantly undergoing transition into graphite, albeit quite slowly; the activation energy for this conversion is nearly equal to the lattice energy of diamond. The fact that diamond was of made of carbon was verified in 1796.

Abraham Gottlob Werner in 1789 coined the term "graphite" (from the Greek *graphein*, "to write") because the material was used commonly in pencils. The structure of graphite was elucidated by John Bernal in 1924 and was henceforth

TABLE 9.1	*Physical Properties of Graphite, Diamond, and C_{60}*		
Physical property	**Graphite**	**Diamond**	**C_{60}**
Crystal structure	ABABAB plane stacking called Bernal stacking In-plane nearest neighbor distance = 1.421 Å ABCABC stacking is rare	Zinc blend, *fcc*, diatomic basis Hexagonal diamond is called Lonsdaleite (Wurzite crystal form)	Fullerene molecules crystallize in an *fcc* structure C_{60}–C_{60} nearest approach is 3.1 Å [5] Octahedral site radius: 2.07 Å
Lattice constant	$a = 2.461$ Å $c = 6.708$ Å Interplanar distance, $d = 3.34$ Å	$a = 3.567$ Å	$a = 14.15$ Å
Bond length	1.42 Å	1.544 Å Lonsdaleite = 1.52 Å	C–C = 1.455 Å C=C = 1.391 Å Cage diameter: 7.11 Å
Bonding	Trigonal bonding network is present in graphite—sp^2 with delocalized π-network Interlayer bonding due to overlap of π-orbitals and not van der Waals forces. Others claim that weak van der Waals forces are responsible for interplanar interaction due to the fourth electron in valence shell [6]	sp^3 hybridization to form tetrahedral bonds	Bonds vary with length and the type of associated polygon. Hybridization is an admixture of sp^2 and sp^3 Bond angles vary from 108 to 110°
Standard heats of formation	0.0 kJ·mol⁻¹	1.67 kJ·mol⁻¹	37.99 kJ·mol⁻¹
Band gap	No band gap, $E_g = 0.0$ eV (~ -0.04 eV)	Wide band gap, $E_g = 5.47$ eV	1.8 eV (1.7 eV)
Atomic density	1.133×10^{23} cm⁻³	1.77×10^{23} cm⁻³ Diamond has the highest atomic density	Molecular density: 1.44×10^{21} cm⁻³
Gravimetric density	2.26 g·cm⁻³	3.52 g·cm⁻³	1.65 g·cm⁻³ (1.72 g·cm⁻³)
Melting point	4200 K (3800 K)	4500 K (3700 K)	Sublimes at 800 K
Electrical conductivity	Semimetal	Insulator	Semiconductor
Thermal conductivity	In plane graphite (∥) has very high conductivity ~2500 W·m⁻¹·K⁻¹ or better	Diamond has the highest thermal conductivity ~2500 W·m⁻¹·K⁻¹	
Hardness	Graphite is a relatively soft material: Mhos hardness 1–2	The hardest material in nature: Mhos hardness ~10	

known commonly as Bernal graphite (consisting of an ABABAB structure). Graphite is essentially a planar material with each carbon atom bonded in a planar configuration to three others by σ-sp^2-bonds. The physical properties of graphite, diamond, and C_{60} are compared in **Table 9.1**. A graphic rendition of the unit cell of graphite is shown in **Figure 9.2** (left). On the right in **Figure 9.2**, the relationship between two planes of graphite, as found in the native material, is illustrated. The lattice parameters, bond length, and interlayer d-spacing are shown in **Figure 9.3**.

FIG. 9.2

The unit cell of graphite (containing two atoms) is shown on the left. On the right, the relationship between two planes of graphite is depicted.

9.0.2 Bonding in Carbon Compounds

Carbon has four electrons with which to form bonds: two $2s$ electrons and two $2p$ electrons; all four are in the $n = 2$ quantum energy state of the atom. Depending on the chemical circumstances, these electrons hybridize to form

FIG. 9.3

Three layers of graphite are shown. The d-spacing between layers is 3.34 Å. The lattice parameter a of the unit cell is equal to 2.46 Å. The C–C bond length is 1.42 Å.

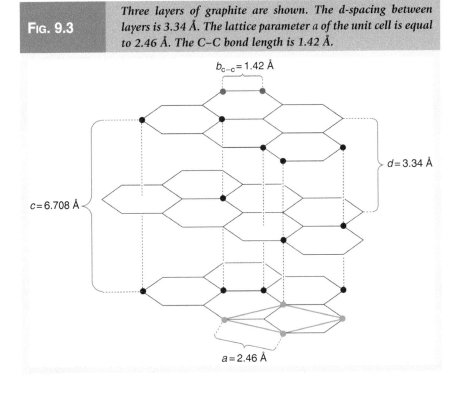

FIG. 9.4

Three kinds of hybrid orbitals of carbon are shown. Tetrahedral carbon has four equivalent sp³ orbitals with which to bond four substituents (e.g., methane, CH₄). The bond angle of sp³ hybrid orbitals is 109.5°. The bond angle of sp² hybrid orbitals is 120° (e.g., ethylene, H₂C=CH₂). The three sp² orbitals per carbon form a single plane. The orientation of the single available p-orbital is perpendicular to the plane of the sp² orbitals and is available to form one π-bond. Finally, the sp hybrid orbital is shown (e.g., acetylene, C₂H₂). The two sp orbitals form a 180° and are therefore linear. The two available p-orbitals are orthogonal to each other and are able to form two π-bonds.

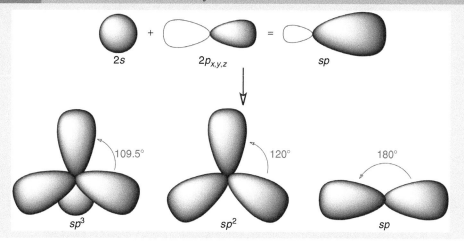

4 sp^3 (no p-orbitals available for bonding), 3 sp^2 (one p-orbital available for bonding), or 2 sp orbitals (two p-orbitals available for bonding). In the sp^2 and sp configurations, the remaining p-electrons go into forming π-bonds (**Fig. 9.4**). Carbon with sp^3 orbitals form tetrahedral structures (bond angle = 109.5°) is like that found in methane or in methylene groups of a hydrocarbon chain; sp^2 orbitals result in planar structures (bond angle = 120°) and are found in compounds like ethylene with π-bonding orbitals perpendicular to the plane of the molecule; and sp structures result in linear molecules (bond angle = 180°) like acetylene with capability to form two orthogonal π-orbitals. Hybrid orbitals are represented by the generalized wavefunction ψ:

$$\psi = s + \chi p_i \qquad (9.1)$$

where χ equals 1 for an sp orbital, 2.5 for an sp^2, and 3.5 for an sp^3. There are gradations of χ in molecules that have strained angles such as in fullerenes and carbon nanotubes. Carbon nanotubes, for example, have more sp^3 character than do unstrained materials like ethylene or graphite. Obviously, having more sp^3 character enhances three-dimensional flexibility in bond angle when compared to the planar sp^2 bond.

9.0.3 The Nano Perspective

Fullerenes were discovered recently and became important enough to award a Nobel Prize to their discoverers. Fullerenes are the perfect nanomaterial. The smallest fullerene, C_{60}, contains 60 carbon atoms formed into the structure of a soccer ball. They are stable (kinetically) and provide great versatility in chemical reactions and applications. The soccer ball shape is indicative of the high level of symmetry of C_{60}. C_{60} is a monodisperse material: All C_{60} are identical. Synthetic batches of C_{60} have 99.99% + purity; no other nanomaterial can make that claim—not quantum dots, not gold clusters, not carbon nanotubes, not diamondoids, or anything else.

Carbon nanotubes are found in the form of semiconductors, conductors, and insulators even though all three kinds have the same relative shape, relative size, and identical composition (only carbon). The different properties arise due to structural differences within the carbon nanotube skeleton. We are hard pressed to find examples of any other material that has such versatility. The concept of the space elevator (an elevator from the Earth's surface into geosynchronous orbit) is indeed quite an intriguing one. The only material that is strong enough of making the space elevator a reality is the carbon nanotube (**Fig. 9.5**).

FIG. 9.5 *The concept of the space elevator is shown. An elevator anchored on Earth (on the ocean surface) at one end and placed in geosynchronous orbit at the other is indeed a "far out" concept that stretches the imagination to its ultimate limit. Carbon nanotubes are the only material that has the necessary strength to pull this off.*

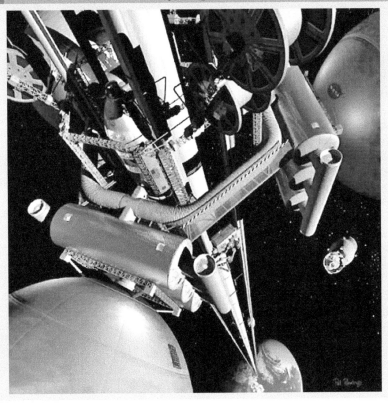

Source: Image courtesy of Michael J. Laine, The Liftport Group. With permission.

Diamondoid materials are also remarkable in the sense that monolayers of carbon in diamond form can be applied to surfaces and are manufactured under relatively mild conditions. Diamondoid materials are made of carbon that contains small amounts of hydrogen, oxygen, nitrogen, or sulfur. Because diamond has excellent physical properties (if hardness, thermal conductivity, and strength are important to you), it is expected to be one of the most valuable construction materials of the future. The strength-to-weight ratio, just like for carbon nanotubes, is expected to be 50 times better than any conventional material in use today. Some nanotechnology pundits already refer to the coming "diamond age" [7]. Applications of diamondoid materials (both current and potential) include: cages for drug and gene delivery, molecular capsules and probes, positional assembly, shape-targeted nanostructures, supramolecular synthesis and host–guest chemistry, fluorescent molecular probes, diamond monolayers, and seed molecules for dendrimer structures. Diamondoids are the only semiconductor materials that show a negative electron affinity [8].

9.1 FULLERENES

Fullerenes (buckyballs) were discovered in 1985 by researchers at Rice University and the University of Essex in England [1]. The general class of fullerenes includes carbon nanotubes as well as buckyballs and derivatives. In this section we focus mainly on the spherical or ellipsoidal forms of fullerenes: C_{60} and its allies. The presence of fullerenes has been detected in the outer bounds of our solar system in various nebulae. Buckyballs form under conditions of extreme temperature and pressure. Large fullerenes consisting of C_{60}, C_{70}, C_{100}, and C_{400} and many species in between have been found in the 4.6 billion-year-old carbonaceous chondrite Allende meteorite in Mexico [9]. There is also evidence that outer space gases are trapped in buckyballs. Australia's Murchison meteorite was found to enclose extraterrestrial helium gas within its cage. Some red giant stars emit a wavelength centered on 21 μm that is caused by interaction with large, complex molecules or solid materials [10]. Fulleranes, a hydrogenated from of fullerenes, are also associated with meteoric masses [11].

Carbon clusters comprise many interesting metastable species [12,13]. Clusters up to 10 atoms are stable in the form of linear chains, whereas 10–30 carbon atoms prefer to form into an annular configuration [14]. P. Eaton in 1964 synthesized a cubic form of carbon known as *cubane* [C_8H_8] [15]. $C_{20}H_{20}$, a dodecahedron, was synthesized in 1983 by L. Paquette [15]. The bond angles in cubane are 90° (highly strained), and in the dodecahedron, bond angles range from 108 to 110°; both deviate slightly from the ideal 109.5° but are characteristic of many "saturated" species [15]. Mass spectrometer data revealed that for carbon number N_C less than 30, clusters ranging from 2 to ca. 30 are well represented [15,16].

Molecular orbital theory calculations showed that linear structures form when N_C is an odd number (with *sp* hybridization) and cage structures like cubane when N_C is even. Linear or ring clusters with N_C = 3, 11, 15, 19, and 23 have tetrahedral bond angles and are therefore more abundant in the mass spectrum [15]. Between 30 and 40 carbons, clusters do not form metastable entities, but above with N_C greater than 40 atoms, enclosed fullerene structures are preferred [1]. The mass spectrum of ionized carbon materials shows a very large peak at $N_C = 60$. Because cage structures appear to possess inherent stability, one would expect that the stable forms of fullerenes also contain an even number of carbon atoms (e.g., C_{60}, C_{70}, C_{76}, C_{78}, C_{80}, C_{84}, C_{90}, etc.). These species are relatively more abundant than neighboring species that have odd numbers of carbon atoms [15].

9.1.1 Fullerene Properties

C_{60} is an extremely stable molecule (in the kinetic sense) with a carbon atom placed at each of the 60 vertices of the structure. C_{60} is not a stable molecule in the thermodynamic sense due to extreme bond strain. The structure is formally known as a truncated icosahedron. C_{60} has 90 edges and 32 faces and, of course, 60 vertices, one at each carbon atom. Of the 32 faces, 12 are pentagons and 20 are hexagons. Euler's theorem of polyhedra relates faces, edges, and vertices of a geometrical solid:

$$f + v = e + 2 \qquad (9.2)$$

where f, v, and e are the number of faces, vertices, and edges of the polyhedron, respectively. In a single layer of graphite, the C–C bond distance is on the average equal to 1.42 Å. However, C_{60} is not like graphite in that it contains pentagonal carbon structures intermingled with hexagons. Pentagons exhibit no double bonds, per se, and hexagons have alternating double bonds. Single bonds (existing on the pentagonal moieties) have bond length equal to 1.46 Å, while double bonds are shorter at 1.40 Å [17]. The diameter of C_{60} is 7.10 Å [1]. Each carbon atom forms a trigonal relationship with three other carbon atoms, just like in graphite. Every carbon in C_{60} is identical (evidenced by ^{13}C NMR spectrum that reveals just one peak at 142.68 ppm).

A more accurate description of the bonds in fullerenes is one that includes some sp^3 character that results in an sp^2–sp^3 admixture. Higher fullerenes are formed by the addition of hexagonal rings around their middles. The limiting case for a fullerene therefore is a carbon nanotube (or vice versa). If carbon nanotubes were to extend from the hemifullerene caps shown in **Figure 9.6,** an *armchair tube* is formed from the cap (a pentagonal at the apex) on the left and a *zigzag tube* is formed from the cap (the hexagonal apex) on the right. There is more discussion about carbon nanotubes to follow.

Fullerenes are resistant to pressure and reclaim their original shape even after experiencing over 3000 atm [18]. Bonding between fullerenes is accomplished via intermolecular interactions (noncovalent) and by covalent bonding. In general, fullerenes interact via weak van der Waals bonds, similar to those experienced between layers of graphite.

We discussed calorimetry briefly in chapter 8 [20]. The difficulty in applying calorimetric techniques to carbon nanotubes is that uniform samples are hard to obtain. This is not an issue with fullerenes, especially the popular ones like C_{60} and C_{70} for which very pure samples can be obtained. The energetic relations among various forms of carbon are graphically illustrated in **Figure 8.8**. The $\Delta H_f^\circ (C_{60})$ and $\Delta H_f (C_{70})$ of these basic fullerenes were found to be 42.5 kJ·mol^{-1}

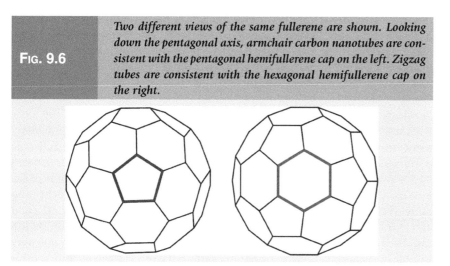

FIG. 9.6 *Two different views of the same fullerene are shown. Looking down the pentagonal axis, armchair carbon nanotubes are consistent with the pentagonal hemifullerene cap on the left. Zigzag tubes are consistent with the hexagonal hemifullerene cap on the right.*

and 40.4 kJ·mol⁻¹ respectively—quite a bit higher than the ΔH_f° (graphite), which is equal to 0.0 kJ·mol⁻¹.

9.1.2 Fullerene Synthesis

Fullerenes are routinely produced in nature by high-energy mechanisms such as lightning strikes and combustion processes. There is evidence that links fullerenes to the Permian extinction well over 250 million years ago. Apparently, fullerenes were produced either upon impact or were already part of the chemical makeup of the suspected asteroid. We discussed some methods to fabricate fullerenes and carbon nanotubes in chapter 4 that included arc-discharge, laser ablation, and flame methods. Electron beam methods produce higher order fullerenes with $N_C > 70$.

A few high-energy processes are capable of producing fullerenes like C_{60}. The electric arc discharge method is one of the most popular and was discussed briefly in chapter 4, "Fabrication Methods." In essence, the surface of a graphite rod is vaporized by exposure to an electric arc discharge under high-current and low-pressure conditions in the presence of an inert gas. If metal catalyst particles are integrated within the graphite target, carbon nanotubes are formed. The products resulting from this procedure usually have a high level of contamination and therefore require significant effort to purify. Fullerene products are collected on water-cooled surfaces (cold fingers) in the reactor. The resulting soot contains ca. 15% fullerene products, mostly C_{60}. Constituents are separated from the parent soot by mass by liquid chromatography (with toluene solvent) and other separation techniques.

Fullerenes and *endohedral fullerenes* (fullerenes that contain atoms, ions, or clusters enclosed within their cavity) are synthesized by laser vaporization. During laser ablation, a graphite target is exposed to extremely high temperatures. The best way to make fullerenes, however, is by combustion of a hydrocarbon fuel under low pressure. The fullerene products produced by the combustion method are significantly cleaner than those produced by arc discharge and laser ablation

(>95%). In addition, the combustion method requires significantly less energy. In this method, the oxidation region of the flame is decoupled from the post-flame region. This gives the operator control over equivalence ratio, temperature, and pressure in the primary zone. In a secondary zone, further residence time is allotted to recycling reactants.

Purification. Fullerenes, C_{60}, can be purified to extremely high levels: +99.99%. For a single stand-alone type of nanomaterial this is quite a remarkable achievement. Innovative, cost-effective methods have been established for fullerene purification. Fractional crystallization in 1,3-diphenylacetone demonstrated 99.5% pure product after three steps and 99.99% following adsorption of the residual C_{70} "contaminant" on charcoal [21]. The temperature-dependent solubility of C_{60} and C_{70} was found to attain a maximum level at 136°C for C_{60} and 41°C for C_{70}. Aggregation of C_{60} and C_{70} provides difficulties in separation but can be controlled by manipulation of the temperature. C_{60} is not very soluble in polar solvents like methanol ($S = 0.0$ mg·mL^{-1}) but is slightly soluble in ethanol ($S = 0.001$ mg·mL^{-1}), isooctane ($S = 0.026$ mg·mL^{-1}), and cyclohexane ($S = 0.036$ mg·mL^{-1}) [22,23]. Solubility is enhanced in solutions containing larger molecules and aromatics, such as benzene and toluene ($S = 2.8$ mg·mL^{-1}), 1,2-dichlorobenzene ($S = 8.5$ mg·mL^{-1}). Even better solubility is obtained in 1-methylnaphthalene ($S = 33.0$ mg·mL^{-1}), 1-phenylnaphthalene ($S = 50.0$ mg·mL^{-1}), and 1-chloronaphthalene ($S = 51.0$ mg·mL^{-1}) [22,23].

9.1.3 *Physical and Chemical Reactions of Fullerenes*

Carbon forms stable bonds in a few preferred geometric structural configurations: tetrahedral (109.5°, hybridization = sp^3, e.g., alkanes), planar (120°, hybridization = sp^2, e.g., alkenes), or linear (180°, hybridization = sp, e.g., alkynes). The preferred geometric configuration for aromatic compounds is also planar. Any deviation from the ideal geometric configuration results in bond strain, higher energy, and enhanced reactivity. Bond strain relief is the driver for fullerene addition chemistry. Any addition to C_{60} results in bond strain relief for all remaining fullerene carbons [24].

Fullerenes are considered to be *soft electrophiles*, which are molecules that are ready to accept electrons from a donor molecule during the first step in a potential chemical reaction. Fullerenes usually are able to accommodate three additional electrons from donors such as hydrogen, methyl groups, and amines. There are three general ways to modify fullerenes: (1) methods that enclose a metal or other small species (e.g., *endohedral* fullerenes); (2) chemical modification of the surface of the fullerene (e.g., to form *exohedral* complexes); and (3) assembly of fullerene one-, two-, and three-dimensional arrays by modification methods 1, 2, or intermolecular interactions. Of course, there do not seem to be any chemical or physical restrictions on combining all three types of fullerene modification methods to form superstructural nanomaterials. *Fullerites* (or fullerides), for example, are crystalline structures of fullerenes composed entirely of C_{60} blocks. The C_{60} form an *fcc* structure held together by weak intermolecular forces with lattice constant $a = 14.17$ Å and the C_{60}–C_{60} bond distance equal to 10.02 Å. Interestingly, the C_{60} molecules comprising the crystal are able to rotate isotropically (freely) at room temperature (**Fig. 9.7**) [17].

FIG. 9.7	*A fulleride crystal is displayed. The crystal shown is an fcc type within which van der Waals gaps separate the constituent fullerenes. Intercalation of metals like lithium between and among fullerenes occurs to form stable complexes.*

Source: Courtesy of Professor Anil K. Rao, Department of Biology, the Metropolitan State College of Denver.

Exohedral Fullerenes. Fullerenes can be modified by chemical methods. Monoadducts form when one type of chemical group is bonded to the surface [25]. An interesting procedure to form monoadducts that involves many types of bonding is shown in **Figure 9.8.** A hierarchical structure is achieved by utilizing all the types of bonding available on a single adduct molecule: thiol linkage to gold surface; lateral hydrogen bonding between adduct molecules; and π–π stacking between fullerenes (**Fig. 9.9**). Combination of multifunctional fullerenes is another approach to forming arrays. In this case, the required functionalities are all located on a single fullerene. For example, four addends are added around the equator of the fullerene and a tethering group is attached to one of the poles [25]. If the four equatorial addends have different length, expanded forms of patterning are possible.

Two- and three-dimensional arrays of fullerenes can be produced by various polymerization methods. For example, dimers result if double bonds between adjacent fullerenes are broken by exposure to laser or UV light and subsequently connected via a cyclobutane derivative—by a [2 + 2] cycloaddition mechanism—forming four-membered carbon rings (**Fig. 9.10**). The cell parameter ($a = 14.17$ Å) of the C_{60} *fcc* lattice is slightly decreased by the formation of these linkages ($a_{rxn} = 14.05$ Å) [26,27].

Rhombohedral, linear orthorhombic, tetrahedral, and other two-dimensional lattice configurations are possible. The arrangement depends on the level of modification, the placement of the functional groups on the fullerene, and length and chemistry of the adduct species (**Fig. 9.11**).

Formation of single adduct (single functional) fullerenes: Each fullerene has 30 π-bonds, many of which are available for supramolecular synthesis, which involves intermolecular interactions. All functionality provided for a chemically modified fullerene in this figure occurs at a single addend attached to the fullerene with covalent bonds. The purpose of the addend is to modify the fullerene with a chemical moiety that provides: (1) an anchor substituent between the fullerene and a metal surface, and (2) a means to interlink with other fullerenes derivatized in the same way or in different ways. The monoadduct contains thiol groups for bonding to a gold surface and carboxylic acid groups to provide intermolecular bonding (via hydrogen bonds) between fullerenes and the π–π interaction capability.

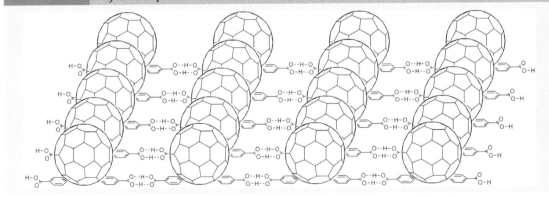

Source: Image redrawn with permission from Glen P. Miller, University of New Hampshire, Center for High-Rate Nanomanufacturing.

There are several types of fulleride–polymer networks. It is relatively easy to bond C_{60} together to form such networks to yield many kinds of one-dimensional, two-dimensional, and layered structures. The Li_4C_{60} polymer serves as an excellent example of a mixed-bonding configuration that yields a close-packed structure.

A two-dimensional array of fullerenes (modified as pictured in Fig. 9.8) on gold substrate is shown. Along the lateral (x) axis, modified fullerenes are held together by hydrogen bonds between carboxylate functionalities. Along the longitudinal (y) axis, fullerenes are attracted to each other via π–π interactions. Molecular models have confirmed the viability of such a system.

Source: Image redrawn with permission from Glen P. Miller, University of New Hampshire, Center for High-Rate Nanomanufacturing.

FIG. 9.10	*C₆₀ dimer formed by [2 + 2] cycloaddition. The structures are formed during prolonged exposure to UV or laser light. The dimers (and higher forms) are insoluble in solvents that buckyballs normally disperse. The dimers revert to individual C_{60} molecules by heat treatment under ambient pressure conditions.*

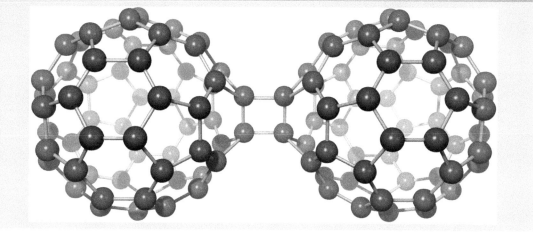

Source:　European Synchrotron Radiation Facility, http://www.esrf.eu/UsersAndScience/Publications/Highlights/2005/Materials/MAT11; Image courtesy of Professor Anil K. Rao, Department of Biology, the Metropolitan State College of Denver. With permission.

XRD of a powder sample revealed that the anisotropic fulleride polymer structure was body centered monoclinic with lattice parameters $a = 9.3267$ Å (involving the C–C lateral bonding scheme); $b = 9.30499$ Å (involving the cycloaddition scheme); $c = 15.03289$ Å and $\beta = 90.949°$. Center-to-center dimensions were found to be approximately 9.33 Å along a and 9.05 Å along the b lattice parameter. Lithium ions reside in pseudotetrahedral and pseudo-octahedral holes formed between the layers [26,27].

On the other hand, Na_2RbC_{60} fullerides are bridged by C–C bonds only. Na_4C_{60} fullerides are bridged with four C–C bonds. Factors that influence the mode of bonding include the charged state of the fullerene and the steric influence of the alkali ion. Two Li^+ actually fit into the pseudo-octahedral site of the Li_4C_{60} fulleride. The charge on the fullerene is less than –4 if Li is applied and approximately equal to –4 if Na is incorporated. All caged carbon structures from fullerenes to nanotubes are excellent electron acceptors.

Endohedral Fullerenes.　Metal atoms inserted within the fullerene cage during energetic high temperature synthesis form endohedral fullerene structures (a.k.a. "dopyballs") and are designated by the notation $X@C_{60}$ where X represents the enclosed metal. Lithium atoms, for example, can be inserted within a fullerene cage by bombardment with 13-eV Li atoms. Radioactive isotope ^{133}Xe gas ions were implanted into fullerenes by an "isotope separator online" (ISOL) instrument [28]. Fullerene targets were made by vacuum evaporation of C_{60} or C_{70} onto Ni foils. Radioactive ^{133}Xe atoms were ionized and accelerated to 40 keV. The Xe ions were separated by mass from the stable ^{129}Xe isotope beforehand. The energetic Xe ions penetrate the fullerene by expanding one of

FIG. 9.11
Generic C_{60} arrays are shown: Top left: rhombohedral; top right; tetrahedral; and bottom: linear orthorhombic polymers. Red lines represent chemical linking groups. Blue squares represent fullerene dimer linkage. Three-dimensional arrays are also possible.

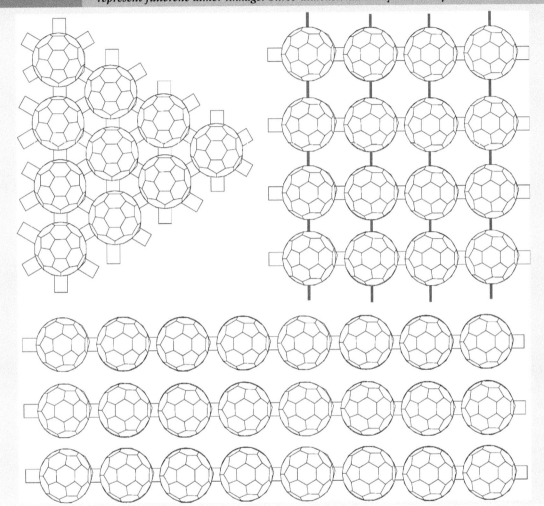

Source: European Synchrotron Radiation Facility, http://www.esrf.eu/UsersAndScience/Publications/Highlights/2005/Materials/MAT11. With permission.

the six-membered rings until the center of the fullerene cavity has been achieved (**Fig. 9.12**). Following implantation, the fullerenes were removed from the Ni target and dissolved in *o*-dichlorobenzene and injected into an HPLC chromatographic column (Cosmosil 5PYE) at a flow rate of 1 mL·min⁻¹. The desired product was detected with UV spectroscopy. These endohedral radioisotope fullerenes (RIF) have application in radiopharmaceutical therapy. Subsequent chemical modification is able to render the RIF hydrophilic with molecular recognition to target cancer cells [28].

The endohedral $Er_3N@C_{60}$, for example, has interesting magneto-topical activity as a solid-state quantum information processing system. This procedure

FIG. 9.12	*Endohedral fullerene. Xenon is inserted into the fullerene cavity by the technique of ion implantation.*

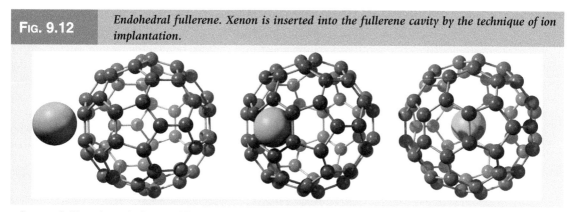

Source: S. Watanabe et al., *Journal of Radioanalytical and Nuclear Chemistry*, 255, 495–498, 2003; Courtesy of Professor Anil K. Rao, Department of Biology, the Metropolitan State College of Denver. With permission.

involves the fabrication of an open-grid endohedral fullerene surface array by template-assisted assembly on an oxide-crystal surface. The procedure applied is as follows: (1) substrate preparation: Ar^+ sputter of 0.5%-wt, Nb-doped $SrTiO_3$ (001) single crystal surface in ultrahigh vacuum; (2) anneal of substrate: $c(2 \times 4)$ reconstruction was generated by annealing the sputtered surface at 1090°C; repeated annealing forms; (3) application of endohedral fullerenes: thermally stable $Er_3N@C_{60}$ is deposited on the surface from the vapor phase at 480°C from a Createc Knudsen cell for 20–30 minutes. A 0.15- to 0.20-monolayer coverage per deposition is achieved by this process. Result no.1: close-packed surface ordering of fullerenes is achieved on a $c(4 \times 2)$ surface reconstruction. Result no.2: the heights of the fullerene islands are ca. 1 nm in the z-direction .

Fulleranes. A less well-known form of fullerenes is the family of fulleranes—hydrogenated versions of the fullerenes. Fulleranes can be prepared from various solvents using Zn–HCl as the reducing agent [11]. The molecule $C_{60}H_{36}$ is not stable in air in the presence of light. Oxidation of this species yields hydroxyl and ketone groups with concomitant cage breakdown. The presence of ozone accelerates the breakdown of the cage. The absorption band of $C_{60}H_{36}$ matches that of several infrared emission bands detected from astrophysical objects such as nebulae [11]. Other fulleranes such as $C_{280}H_{120}$, $C_{70}H_{60}$, $C_{100}H_{60}$, $C_{120}H_{72}$, and $C_{336}H_{144}$ were modeled by computer simulations. Icosahedral fulleranes such as $C_{80}H_{80}$ and $C_{180}H_{180}$ have shown remarkable stability [29].

9.2 CARBON NANOTUBES

Nanomaterials are remarkable materials. Carbon nanotubes, especially single-walled carbon nanotubes, are one of the most remarkable of nanomaterials. Their physics and chemistry are both well understood. Their synthesis is closing the gap between short tubes and long tubes, polydisperse and monodisperse batches, chiral and achiral forms, metallic and semiconducting, and pure and highly pure. Their potential is enormous, although it remains to be seen how many commercially viable applications are developed over the next decade.

Their electrical, mechanical, chemical, and thermal properties are, in no uncertain terms, quite fantastic. The discovery of fullerenes, in addition to adding another allotrope to known forms of carbon, ignited the imagination. We now present our discussion on carbon nanotubes—elongated versions of the fullerenes.

We do not focus our effort on multiwalled tubes in this section although manufacture of multiwalled nanotubes (MWNTs) has achieved the status of economic viability—mostly in the capacity of fillers for polymeric matrix materials. The first nanotubes discovered were MWNTs; their discovery preceded that of SWNTs by several years.

9.2.1 Structure of Single-Walled Carbon Nanotubes

Two types of high symmetry SWNTs are considered to be archetypal: the zigzag SWNT and the armchair SWNT. All other kinds of SWNTs are called *chiral* nanotubes and have structure somewhere in between these two limits. Each chiral nanotube, as the term implies, has a mirror image. The two archetypal tube types and a generic chiral form are shown in **Figure 9.13a**. In the figure, a small planar segment of graphite is laid out to form a plane—an unrolled carbon nanotube. Just like graphite, carbon atoms exist in a hexagonal configuration. In zigzag tubes, hexagons of carbon are arranged around the circumference so that an apex of the hexagon is oriented parallel to the longitudinal axis (the long axis) of the SWNT (shown at the right in the figure). Armchair SWNTs, on the

FIG. 9.13   *(a) An unrolled carbon nanotube (called a graphene sheet) is displayed. Three types of single-walled carbon nanotubes are enclosed in red rectangles. Bottom right: a zigzag tube SWNT in which the hexagons point along the long axis of the tube; top left: an armchair tube SWNT in which the flat side of the hexagon lies along the long axis of the tube; and bottom left: a chiral nanotube in which the configuration lies between the two extremes. (b) Single-walled carbon nanotubes are shown schematically from the perspective of the tube axis. Left: armchair SWNT; middle: chiral SWNT; and right: zigzag SWNT. The vector renditions represent lines through the long axis (the pointing end) of the hexagones.*

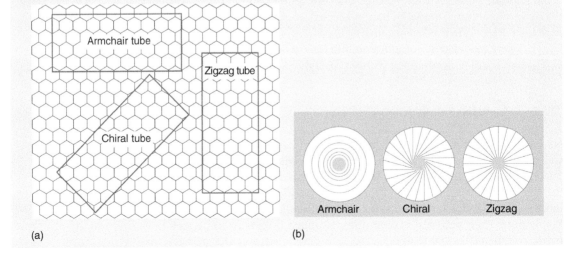

(a)   (b)

other hand, have the apex oriented along the circumference of the tube and a C–C bond oriented along the longitudinal axis of the tube. Another way to visualize the two is to imagine that the tubes are derived from the hemispherical fullerenes shown in **Figure 9.6.** The end cap of a zigzag tube corresponds to the hemifullerene depicted on the right in the image (e.g., the hexagonal derivative). The cap of an armchair tube, on the other hand, corresponds to the fullerene on the left (e.g., the pentagonal derivative). Front-end views of the three basic kinds of fullerenes are represented schematically in **Figure 9.13b.** The physics of SWNTs is extremely rich and the mathematical description of SWNTs is quite well defined.

Vector Notation for SWNT Structures. Information for this section was gleaned from excellent textual and Internet sources—in particular, *Physical Properties of Carbon Nanotubes* by R. Saito, G. Dresselhaus, and M. S. Dresselhaus (Imperial College Press, London, 2004) and *Carbon Nanotubes and Related Structures* by P. J. F. Harris (Cambridge University Press, Cambridge, 2001). Most nanotubes have chiral form [30]. Specification of the structure is accomplished via the chiral vector **C**. The origin of the vector intersects two equivalent points on the graphene sheet. There are many ways to produce degeneracies (folding the sheet in different ways that produce the same nanotube). Just as in group theory, there are irreducible representations. The sector bounded by two chiral vectors is shown in **Figure 9.14** and enclose the irreducible region. This wedge is 1/12 of the graphene lattice [30].

FIG. 9.14	*Illustration of nanotube structure and the chiral vector. The dark arrows represent the tube axes of zigzag and armchair tubes. The arrow in the middle represents the tube axis for chiral tubes, the vector sum of the others.*

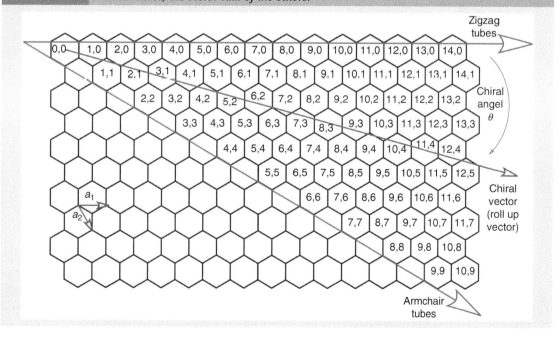

The chiral vector is described by

$$C = n\mathbf{a}_1 + m\mathbf{a}_2 \tag{9.3}$$

where $\mathbf{a}_1$ and $\mathbf{a}_2$ are the unit base cell vectors of graphite (e.g., in the form of the graphene sheet) and $n \geq m$ [30]. The unit cell of the graphene sheet is not the same as the translational unit cell of the nanotube along the tube axis (more about that later). The magnitude of the unit cell vectors $\mathbf{a}_1$ and $\mathbf{a}_2$ of graphite is

$$|a_1| = |a_2| = a = 0.246 \text{ nm} \tag{9.4}$$

The measured C–C bond length is 0.142 nm. Derivation of the bond length from the lattice constant is given by

$$b = \frac{a}{\sqrt{3}} = 0.142 \text{ nm} \tag{9.5}$$

The diameter of the nanotube is also given by the ratio of the circumference of the tube, L, to π:

$$d_{\text{SWNT}} = \frac{L}{\pi} = a\sqrt{n^2 + nm + m^2} \tag{9.6}$$

or, in terms of n and m and b, is given by

$$d_{\text{SWNT}} = \frac{|C|}{\pi} = \frac{b}{\pi}\sqrt{3(n^2 + mn + m^2)} \tag{9.7}$$

where b is the C–C bond length in graphite as before. These equations assume that the C–C bond length is 0.142 nm, when in actuality it is longer.

Because there is hexagonal symmetry, the only tubes that need to be considered lie within the range $0 \leq |m| \leq n$ [30–33]. If $m = 0$, all tubes are zigzag tubes. If $m = n$, all tubes are armchair tubes. Any other combination of m and n yields chiral tubes. The range of the chiral angle θ is $0 \leq \theta < 30°$. The chiral angle for zigzag tubes is $0°$, and the chiral angle for armchair tubes is $30°$. The chiral angle θ is given by

$$\cos\theta = \frac{2n + m}{2\sqrt{n^2 + nm + m^2}} \tag{9.8}$$

With some trigonometric rearrangement, another common form is

$$\theta = \sin^{-1}\frac{\sqrt{3}m}{2\sqrt{n^2 + nm + m^2}} \tag{9.9}$$

The *translation vector* $\mathbf{T}$ is the unit vector of the carbon nanotube. $\mathbf{T}$ is parallel to the tube axis and normal to the chiral vector. The translational unit cell for all nanotubes is in the form of a cylinder [30]. Translational unit cells for zigzag and armchair tubes are shown in **Figure 9.15**. For chiral tubes, the unit cell will be much larger than for either the armchair or the zigzag tube.

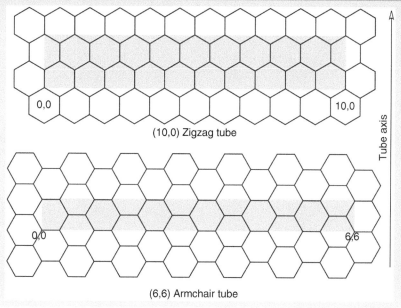

FIG. 9.15

The translational unit cell of a (10,0) zigzag tube is shown above. The width of the cell is equal to $\sqrt{3}a$, where a is the unit vector for the graphene lattice. The translational unit cell for a (6,6) armchair tube is shown below. The width of this unit cell is equal to the unit vector a of the graphene lattice [30]. Chiral tubes that have lower symmetry are characterized by larger cells than the two depicted in the figure [30]. The magnitude of the chiral vector is the circumference of the tube. The magnitude of the translational vector is the width of the cell. The chiral vector (along the circumference) and the translational vector (along the tube axis) are orthogonal to one another.

(10,0) Zigzag tube

(6,6) Armchair tube

Tube axis

Vector Notation and Electrical Conductivity. Not all carbon nanotubes conduct electricity. Some kinds are semiconducting. The distinction between the two kinds is found in the vector notation. Tubes are conducting if $(m - n)$ or $(n - m)$ is a multiple of 3: $n - m = 3q$, where q is an integer. Therefore, all armchair tubes are metallic because $n - m = 0$, all zigzag tubes with n a multiple of 3 are conducting—for example, (6,0), (9,0,), (12,0), etc.—and all chiral tubes that fall into the general category of $n - m = 3q$ are conducting as well (e.g., (6,3), (9,6), (12,9), etc.). In **Figure 9.16,** an unrolled nanotube graphene grid depicting the kinds of nanotubes is utilized again to show which tubes are semiconducting or metallic.

How to Make a Carbon Nanotube. A good exercise for students is to print out a hexagon matrix of carbon atoms onto a piece of transparency film. Print a graphene lattice on the transparency and fold the tube into various configurations. A sample is shown in example 9.1 (**Fig. 9.17**).

For more information concerning vector relationships and symmetry in carbon nanotubes, consult the text *Physical Properties of Carbon Nanotubes* by R. Saito, G. Dresselhaus, and M. S. Dresselhaus (Imperial College Press, London, 2004).

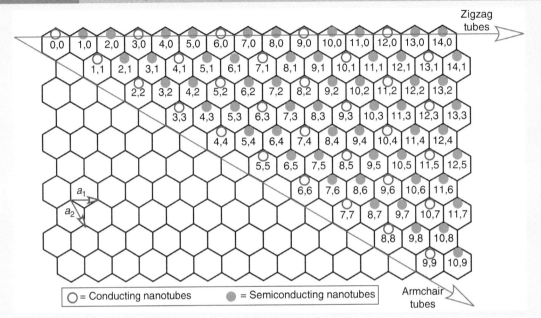

Fig. 9.16 SWNTs are either electrically semiconducting or electrically conducting (metallic). Semiconducting SWNTs are represented by solid green dots and conducting tubes by hollow purple dots. Tubes are conducting if $(m-n)$ or $(n-m)$ is a multiple of 3: $n-m=3q$, where q is an integer. Therefore, all armchair tubes are metallic, all zigzag tubes with n a multiple of 3 are conducting, and all chiral tubes that fall into the general category of $n-m=3q$ are conducting as well.

9.2.2 Physical Properties of Single-Walled Carbon Nanotubes

We have been saying all along that SWNTs have some incredible physical properties. We now discuss a few of the well-known ones.

Mechanical Properties.   CNTs in general are some of the strongest and stiffest materials known to science. As opposed to other materials that consist of melded

---

EXAMPLE 9.1   *Making a Carbon Nanotube*

The best way to understand the geometrical relationships within a carbon nanotube is to construct one. Start by printing out a graphene sheet on a clear 8.5-× 11-in. slide (transparency film, used by overhead projectors and very rare nowadays). A good number is ca. 15 hexagons to fill the width of your page.

The simplest tube to make is a zigzag tube. Let us start with the (10,0) tube. Take the (0,0) atom and superimpose it on the top of the (10,0) atom. You now have a (10,0) zigzag tube. A (5,5) armchair tube is also constructed in *Figure 9.17*.

Chiral tubes are made in a similar fashion. Notice that the zigzag and armchair tubes have a high degree of symmetry. You will also notice that when the sheet is rolled up, the ends of the chiral vector should intersect. The magnitude of the chiral vector then is the circumference of the newly formed tube.

| FIG. 9.17 | *Making (10,0) and a (5,5) carbon nanotubes by folding a graphene transparency.* |

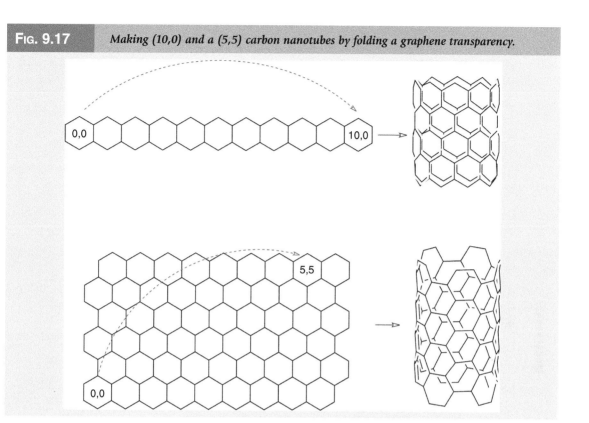

grains and micron-sized substructure, the constituents holding nanotubes together are carbon bonds—the strongest in nature. The breaking point therefore of the C–C or C=C bond must be an enormous number. The theoretical breaking point of carbon nanotubes then must be proportionally enormous and a function of the number of C–C bonds that contribute to the circumferential structure (the translational unit cell). The covalent sp^2–sp^3 bonds that hold carbons in place in a tube are the basis for the strength of the material. A multiwalled carbon nanotube demonstrated experimental tensile strength on the order of 63 GPa compared to high-carbon steel that measured in at a mere 1.2 GPa [34].

The specific strength of carbon nanotubes is on the order of 48.5×10^3 kN · m · kg^{-1}. The specific strength of steel is 1.54×10^2 kN · m · kg^{-1} [35]. The combination of high tensile strength with low density in a material is an engineer's dream—definitely those of aerospace engineers. The tubes undergo plastic deformation under extensive tensile stresses. Under compression, tubes are not very strong due to their hollow structure and they tend to buckle (**Table 9.2**).

A good exercise is to calculate the diameter of a megabundle of SWNTs that would be required to pick up the Empire State Building (weight = 365,000 tons). Information about the type of nanotube, the geometrical packing area, and the density of a bundle of SWNTs would be required. For the sake of simplicity, assume the tubes are zigzag (10,10) tubes. Under mechanical tensile stress, the tube undergoes ~0.35% axial stretch and ~0.45% radial stretch [37,38].

TABLE 9.2	*Mechanical Properties of Materials*			
Material	**Young's modulus (GPa)**	**Tensile strength (GPa)**	**Density (g · cm⁻³)**	
SWNTs	1054	150	1.33	
MWNTs	1200	150	2.6	
Steel	208	0.4	7.8	
Epoxy	3.5	0.005	1.25	
Wood	16	0.008	0.6	

Source: Basic properties of carbon nanotubes, Applied Nanotechnologies, Inc., applied-nanotech. com/cntproperties.htm#Mechanical%20Properties (2005).

EXAMPLE 9.2	*Tensile Strength of Carbon–Carbon Bonds*

(a) What is the theoretical tensile strength σ_S in gigapascals (1 GPa = 10^9 N · m²) of carbon–carbon bonds of different order [Gilkes]? Assume that all bonds stretch to 2.5× their respective equilibrium length before breaking.

(b) How much mass under Earth's gravitational pull are these bonds able to hold without breaking?

Type bond	Bond order	Bond length, L (nm)	Bond dissociation energy, E_{BDE} (kJ · mol⁻¹)	Bond break length, $d = 2.5L - L$
C–C	1.0	0.154	348	+0.231
C≐C	~1.33	0.142	480[a]	+0.213
C=C	2.0	0.134	614	+0.201
C≡C	3.0	0.120	839	+0.180

[a] Hypothetical value based on a bond order ≈1.33 of graphite, bond length ≈0.142 nm.

Solution:

The bond dissociation energy for one C–C bond: $E_{Carbon\ Bond} = \left(\dfrac{E_{BDE}}{N} \right) \left(\dfrac{10^3\ J}{kJ} \right)$ (9.10)

where N is Avogadro's number

The force required to break one C–C bond: $F_{Carbon\ Bond} = \left(\dfrac{E_{Carbon\ Bond}}{d} \right)$ (9.11)

The covalent radius r_C of a carbon atom is 0.077 nm—the cross-sectional area A_C:

$$A_C = \pi r_C^2 = (3.1416)(0.077 \times 10^{-9}\ m)^2 = 1.86 \times 10^{-20}\ m^2$$
     (9.12)

Tensile strength σ_S (1 Pa = N · m⁻²) of one C–C bond: $\sigma_{Carbon\ Bond} = \dfrac{F_{Carbon\ Bond}}{A_C} \left(\dfrac{1\,GPa}{10^9\,Pa} \right)$.

The maximum force F_{Max} that one C–C bond is able to hold without breaking is the product of the tensile strength and the area of the material A_M (e.g., the diameter of a cable)

If it is just for one C–C bond, then $A_M = A_C$: $F_{Max} = A_M \sigma_S = A_C \sigma_{Carbon\ Bond}$ (9.13)

EXAMPLE 9.2 (CONTD.)	*Tensile Strength of Carbon–Carbon Bonds*

The maximum mass m_{Max} one C–C is able to hold without breaking: $m_{Max} = \dfrac{F_{Carbon\ Bond}}{g}$ (9.14)

Worked example for the single C–C bond:

$$E_{C-C} = \left(\frac{348\ \text{kJ} \cdot \text{mol}^{-1}}{6.022 \times 10^{23}\ \text{bonds} \cdot \text{mol}^{-1}}\right)\left(\frac{10^3\ \text{J}}{\text{kJ}}\right) = 5.78 \times 10^{-19}\ \text{J} \cdot \text{bond}^{-1};$$

$$F_{C-C} = \left(\frac{5.78 \times 10^{-19}\ \text{J} \cdot \text{bond}^{-1}}{0.231 \times 10^{-9}\ \text{m}}\right) = 2.50 \times 10^{-9}\ \text{N}$$

$$\sigma_{C-C} = \frac{2.50 \times 10^{-9}\ \text{N}}{1.86 \times 10^{-20}\ \text{m}^2}\left(\frac{1\ \text{GPa}}{10^9\ \text{Pa}}\right) = 134\ \text{GPa}; \quad F_{Max} = (1.86 \times 10^{-20}\ \text{m}^2)(134\ \text{GPa})\left(\frac{10^9\ \text{Pa}}{\text{GPa}}\right) = 2.50 \times 10^{-9}\ \text{N}$$

$$F_{Max} = (1.86 \times 10^{-20}\ \text{m}^2)(134\ \text{GPa})\left(\frac{10^9\ \text{Pa}}{\text{GPa}}\right) = 2.50 \times 10^{-9}\ \text{N}; \quad m_{Max} = \frac{F}{g} = \frac{2.50 \times 10^{-9}\ \text{N}}{9.801\ \text{m} \cdot \text{s}^{-2}} = 2.54 \times 10^{-10}\ \text{kg}$$

Type of carbon bond	Bond dissociation energy $(J \times 10^{-19})$	Force, $F (N \times 10^{-9})$	Tensile strength, σ_s (GPa)	(b) Mass, m_{Max} $(kg \times 10^{-10})$
C–C	5.78	2.50	134	2.54
C≐C	7.97	3.74	201	3.82
C=C	10.2	5.07	272	5.18
C≡C	13.9	7.74	416	7.90

Clearly, the tensile strength increases with the bond strength. Therefore, theoretical limit of the tensile strength of one of the strongest bonds in nature is quite high. If the mass of two carbons in the bond is 1.99×10^{-26} kg, these bonds can hold about 10^{15}–10^{16} times their own mass.

Electrical Conduction. Nanotubes are one-dimensional nanomaterials. Many refer to them as *pseudo-one-dimensional nanomaterials*. Due to confinement, electrical conduction is considered to be quantized, and because the mean-free path of electrons is on the order of the dimensions of nanotubes, electrical conduction is also considered to be ballistic. Because resistivity in nanotubes is constant, nanotubes are excellent materials for high current applications [39]. S. Frank et al. in 1998 showed, by scanning probe microscopic methods, that jumps in conductance were exhibited by SWNTs in contact with mercury [40]. D. Tománek et al. confirmed these results in 1999 [41].

Metallic single-walled carbon nanotubes, in theory, should be able to carry electric current density greater than 1000× that of silver or copper [42].

Thermal Properties. Along the tube axis, carbon nanotubes have excellent thermal conductivity to the tune of ~6000 W · m^{-1} · K^{-1} at room temperature [42]. On the other hand, copper, an excellent conductor of heat, is valued at 385 W · m^{-1} · K^{-1}

[42]. Along the diameter of the tube, however, carbon nanotubes are insulating. The thermal stability of carbon nanotubes is very high (ca. 3100 K in vacuum). In the presence of oxygen, carbon nanotubes are easily oxidized at ~900 K.

Optical Properties.   Metallic tubes, as for any other metal, have no band gap (0.0 *e*V). Metallic carbon nanotubes, as we found out earlier, arise when the condition $n - m = 3q$ is satisfied. Semiconductor tubes, on the other hand, do have a bandgap that is a function of the diameter, another example of a nanomaterial property that relies on size . The bandgap ranges in energy from 0.4 to 0.7 *e*V [39]:

$$E_g = \frac{2wb}{d_{\text{SWNT}}} \tag{9.15}$$

where E_g is the band gap in electron volts, w is the tight-binding overlap energy (2.7 ± 0.1 *e*V), and b is the bond length as before.

The difference is best illustrated in a *density of states* (DoS) plot. DoS is obtained by scanning tunneling spectroscopy (STS). STS is obtained by an STM parked over a predesignated region of the sample while the bias voltage transmitting from the tip is scanned from plus to minus. The DoS is a function of the number of quantum states in the energy range between E and $E + dE$ divided by the product of dE and the volume of the sample. Examples of generic DoS plots of a metallic and a semiconducting nanotube are shown in **Figure 9.18.** In a metal, the Fermi level is populated and therefore the DoS shows a positive population in that area. For a semiconductor, the Fermi level (technically, the chemical potential) is empty (the bandgap), and the DoS in that region has zero population (no states). As energy is increased on either side, peaks called van Hove singularities appear. These regions are responsible for the optical properties of carbon nanotubes (e.g., transitions between the van Hove singularities) [39].

FIG. 9.18   *Optical properties of single-walled carbon nanotubes. Van Hove singularities are shown for (a) (9,9) metallic conducting tubes and (b) (11,7) semiconducting tubes. Note that there is no band gap in the metallic tube (all states populated). The semiconducting tube has the region in the center with no populated states. Optical transitions occur between the van Hove peaks.*

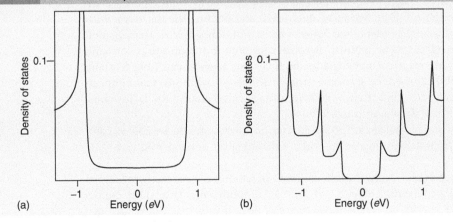

(a)   (b)

Source: www.pa.msu.edu/cmp/csc/nanotube.html

| FIG. 9.19 | *A generic Raman spectrum of single-walled carbon nanotubes is shown. The spectrum is unique to SWNTs because of the radial breathing modes found at lower energy. The location of the RBMs is a function of the SWNT diameter; the larger the diameter is, the lower energy the breathing mode has. The intense peak at ca. 1600 is the tangential E_{2g} mode (a.k.a. the G-band) that reflects the electronic properties of the SWNT. The D-band is an indication of the level of disorder in the batch.* |

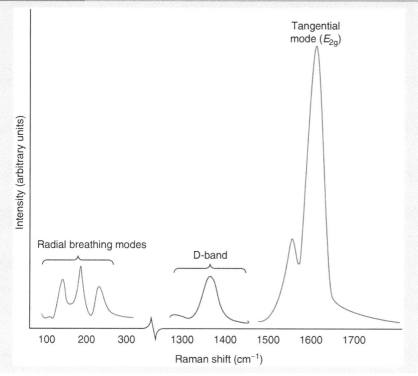

Raman Spectra. Raman scattering spectra provide a clear and unambiguous indication of the presence of SWNTs. Due to unique symmetrical phonon modes of SWNTs, such as the radial breathing mode (RBM) and the tangential mode, Raman spectra acquired from SWNTs are diagnostic. Three features are apparent in Raman spectra of SWNTs in **Figure 9.19**: (1) the intense band on the right in the figure corresponds to tangential symmetric vibrations (the G-band or E_{2g} tangential mode) around 1590 cm^{-1} and is related to the strain in the tubes and semiconducting properties; (2) the D-band (ca. 1340 cm^{-1}) in the middle represents the level of disorder in the tube/bundle; and (3) the radial breathing modes (140–300 cm^{-1}) are indicative of the diameter of the SWNTs. Another band, not shown, is the G′ band, an overtone of the D-band, that provides information about the electronic structure of the nanotube (e.g., whether it is metallic vs. semiconducting).

Carbon Nanotube Bundles. A common example put forward by nanotubers is that of the 1-nm diameter nanotube (inner diameter). If the wall thickness is assumed to be 0.34 nm, then the diameter of the SWNT overall is 1.68 nm.

TABLE 9.3	SWNT Bundle Parameters	
Nanotube index (n, m)	**Lattice parameter, a_{SWNT} (nm)**	**Density (cm^{-3})**
10,10	1.678	1.33
17,0	1.652	1.34
12,6	1.652	1.40

Source: T. A. Adams and D. Tomànek, Physical properties of carbon nanotubes, The Nanotube Site, pa.msu.edu/cmp/csc/nanotube.html (visited 2007).

This is approximately a (10,10) tube; diameter = 1.65 nm. Carbon nanotubes usually exist in bundles. Due to van der Waals attractions between crystalline nanotubes, similar carbon nanotubes are drawn to each other [43]. Tubes are bundled, if viewed from the cross-section, in a close-packed two-dimensional triangular lattice. The lattice constant, a_{SWNT}, is estimated to be 1.7 nm [44]. The lattice parameter and densities of sample nanotubes are shown in **Table 9.3** [39].

Spacing between tubes in a bundle is close to the value displayed by graphite: ca. 0.34 nm. Multiwalled nanotubes exhibit the same general value for intertube spacing. It turns out that spacing between tubes is dependent on the chirality. Armchair tube bundles show intertube spacing of 0.338 nm; zigzag tube spacing is on the order of 0.341 nm and chiral tubes with index of $(2n, m)$ have spacing equal to 0.339 nm. The lattice parameter can be estimated by [39]

$$a_{SWNT} = d_{SWNT} + 0.34 \text{ nm} \qquad (9.16)$$

where d_{SWNT} is the diameter of the tube as before. The van der Waals gap between SWNTs in bundles is discussed further in chapter 10.

9.2.3 Synthesis of Carbon Nanotubes

The best means by far to grow SWNTs is by chemical vapor deposition. High-energy methods such as laser ablation and electric arc are energy intensive and therefore make scale-up problematic. Extreme temperature conditions (1000–2000°C or greater) make it impossible to exert any control over tube purity, uniformity and orientation. The product is usually in the form of a tangled mass with nanotubes projecting in all directions. Flame methods to produce MWNTs, on the other hand, have achieved a certain level of sophistication with regard to bulk processing and purity. MWNTs have found a niche as fillers in the polymeric materials industry and have become an economically viable process. Flame synthesis is a subset of CVD fabrication methods.

Chemical vapor deposition was discussed briefly in chapter 4 and we will not elaborate much more on the process itself. Instead we will focus on recent developments that have enhanced CVD methods to the point that it has become the "method of choice" for researchers. Although its commercial potential is still in its infancy, the future bodes well for CVD methods—especially for fabrication of SWNTs. Commonly used carbon sources include methane [45], acetylene [46], ethylene [2,3], alcohols [47], CO [48], propylene [45], hexane [49], benzene [50], and others.

Synthesis of carbon nanotubes, especially SWNTs, is remarkably easy; it is ironic that so many nanomaterial synthesis processes are relatively simple, cost effective, and straightforward with high throughput. The catalyst is generated by any number of techniques. Physical methods include RF sputter and thermal evaporation; chemical methods include aqueous incipient wetness impregnation (AIWI) and template synthesis [45]. Substrates like silicon, silicon oxide, alumina, and mica serve as the support for catalyst nanoparticles.

Process Improvements. In general, current CVD methods produce CNTs that are polydisperse with regard to diameter, chirality, and length; randomly oriented; defect laden; and impure, for the most part. Nanotube length has been limited to a few hundred microns, although nanotubes in the millimeter to centimeter range have been reported [49]. However, mechanical properties of SWNT bundles 20 cm in length failed to live up to expectations [49]. Although fabricating a 20-cm bundle is quite an achievement, researchers concluded that the strands were composed of shorter nanotubes, hence the poor mechanical properties.

The work of Hata et al. demonstrated highly pure 2.5-mm SWNTs made from ethylene and water vapor in the CVD atmosphere [51]. The function of the water vapor is to extend catalyst lifetime by oxidizing amorphous carbon by-products formed during the CVD process. The purity of the SWNT product was 99.8% and the SWNT/catalyst ratio was reported to exceed 50,000 [51]. The application of water to scrub catalyst particles was reported by Cao et al. in 2001 [52]. Other mild oxidizers such as oxygen and CO_2 have also been applied during the CVD process [53,54].

Other improvements have been made over the past few years. The use of alternate carbon sources was reported by Maruyama et al. [46]. High-quality SWNTs were produced at 550°C, a very low temperature for a CVD process. Alcohols are partially oxygenated hydrocarbons and may provide a built-in means of scrubbing catalyst particles and removing unwanted CVD by-products. The use of ohmically heated substrates has demonstrated an efficient and cost-effective means of forming CNTs by CVD [55]. MWNTs grown from Ni, Co, or Fe catalyst particles were supported on resistively heated carbon paper. Aligned two-dimensional arrays of CNTs can be produced with the application of an electric field [56,57]. CNTs formed in RF plasmas are aligned by the applied field and thereby it is possible to fabricate aligned vertical arrays of carbon nanotubes in this way.

Seidel et al. showed that SWNTs are efficiently synthesized between catalyst-covered Mo pads [58]. The purpose was to develop new methods to fabricate electronic device circuitry. Evaporation of a thin layer of alumina anchors the catalyst particles at predetermined locations on the substrate. The thin alumina layer also prevents the diffusion of the catalyst into the Mo substrate [58,59]—a tribute to the versatility of nanoengineering! G. S. Duesberg et al. showed growth of single strands of MWNTs from lithographically fabricated nanoholes in SiO_2 [60]. The diameter of the catalyst particle and hence the diameter of the MWNT were directed by the diameter of the nanohole—a template synthesis process. Catalysts were implanted into the nanoholes by "spin-on" deposition or by application of a high-precision ion coater. SWNTs were also grown in nanoholes [60].

Although technology is not stressed in this text, many recent developments are demonstrating the incredible utility of carbon nanotubes. Two-centimeter-long nanotube aspect ratios exceeding 900,000 were produced by V. Shanov and

M. Schultz of the University of Cincinnati [61]. Apparently an innovative means to keep the catalyst alive during the CVD process was discovered (details not available).

A nanotube-based synthetic gecko tape was developed by Ge et al. [62]. Micropatterned carbon nanotube arrays were transferred onto a flexible polymer tape patterned after gecko setae (chapter 13). The tape supported a shear stress of 36 N · cm^{-2}, approximately four times better than that of the gecko foot and it was able to stick to many different kinds of surfaces, even Teflon [62]. The setae were represented by nanotube bundles and the *spatulae* (on the order of 200 nm in geckos) were represented by individual carbon nanotubes. The stickiness is due to van der Waals forces and offers a means to provide reversible adhesion that also has the capability to conduct electricity [62].

Purification and Separation.   More often than not, carbon CVD products require purification and separation in order to extract nanotubes. Reflux of sooty products in 10% HNO$_3$ at 150°C for 16 h removes most of the amorphous carbon and graphitic debris [30] by oxidation and then, depending on the type of catalyst, most of the metal. The process is not particularly effective for graphite encapsulated metal catalysts. Following rinsing in surfactants and washing in toluene, the nanotubes undergo further graphitization at temperatures exceeding 1500°C to remove structural defects (heptagons and pentagons). Other purification methods apply equally harsh conditions.

Removal of the protective fullerene cap on nanotubes is also desired in many applications (e.g., hydrogen storage and functionalization). Defect-rich nanotubes in particular are easy to oxidize, albeit at great cost: Most of the product is lost in the process [63].

Purification is just the first part of the process. Separation is the other. To date, no one has been able to synthesize exactly one kind of nanotube, although exclusive fabrication of SWNTs, MWNTs, or DWNTs (double-walled carbon nanotubes) is possible by controlling reaction conditions (pressure, gas composition, catalyst composition and size, and temperature) [38]. Most as-formed product requires postsynthesis separation because, in addition to numerous carbon byproducts, the batch includes a mixture of tubes with polydisperse diameter, length, and chirality and both semiconducting and metallic tubes.

Many types of separation procedures have been applied to CNTs—some effective and some not effective [38]: chromatography (size exclusion, HPLC, gel permeation, and ion exchange); electrophoresis (capillary, ac dielectrophoresis, gel, dielectric field-flow fractionation); fluid-based methods (flow-field fractionation [FFF], nematic liquid crystal extraction); destructive methods (current-induced oxidation, fluorination, and annealing); chemical methods (amino acid–amine adsorption, photoelectrochemistry, density-gradient centrifugation). Capillary electrophoresis has been applied to separate CNTs based on length [21]. Flow-field fractionation [64] has proven to be an effective means of separating CNTs. One means of separating semiconducting tubes from conducting (metallic) tubes is by passing a large current through an SWNT mass [65]. This method, however, results in the destruction of the metallic tubes due to defects caused by the large current that lead to exclusive oxidation of the conducting tubes [65]. Hongie Dai et al. of Stanford utilized low-temperature methane plasma to selectively destroy metallic tubes and nanotubes with diameter in the range of 1.0–1.3 nm [38,68].

Researchers at Rice University recently developed a means to separate semiconducting nanotubes from metallic ones based on size [67]. The separation procedure is based on the diameter-dependent dielectric constant of individual nanotubes. Dissolved nanotubes were pumped into an electrified chamber in which metallic nanotubes became trapped. Semiconducting tubes, depending on their chirality and size, floated at different levels in the chamber. The smaller diameter tubes possessed larger dielectric constants and were found lower in the system. As the speed of the flow was varied, with upper level currents traveling faster, the Rice scientists were able to separate tubes [67].

9.2.4 Growth Mechanisms

R. E. Smalley et al. proposed a straightforward explanation of nanotube growth [68]. The theory explains the size and structure of two kinds of graphitic materials: SWNTs and onions (graphite encapsulated catalysts) (**Fig. 9.20**). The model,

FIG. 9.20 *The yarmulke mechanism of SWNT growth is depicted. It is energetically better for the growing fullerene to remain in the form of a tube rather than spread out over the particle surface if the curvature of the particle is extreme (such as the case for very small particles)— the trade-off between dangling bonds and bond strain. Encapsulation of larger catalyst particles is depicted below. Depending on process conditions, large particles are either encapsulated or grow multiwalled tubes. The onion configuration is schematically depicted at the bottom right of the figure. The structure resembles graphite with interlayer spacing on the order of 0.34 nm. Once a particle is encapsulated it loses its catalytic function.*

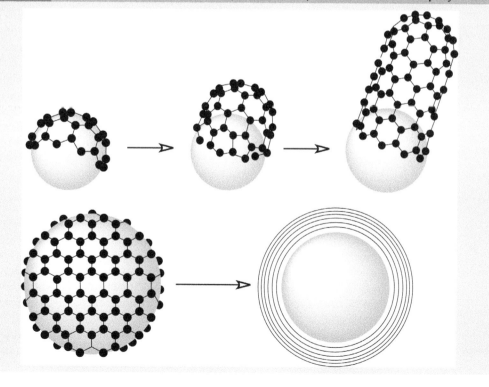

called the *yarmulke* mechanism, is based on the formation of a hemispherical fullerene cap on the surface of the iron nanoparticle. What happens after the formation of the cap depends on the size of the catalyst particle. Experimental results demonstrated the gas-phase synthesis of SWNTs from iron catalysts formed by the decomposition of ferrocene. Following decomposition and nucleation, the catalyst grows by collision with other clusters. Depending on reaction conditions, clusters in the range of 0.7–1.4 nm are formed (50–100 iron atoms). CO gas, the carbon source, disproportionates on the cluster surface according to the Boudouard reaction:

$$CO + CO \xrightarrow{\text{Catalyst}} C_{(s)} + CO_2 \qquad (9.17)$$

Classical thermodynamic principles can be extended to explain the energy associated with graphitic structures. A single-walled carbon nanotube is nothing more than a sheet of graphite (called a graphene sheet) rolled to make a cylinder. Multiwalled nanotubes are a series of concentric SWNTs of increasing diameter. However an excess amount of energy is required to get this rolling accomplished. The trade-off in energy is a balance between forming an open structured planar sheet of graphite of limited size (more accurately, a graphene sheet that has numerous dangling bonds) and forming a closed structure with bond strain but with no dangling bonds. As we find out, the size of the catalyst particle has great influence over these two parameters.

The growth mechanisms of carbon nanotubes were discussed briefly in chapter 8. In this chapter, we will touch upon on the distinctions between base growth and tip growth and other factors involved in CNT growth. The types of synthesis process—e.g., CVD, arc discharge, laser ablations, and PECVD (plasma-enhanced CVD), affect the growth mechanism of CNTs. CVD is a cleaner process than PECVD, which contains reactive radicals and atomic hydrogen in the growth environment [68]. The types of catalyst applied (Fe vs. Fe_2O_3), the nature of the support material (silica, alumina), the carbon source gas (methane, acetylene, and ethylene), and the CVD conditions also play important roles in CNT growth (type of gases, partial pressures, temperature, and flow rates). The formation of chemical intermediates like metal carbides must also be factored into the growth equation.

The base growth mechanism is typical of SWNTs. In this case, the catalyst particle remains anchored to the oxide support material, usually aluminum oxide. MWNTs can be grown by tip or base growth, depending on the substrate (MWNTs can also be grown without the participation of metal catalysts). If silica-type substrates are utilized, the tip growth mechanism is favored. The strength of the metal oxide support interaction is thought to play a major role in determining which kind of growth is expressed—e.g., alumina forms stronger interactions with catalyst particles than does silica. In situ analysis of CNT growth was accomplished with TEM methods [69]. CNTs were synthesized directly on Ni catalyst particles deposited directly on a copper TEM grid [69]. Results showed that tip, base, and intermediate growth modes (where the catalyst particle is located in the middle of a growing nanotube) occurred simultaneously on the grid.

Limitations to nanotube growth are due to several factors. According to one theory, whole nanotubes need to slide on the surface of the substrate [70]. Once the tube becomes longer, van der Waals forces between the nanotube and the substrate become too strong and overcome the force required to move the nanotube during growth. Thus, growth is terminated. This apparent difficulty can be overcome by application of an electric field (from RF sources). In this case, the tubes would align with the field during growth. This is not a problem with the tip growth mechanism. Another theory proposes that mass transport limitations account for termination of nanotube growth. The flow rate of carbon source gases to the catalyst becomes restricted as the nanotube forest is grown [70]. Also, SEM and AFM showed that most tubes nucleate via base growth and thereby the restrictions placed on mass transport apply [70].

In general, carbon source materials like methane or acetylene adsorb onto the surface of the catalyst and decompose. The use of saturated hydrocarbons like methane and propane initialize endothermic conditions whereas the use of unsaturated gases like acetylene or ethylene produce exothermic conditions— e.g., $C_2H_2 \rightarrow 2C_{graphitic} + H_2$, $\Delta H_{rxn} = +262.8$ kJ·mol^{-1} [71]. The liberated carbon diffuses into the metal catalyst or upon its surface. The activation energy for surface diffusion is obviously smaller than diffusion through the bulk of the metal and is therefore the preferred path, at least in the formation of SWNTs. The surface diffusion argument, however, does not adequately explain the growth of MWNTs [71]. Because certain amount of carbon must be dissolved in the catalyst before nanotube growth is possible, other methods of carbon transport may also affect CNT growth.

Klinke et al. applied thermodynamic calculations and finite element analysis to CNT growth based on hydrocarbon precursors at elevated temperatures [71]. Temperature gradients were thought to drive the diffusion of carbon during the growth phase but more recently, a concentration gradient is more likely the cause of the carbon flow with regard to the catalyst [71]. Carbon precipitates at the part of the catalyst known as the depletion zone [71]. The temperature however plays a major role in activating the diffusion process. Although much progress has been achieved, there is much about CNT growth mechanisms in general that remain to be clarified.

9.2.5 Chemical Modification of Carbon Nanotubes

This field of nanochemistry has burgeoned in recent years, especially as chemically modified nanotubes have become increasingly significant components in nanocomposites. The applications of carbon nanotubes in polymers are very broad—electrical conductors and heat transfer conduits to structural elements. The extreme aspect ratio of SWNTs, ranging from 1000 to 10,000 and more, makes their use as fillers in polymeric matrices very attractive to nanocomposite technology. Chemical challenges facing nanotube chemistry include dispersion and solubility in the polymer, a fact exacerbated by bundling. Dispersion and dissolution is problematic with SWNTs due to strong intertube attractions (bundling) and the fact that they are large molecules. Interfacial adhesion between the polymer and the "smooth" nanotubes is also problematic. In order to reinforce, enhance, or add properties to composite materials, carbon nanotubes need to be modified chemically. Although physical techniques such as sonication are able to temporarily disperse CNTs, once the power is turned off, the tubes reagglomerate into bundles. SWNTs, however, require special treatment to enhance

FIG. 9.21   *Two growth mechanisms of multiwalled carbon nanotubes are shown. On the left, the base-growth mechanism is depicted. Catalyst particles that are adsorbed onto alumina support materials facilitate base growth of carbon nanotubes. Tip growth, depicted on the right in the figure, is the case in which catalyst particles do not adhere to the surface of the substrate and occurs on silica surfaces. Chemisorption between catalyst particles and alumina appears to be strong enough to anchor the catalyst to the surface. The carbon source shown above is methane. Methane decomposes following adsorption onto the surface of the metal catalyst.*

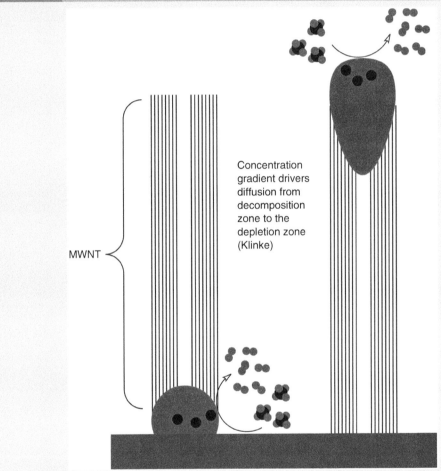

MWNT

Concentration gradient drivers diffusion from decomposition zone to the depletion zone (Klinke)

their chemical compatibility, dispersion, dissolution, and interfacial adhesion properties. Although significant strides have been made in recent years, fabrication of CNTs with uniform length, diameter, and chirality; inexpensive mass production; and purity plague advanced nanotube development.

SWNTS are all surface, both inside and outside. Every single atom is a surface atom. Carbon nanotubes are supermolecules that are held together by some of the strongest bonds known to nature, the carbon–carbon bond; there are no dangling bonds since the ends are technically capped by hemispherical fullerenes.

One would expect that SWNTs in particular would be extremely reactive due to high surface energy. However, eventhough carbon nanotubes have no bulk atoms, they are relatively inert—not because of thermodynamic considerations but rather because they have no chemical functional groups. It is possible, however, to disturb the conjugated system of CNTs by attack of strong acids such as HNO_3 and/or H_2SO_4 to form carboxylates [24]. Applications of heat and/or ultrasonication along with acid treatment result in the formation of dangling bonds that render the tubes reactive [24]. The apparent stability of SWNTs is demonstrated by the low degree of functionalization, even under such harsh conditions.

Under the right conditions, the intrinsic curvature of chemical bonds (thermodynamic properties) making up nanotubes provides the basis for chemical modification. Modification of the outer surface of SWNTs reduces intertube attraction because the perfect Van der Waals interactions no longer overlap exactly. Chemical functionalization also enhances interaction between the SWNT or MWNT and a potential polymer matrix material. The design of modified nanotubes to suit a specific purpose is a high-priority research endeavor.

Single-walled carbon nanotubes are made of two components: the high-aspect-ratio tube and the end caps. The end caps, hemispherical fullerenes, naturally exhibit fullerene behavior and are therefore more susceptible to oxidation and addition reactions. There are two generic types of chemical modification that can be performed on nanotubes: covalent modification and noncovalent modification [72,73].

Covalent chemical modification involves breaking old covalent bonds and forming new covalent bonds. Processes such as oxidation and fluorination are examples of covalent modification. Noncovalent modification implies that carbon nanotubes are functionalized by means of intermolecular bonding processes, some forms of which were discussed in previous sections of this chapter. By its very nature, covalent modification perturbs the structure of the nanotube and introduces defects into the carbon nanotube that disrupt the aromatic structure. Defects alter the mechanical strength, electrical conductivity, and thermal conductivity of the nanotube. On the flip side, depending on the chemical nature of the modifying group, the nanotube is capable of linking to its host polymeric matrix. Such cross-linking, always an important aspect of polymer design, serves to strengthen the polymer as a whole. A balance must be achieved between the degree of chemical modification of the nanotube and the degree of enhancement of the strength of the polymer as a whole: What is the load-transfer efficiency from the polymer to the nanotube [73,74]?

Noncovalent types of modification are governed by thermodynamic parameters. This form of chemical functionalization is accomplished by means of intermolecular attractions between an adduct molecule and the nanotube. In this way the backbone structure of the nanotube is not altered and properties such as mechanical strength, electrical conductivity, and thermal conductivity are compromised, relatively speaking. The problems of solubility, dispersion, and separation of native tubes are addressed by noncovalent chemical functionalization. As a result, subsequent processing of nanotubes with the desired outcome is more likely.

The potential routes to nanotube chemical modification are summarized in **Table 9.4.**

TABLE 9.4	*Chemical Modification of Carbon Nanotubes*	
Major category	**Type**	**Comments/examples**
Noncovalent methods	$\pi-\pi$ Stacking	Increased solubility in aromatic solvents, aromatic surfactants, and aromatic polymers Immobilization of chemical and biological molecules by polynuclear aromatic compounds (anthracene, pyrene, etc.) Increase solubility of SWNT in aqueous media by application of ionic pyrenes [75] Immobilize enzymes on a surface via pyrene-derivative modified CNTs
	Polymer wrapping	Poly(styrene sulfonate), polyvinyl sulfate, polyvinyl alcohol, polyethylene glycol, and many others wrapping around SWNT driven by thermodynamic propensity to eliminate interface between hydrophobic surface of the SWNT and aqueous medium [76]
	Surfactants	Application of surfactants aids in dispersion but forms no permanent structures
Covalent methods	Side wall chemistry	Electrophilic nature of SWNT electronic structure
	End and defect chemistry	Most reactive region of SWNT (highest bond strain)
	Grafting to	Polymers are grown first and then grafted to the SWNT. Poly(styrene), PS, with nitride functionality was synthesized by atom transfer free radical polymerization (ATRP) followed by end-group transformation (PS–N_3) and then grafted onto the SWNT [77]
	Grafting from	Polymers are grown from immobilized groups on SWNTs. Immobilization of initiators onto tubes followed by surface-initiated polymerization (e.g., formation of polymer brushes) [77,78]. ATRP styrene with 2-bromopropionate immobilized on SWNT. Methly-2-bromopropionate was added as initiator. The PS was attached covalently to the SWNT
	Fluorination	Fluorination destroys electrical conductivity of nanotubes [79]. Oxidation first, then fluorination under F gas to produce routes to further chemical derivatization [79]
	Carboxylation	Addition of carboxylic groups + sonication shortens the length of the nanotubes and thereby enhances dispersion
	Electrochemical reduction	Reduction of aryldiazonium salts [80]

Noncovalent Modification. This form of chemical derivatization is relatively nonspecific and is based on intermolecular interactions that do not involve strong bonds. The electrical conductance of SWNTs, for example, relies strongly on its immediate environment and modification by noncovalent methods tend to preserve the basic structure although electronic properties (e.g., the bandgap) may be influenced [72].

R. E. Smalley et al. reported in 2001 about the derivatization of SWNTs by a process called *polymer wrapping* [76]. The process entailed association of the SWNT with water-soluble linear polymers such as polyvinyl pyrrolidone (PVP) and polystyrene sulfonate (PSS). The process was successful in disrupting the hydrophobicity of the nanotube as well as the intertube attraction to form bundles [76]. Unwrapping of the nanotubes occurred by changing the supporting solvent. Chemical modification by this process also allowed for analysis by solution-phase techniques such as chromatography and electrophoresis [76].

According to the authors, a thermodynamic driving force strives to eliminate the hydrophobic interface between the SWNT and the water solvent [76].

Although noncovalent modification technically does not affect the structure of nanotubes, it is capable of influencing the bandgap of the material [81]. Metallotetraphenyl porphyrins are electron donors; SWNTs are electrophiles. In one experiment, the wavelength and absorption intensity of the SWNT were red shifted and increased, respectively, relative to the undoped state [81]. Electron donation into the *p*-doped valence band increased the electronic density while simultaneously decreasing the transition energy of the unoccupied conduction band [81].

A. Noy et al. accomplished layer-by-layer electrostatic assembly of polyelectrolyte nanoshells on individual suspended carbon nanotubes [82]. The goal of the study was to develop a robust strategy for noncovalent modification. The purpose of the nanotube was to serve as a bridge and template. The first step involved exposing the CNTs to pyrene derivative (1,3,6,8-pyrenetetrasulfonic acid tetrasodium salt and 1-pyrenepropylamine hydrochloride or PyrNH$_3$) followed by layer-by-layer deposition of polyelectrolyte macro-ions. The polymers used in the study were poly(diallyldimethylammonium chloride (PDDA) and polystyrene sulfonate sodium salt (PSS). The entire process was based on stepwise self-assembly among the constituents. A generic version of carbon nanotube decorated with pyrene groups is shown in **Figure 9.22**.

Covalent Modification.   Covalent modification strategies rely on the curvature of the tube and on the electronic properties [81]. Functionalization of SWNTs to

FIG. 9.22   *The planar pyrene molecule reacts with nanotube surfaces by π–π interactions—intermolecular interactions that are not as strong as covalent bonds. They serve as nonspecific anchoring constituents. Reactive groups attached to the pyrene enable further chemistry to take place.*

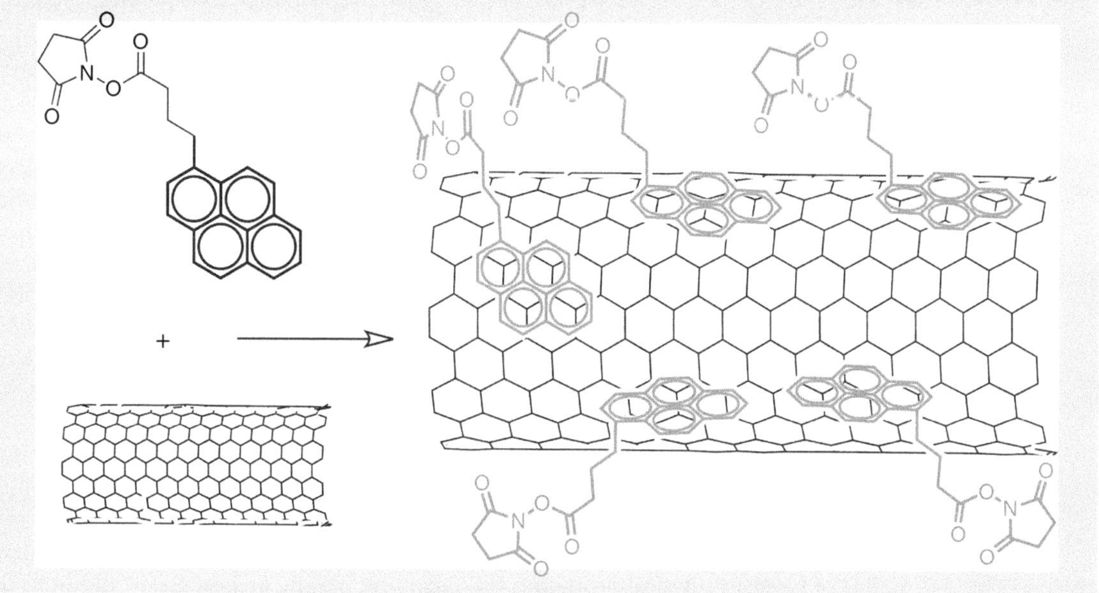

generate $-CO_2H$ decorations takes place in a rather severe chemical environment. SWNTs are first refluxed in 2–3 M nitric acid for 12–48 h at 115°C. Chemical modification can cease at this point if carboxylic acid groups are the desired decoration. The product is washed and dried. If further modification is required, carboxylated SWNTs suspended in dimethylformamide (DMF) are exposed to thionylchloride ($SOCl_2$). This process transforms the $-CO_2H$ group into a $-COCl$ chloric acid derivative. This transformation renders the SWNTs reactive to further modification by long-chain amines such as dodecylamine ($C_{12}H_{27}N$, a.k.a. *f12*) and octadecylamine ($C_{18}H_{39}N$, a.k.a. *f18*). A surfactant, NaDDBS (sodium dodecylbenzenesulfonic acid), was used to disperse the mixtures [83].

Evidence of carboxylation is provided by FTIR and Raman analysis. The presence of sp^3-hybridized carbon, a consequence of carboxylation, is indicated by the increase in the Raman disorder mode (D-band) at 1292 cm^{-1}. The G-mode, also called the tangential mode, is shifted to higher energy, an indication of the electron withdrawing affect of the carboxylate groups [83]. FTIR analysis shows the diagnostic carbonyl stretch located at 1727 cm^{-1}. Following treatment of the carboxylate moiety to form the chloric acid derivative, an amide linkage is formed between the *f12* or *f18* and the SWNT [83]. One means of measuring the degree of functionalization, measured as the percentage of carbon atoms that are actually derivatized, is by TGA. Alkyl chain mass losses account for up to 48% of the total mass of the derivatized SWNT system. This yields a grafting density of approximately four alkyl chains per nanometer of a (10,10) SWNT. Overall, chemical modification of the SWNT surface was less than 5% [83].

In another application, SWNTs were first carboxylated in a sulfuric-nitric acid (3:1%-vol) and then sonicated for 3 h at ambient temperature [74]. Following dilution with distilled water (1:5), the mixture was forced through a PTFE filter (10-μm pore size) and washed until no residual acid remained. The success of carboxylation was verified by FTIR [74]. Di-epoxide-terminated molecules were then attached to the CO_2H-modified SWNTs by adding EPON 828 and subsequent sonication for 1 h. KOH was added as a catalyst at 70°C. After washing and filtering, the derivatized SWNTs were analyzed by TGA (to quantify the amount of attached molecules) and FTIR (to verify the presence of epoxides). The chemical behavior of the epoxide-modified carbon nanotubes was different from the carboxylated nanotubes; this was confirmed by quantitative titrations that showed carboxyl terminals were consumed during the modification track to form epoxides. The use of an epoxide active group allows for further interaction with an epoxide-based bulk polymer with load transfer transiting through covalent bonds between the bulk polymer and the nanotubes. In thermoplastic polymers containing derivatized nanotubes, load transfer occurs by way of van der Waals interactions. If the groups attached to the nanotubes are long enough, the authors claim that load transfer efficiency characteristics are expected to improve significantly [74] (**Fig. 9.23**).

9.3　Diamondoid Nanomaterials

Diamonds are a very costly material, one of the most beautiful, and have a slew of fantastic physical properties [6]. Diamond has the highest elastic modulus of

FIG. 9.23 Covalent surface modification involves making bonds between the carbon framework of the nanotube and the modifying moiety. A multiwalled carbon nanotube underwent carboxylation by application of a sulfuric-nitric acid (3:1) mixture while under sonication. Following rinsing and purification, the carboxylated MWNTs were reacted with di-epoxide-terminated molecules (Epon 828) in the presence of KOH catalyst. The terminal epoxide is available to undergo further chemical reactions. If the derivatized nanotubes are immersed into a bulk epoxy polymer matrix, strong links with the bulk polymeric material are formed. Such linking allows for efficient load transfer from the bulk polymer to the MWNTs. In this way, the MWNTs are able to enhance the mechanical properties of the polymer composite material.

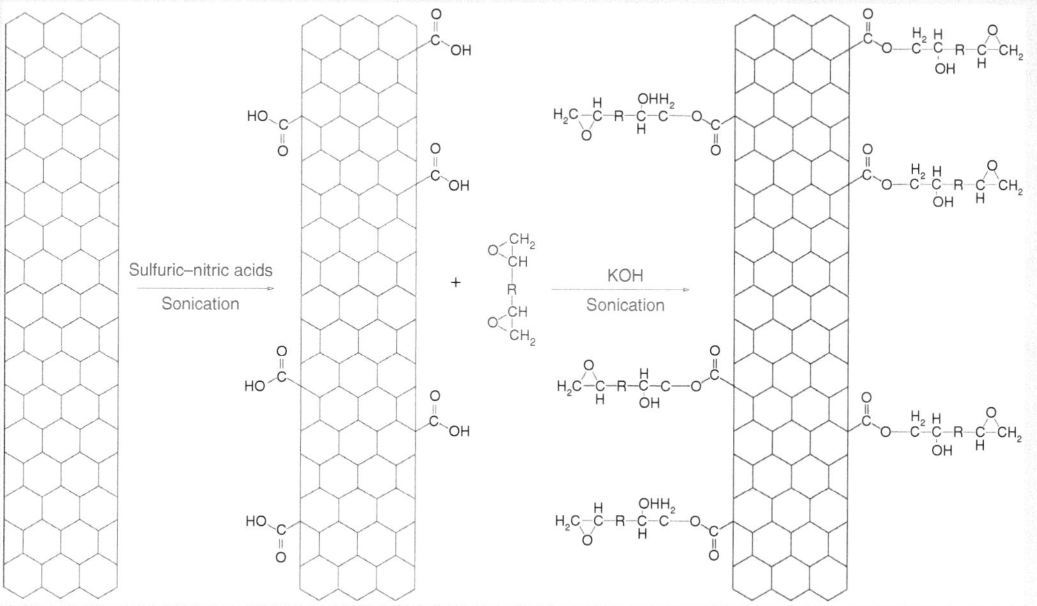

Source: Image redrawn with permission from Linda S. Schadler, Materials and Science Engineering, Rensselaer Polytechnic Institute, Troy, New York.

any material (1050 GPa), one of the highest electrical resistivities ($\sim 10^{16}\ \Omega \cdot cm$), high thermal conductivity ($2000\ W \cdot m^{-1} \cdot K^{-1}$), and low thermal expansion ($1.2 \times 10^{-6}\ K^{-1}$). Diamond is for the most part resistant to chemicals; transparent to visible, IR, and microwave radiation; and insulating (high band gap) [6]. The fabrication of diamonds has always been one of the holy grails of science. The feat was accomplished in 1954 by the General Electric Corporation; diamonds were formed under high-temperature and -pressure conditions (HPHT) [6]. These diamonds have numerous defects but are useful in industrial processes. Bulk diamonds are metastable forms of carbon. Diamond nanoparticles and thin films are also metastable, perhaps more so.

Diamondoids were first isolated in 1933 from Czechoslovakian crude oil and named adamantane (from the Latin *adamentum,* "the hardest metal," and the Greek *adamas,* "unconquerable"). Deposition of diamondoids is an undesirable occurrence during the oil refining process and subsequent transportation of

TABLE 9.5	*Comparison of Diamond, Diamondoid, and Amorphous Carbons*		
Physical property	**Diamond**	**CVD diamondoid**	**Amorphous carbon (diamond-like)**
Crystal structure	Cubic, $a = 3.567$ Faceted crystals	Cubic, $a = 3.561$ Faceted crystals	Mixed sp^2 and sp^3 Smooth to rough Domains of microcrystalline diamonds
Hardness (H_v)	7000–10,000	3000–12,000	900–1200
Density ($g \cdot cm^{-3}$)	3.51	2.8–3.5	1.2–2.6
Electrical resistivity ($W \cdot cm$)	~10^{16}	~10^{13}	10^6–10^{14}
Thermal conductivity ($W \cdot m^{-1} \cdot K^{-1}$)	2000	1100 (2100)[a]	(1700)[a]
Thermal coefficient of expansion (K^{-1})	1.2×10^{-6}	1×10^{-6}–2.0×10^{-6} [86]	—

Sources: M. Ohring, *Materials science of thin films: Deposition and structure*, 2nd ed., Academic Press, San Diego, 2002; R. C. DeVries, *Annual Review of Materials Science*, 7, 161, 1987; H. C. Tsai and D. B. Bogy, *Journal of Vacuum Science and Technology*, A5, 3287, 1987.
[a] J. Norley, *Electronics Cooling*, 7, 50–51 (2001).

products like natural gas, gas, and petroleum products. Diamonds are made entirely of carbon that crystallizes in two forms: diamond and Lonsdaleite.

Diamondoids are materials that resemble diamond with regard to hardness, strength, stiffness, structure, and density. Diamondoids are covalently bonded sp^3 tetrahedral materials consisting of light elements with valence of three or greater. Diamond and sapphire are examples of diamondoid materials, but the definition is sometimes extended to sp^2-hybridized structures that form in planes like graphite. Therefore, fullerenes and nanotubes are sometimes referred to as diamondoid materials. We will not include them in the general category of diamondoid materials (**Table 9.5**).

Another application of diamondoid material thin films is to provide a surface with the capability of radiation resistance (called radiation hardness)—a required feature for the development of radiation detectors and linear colliders that undergo severe radiation environments.

9.3.1 Diamondoids

Adamantane is a diamond-like single cage structure with chemical formula [$C_{10}H_{16}$]. Two or more linked cage structures are called diamantanes, triamantanes, and, in general, polyamantanes—up to a dozen or so adamantane groups that include isotetraamantane (18 faces and four isomers) [$C_{22}H_{28}$], cyclohexamantane [$C_{26}H_{30}$], and superadamantane [$C_{35}H_{36}$] (**Fig. 9.24**). These structures are similar to diamond and have hydrogen terminal groups that stabilize the surface (just like diamond).

We are aware that nanomaterials have size-dependent properties. Ideally, we all want to study the quantum confinement properties of ideal materials. In reality, we usually deal with materials that are polydisperse or charged with nonuniform surface reconstruction or termination [8,87,88]. Diamondoid materials offer the opportunity to study electronic properties of the smallest possible single cage unit of the diamond lattice. Each dangling bond is terminated with hydrogen

FIG. 9.24 *Left: adamantane; middle: triamantane; and right: pentamantane that has been chemically modified with a thiol group—at the locus of bridging groups. Although these structures are considered to be diamondoids, they are really just simple hydrocarbons that are able to form into cages. The thiol group is able to bind the adamantane to a gold surface (or a gold nanoparticle) forming a diamond-like layer.*

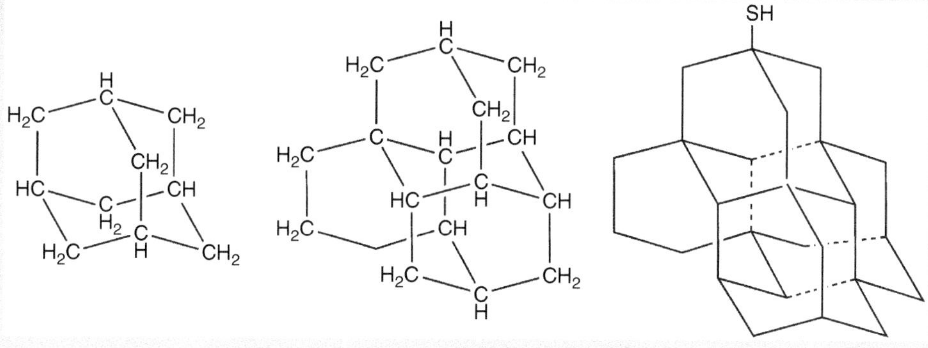

and each diamondoid is capable of adding another cage to form semiconductor clusters [8]. Bostedt et al. in 2004 monitored the size-dependent changes in electrical properties of individual diamond clusters [8]. Diamondoids (adamantane to hexamantane) were brought into the gas phase and analyzed by carbon K-edge x-ray adsorption spectroscopy. Data revealed that diamondoids absorb x-rays at lower energies and that the terminal hydrogens influence the density of states of the nanoparticles. Adamantane absorption resembles the x-ray profile of cyclohexane while hexamantane begins to show the characteristic profile of bulk diamond [8].

9.3.2 Thin Diamond Films (and Other Ultrahard Substances)

There are three hard substances known to science: diamond, boron–nitride, and carbon nitride. Interestingly, silicon carbide is not on this list. All three can be deposited by CVD. The deposition of perfect diamond lattices by CVD has been accomplished in the past couple of decades. In general, hydrocarbon sources such as CH_4 are activated over a solid substrate surface (720–1200°C) by application of a hot filament or electrical methods (DC arc discharge, RF, or microwave). In this way, true diamond lattice structures are formed—or diamond-like structures that are mostly amorphous with microcrystalline phase inclusions [86,89]. Room temperature thermal conductivity of such diamond-like carbon thin films is 1700 $W \cdot m^{-1} \cdot K^{-1}$. Values for CVD diamond as high as 2100 $W \cdot m^{-1} \cdot K^{-1}$ have been achieved [86,89].

CVD of metastable diamond occurs over a narrow range where there is a very small free energy difference between diamond and graphite: 2.1 $kJ \cdot mol^{-1}$ (**Fig. 9.25**) [6]. Kinetic factors can predominate in this small domain and the probability that both phases of carbon are produced is finite [6]. Kinetic control can be inserted to prevent the formation of graphite or its removal as it forms. This is accomplished by generating a superequilibrium (supersaturation) of atomic

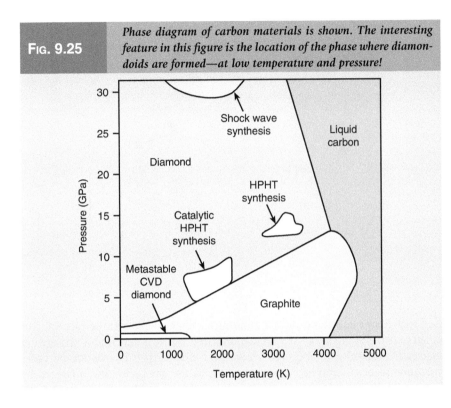

FIG. 9.25   *Phase diagram of carbon materials is shown. The interesting feature in this figure is the location of the phase where diamondoids are formed—at low temperature and pressure!*

hydrogen radicals [H·]. In this way, the atomic hydrogen radical prevents the formation of graphite by dissolving it during its nucleation [6]. Typical conditions to form microcrystalline diamond films are as follows: 0.5–2% CH_4 in H_2 at 50–100 torr; 2.45-GHz microwave plasma or hot filament; and a roughened silica substrate held between 800 and 900°C. Polycrystalline diamond was produced at a rate of 2 $\mu m \cdot h^{-1}$ with grain size equal to 1–10 μm. Nanocrystalline films are characterized by diamond grain size between 5 and 50 nm. CVD conditions are altered by significant dilution of reactants achieved by the presence of a 90% argon fraction.

9.3.3 Chemical Modification of CVD Diamond

The growth of covalently bonded nitrophenyl layers on atomically smooth boron-doped crystalline diamond surfaces was successfully accomplished by Uetsuka et al. in 2006 [90]. The purpose was to fabricate a biosensor that has long-term chemical stability and inherent biocompatibility. Chemical modification of diamond surfaces, because of its remarkable properties, is problematic, although reactions with atomic hydrogen, fluorine, and chlorine have been reported [90]. In 2000, Takahashi et al. demonstrated that covalent layers were possible on diamond surfaces by photochemical chlorination–amination–carboxylation of hydrogen terminated CVD diamond surfaces. Subsequent photochemical methods were able to modify CVD diamond with alkenes as well [90,91]. Attachment to ultrananocrystalline diamond (UNCD) has also been accomplished [90,92].

Boron-doped single-crystal diamond films (1 μm thickness) were formed homoepitaxially on diamond substrates (900°C) by microwave plasma-assisted (1200 W) CVD at 50 torr in 0.6% CH_4 in H_2 in the presence of B_2H_6 [90]. The boron concentration in carbon was 1.6×10^4 ppm. Hydrogen termination was achieved by application of pure hydrogen plasma for 5 min. One goal of the preparation was to form oxygen-terminated groups. This was accomplished by exposing the as-formed diamond film to oxygen plasma (13.56 MHz, 300 W) at 20 torr for 2.5 min. The degree of success was determined by wettability experiments; the contact angle θ approached 0° upon rendering of the surface to a hydrophilic state. Prior to this step, the contact angle was >94°, suggesting strong hydrophobicity of the H-terminated surface. The resulting surface had surface roughness < 1 Å. Nitrophenyl groups were attached by reduction of 4-nitrobenzene diazonium tetrafluoroborate. Attachment was initiated by cyclic voltammetry or constant potential attachment (**Fig. 9.26**).

In another study (R. J. Hamers et al.), researchers were able to attach DNA to CVD-formed diamond films [94,95]. DNA or other biological entities such as antibodies are chemically stable when they are bonded covalently to diamond surfaces. The use of electrical signals to direct chemical modification of diamond surfaces has become a very effective means of generating electrodes that have biosensing capability. The procedure to form the biological active electrodes is as follows: a Si wafer was coated with 300 nm of silicon nitride. Photolithographic masks were used to transfer a pattern of Ti–Mo derivatized contacts to the surface. Nanocrystalline diamond powder was introduced followed by lift-off of the photoresist. Nanocrystalline diamond was applied on Ti–Mo contacts in a 2.45-GHz microwave plasma reactor that contained methane, hydrogen, and B_2H_6. Grain size by this method was 20–100 nm. The film was continuous and pinhole free. The surface layer was H-terminated as discussed before. Forty millimolars of 4-nitrobenzene diazonium tetrafluoroborate in 1% sodium dodecyl sulfate (SDS) surfactant produced aryl-nitro groups attached to the surface. Reduction of the nitro groups to primary amines proceeded via cyclic voltammetry, followed by reaction with bifunctional linking groups that

FIG. 9.26	*Reduction of 4-nitrobenzene diazonium tetrafluoroborate to form a reactive radical and subsequent attachment to a CVD diamond film is depicted. The nitro-moiety allows for further attachment of functional groups such as DNA or the creation of additional layers.*

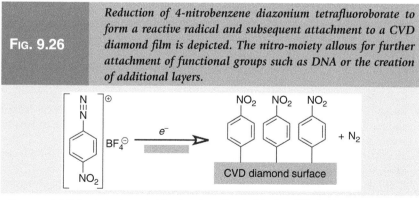

Source: J. F. Bahr et al., *Journal of the American Chemical Society*, 123, 6536–6542, 2001. With permission.

Attachment of DNA to a CVD diamond nanocrystalline thin film [94]. These DNA-modified diamond films show extraordinary stability under conditions that would denature natural DNA (elevated temperatures, repeated hybridization–denaturation cycles).

Source: Image redrawn with permission; R. J. Hamers, Department of Chemistry, University of Wisconsin.

are able to covalently attach to DNA—a heterobifunctional cross-linking molecule called sulfo-succinimidyl 4-(*N*-maleimidomethyl)cyclohexan-1-carboxylate (or SSMCC) [94] (**Fig. 9.27**).

References

1. H. W. Kroto, J. R. Heath, S. C. O'Brien, R. F. Curl, and R. E. Smalley, C60: buckminster fullerene, *Nature*, 318, 162 (1985).
2. S. Iijima, Helical microtubules of graphitic carbon, *Nature*, 354, 56 (1991).
3. S. Iijima and T. Ichihashi, Single-shell carbon nanotubes of 1-nm diameter, *Nature*, 363, 603 (1993).
4. D. S. Bethune, C. H. Kiang, M. S. De Vries, G. Gorman, R. Savoy, J. Vasquez, and R. Beyers, Cobalt-catalyzed growth of carbon nanotubes with single-atomic-layer walls, *Nature*, 363, 605 (1993).
5. Properties of carbon and C_{60}. Data compiled by the CSC (Creative Science Center) and the Sussex Fullerene Group, University of Sussex, Falmer, Brighton, East Sussex; www.creative-science.org.uk/propc60.html
6. M. Ohring, *Materials science of thin films: Deposition and structure*, 2nd ed., Academic Press, San Diego (2002).
7. H. Kluytmans, Nanotechnology and the arrival of the diamond age, www.dse.nl/~hkl/e_nano1.htm (1997).
8. C. Bostedt, Electronic properties of diamond clusters (diamondoids, nanodiamonds), www.physik.tu-berlin.de/cluster/diamondoids.html (2004).

9. L. Becker and T. E. Bunch, Fullerenes, fulleranes and PAHs in the allende meteorite, *Meteoritics*, 32, 479–487 (1994).

10. A. Hellemans, Labs hold the key to the 21-micrometer mystery, *Science*, 284, 1113 (1999).

11. F. Cataldo, Fullerane, the hydrogenated fullerene: Properties and astrochemical considerations, *Fullerenes, Nanotubes and Carbon Nanostructures*, 11, 295–316 (2003).

12. P. Mauron, *Growth mechanisms and structure of carbon nanotubes*, inaugural dissertation, Universität Freiburg, Department of Physics (2003).

13. E. A. Rohlfing, D. M. Cox, and A. Kaldor, Production and characterization of supersonic cluster beams, *Journal of Physical Chemistry*, 81, 3322–3330 (1984).

14. H. S. Carman and R. N. Compton, Electron attachment to c_n clusters (n#30), *Journal of Physical Chemistry*, 98, 2473–2476 (1993).

15. C. P. Poole and F. J. Owens, *Introduction to nanotechnology*, Wiley-Interscience, John Wiley & Sons, Inc., Hoboken, NJ(2003).

16. S. Sugano and H. Koizuni. *Microcluster physics*, Springer-Verlag, Heidelberg (1988).

17. G. Cao, *Nanostructures & nanomaterials: Synthesis, properties & applications*, Imperial College Press, London (2004).

18. P. Holister, C. Román Vas, and T. Harper, Fullerenes, Technology White Papers, NR.7, Científica, www.scientifica.com (2003).

19. H.-D. Beckhaus, C. Rüchardt, M. Kao, F. Diedrich, and C. S. Foote, The stability of buckminsterfullerene (C_{60}): Experimental determination of the heat of formation, *Angewandte Chemie International Edition in English*, 31, 63–64 (1992).

20. H.-D. Beckhaus, S. Verevkin, C. Rüchardt, F. Diedrich, C. Thilgen, H.-U. ter Meer, H. Mohin, and W. Müller, C_{70} is more stable than C_{60}: Experimental determination of the heat of formation of C_{70}, *Angewandte Chemie International Edition in English*, 33, 996–998 (1994).

21. R. J. Doome, A. Fonseca, H. Richter, J. B. Nagy, P. A. Thiry, and A. A. Lucas, Purification of C_{60} by fractional crystallization, *Journal of Physics and Chemistry of Solids*, 58, 1839–1843 (1997).

22. N. Sivaraman, R. Dhamodaran, I. Kalliappan, T. G. Srinivasan, P. R. Vasudeva Rao, and C. K. Matthews, Solubility of C_{60} in organic solvents, *Journal of Organic Chemistry*, 57, 6077–6079 (1992).

23. R. S. Ruoff, D. S. Tse, R. Malhotra, and D. Lorents, Solubility of fullerene (C_{60}) in a variety of solvents, *Journal of Physical Chemistry*, 97, 3379–3383 (1993).

24. S. Niyogi, M. A. Harmon, H. Hu, B. Zhao, P. Bhowmik, R. Sen, M. E. Itkis, and R. C. Haddon, Chemistry of single-walled carbon nanotubes, *Accounts of Chemical Research*, 35, 1105–1113 (2002).

25. M. Jazdzyk, W. Jia, and G. P. Miller, *Patterning surfaces with functionalized fullerenes*, Poster, Emerging Technologies and the Chemical Sciences, NERM 2006, American Chemical Society (2006).

26. Mixed interfullerene bonding motifs in C60-based polymers, European Synchrotron Radiation Facility, http://www.esrf.eu/UsersAndScience/Publications/Highlights/2005/ (2007).

27. S. Margadonna, D. Pontiroli, M. Belli, T. Shiroka, M. Riccò, and M. Brunelli, Li_4C_{60}: A polymeric fulleride with a two-dimensional architecture and mixed interfullerene bonding motifs, *Journal of the American Chemical Society*, 126, 15032–15033 (2004).

28. S. Watanabe, N. S. Ishioka, T. Sekine, A. Osa, M. Koizumi, H. Shimomura, and H. K. Yoshikawa Muramatsu, Production of endohedral fullerene—[133]Xe by ion implantation, *Journal of Radioanalytical and Nuclear Chemistry*, 255, 495–498 (2003).

29. M. Linnolahti, A. J. Karttunen, and T. A. Pakkane, Remarkably stable fulleranes: $C_{80}H_{80}$ and $C_{180}H_{180}$. *ChemPhysChem*, 7, 1661–1663 (2006).

30. P. J. F. Harris, *Carbon nanotubes and related structures: New materials for the twenty-first century*, Cambridge University Press, Cambridge (2001).

31. M. S. Dresselhaus, G. Dresselhaus, and P. C. Eklund, *Science of fullerenes and carbon nanotubes*, Academic Press, San Diego (1996).

32. M. S. Dresselhaus, G. Dresselhaus, and R. Saito, Physics of carbon nanotubes, *Carbon*, 33, 883 (1995).

33. R. Saito, M. S. Dresselhaus, and G. Dresselhaus, *Physical properties of carbon nanotubes*, Imperial College Press, London (1998).

34. M.-F. Yu, Strength and breaking mechanism of multiwalled carbon nanotubes under tensile load, *Science*, 287, 637–640 (2000).

35. P. G. Collins, M. S. Arnold, and P. Avouris, Engineering carbon nanotubes and nanotube circuits using electrical breakdown, *Science*, 292, 706–709 (2001).

36. Basic properties of carbon nanotubes, Applied Nanotechnologies, Inc., applied-nanotech.com/cntproperties.htm#Mechanical%20Properties (2005).

37. A.N. Kolmogorov, V.H. Crespi, M.H. Schleier-Smith, J.C. Ellenbogen, Nanotube-substrate interactions: Distinguishing carbon nanotubes by the helical angle, *Physics Review Letters*, 92, 085503 (2004).

38. M. D. Taczak, Controlling the structure and properties of carbon nanotubes, MITRE Corporation, the MITRE Nanosystems Group, www.mitre.org/tech/nanotech (2007).

39. T. A. Adams and D. Tománek. Physical properties of carbon nanotubes, The Nanotube Site, pa.msu.edu/cmp/csc/nanotube.html (visit 2007).

40. S. Frank, P. Poncharal, Z. L. Wang, and W. A. de Heer, Carbon nanotube quantum resistors, *Science*, 280, 1744–1746 (1998).

41. S. Sanvito, Y.-K. Kwon, D. Tománek, and C. J. Lambert, Fractional quantum conductance in carbon nanotubes, *Physics Review Letters*, 84, 1974 (2000).

42. P. G. Collins and P. Avouris, Nanotubes for electronics, *Scientific American*, December, 62–69 (2000).

43. A. Thess, R. Lee, P. Nikolaev, H. Dai, P. Petit, J. Robert, C. Xu, Y. H. Lee, S. G. Kim, A. G. Rinzler, D. T. Colbert, G. Scuseria, D. Tománek, J. E. Fischer, and R. E. Smalley, Crystalline ropes of metallic carbon nanotubes, *Science*, 273, 483 (1996).

44. G. Gao, T. Cagin, and W. A. Goddard, III, Energetics, structure, mechanical and vibrational properties of single walled carbon nanotubes, *Nanotechnology*, 9, 184–191 (1998).

45. G. L. Hornyak, L. Grigorian, A. C. Dillon, P. A. Parilla, and M. J. Heben, A temperature window for chemical vapor deposition growth of single walled carbon nanotubes, *Journal of Physical Chemistry B*, 106, 2821–2825 (2002).

46. S. Maruyama, Y. Murakami, Y. Miyauchi, and S. Chiashi, Generation of single-walled carbon nanotubes from alcohol and generation mechanism by molecular dynamic simulations, *Journal of Nanoscience and Nanotechnology*, 4, 360 (2004).

47. L. Ci, S. Xie, D. Tang, X. Yan, Y. Li, Z. Liu, X. Zhou, W. Zhou, and G. Wang, Controllable growth of single wall carbon nanotubes by pyrolyzing acetylene on the floating iron catalysts, *Chemical Physics Letters*, 349, 191 (2001).

48. H. Dai, A. G. Rinzler, P. Nikolaev, A. Thess, D. T. Colbert, and R. E. Smalley, Single-wall nanotubes produced by metal-catalyzed disproportionation of carbon monoxide, *Chemical Physics Letters*, 260, 471 (1996).

49. H. W. Zhu, C. L. Xu, D. H. Wu, B. Q. Wei, R. Vajtai, and P. M. Ajayan, Direct synthesis of long single-walled carbon nanotube strands, *Science*, 296, 884–886 (2002).

50. N. Hatta and K. Murata, Very long graphitic nano-tubules synthesized by plasma-decomposition of benzene, *Chemical Physics Letters*, 217, 398 (1994).

51. K. Hata, D.N. Futaba, K. Mizuno, T. Namai, M. Yumra, and S. Iijima, Water-assisted highly efficient synthesis of impurity-free single-walled carbon nanotubes, *Science*, 306, 1362 (2004).

52. A. Cao, X. Zhang, C. Xu, J. Liang, D. Whu, and B. Wei, Aligned carbon nanotube growth under oxidative ambient, *Journal of Materials Research*, 16, 3107 (2001).

53. S. C. Tsang, P. J. F. Harris, and M. L. H. Green, Thinning and opening of carbon nanotubes by oxidation using carbondioxide, *Nature*, 362, 520–522 (1993).

54. P. M. Ajayan, T. W. Ebbesen, T. Ichihasi, S. Iijima, K. Tanigaki, and H. Hiura, Opening carbon nanotubes with oxygen and implication for filling, *Nature*, 362, 522–523 (1993).

55. O. Smiljanic, T. Dellero, A. Serventi. G. Lebrun, B. L. Stansfield, J. P. Dodelet, M. Trudeau, and S. Desilets, Growth of carbon nanotubes on ohmically heated carbon paper, *Chemical Physics Letters*, 342, 503 (2001).

56. Y. Zhang, Electric field directed growth of single-walled carbon nanotubes, *Applied Physics Letters*, 79, 3155 (2001).

57. T. Ono, E. Oesterschulze, G. Georgiev, A. Georgiev, and R. Kassing, Field-assisted assembly and alignment of carbon nanofibres, *Nanotechnology*, 14, 37 (2003).

58. R. V. Seidel, *Carbon nanotube devices*, Ph.D. dissertation, Technischen Universität Dresden (2004).

59. R. Seidel, M. Liebau, G. S. Duesberg, F. Kreupl, E. Unger, A. P. Graham, W. Hoenlein, and W. Pompe, In-situ contacted single-walled carbon nanotubes and contact improvement by electroless deposition, *Nano Letters*, 7, 965–968 (2003).

60. G. S. Duesberg, A. P. Graham, M. Liebau, R. Seidel, E. Unger, F. Kreupl, and W. Hoenlein, Growth of isolated carbon nanotubes with lithographically defined diameter and location, *Nano Letters*, 3, 257–259 (2003).

61. V. Shanov and M. Schultz, The longest carbon nanotubes you've ever seen, *Nanotechnology Now*; www.nanotech-now.com/news.cgi?story_id=22540; Posted May 10 (2007). The breakthrough was presented in April 2007 at the *Single Wall Carbon Nanotube Nucleation and Growth Mechanisms* workshop organized by NASA and Rice University.

62. L. Ge, S. Sethi, L. Ci, P. M. Ajayan, and A. Dhinojwala, Carbon nanotube-based synthetic gecko tapes, *Proceedings of the National Academy of Science USA*, 104, 10792–10795 (2007).

63. T. W. Ebbesen, P. M. Ajayan, H. Hiura, and K. Tanigaki, Purification of carbon nanotubes, *Nature*, 367, 519 (1994).

64. N. Tagmatarchis, A. Zatton, P. Rescgiglian, and M. Prato, Separation and purification of functionalized water-soluble multi-walled carbon nanotubes by field-flow fractionation, *Carbon*, 43, 19984–19989 (2004).

65. P. G. Collins, M. S. Arnold, and P. Avouris, Engineering carbon nanotubes and nanotube circuits using electrical breakdown, *Science*, 292, 706–709 (2001).

66. G. Zhang, P. Qi, X. Wang, Y. Lu, X. Li, R. Tu, S. Bangsaruntip, D. Mann, L. Zhang, and H. Dai, Selective etching of metallic carbon nanotubes by gas-phase reaction, *Science*, 314, 974–977 (2006).

67. Rice Develops Method to Separate Nanotubes Based on Size, brightsurf.com/news/headlines/25097/Rice_develops_first_method_to_sort_nanotubes_by_size.html; based on information acquired from H. Schmidt, Carbon Nanotechnology Laboratory at Rice University (2007).

68. M. Meyyappan, Growth: CVD and PECVD, Chapter 4 in *Carbon nanotubes: Science and applications*, M. Meyyappan, ed., CRC Press, Boca Raton (2005).

69. S. Huang and J. Liu, Direct growth of single walled carbon nanotubes on flat substrates for nanoscale electronic applications, Chapter 4 in *Applied physics of carbon nanotubes: Fundamentals of theory, optics and transport devices*, S. V. Rotkin, S. Subramoney, eds., Springer-Verlag, New York (2005).

70. S.-J. Eum, H.-K. Kang, and C.-W. Yang, Electron microscopy investigation at the initial growth stage of carbon nanotubes, *Journal of the Korean Physical Society*, 42, S727–S731 (2003).

71. C. Klinke, J.-M. Bonard, and K. Kern, Thermodynamic calculations on the catalytic growth of multiwall carbon nanotubes, *Physics Review B*, 71, 035403:1–7 (2005).

72. R. J. Chen, S. Bangsaruntip, K. A. Drouvalakis, N. W. S. Kam, M. Shim, Y. Li, W. Kim, P. J. Utz, and H. Dai, Noncovalent functionalization of carbon nanotubes for highly specific biosensors, *Proceedings of the National Academy of Science*, 100, 4984–4989 (2003).

73. P. Liu, Modifications of carbon nanotubes with polymers, *European Polymer Journal*, 41, 2693 (2005).

74. A. Eitan, K. Jiang, D. Dukes, R. Andrews, and L. S. Schadler, Surface modification of multi-walled carbon nanotubes: Toward tailoring of the interface in polymer composites, *Chemistry of Materials*, 15, 3198–3201 (2003).

75. H. Palmoniemi, T. Aarital, T. Laiho, H. Liuke, J. Lukkari, and K. Haapakka, Noncovalent functionalization of single-wall carbon nanotubes to improve water-solubility, University of Turku, Department of Chemistry, Poster, Segovia, Abstract_Poster_PaloniemiHanna.pdf (2004).

76. M. J. O'Connell, P. Boul, L. M. Ericson, C. Huffman, Y. Wang, E. Haroz, C. Kuper, J. Tour, K. D. Ausman, and R. E. Smalley, Reversible water-solubilization of single-walled carbon nanotubes by polymer wrapping, *Chemical Physics Letters*, 342, 265–271 (2001).

77. S. Qin, W. T. Ford, D. E. Resasco, and J. E. Herrera, Functionalization of single-walled carbon nanotubes with polystyrene via grafting to and grafting from methods, *Macromolecules*, 37, 752–757 (2004).

78. X. Lou, C. Detrembleur, V. Sciannamea, C. Pagnoulle, and R. Jérôme, Grafting of alkoxamine end-capped (co)polymers onto multi-walled carbon nanotubes, *Polymer*, 45, 6097–6102 (2004).

79. H. Aziz, Routes to carbon nanotube solubilization and applications, Department of Chemistry, Duke University: www.lib.duke.edu/chem/chem110/papers/ Hamza%20Aziz.thm (2003).

80. J. F. Bahr, J. Yang, D. V. Kosynkin, M. J. Bronikowski, R. E. Smalley, and J. M. Tour, Functionalization of carbon nanotubes by electrochemical reduction of aryl diazonium salts: A bucky paper electrode, *Journal of the American Chemical Society*, 123, 6536–6542 (2001).

81. D. R. Kauffman, O. Kuzmych, and A. Star, Modification of the semiconducting single walled carbon nanotube valence band through non-covalent attachment of tetraphenyl metalloporphyrins, Department of Chemistry, University of Pittsburgh, www.nanofab.ece.cmu.edu/avs/abstracts/DouglasKauffman.doc

82. A. B. Artyukhin, O. Bakajin, P. Stroeve, and A. Noy, Layer-by-layer electrostatic self-assembly of polyelectrolyte nanoshells on individual carbon nanotube templates, *Langmuir*, 20, 1442–1448 (2004).

83. R. Haggenmueller, F. Du, J. E. Fischer, and K. I. Winey, Interfacial in situ polymerization of single wall carbon nanotube/nylon 6,6 nanocomposites, *Polymer*, 47, 2381–2388 (2006).

84. R.C. DeVries, Synthesis of diamond under metastable conditions, *Annual Review of Materials Science*, 17, 161–187 (1987).

85. H. C. Tsai and D. B. Bogy, Characterization of diamond like carbon films and their application as overcoats on thin film media for magnetic recording, *Journal of Vacuum Science and Technology*, A5, 3287–3312 (1987).

86. J. Norley, The role of natural graphite in electronics cooling, *Electronics Cooling*, 7, 50–51 (2001).

87. T. M. Willey, C. Bostedt, T. van Buuren, J. E. Dahl, S. G. Liu, R. M. K. Carlson, L. J. Terminello, and T. Möller, Molecular limits to the quantum confinement model in diamond clusters, *Physics Review Letters*, 95, 113401 (2005).

88. J. E. Dahl, S. G. Liu, and R. M. K. Carlson, Isolation and structure of diamondoids, nanometer-sized diamond molecules, *Science*, 299, 96 (2003).

89. J. Rantala, Diamonds are a thermal designer's best friend, *Electronics Cooling*, electronics-cooling.com/articles/2002/2002_february_techdata.php (2002).

90. H. Uetsuka, D. Shin, N. Tokkuda, K. Saeki, and C. E. Nebel, Electrochemical grafting of boron-doped single-crystalline chemical vapor deposition diamond with nitrophenyl molecules, *Langmuir,* 23, 3466–3472 (2007).

91. W. Yang, S. E. Baker, J. E. Butler, C.-S. Lee, J. N. Russwell, L. Shang, B. Sun, and R. J. Hamers, Electrically addressable biomolecular functionalization of conductive nanocrystalline diamond films, *Chemistry of Materials,* 17, 938–940 (2005).

92. J. Wang, M. A. Firestone, O. Auciello, and J. A. Carlisle, Diamond films by electrochemical reduction of aryldiazonium salts, *Langmuir,* 20, 11450 (2004).

93. J. F. Bahr, J. Yang, D. V. Kosynkin, M. J. Bronikowski, R. E. Smalley, and J. M. Tour, Functionalization of carbon nanotubes by electrochemical reduction of aryl diazonium salts: A bucky paper electrode, *Journal of the American Chemical Society,* 123, 6536–6542 (2001).

94. W. Yang, O. Auciello, J. E. Butler, W. Cai, J. A. Carlisle, J. E. Gerbi, D. M. Gruen, T. Knickerbocker, T. L. Lasseter, J. N. Russell, L. M. Smith, and R. J. Hamers, DNA-modified nanocrystalline diamond thin-films as stable, biologically active substrates, *Nature Materials,* 1, 253–257 (2002).

Problems

9.1 Carbon is fundamental to life. Why? Why would silicon not be as good?

9.2 Calculate the diameter of the following SWNTs:

 a. (6,6)
 b. (10,10)
 c. (12,12)
 d. (12,0)
 e. (9,3)
 f. (10,2)
 g. (11,7)

9.3 Which SWNTs are conducting? Which ones are semiconducting?

 a. (4,3)
 b. (4,4)
 c. (11,5)
 d. (5,1)
 e. (9,0)
 f. (9,9)
 g. (12,4)

9.4 Calculate the tensile strength of a (12,12) carbon nanotube. How many are required to pick up the Empire State building? Use the dissociation energy of a 1.5 C–C bond (between energy between that of a C–C and a C=C bond). The Empire State building weighs 365,000 tons.

9.5 Explain the van Hove singularity in more detail.

9.6 Explain the differences between Raman and infrared techniques.

9.7 Name the fundamental Raman modes of SWNTs and describe briefly their energy and source in the nanotube.

9.8 Please research the following: The translation vector that describes SWNTs is given by

$$T = t_1 a_1 + t_2 a_2$$

where

$$t_1 = \frac{2m+n}{d_H} \quad \text{and} \quad t_2 = \frac{-(2n+m)}{d_H}$$

where d_H is the greatest common divisor.

 a. Show that $T = \frac{\sqrt{3}C}{d_H}$ if $n - m \neq 3rd_H$.

 b. Draw the unit cell for a (5,5) SWNT. Determine the area of a (5,5) nanotube unit cell (use: area in square nanometers $= |C \times T|$)

 c. Draw an image of the graphene sheet representing this tube.

9.9 Using a transparency, construct a nano-tube that is

a. (10,10)
b. (5,5)
c. (5,0)
d. (5.3)
e. (10,0)

Scotch tape the ends of the transparency together to hold the tube together.

9.10 Do some research and find an expression that reveals the number of carbon atoms per unit cell.

9.11 Determine the chiral angle in the following nanotubes:

a. (4,3)
b. (4,4)
c. (11,5)
d. (5,1)
e. (9,0)
f. (9,9)
g. (12,4)
h. (10,10)
i. (5,5)
j. (5,0)
k. (5.3)
l. (10,0)

9.12 Some carbon nanotubes are insulating. Why do you think this could be?

9.13 SWNTs are often found in bundles. Why do you think this occurs?

9.14 How many peaks do you expect in a ^{13}C NMR spectrum of C_{60}? Of C_{70}?

9.15 Explain briefly the mechanism of formation of SWNTs.

9.16 Why is *base-growth* mechanism favored over *tip-growth* from catalysts on alumina substrates?

9.17 Would it be possible to determine nanotube diameter from Raman spectra? Explain your answer.

9.18 Can you think of any other pure material (one element) that can exist as a conductor, a semiconductor, or an insulator based solely on its structure and size?

9.19 Boron–nitrogen materials also form solid sp^2 structures. How are they different from or similar to graphene? Boron ($3\ e^-$) and nitrogen ($5\ e^-$) are on the left and right side of carbon, respectively, in the periodic table.

9.20 Check Euler's rule of polyhedra, $f + v = e + 2$, for a tetrahedron (where f = faces, v = vertices, e = edges); for an octahedron; and for a cuboctahedron. In addition to metal clusters, the formula also works well for sp^2-bonded carbon materials. Does it apply to C_{60}? How about C_{70}?

9.21 Theorize as to why small carbon clusters (~20 atoms or less) prefer to exist in chains. (Consult M. Di Ventra et al., *Introduction to Nanoscale Science and Technology*, Kluwer Academic Press, the Netherlands, 2004.)

9.22 How are the energetics of fullerenes related to curvature of the nanostructure? Is there a relation to $1/R$ or $1/R^2$ (where R is the curvature)? (Consult M. Di Ventra et al., *Introduction to Nanoscale Science and Technology*, Kluwer Academic Press, the Netherlands, 2004.)

CHEMICAL INTERACTIONS AT THE NANOSCALE

Even the formal justification of the electron-pair bond in the simplest cases ...
requires a formidable array of symbols and equations.

LINUS PAULING, 1931

Chapter 10

THREADS

The chemistry division consists of four chapters. We started with chapter 9, "Carbon-Based Nano-materials." These materials form an extremely important class of materials and consequently the topic was awarded a special chapter. Materials based on carbon contribute to zero dimension (fullerenes), one dimension (carbon nanotubes), two dimensions (diamond thin films), and three dimensions (hierarchical structures). Carbon contributes to a wide class of applications including conductors, semiconductors, insulators, and structural components. The importance of nanocarbon materials in biology, biotechnology, and medical technology cannot be understated.

In this chapter, we present and discuss many kinds of intermolecular chemical interactions. Although the chapter is all about chemistry, the topic unavoidably includes a hefty dose of physics. Supramolecular chemistry is the focus of chapter 11.

All of supramolecular chemistry takes place at the nanoscale. Chapter 12 delves into chemical modification of nanomaterials, template synthesis, and other nanorelevant topics. With chapter 12, we conclude the chemistry division of the text.

Since numerous subdisciplines of chemistry infringe upon the domains of physics at one end and biology at the other, we will try our best to stay within the boundaries of synthesis-oriented topics as they apply to nanomaterials. Chapters 13 and 14 are allocated to the biological nanoscience division; they focus on natural nanomaterials and biochemical nanoscience, respectively.

It is time to refresh our memories once again concerning chapter 2 on the societal implications of nano. As you proceed through this chapter and the rest of the text, always ask the question, "What are the consequences—good, bad, or indifferent—of these materials?"

10.0   BONDING CONSIDERATIONS AT THE NANOSCALE

Chemical and physical interactions between atoms and molecules comprise two sides of the same coin. This apparent dichotomy is once again a product of convenience, generated to feed our instinct to catalog. Two atoms held together by attractive forces form a chemical bond that in turn yields a molecule (from the Latin *moles*, "small unit of mass" or "mass barrier" based on the Greek *molos*, "exertion"). Attractive forces between atoms form strong *intramolecular* bonds as a result of chemical reactions. We have briefly introduced types of *intramolecular* bonding such as the covalent bond, the ionic bond, and other types like the metallic bond in chapter 5. We now expand our discussion to the *intermolecular bond*—a type of interaction that exists between two or more molecules.

We define physical processes as those that involve no change in the chemical structure of molecules. We define chemical processes as those that do involve changes in the chemical structure of molecules. If we apply these definitions, in the strictest sense, to intermolecular bonds, they should be classified as a physical interaction because the structure (chemical nature) of the precursor molecules is not altered (significantly). However, any type of bonding, regardless of how strong or weak, causes perturbations to the integrated molecular orbital of any preexisting molecular system—perturbations that can be construed to be chemical in nature. Although no bonds are made or broken (except in the case of disulfide linkages), is the folding of a protein a chemical or a physical process? The result of the folding yields a molecule, albeit a quite large one, that is bestowed with altered chemical properties.

Interactions between atoms or between molecules, regardless of our need to classify them, belong to a continuum of bonding. It is just a matter of degree with regard to bond strength and the type of bond that is formed between specific chemical species that allows us to classify them as one type or another. All bonding is due to some kind of electrostatic attraction or repulsion or a combination of both and therefore belongs to the "electrostatic bonding continuum." Although bonding forces involved in forming intramolecular and intermolecular bonds have many different labels and are assigned numerous levels of strength, they all fall under the generalized umbrella of Coulombic (electrostatic) interactions. Electrons and their relationship with positively charged nuclei account for all types of bonding that occur between and among atoms and molecules. The distinction, therefore, between chemical (intramolecular) and physical (intermolecular) interactions is somewhat arbitrary but certainly one that helps us to derive a perspective and to seek an understanding of macromolecular systems. In an analogous manner, the distinction between chemisorption and physisorption is also worth a second look with respect to this context. In the former, a chemical bond is formed between an atom or molecule and a substrate; in the latter, a physical interaction is responsible for the attachment. Physisorption, therefore, can be considered to be one of the weakest types of chemical bonding.

Two or more molecules that are held together by attractive forces other than the previously mentioned "big three" also result in a generalized chemical bond. These bonds assume many forms and range in strength from almost nothing to approximately kT to several tens or more kilojoules per mole, but they are usually much weaker than those found in ionic, covalent, and metallic materials. The formation of covalent bonds, for example, requires energy between 80 and 400 times kT (200–1000 $kJ \cdot mol^{-1}$) [1]. These bonds are generally stable to fluctuations in their environment. Intermolecular bonding, as we shall soon discover, falls under another set of parameters and can be significantly influenced by the immediate environment.

Molecules held together by intermolecular forces are characterized by the following generalized parameters:

- Primary structure of molecules that are bound to each other by intermolecular bonds is not altered (e.g., the covalent backbone of each molecule remains unchanged although the molecule exists in a newly combined state).
- Strength of intermolecular bonds is relatively weak compared to the big three.
- Intermolecular bonds are generally formed under thermodynamically controlled conditions.
- Solvent is important during synthesis of macromolecules and their subsequent maintenance.
- Both enthalpic and entropic conditions are important to overall ΔG.
- Although individually relatively weak, the energy of intermolecular bonds summed over the whole molecule or system can be quite formidable.

FIG. 10.1

The ancients believed that solids were held together by hooks—a concept that would endure into the nineteenth century. Water was a smooth material that had no hooks. Atoms found in air were soft. (J. Pfeffer and S. Nir, Modern Physics: An Introduction Text, World Scientific Publishing Company, 183, 2001.) Descartes and others rejuvenated the microscopic "hook and eye" theory.

10.0.1   *Background*

The concept of chemical bonding dates back to ancient times when Asklepiades of Prusa (ca. 100 B.C.) theorized about clusters of atoms [2]. In ancient Rome, Lucretius speculated in his *De Rerum Natura* (*On the Nature of Things*) that atoms were tiny spheres held together by fishhooks and that atoms bonded when these hooks became entangled [2]. Robert Boyle in 1661 (*The Sceptical Chymist*) rejuvenated the idea of the cluster and that chemical reactions resulted in rearrangement of the clusters. His corpuscular theory of matter helped construct the platform of modern chemistry. Later, in the early seventeenth century, René Descartes proposed that "some atoms were furnished with hook-like projections, and others, with eye-like ones" (**Fig. 10.1**). He affirmed that two atoms combined when the "hook of one got caught in the eye of the other."

The idea of "bonded combinations of atoms" was broadened into the concept of chemical affinity in 1718 by French chemist E. F. Geoffroy, who developed one of the first periodic tables: the Affinity Table (**Fig. 10.2**).

William Higgins in 1789 proposed the concept of *valency* based on "ultimate particles" that had an associated strength of force between them. It was a few years later in 1803 that John Dalton proposed the law of simple proportions and established the basics for modernistic atomic theory (**Fig. 10.3**).

In 1811, a paper by Amadeo Avogadro titled "Essay on Determining the Relative Masses of the Elementary Molecules of Bodies" was published. In it Avogadro claimed that "atoms are united by attractions to form a single molecule." M.-A.-A. Guadin of France correctly proposed molecular formulae (e.g., H_2O) in 1833. Friedrich Kekule in 1857 claimed that carbon was tetrahedrally bonded to four other constituents. In 1861, Joseph Loschmidt in his book *Chemischen Studien I* introduced the double-bonded carbon structure. The first stick-and-ball model of molecules was fabricated by August W. von Hofmann in 1865. Alexander C. Brown expanded the concept of valency. Emil Fischer in 1894 introduced the idea of "lock-and-key," specifically within the context of enzyme action. In 1916, G. N. Lewis generated the first *Lewis structures* and the *octet rule* by using dots to represent shared electrons.

It was not until 1931, following the discovery of the electron and the development of quantum mechanics, that Linus Pauling and colleagues established the modern understanding of the chemical bond, reported in his landmark work titled *The Nature of the Chemical Bond* [3]. Hybridization theory and

Fig. 10.2 *E. F. Geoffroy of France in 1718 developed a theory about chemical affinity. He created a table that described "certain alchemical forces that draw components together." It never ceases to amaze how early scientists were able to speculate about the fundamental nature of matter without the ability to see atoms.*

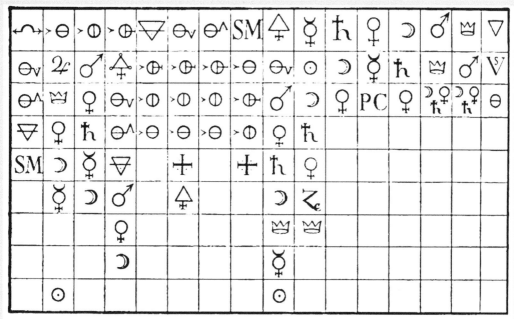

Esprits acides
Acide du sel marin
Acide nitreux
Acide vitriolique
Sel alcali fixe
Sel alcali volatil

Terre absorbante
Substances metalliques
Mercure
Regule d'Antimoine
Or
Argent

Cuivre
Fer
Plomb
Etain
Zinc
Pierre calaminaire

Soufre mineral   [Principe
Principe huileux ou Soufre
Esprit de vinaigre.
Eau.
Sel.   [dents
Esprit de vin et Esprits ar.

Fig. 10.3 *John Dalton also believed in the "hook theory" of atomic combination. Regardless of his understanding of the exact mechanism of bonding, hooks, or others, he was able to compile an impressive table that describes the various combinations of atoms to form common molecules.*

Hydrogen
Nitrogen
Carbon
Oxygen
Sulfur
Phosphorus
Alumina
Soda
Potash
Copper
Lead

Water
Ammonia
Olefiant gas
Carbonic oxide
Carbonic acid

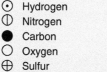

Sulfuric acid

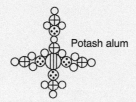

Potash alum

electronegativity were just a few of the phenomena described in this fundamental work. Details concerning the nature of the *intermolecular* bond also emerged about this time. K. L. Wolf in 1937 described the hydrogen bond, the first intermolecular bond to be designated as such, in *supermolecular* systems. In 1953, Watson and Crick's (and Franklin's) landmark work illuminating the structure of DNA led to a new age in supramolecular chemistry. The ingenuity exhibited by all these scientists is certainly inspirational if not outright astounding. Most of the bonding models were developed without knowledge of the electron and, for the first few thousand years, without the benefit of any clear proof of the existence of the atom itself.

One can just imagine the exhilaration experienced by the first person to see the atom—a distinction that goes to Professor Erwin Mueller (the inventor of the field-emission, atomic probe, and field-ion microscopes) of Penn State University in 1955. He used an electron microscope that he developed: "It was a sticky day in August 1955 that I became the first person to see an atom."

The first truly atomic-scale resolution image of an atom was achieved by University of Chicago graduate student Joe Wall in 1969 with an STEM (Discover Magazine.com/2007). Of course, as we know so well by now, the development of scanning tunneling microscopy (STM) allowed for manipulation of single atoms and a close-up look at crystal structure and molecular bonding.

10.0.2 *Intramolecular versus Intermolecular Bonding*

We proceed now to reinforce our previously developed dichotomy in bonding: the differences between the intramolecular bond and the intermolecular bond. The three major kinds of intramolecular bonds are summarized in **Table 10.1**.

TABLE 10.1	*The "Big Three" of Intramolecular Bonding*	
Type of bond	**Comments/examples**	**Bond strength (kJ · mol^{-1})**
Ionic bonds	Nondirectional bonds between ions of opposite electronic charge Bond strength is proportional to the magnitude of the charges and inversely proportional to distance between them	Lattice energies (calculated and thermochemical)
	NaF	910–923
	NaCl	769–786
	KF	808–821
	KCl	701–715
	CsF	744–740
	CsCl	657–659
	Al_2O_3	15,916
	Ga_2O_3	15,590–15,220
	Ti_2O_3	14,702
	Na_2O	2,481
	MgO	3,795–3,791
	CaO	3,414–3,401
	$NaNO_3$	755–756

TABLE 10.1 (CONTD.)	The "Big Three" of Intramolecular Bonding	
Type of bond	**Comments/examples**	**Bond strength (kJ · mol⁻¹)**
Covalent bonds	Highly directional (with well-defined angles), hybrid bonds and sharing of outer-orbital electrons	Strong bonds ranging from 100 to 1000 kJ · mol⁻¹
	H–H	436
	O–H	463
	C–H	412
	C–C	348
	C=C	690
	C≡C	839
	C–O	360
	C=O	690
	C≡O	1072
	C–Cl	338
	C–Br	276
	C–I	238
	C–N	305
	N≡N	942
Metallic bonds	Nondirectional delocalized electrons exit as "electron gas" based on free electrons of metal. Transition metals are the most common metals	Energy of cohesion, ΔH_{sub}
	Au (*fcc*)	368
	Ag (*fcc*)	284
	Cu (*fcc*)	336
	Al (*bcc*)	327
	Hg (*rhomb*)	60.3
	Os (*hexagonal*)	788
	W (*bcc*)	848
	Mo (*bcc*)	718
	Fe (*tetragonal*)	413
	Pb (*fcc*)	196
	Mn (*hexagonal*)	282
	Po (*monoclinic*)	145
	Pt (*fcc*)	564

In order for atoms to stay in close proximity, the strength of the bond that holds them together must be able to overcome the thermal kinetic energy of an average molecule

$$KE_{Molecule} > \frac{3}{2}k_B T \qquad (10.1)$$

where k_B is the Bolztmann constant (1.3806×10^{-23} J · K⁻¹) and T is the temperature in Kelvin. At room temperature, the kinetic energy of an average molecule is ca. 3.72 kJ·mol⁻¹. If bond strength is below this value, atoms run the risk of vibrating off the bond and going on separate trajectories. It is clear from **Table 10.1** that intramolecular bonds are quiet strong and run no risk of dissociation at $k_B T$. However, many ionic compounds, however, seemingly dissolve in water quite easily. What is the basis of this apparent paradox? The heat of solvation, whether exo- or endothermic, is the difference between the lattice enthalpy and the enthalpy of solvation. Obviously, other factors besides thermal energy are at work. For example, enough free energy is gained by the interaction of the ions in a crystal with the molecules of the solvent to allow for dissolution of the solid.

Although covalent interactions are always part of the bonding scenario, they take a backstage role for now as we direct our attention to intermolecular attractions that combine to form the primary actors (attractors) of this chapter and the next: supramolecular chemistry. Chemistry of the nanoscale is very much about exploitation of weak interactions: some types that exist below the kinetic energy of the average molecule. So long as temperature, pH, and pressure remain relatively constant (made possible by confined or insulated systems and proper choice of environment), supramolecular structures, including the major subset of biomacromolecules, remain in stable form. Intra- and intermolecular bonding are compared side by side in **Table 10.2**.

10.0.3 Types of Intermolecular Bonding

In this section we set the table of intermolecular bonding by providing a brief overview. All intermolecular interactions arise from the relationships between

TABLE 10.2	*Intermolecular—Intramolecular Bond Comparison*	
Parameter	**Intermolecular**	**Intramolecular**
Molecular structure of constituents	Molecular structure remains relatively intact after intermolecular bonding episode. Example: amino acids linked by hydrogen bonds that are involved in the 2° (folding) structure of a protein. The 1° structure of the protein is not altered	New molecules formed with chemical properties that may be radically different from precursors. Amino acids linked by condensation reactions form the 1° structure of a protein
Bond strength	Bond strength in general ranges from nearly nothing to less than 100 kJ·mol⁻¹	The bond strength of the CO triple bond is 1072 kJ·mol⁻¹
Reaction type	Thermodynamic control	Kinetic control
Importance of solvent and immediate environment	Extremely important during formation and maintenance (stability) of macromolecular systems (e.g., biological systems)	Solvent important during synthesis to stabilize intermediates and affect solubility
ΔG	Both ΔH and $T\Delta S$ are important	ΔH is important
Overall bond strength	Even weak bonds between molecules summed over all molecules can influence physical properties (e.g., the hydrogen bond in water). The gecko's ability to cling to a ceiling is another example of overall bond strength	Overall enthalpy is a straightforward extensive property of the material
Self-repair	Yes	No
Degrees of freedom	Greater flexibility	Limited

and among the electrons, the nuclei, and the solvent if in liquid phase. The basic types of attractive intermolecular interactions are listed and discussed below.

Ion–Ion and Ion–Dipole Interactions. Ion–ion interactions are purely electrostatic in nature. Apart from those found in ionic crystals, ion–ion bonds are technically characterized as individually weak. For example, the interaction between a charged species (e.g., an anionic macromolecule) and a cation (either in atomic or molecular form) falls into this category. Obviously, ion–ion reactions that are relevant to nanoscience take place in solvents, although gas phase ion–ion reactions also exist. Electron transfer from one atom to another is typical in such reactions, although subtle exchanges between (+) and (−) charges are also observed. To be purely intermolecular, no electron transfer should occur (e.g., the discrete character of the molecule is retained during the interaction). These kinds of interactions are generally classified as nondirectional (delocalized), nonspecific, and lacking unique stoichiometry. Although many ion–ion configurations incorporate other selectivity criteria (e.g., size exclusion), in general, ion–ion interaction reactions are difficult to assimilate into a molecular design process. Ion–dipole interactions are also considered to be purely electrostatic (Coulombic); the difference arises only in the strength of the charge on the dipole. The strength of ion–dipole interactions is reduced compared to ion–ion interactions. The strongest interaction occurs when the proper alignment of ion to dipole occurs.

Van der Waals Interactions. Van der Waals interactions include a wide array of electrostatic interactions based on polarity and electric charge. The *dipole–dipole* electrostatic interaction (the Keesom van der Waals interaction) is based on Coulombic attraction between two permanent dipoles, each characterized by a unique dipole moment μ. The angle of approach is a critical factor that influences the strength of the intermolecular bond. Rotations around bonds influence the strength of the dipole interaction and can range from attractive to repulsive. A unique subset of dipole–dipole interactions is the hydrogen bond—a category that deserves its own section in this text.

Polarization interactions contribute two major subsets of the van der Waals interaction category: *ion-induced dipole, dipole-induced dipole,* and *induced dipole– induced dipole* interactions. Ion-induced dipole interactions are technically not considered to be a member of the general class of van der Waals forces. Dipole-dipole interactions occur in two ways: head on head, similar in form to the hydrogen bond, and in the case where the polar ends of molecules overlap simultaneously. Dipole-induced dipole interactions (the Debye form of the van der Waals interaction caused by the formation of instantaneous dipoles) are caused by polarization of a previously nonpolar molecule by a nearby dipolar molecule. The last category, induced dipole–induced dipole (London dispersion forces), is the most common form of van der Waals interactions. London dispersion forces are characterized by charge fluctuations (e.g., dynamic electron cloud fluctuations of participating molecules). All electron clouds of all elements and molecules are subject to polarization from outside influences. Once polarized, molecules are able to remain in combined form due to these weak electrostatic interactions, depending on the strength of perturbations arising from the local environment.

Hydrogen Bonds and Allies.   Hydrogen bonds are dipole–dipole interactions in which a relatively strong bond is formed between substituents. It occurs between two molecules that contain highly electronegative (and small) atoms like O and N and a shared hydrogen. H-bonds are directional and can therefore be quite specific (e.g., the interaction between nucleotide base pairs to form DNA). Their utility in supramolecular design, unlike ion–ion, ion–dipole, or other forms of dipole–dipole interactions, is highly valued. Cα–H-bonds and halogen bonds, are similar in character to hydrogen bonds and will be discussed later in the text.

Hydrophobic Interactions.   Hydrophobic interactions occur between nonpolar molecules in aqueous solvents. Nonpolar molecules do not form hydrogen bonds with water but are nonetheless "organized" by water molecules (hydrated). Additional hydrophobic constituents in the solution cause release of organized water (entropy driven) and eventual coalescence of nonpolar constituents. Hydrophobic interactions are not classified as true bonding. They are nondirectional and nonspecific. Although extremely important in biology, hydrophobic interactions are difficult to account for in the molecular design process. *Attractive entropic elasticity* is a characteristic of macromolecular systems that accounts for an attractive–recoiling force produced by deformation due to extension of the molecules in a polymer.

Repulsive Forces.   *Steric repulsion* occurs from overlap of electron clouds and is based on the Pauli exclusion principle (no two electrons can have the same quantum number). Repulsion forces arising from the (+) charge of nuclei also contribute to electrostatic repulsion between two atoms or molecules. Repulsive interactions are described by Lennard–Jones potential functions and quantum mechanics. *Born repulsion* is the electrostatic repulsion between two entities with the same electrical charge. Subtle forms of repulsion also occur between two hydrophilic molecules. In particular, *hydrophilic repulsion* occurs in aqueous solutions between polar molecules to maximize the formation of hydrogen bonds with water. *Repulsive entropic elasticity* is the opposite of attractive entropic elasticity. In this case, repulsive forces arise due to compression of polymeric or other macromolecular systems. *Repulsive osmotic forces* arise from diffuse counter-ion electrical double layers between two charged surfaces and from steric repulsion in polymers originating from solvent translational entropy and interchain excluded volume.

10.0.4   *The Nano Perspective*

All types of bonding are important in nanomaterials. In chapter 6 we probed the various methods by which nanomaterials seek stabilization. Distortion of the lattice structure of metals held together by metallic bonds, for example, is an effective means of providing relief for the high energy and strain of the surface. Agglomeration, of course, is a straightforward means of alleviating instability. The adsorption (chemisorption or physisorption) of chemical moieties is fundamentally one of the most effective means of stabilizing nanomaterials. Chemisorbed ligands formed by dative bonds to the surfaces of nanometals serve to stabilize clusters. All systems seek the condition of minimal energy. Intramolecular metallic bonds hold together the Au-55 cluster, but intermolecular bonds stabilize the cluster through the action of ligands. Ionic bonds hold together TiO_2 nanoparticles, but dye molecules attached to the surface via intermolecular bonds serve to stabilize and functionalize. Carbon nanotubes, although kinetically stable, are held together by strong covalent bonds but can be chemically modified by covalent, ionic, and intermolecular reactions such as π-interactions.

Intermolecular bonds are well represented by nanoscale materials that we shall now designate as *soft matter* [4]. The term soft matter applies to materials like polymers, colloids, amphiphiles, and liquid crystals and, of course, almost all the biological materials. Colloid chemistry is an often neglected form of soft chemistry [4]. Well, not anymore! The soft matter of nature exhibits a level of bonding expertise that is by far the most impressive. The functionality of nanoscale biological systems relies on the timely manipulation of numerous forms of chemical bonding—often simultaneously. Supramolecular chemistry, our chemistry of the nanoscale, is discussed in chapter 11. Supramolecular chemistry is chemistry beyond the single molecule and is the chemistry of the intermolecular bond [5–7].

Several tables listing the types of intermolecular interactions follow. Please view these lists somewhat open mindedly and consider that the distinctions between and among all types of bonding are fuzzy at best. Numerous opportunities will become available throughout the text to sort out and evaluate unique bonding character. We certainly understand that all intermolecular bonds are derived from generalized Coulombic interactions and that there is significant intermingling of one type of bonding category into another type—especially in cooperative interactions. It is all a matter of degree, albeit one that forms yet another continuum comprising bonding in matter.

We also ask that the relationship of bonding to the perspective of nanomaterials be placed at the back of your mind as well as within the frontal lobes during all the forthcoming discussion. Bonding between and among entities that are larger than atoms or molecules, such as between materials with nanoscale proportions, is the sum total of the individual "bonding elements" integrated across the entire volume and surface of the constituents. The sum total energy in these cases may be quite large.

10.1 ELECTROSTATIC INTERACTIONS

Although all bonding is electrostatic, we especially direct our attention to interactions in which there is a clear case for physical charge separation between entities that have an extra electron or two or three or are missing electrons. Permanent dipoles are not charged but exhibit charge separation within their structure. The permanent dipole is due to the difference in electronegativity of its atomic constituents. All chemical bonding is due to interactions between electronic charges that arise from manifestations of Coulomb's law, given below in its force and energy forms:

| FIG. 10.4 | *A commemorative stamp issued by the French postal service to honor the work of Charles-Augustin Coulomb is depicted. He is best known for the formulation of Coulomb's law—that "the force between two electrical charges is proportional to the product of the charges and inversely proportional to the square of the distance between them" (Geocities.com).* |

$$F = \frac{1}{4\pi\varepsilon_o\varepsilon_r}\left(\frac{q_1q_2}{r^2}\right) \qquad (10.2)$$

$$E = \frac{1}{4\pi\varepsilon_o\varepsilon_r}\left(\frac{q_1q_2}{r}\right) \qquad (10.3)$$

where F is the force experienced between two elementary charges q, ε_o and ε_r are the dielectric permittivity of vacuum and medium, respectively, and r is the distance between the two charges. E is the potential energy between two point charges and is inversely proportional to the distance between them. The value of the vacuum permittivity constant ε_o is equal to 8.854×10^{-12} C$^2\cdot$J$^{-1}\cdot$m^{-1} and that of the elementary charge is 1.6022×10^{-19} C. Ion–ion, ion–dipole, and dipole–dipole interactions are electrostatic interactions between atoms and/or molecules that possess relatively permanent electrostatic charges. Charles-Augustin Coulomb is pictured in **Figure 10.4.**

Repulsive Interactions. Lest we place too little value on repulsive interactions, we shall provide a cursory overview of a few kinds of electrostatic repulsive interactions that exist in materials. Without repulsive interactions, all matter would collapse and there would be no investigators and nothing to investigate. Pauli exclusion is responsible for repulsive forces. Repulsive interactions provide the balance to attractive interactions at the distance of equilibrium position. Interaction potentials function based on distance between two charged species and like the *Lennard–Jones 6–12* potential, mentioned in chapter 6, incorporate both repulsive and attractive forces that additively combine to generate the curve shown in **Table 10.3.** In 1903, Gustav Mie developed an expression for a generalized interaction pair potential (before definitive evidence for atoms and molecules existed). The overall energy potential V_{total} (or U_{total}) is

TABLE 10.3	*Repulsive Interactions*	
Description	**Example**	**Repulsion energy (kJ·mol⁻¹)**
Steric repulsion	Steric repulsion results from filled orbitals that cannot be involved in bonding; such excluded volume contributes to a kinetic effect Reduces strength of interaction important in *lock and key* scenarios [8] Molecular crowding in cells Equilibrium constants inside small cavities or cells are different than of standard equilibrium constants determined in a beaker	Variable
Steric hindrance (steric resistance) Steric protection Steric shielding (of charged groups)	 *The staggered version of ethane (left) is more stable than the eclipsed version (right). Unwanted reactions can be prevented by steric protection (t-butyl groups are often used).*	12.5 kJ·mol⁻¹
Lennard–Jones 6–12 potential	 *The Lennard–Jones potential. The green line represents the average of the attraction (red) and repulsion (blue) curves. Circles represent atomic pairs. Repulsion due to Pauli exclusion and repulsive forces, therefore, is close range. Potential energy is inversely proportional to r¹². Repulsion potentials are positive and attraction potentials are negative.*	Interaction energy is a function of distance

$$V_{total} = -\frac{C_1}{r^n} + \frac{C_2}{r^m} \qquad (10.4)$$

where C_1 and C_2 are constants and r is the interatomic or intermolecular distance. You should be able to recognize the form of equation (10.4) and that exponents n and m are similar in form to the Lennard–Jones 6–12 exponents that we encountered in chapter 6. The Lennard–Jones potential is shown in **Table 10.3.**

Other electrostatic repulsive species exist that are equally as important, especially with regard to macromolecules and biological molecules. The steric (repulsive) character of rotamers such as ethane requires about 12 kJ·mol^{-1} to overcome the barrier to rotation. Steric hindrance in the form of t-butyl groups is able to protect chemically active species (**Table 10.3**).

10.1.1   Ion Pair Interactions

Ion–ion interactions are nondirectional. The strength of the interaction, the bond energy, is directly proportional to the magnitude of the charges involved and indirectly proportional to the distance between them. Expressions that describe the ion-attractive type interaction are extensions of Coulomb's law. Ionic solids are characterized by their lattice energy. The lattice energy is the change in enthalpy between an ionic solid and its ionic forms in the gas phase. The lattice energy is a good approximation of the strength of the ionic bond in the solid. An estimate of lattice energy for ionic solids is given by

$$E_a = \frac{\mathcal{M}\, z_1 z_2 e^2}{4\pi\varepsilon_o d_o} \qquad (10.5)$$

where $\mathcal{M}$ is the Madelung constant for the crystal, z_1 and z_2 are the respective integral charges on the ions, d_o is the equilibrium interatomic spacing, and e and ε_o assume their usual values. The Madelung constant $\mathcal{M}$ is a scaling constant that factors in the geometric arrangement of ions in a crystal to the nth level of coordination—nearest neighbors and beyond. Sample Madelung constants for a few crystals are NaCl (1.748), CsCl (1.763), CaF$_2$ (2.519), TiO$_2$ (2.408), and Al$_2$O$_3$ (4.172). A better approximation of E_a is obtained if a repulsive expression (r) is added to the lattice energy approximation. The total energy E_L of the lattice then is given by

$$E_L = E_a + E_r \qquad (10.6)$$

$$E_a = \frac{\mathcal{M} z_1 z_2 e^2}{4\pi\varepsilon_o d_o}\left(1 - \frac{d^*}{d_o}\right) \qquad (10.7)$$

where d_o is the equilibrium spacing as before and d^* is typically equal to 0.345 Å. The ion–ion energy depicted in this form of the equation yields energy in terms of electron volts (of the single ion pair interaction). It is difficult to determine lattice energy solely by experimental means; however, these calculations yield fairly good approximations of bond strength in ionic compounds.

The sodium chloride crystal is the universal example of a material based on ion–ion interactions. Each sodium cation is surrounded by six adjacent chloride anions in a face-centered cubic close-packed configuration. Each chloride, in a

FIG. 10.5	*The dissolution of NaCl is shown. Dissolution is driven by the ability of water to stabilize sodium cations and chloride anions. Hydration spheres with specific structure are able to solvate the ions.*

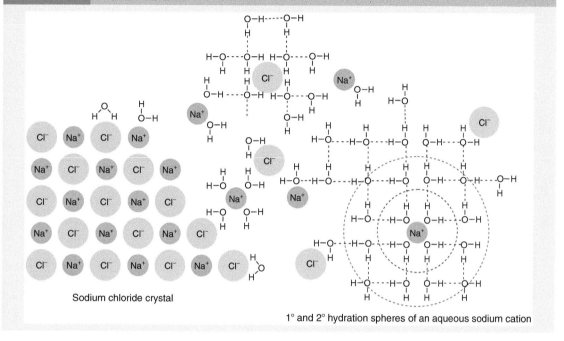

Sodium chloride crystal

1° and 2° hydration spheres of an aqueous sodium cation

complementary fashion, is coordinated by six sodium cations. The dissolution of a NaCl crystal is shown in **Figure 10.5**. Sodium chloride, although possessing relatively high lattice energy, will dissolve readily in water.

NaCl is technically not a supramolecular compound because it possesses only strong bonds and its precursors are all atoms. It is, however, an example of the simplest kind of donor–acceptor relationship—one that exhibits self-organizational behavior [9]. Ionic chelation compounds such as *tris*-diazobicyclooctane and cyanides serve as better examples of supramolecular ionic compounds that are able to complex metal anionic clusters such as $[Fe(CN)_6]^{3-}$ [9]. Other well-known chelating agents such as EDTA (ethylene-diaminetetraacetic acid) sport two lone pairs and four negative charges (Y^{4-} species), depending on the solution pH, and are readily able to complex metal cations. The coordination of biological metals (cofactors) by organic supramolecular moieties relies on ion–ion interactions as well as other types of bonding species (e.g., salt bridges). Ionic compounds—inorganic, metallic, and organic—all contribute in a profound way to supramolecular and biological systems (**Table 10.4**).

10.1.2 Solvent Effects

Although lattice energies of ionic solids are relatively high (**Table 10.1**), many ionic compounds dissolve readily in water. We now address the paradoxical statement proposed earlier. If we inspect a Born–Haber style cycle for NaCl, starting from ions in their gaseous state, we see that the lattice enthalpy for NaCl

TABLE 10.4   *Ion Pair Interactions*

Description	Example	Bond strength (kJ · mol⁻¹)
Ion–ion interactions $$E_{Ion-ion} = \frac{q_1 q_2 e^2}{4\pi \varepsilon_0 \varepsilon_r r} \quad (10.8)$$	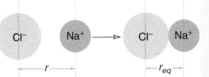	50–300 (100–350) Usually >190

Ion–ion attractions are responsible for the formation of NaCl crystals. Long-range attractive or repulsive forces have energy proportional to $1/r^2$; repulsive energy is proportional to $1/r$ (Table 10.3). Ion attractions are nondirectional and depend on the dielectric constant of the medium. The local minimum depicted in Table 10.3 corresponds to the equilibrium distance between the two charges.

| Salt bridge | 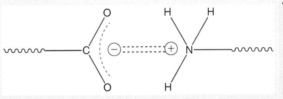 | 4–25 |

Strength of salt bridges can be as high as 25 kJ·mol⁻¹. Salt bridges are commonly found in protein complexes, especially in the active pocket of enzymes (usually between +Arg, +Lys, +His, and –Asp, –Glu).

EXAMPLE 10.1   *Lattice Energy of Ion–Ion Compounds*

(a) Calculate the lattice energy (ΔH_L°) of the NaCl crystal, at 25°C and standard pressure, from the Born–Haber cycle method ($\Delta H_f^\circ = -411$ kJ · mol⁻¹; $\Delta H_{sub}^\circ = +108$ kJ · mol⁻¹; BE(Cl₂) = +242 kJ · mol⁻¹; $I_{Na} = 5.13$ eV; $E_{Cl} = 3.68$ eV; the Faraday constant, $\mathcal{F} = 96{,}484.56$ C · mol⁻¹.

(b) Calculate the answer using e–e potential energy functions and the Madelung constant (equation 10.7) and compare to the value derived in (a).

Solutions:

(a)
$$\Delta H_L^\circ = -\Delta H_f^\circ + \Delta H_{sub}^\circ + \Delta H_{diss}^\circ + \Delta H_{ion}^\circ, Na + \Delta H_{e-aff,Cl}^\circ$$
$$= -\Delta H_f^\circ + \Delta H_{sub}^\circ + [BE(Cl_2)/2] + I\mathcal{F} + E\mathcal{F} \qquad (10.9)$$

The enthalpy of formation of NaCl is negative and the enthalpy of sublimation is positive. The energy of the Cl–Cl bond is 242 kJ·mol⁻¹, the ionization energy associated with Na → Na⁺ + e⁻ is 5.14 eV; the electron affinity E of Cl° + e⁻ → Cl⁻ is 3.68 eV.

$\Delta H_L^\circ = -(-411 \text{ kJ·mol}^{-1}) + (108 \text{ kJ·mol}^{-1}) + [242/2 \text{ kJ·mol}^{-1}] + [(5.13 \text{ eV})(98{,}484.56 \text{ C·mol}^{-1})/1000]$
$- [(3.68 \text{ eV})(96{,}484.56 \text{ C·mol}^{-1})/1000]; \Delta H_L^\circ$
$= [411 + 108 + 121 + 495 - 355] \text{ kJ·mol}^{-1} \approx 780 \text{ kJ·mol}^{-1}$

EXAMPLE 10.1 (CONTD.) *Lattice Energy of Ion–Ion Compounds*

(b) *In general, the potential energy function, PE, is given by Coulomb's law: PE $\propto [q_1 q_2/d]$ where q is equal to e (1.6022 × 10⁻¹⁹ C) and d is the distance between the two charges. The total electronic interaction of a single Na^+ and the chlorides in the crystal is mediated by the Madelung constant M that equals 1.748 for NaCl. Directionality is not a factor.*

The NaCl crystal comprises two intermingled fcc structures. The distance from Na^+ to Cl^- ions from all possible directions within its realm of coordination comprises the Madelung constant used below.

The sum of the ionic radii, d, of the closest Na–Cl pair is [0.97 + 1.81 Å] = 2.78 Å, but the actual lattice constant of NaCl is 2.825 Å. Note that d is equal to 1/2 the lattice constant a of a NaCl crystal (a = 5.65 Å).

The ionic potential energy relationship for the NaCl crystal is

$$\Delta H_L = \frac{\mathcal{N}_A \, \mathcal{M} \, z_{Na} z_{Cl} \, e^2}{4\pi\varepsilon_o d_o}\left(1 - \frac{d^*}{d_o}\right)$$

where $\mathcal{N}_A$ is the Avogadro number, $\mathcal{M}$ is the Madelung constant for NaCl equal to 1.748, and d^* is a constant equal to 0.345 Å. The expression $[1 - d^*/d]$ is the repulsion term.

$$\Delta H_L = \frac{(6.022\times10^{23})(1.748)(+1)(-1)(1.6022\times10^{-19}\,C)^2}{4\pi(8.854\times10^{-12}\,C^2\cdot J^{-1}\cdot m^{-1})(2.78\times10^{-10}\,m)}\left(1 - \frac{0.345\,n}{2.78\,n}\right)$$

$$= -\frac{(1.389\times10^{-4}\,J\cdot m\cdot mol^{-1})(1.748)}{(2.78\times10^{-10}\,m)}(0.876)(1\,kJ\cdot1000\,J^{-1}) = 765\,kJ\cdot mol^{-1}$$

Please recall that the lattice energy is altered for compounds that have high surface area to volume ratio, such as for nanomaterials.

---

is 788 kJ · mol⁻¹. For this solid to dissolve into its constituent ions, there must be quite a hefty investment in energy. This energy comes from the formation of new bonds formed by hydration of the ion (ΔH_{Hyd}). The formation of new coordinative water bonds to Na^+ and Cl^- species is 784 kJ · mol⁻¹ (other estimates claim that the hydration of ions from the gaseous state, ΔH_{Hyd}, is $Na^+ = -406$ kJ · mol⁻¹ and ΔH_{Hyd}, $Cl^- = -363$ kJ · mol⁻¹ = 769 kJ · mol⁻¹). Based on enthalpic considerations alone, ΔH_{Sol} is equal to +3.88 kJ · mol⁻¹, an endothermic discrepancy between the lattice energy and the energy of hydration—certainly not enough to drive the dissolution of NaCl in water. What other factor must we consider?

$$\Delta H_L \text{ (always endothermic)} + \Delta H_{Hyd} \text{ (always exothermic)}$$
$$= \Delta H_{Sol} \text{ (endo- or exothermic)} \tag{10.10}$$

If we consider all of the entropic trade-offs in the process of dissolution and hydration, we find that the overall free energy, ΔG, of the dissolution reaction is negative because most dissolution reactions involving ionic compounds occur spontaneously. There are three entropic factors to consider: (1) completely replacing the crystalline order of the lattice by independent ions decreases overall order and ΔS is positive; (2) rearranging the host solvent (creating cavities) increases order and ΔS yields a negative contribution to the overall entropy; and (3) incorporating the ions within relatively ordered hydration cages (hydration spheres) causes more order and ΔS is also negative in this case. For the dissolution of NaCl in water, the value of the entropy for each step is $+240 \ \text{J} \cdot \text{mol}^{-1} \cdot \text{K}^{-1}$, $-20 \ \text{J} \cdot \text{mol}^{-1} \cdot \text{K}^{-1}$, and $-180 \ \text{J} \cdot \text{mol}^{-1} \cdot \text{K}^{-1}$, respectively. The overall free energy, ΔG, is found from

$$\Delta G = \Sigma(\Delta H) - T\Sigma(\Delta S) \tag{10.11}$$

and

$$\Delta G = (+788 - 784) \ \text{kJ} \cdot \text{mol}^{-1} - [298 \ \text{K} \ (+0.240 - 0.020 \\ - 0.180) \ \text{kJ} \cdot \text{mol}^{-1} \cdot \text{K}^{-1}] \approx -8 \ \text{kJ} \cdot \text{mol}^{-1}.$$

The overall free energy is less than zero and the dissolution process occurs spontaneously at 298 K. Because ΔH_{Sol} is positive, the solution becomes cooler during dissolution of NaCl.

The purpose of this exercise is to bring attention to solvent effects. The solvent, as we shall find out, plays a major role in nanochemical processes. Dissolution, dispersion, stabilization, surfactants, ligand exchange reactions, and hydrolysis all figure into the grand scheme of supramolecular chemistry. The ion–ion example is perhaps one of the easiest concepts to visualize. Unfortunately, it only gets more complex from this point.

The Dielectric Constant.   The dielectric constant is based on the phenomenon of permittivity. Permittivity is an intensive quantity that gauges the interaction of an electric field within a medium. The dielectric constant ε_r of a specific material is the ratio of the permittivity within a material to the permittivity of free space:

$$\varepsilon_r = \frac{\varepsilon_s}{\varepsilon_o} \tag{10.12}$$

where ε_s is the specific permittivity of a material. The dielectric constant of air is approximately 1. The dielectric constant of water is 78.85. The relative polarity (or nonpolarity) of a solvent is scaled by the dielectric constant. The strength of ion–ion attractions, as well as any other type of electrostatic interaction, is mitigated by the medium enclosing the charges. As we will learn later, the design of host–guest complexes must take into account not only the type of bonding interaction but also the immediate environmental conditions, such as temperature and solvent.

Dielectric Constant and Solvents.   In biology, ionic bonds serve an important function. The tertiary (3°) structure of proteins is stabilized by the participation of ionic bonds to form salt bridges between the carboxylate (–COO⁻) moiety of aspartate or glutamate and the amine (–NH₃⁺) group of lysine. In the enzyme aspartate aminotransferase, the salt bridge between Lys-68 and Glu-265 plays a vital role in the kinetic mechanisms of the enzyme action [10]. The surfaces of proteins are usually enveloped in water (e.g., polar groups on the protein face outward into the aqueous medium). The interior environment of proteins, on the other hand, especially the active sites, is often hydrophobic in nature. Proteins are able to sequester single molecules of water in the confined space of an active pocket and utilize water's polar character to influence substrate-specific reactions—all within the hydrophobic pocket of an enzyme. The dielectric nature of water, or any physiological fluid, is an important factor in enzyme function and protein structure.

10.1.3 Ion–Dipole and Dipole–Dipole Interactions

Ion–Dipole Interactions.   Ion–dipole interactions comprise the next level of complexity along the electrostatic interaction continuum as weaker interactions are accompanied by greater complexity of mathematical expressions. The strength of ion–dipole interactions is significantly less than that of ion–ion interactions and ranges from 50 to 200 kJ · mol^{-1} [9]. A dipole (from the Greek *dyo*, "two" + *polos*, "pivot") is exactly what the term implies—that regions of opposite charge (poles) exist within a single molecule.

In the case of the dipole, we need to introduce another electrostatic term known as the *dipole moment*. The measure of dipole strength is the electric dipole moment, μ—the product of charge (in coulombs) and the distance between the charges (meters). The *debye* is the unit of measure for the dipole moment—1 debye, D = 3.336 × 10^{-30} C · m and is symbolized by $\leftrightarrow$ with the positive charge stationed at the start of the arrow. Permanent dipoles are formed when a molecule is composed of two or more atoms that have substantially different electronegativity. The presence of a molecular dipole depends on the symmetry of the molecule and the electronegative nature of its constituent atoms. For example, carbon tetrachloride (CCl_4) does not have a dipole moment even though four very electronegative chlorine atoms are attached to a significantly less electronegative carbon atom. Chloroform ($CHCl_3$), on the other hand, does possess a dipole moment. Quadrupole, octupole, and higher orders of charge distribution also are found within molecules.

An ion–dipole interaction is an interaction between a charged ion, in elemental or molecular form, and a polar molecule (e.g., one that contains a dipole). The implicit existence of ion–dipole interactions was mentioned in the previous section during our discussion of solvent effects. Na^+ coordinated by water molecules serves as an excellent example of an ion–dipole interaction. The ion–dipole attraction between water and Na^+ contributes to the overall energy required to overcome the lattice energy of a NaCl crystal. The energy between ion–dipole attractions is inversely proportional to the square of the distance between them, r^2. The coordination of Na^+ or K^+ by crown ether complexes is a good example of an ion–dipole interaction found in supramolecular organic and biochemical systems. The energy of an ion–dipole attraction is given by the following expression:

$$E_{Ion-dipole} = \frac{ze\mu\cos\theta}{4\pi\varepsilon_0\varepsilon_r r^2} \tag{10.13}$$

where z is the unitless magnitude of the charge on the ion, μ is the dipole moment in D, $\cos\theta$ is the angle of approach of the ion to the center of the dipole (ranges from 0° to 180°). Obviously, the strongest interaction occurs when θ equals zero.

In biology, ion–dipole interactions are extremely important. Metal cations are coordinated in the center of porphyrin ring complexes and include the binding of Fe^{2+} or Mg^{2+} by heme, cytochrome-C, and chlorophyll. A cobalt cation is bound within the corrin ring system of vitamin B_{12}. These kinds of interactions are classified as dative bonds—bonds that coordinate metal cationic species within an organic framework. The distinction between metal cation–dipole interactions and dative or coordinative bonds is rather blurry. In some instances, the metal cation serves as a template around which a supramolecular species is synthesized.

Another form of ion interactions involves negative ions (anions). Anion binding is prevalent in biological systems. This is important because nearly two thirds of all enzyme substrates are anions. An ion–dipole interaction between a phosphate moiety and a *dipolar group transfer receptor* provides overall electrostatic balance [11]. Anion recognition is a fundamental mechanism in the development of antibacterial agents. Anions form nondirectional interactions with neutral, positively polarized, and cationic groups. Since anions are heavily solvated, the binding energy of any potential host must overcome this high solvation energy barrier—another factor to consider in supramolecular host–guest system design.

The action of ionophores depends on an ion–dipole mechanism. An ionophore is a lipid-soluble structure, channel, or carrier that transports metal ions through natural or synthetic lipid membranes. They range in size from small

---

EXAMPLE 10.2   *Ion–Ion Interaction: Effects of the Liquid Medium Ion–Dipole Interactions*

What is the ion–ion potential energy of Na^+ and Cl^- ions that are 1 nm apart in water? in methanol? in trimethyl amine? ε_r of water, methanol, and trimethyl amine at 25°C are 78.85, 32.63, and 2.44, respectively.

Solutions:
Permittivity of free space, $\varepsilon_0 = 8.854 \times 10^{-12}\ C^2 \cdot J^{-1} \cdot m^{-1}$, $e = 1.6022 \times 10^{-19}\ C$

For water:

$$= \frac{(+1)(-1)(1.6022 \times 10^{-19}\,C)^2}{(4\pi)(8.854 \times 10^{-12}\,C^{-2}J^{-1}m^{-1})(78.85)(1 \times 10^{-9}\,m)}$$

$$= -\frac{(8.988 \times 10^9\,J \cdot m \cdot C^{-2})(2.567 \times 10^{-38}\,C^2)}{(78.85)(1 \times 10^{-9}\,m)} = -2.926 \times 10^{-21}\,J$$

$$= -(2.926 \times 10^{-21}\,J \cdot bond^{-1})(6.022 \times 10^{23}\,bonds \cdot mol^{-1})(1\,kJ \cdot 1000\,J^{-1}) = -1.76\,kJ \cdot mol^{-1}$$

For methanol:

$$= -\frac{(8.988 \times 10^9\,J \cdot m \cdot C^{-2})(2.567 \times 10^{-38}\,C^2)}{(32.63)(1 \times 10^{-9}\,m)} = -7.07 \times 10^{-21}\,J$$

$$= -(7.07 \times 10^{-21}\,J \cdot bond^{-1})(6.022 \times 10^{23}\,bonds \cdot mol^{-1})(1\,kJ \cdot 1000\,J^{-1}) = -4.26\,kJ \cdot mol^{-1}$$

For trimethyl amine:

$$= -\frac{(8.988 \times 10^9\,J \cdot m \cdot C^{-2})(2.567 \times 10^{-38}\,C^2)}{(2.44)(1 \times 10^{-9}\,m)} = -9.46 \times 10^{-20}\,J$$

$$= -(9.46 \times 10^{-20}\,J \cdot bond^{-1})(6.022 \times 10^{23}\,bonds \cdot mol^{-1})(1\,kJ \cdot 1000J^{-1}) = -56.9\,kJ \cdot mol^{-1}$$

EXAMPLE 10.3	*Ion–Dipole Interactions*

Calculate the interaction energy between a potassium cation and urea in carbon tetrachloride. The dipole moment, μ, of urea is equal to 4.56 D. Assume that the ion is 2-nm radius from the urea at an angle of $45°$ and that the solvent is chloroform ($\varepsilon_r = 2.238$ for carbon tetrachloride). What would the interaction energy be if the K^+ cation were in line with the carbonyl group of the urea?

Solution:

$$E_{Ion-dipole} = \frac{ze\mu\cos\theta}{4\pi\varepsilon_o\varepsilon_r r^2}$$

$$= \left[\frac{(+1)(1.6022\times10^{-19}\,C)(4.56\times3.336\times10^{-30}\,C\cdot m)\cos 45}{4\pi(8.854\times10^{-12}\,C^2\cdot J^{-1}\cdot m^{-1})(2.238)(2\times10^{-9}\,m)^2}\right]\left(\frac{6.022\times10^{23}}{mol}\right)\left(\frac{1\,kJ}{1000\,J}\right)$$

$$= -1.04\ kJ\cdot mol^{-1}$$

The minus sign indicates that all attractive interactions are negative. If the cation interacted head-on with the urea, the interaction energy would be greater ($\cos 0° = 1$) = -1.47 kJ · mol^{-1}.

molecules that complex the metal ion and shield its charge to larger structures that are able to sequester the ion completely. Structures called *ionophore channel-formers* insert hydrophilic pores into membranes.

Dipole–Dipole Interactions. Dipole–dipole interactions are weaker forms of electrostatic interactions and therefore exhibit yet more mathematical complexity:

$$E_{Dipole-dipole} = -\frac{\mu_A\mu_B}{4\pi\varepsilon_o\varepsilon_r r^3}\left(2\cos\theta_A\cos\theta_B - \sin\theta_A\sin\theta_B\cos\phi\right) \tag{10.14}$$

If dipoles react in a head-on configuration, θ is equal to zero and the expression simplifies to

$$E_{Dipole-dipole} = -2\left(\frac{\mu_A\mu_B}{4\pi\varepsilon_o\varepsilon_r r^3}\right) \tag{10.15}$$

Please refer to **Table 10.5** for an illustration of the dipole–dipole interaction. In dipole–dipole interactions, the energy of interaction is inversely proportional to the cube of the distance between the dipoles, r^3. Dipole–dipole interactions also involve rotations (ϕ). Rotation of dipoles acts to diminish the overall strength of interactions in a solution due to averaging as the system alternates between attractive and repulsive domains (**Table 10.5**).

The dipole–dipole interaction is the basis of the hydrogen bond. It is without question one of the most important types of bonding—whether inside or outside the biological domain. From the macroscopic perspective, hydrogen bonding in ice makes life possible under frozen lakes. From the nanoscopic perspective, hydrogen bonds act collectively to impart structure to DNA. The hydrogen bonds between and among water molecules are strong and flexible enough to

TABLE 10.5	*Ion–Dipole and Dipole–Dipole Interactions*	
Description	**Example**	**Bond strength (kJ · mol⁻¹)**
Ion–dipole	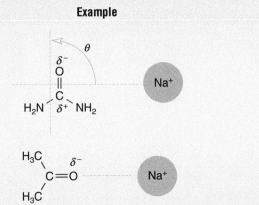	50–200 (40–120)

Ion–dipole interactions. Two approaches of the sodium cation are shown. Ion–dipole reactions can be attractive or repulsive. The strength of the interaction has $1/r^2$ dependence.

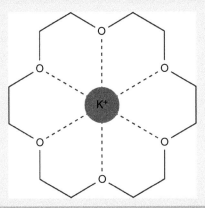

Crown ether ethers also are able to undergo ion–dipole interactions. A potassium crown ether [K(18-Crown-6)] molecular complex is displayed.

Dipole–dipole	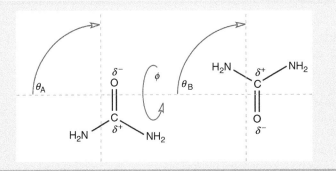	5–50 (5–40)

A strong interaction between two dipoles requires the proper alignment.

TABLE 10.5 (CONTD.)	Ion–Dipole and Dipole–Dipole Interactions	
Description	**Example**	**Bond strength (kJ · mol⁻¹)**
		5–50 (5–40)

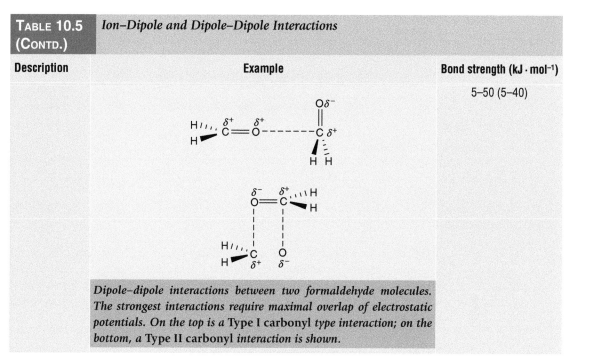

Dipole–dipole interactions between two formaldehyde molecules. The strongest interactions require maximal overlap of electrostatic potentials. On the top is a Type I carbonyl type interaction; on the bottom, a Type II carbonyl interaction is shown.

enable living systems to adjust to prevailing circumstances. More about hydrogen bonding is presented in section 10.2.

10.1.4 Dative Bonds

The stabilization of metal ions by ligands that donate electron pairs to vacant metal outer orbitals is called dative bonding. The dative bond (from the Latin

---

EXAMPLE 10.4 **Dipole–Dipole Interactions**

Apply the same conditions that were given in example 10.3 to the following dipole–dipole example. Assume that the head-on dipole configuration is intact. What is the value at 0.4 nm approach?

Solution:

$$E_{Dipole-dipole} = -2\left(\frac{\mu_A \mu_B}{4\pi\varepsilon_o\varepsilon_r r^3}\right)$$

$$= -2\left[\frac{\left(4.56 \times 3.336 \times 10^{-30}\,C\cdot m\right)^2}{4\pi\left(8.854 \times 10^{-12}\,C^2\cdot J^{-1}\cdot m^{-1}\right)(2.238)\left(2 \times 10^{-9}\,m\right)}\right]\left(\frac{6.022 \times 10^{23}\cdot mol^{-1}}{1\,kJ\cdot 1000\,J^{-1}}\right)$$

$$= -0.031\,kJ\cdot mol^{-1}$$

This is a much smaller value than that for the ion–dipole interaction of example 10.3. The minus sign indicates that all attractive interactions are negative.
At 0.4 nm, the dipole–dipole interaction is much stronger: $E_{Dipole-Dipole} = -1.96\,kJ\cdot mol^{-1}$.

The first metal coordination complex was synthesized by Alfred Werner in 1913. Six-coordinate complexes assume different colors depending on the ligands: $[Co(NH_3)_6]^{3+}$ = yellow-orange; $[Co(NH_3)_5Cl]^{2+}$ = purple; $[Co(NH_3)_5H_2O]^{3+}$ = red; and $[Co(NH_3)_4Cl_2]^+$ = green. What is the point of this? Ligands influence the physical (and chemical) properties of complexes.

datives, "to give, donate") involves a metal (usually a transition metal) that forms neutral, cationic, or anionic complexes. The coordination of metal ions can vary from 2 to 12 ligands, but the usual number of coordinating ligands is 6. Complexation of metals is relevant in biology, pharmacology, environmental chemistry, and supramolecular chemistry.

The process of binding metals to form macrocyclic rings is called chelation (from the Greek *chela,* "claw"). Morgan and Drew coined the term "chelate" in 1920 [12]. Alfred Werner invented the field of coordination chemistry, for which he won the Nobel Prize in 1913. He is known for the reaction depicted in **Figure 10.6.**

A dative bond is like a covalent bond in which electrons are shared, but in the case of dative bonding, both electrons are usually donated by one atom. Dative bonds are also known as "coordination or dipolar bonds" (ion–dipole bonds). Once again we are up against the chemical bonding continuum, trying to define a type of bonding that can be characterized in several ways. Coordination bonding comes in many forms: ionic, covalent, dative, or mixed. The major types of metal–ligand binding are listed in **Table 10.6.** The binding (stabilization) of metal clusters is a purely nano phenomenon. As we found out in chapter 6, clusters are inherently unstable due to the high associated surface energy. Ligands provide avenues to coordinate (and therefore stabilize) surface atoms or individual metal cations—both are deficient with regard to "nearest neighbors." The use of ligands to stabilize clusters also opens new routes to hierarchical nanomaterial synthesis.

Ligand-Stabilized Clusters. One of the significant achievements of nanocluster science is the stabilization of metal complexes like Au_{55} with organic ligands to

---

EXAMPLE 10.5 *Binding Constant of Zn²⁺*

Calculate ΔG of formation of the EDTA–Co²⁺ coordination complex at 25°C in a buffered solution containing EDTA and Co²⁺ if the overall formation constant K_f of the EDTA–Co²⁺ complex is equal to 2.0×10^{16}.

Solution:

$$\Delta G = -RT\,lnK_f = -[8.314 \times 10^{-3}\ kJ \cdot mol^{-1} \cdot K^{-1}\,(298\ K)\,ln\,(2.0 \times 10^{16}) = -93\ kJ \cdot mol^{-1}$$

TABLE 10.6	*Metal–Ligand Interactions*
Description	**Example**
Dative bonding	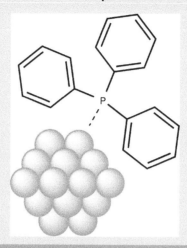

Au$_{55}$(PPh$_3$)$_{12}$Cl$_6$ cluster is an example of a ligand-stabilized system [13–16]. The size of the constituents is not in scale and the Au-ligand structure is not exactly displayed; there are actually 12 phosphine ligands surrounding the gold cluster. Coordination of metals by ligands occurs by donation of a lone pair of electrons—e.g., dative bonds.

Metal chelation

$$CO_2H—CH_2 \quad H_2 \quad H_2 \quad \overset{H_2}{\underset{H_2C—CO_2H}{N}}—\overset{C}{C}-CO_2H$$

$$HO_2C—CH_2 \quad N—C-C—N$$

Ethylenediaminetetraacetic acid
EDTA

$$CO^{2+} \mid pH > 10$$

EDTA is one of the best chelation agents in industry and research. It has six ligands with which to bind metal cations. The mechanism is based on Lewis acid–base reactions (dative bonds).

(*continued*)

TABLE 10.6 (CONTD.)	*Metal–Ligand Interactions*
Description	**Example**
Typical metal coordination by monodentate ligands	

A hexaaquocobalt(III) complex is depicted. The water molecules form dative bonds to the Co^{3+} metal cation via lone pairs on oxygen atoms.

Agostic interactions (α, β, γ C–H bonding)	

Agostic bonds involve the attraction of a C–H bond to a metal. Their structure (top) was elucidated in 1983 by M. Brookhart and M. L. H. Green (M. Brookhart and M. L. H. Green, Journal of Organometallic Chemistry, 250, 395–408, 1983.) M = metal; X = H, halogens or other substituents. Three kinds of organometallic titanium complexes are shown: left: α-agostic; middle: β-agostic; and right: γ-agostic. (I. Vidal et al., Organometals, 25, 5638–5647, 2006. With permission.)

form the $Au_{55}(PPh_3)_{12}Cl_6$ cluster [13–16]. More recently, a tetrahedral $Au_{20}(PPh_3)_8$ was rendered stable when coordinated to eight PPh_3 (triphenyl phosphine) ligands. Following collision-induced dissociation, stable $Au_{20}(PPh_3)_4^{2+}$ species were formed that had one PPh_3 group bonded to each corner of a tetrahedron [18]. The PPh_3 is zero valent and forms dative bonds to the gold cluster. Batches of ligand-stabilized clusters can be dried, packaged, and shipped to far away destinations without fear of decomposition or agglomeration. In terms of molecular design, dative bonds are expected to contribute significantly to molecular manufacturing processes in the near future [19]. B. Olenyuk et al. state [19]

A recent novel synthetic protocol in the construction of organized nanoscopic assemblies from multiple building blocks in a single step, namely self-assembly, relies on critical information about the shape and properties of the resulting structure being preprogrammed into each individual building block. Although this approach was initiated by artificial mimicking of natural receptors that utilize weak hydrogen bonds, it has now resulted in an entirely different "unnatural" strategy, molecular architecture that employs transition metals and dative bonds to achieve structurally well defined, highly ordered assemblages.

This approach relies on the fact that fewer metal–ligand bonds may be used in place of several hydrogen bonds owing to their greater strength. Another advantage lies in the existence of a large variety of transition metals with different co-ordination numbers [sic], thus facilitating the building of diverse nanoscopic entities, with tremendous variations in shapes and sizes.

Binding modes in coordination complexes goes beyond donation of neutral lone electron pairs. Ionic constituents, primarily in the form of anions, also form complexes with metals. The mechanism of chelation involves the use of both types of ligand donors (ionic and neutral), sometimes simultaneously to bind the same metal cation. Dative bonds are, of course, important to nanoscience, and the formation of well-ordered organic monolayers has great potential for use in sensors and other devices. For example, the adsorption of pyrimidine onto a Ge(100) surface was found to have a "double dative bond configuration"— confirmed by STM, TPD, and density functional theory (DFT) [20]. Specifically, dative bonds are formed between the nitrogen of pyrimidine (Lewis base) and a Ge-dimer (the *buckle-down* atom). In this configuration, electron transfer occurs from one member of the Ge-dimer to the other and is accomplished without the loss of aromaticity in the rings of the pyrimidine molecules [20].

Chelates. Chelation is a specialized form of metal coordination binding that involves a molecule with at least two binding sites or ligands (from the Latin *ligamentum,* "to bind, tie"). Monodentate (from the Latin *dentis,* "tooth"), bidentate to multidentate chelates are named according to the number of ligands that a single molecule can provide to bind a metal cation. The more ligands that a molecule can provide to bind a metal or cation, the better is the entropic trade-off and the more spontaneous the binding becomes. Hence, one of the driving forces behind the chelate effect is the increase in entropy afforded by polydentate ligands that release monodentate counterparts into solution. On the other hand, the enthalpy of complexation for a polydentate chelate is about the same as that of an equivalent group of monodentate ligands.

The exothermic character and high formation (complexation) constant of chelate binding processes is due in part to the *chelate effect*—the enhanced stability exhibited by metal–ligand complexes that contain chelate rings (e.g., polydentate ligands) compared to similar configurations that contain fewer or no rings (e.g., monodentate ligands) [8]. The magnitude of the effect depends on the free energy change, the basicity of the ligand, the *p*H of the solution, and the type and concentration of the chelate and substrate [21]. Multidentate chelate complexes remain stable even under extremely dilute conditions, whereas monodentate ligand complexes are able to dissociate under dilute conditions.

The coordination sphere of a metal–ligand complex is composed of the number of ligands directly attached to the metal. These are called inner-sphere complexes. Complex cations, however, are able to associate with anions without displacement of inner-sphere ligands. These are called outer-sphere complexes and are characterized by short lifetimes and nonspecific geometry. Coordination complexes consisting

---

EXAMPLE 10.6   *Free Energy of Chelation of EDTA*

Calculate ΔG of formation of the EDTA–Co^{2+} coordination complex at $25\,°C$ in a buffered solution containing EDTA and Co^{2+} if the overall formation constant β of the EDTA–Co^{2+} complex is equal to 2.0×10^{16}.

Solution:

$$\Delta G = -RT\,ln\beta = -[8.314 \times 10^{-3}\ kJ \cdot mol^{-1} \cdot K^{-1}\ (298\ K)\ ln\ (2.0 \times 10^{16}) = -93\ kJ \cdot mol^{-1}$$

---

of multidentate ligands invest a large covalent contribution to the inner-sphere structure. These species are long-lived and have a definitive geometry.

The free energy of complexation is given by $\Delta G = -RT \ln \beta$ where β is the overall equilibrium complexation constant). The stability constant for adding a succession of monodentate ligands is as follows [8,9]:

$$ML_{n-1} + L \rightarrow ML_n:\ K_n = \frac{[ML_n]}{[M_{n-1}][L]} \tag{10.16}$$

and for the cumulative addition of ligands:

$$ML + nL \rightarrow ML_n:\ \beta_n = \frac{[ML_n]}{[M][L]^n} = \prod_1^n K_n, \tag{10.17}$$

where $\beta_n = (K_1)\,(K_2)\,(K_3)\,(K_4)\,...\,(K_n)$

EXAMPLE 10.7   *Free Energy and Formation Constants of Chelation Reactions*

Calculate ΔG and the formation constants K_f (and log K_f) for the following coordination reactions at 298 K and discuss the reasons for the differences. M = metal cation. In reaction a, four monodentate ligands coordinate a single metal cation. In reaction b, two bidentate ligands coordinate a single cation.

   a. $M^{2+}_{(aq)} + 4NH_{3(aq)} \rightarrow [M(NH_3)_4]^{2+}_{(aq)}$ $\Delta H° = -53\ kJ \cdot mol^{-1}$, $\Delta S° = -42\ J \cdot mol^{-1} \cdot K^{-1}$
   b. $M^{2+}_{(aq)} + 2(NH_2-CH_2-CH_2-NH_2)_{(aq)} \rightarrow [M(NH_2-CH_2-CH_2-NH_2)_2]^{2+}_{(aq)}$
     $\Delta H° = -56\ kJ \cdot mol^{-1}$, $\Delta S° = 10\ J \cdot mol^{-1} \cdot K^{-1}$

Solution:
In all cases, use $\Delta G° = \Delta H° - T\Delta S°$ and $\Delta G° = -RT\,lnK_f$ (where $logK_f = lnK_f/2.303$)

   a. $\Delta G = -53\ kJ \cdot mol^{-1} - (298\ K)(-0.042\ kJ \cdot mol^{-1} \cdot K^{-1})$
      $\Delta G = -41\ kJ \cdot mol^{-1}$
      $lnK_f = \Delta G/[-RT] = -41\ kJ \cdot mol^{-1}/[-0.00834\ kJ \cdot mol^{-1} \cdot K^{-1} \times 298\ K] = 16.5$
      $logK_f = lnK_f/2.303 = 7.2$
      $K_f = 1.5 \times 10^7\ M^{-1}$
   b. $\Delta G = -56\ kJ \cdot mol^{-1} - (298\ K)(+0.010\ kJ \cdot mol^{-1} \cdot K^{-1})$
      $\Delta G = -59\ kJ \cdot mol^{-1}$
      $lnK_f = \Delta G/[-RT] = -59\ kJ \cdot mol^{-1}/[-0.00834\ kJ \cdot mol^{-1} \cdot K^{-1} \times 298\ K] = 23.8$
      $logK_f = lnK_f/2.303 = 10.3$
      $K_f = 2.3 \times 10^{10}\ M^{-1}$

There is a three order of magnitude increase in the complexation constant of the metal dication by the bidentate ligand. The difference is due to the ca. +52 $J \cdot mol^{-1} \cdot K^{-1}$ increase in entropy afforded by the bidentate chelation agent.

Agostic Interactions.   Agostic (from the Greek "to hold close") interactions involve weak bonding interactions between transition metals (usually second row metals) and the alkyl hydrogen of a C–H bond; specifically, the alkyl hydrogen is "held close" to the metal atom. The strength of the interaction is about 42–63 kJ·mol^{-1} [17]. In 1983, M. Green and M. L. H. Brookhart introduced the term *agostic* to the chemical bond community. Other forms involving transition metals and groups with more polar character than alky groups, more polar groups such as R-S–H and R-N–H, are also included into the category of agostic bonding [17]. Although its exact nature is still not clear, agostic bonds have been described as either van der Waals type or "three-center-two-electron" bonds. Agostic bonds are not a special case of hydrogen bonds but should be considered more along the line of a real bond.

10.1.5   π-Interactions

π-interactions are another important type of electrostatic interaction (energy 0–50 kJ·mol^{-1}) [9]. For example, DNA, in addition to being held together by hydrogen bonds, is held together by π-interactions between aromatic constituents of the nucleotide bases. Interplanar π–π interactions are responsible for the three-dimensional structure of graphite and for its lubrication properties. Intercalation phenomena, aromatic packing in crystals, the 3° structure of proteins, complexation in host–guest configurations, and stacking of porphyrins are all dependent on π–π interactions [22,23]. DNA, graphite, and molecular tweezers all rely on intercalation to provide additional functionality. The basic forms of π-interactions are shown in **Table 10.7.**

C. A. Hunter et al. in 1990 proposed that aromatic interactions rely heavily on geometrical factors in order for bonding to occur [22]. The σ-framework is an important component in facilitating π–π interactions because attractions between π–σ systems (via van der Waals forces) are strong enough to overcome direct π–π repulsive forces. The authors claim that the geometrical point of intermolecular contact rather than the overall molecular oxidation–reduction potential is what dominates the interaction. If true, this implies that conventional electron-donor~electron-acceptor models are not applicable to π–π interactions [22,23]. The Hunter–Saunders model is based on competing π–π and van der Waals electrostatic interactions. Van der Waals attractions arise between the electrons of the π-system and the positive-leaning σ-framework of another aromatic molecule [9,22,23]. Stacking of porphyrins in nanostructured crystals happens by means of a coplanar-offset geometry. This example demonstrates how cooperative interactions are most likely responsible for binding in complex systems. Molecular tweezers that consist of two aromatic surfaces are able to sandwich adenine with a "high degree of face-to-face complementarity" [25]. The binding energy for this complex is on the order of 25.1 kJ·mol^{-1} in deuterated chloroform [25].

Neutral Guests.   The binding of neutral guests raises certain issues with regard to selectivity and stability [9]. However, hydrogen bonding, dative bonding, hydrophobic effects, and π–π interactions and mixed forms are all ways of pulling neutral molecules into a reactive zone. The basic driver in chemical binding is to maximize the amount of orbital overlap. The electronic configuration of an

TABLE 10.7	π-Interactions	
Description	**Example**	**Bond Strength (kJ · mol⁻¹)**

Description	Example	Bond Strength (kJ · mol⁻¹)
π–π Interactions		0–50 (10–15 Face) (15–20 Edge)
	Attraction between the electron-rich interior of the aromatic ring with the electron-poor exterior is the most likely configuration. The interaction is considered to be electrostatic and van der Waals. The configuration in the middle is least likely. The interaction found in graphite is shown on the right.	
d-Block–π-interactions	Ferrocene $Fe(C_5H_5)_2$	5–80
	Coordination of iron between two cyclopentadyl rings yields the compound ferrocene—one of the most important in the synthesis of carbon nanotubes. The preparation and organometallic chemistry of ferrocene was established in 1951 (G. Wilkinson, Journal of Organometalic Chemistry, 100, 273–278, 1975. With permission.)	
Cation–π-interactions		
	Metal addition to alkenes to form strong covalent bonds—a superdative bond of sorts.	

unsuitable bonding partner, like a neutral molecule, can be altered by a cooperative effort. The binding of *p*-benzoquinone with hydrogen bonds and π–π interactions demonstrates the concept of cooperativity between (or among) different types of molecules. A π–π interaction acting in concert with a hydrogen bond is able to alter the electronic properties of the neutral guest and render it bondable by changing its polarity or tweaking its symmetry [9]. This example demonstrates how such intermolecular interactions can influence the chemical nature of the guest molecule [8]. The best examples of cooperative bonding exist in the active pockets of enzymes.

10.2 HYDROGEN BONDING

Of all the kinds of intermolecular bonding, chemists and biologists alike are most familiar with the hydrogen bond (H-bonds), a special subset of dipole–dipole interactions. H-bonds are mostly directional bonds that consist of a shared hydrogen atom between small electronegative elements (usually O, N, F, Cl) that possess at least one lone pair of electrons. Carbon, however, is another element capable of forming hydrogen bonds, especially if an electron-withdrawing group is nearby on its backbone structure [9,26]. Physical properties like boiling point, melting point, and surface tension are influenced by hydrogen bonds. For example, the boiling point of Group V, VI, and VII hydrides is anomalously higher for NH_3, H_2O, and HF than their heavier counterparts. **Table 10.8** lists some of the relevant features of hydrogen bonds and those of other compounds that make hydrogen bond-like bonds.

10.2.1 Standard Hydrogen Bonds

Although formed from dipole–dipole interactions, the bond energy of H-bonds can be much stronger than those of standard dipole–dipole bonds: ~120 versus ~50 kJ · mol^{-1}, respectively. Hydrogen bond length ranges from 2.5 to 3.5 Å. This is somewhat more than the length of a normal covalent bond between hydrogen and other electronegative atoms (<2.0 Å). H-bond length depends on the electronegative constituents of the molecules involved. For example, an H-bond between water and chlorine, a larger (softer) and therefore less electronegative atom, is generally longer and weaker than those between water and fluorine [9]. The strength of hydrogen bonds also depends on the immediate environment: the medium (type of liquid or gas phase), chemical composition, pH, and temperature. The hydrogen bond is responsible for the physical properties of water, the binding in DNA, and protein folding to form 2° structures.

Even though the hydrogen bond is relatively weak compared to covalent bonds, the summation of all hydrogen bonds over the volume and surface of a system results in compounds or phases that are quite stable. In the liquid state, each water molecule is involved in an average of 3–3.5 hydrogen bonds—even as hydrogen bonds in water are formed and broken every 10^{-12} s. Molecules of water pack together to form short-lived hydrogen-bonded tetrahedral lattice transients called *flickering clusters* $[8H_2O \leftrightarrow 2(H_2O)_4]$ caused by fluctuations in thermal energy. In addition to hydrogen bonding interactions, water molecules experience dispersion and polarization. Polarization effects in water mutually act to strengthen the hydrogen bond—an example of another cooperative effect.

For the pure octomeric form of water, ΔH_{vap} is equal to 44.3 kJ · mol^{-1}. Many kinds of interactions contribute to the heat of vaporization: hydrogen bonding (35 kJ · mol^{-1}), dipole interactions (1.7 kJ · mol^{-1}), van der Waals interactions (5.0 kJ · mol^{-1}), and the RT (or k_BT) contribution (2.5 kJ · mol^{-1}) [30]. In ice, each water molecule is coordinated by four hydrogen bonds to form a tetrahedral

TABLE 10.8	*Hydrogen Bonding and Allies*	
Description	**Example**	**Bond strength (kJ · mol⁻¹)**

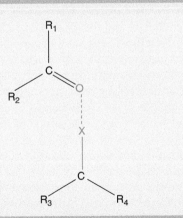

Standard hydrogen bonding example shown with bond strength:

4–120 [9]
(15–40 strong, 5–15 moderate, <5 weak)

Hydrogen atoms are shared between electronegative oxygen atoms. H-bonds are directional; bond character is electrostatic (90%) + some covalent (10%) and has short-range attraction (0.25–0.35 nm). (O. Markovitch and N. Agmon, Journal of Physical Chemistry A, 111, 2253–2256, 2007. With permission.)

O–·····HO	21
O–H······N	29
N–H······N	13

C–α–H······O/N
C–α–H······π
Hydrogen bonds

The archetypical weak hydrogen bond of weak acid–strong base or soft acid–hard base is one kind. Hydrogens also donated to π-acceptors like C=C and phenyl rings are included in this category

Cα–hydrogen bond in amino acids. These bonds are commonly found in protein structures. (S. Aravinda et al., Biochemical and Biophysical Research Communications, 272, 933–936, 2000. With permission.)

Halogen bonding

Weak to strong, just like with H-bonds
Some bond strengths exceed H-bonds
Bond strength dependent on environment

Halogen bonding is analogous to hydrogen bonding. The substituent X is F, Cl, Br, or I or oxygen-containing groups like carbonyls (pictured), carboxylates, hydroxyls, phosphates, or sulfates. (P. Auffinger et al., Proceedings of the National Academy of Sciences, USA, 101, 16789–16794, 2004. With permission.)

FIG. 10.7 *Various orientations of hydrogen bonds are depicted. In ice, each water molecule is coordinated to form a tetrahedral structure consisting of two shared bonds and two donated bonds.*

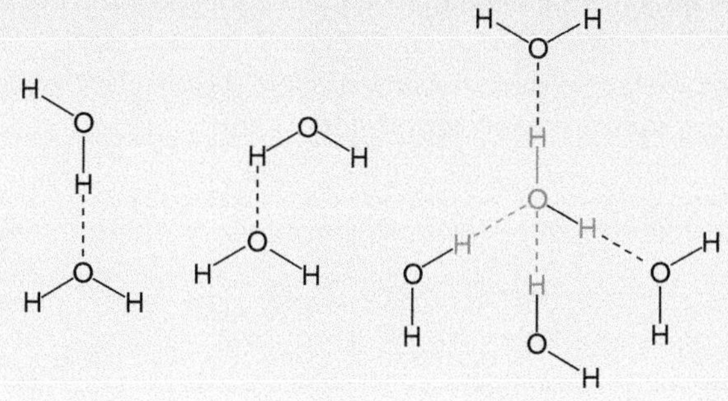

geometry (two shared, two donated). Such factors make H-bonding by itself a cooperative effort [31]. The structure of water is a function of the concentration of ions and other constituents. Hydration shells are another aspect of organization in water. Hydration around an ion changes the structure of the hydrogen bond as H_2O reorients itself to accommodate the ion. Depending on the charge of the ion (+ or −), water will direct its opposite polarized charge to create the hydration sphere. Various orientations of hydrogen bonds in water are shown in **Figure 10.7**.

The structure of water in confined spaces, as in the active pockets of enzymes and in carbon nanotubes, is a vital area of current research. Electrodes consisting of nanolayers of chemically modified with osmium bipyridine, poly(allylamine) and poly(vinyl)sulfonate were found to alter the hydration number of certain ions [32]. Molecular dynamic simulations have shown that the hydration number of Na^+ (4.5 in bulk water) is reduced to 3.5 when confined in 0.90-nm diameter nanotubes [33]. In 0.82-nm diameter tubes, the hydration number is reduced even further [33]. Solubility of ions in water is a function of the ability to create hydration spheres around the ion. If space is restricted, the solubility of ions goes down. The solubility of NaCl decreases by a factor of 2 if the solvent is confined in 0.8-nm diameter pores [34]. Confinement effects are able to influence the course of chemical reactions—the case once again in the active pockets of enzymes.

Hydrogen bonding also plays a role in solids. Approximately 20–25% of the lattice energy of dihydrogen phosphate salts is due to hydrogen bonds [36]. The hydrogen bond also contributes to the packing and crystal geometry of proteins. Supramolecular chemistry (chapter 11) is highly dependent on the hydrogen bond. Large organic supramolecules based on melamine and cyanuric acid (**Fig. 10.8**) or extremely large supramolecules like DNA all rely on hydrogen bonding to maintain the integrity of their structure.

Just as with the naming of disciplines and subdisciplines of chemistry, boundaries that distinguish between and among the types of H-bonding are also fuzzy. Hydrogen bonds have both covalent and ionic character, are strongly directional, or have angular components. Bonds can be short or long. Because of these and other effects imparted by the solvent and other solutes, a complete description of the bonding scenario is difficult to achieve (**Table 10.9**).

FIG. 10.8 *Hydrogen bonding is important in supramolecular complexes such as the melamine–cyanuric acid complex shown. It is clear from the image that hydrogen bonds are able to direct supramolecular geometry with high specificity. The same is true for DNA; we are all quite familiar with its specificity and overall structure. The rosette structure on the left is a kind of supramolecular structure that creates a cavity in the center. A ribbon is formed by control of the size of the R-groups. Such nanomaterials are capable of self-assembly due to the complementarity character of hydrogen bonds.*

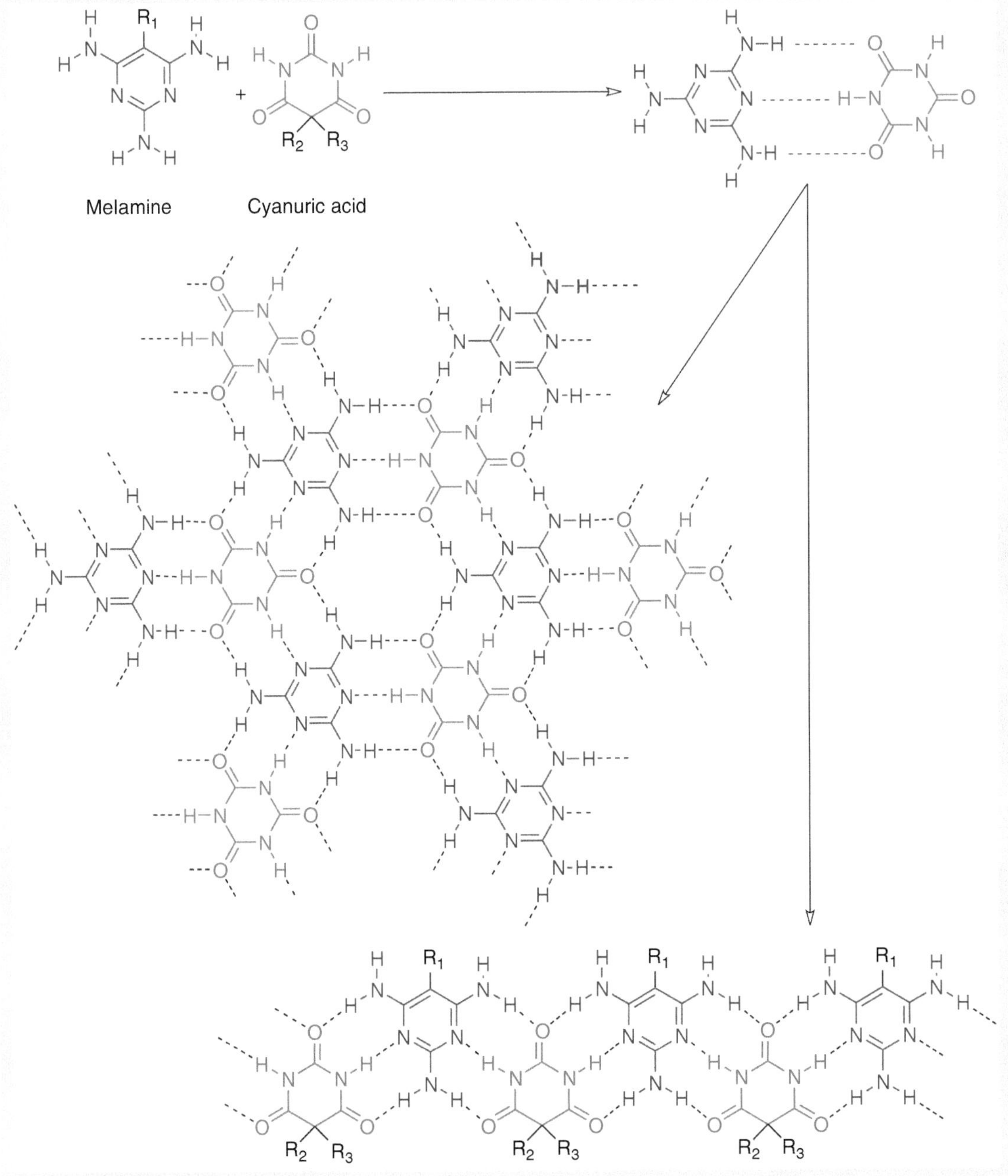

	Hydrogen bond strength	Covalent bond strength	Energy ratio %
Molecule	$(H_nX\text{--}H)$ $(kJ \cdot mol^{-1})$	$(H_nX\cdots\cdots H\text{--}XH)$ $(kJ \cdot mol^{-1})$	hydrogen/covalent
H_2S	7	363	1.9
NH_3	17	386	4.4
H_2O	22 (Pure water: 10.2) [35]	464	4.7
HF	29	565	5.1

TABLE 10.9 *Hydrogen Bond Energies of Some Common Molecules*

Note: X is S, N, O, or F.

Technically, the hydrogen bond is the result of an insipient proton transfer reaction [37]. Dissociation energies of hydrogen bonds range from 0.8 to 170 kJ·mol^{-1} depending on the ambient conditions [37]. The hydrogen bond is characterized by broad transitions that merge continuously with covalent bonds, ionic interactions, van der Waals phenomena, and π-cation interactions. Interestingly, it was shown by simulation that the strength of hydrogen bonds is stronger in ordered clathrate solvation shells of nonpolar groups (18.5 vs. 17.4 kJ·mol^{-1}) and that the correlation time of hydrogen bonds persisted significantly longer near hydrophobic groups [31]. If cations like Na$^+$ are nearby, hydrogen bonds in the hydration sphere are weakened by the strong electric field of the cation. The Na$^+$ cation is capable of causing perturbations to the alignment of the water dipoles around hydrophobic groups, distorting the hydrogen bond [31]. Others claim that water at hydrophobic surfaces demonstrates weaker hydrogen bonding but stronger orientation effects [38]. Selective probing of the molecular structure of CCl_4–H_2O and hydrocarbon–H_2O interfaces with vibrational methods found that H-bonding was weakened near hydrophobic surfaces [38].

The Born–Haber method of determining the bond strength is an approximate method at best because it is not able to discriminate among the numerous types of bonding that may be involved. Many methods have been applied to determine the bond strength of H-bonds. For example, the hydrogen bond strength was determined for a fluorescent solute, neutral red, in water by measuring the temperature dependence of the ratio of the intensity of the dye at 625 nm in water to its intensity in benzene (at 530 nm). The H-bond energy was determined to be 10.1 kJ·mol^{-1} [35].

Analytical Methods. Infrared methods such as FTIR rely on shifts (in reciprocal centimeters) and the width of the –OH stretching bands to detect H-bonds. Hydrogen bonding in general results in decreased frequency and a broadening of the absorption band [39]. H-bonding in alcohols, for example, exhibits this phenomenon. FTIR spectra of alcohols reveal two types of –OH stretching frequencies: a "free" hydroxyl stretch (no hydrogen bonding) located at ca. 3620–3640 cm^{-1} and a broader bound hydroxyl stretch (with hydrogen bonding) at 3250–3450 cm^{-1} [40]. Polarized infrared absorption spectra acquired from silicate minerals (pectolite and serandite) showed that H-bonds are short, strong, and asymmetric [41]. FTIR and XRD were applied in tandem to analyze the active-site structure role of Tyr^{34} residue of manganese superoxide dismutase. Results indicate that the phenolic hydroxyl of Tyr^{34} is hydrogen bonded and acts as a proton donor to an adjacent H_2O in the active pocket [42].

EXAMPLE 10.8   *Calculation of the Hydrogen Bond Strength of a Crystalline Solid*

A hypothetical material that contains a small electronegative anion, X^- forms a crystalline solid with an organic carbonyl, R, of the formula MXR where M is some metal. Using the Born–Haber cycle, calculate the strength of the hydrogen bond if:

$\Delta H_L \, MX \cdot R = +725 \text{ kJ} \cdot \text{mol}^{-1}$
$\Delta H_{Sol}, MX \cdot R = -2.9 \text{ kJ} \cdot \text{mol}^{-1}$
$\Delta H_{Vap} R = +19.7 \text{ kJ} \cdot \text{mol}^{-1}$
$\Delta H_L MX = +798 \text{ kJ} \cdot \text{mol}^{-1}$
$\Delta H_{Sol}, MX = +34 \text{ kJ} \cdot \text{mol}^{-1}$

Solution:
Using the Born–Haber cycle:

$MX_{(s)} + R_{(l)} \rightarrow M^+_{(g)} + X^-_{(g)} + R_{(l)}$ $\Delta H_L = +798 \text{ kJ} \cdot \text{mol}^{-1}$ *[lattice energy of $MX_{(s)}$]*

$R_{(l)} \rightarrow R_{(g)}$ $\Delta H_{vap} = +19.7 \text{ kJ} \cdot \text{mol}^{-1}$ *[energy of vaporization of $R_{(l)}$]*

$MX_{(s)} + R_{(l)} \rightarrow M^+_{(g)} + X^-_{(g)} + R_{(g)}$ $\Delta H_{rxn} = +817.7 \text{ kJ} \cdot \text{mol}^{-1}$ *[overall energy on left side]*

$MX_{(s)} + R_{(l)} \rightarrow M^+_{(l)} + X^-_{(l)} + R_{(l)}$ $\Delta H_{sol} = +34 \text{ kJ} \cdot \text{mol}^{-1}$ *[solvation energy of $MX_{(s)}$]*

$M^+_{(l)} + X^-_{(l)} + R_{(l)} \rightarrow MX \cdot R_{(s)}$ $\Delta H_{sol} = -2.9 \text{ kJ} \cdot \text{mol}^{-1}$ *[solvation energy of $MX \cdot R_{(s)}$]*

$MX \cdot R_{(s)} \rightarrow M^+_{(g)} + X^- \cdots\cdots R_{(g)}$ $\Delta H_L = 725 \text{ kJ} \cdot \text{mol}^{-1}$ *[lattice energy of $MX \cdot R_{(s)}$]*

$M^+_{(g)} + X^- \cdots\cdots R_{(g)} \rightarrow M^+_{(g)} + X^-_{(g)} + R_{(g)}$ $\Delta H_{H\text{-}bond} = H\text{-}B \text{ kJ} \cdot \text{mol}^{-1}$ *[unknown energy of H–bond]*

$MX_{(s)} + R_{(l)} \rightarrow M^+_{(g)} + X^-_{(g)} + R_{(g)}$ $\Delta H_{rxn} = +761.9 \text{ kJ} \cdot \text{mol}^{-1} + H\text{-}B$ *[overall energy on right side]*

Balance
817.7 = 761.9 + H–bond
H–bond energy = 55.8 kJ · mol⁻¹ $\Delta H_{H\text{-}bond} = -55.8 \text{ kJ} \cdot \text{mol}^{-1}$

Glass transition temperature analysis by differential scanning calorimetry (DSC) was used to determine the behavior and extent of hydrogen bonding in phenolic polymers. The thermodynamic competition between inter- and intramolecular hydrogen bonding was evaluated by molecular mechanics modeling [43]. Analysis of water by NMR methods revealed an up-field (lower frequency) chemical shift of the proton involved in hydrogen bonding, and the more hydrogen bonding there was, the greater the up-field shift became. At higher temperatures (100°C), the shift was even greater; better shielding occurred due to the reduced strength of the hydrogen bond [43].

Molecular modeling methods have proven to be extremely useful. Ab initio molecular orbital theory methods were employed to determine the stability of hydrogen bonds in methyl substituted adenine and uracil [44]. The results agreed quite well with experimentally determined binding constants [44]. Bond strength, depending on the substitution constituent, ranged from 40 to 60 kJ · mol⁻¹.

10.2.2 C-α-H···O Hydrogen Bonds

Weaker forms of hydrogen bonding are exhibited by C-α-H · · · · · · ·O compounds [45]. These are of the form C–H · · · · · ·O/N or C–H · · · · · ·π. These weaker forms of hydrogen bonding were all discovered in the 1930s [45]. Validation of their existence was accomplished in the 1960s [45]. By the time the 1990s rolled around, the C–H · · · · ·O and π-types of hydrogen bonds were fully accepted—even by the crystallographers who grudgingly released their stamp of approval. Intermolecular O–H · · · · ·π hydrogen bonds were discovered in solutions of R–O–H donors in solvents with π-acceptors like benzene and toluene. This weak form of the hydrogen bond plays a role in supramolecular chemistry. Quinones in particular, like 1,4-benzoquinone, can be formed in a planar configuration that depends on an extensive network of C–H · · · · ·O bonds [45]. Weak H-bonds also exist in purine and pyrimidine interactions and in collagen and polyglycine [45].

In 2000, researchers detected an unusual C-terminal conformation that was due to a strong C–H · · · · ·O bond between Ala(4)-CαH and D-leu(9)CO [28]. Ab initio quantum calculations validated the presence of C–H · · ·O hydrogen binding and showed that peptide CαH groups can be good hydrogen donors to water molecules with the binding energy in the range of 8–11 kJ·mol^{-1} for nonpolar and polar amino acids, respectively [46]. This relatively new type of hydrogen bond is considered to be important in many kinds of molecular complexes and crystals [47,48]. C–H · · · · ·O bonds are important to biological systems with nontrivial bond energies. **Table 10.10** lists simulated FTIR and NMR results for several amino acids as proton donors to water [48].

10.2.3 Halogen Bonds

An interesting analogy to H-bonds is the halogen bond. The interactions are fundamentally electrostatic but polarization, dispersion, and charge transfer also contribute to the bond—just like with hydrogen bonds [29]. The Br · · · · ·O distance is fairly long, ~3 Å, 12% shorter than the van der Waals radii sum (data acquired from an aldose reductase complex with halogen inhibitor) [14]. In

TABLE 10.10	*Interaction of Proton Donors with Water*			
Proton donor	**Interaction energy ΔE (kJ·mol^{-1})**	**FTIR stretch shift: C–H Δν (cm^{-1})**	**NMR isotropic shift of bridging proton Δσ$_{iso}$ (ppm)**	**Anisotropic shift of bridging proton Δσ$_{aniso}$ (ppm)**
H–OH	−18.9	−31 (O–H)	−2.6	10.9
F$_2$HCH	−10.6	26	−1.3	6.1
Gly [H]	−10.6	14	−1.4	5.9
Ala [CH$_3$]	−8.8	51	−1.4	6.8
Val [CH(CH$_3$)$_2$]	−8.4	56	−1.5	6.7
Ser [CH$_2$OH]	−9.6	22	−1.7	7.5
Cys [CH$_2$SH]	−7.9	51	−1.6	7.7
Lys$^+$ [(CH$_2$)$_4$NH$_3^+$]	−20.6	6	−1.7	6.7
Asp$^-$ [CH$_2$COO$^-$]	+5.98	70	−1.5	6.7

other compounds, bond lengths on the order of 80% of the van der Waals (*vdw*)-radii sum are not uncommon [14]. The thyroid hormone utilizes halogen bonds in molecular recognition. Halogen bonds are short-contact charge-transfer bonds that involve a negative charge from oxygen, nitrogen, or sulfur (well-known Lewis bases) and a polarizable halogen such as bromine (Lewis acid) [29]. The most celebrated naturally occurring halogen-bonded macromolecules are the class of thyroid hormones. Short $I \cdots O$ contacts between tetraiodothyroxine and its transport protein transthyretin play a role in recognition. There are greater than 3500 halogenated metabolites, including antibiotics [29]. The authors of this paper state:

> It is stunning to see how Nature has exploited all possible intermolecular interactions, even the most "exotic" ones, to design very specific and efficient recognition systems.
>
> Halogen atoms can be involved in electrostatic-type interactions that are strong enough to compete with hydrogen bonds, in addition to their better documented abilities to serve as electron withdrawing substituents or their supposed "hydrophilic" properties, would contribute to the design of ligands by providing a framework for the use of this interatomic interaction.

This type of bond has already been utilized to form supramolecular assemblies. It is capable of inducing conformational perturbations and therefore, with proper preorganization, design parameters of supramolecular assemblies can be expanded.

10.2.4   Hydrogen Bonds and Living Things

The effect of hydrogen bonding on living systems is fundamental to life itself [49]. The body temperature of humans is 36.1–37.8°C. Reduction in H-bond strength would bring about the following ramifications (at body temperature): Reduction in hydrogen bond strength (by 29%) would cause water to boil at 37°C; by 18%, proteins would denature; and by 11%, the potassium ion becomes *kosmotropic* (or *chaotropic*, the ability to destabilize H-bonding and hydrophobic interactions—the more negative ΔG_{hyd}, the more kosmotropic the salt becomes); and a decrease in H-bond strength by 7% would result in increased pK_w by a factor of 3; by 5%, CO_2 would be 70% less soluble.

Increase in H-bond strength would effect the following changes (at body temperature): an increase of 2% would cause significant metabolic changes; by 3%, viscosity increases by 23% and diffusivity is reduced by 19%; by 5%, O_2 would be 270% more soluble; by 5%, CO_2 would be 440% more soluble; by 7%, pK_w would decrease by 1.7; by 11%, Na^+ becomes chaotropic; by 18%, water freezes; and by 51%, many proteins cold denature [48]. There is, of course, a wealth of information about the hydrogen bond but due to the limited space, we must by necessity move on to the next category of bonding. We recommend that you independently pursue more information about the hydrogen bond and its importance in chemistry and biology—and to skating on frozen lake surfaces [50].

10.3   VAN DER WAALS ATTRACTIONS

In 1873, J. D. van der Waals (**Fig. 10.9**), a Dutch scientist, noticed that the pressure of a real gas was lower than the value predicted by the ideal gas relation. He

FIG. 10.9

Johannes Diderik van der Waals was awarded the Nobel Prize in physics in 1910 for his contributions to understanding equations of state for gases and liquids—the continuity of the gas and liquid state. Indeed, van der Waals showed in 1893 that there is a material continuum among the states of matter. In his famous equation, he showed that van der Waals forces need to be considered in describing the physical state of a gas (e.g., its deviation from ideality). In his law of corresponding states he developed a general form of the equation of states. Van der Waals died in 1923.

surmised that this was due to attractive and repulsive forces between atoms and molecules of the gas—a condition not addressed by the ideal gas laws. Van der Waals was awarded the Nobel Prize in physics in 1910 for his work on the equation of state for gases and liquids. His equation of state for homogeneous substances consists of the usual pressure, volume, and temperature correlations, but also includes constant factors, characteristics of a specific gas, that offer corrections to the ideal gas law (e.g., nonideality due to deviations from the action of intermolecular forces) and that atoms and molecules have nonzero volume. These weak forces are known as van der Waals forces and they constitute an important category of intermolecular attractions, especially if summed over many molecules. The generalized van der Waals equation is an expanded version of the well-known $PV = nRT$ form. The term n is the number of moles of gas, and subtraction of the product nb from V corrects for intermolecular repulsion [51].

$$\left[P + \left(\frac{an^2}{V^2}\right)\right](V - nb) = nRT \qquad (10.18)$$

The term an^2/V^2 is a measure of the intermolecular attraction in the system [51]. Familiar forms of this equation employ a molar volume V_m term. The constants a (units of $L^2 \cdot atm \cdot mol^{-2}$) and b (units of $L \cdot mol^{-1}$) are typical of a specific gas, and the equation applies appropriately to spherical atoms or molecules. Van der Waals attractions arise due to the polarization of an electron cloud of an atom or molecule by a nearby nucleus that results in an electrostatic attraction that is <5 $kJ \cdot mol^{-1}$ in strength [9]. VDW interactions are weak intermolecular or particulate forces with $1/r^6$ dependence. Transient fluctuating dipoles are VDW interactions that contribute to stabilize hydrophobic guests in hydrophobic pockets [8].

10.3.1   Contributions to the van der Waals Interaction

Van der Waals forces can be broken down into three contributions: (1) the Keesom-*vdw* attraction is a dipole–dipole angle-averaged orientation that involves electrostatic interactions between charges such as ions, dipoles, quadrupoles, and other permanent multipoles; (2) the Debye-*vdw* interaction involves the process of attraction by induction (polarization). In this case, a molecule with a permanent charge or dipole induces a dipole (or multipole) in another molecule within proximity; and (3) the final type of *vdw* attraction is the London-*vdw* (or dispersion) force that relies on transient dipoles for interaction. London dispersion forces are the only attraction force experienced by the inert gases. Except for the case of the inert gases, all of these *vdw* forces depend on the shape (iso- or anisotropic charge distribution) and orientation of the participating molecules (**Table 10.11**).

The Keesom type of *vdw* interaction involves electrostatic forces that can be attractive or repulsive depending on the rotational configuration of the participating molecules. Due to thermal motion, electrostatic forces based on the Keesom form are thermally averaged due to molecular rotations that limit the strength of the interaction. Keesom and Debye interactions abide by classical electrostatic (Coulombic) and Boltzmann distribution laws [52]. London dispersion phenomena require quantum mechanical explanations [52]. The Keesom form for two dissimilar molecules is given by

$$V_P(r) = -\frac{\mu_1^2 \mu_2^2}{3(4\pi\varepsilon_o)^2 k_B T r_{12}^6} = \frac{C_P}{r^6} \tag{10.19}$$

where μ is the dipole moment of the species and C_P is the Keesom dipolar orientation interaction coefficient. The Debye dipole–induced dipole interaction is expressed as

$$V_I(r) = -\frac{\mu_1^2 \alpha_2 + \mu_2^2 \alpha_1}{(4\pi\varepsilon_o)^2 r^6} = \frac{C_I}{r^6} \tag{10.20}$$

where α is the polarizability of the species and C_I is the Debye-induced interaction coefficient.

London dispersion is represented by

$$V_d(r) = -\frac{3\alpha_1\alpha_2}{2(4\pi\varepsilon_o)^2 r^6}\left(\frac{I_1 I_2}{I_1 + I_2}\right) = \frac{C_d}{r^6} \tag{10.21}$$

TABLE 10.11	*Van der Waals Interactions*	
Description	**Example**	**Bond strength (kJ · mol⁻¹)**
Keesom forces	*Keesom type van der Waals alignment. Permanent dipole–permanent dipole interactions are capable of repulsive interactions upon rotation.*	0.4–4.0
Debye forces	*Debye type van der Waals permanent dipole–induced dipole interaction.*	(<5)
London dispersion forces	*London type (dispersion) van der Waals induced dipole–induced dipole (above) and transitory induced dipole (below) interactions.*	<<5
Hamaker constant	Hamaker constants are combinations of all van der Waals forces that are used to evaluate colloid–colloid, colloid–surface, and similar interactions. The Hamaker constant is shape dependent and specific for materials. Most Hamaker constants are on the order of 10^{-19}–10^{-20}J.	10^{-19}–10^{-20} J

where I_n represents the ionization potential of the species. The link to quantum mechanics is made through the ionization potential term. The total van der Waals interaction is the sum of the three contributions:

$$V_W(r) = \frac{C_W}{r^6} = \frac{C_P + C_1 + C_d}{r^6} \qquad (10.22)$$

The *vdw* attractions between colloids and other solubilized materials are summarized in the Hamaker constant A. Proper shape-dependent terms are incorporated into the expression of the Hamaker constant. A complete derivation of the Hamaker constant is beyond the scope of this text. A generalized relation appropriate for spherical particles is [53,54]

$$V_W(r) = \frac{A}{6}\left[\left(\frac{2R^2}{d^2 + 4Rd}\right) + \left(\frac{2R^2}{d^2 + 4Rd + 4d^2}\right) + \ln\left(\frac{d^2 + 4Rd}{d^2 + 4Rd + 4d^2}\right)\right] \qquad (10.23)$$

All Hamaker constants range from 10^{-19} to 10^{-20} J. The Hamaker constants of water (4.4×10^{-20} J), toluene (6.3×10^{-20} J), metals ($25–40 \times 10^{-20}$ J), quartz (8.7×10^{-20} J), and CCl_4 (5.5×10^{-20} J) serve as a few examples [52].

Of course, we are not interested per se in the attractions between gases but rather of molecules in solution, especially large molecules. Depending on the size, shape, proximity, and orientation of one molecule to another, physical properties are expected to be affected, most notably the boiling point of liquids and the stability of monolayers. Since all atoms and molecules are impacted by van der Waals forces, it is considered to be a universal factor in intermolecular bonding. In nature, the humble gecko (chapter 13) has mastered the ability to cling to ceilings, whether wet or dry, due to capillary forces or van der Waals forces, respectively. At the nanoscale, van der Waals forces assume a greater level of importance than they do at the bulk scale; small molecules or clusters are influenced greatly by van der Waals forces and must be considered in the design of devices.

The Lennard–Jones potential relationship gives us a good feel for van der Waals interactions and it serves as a good first approximation of the potential energy between any two atoms or molecules that approach each other. Van der Waals energy is inversely dependent on the sixth power of the distance between two molecules (e.g., *vdw* $\propto r^{-6}$). The same dependency is exhibited by the attractive force component in the Lennard–Jones potential treatment. The van der Waals radius of atoms can be calculated from the generalized *Lennard–Jones 6–12* equation. Some van der Waals radii are listed in **Table 10.12**. Reading towards the right in the periodic table, *vdw* radii decrease. As we proceed down a group, *vdw* radii in general increase in magnitude. As a first approximation for molecular dynamic simulations, atoms cannot approach each other closer than their *vdw* radius allows. The van der Waals radius is the equilibrium distance-closest approach between two atoms or molecules. Generic hard-sphere van der Waals radii are shown in **Figure 10.10**. Space-filling models generated by computer programs do a great job in approximating the van der Waals surface. Selected examples of space-filling models of large molecules are shown in **Figure 10.11a** and **b.**

TABLE 10.12	Van der Waals Radii of Some Common Elements		
Atom	**Radius (Å)**	**Atom**	**Radius (Å)**
H	1.20	Te	2.06
C	1.70	As	1.85
N	1.55	Sb	2.2
O	1.52	Cl	1.75
F	1.47	Br	1.85
S	1.85	I	1.98
Se	1.90	P	1.80
Si	2.10	Ni	1.63
Na	2.27	Pb	2.02
K	2.75	Xe	2.16
Au	1.66	Zn	1.39

Source: A. Bondi, *Journal of Physical Chemistry*, 68, 441–451, 1964.

10.3.2 Van der Waals Radius

Van der Waals forces involve the interaction between closed-shell molecules [55]. They include interactions between partially distributed electrical charges of polar molecules and the repulsive interactions (Pauli exclusion) as molecules approach within the van der Waals radius. Although repulsive forces are indeed important (very important), we focus primarily upon the attractive variety of van der Waals interactions. The induced dipole is the fundamental electrostatic entity involved in van der Waals interactions, and there are many forms of van der Waals forces. Strictly speaking, we will refer to the generalized van der Waals force that arises from the polarization of the electronic space of a molecule by a nearby nucleus that results in fluctuating dipoles, multipoles, octupoles, etc. [9].

10.3.3 Physical Property Dependence

Van der Waals forces are roughly proportional to the number of atoms, and hence the number of electrons, incorporated into a molecular structure. As we

FIG. 10.10 *A hard-sphere model of the van der Waals radii of two atoms in near proximity is shown. The equilibrium distance corresponds to a local energy minimum in a Lennard–Jones type potential diagram.*

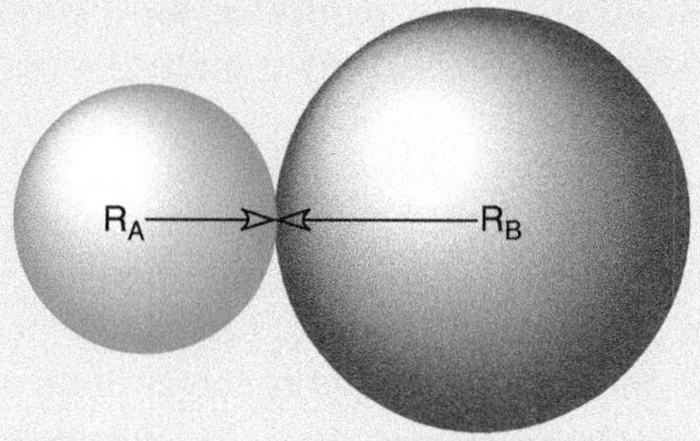

Fig. 10.11	*The structure of vitamin A is shown by two illustrations: (a) by a succinct chemical formula and (b) in the form of a space-filling model. Computer models are also adept at showing the van der Waals surface—a somewhat better rendition of the molecular orbital that envelopes the entire molecule.*

(a)

(b)

demonstrated earlier with the boiling point of small hydrides, the boiling points of linear methane series (paraffin series) alkanes (C_nH_{2n+2}) serve as an excellent example of how van der Waals interactions affect physical properties [57]. Since these molecules have extremely limited polarity, van der Waals attractions are expected to contribute to most of the interbonding potential energy. The boiling point of *n*-alkanes will be higher than their branched-chain counterparts due to the greater common surface area (better fit) shared among linear molecules. Adjacent linear aliphatic chains, in particular, form strong *vdw* interactions. Lateral bonding via *vdw* interactions in self-assembled monolayers serves to give structural integrity to the layers.

Molecular shape also plays a role in the melting point (again related to van der Waals forces) of materials. Specifically, even numbered alkanes pack better than odd ones. Van der Waals interactions are therefore maximized in such even-carbon alkanes. The estimated energy per added methylene group boils down to about 3.5 kJ · mol^{-1}, a value near the energy of thermal motion at room temperature. Others estimate ΔH°_{vap} for the homologous series of alkanes to be 4.9 kJ · mol^{-1} per added methylene group [58] (**Table 10.13**).

EXAMPLE 10.9 *Van der Waals Radius Calculation*

Calculate the VDW radius, r_{VDW}, for the water molecule and compare to that of carbon tetrachloride. What is a major assumption of this equation and what are the major limitations of this method? Which is a better approximation?

Solution:
Use the following relationship:

$$\frac{4}{3}\pi r_{VDW}^3 = f_P\left(\frac{M}{\mathcal{N}_A\rho}\right)$$

(10.24)

where M is the molecular weight, $\mathcal{N}_A$ is the Avogadro number, ρ is the density of water, and f_P is an estimated packing fraction of liquid water (equal to 0.7405 assuming 12 nearest neighbors—chapter 5).
The equation assumes a rather hard-sphere model and close packing.

For water:

$$r_{VDW} = \sqrt[3]{0.7405\left(\frac{3}{4\pi}\right)\left(\frac{M}{\mathcal{N}_A\rho}\right)} = \sqrt[3]{0.7405\left(\frac{3}{4\pi}\right)\left(\frac{18\,\text{g}\cdot\text{mol}^{-1}}{6.022\times10^{23}\,\text{mol}^{-1}\times1.00\,\text{g}\cdot\text{cm}^3}\right)} = 1.74\,\text{Å}$$

For carbon tetrachloride:

$$r_{VDW} = \sqrt[3]{0.7405\left(\frac{3}{4\pi}\right)\left(\frac{M}{\mathcal{N}_A\rho}\right)} = \sqrt[3]{0.7405\left(\frac{3}{4\pi}\right)\left(\frac{153.82\,\text{g}\cdot\text{mol}^{-1}}{6.022\times10^{23}\,\text{mol}^{-1}\times1.5867\,\text{g}\cdot\text{cm}^3}\right)} = 3.05\,\text{Å}$$

CCl_4 is more purely tetrahedral than water and by default better approximates a sphere. In both cases the VDW radius is overestimated by ca. 30%. The molecular diameter of water is about 2.75 Å and from the preceding equation, the derived molecular diameter is 3.48 Å. Also recall that both water and carbon tetrachloride are bonded with strong covalent bonds. This also helps to explain the serious deviation from experimental results.
Better packing parameters are available for water (0.63 → 1.65 Å) and carbon tetrachloride (0.53 → 2.74 Å).

TABLE 10.13 *Effect of Size on Physical Properties of n-Alkanes*

n-Alkane	Molecular weight	Melting point	Boiling point	ΔH_{vap}
Methane	16.04	−182	−164	8.91
Ethane	30.07	−183.3	−88.6	15.6
Propane	44.11	−189.7	−42.1	19.0
Butane	58.12	−138.4	−0.5	24.3
Pentane	72.15	−130	36.1	27.6
Hexane	86.18	−95	69	31.9
Heptane	100.21	−90.6	98.4	37.4
Octane	114.23	−56.8	125.7	38.6
Nonane	126.26	−51	150.8	43.8
Decane	142.29	−29.7	174.1	45.7
Undecane	156.32	−25.6	196	48.0
Dodecane	170.34	−9.6	216.3	49.6
Tridecane	184.37	−5.5	234.4	54.4
Tetradecane	198.40	5.9	253.7	57.5
Pentadecane	212.42	10	270.6	61.2

We understand that van der Waals forces are relatively weak and only exert their influence at very short distances (proportional to r^6) for closed-shell molecules.

Van der Waals forces are active in stabilizing the lateral attractions experienced by nitrogen molecules in BET analysis. They are not so important arguably, with regard to supramolecular design because of their inherent lack of directionality [9]. Van der Waals interactions are a factor, however, in molecular cavities that trap small organic molecules such as the van der Waals inclusion complex toluene by *p-tert*-butylcalix[4]arene [9]. In ever-complex supramolecular clusters, it would be nearly impossible to predict the sum total of all van der Waals forces. For other systems, the van der Waals contributions are well known. Van der Waals interactions bundled into the Hamaker constant serve as a valuable tool in predicting the physical properties of colloidal systems. Discussion of the shape-dependent forms of the Hamaker constant is beyond the scope of this text. An example of a van der Waals inclusion complex molecule is shown in **Figure 10.12**.

Van der Waals forces, although not of the highest priority in supramolecular design, are important in polymer chemistry. Van der Waals interactions are important in the chemistry of macromolecules and self-assembled molecular assembly, but they also are important in the stabilization of fullerene clusters (and carbon nanotube bundles). Both fullerenes and nanotubes are formed in an uncharged state. The surfaces of both are kinetically stable. Single-walled carbon nanotubes in particular are extremely difficult to dissolve, especially if they are bundled (e.g., they behave like a *supermolecule*). Large molecules in general are more difficult to solvate than smaller ones. A bundle of SWNTs may

Fig. 10.12

A van der Waals inclusion complex (p-tert-butylcalix[4]arene) is depicted. This guest molecule is able to accommodate, for example, toluene within its cavity. Van der Waals attractions are nondirectional and therefore a high level of specificity between guest and host is not expected.

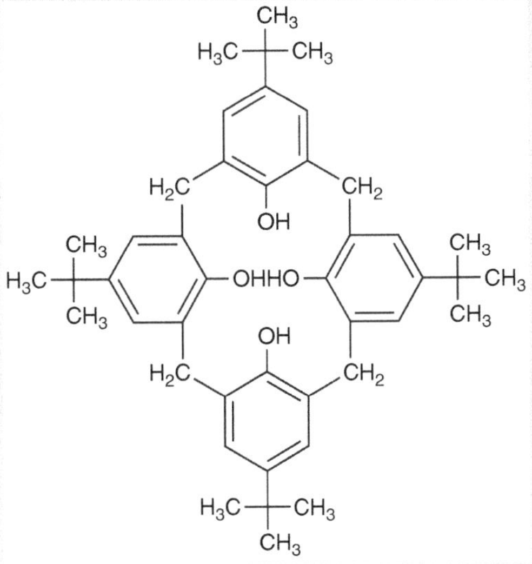

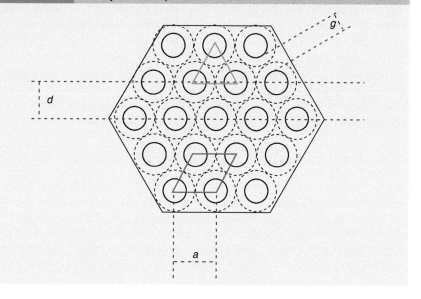

FIG. 10.13 *Van der Waals gaps exist between and among single-walled carbon nanotubes within a bundle. It is within these interstitial spaces that intercalation occurs. In hydrogen storage by SWNTs, it is thought that hydrogen is sequestered within the interstices formed as a result of the van der Waals gap. The phenomenon of hydrogen storage, although showing much promise at first, has yet to be a proven.*

consist of tens to hundreds of carbon nanotubes that are many microns in length (even millimeters and centimeters in length). Trying to disperse the bundle into individual nanotubes is a formidable task that requires extra energy to accomplish. A phenomenon known as the *van der Waals gap* is commonly applied to the distance between fullerenes in the solid state and bundled SWNTs. The van der Waals gap is equal to 3.15 Å in both cases. Fullerenes crystallize in a three-dimensional *fcc* close-packed structure while SWNTs bundle in a two-dimensional trigonal lattice (**Fig. 10.13**). The van der Waals gaps within the nanotube bundles serve as sites for intercalation.

10.4 HYDROPHOBIC EFFECT

The hydrophobic effect is one of the least understood but a very significant noncovalent interaction—especially for biological materials. Entropy-driven processes play a major role in forming hydrophobic-coupled structures.

10.4.1 Background

G. S. Hartley first described the hydrophobic effect in 1936:

> The antipathy of the paraffin chain in water is, however, frequently misunderstood. There is no question of actual repulsion between individual water

molecules and paraffin chains; nor is there any very strong attraction of paraffin chains for one another. There is however, a very strong attraction of water molecules for one another in comparison with which the paraffin-paraffin or paraffin-water attractions are slight.

Tanford in 1973 coined the phrase hydrophobic effect [59,60]. A little bit later, Hildebrand asked if there was such a thing as a hydrophobic effect [61]. Tanford essentially agreed with Hartley's observation given previously. He stated that the attraction between water and itself was the reason for the unfavorable interfacial free energy between water and a hydrocarbon. Hydrophobic interactions play a major role in the formation of micelles, membranes, DNA, and folding and interaction in globular proteins and molecular recognition [62]. Regardless of all the research that has been conducted to unravel the mysteries of the hydrophobic effect, it is only recently that molecular modeling methods have shed some light on the molecular scale understanding of the phenomenon. An example of a hydrophobic interaction is shown in **Figure 10.14**.

10.4.2   *Water and the Hydrophobic Effect*

In pure water, entropy is maximized by its unique hydrogen-bonded structure and electrostatic relations. If a hydrophobic molecule or substance is added to the water, the structure of the water is disrupted and the overall entropy is decreased by creation of a cavity. This is an unfavorable state. This cavity lacks the electrostatic potential to interact with the water molecules. If many such cavities exist, a high surface area of such individual hydrophobic entities is created—not desirable. The water counteracts this by forming cage structures around the

FIG. 10.14   *A hydrophobic interaction with host–guest specificity is depicted. When a hydrophobic surface is exposed to an aqueous medium, water molecules form an organized layer on the surface. Water molecules undergo tangential organization of dipoles on the hydrophobic surface. (D.T. Bowron, J. L. Finney, and A. K. Soper, Journal of Physical Chemistry, 102, 3551–3563, 1998.) Displacement of the water molecules by another hydrophobic entity raises the entropy of the systems—the increase in entropy gained by the removal of hydrophobic surface area from ordered solvating water. This phenomenon and water's intrinsic attraction to other water molecules act as drivers in hydrophobic organization. Hydrophobic interactions are one of the most important noncovalent intermolecular forces in the folding of linear polypeptides.*

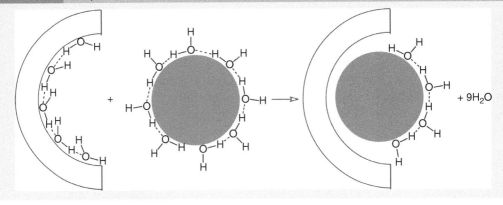

TABLE 10.14 *Hydrophobic Effects*

Description	Example	Bond strength (kJ·mol⁻¹)
Micelles Lipid bilayers Hydrophobic pockets	Micelles Lipid bilayers	0–0.2

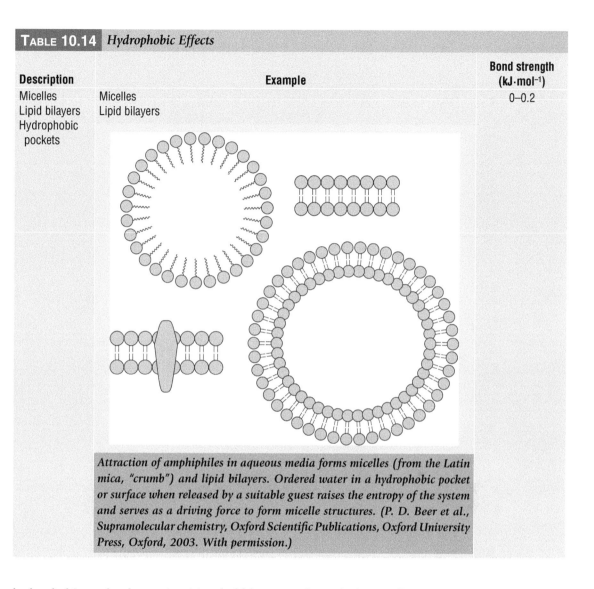

Attraction of amphiphiles in aqueous media forms micelles (from the Latin mica, "crumb") and lipid bilayers. Ordered water in a hydrophobic pocket or surface when released by a suitable guest raises the entropy of the system and serves as a driving force to form micelle structures. (P. D. Beer et al., Supramolecular chemistry, Oxford Scientific Publications, Oxford University Press, Oxford, 2003. With permission.)

hydrophobic molecules or inspiring bubbles to coalesce (reduce collective surface area). This action lowers the overall surface area (**Table 10.14**).

Recent work has shown that small and large hydrophobic solutes are hydrated in different ways [62]. Pressure, temperature, and additives all affect the hydrophobic interaction, regardless of the size of the hydrophobic solutes. However, the behavior of the solute–water system is a function of the length scale of the hydrophobic moiety [62] (e.g., small solutes have different hydration thermodynamics than do large solutes). For example, small solutes exhibit entropic effects while large solutes exhibit enthalpic effects. Apparently, according to simulations, there is a crossover state of nanoscopic size that consists of both molecular and macroscopic hydration thermodynamics [62]. This is important because biological and supramolecular assemblies include the integration of molecules as small as amino acids and as large as proteins. Chaotropic substances such as urea and guanidinium chloride increase the solubility

TABLE 10.15   *Hydrophobic Character Scales of Amino Acids*

Amino acid	Mol. wt.	[64][a]	[65][b]	[66][c]	[67][d]	Amino acid	Mol. wt.	[64][a]	[65][b]	[66][c]	[67][d]
		Rating						Rating			
ALA[1]	89	1.80	0.42	8.11	-2.09	LEU[1]	131	3.80	1.80	9.54	-7.45
ARG[2]	174	-4.50	-1.56	-83.3	12.6	LYS[2]	146	-3.90	-2.03	-39.8	12.6
ASN[3]	132	-3.50	-1.03	-40.5	0.837	MET[1]	149	1.90	1.18	-6.19	-5.44
ASP[4]	133	-3.50	-0.51	-45.8	12.6	PHE[1]	165	2.80	1.74	-3.18	-10.5
CYS[5]	121	2.50	0.84	-5.19	-4.18	PRO[1]	115	-1.60	0.86	0.00	0.00
GLN[3]	146	-3.50	-0.96	-39.2	0.837	SER[3]	105	-0.80	-0.64	-21.2	1.26
GLU[4]	147	-3.50	-0.37	-42.7	12.6	THR[3]	119	-0.70	-0.26	-20.4	-1.67
GLY[1]	75	-0.40	0.00	10.0	0.00	TRP[5]	204	-0.90	1.46	-24.6	-14.2
HIS[2]	155	-3.20	-2.28	-43.0	-2.09	TYR[3]	181	-1.30	0.51	-25.6	-9.62
ILE[1]	131	4.50	1.81	9.00	-7.45	VAL[1]	117	4.20	1.34	8.32	-6.28

Note: [1] neutral, nonpolar; [2] basic, polar; [3] neutral, polar; [4] acidic, polar; [5] neutral, slightly polar.

a   Value determined by hydropathicity.

b   Value determined by HPLC.

c   Value determined by hydration potential (kJ·mol^{-1}).

d   Value based on hydrophilicity.

of hydrophobic particles in aqueous media. These materials have the capability to destabilize aggregations of nonpolar solute particles and micelles and to denature proteins [63].

10.4.3 Amino Acids and Proteins

There are numerous *hydrophobicity scales* that catalog amino acids with regard to their hydrophobic and, conversely hydrophilic, natures (**Table 10.15**). The purpose of the scales is to be able to predict or determine the hydrophobicity of proteins assembled of those amino acids. The hydrophobicity values for the scale of Kyte and Doolittle (1982) were derived from experimentally measured amino acid solubility in various organic solvents [64]. More detail will be afforded to the topic of amino acids and proteins in chapter 14.

The hydrophobicity of proteins can then be acquired by compiling an average hydrophobicity over all amino acids that comprise the protein [68]. Advanced calculations of *vdw* volume are used routinely in developing drug compounds [69]. The hydrophobic effect and cooperativity are figured in together in the calculation of binding constants [70].

There is no question about the importance of van der Waals interactions in all materials—especially in nanomaterials like colloids, proteins, and pharmaceuticals. Van der Waals forces are not the most significant interaction at the macroscale (unless you are a gecko), but they assume huge importance at the nanoscale. If only van der Waals himself could see how his fundamental concept has been developed [71].

References

1. H. Y. Erbil, *Surface chemistry of solid and liquid interfaces*, Blackwell Publishing, Oxford (2006).
2. Chemical bond—History, *Science Encyclopedia*, Vol. 2, http://science.jrank.org/collection/2/Gale-Encyclopedia-Science.html (2007).
3. L. Pauling, *The nature of the chemical bond*, Cornell University Press, Ithaca, NY (1960).
4. I. W. Hamley, *Introduction to soft matter: Polymers, colloids, amphiphiles and liquid crystals*, John Wiley & Sons, Ltd., West Sussex (2000).
5. J.-M. Lehn, *Supramolecular chemistry*, Weinheim, Germany: VCH (1995).
6. J.-M. Lehn, *Organic chemistry: Its language and its state of the art*, Proc. Centenary Geneva Conf., M.V. Kisakürek, VHCA, Basel; VCH, Weinheim (1993).
7. J.-M. Lehn, Cryptates: Inclusion complexes of macrocyclic receptor molecules, *Pure Applied Chemistry*, 50, 871–892 (1978).
8. P. D. Beer, P. A. Gale, and D. K. Smith, *Supramolecular chemistry*, Oxford Scientific Publications, Oxford University Press, Oxford (2003).
9. J. W. Steed and J. L. Atwood, *Supramolecular chemistry*, John Wiley & Sons, Ltd., Chichester (2000).
10. E. Deu, K. A. Koch, and J. F. Kirsch, The role of the conserved Lys68* Intersubunit salt bridge in aspartate aminotransferase kinetics: Multiple forced covariant amino acid substitutions in natural variants, *Protein Science*, 11, 1062–1072 (2002)
11. N. K. Vyas, M. N. Vyas, and F. A. Quiocho, Crystal structure of M tuberculosis ASB phosphate transport receptor: Specificity and charge compensation dominated by ion–dipole interactions, *Structure*, 11, 756–774 (2003).

12. G. T. Morgan and H. D. K. Drew, Researches on residual affinity and coordination. Part II, acetyl acetones of selenium and tellurium, *Journal of the Chemical Society*, 117, 1456–1465 (1920).

13. G. Schmid, Large clusters and colloids: Metals in the embryonic state, *Chemical Reviews*, 92, 1709–1727 (1992).

14. G. Schmid, ed., *Clusters and colloids: From theory to applications*, VCH, New York (1994).

15. G. Schmid and G. L. Hornyak, Metal clusters: New perspectives in future nano-electronics, *Current Opinion in Solar State Materials Science*, 2, 204 (1997).

16. G. L. Hornyak, M. Kroell, R. Pugin, T. Sawitowski, G. Schmid, J-O. Bovin, H. Hofmeister, and S. Hopfe, Gold clusters and colloids in alumina membranes, *Chemistry: A European Journal*, 3, 1951 (1997).

17. I. Vidal, S. Mechor, I. Alkorta, J. Elguero, M. R. Sundberg, and J. A. Dobado, On the existence of α-agostic bonds: Bonding analyses of titanium alkyl complexes, *Organometals*, 25, 5638–5647 (2006).

18. H.-F. Zhang, M. Stender, R. Zhang, J. Li, and L.-S. Wang, Toward the solution of the tetrahedral Au_{20} cluster, *Journal of Physical Chemistry B*, 108, 12259–12263 (2004).

19. B. Olenyuk, A. Fechtenkötter, and P. J. Stang, Molecular architecture of cyclic nanostructures: Use of co-ordination chemistry in the building of supermolecules with predefined geometric shapes, *Journal of the Chemical Society, Dalton Transactions*, 1707–1728 (1998).

20. J. Y. Lee, S. J. Jung, S. Hong, and S. Kim, Double dative bond configuration: Pyrimidine on Ge(100), *Journal of Physical Chemistry B*, 109, 348–351 (2005).

21. J. J. da Silva and R. Frausto, The chelate effect redefined, *Journal of Chemical Education*, 60, 390–392 (1993).

22. C. A. Hunter and J. K. M. Saunders, The nature of π–π interactions, *Journal of the American Chemical Society*, 112, 5525–5534 (1990).

23. P. A. Brooksby, C. A. Hunter, A. J. McQuillan, D. H. Purvis, A. E. Rowan, R. J. Shannon, and R. Walsh, Supramolecular activation of *p*-benzoquinone, *Angewandte Chemie International Edition*, 33, 2489–2491 (1994).

24. G. Wilkinson, The iron sandwich. A recollection of the first four months, *Journal of Organometallic Chemistry*, 100, 273–278 (1975).

25. M. Kamieth, U. Burkert, P. S. Corbin, S. J. Dell, S.C. Zimmerman, and F.-G. Klärner, Molecular tweezers as synthetic receptors: Molecular recognition of electron-deficient aromatic substrates by chemically bonded stationary phases, *European Journal of Organic Chemistry*, 1999, 2741–2749 (1999).

26. G. A. Jeffrey, *An introduction to hydrogen bonding*, Oxford University Press, Oxford (1997).

27. O. Markovitch and N. Agmon, Structure and energetics of the hydronium hydration shells, *Journal of Physical Chemistry A*, 111, 2253–2256 (2007).

28. S. Aravinda, N. Shamala, A. Pramanik, C. Das, and P. Balaram, An unusual C–H$\cdots$O hydrogen bond mediated reversal of polypeptide chain direction in an synthetic peptide helix, *Biochemical and Biophysical Research Communications*, 272, 933–936 (2000).

29. P. Auffinger, F. A. Hays, E. Westhof, and P. S. Ho, Halogen bonds in biological molecules, *Proceedings of the National Academy of Science, USA*, 101, 16789–16794 (2004).

30. J. Barciszewski, J. Jurczak, S. Porowski, T. Specht, and V. A. Erdmann, The role of water structure in conformational changes of nucleic acids in ambient and high pressure conditions, *European Journal of Biochemistry*, 260, 293–307 (1999).

31. H. Xu, H. A. Stern, and B. J. Berne, Can water polarizability be ignored in hydrogen bond kinetics? *Journal of Physical Chemistry*, 106, 2054–2060 (2002).

32. M. Tagliazucchi, D. Grumelli, and E. J. Calvo, Nanostructured modified electrodes: Role of ions and solvent flux in redox active polyelectrolyte multilayer films, *Physical Chemistry Chemical Physics*, 8, 5086–5095 (2006).

33. H. Liu, S. Murad, and C. J. Jameson, Ion permeation dynamics in carbon nanotubes, *Journal of Chemical Physics*, 125, 084713–084726 (2006).

34. A. A. Malani, Structure and dynamics of water under confinement, presentation, Indian Institute of Science, Bangalore (viewed 2007).

35. M. K. Singh and G. E. Walrafen, New method for determining H-bond energy: Fluorescence from neutral red in water compared to fluorescence from neutral red in benzene, *Journal of Solution Chemistry*, 34, 579–583 (2005).

36. C. B. Aakeröy, K. R. Seddon, and M. Leslie, Hydrogen-bonding contributions to the lattice energy of salts for second harmonic generation, *Structural Chemistry*, 3, 63–65 (1992).

37. T. Steiner, The hydrogen bond in the solid state, *Angewandte Chemie International Edition*, 41, 48–76 (2002).

38. L. F. Scatena, M. G. Brown, and G. L. Richmond, Water at hydrophobic surfaces: Weak hydrogen bonding and strong orientation effects, *Science*, 292, 908–912 (2001).

39. R. S. Drago, *Physical methods for chemists*, 2nd ed., Saunders College Publishing, Hartcourt-Brace-Jovanovich College Publishers, New York (1992).

40. A. Streitwieser and C. H. Heathcock, *Introduction to organic chemistry*, 3rd ed., Macmillan Publishing Company, New York (1985).

41. V. M. F. Hammer, E. Libowitsky, and G. R. Rossman, Single-crystal spectroscopy of very strong hydrogen bonds in pectolite, $NaCa_2 Si_3O_8(OH)$, and serandite, $NaMn_2 Si_3O_8(OH)$, *American Mineralogist*, 83, 569–576 (1998).

42. I. Ayala, J. J. P. Perry, J. Szczepanski, J. A. Tainer, M. T. Vala, H. S. Nick, and D. N. Silverman, Hydrogen bonding in human manganese superoxide dismutase containing 3-fluorotyrosine, *Biophysical Journal*, 89, 4171–4179 (2005).

43. C. L. Aronson, D. Beloskur, I. S. Frampton, J. McKie, and B. Burland, The effect of macromolecular architecture on functional group accessibility: Hydrogen bonding in blends containing phenolic photoresist polymers, *Polymer Bulletin*, 53, 413–424 (2005).

44. S.-I. Kawahara, K. Taira, M. Sekine, and T. Uchimaru, Evaluation of the hydrogen bond energy of base pairs formed between substituted 9-methyladenine derivatives and 1-methyluracil by use of molecular orbital theory, *Nucleic Acids Symposium Series*, 44, 237–238 (2000).

45. G. R. Desiraju and T. Steiner, *The weak hydrogen bond*, Oxford University Press, Oxford (2001).

46. S. Scheiner, T. Kar, and Y. Gu, Strength of the $C\alpha H \cdots O$ hydrogen bond of amino acid residues, *Journal of Biological Chemistry*, 276, 9832–9837 (2001).

47. G. R. Desiraju, Crystal gazing: Structure prediction and polymorphism, *Science*, 278, 404–405 (1997).

48. T. Steiner, The influence of C-H$\cdots$O interactions on the conformation of methyl groups quantified from neutron diffraction data, *Journal of Physical Chemistry A*, 104, 433–435 (2000).

49. M. F. Chaplin, Water's hydrogen bond strength. In *Water of life: Counterfactual chemistry and fine-tuning in biochemistry*, R. M. Lynden-Bell, S. C. Morris, J. D. Barrow, J. L. Finney, and C. L. Harper, eds., manuscript in preparation (2007).

50. A. H. Nissan, The hydrogen bond strength of ice, *Nature*, 178, 1411–1412 (1956).

51. I. A. Levine, *Physical chemistry*, 3rd ed., McGraw–Hill Book Co., New York (1988).

52. H. Y. Erbil, *Surface chemistry of solid and liquid interfaces*, Blackwell Publishing, Oxford (2006).

53. G. Cao, *Nanostructures & nanomaterials: Synthesis, properties and applications*, Imperial College Press, Singapore (2004).

54. P. C. Hiemnz, *Principles of colloid and surface chemistry*, Marcel Dekker, New York (1977).

55. P. W. Atkins, *Physical chemistry*, 4th ed., W. H. Freeman & Co., New York (1990).

56. A. Bondi, van der Waals volumes and radii, *Journal of Physical Chemistry*, 68, 441–451 (1964).
57. R. C. Wiest, ed., *CRC handbook of chemistry and physics*, CRC Press, Boca Raton, FL (1985).
58. J. D. Dunitz and A. Gavezzotti, Attractions and repulsions in molecular crystals: What can be learned from the crystal structure of condensed ring aromatic hydrocarbons? *Accounts in Chemical Research*, 32, 677–684 (1999).
59. C. Tanford, *The hydrophobic effect*, Wiley, New York (1973).
60. C. Tanford, Interfacial free energy and the hydrophobic effect, *Proceedings of the National Academy of Science, USA*, 76, 4175–4176 (1979).
61. J. H. Hildebrand, Is there a hydrophobic effect? *Proceedings of the National Academy of Science, USA*, 76, 194 (1979).
62. S. Rajamani, T. M. Truskett, and S. Garde, Hydrophobic hydration from small to large length scales: Understanding and manipulating the crossover, *Proceedings of the National Academy of Science, USA*, 102, 9475–9480 (2005).
63. S. Moelbert and P. De Los Rios, Chaotropic effect and preferential binding in a hydrophobic interaction model, *Journal of Chemical Physics*, 119, 7988–8001 (2003).
64. J. Kyte and R. F. Doolittle, A simple method for displaying the hydropathic character of a protein, *Journal of Molecular Biology*, 157, 105–132 (1982).
65. R. Cowan and R. G. Whittaker, Hydrophobicity indices for amino acid residues as determined by HPLC, *Peptide Research*, 3, 75–80 (1990).
66. D. Ring, Y. Wolman, N. Friedmann, and S. L. Miller, Prebiotic synthesis of hydrophobic and protein amino acids, *Proceedings of the National Academy of Science, USA*, 69, 765–768 (1972).
67. T. P. Hopp and K. R. Woods, Prediction of protein antigenic determinants from amino acid sequences, *Proceedings of the National Academy of Science, USA*, 78, 3824–3828 (1981).
68. C. M. Roth, B. L. Neal, and A. M. Lenhoff, van der Waals interactions involving proteins, *Biophysics Journal*, 70, 977–987 (1996).
69. Y. H. Zhao, M. H. Abraham, and A. M. Zissimos, Fast calculation of van der Waals volume as a sum of atomic and bond contributions and its application to drug compounds, *Journal of Organic Chemistry*, 68, 7368–7373 (2003).
70. D. H. Williams and B. Bardsley, Estimating binding constants—The hydrophobic effect and cooperativity, *Perspectives in Drug Discovery and Design*, 17, 43–59 (2004).
71. J. D. van der Waals, *Over de Continuiteit van den Gas—en. Vloeistoftoestand*, thesis, University of Leiden (1873).
72. J. Pfeffer and S. Nir, *Modern physics: An introduction text*, World Scientific Publishing Company, Hackensack, NJ, 183 (2001).
73. M. Brookhart and M. L. H. Green, Carbon–hydrogen transition metal bonds, *Journal of Organometallic Chemistry*, 250, 395–408 (1983).
74. D. T. Bowron, J. L. Finney, and A. K. Soper, Structural investigation of solute–solute interactions in aqueous solutions of tertiary butanol, *Journal of Physical Chemistry*, 102, 3551–3563 (1998).

Problems

10.1 What is the ion–ion potential interaction energy for Na^+ and Cl^- if the ions are in water at 25°C at a distance of 3 nm? 2 nm? 1 nm?

Permittivity of free space, $\varepsilon_o = 8.854 \times 10^{-12}$ $C^2 \cdot J^{-1} \cdot m^{-1}$, $e = 1.6022 \times 10^{-19}$ C, $\varepsilon_{r'H_2O} = 78.85$.

10.2 How close must two attractive ions be in order to satisfy the 50–300 kJ·mol⁻¹ range?

10.3 What ways can you think of to alleviate entropic factors in systems that have increased level of organization?

a. Hydrophobic entities in water?

b. Micelles?

10.4 The dipole of diethyl ketone is in line with a potassium cation at a distance of 0.5 nm. What is the energy of the interaction in water ($\varepsilon_r = 78.85$)? In methanol ($\varepsilon_r = 32.63$)? In trimethyl amine ($\varepsilon_r = 2.44$) at 298 K?

10.5 Which molecules have a permanent dipole moment? Why?

   a. CH_4
   b. CO
   c. N_2
   d. H_2O
   e. CO_2
   f. *n*-Heptane
   g. HF

10.6 List, in order of lowest to highest boiling point, the following hydrides and explain.

   a. HF, CH_4, H_2O, NH_3
   b. H_2S, H_2Se, H_2Te, H_2O

10.7 Molecular shape is also an important consideration in van der Waals bonding. How would the boiling point of a solution filled with nonhomogeneously shaped molecules compare to one that is homogeneous?

10.8 How would the melting and boiling points of *n*-butane and methylpropane compare? Is there any correlation between mp, bp, and thermodynamic stability? The $\Delta H°$ of the reaction $CH_3CH_2CH_2CH_3 \rightarrow (CH_3)_2CH_2CH_3$ is -8.58 kJ·mol^{-1}.

10.9 What kinds of interactions do you expect between the following ion or molecular pairs? List them in order of increasing strength:

   a. Dichloromethane and methanol
   b. Octane and heptane
   c. Bromide anion and octane
   d. Na+ and water
   e. Dimethyl ketone and heptane
   f. Ethanol and water

10.10 Calculate the cohesion energy (binding energy) of a KBr crystal from a Born–Haber cycle. How does the value compare to that of NaCl? Why the difference?

Experience the Born–Haber cycle for KBr by writing out all the equations involved. Feel free to use references.

10.11 Calculate the density of a fullerene crystal made of C_{60} that is fcc-close-packed structure. What is the packing fraction of the solid-state form of the C_{60}?

10.12 Rank the following ionic compounds in order of decreasing lattice energy and explain the trend:

   a. $NaCl — NaF — NaBr — Na_2O — NaI — NaNO_3$
   b. $Al_2O_3 — MgO — KCl — KBr — KI$

10.13 Using your newly acquired knowledge of chemical bonding, why does acetone have a lower surface tension than water?

10.14 Calculate ΔG and the formation constants K_f (and $\log K_f$) for the following coordination reactions at 298 K and discuss the reasons for the differences. M = metal cation:

   a. $M^{2+}_{(aq)} + 4NH_{3(aq)} \rightarrow [M(NH_3)_4]^{2+}_{(aq)}$
      $\Delta H° = -53$ kJ·mol^{-1}, $\Delta S° = -42$ J·mol^{-1}·K^{-1}
   b. $M^{2+}_{(aq)} + 4CH_3–NH_{2(aq)} \rightarrow [M(CH_3–NH_2)_4]^{2+}_{(aq)}$
      $\Delta H° = -57$ kJ·mol^{-1}, $\Delta S° = -67$ J·mol^{-1}·K^{-1}
   c. $M^{2+}_{(aq)} + 2(NH_2–CH_2–CH_2–NH_2)_{(aq)} \rightarrow [M(NH_2–CH_2–CH_2–NH_2)_2]^{2+}_{(aq)}$
      $\Delta H° = -56$ kJ·mol^{-1}, $\Delta S° = 10$ J·mol^{-1}·K^{-1}

10.15 Calculate ΔG and the formation constant K_f (and $\log K_f$) for each of the following coordination reactions at 298 K in water and discuss the reasons for the differences. M = metal cation. How does the value for ethylenediamine (en) compare to the answers you derived in problem 11.3? What would account for the difference in the K_f between the metal chelated in problem 10.14?

   a. $[M(H_2O)_6]^{2+}_{(aq)} + 2NH_{3(aq)} \rightarrow [M(H_2O)_4(NH_3)_2]^{2+}_{(aq)} + 2H_2O$
      $\Delta H° = -46$ kJ·mol^{-1}, $\Delta S° = -8$ J·mol^{-1}·K^{-1}
   b. $[M(H_2O)_6]^{2+}_{(aq)} + (NH_2–CH_2–CH_2–NH_2)_{(aq)} \rightarrow [M(H_2O)_4(NH_2–CH_2–CH_2–NH_2)]^{2+}_{(aq)} + 2H_2O$
      $\Delta H° = -54$ kJ·mol^{-1}, $\Delta S° = 23$ J·mol^{-1}·K^{-1}

10.16 Answer the following questions with short sentences:

a. How do enzymes function? Give an example and describe the chemical processes (and bonding) involved.

b. What is a cofactor and what is its purpose? How is it bound? What is a coenzyme and what is its purpose. How is it bound? Give an example of each.

c. What is a heme group and how is it coordinated? What is the name of the generic type of bioorganic compound that coordinates Fe in living systems? How does it bind oxygen?

d. What is a vitamin? Please supply an example and describe its generic structure.

e. Of what is the tobacco mosaic virus composed? Is it a supramolecule? What kinds of bonds hold it together?

f. Would other metals besides magnesium function properly in the chlorophyll supramolecule? Why or why not?

10.17 Calculate the lattice energy of a KCl crystal by the Born–Haber method if $\Delta H^\circ_{f,KCl} = -438$ kJ·mol^{-1}, $\Delta H_{sub,K} = +89$ kJ·mol^{-1}, $\Delta H_{ion,K} = +425$ kJ·mol^{-1}, $\Delta H_{diss,Cl2} = +244$ kJ·mol^{-1}, $\Delta H_{e-aff,Cl} = +355$ kJ·mol^{-1}. How does the value compare to that found for NaCl? Why the difference?

10.18 In a nanotube bundle of uniform SWNTs, what force holds the bundle in place? What factors contribute to the integrity of the bundle?

10.19 Are all interactions between atoms and molecules electrostatic in nature?

10.20 We know that hydrogen bonds when taken collectively are quite strong. Is it possible to calculate the strength of H-bond from its ability to rupture strong containers when water freezes? How would you go about setting a value in this problem?

10.21 Hydrogen bonds are fundamental to life. Give as many examples as possible.

10.22 A linear amphiphile has a cross-sectional area of 50 Å^2. How much area of water surface, assuming tightest packing, would 8.0×10^{-2} mol of this species cover? How many moles would be required to cover Lake Michigan?

SUPRAMOLECULAR CHEMISTRY

Chemistry is an ever expanding "universe" at the microscopic level requiring mastery of the invisible ... at ever increasing levels of complexity.

NICHOLAS J. TURRO,
Columbia University, 2005

Chapter 11

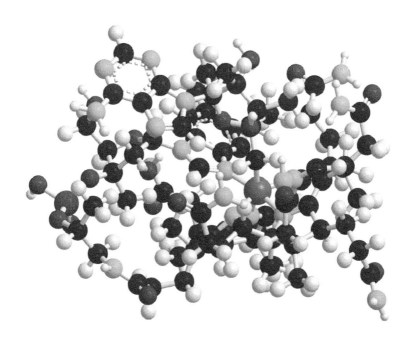

Fig. 11.0 *Nicholas Turro of Columbia University, renowned supramolecular chemist.*

Source: Image courtesy of Dr. Nicholas Turro.

THREADS

Chapter 11 is about chemistry and synthesis at the nanoscale. How different would you expect the subject matter in this chapter to be from any other found in traditional chemistry textbooks? Chemistry, after all, is just chemistry. Traditional chemistry has always involved some semblance of the nanoscale, the best example of which is given by colloid chemistry. Chemistry at the nanoscale involves every kind of traditional chemistry. As we have stated before, human beings have dabbled in the chemistry of nanoscale materials for the past millennia or so, but we have had limited success in controlling the size, shape and orientation of nanomaterials—atleast until now. In the latter part of the twentieth century to the present, major breakthroughs have been accomplished in that regard. The key ingredient is to keep nanomaterials from morphing into larger materials or decomposing into smaller ones.

Supramolecular chemistry is a new field. Biochemistry is an old field. Colloid chemistry is an old field. Chemical functionalization of nanomaterials is new because the substrates are relatively new. Polymerization is an old technique. Nanocomposites, quantum dots, and carbon nanotubes are new materials. C, N, H, O, F, S, and P are old materials.

Our understanding of chemical bonding has expanded significantly in the past few years, and we plan deliberately, consciously, subliminally, and subconsciously (hopefully, not unconsciously) to make reference to chemical bonding and intermolecular interactions throughout the chapter; it is unavoidable to do otherwise. Chapter 12 takes on more chemistry topics: cluster synthesis, self-assembled monolayers, template synthesis, and more chemical modification of nanomaterials.

We can never, especially at this stage, sidestep our analogies to nature and learn from this ultimate tutor. Nature has incorporated, by the most ingenious of means, the entire continuum of chemical bonding into its wondrous biological products with the utmost in efficiency and under the mildest synthesis conditions imaginable—just how perfect is that? From nature we continue to learn how this is accomplished so that we can construct our own version of macro-, super-, and supramolecules. In chapters 13 and 14, we explore nature's world in more detail. The closer we get to biological topics, the more relevant societal issues seem to become. Keep them in mind.

11.0 CHEMISTRY OF NANOMATERIALS

Every discipline of chemistry applies with equal validity to nanoscience. Elements of organic, inorganic, organometallic, coordination, physical, bio, surface and interfacial, colloidal, electro, and even cosmo chemistry have ties to nanoscience. This chapter is about chemistry of the nanoscale and focuses on molecules and synthesis and in particular supermolecules, supramolecules, and their design. The focus is not so much upon properties and phenomena as in previous chapters, although we do not pledge to stay clear of such topics. However, we have limited time and limited space and must settle for this precursory format.

Chemistry of the nanoscale is all of traditional chemistry plus chemistry that is conducted on nanomaterial synthesis and the subsequent chemical derivatization on nanomaterial substrates. The chemistry involved in building hierarchical structures is also presented and discussed briefly. Nanoscale chemistry is a convergence of disciplines at the molecular level that includes an infusion of biochemistry and chemical engineering to create designer molecules. This chapter could easily be expanded into additional chapters that address supramolecular chemistry in more detail, self-assembly, and chemical functionalization. Unfortunately, we must be rather succinct in our presentation.

We start our journey with a generalized introduction to nanoscale chemistry, list types of chemical synthesis, and discuss thermodynamic versus kinetic control of reactions. Then we tackle the burgeoning field of supramolecular chemistry. Supramolecular chemistry is based on complementary synthesis of thermodynamically stable compounds that are held together by relatively weak intermolecular interactions. We have just had a taste of intermolecular interactions in the previous chapter, and we understand, at least at the introductory level, how the different types of chemical bonds work together to form macromolecular systems.

First, supramolecular chemistry is the chemistry of the intermolecular chemical bond. Secondly, reactions are designed to synthesize supramolecules proceed via thermodynamic control rather than kinetic control, a distinction that leads to thermodynamically stable products. Lastly, factors such as solvent effects and entropic considerations fundamentally have greater impact on the course of a reaction and overall product stability. We have had a small dose of each of these topics in previous chapters. There is no question that this exciting new field has roots in biology and organic chemistry, and from that unifying principle, we shall proceed to take our first steps.

In section 11.3, we provide a few outstanding examples of hierarchical supramolecular fabrication to demonstrate the interplay between intramolecular and intermolecular bonding. Traditionally, the chemistry of self-assembly is a subset of supramolecular chemistry but one that is mostly restricted to two-dimensional synthesis strategies (e.g., one monolayer at a time). The synthesis of self-assembled monolayers, however, does not rely exclusively on intermolecular bonds and in that regard does not fall completely under the supramolecular umbrella. Chemical functionalization is addressed more fully in the next chapter. Chemical functionalization involves the chemical modification of nanomaterials or the application of nanomaterials as building blocks with which to form three-dimensional hierarchical structures.

11.0.1 Background

German chemist Friedrich Wöhler synthesized urea (carbonyl diamide $[CO(NH_2)_2]$) in 1828. This accomplishment was the first experimental debunking of the *Vital Force* school of thought by demonstrating that organic compounds can be synthesized from inorganic starting materials (e.g., that synthesis of biological molecules is no longer exclusive to nature). The beginnings of *molecular chemistry* are attributed to the work of Wöhler. *Macromolecular chemistry* is the chemistry of larger molecules and has its roots in the early part of the twentieth century. Hermann Staudinger received the Nobel Prize in chemistry for his discoveries in the new field of macromolecular chemistry. He also was fundamental in laying the foundation for polymer chemistry by stating that "polymers were comprised of long-chains of repeating molecular units."

In 1894, Emil Fischer, another Nobel laureate, proposed the rigid enzyme *lock and key* model for enzyme action. The *lock* in the analogy is the enzyme and the *key* represents the substrate, usually a small molecule. He also contributed to understanding of the synthesis of sugars and purines. Paul Ehrlich developed the first antibiotic and popularized the terms *chemotherapy, magic bullet,* and *blood–brain barrier* and predicted the term *autoimmunity.* He investigated the affinity of chemical substances for use against antigens and laid down much of the foundation of modern immunology. In 1908, Paul Ehrlich shared the Nobel Prize in medicine with Ilya I. Mechnikov for their work on immunity. In the same period, Alfred Werner invented the field of coordination chemistry, for which he won the Nobel Prize for chemistry in 1913.

Irving Langmuir and Katherine Blodgett, mentioned briefly in the introduction and sporadically throughout the text, developed the science of self-assembled monolayers in 1938. Not surprisingly, the process is named after them; the Langmuir–Blodgett process involves the synthesis of monomolecular layers on the surfaces of water, metal, or glass. Blodgett later went on to develop a coating on glass that consisted of 44 monolayers of barium stearate. The result was a glass that is 99% nonreflective, appropriately named "invisible glass"—one of the first deliberate applications of nanomaterials. The development of invisible glass helped to mitigate distortion in eyeglasses and other optical equipment. Katherine Blodgett also used interference colors to measure the thickness of each layer as it was applied. In retrospect, Katherine Blodgett should have been awarded a Nobel Prize for the breakthroughs she ingeniously delivered to the world of science.

James D. Watson, Francis H. C. Crick, and Rosalind Franklin uncovered the structure of DNA in 1953. Watson and Crick received the Nobel Prize for their work in 1962. Franklin did not win the prize due to her untimely death in 1958 of cancer, coupled with the policy that the Nobel Prize is not awarded posthumously. DNA (aside from its fundamental role in life) is perhaps the most celebrated and vitally important supramolecule in the brief history of the field. DNA has also become one of the most interesting and useful nanomaterials.

Supramolecular chemistry officially began in the second half of the twentieth century with the work of Woodward, Pederson, Cram, and Lehn. Robert Woodward was awarded the Nobel Prize in chemistry in 1965 for his work in organic synthesis. In 1987, Charles Pedersen (crown ethers), Donald Cram (host–guest chemistry), and Jean-Marie Lehn (cryptands) were awarded the Nobel Prize in chemistry for development of complexes that demonstrated high selectivity and structural specificity. They are considered to be the founders of modern supramolecular chemistry. More recently, Nicholas Turro of Columbia University is widely recognized as a pioneer in the field of supramolecular chemistry (a.k.a. host–guest chemistry) [1,2].

Although the existence of fullerenes and their derivatives (the carbon nanotubes) was unveiled in the latter part of the twentieth century, they cannot technically be considered as supramolecules, at least not according to our upcoming definition. Rather, these large molecular structures must be designated as *supermolecules*: big versions of the same repeating small molecular (or atomic) unit that are held together throughout with strong covalent bonds. Colloids, sodium chloride, and gold clusters are not supramolecules either. These three materials are relatively homogeneous and are held together with covalent, ionic, or metallic bonds, respectively. The distinction, however, is another boundary that we shall certainly violate.

Colloid and surface scientists were the first researchers to be actively involved in supramolecularity and self-assembly [3]. Whereas colloids are not considered to be *designed* (not a fair assessment according to some), supramolecular structures are designed to assemble in a specific way that yields unique structure and function [3]. Colloid chemists and surface scientists laid down much of the foundation for nanoscience and deserve all the credit that is due. It was their discovery and work that established much of the theory and practice (supramolecular chemistry, self-assembly, the Hamaker constant to nucleation, and other relevant phenomena) with which we now are so familiar.

Self-assembly is a phenomenon we that is self-explanatory. The phrase implies that things, components, constituents, molecules, etc. all come together spontaneously without the input of energy or design. Some input energy is usually required, of course, but it is expressed at new levels of subtlety. The driving energy may be sequestered in the surface of a nanoparticle, in the form of a molecular recognition couple or within an excited state of a molecule. Rest assured, it is there. From what we have gleaned in previous chapters, we know that the making and breaking of relatively weak intermolecular bonds are controlled by entropic trade-offs and small ΔH_{Rxn}, thus giving the appearance of self-assembly. The following statement embodies the importance of self-assembly to nanoscience [3]:

"Self-assemblies are not merely beautiful structures (although this is certainly a source of their appeal); they allow alternative solutions to problems that have hitherto been addressed only at the single-molecule level." This is certainly well stated. We now have developed new routes to synthesize materials that were essentially "unsynthesizable" in the past. These new routes go through the zone of nanoscience.

Chemical functionalization techniques are not new. Many, if not all, synthesis techniques to produce and modify nanomaterials have roots in inorganic, organic, polymer, colloid, and biochemistry. For example, carbon nanotubes, in addition to DNA, are some of the most remarkable materials known to nanoscience, or science in general for that matter. Neat carbon nanotubes without chemical modification do not enhance the mechanical properties of nanocomposites. Scientists had to find ways to incorporate them into polymer matrices and other technological materials without loss of their remarkable properties. Various chemical modification schemes have been developed to do just that. Attachment of proteins to solid surfaces and modifying targeted groups is the basis of biological sensor technology.

In chapters 5–8 we addressed physical properties and phenomena, thermodynamics, and chemical bonding without much fanfare about synthesis. It is time for synthesis, from the bottom up, to carry the baton of nanoscience in the forward direction. Bottom-up chemical routes are becoming increasingly practical as better understanding of the forces that drive the assembly of nanomaterials is acquired. Although top-down fabrication will always be important and bottom-up chemical methods are still in the minority, bottom-up chemical synthesis methods are gaining ground fast. Atoms and molecules can be added batch-wise to surfaces to stabilize nanoparticles and to create entities with highly specific structure and functionality. Nature does, after all, accomplish wondrous things in this way. The importance of chemical methods cannot be understated.

Supramolecular chemistry is chemistry "beyond the molecule" [3–6]. It is the chemistry of the intermolecular bond [3–6]. The term *supramolecular chemistry* was introduced in 1978 by Nobel laureate Jean-Marie Lehn [6]: "Just as there is a field of molecular chemistry based on the covalent bond, there is a field of supramolecular chemistry, the chemistry of molecular assemblies and of the intermolecular bond."

11.0.2 *Types of Chemical Synthesis*

Although every type of chemical synthesis approach applies with validity to nanomaterials, nanoscale synthesis must be considered to be a specialized field—one that takes seriously the effects of small dimensions and the importance of forces that are considered to be relatively weak on larger scales. Although catalysts, colored glasses, and photoemulsions have been studied and used in industrial processes for centuries, the conscious and deliberate use of nanomaterials as precursor materials, substrates, or moderators in reactions to synthesize new materials is a relatively new undertaking. Today, many routes to nanomaterials and hierarchical structures go through the supramolecule or nanomaterial phase. Just as with any table that involves a catalog, the divisions, although not entirely arbitrary, are susceptible to interpretation. For the purposes of the text, consider them to be pedagogically oriented and susceptible to change over time. **Table 11.1** is directed towards students who have not had the blessing of a rigorous background in chemistry.

The purpose of this table is to help to construct a perspective about the vast field of nanochemistry. It is also provided to illustrate the extreme interdisciplinary nature of nanoscience. As you proceed in this text, try to correlate the subject matter with one or more of the disciplines listed in the table. Once again, it was Nobel Prize winner Jean-Marie Lehn who said in 1993 [5]:

> Definitions have a clear, precise core but often fuzzy borders, where interpenetration between areas takes place. These fuzzy regions in fact play a positive role

TABLE 11.1	*Chemical Synthesis—All Bottom-Up Methods*	
Type of synthesis	**Discipline[a]**	**Comments**
Traditional synthesis-oriented fields of chemistry	Inorganic chemistry Coordination chemistry	Noncarbon chemistry. Basic coordination chemistry contributed to the founding of supramolecular chemistry. The use of ligands to stabilize metals also applies in nanoscience.
	Organometallic chemistry	Organometallic chemistry is a hybrid field. Supramolecular chemistry in nature and in the laboratory involve organometallic complexes that have functionality.
	Organic chemistry	The chemistry of carbon compounds. Synthesis of naturally occurring compounds by organic methods began over 100 years ago. For the most part, organic chemistry involves breaking and making covalent bonds and kinetic control of the reaction. Organic chemical synthesis is used to make precursors to supramolecular species.
	Polymer chemistry	Polymer chemistry is an offshoot of organic chemistry. It is the use of monomeric constituents to form a bulk material with specific properties. Nanoscale polymeric materials contribute to drug delivery systems. The field of nanocomposites is an offshoot of polymer chemistry.
	Biochemistry	Bottom-up synthesis with roots in biology is the primary focus of biomimetic technology. Supramolecular chemistry owes much of its existence to biochemistry and models many of its reactions on biochemical phenomena.
Specialized fields of traditional chemistry	Colloid chemistry	From several nanometers to several microns, colloids occupy a broad stage in science and nanoscience. Concepts of surface area and surface energy were first developed in colloid chemistry. Colloid chemists do not get the credit that is due them.
	Surface science	Surface science is the study of the physical chemistry of surfaces and interfaces, primarily those involving a solid–liquid or solid–gas interface. Surface science of nanomaterials is a new field.
	Catalysis	Although micro- and nanoscale catalysts have been on the scene for decades (centuries), the use of surfactant-stabilized catalysts of extremely small size is a relatively new development for hydrogenation and oxidation reactions. Materials not thought to be good catalysts, such as gold, have proven to be otherwise.
Recently developed fields of chemistry	Supramolecular chemistry	Supramolecular chemistry is the convergence of organic, inorganic, polymer, and biochemistries. It is the chemistry of more than one molecule—chemistry beyond the atom or molecule. It is the chemistry of the intermolecular bond.
	Metallo-supramolecular chemistry	Developed in the 1990s, metallo-supramolecular chemistry involves metal-directed synthesis and metal-directed self-assembly in the formation of discrete to infinite structures.
	Chemical functionalization of nanomaterials	Nanomaterials are the starting materials. Chemical modification of nanomaterials offers stability and function. Due to the high surface energy of nanomaterials, reactivity is the rule rather than the exception. Functionalization allows for carbon nanotubes to be inserted into polymers and in the stabilization of nanometals.
	Template synthesis	Kinetic confinement of nanoscale reactions to directly produce nanomaterials that are confined in small spaces. The use of porous alumina membranes, zeolites, block copolymers, and micelles to form nanomaterials that would be impossible to form by other methods is gaining momentum.
	Physical–chemical synthesis methods	ALD, MBE, and CVD are physical–chemical methods to form nanomaterials—for the most part, 2-D nanomaterials, although 0-D and 1-D materials are also possible by sputtering, evaporation, and templates.

(*continued*)

TABLE 11.1 (CONTD.)	*Chemical Synthesis—All Bottom-Up Methods*	
Type of synthesis	**Discipline[a]**	**Comments**
	Nanomaterial chemistry[b]	Use of catalysts, colloids, carbon nanotubes, or other nanomaterials to form new nanomaterials with greater complexity, hierarchical nanostructured materials, or bulk materials with nanoscale components
	Nanochemical engineering	Serious challenges face the scale up of nanomaterials to industrial proportions. Experiment repeatability from lab to lab let alone to the production line has not been realized in bottom-up methods. The interdisciplinary nature of nanotechnology offers its own set of issues.
	Chemistry in nanomedicine	The pharmaceutical industry has always had a chemical, supramolecular basis. Now more than ever, nanomaterials have made inroads into medical imaging, cancer therapy, structural prosthetics, Alzheimer's disease mitigation, and numerous other applications.
Biochemical nanoscience	Bionanotechnology	Biomimetics and bionano are already big industries. The pharma industry has been nano for decades. Trying to mimic nature's incredible arsenal of nanoscale materials and phenomena is burgeoning.

[a] There are many more subdisciplines, of course.
[b] Nanomaterials chemistry in the context of this table implies the use of nanoscale materials as a starting material or substrate in a chemical reaction or process.

since it is often there that mutual fertilization between areas may occur. This is certainly also true for the case at hand, the case of supramolecular chemistry and its language.

11.0.3 Thermodynamic versus Kinetic Control and Selectivity

There are two general types of chemical synthesis: (1) those that are conducted under kinetic control conditions and (2) those that are conducted under thermodynamic control conditions. Once again we create a boundary for the purposes of understanding. In actuality, both factors are always present in any reaction and in every kind of synthesis. The important aspect to remember is that we are able to manipulate experimental conditions to favor a preferred product whether it is the kinetic product or whether it is the thermodynamic product.

According to IUPAC (the International Union of Pure and Applied Chemistry), thermodynamic control of product composition indicates that "… conditions that lead to reaction products in a proportion governed by the equilibrium constant for their interconversion of reaction intermediates formed in or after the rate-limiting step." And, according to IUPAC, kinetic control of product composition indicates that "… conditions (including reaction times) that lead to reaction products in a proportion governed by the relative rates of the parallel (forward) reactions in which the products are formed, rather than by the respective overall equilibrium constants."

Table 11.2 lists some of the differences between the two aspects of chemical reaction control. The universal time-honored and often used example of

TABLE 11.2	*Thermodynamic versus Kinetic Control*	
Parameter	**Kinetic control**	**Thermodynamic control**
Field of use	Organic synthesis	Supramolecular synthesis
General descriptor	Rate of product formation	Relative stability of products
Strength of bonds	Strong	Strong or weak
Types of bonds	Covalent, ionic, metallic; the "big three"	Interactions: covalent + hydrogen bonding, van der Waals, hydrophobic, $\pi-\pi$
Driving force	Product with lowest activation energy (competing rates of product formation) Reverse reactions not favored	Formation of thermodynamically stable products (equilibrium thermodynamics) Reverse reactions are possible (if allowed)
Thermodynamic parameters	Although entropy and enthalpy are important, kinetic methods can be made to overcome thermodynamic disadvantages	Entropy and enthalpy important in nanoscale chemistry and are used to drive reactions
Reversibility	Generally irreversible for kinetic product to predominate	Generally reversible Thermodynamic product predominates
Equilibrium constant	The equilibrium constant is the same for both. Product proportion governed by kinetic factors ("Le Chatalier's principle" manipulation, confinement, and other kinetic tricks)	Reaction product proportion governed by equilibrium constant
Types of reactants	Covalent, ionic, and metallic substrates	The same three plus supramolecules and nanomaterials
Stability of intermediates	More stable transition state with lower activation energy resulting in faster reactions	Less stable transition states with higher activation energy in covalent reactions. Self-assembly process occurs at low temperature to form thermodynamic products.
Stability of products	Products energetically less stable Products are kinetically stable (no path available for decomposition or reaction)	Products energetically more stable
Purity of products	Kinetic products as a rule require purification (there are exceptions, of course)	Thermodynamic nanomaterials are monodisperse and pure.
Number of products	Numerous products are possible	Thermodynamically favored (usually one product favored)
Molecular recognition	Molecular recognition requires a long gestation period. Kinetic reactions by definition usually occur quickly	Allows manipulation of free energy of product. Desired compound will dominate the mixture at the expense of undesired products
Temperature	Favors reaction path of lowest activation energy—hence lowest temperature	Reactions proceed at higher temperatures in covalent chemistry; low temperature (over long periods of time) in intermolecular domain.
Biology	Interestingly, kinetic reactions are prominent in the biological world—living things' need for reactions to happen quickly. Energy to overcome activation barriers provided from external sources	Thermodynamic factors were probably more important in the prebiotic world. Underlying thermodynamic self-assembly and other reactions are governed by kinetic factors.

thermodynamic versus kinetic control is the addition of hydrogen bromide (HBr) to the double bond of butadiene. The 1,2-addition to form 3-bromo-1-butene (kinetic product) is faster than the 1,4-addition to form 1-bromo-2-butene (thermodynamic product) (**Fig. 11.1a**). The reaction coordinate for both types is depicted in **Figure 11.1b**.

Fig. 11.1

(a) The addition of HBr to the alkene 1,3-butadiene is pictured. Two products are formed: On the top, the kinetic product is formed faster by the 1,2-addition; on the bottom, the thermodynamic product is formed by the slower (thermodynamically stable) 1,4-addition. 1-Bromo-2-butene is the thermodynamically favored product (blue curve) with higher activation energy and overall more negative ΔG. (b) Graphical comparison of thermodynamic versus kinetic control of a reaction. Kinetic reactions are essentially irreversible and are controlled by the rates of formation of constituents. The transition state has lower activation energy. Thermodynamic control implies that equilibrium conditions are in place. Right: The universal example of kinetic versus thermodynamic control is depicted. It is the reaction of 1,3-butadiene with HBr. The thermodynamic product 1-bromo-2-butene is depicted in red. The overall change in free energy is depicted on the right. The thermodynamic product is expected to be more stable than the kinetic product. Left: The transition state, K, is formed at a lower energy than the transition state, T, of the thermodynamic product.

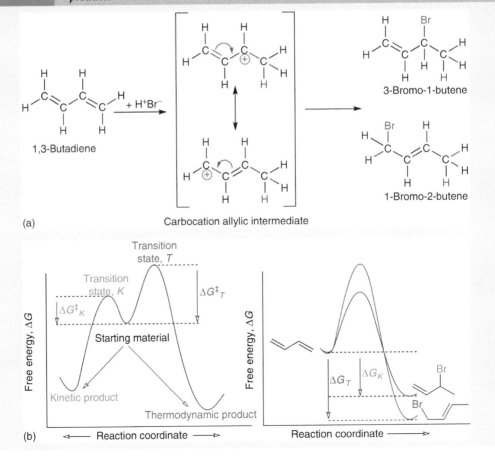

At low temperatures, the less stable kinetic product is favored because the intermediate in the reaction goes through a pathway that requires lower energy (e.g., lower activation energy). At elevated temperatures, there is enough energy available for the reaction to proceed through another intermediate: the one with higher activation energy that leads to the thermodynamically stable product. With enough time or if the temperature is raised, the kinetic product is transformed eventually into the thermodynamic product. Therefore, at higher temperatures and longer reaction times, the thermodynamic product is favored. The example of the transformation of diamond into graphite illustrates this process. There is a distinction to be made at this point between reactions like the one discussed in **Figure 11.1** and those that we consider for supramolecular chemistry. Specifically, in the HBr reaction, strong covalent bonds (intramolecular) are made and broken. In supramolecular chemistry, weak intermolecular bonds are made and broken and they fall mostly under the umbrella of thermodynamics.

A phenomenon called the *template effect* is a good example of a kinetic parameter that is able to influence the outcome of a reaction by control of the steric pathway. Templates in general exert kinetic control over a chemical reaction. In actuality, the template effect was fundamental in giving supramolecular chemistry its start; supramolecular chemistry may never have existed had it not been for the participation of some good kinetically based luck [7]. A kinetic pathway due to the template effect was responsible for the formation of the historic crown ether, dibenzol[18]crown-6, synthesized by Nobel laureate Charles Cram. Although thermodynamically less stable, the crown ether was favored over the thermodynamically more stable polycondensation product (**Fig. 11.2**) [7].

The *kinetic template effect* in this historic reaction was due to the ability of the K^+ cation to organize the assembly of a stable intermediate into an octahedral geometry. By this way, the templating action of the potassium ion was able to place reactive moieties on either end of the precursor molecule into close proximity [7]. Often you will here how a nanomaterial is *metastable* or *kinetically stable*. These terms imply that the material is not thermodynamically stable but that it is able to exist nonetheless. Kinetically metastable materials will exist when: (1) there are no other molecules around with which to react; (2) the environmental conditions are able to support their stability (e.g., low temperature); (3) the space remains confined; or (4) it becomes chemically stabilized by another reaction (e.g., a shell of ligands). The template effect witnessed in confined spaces is able to bring about reactions that normally would not occur in other environs, somewhat akin to catalytic processes that serve to reduce the activation energy of a reaction by forming kinetically stable intermediates.

In the case of the *thermodynamic template effect*, the reaction equilibrium shifts to favor a metal-stabilized product with high yield [8]. Without the presence of the templating metal, other products would participate in the product mixture. The metal then is able to exclude other reactions from occurring by forming a thermodynamically stable product. A metal cation's ability to "select" the best ligand to form an energetically favored complex in a reaction mixture alters the equilibrium in favor of the thermodynamic product [7].

Thermodynamic factors play a major role in selectivity, molecular recognition, self-assembly, and self-replication. Several processes consist of numerous steps involving both thermodynamic and kinetic factors. For example, thermodynamic

FIG. 11.2   *Synthesis of the crown ether [18] crown-6 by the template action of a potassium cation. In the presence of KOH, the oxygen atoms of the precursors form dative bonds with the potassium cation. Notice that in this configuration, the distal ends of the linear molecules are placed within close proximity of each other—a by-product of the kinetic templating effect of the metal cation [7]. In the presence of an organic base, no direction occurs and the polymer condensate consisting of precursors is formed. The kinetic pathway leads to the formation of one of the first crown ethers. Conducting this reaction in the condition of high dilution enhances the formation of the macrocycle rather than the polymer.*

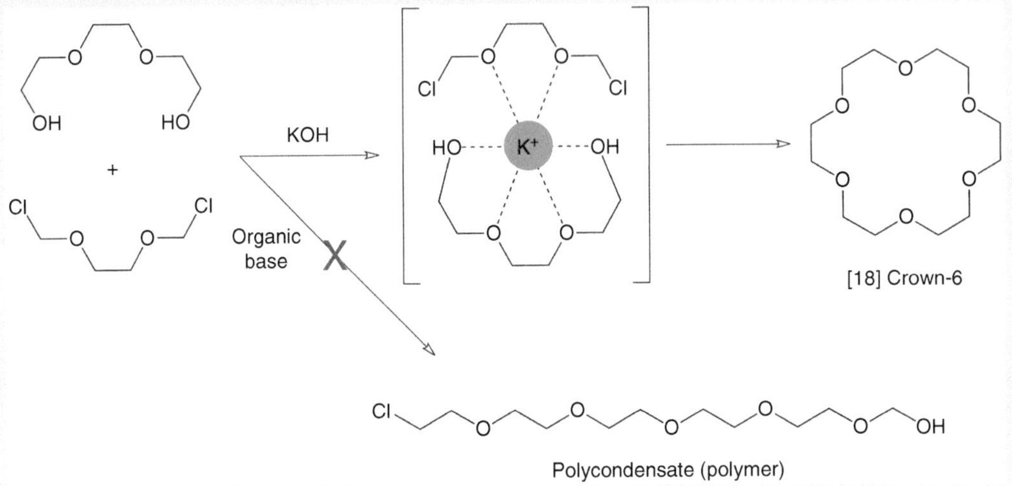

self-assembly may initially involve $\pi-\pi$ stacking, while subsequent covalent reactions involve kinetic parameters [7]. Another example involves hydrogen bonding—an intermolecular attraction characteristic of thermodynamic self-replicating systems—to form precursor molecules that are covalently modified later in the process. The action of enzymes involves both thermodynamic and kinetic controls. A potential substrate is lured into an active pocket by means of intermolecular factors and then modified under kinetic conditions to form a completely new molecule. Ripping a hydrogen atom from a substrate during an oxidation process taking place in the active pocket of an enzyme requires the kinetic confinement of the substrate and the application of strain to the bond in question. Once all factors are in place, the hydrogen is removed under physiological conditions.

11.0.4   *Introduction to Supramolecular Design*

Selectivity in the host–guest process is an important aspect of supramolecular chemistry and other chemistry of the nanoscale. As chemists, we design reactions in ways that favor specific products—preferably in high yield. Such prudent strategy saves time and expense in necessary post facto purification, further processing, and proper functioning. Selectivity is highly desired in any chemical reaction, whether it is chromatography, polymerization, or enzyme–substrate interactions. With regard to supramolecular chemistry (and biochemistry), selectivity is the ability of a potential host molecule to distinguish between and among potential guest substrates and be able to bind that guest molecule for some predetermined length of time. The host must prefer a certain substrate or class of substrates over others in order to accomplish any meaningful function.

Distinguishing between a structural (static) element and a functional (dynamic) element is important to any kind of molecular design. For example, if one plans to design a structural element that fits into a larger supramolecular system, it would be prudent to consider a reaction that forms a complex with a high binding constant. On the other hand, if the purpose of your supramolecular system is more centered on chemical processing, such as turning over a substrate in an enzymatic system, a prohibitively high binding constant may not be what you want. The rate of complexation of a potential guest influences the outcome of a host–guest relationship. Biological enzymes, for example, depend on kinetically selective conditions. The formation of an enzyme–substrate complex that has a prohibitively large binding constant would slow down the rate of the primary reaction, which is certainly not desirable in metabolic processes [7].

The ability to accomplish a chemical task at a facilitative, moderate, or deliberate pace relies on both the kinetic and thermodynamic parameters of the chemical process. Depending on the application, whether it has a metabolic, catalytic, or structural mission, the timescale of the process, its product quality, its yield and rate of turnover mean the difference in investment of time, energy, capital, and personnel. In the case of nature, life and death itself depend on host–guest relationship parameters. The best example of the criticality involved in such metabolic processes is the binding of oxygen by the heme complex (e.g., life based on carbon monoxide simply would not work under the current metabolic hierarchy).

11.0.5 The Nano Perspective

The material presented in this chapter falls under the category of *soft matter* as opposed to hard matter that consists of, for example, minerals and metals. Soft matter lacks long-range crystalline order and is held together by intermolecular bonds. Members of the class of soft matter include living tissue, polymers, micelles, colloids, amphiphiles, liquid crystals, gels, and supramolecules [9,10]. The strength of many kinds of intermolecular attractions between and among components of soft matter is on the order of magnitude of $k_B T$ where k_B is the Boltzmann constant. All of the materials considered to be soft matter get their start at the nanoscale. A comparison between soft and hard matter is given in **Table 11.3**. However, hard matter makes great for substrate materials upon which soft matter can be manipulated.

We spoke briefly about the threshold between single- and multicellular organisms in chapter 5 and what roles increased surface area and size may have contributed to crossing that threshold. Before the existence of eukaryotic cells there was the prokaryote and before that the protocell—a cell that is not considered to be life per se but possesses some of the rudimentary characters of living things. The protocell is thought to have existed within a thermodynamic envelope that was separate from its environment and within which it was able to replicate itself (grow?) by exploiting nutrients and energy sources that existed outside its

TABLE 11.3	*Soft Matter versus Hard Matter*	
Parameter	**Soft matter**	**Hard matter**
Examples	Assembly of large molecules [9]: micelles, colloids, amphiphiles, liquid crystals, self-assembled monolayers, polymers, gels, sols, supramolecules, biological materials, composites	Assembly of atoms and small molecules [9]: metals, minerals, ionic solids, semiconductors, ceramics, composites (concrete), ice, graphite, carbon nanotubes, quantum dots, and fibers
Bonding and energy	Intermolecular interactions on the order of $1–10\ k_BT$ Energy proportional to k_BT; entropy measured in units of k_B	Intramolecular interactions of covalent, metal, and ionic bonds Energy $\gg k_BT$
Structure	Micro- and nanostructure between 1 nm and 1 μm. Although nanomaterials self-purify, long-range order is absent in 2-D soft matter nanomaterials. Order is achieved on the mesoscopic scale	Contains micro- and nanostructure but may be homogeneous on the bulk scale Crystalline materials have long-range order.
Reaction control	Thermodynamic and kinetic	Thermodynamic and kinetic
Behavior	Viscoelastic	Rigid structures prevalent

Sources: J. G. E. M. Fraaijie, Soft matter. Presentation ISM-Series, University of Leiden, wwwchem.leidenuniv.nl/scm/Presentations/ISM/ISM09/ISM09.pdf (viewed 2007); I. W. Hamley, *Introduction to soft matter*, John Wiley & Sons, Ltd., Chichester (2000).

envelope [11]. The protocell required a minimal lipid structure to form an exclusion membrane and the vesicles within (a.k.a. the liposomes). These structures must have formed spontaneously (negative ΔG) via self-assembly by the action of large molecules (amphiphiles) that have hydrophobic and hydrophilic ends. How fortuitous was this scenario? Transfer of nutrients and energy perhaps first occurred in such structures.

The theory of Morowitz, Heinz, and Deamer states that self-assembled membrane vesicles made of rare organic amphiphiles were responsible for the beginnings of life. A prebiotic foundation for the photosynthetic process was most likely established by an amphiphilic pigment molecule that became incorporated into a lipid bilayer [11]. It is not as far-fetched as one might think to imagine a system in which a source of energy was spontaneously incorporated into a system that required energy, thus developing a relationship between structure and function in living things. For example, why would any prebiotic structure require energy in the first place? What driving force is there that requires materials to seek energy? Is the answer "to make more of themselves"? we then ask, why make "more of themselves" in the first place?

Proteinoid microspheres (protein, from the Greek *protos*, "first quality"), formed from a mixture of amino acids at ca.150°C, have been shown to spontaneously assemble into spherical structures (in the laboratory) [12]. This work was based on Stanley L. Miller's landmark paper in 1953 that described the synthesis of primitive α-amino acids under prebiotic conditions [13,14]. DNA and RNA are also supersized molecules that have built-in self-assembly mechanisms. Initially, there may have been a non-DNA–RNA means to transfer information—a fact contrary to the "central dogma" of biology. Prions, for example, are considered (arguably) to replicate (reproduce?) without the aid of nucleic acids. Prions have the ability to influence (infect?) the normal form of the amyloid protein with the identical amino acid sequence by initiating refolding into an abnormal shape that eventually sets up a chain reaction (self-assembly?) [15–17]. Interestingly, prions are able to resist proteases better than normal counterparts as well as higher levels of heat and radiation—conditions that may have existed in early Earth history. Prions may be the most primitive form of protein. Certainly these are interesting aspects to ponder. Prions have been determined to be the active agent in the cause of Alzheimer's disease.

The message implied in the previous paragraph is that supramolecular nanoscale materials were made of natural inorganic and organic elements and not necessarily from biological sources. Later, biological forces were able to exploit processes that require little input of energy, such as the molecular self-assembly process, to form supramolecular molecules and structures. Whether DNA came first or some kind of prion preceded the advent of DNA is a debate better left to *paleobiologists, paleochemists,* and perhaps nanoscientists to sort out in coming years. The rest is history and we anticipate many exciting developments in this extremely relevant field in the very near future. The *nano perspective* of supramolecular chemistry is one of great size and breadth. It is a field that will render unimaginable impact upon our technological world. The contributions of pharmaceuticals have already improved the quality of life for humans. What wonders will we be able to contribute to civilization and with equal importance—give back to the natural world?

11.1　Supramolecular Chemistry

Supramolecular chemistry (from the Latin *supra,* "above, beyond," + chemistry) has its roots in organic, colloid, coordination, polymer, and biochemistries. It is defined in a general sense as any chemistry that addresses synthesis "beyond that of the atom or molecule" and is widely known as the "chemistry of the intermolecular bond" [18–20]. The study of materials that have two or more molecules, in the narrowest sense, conforms to this definition. This relatively new field of chemistry involves design, synthesis, and characterization of large molecules. It is the chemistry of molecular assemblies and the intermolecular bond—a generalized coordination chemistry that goes beyond the simple binding by ligands of transition metals. An important aspect of supramolecular chemistry is that molecules that are part of a supramolecular complex retain their original structure without too much modification, although their chemical nature may be modified.

11.1.1　The Host–Guest Relationship

Supramolecular chemistry is indeed very much about the *lock and key* (or guest–host) hypothesis first proposed by Emil Fischer in 1894. Within the context of nanoscience, supramolecular chemistry focuses on noncovalent interactions between molecules. Fabricating a *supermolecule,* on the other hand, is akin to making the same molecule that is held together with strong bonds simply bigger. Many nanomaterials are supermolecules. Many kinds of nanomaterials are supramolecules. This is not to say that nanogold has different properties apart from the bulk and atomic forms because it certainly does; it just is not due to supramolecular chemistry. Another strongly bonded material, NaCl, is capable of extremely primitive "atomic recognition" and "self-assembly," but it is not the same self-assembly characteristic of soft-matter materials. Carbon nanotubes are precursors in supramolecular chemical reactions, but are not considered to be supramolecules, although they are certainly supermolecules. In the carbon nanotube, there are no precursor molecules, just carbon atoms that are covalently bonded (very strongly) to form a very long molecule—a supermolecule. Although nomenclature and categories are convenient tools, distinguishing between terms *supra* and *super* should not provide any cause for consternation.

Supramolecular chemistry is focused on the relationship between a host molecule and a potential guest atom, molecule, or substrate. **Figure 11.3** describes in minimal detail the fundamental concept of supramolecular chemistry: the host–guest relationship.

Aside from the obvious conclusion that supramolecular structures can be rather large, each molecule has the ability to accommodate other complementary pairs until the precursor material is exhausted—a version of the *ying–yang* scenario of the molecular world. This general principle forms the basis for

FIG. 11.3

A hypothetical host–guest interaction leading to a supramolecular structure is displayed. The "guest" hexagons (blue) "fit" perfectly in the grooves formed by a complementary "host" structure (red). The process continues as long as material transport is unhindered, precursor supply is abundant, or the complex remains soluble. Structures such as the one depicted come together due to two factors: molecular recognition and molecular self-assembly. Host–guest chemistry is about weak intermolecular forces that hold a guest and host together. Host and guest molecules, individually, are held together by covalent bonds. The bond to the substrate is often of covalent character. Each individually is a molecule. Together they comprise a supramolecule.

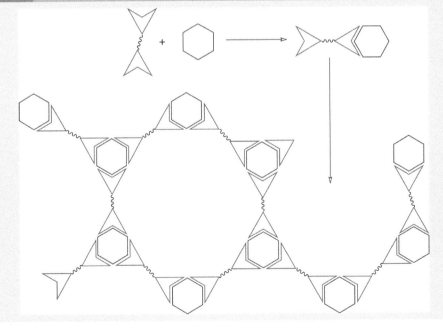

coordination compounds (that involve a ligand and a metal), the enzyme–substrate specificity, and the antibody–antigen relationship that is fundamental to the immune system. *Complementarity* is perhaps the best term to describe the host–guest concept [8]. Hosts can be biological, organic, or inorganic; guests are usually smaller entities like ions, metals, and neutral chemicals. Hosts can be anionic, cationic, or neutral or combinations of anionic–neutral or cationic–neutral and even zwitterionic. Substrates are also found in anionic, cationic or neutral states.

11.1.2   Molecular Recognition

There is no better way to introduce a subject than by quoting one of the founders of the field—Jean-Marie Lehn:

> Molecular recognition is defined by the energy and the information involved in binding and selection of a substrate(s) by a given receptor molecule; it may also involve a specific function. Mere binding is not recognition, although it is often taken as such. One may say that recognition is binding with a purpose. It implies a pattern recognition process through a structurally well-defined set of intermolecular interactions. Binding of σ to ρ forms a complex or supermolecule characterized by its (thermodynamic and kinetic) stability and selectivity, i.e., by the amount of energy and of information brought into operation.

According to this quotation, we understand that there is something special about this field, above and beyond the normal boundaries of chemistry with which we are so familiar. Supramolecular chemistry implies purpose, design, and engineering, as well as an inherent beauty. As a case in point, DNA, the premier supramolecule, is all about information—a molecule with a purpose. It is no wonder with the advent and development of supramolecular chemistry that we are ready to knock on nature's door itself—ready to insert a key, albeit still on the small scale, into its mysterious lock. "Binding with a purpose!" Molecular recognition, after all, provides the foundation for replication mechanisms.

Molecular recognition represents a process that defines specific interactions between two or more molecules (or supramolecules) by means of noncovalent forces. Host–guest chemistry, and hence supramolecular chemistry, in general is based on molecular recognition. However, recognition in its broadest sense is responsible for colligative properties such as melting point, freezing point, and boiling point—relationships that arise between and among atomic or molecular constituents of a material. Atoms are able to recognize each other and bond with each other "knowing" that the minimal energy state will result from such binding. But there is no information or function associated with such colligative properties, only "purposeless" stochastic interactions that are driven towards a state of minimal energy. Supramolecular chemistry, in another broad sense, is chemistry with a purpose.

Moving on from sodium chloride, the *fitting* together of molecules implies a higher level of recognition called molecular recognition. Is *information* exchanged in all molecular recognition processes? If we label one bonding episode as "1" and a nonbonding episode as "0," then our answer is in the affirmative. It would be a stretch, however, to claim that any sort of information exchange is actually involved. We know that DNA is capable of undergoing supramolecular interactions that code for information transfer processes. Although no information may be exchanged per se in many supramolecular reactions, the foundations for information exchange are there to be exploited when the conditions are appropriately defined.

Supramolecules come together to form larger entities with the same goal that drives metals to agglomerate: achieving that state of lower energy. Because supramolecular reactions involve intermolecular interactions, the reactions are reversible. Reversibility (a condition characterized by small reaction constants, K_{rxn}) is a fundamental property of biological phenomena. If RNA coding were not reversible, then there would be no protein synthesis, at least in the way with which we are familiar.

The aim of early molecular recognition processes was to bind metal cations [21]. C. J. Pedersen in 1967 helped to launch the science of molecular recognition after his synthesis of crown ethers [22,23]. Crown ethers are able to selectively bind alkali metals [22,23]. The binding of metals by chelates is a generalized form of molecular recognition within the general bounds of the definition given in this chapter. EDTA, for example, is able to bind many kinds of metals. Is this a case of molecular recognition? The answer is, "Yes, it is, but in a less specific way." Other chelating agents offer better specificity. Iron is coordinated in heme by a porphyrin ring to form a near planar complex. Is the binding of iron by porphyrin a true case of molecular recognition or more like recognition by a molecular species of a cation?

The biological realm offers numerous examples of specific molecular interactions. DNA, RNA, structural proteins, antibodies, cellular receptors, and enzymes are a few examples of materials that combine by the action of noncovalent bonds. This is not to say that DNA, RNA, and proteins are without covalent bonds. Covalent bonds are well represented in the backbone structures of each material. However, higher order noncovalent interactions also occur between DNA and protein (histones), RNA and ribosomes, sugar and lectin, antibody and antigen, and between magnesium or iron and porphyrins (chlorophyll and heme, respectively). Molecular recognition forms the fundamental basis of immunology and of life itself.

Taking back the clock to the timeless period of prebiotic protocellular matter, insipient stages of molecular recognition must have been linked, albeit tenuously, to some specific function or just some statistically based process that found its way to achieve the state of lowest energy and stability. The purpose of life is to replicate. The function of molecular recognition is to make life possible. Just think for a moment about the attraction that results in romance and the chemistry in relationships—in actuality nothing more than a glorified version of molecular recognition with the underlying end goal of replication.

Types of Molecular Recognition.   There are two basic kinds of molecular recognition. *Static* molecular recognition is similar to the *lock and key* hypothesis proposed by Emil Fischer in the later nineteenth century. It involves complexation between one host and one substrate: a one-on-one relationship. *Dynamic molecular recognition*, based on dynamic equilibrium, is more complex and involves more hosts (or sites) and more guests. In one host, several binding sites may exist simultaneously. If a guest is bound by one site, the thermodynamic propensity to associate by another site may be increased (*positive allosteric effect*) or decreased (*negative allosteric effect*). One of the consequences of dynamic molecular recognition is the capability to regulate substrate binding, a useful characteristic of sensors [24,25]. Unless we are dealing with an isolated host and guest, all integrated supramolecular systems of two or more molecules exert influence on one another and the whole.

Preorganization and Complementarity.   Right along with the terms molecular recognition, host–guest, and lock and key are the terms *preorganization* and *complementarity*. Preorganization is an engineering term—a design term involved in the selection of host and guest so that a thermodynamically stable product is formed during a process. The term complementarity describes a state in which one component complements the structure, the electronic character, and function of another component. It is the "hand-in-glove" term fundamental to supramolecular chemical compounds and synthetic procedures. Complementarity in physics implies that the particle–wave duality of the electron explains certain phenomena associated with electrons. Complementarity in chemistry means a good fit. The best example of a good fit is that between an enzyme and its substrate. The most commonly used example is that of the enzyme carboxypeptidase-A (CPA) (**Fig. 11.4**). We shall conform to the trend and use it ourselves to illustrate this important point.

CPA is a peptidyl-L-amino-acid hydrolase (metalloexopeptidase) that exists in the small intestine and is made up of 307 amino acid residues. Its job (function, purpose) is to cleave the terminal amino acid from the carboxyl terminal (C-terminus) with the assistance of the metal cofactor zinc. A single activated water molecule and the zinc dication work together to hydrolyze the peptide bond. The hydrophobic pocket is there to provide the proper chemical environment for groups such as the phenyl depicted in **Figure 11.4.** The *active site* of CPA pictured in **Figure 11.4** is a static version frozen in time. The process is actually quite dynamic. CPA is formed in the pancreas and once in the small intestine it is activated by trypsin, another peptidase, to open its active site. The entire process becomes even more remarkable when the turnover rate is considered (in the range of 100–1000 substrates per second).

The DNA double helix is sine qua non when it comes to biological complementarity—a complementarity based on hydrogen bonds between heterocyclic nucleotide base purines (adenine and guanine) and pyrimidines (cytosine and thymine) (**Fig. 11.5a**). The backbone of the DNA consists of covalent-bonded phosphate and deoxyribose sugar molecules that provide the framework (substrate) for linked nucleotides. The primary structure is composed of a nucleotide base-pair sequence that is able to code information in groups of three that consist of various permutations of the four base pairs adenine, guanine, cytosine, and thymine (ACGT). "A" only binds with "T" and "C" only binds with "G." The coupled base pair trios run up and down the length of the double helix to form a DNA strand that makes up the chromosome—indeed, a hierarchical extension of complementarity. Informational exchange is linked to the order of the base pairs. Information coded by DNA exists at a very high and sophisticated level of complementarity.

Therefore, beginning with a simple complementary molecular recognition system consisting of mainly hydrogen bonds, an intricate information data system is constructed. Overall, complementary hydrogen bonding, enthalpic π–π stacking interactions, and entropy-driven hydrophobic interactions

FIG. 11.4

A schematic rendition of carboxypeptidase-A (CPA) and substrate (in blue). The primary function of the enzyme carboxypeptidase-A (CPA) is to hydrolyze the carboxy-terminal (C-terminus) peptide bond of proteins. By this action, the last amino acid in a peptide is removed. The interplay between and among hydrophobic pocket, hydrogen bonds, salt bridge, anionic and cationic moieties, metal cofactor, and a water molecule is certainly mind boggling. The action of CPA is considered to be nonselective (e.g., it acts on nearly all C-terminus peptides). The products of the reaction depicted in the figure are the amino acid phenylalanine and the rest of the protein. The squiggly red line indicates the site where hydrolysis of the peptide bond takes place. Carboxypeptidases A_1, A_2, and B are made in the pancreases and are involved in the breakdown of food stuffs in the small intestine. CPs also assist in the synthesis of other proteins like insulin and many other processes. CPA is a metallo-CP. CPs that prefer amino acids that contain phenyl groups (shown here) or branched hydrocarbons are know as CPA (where A stands for aromatic or aliphatic). CPA is converted into its active form by another enzyme called enteropeptidase. By this mechanism, the digestion of CPA is prevented.

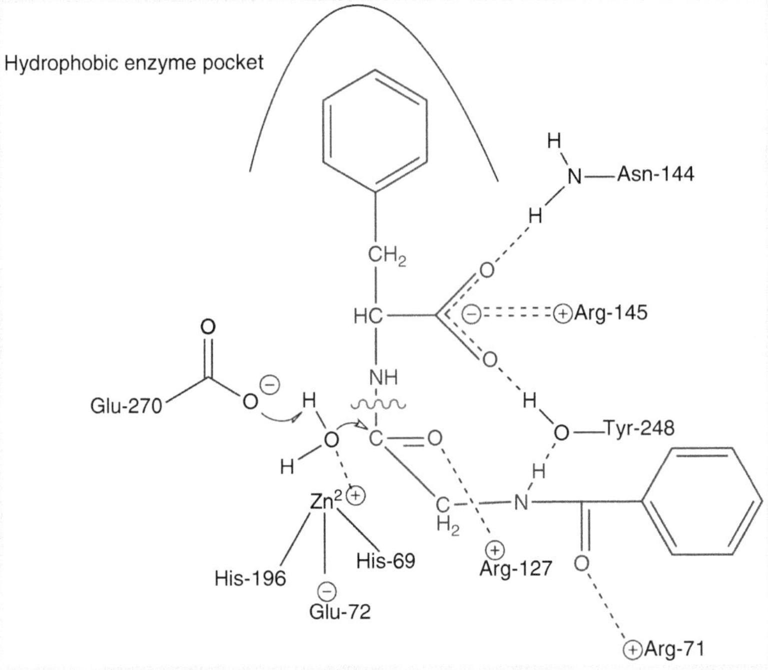

contribute to the self-assembly of DNA into the α-helix configuration (**Fig. 11.5b**) [26].

Molecular Recognition in Nature. Metal binding by macrocyclic biological molecules is another type of molecular recognition exhibited by living systems. This kind of binding is classified as molecular recognition between a metal cation and an organic chemical. It is the least complex form of molecular recognition. Potassium in the form of a monovalent cation electrolyte is responsible for maintaining the membrane potential of cells and also serves as a cofactor for some enzymes (pyruvate kinase, Na–K-ATPase). Aside from its apparent complementary relationship with crown ethers, potassium also has a special relationship with a biological macrocyclic molecule called valinomycin $[C_{54}H_{90}N_6O_{18}]$ (**Fig. 11.6a**). Valinomycin is a depsipeptide that consists of six alternating amino acid and organic acid precursors (**Fig. 11.6b**). It is a natural compound that also happens to be an effective antibiotic. Valinomycin specializes in the transport of the K+ cation (and Rb and Cs cations) and acts specifically to increase the permeability of membranes to potassium cations [27]. The antibiotic function arises from its ability to alter the permeability of K+ across cellular membranes in bacteria, mitochondria, and erythrocytes. Valinomycin was isolated in 1955 by H. Brockmann and G. Schmidt-Kastner [28].

Valinomycin synthesis, structure, and function are summarized in four steps: (1) Covalently bonded precursors combine to form a repeating unit (a monomer) consisting of four precursors (**Fig. 11.6b**); (2) three monomer units are covalently bonded to form a macrocyclic compound; (3) hydrogen bond interactions stabilize a conformation that allows tight binding of the potassium cation (e.g., the ring is puckered to form a three-dimensional octahedral enclosure around the potassium cation); and (4) after capture and a change in its conformation, a hydrophobic perimeter is generated that enables the valinomycin, with its sequestered cationic cargo, to become soluble within the lipid of a membrane, thereby allowing passage. This is an excellent example of a simple function at the nanoscale—a function that is critical to life. No information is coded into this process. Thermodynamic (and perhaps kinetic) driving forces are the only factors involved in its function. However, living things created the environment within which valinomycin operates—an environment that did

FIG. 11.5

(a) DNA consists of complementary sets of purine–pyrimidine base pairs connected to a phosphate–deoxyribose backbone—all formed into a double-helix structure. The purine adenine pairs only with the pyrimidine thymine; the purine guanine pairs only with the pyrimidine cytosine. Molecular recognition is achieved with highly specified hydrogen bonding. This is a remarkable complementary situation that is able to store data, regulate cellular function, and make replicas.

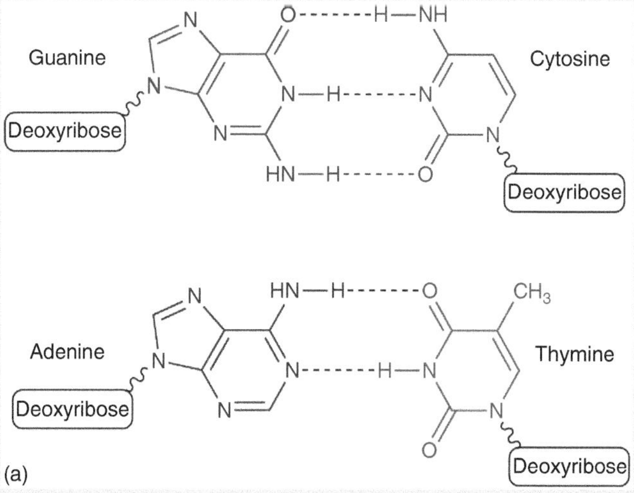

(a)

(continued)

Fig. 11.5 (Contd.)

(b) The double helix is the result of straightforward hydrogen bonding between purine and pyrimidine nucleotide base pairs, π-stacking interactions between the aromatic constituents of the nucleotides, and complex hydrophobic factors. Red: thymine; blue: adenine; green: guanine; purple: cytosine. The code in this segment is TAGCT on the left strand. Additional carbons in the backbone and the sugar are not shown for purposes of clarity. Covalent bonds hold the nucleotides to the backbone structure. Hydrogen bonds between the nucleotides are unzipped during replication. It is difficult to show on this image a rigorous rendition of the π–π interactions that occur between the linkage groups.

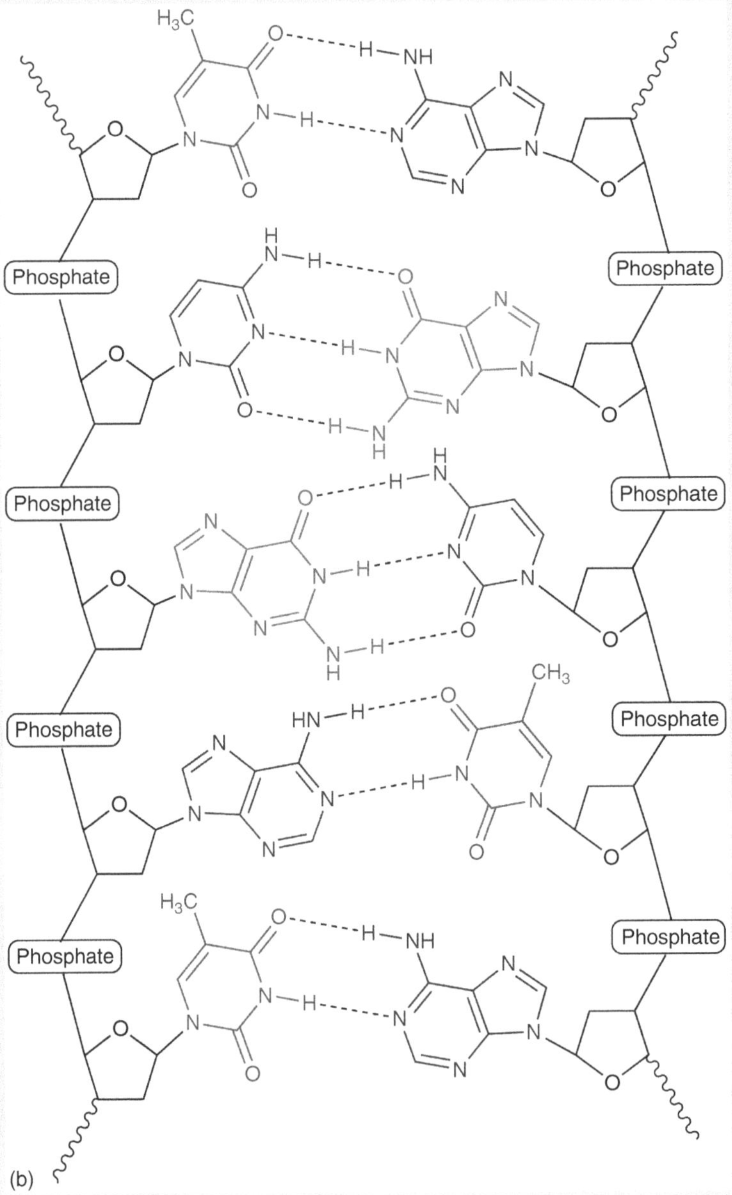

(b)

FIG. 11.6 *(a) The biological macrocycle valinomycin is a mobile carrier of K⁺ across phospholipid bilayer membranes. Hydrogen bonding between –[N–H···O=C]– enhances the folding of the ionophore into an octahedral shape around K⁺. Such action is termed preorganization. Valinomycin's antibiotic potential relies on its ability to disrupt the transport of potassium across bacterial membranes. (b) Precursor molecules of valinomycin (from left to right): orange = lactic acid; blue = valine; green = hydroxy-isovaleric acid; red = valine. These precursors are bonded together with strong covalent bonds, thus synthesizing a supermolecule. It is only the metal cation guest that is bonded with labile intermolecular bonds.*

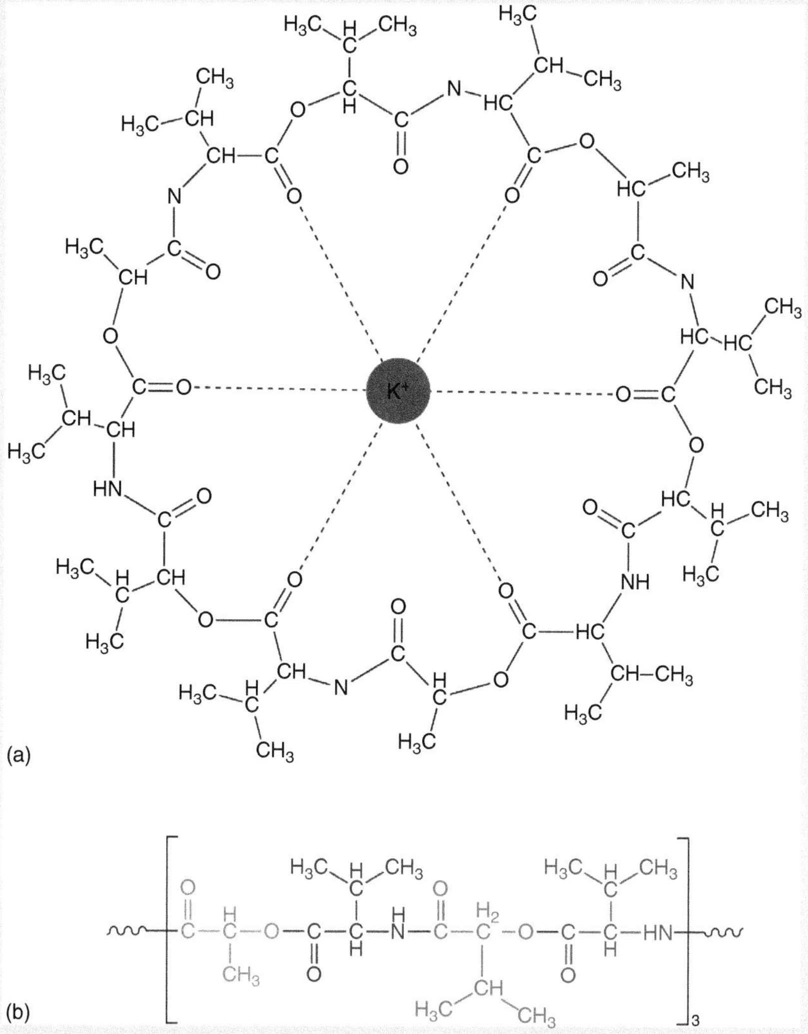

Source: F.F. Nachtigall et al., *Journal of the Brazilian Chemical Society*, 13, 295–299, 2002.

require information, translation, and synthesis, which all existed in a state of lowest energy specially designed for valinomycin to accomplish its task. Perhaps the earliest forms of "living things" did not rely on coded information. If true, this makes the valinomycin–potassium cation relationship one of the oldest in the life evolutionary timeline. Nature exploited the special relationship between valinomycin and the potassium cation and created an environment for it within which to function. Once inside the cell, the potassium is released in another environment designed just to accomplish that special task. Human science has found a way to apply valinomycin in the capacity as an ion-exchange component in potassium selective electrodes.

A more advanced form of molecular recognition is illustrated by the humble silkworm moth, *Bombyx mori*. It is a form of molecular recognition between a molecule and a protein receptor. Although still rather simplistic, in some ways this relationship can be considered as the ultimate expression of molecular recognition in terms of single molecule detection. The substrate molecule is a sex pheromone called bombykol—another type of naturally occurring supramolecule precursor. Pheromones are examples of precursor molecules that participate in extremely sensitive and selective reactions.

Bombykol [$C_{16}H_{30}O$] (**Fig. 11.7**) is released by the female silkworm moth and is detected by pheromone-binding protein receptors that exist in the nano- and microstructure of the *sensilla* of the male moth [29,30]. Sensilla are capable of detection down to a single molecule of bombykol [29,30]. Nature has achieved the level of single molecule detection! Considering the extreme distances between the pheromone and its molecular recognition receptor (often measured in kilometers) and the dilution factor involved, this process is nothing short of remarkable.

Jean-Henri Fabre, a French naturalist, discovered pheromones in 1870 when a female peacock moth attracted numerous male suitors within moments after emerging from its cocoon. Adolf Butenandt, a German biochemist, is credited with naming the silkworm moth pheromone bombykol in 1959. The word "perfume" has its roots in the word pheromone (from the Greek *pherein*, "to carry," + *hormon*, to "impel, urge on"). Biological entities such as cells are considered to be a *collection of sophisticated nanomachines* that self-assemble to interact with complex chemical and physical networks [31]. The example of bombykol is just one great example of natural molecular recognition mediated by nanostructures that are integrated into complex chemical and physical networks. Perhaps the highest form of "dynamic molecular recognition" is exhibited by the immune system. The immune system is an example of a dynamic molecular recognition system that is able to scan antigenic invaders, construct a chemical surface that is specific to the antigenic surface and subsequently neutralize it,

Fig. 11.7

Bombykol is a relatively simple molecule; it is not very much like a supramolecule at all but more like a large precursor molecule. It obviously plays the role of the guest to the host protein receptors of the sensilla of the male moth.

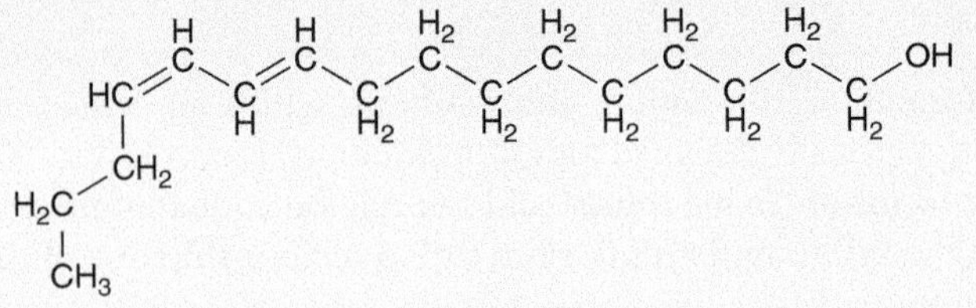

and make it soluble to hydrolysis or excretion. We know that it works quite well most of the time.

Molecular recognition is one of the greatest challenges faced by the developers of pharmaceuticals. The overall objective is to create a drug with selective action, rather than one that has systemic effects. Drugs that are able to target only sick cells via molecular recognition would certainly spare the host the agony of systemic chemotherapy treatment. Cancerous liver cells, for example, have abundant folic acid receptors on the surface. Recently Z. Tang et al. have proposed a system based on *aptamers* that are able to characterize cancer cells based on molecular recognition [32]. They describe a means to synthesize molecular probes for specific recognition of cancer cells. Aptamers are short single strands of DNA or RNA (oligonucleotides) or peptides with capability to bind target molecules. Aptamers are created by statistically based techniques from a large pool of candidates and are touted to be the "chemist's antibody" that can function as molecular probes for many kinds of biochemical applications [32]. "The strategies used here will be highly useful for aptamer selection against complex target samples in order to generate a large number of aptamers in a variety of biomedical and biotechnological applications, thus paving the way for molecular diagnosis, therapy, and biomarker discovery."

Aptamers based on nucleic acids have the capability of molecular recognition similar to that of antibodies, but "antibodies" that can be easily modified chemically to extend lifetime in fluids and bioavailability [33]. Standard antibodies are generally not modifiable because they lose specificity during the process. Antiviral aptamers are actively being developed as a mode of therapy against tenacious virus systems [33]. Small RNA is capable of folding into three-dimensional structures that are able to bind to proteins and hence deactivate them in a manner similar to other protein antagonists [34]. Natural decoy RNAs in the recent past have been shown to slow down the replication of the AIDS virus [34,35]. An aptamer product that targets vascular endothelial growth factor (cause of age-related macular degeneration) is in Phase II/III clinical trials. Another aptamer that targets thrombin is used as an anticoagulent that is reversible upon application of an antidote [34].

11.1.3 Synthetic Supramolecular Host Species

The number of supramolecules is vast and it would be impossible to list them all. The number of supramolecular precursor species is equally as vast. We will catalog only a few of the well-known supramolecular precursor organic host species. Biological supramolecular building blocks and complexes are numerous as well. Supramolecular building blocks (precursors) are covalently bonded macromolecules that either act separately, as in the case of ionophores, or in concert with others, as in the case of proteins, coenzymes, chlorophyll, columnar structures, and DNA.

The shape of substrates is also an important consideration, especially to the supramolecular receptor design engineer (and chemist!). Substrates assume varied geometries: spherical, linear, trigonal-planar, tetrahedral, and octahedral, to name just a few. Recognition between guest and host occurs by means of the many intermolecular interactions discussed in the previous chapter: electrostatic, hydrogen bonding, Lewis acid–base relations, and hydrophobicity. Inorganic anion substrates include atomic species such as fluoride, chloride, bromide, and iodide; linear molecules such as nitride, cyanide, thiocyanide, and hydroxide; trigonal species like carbonates and nitrates; tetrahedral moieties that include phosphates, sulfates, vanadates, molybdates, and manganates; and octahedral entities such as ferricyanide and other octahedrally coordinated metal cyanides [8]. Approximately 70–75% of biological substrates are anions. To become familiar with the multitude of organic species, the supramolecular precursors are presented in **Table 11.4** in alphabetical order. There is no way we can discuss each and every one in great detail, but a few selected hosts are provided with a brief discussion.

Host–guest complexes come in a variety of forms [4,7,8]. There are several kinds of host–guest relationships: (1) *encapsulation*, in which the guest molecule is sequestered within the host; (2) *nesting*, where the guest sits on the surface of the host; (3) *perching*, in which the guest is associated with the edge of the host; (4) *nonpolar surface interaction* in which the guest is placed near the host; (5) the *sandwich* relationship, akin to that of the iron in ferrocene; and (6) the *wrapping* configuration, in which the guest is surrounded by the host with an exit portal available for decomplexation [7]. At higher levels of complexity, host materials are able to complex with other host materials to form superhosts or bind with complex guests. Some hosts are able to bind more than one guest. Binding in hosts often occurs by means of dative style bonds (Lewis acid–base relationships) if the guests are metal cations, but all kinds of intermolecular interactions are capable of binding guest species. Cooperative binding alliances work in concert to attract and then hold guest species and release them when needed.

Hosts are able to accommodate neutral species, anions, and cations or, in some cases, both anions and cations, simultaneously, or zwitterions. In addition to atomic species and molecular anions, substrates like alcohols, amino acids, and peptides are able to fill the role of guest species in enzymatic reactions. Supramolecular design requires diligent analysis of potential host–guest relationships. In order to achieve the level of a tangible nanomaterial, more than a simple dative bond with a metal is often required. A nanomaterial with supramolecular ancestry is the sum of all its host–guest reactions. There are few generalized categories of host–guest relationships. They consist of: (1) the same kind (homogeneous) in which the nanomaterial is assembled in one step; (2) the same kind, but in which the nanomaterial is assembled in multiple steps by the same process; (3) multiple tandem steps consisting of different processes (different host–guest relationships); or (4) simultaneous reactions consisting of heterogeneous processes akin to what was most likely found in the primal soup.

With regard to nomenclature the suffix "and" indicates the synthetic host species (e.g., cryptand, carcerand, and catenand). The suffix "ate" (also "plexes") indicates that the host species exists in a complexed form (e.g., cryptate, encarcerate, and catenate or caviplexes, carceplexes, and spheraplexes). As one can see from the table, the list of supramolecular hosts is quite imposing. We will discuss a few of them in more detail. Several host species are depicted in **Figure 11.8a–p**.

Chelates.   Chelates and the chelate effect were introduced and defined in the last chapter. Chelates represent a broad range of compounds that are vital to biology and extremely useful in supramolecular assembly. Chelates are able to

TABLE 11.4	*Catalog of Supramolecular Host Species A–Z*		
Host species	**Description**	**Species**	**Description**
Amphiphiles Surfactants	Molecules with hydrophilic and hydrophobic groups (e.g., soaps, phospholipids). Surfactants are able to self-assemble in aqueous solutions and change the properties of water (surface tension).	Intercalates	Layered solids, intercalates can be considered as clathrates. Examples include graphite, cationic/anionic clay minerals, and metal phosphates. Organic urea clathrates form solid-state clathrates. Urea clathrates form helical host channels.
Antennae complexes	Complexes capable of harvesting light (e.g., metal polypridyls)	Ionophores	Organic ion carriers in biological systems that provide binding and shielding to metal cations—K^+ in particular. The purpose of ionophores is to transport metal ions through membranes.
Calixarenes	Macrocycle with –OH centers + phenyl rings (e.g., *p-t*-butly-calix[4]arene). Useful in cation complexation. Calixarenes are cavitands.	Katapinands	Macrobicyclic cavity hosts that complex halides. One of the first examples of anion binding
Carcerands Carceplexes Hemicarcerands	Permanently trap guest species unless covalent matrix of host ruptures. Hemicarcerands have a measurable activation barrier for releasing guest. Contain a concave inner surface—a molecular capsule. Prominent in neutral guest binding with *vdw* forces	Lariat ethers	A crown ether with appendages that enhance metal cation complexation by providing another dimension to binding
Catenanes Catenands Catenates	A host consisting of two or more rings that are interlocked in a mechanical way (e.g., [2]catenand). Although linked mechanically, the rings are not covalently bonded. Complexes are stabilized by H-bonds/charge transfer $\pi-\pi$ interactions.	Lewis acid hosts	Lewis acid hosts are capable of binding anions. Organoboron compounds are electron deficient. These hosts are opposite to chelates that are Lewis bases. Binding usually occurs at the periphery of the molecule or between two molecules.
Cavitands Cavitates Caviplexes	Host with intramolecular cavities. Host–guest complexes are stable in both liquid and solid state. A molecular container with an enforced concave surface. Calixarenes are cavitands. Cavitands are neutral molecules; cyclodextrins are cavitands.	Ligands	A ligand is an organic molecule that is able to complex with a metal. The Au–triphenylphosphine is an example of a metal–ligand complex. Water is a monodentate inorganic ligand that is able to form complexes with metals.
Chelates	The binding of a single metal atom with two or more binding sites on the same ligand. In the general sense, most hosts that bind metals can be considered to be chelates. Chelates can be charged or neutral. Recognition is by Lewis acid–base interaction.	Micelles Lipid bilayers Reversed micelles	Micelles (3-D) and lipid bilayers (2-D) are made of amphiphilic molecules that are able to self-assemble in aqueous solutions. Reversed micelles self-assemble in hydrophobic solutions.

(*continued*)

TABLE 11.4 (CONTD.)	*Catalog of Supramolecular Host Species A–Z*		
Host species	**Description**	**Species**	**Description**
Clathrands Clathrates	Hosts with extramolecular cavities— between two or more host molecules. Water forms polyhedral networks (hydrates) around dissolved species. Zeolites are inorganic solid-state clathrates. Clathrates are a broad category that includes intercalates, helical tubulands, inclusion polymers and zeolites, and more.	Molecular clefts Molecular tweezers	Two guest-binding domains are positioned to form a binding site.
Coordination polymers	Metal–ligand directional coordination reactions to produce crystalline poly-meric architecture. Guest bound by Hoffman type clathrate	Molecular knots	Molecular knots are intricate forms of catenanes formed with multiple metal helicates and other substructures.
Corands Corates Coraplexes Azacorand	Crown ethers belong to this general class. A corand is a closed monocyclic hetero-ring. If containing nitrogen and/or oxygen it is called an azacorand or azacrown.	Molecular squares Molecular boxes	Squares, triangles, and cubes with metal coordination at corners and linked with aromatic species. Molecular boxes, etc. are assembled with metal coordination.
Crown ethers Anticrowns Azacrowns Heterocrowns	The roots of supramolecular chemistry are traced to the crown ethers of C. J. Pedersen. Crown ethers are oxygen-containing heterocycle corands. They are 2-D heterocycles that contain oxygen. Azacrowns contain nitrogen. Crowns are suitable for cation binding.	Podands Podates	Open chain hosts. They are acyclic hosts with pendant binding sites. Open crown ethers form podants. Because they are open structures, they do not have the affinity for cations as do the crown ethers due to unfavor-able enthalpic and entropic effects.
Cryptands Cryptates Hetero-cryptands Cryptophanes	Cryptands are 3-D bicyclic or oligocyclic systems. Crown ethers that have 3-D components are cryptands. Cryptands have high stability constants because the cryptand cavity is poorly solvated and there is less reorganization required to bind metals.	Rosettes	Hydrogen bonded structures consisting of melamine and cyanuric acid form planar circular structure with a central cavity.
Cyclophanes Paracyclo-phanes	A cyclophane is an organic host molecule with a bridged aromatic ring. Cyclo-phanes contain multiple bridged benzene rings. The bridges are usually in the form of an aliphatic chain. Cyclophanes have parallel aromatic walls with a well-defined cavity. Cyclophanes are good hosts to hydrophobic apolar guests.	Porphyrins	Porphyrins are classical coordination compounds consisting of planar pyrrole rings that bind by classical chelate and macrocycle effects. They are tetrapyrrole macrocycles. Binding of Fe by heme is a well-known example.
Dendrimers	Dendrimers are cascade molecules with highly branched 3-D structure. The dendritic core is porous and capable of hosting a variety of materials. By *site isolation*, a dendrimer can be constructed from a core guest. *Guest inclusion* is the case in which the dendrimer is already formed and a guest is invited in.	Rotaxanes Pseudorotax-anes	Rotaxanes are long molecules that exist in the center of macrocyclic rings in needle–thread fashion. These are permanent structures, especially if the linear molecule is terminated with a bulky group. Pseudorotaxanes allow the central molecule to slip out. Rotaxanes are similar to catenanes.

TABLE 11.4 (CONTD.)	*Catalog of Supramolecular Host Species A–Z*		
Host species	**Description**	**Species**	**Description**
Fullerenes	The cavity within a fullerene is a perfect place to sequester metal or other species. Metals are introduced by ion implantations and chemical techniques. Space inside carbon cage ranges from 0.4 to 1.0 nm for C_{60}–C_{240}.	Schiff's bases	Macrocycles formed from Schiff's condensation (amine + aldehyde–water = imine) to complex metals
Helicands Helicates Helical tubulands	These clathrates can coordinate metal ions or anions like Cl^-. They are usually made of organic hosts. Helicate assembly occurs with templated metal assist of organic threads that resemble DNA—transition metal-directed assembly. Only helices of strands of identical length would self-assemble. The metal ion sits in the center of periodic twists in the helix.	Sepulchrates Sarccophagines	Sepulchrates are cryptands which are made from cobalt(III) template synthesis of *tris*(chelate) complexes of bidentate ligands. As with other cryptands, the metal is sequestered. Sarcophagines contain C-bridged species.
Hoffman inclusion compounds	Belong to the family of clathrates. Lattice voids in inorganic coordination compounds with general formula: $M(NH_3)_2M'(CN)_4 \cdot 2G$ where M = Zn, Cd; M' = Ni, Pd, Pt and G = small aromatic molelcule	Siderophores Siderands	Siderophores are naturally occurring ligands like myobactin and enterobactin that mobilize iron. Siderands are synthetic analogs. Siderophores form a 6-coordinate complex with Fe.
Hourglass inclusions	Formed by the nonstoichiometric inclusion of water-soluble organic dye molecules within an organic lattice like K_2SO_4. The colored and colorless regions overall look like an hourglass.	Spherands Spheraplexes Hemispherands	Spherands are macroscyclic cation hosts. They are based on *p*-methylanisole. Crowns and cryptands are flexible. Spherands are rigid. Due to this, spherands have high selectivity.
Hybrid hosts	Hybrid hosts form by combining two or more kinds of hosts: crown ethers, cryptands, spherands and podands, etc. A wide variety of hosts are created by mixing host types.	Speleands Speleates	A speleand is an amphiphilic receptor that combines two or more forms of guest recognition. A synergistic enhancement of binding results. Polar or charged sites can be combined with a hydrophobic residue. The active sites of enzymes certainly display speleand character.
Hydroquinone Phenol cages Hexahosts	Hydroquinone and phenol form hexagonal cyclic hydrogen-bonded rings with bulk substituted groups alternating in up–down pattern. Cavities are formed between rings. These are a kind of clathrate.	Zwitterions	Neutral molecules with both a positive and a negative charge. Anionic binding proteins and enzymes are zwitterionic.

Sources: J.-M. Lehn, *Supramolecular chemistry: Concepts and perspectives*, VCH Verlagsgesellscahft GmbH, Weinheim (1995); J. W. Steed and J. L. Atwood, *Supramolecular chemistry*, John Wiley & Sons, Ltd., Chichester (2000); P. D. Beer, P. A. Gale, and D. K. Smith, *Supramolecular chemistry*, Oxford Chemistry Primers, Oxford Science Publications, Oxford University Press, Oxford (2003).

Fig. 11.8

(a) Chelate: [Ru(2,2'-bipyridyl)₃]²⁺ hexadentate chelate; (b) crown ether corand: [24]crown-8; (c) azacrown ether analog; (d) MCM-41 zeolite pore structure: 1.5–10 nm, surface area: 400–1700 m²·g; pore volume limit: 1.1 cm³·g; pentasil zeolite ZSM-5 0.54 × 0.56 nm; (e) cryptands are three-dimensional analogs of the crown ethers. A [2.2.2]cryptand is on the top and a [2.2.1]cryptand is on the bottom; (f) benzol[18]crown-6crown ether; (g) fullerene hosting a xenon gas atom; although it is a host molecule, would you consider a fullerene a supramolecule? (h) A podand: pentaethyleneglycol dimethylether if R = Me (EG5); (i) p-tert-butylcalix[6]arene, a cyclophane, is pictured on the bottom; a spherand is pictured on top; (j) [2]catenane; (k) α-cyclodextrin [6-sugar ring]; (l) dendrimer made from amine groups—poly(propyleneimine); (m) trimesic acid (1,3,5-benzentricarboxylic acid) tubuland network precursor; (n) helicate precursor 2,2':6',2":6",2'":6'",2"":6"",2""'-sexipyridine; (o) and (p) molecular squares [(en)Pd(4,4'-bipy)]₄(NO₃)₈, and [Cd(4,4'-bipy)]₄(NO₃)₈.

complex with metal ions to form very stable structures and bind metals on the basis of Lewis acid–base recognition—dative coordinative bonds that collectively contribute to form a complex (**Fig. 11.8a**). In the broadest sense, many host species can be considered to be glorified chelates.

Clathrates.   Clathrates (and intercalates) form the largest and most diverse class of host materials. Inorganic, organic, biological, mineral clays, graphite, and polymers all have representative clathrate compounds. Clathrates (from the Greek *klethra*, "bars") are chemical substances that consist of a lattice that is able to sequester or trap guest species. A clathrate hydrate (a gas hydrate) is pictured in **Figure 11.9**.

Crown Ethers.   Supramolecular chemistry began with the crown ether (**Fig. 11.8b, c, f**). Crown ethers have oxygen atoms that link aliphatic ethyl bridges to form a circular macrocycle (heterocycle). Crowns that contain nitrogen-bearing ligands are called azacrowns. Crown ethers belong to a class of host molecules called *corands*. The cavity diameter of crown ethers can be tailor fitted to accommodate

FIG. 11.9

A clathrate consisting of a water cage enclosing one methane molecule is depicted. This structure is called a methane clathrate, methane hydrate, or even methane ice. Its physical form is that of solid water that contains a significant amount of methane. Naturally occurring methane clathrates are found under sediments on the ocean floor. The clathrate remains stable up to 18°C. There is usually 1 mol of CH_4 per 5.8 mol of water with a density of 0.9 g·cm^{-3}. The basic structure consists of 20 water molecules to form a dodecahedral shape. If oceans were to warm further, the release of methane would significantly contribute methane to the atmosphere and further exacerbate global warming.

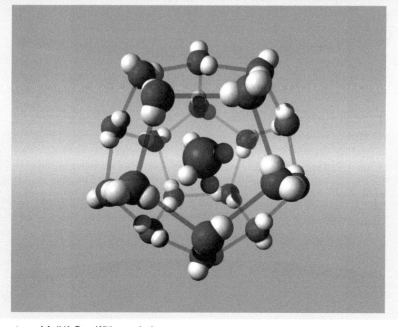

Source: Image courtesy of Anil K. Rao. With permission.

cations of differing sizes. The [12]crown-4 (four oxygens with four aliphatic ethyl group bridges) has a cavity size of 1.2–1.5 Å, perfect for a Li$^+$ cation. The [21] crown-7 possesses a cavity that is 3.4–4.3 Å, perfect for the accommodation of a Cs$^+$ cation [8].

Cryptands.   Cryptands are cage-like bicyclic molecules that are three-dimensional analogs of the crown ethers [8]. Mixed cryptands contain other atoms besides oxygen, usually nitrogen (**Fig. 11.8e**). Amine-based cryptands have an affinity for binding the ammonium cation, alkaline earths, and lanthanides. The cavity of cryptands, as the name implies, is deep within the molecule. Therefore, once a metal is bound within the cavity, it tends to remain there.

Cyclophanes.   Cyclophanes consist of benzene rings held together by aliphatic or amine bridges (**Fig. 11.10**). A positive charge imparted to cyclophanes is able to enhance its selectivity towards anionic-hydrophobic guests. In this way, the positive charge attracts the anion and the hydrophobic pocket provides for the binding specificity [8]. The size of the cavity depends on the number of benzene rings in the macrocycle.

Dendrimers.   Dendrimers are incredible molecules with a fractal-like quality very much like the roots or branches of a tree [36]. A fractal is a structure that does not exist in integer space but rather a space governed by fractions—the dimensionality of a fraction [36]. It is a structure that is subdivided but one in which each subdivision resembles the structure as a whole—a property known as *self-similarity*. Dendrimers are cascade molecules with highly branched three-dimensional structure and are often spherical in shape (**Fig. 11.8l**) [36]. The subunits of dendrimers are relatively simple amines or other branchable molecules. There are three general ways to synthesize dendrimers: divergent, convergent, or by site isolation. Site isolation technique begins with a core "guest" from which the dendrimer is constructed. In divergent methods, the dendrimer is assembled from the center to the periphery. In the convergent method, the opposite strategy is employed. Each layer (or generation) is built step by step. The dendritic core is porous and capable of hosting a variety of materials. For

| **Fig. 11.10** | *Cyclophanes are organic host molecules that contain bridged aromatic rings that may contain a cavity [7]. A [2,2]-cyclophane is depicted. With appropriate positioning (preorganization), cyclophanes are able to contribute steric and electrostatic complementarity to a potential guest [7]. Calixarenes and hemicarcerands are also cyclophanes. Cyclophanes are able to bind neutral molecules and cations.* |

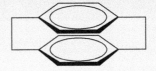

Fig. 11.11

A dendrimeric synthesis scheme is depicted. 1-Bromopentane undergoes nucleophilic substitution with triethyl sodiomethanetricarboxylate. Lithium aluminum hydride is then applied to reduce the ester groups to alcohols. Functionalization of the terminal groups into good leaving groups (tosylates) prepares the process for another generation of growth. The process continues until a dendrimer is synthesized.

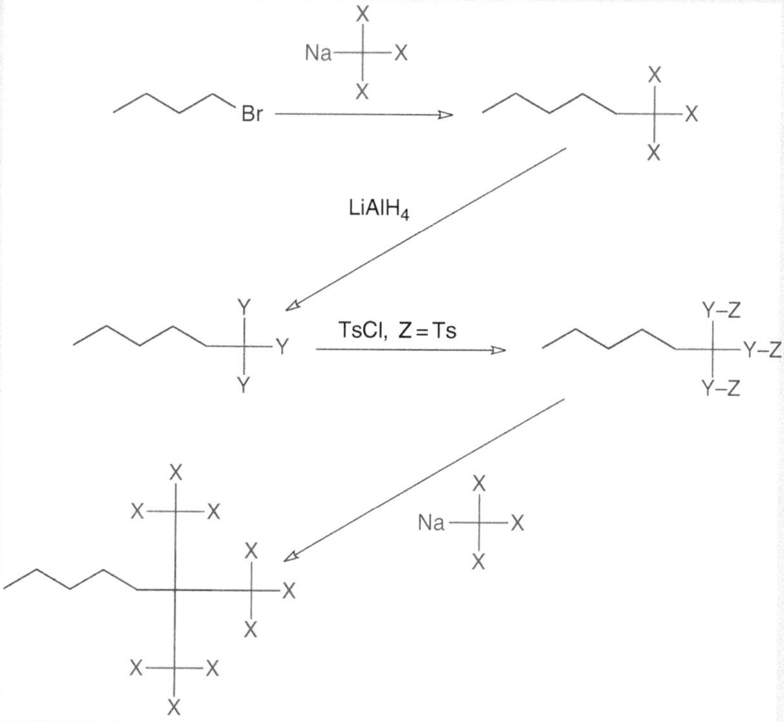

example, a dendrimer can be constructed from polyethylene glycol arms attached to an iron–porphyrin core to form a dendritic water-soluble heme [7].

The first dendrimer (the *Newkome* dendrimer) was synthesized in 1985. Dendrimers are synthesized by a series of repetitive growth-activation steps (**Fig. 11.11**). First, 1-bromopentane [$CH_3(CH_2)_3CH_2$-Br] underwent nucleophilic substitution by triethyl sodiomethanetricarboxylate [$(CO_2Et)_3Na$] in DMF (dimethylformadide) and benzene at 80°C. Lithium aluminum hydride [$LiAlH_4$] in diethlyether was used to reduce the ester groups to alcohols. The terminal groups were then transformed into tosylates (good leaving groups) by the application of tosyl chloride for further activation to complete the first generation of the dendrimer. Nucleophilic substitution was then performed again. The process was repeated for each generation of the dendrimer structure.

Recently, a Swedish research team headed by Michael Malkoch found a way to eliminate half the number of steps involved in dendrimer synthesis [37,38]. In the new method, two different monomers are used alternatively without the need for an activation step. B. Sharpless, of the Scripps Research Institute in La Jolla, stated in 2007 that "Malkoch and colleagues have provided a beautiful new dendrimer synthesis. It is simple and efficient, and to the best of my knowledge, represents the state of the art."

11.1.4   Surfactants and Micelles

Surfactants and Micelles. Surfactants are surface-active agents. They are part of the larger class of *soft matter.* The cleansing characteristic of soap is based on surfactants. The hydrophobic ends of the surfactants form an interface with insoluble dirt particles while the hydrophilic ends interface with the water; this renders the particle water soluble and cleansing ensues. The addition of surfactants also reduces the surface tension of water. **Table 11.5** lists the major categories of surfactants. *Amphiphiles* are molecules that have a polar moiety attached to one end and a nonpolar group at the other. When placed in water, self-organization of the amphiphiles into low-energy forms takes place as the hydrophilic ends retain contact with the aqueous phase and the nonpolar ends agglomerate by creating a middle where water is absent as is the case for micelles. At a gas–liquid interface, the polar ends of amphiphiles stay in the water while the tails interface with the gas (usually air), thereby forming a film across the surface. Depending on the chemical nature of the amphiphile, many different structures are possible, from spheres to cylinders to planar bilayered structures, giving supramolecular design using amphiphiles great flexibility.

The simplest amphiphilic structure is the micelle. Although micelles are homogeneous in the sense that they consist of the same molecule, they must be considered as supramolecular structures. Why? The molecular components of micelles are held together by intermolecular forces. Assuming that all amphiphiles in a solution are identical, predictions about their structure are made by application of the *critical packing parameter* [37]. The *cpp* is given by

$$cpp = \frac{V}{A_{cap}L_{cap}} \tag{11.1}$$

where V is the volume of the amphiphile, A_{cap} is the area of the head-group, and L_{cap} is the length of the head-group. Packing shapes vary from cone like to cylinder like to wedge like. For single tail surfactants that have relatively large head-groups (e.g., sodium dodecyl sulfate, SDS), *cpp* is equal to 1/3 and the packing shape is conical with the tails petering out into a point. Spherical micelles conform to these criteria as well. Single-tailed amphiphiles with small head-groups (e.g., hexadecyltrimethylammonium bromide, CTAB) pack into a truncated cone with *cpp* between 1/3 and 1/2 and show incipient cylindrical tendencies. Double-tailed amphiphiles with large heads (e.g., phoshpatidic acid) form truncated cones (*cpp* = 1/2–1) that in turn form flexible bilayers. Double-tailed amphiphiles with small heads (e.g., phosphatidyl serine) form cylinders (*cpp* = 1) that settle eventually into planar bilayers (like membranes) or into inverted truncated cones (wedges). In the cases where *cpp* > 1, inverted micelles (reverse micelles) are formed under the proper solution conditions (e.g., phosphatidyl ethanol amine) [37].

Each micellular system is characterized by a unique *critical micelle concentration* (CMC). Surfactants exist as independent molecules up until a limiting condition at the point of the CMC. It is one of the most important surfactant

TABLE 11.5	*Types of Surfactants*	
Type of surfactant	**Comments**	**Example**
Anionic	An example of an amphiphilic molecule with hydrophobic tails and hydrophilic head groups	*Sodium laurate $[CH_3(CH_2)_{10}(CO_2)Na]$ anionic surfactant commonly found in shampoo. (Image redrawn with permission from Tapani Viitala, KSV Instruments, Ltd.)*
	Double-tailed anionic surfactant	*Dipalmitoyl phospahtidyl serine is a double-tailed anionic surfactant.*
Cationic	Used as a topical antiseptic against bacteria and fungi. It is also used in a buffer solution for extracting DNA.	*Cetrimonium bromide is a cationic surfactant.*
Non-ionic	Non-ionic surfactants are used in detergents that are able to dissolve proteins. Also act as emulsifiers in beverages and food products	*Triton X-100 $[C_{14}H_{22}O(C_2H_4O)_n]$ is a non-ionic surfactant. The hydrophilic contribution is from the poly(ethylene oxide) or from a sugar. Nonionic surfactants are used in detergents and are able to solubilize membrane proteins.*
Zwitter-ionic	Zwitterions have both acidic and basic functional groups on the same molecule.	*HEPES[4-(2-hydroxyethyl)-1-piperazineethanesulfonic acid] is a zwitterionic surfactant that contains both anionic and cationic groups. The molecule exists in the zwitterionic form at certain pH.*

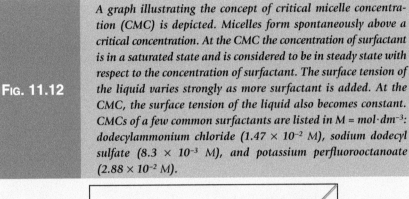

FIG. 11.12 *A graph illustrating the concept of critical micelle concentration (CMC) is depicted. Micelles form spontaneously above a critical concentration. At the CMC the concentration of surfactant is in a saturated state and is considered to be in steady state with respect to the concentration of surfactant. The surface tension of the liquid varies strongly as more surfactant is added. At the CMC, the surface tension of the liquid also becomes constant. CMCs of a few common surfactants are listed in M = mol·dm⁻³: dodecylammonium chloride (1.47 × 10⁻² M), sodium dodecyl sulfate (8.3 × 10⁻³ M), and potassium perfluorooctanoate (2.88 × 10⁻² M).*

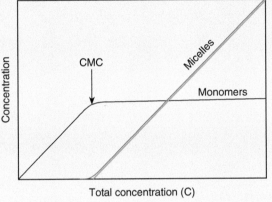

Source: Image redrawn with permission from Tapani Viitala, KSV Instruments, Ltd.

properties to chemical engineers. In plots describing the concentration of surfactant and micelles, micelles form when a critical concentration is exceeded. At this point the concentration of the micelle increases linearly, but the concentration of surfactant monomers achieves a steady-state concentration (**Fig. 11.12**). The CMC of anionic surfactants is $\sim 10^{-2}$ M and that for nonionic surfactants $\sim 10^{-4}$ M. For SDS, CMC = 8.3×10^{-3} M. This means that, at this molarity, the concentration of SDS plateaus and solution dynamics is dominated by micelle formation.

11.1.5 *Biological Supramolecular Host Species*

We are all quite familiar with nature's arsenal of biological host molecules and host nanomaterials. Chapters 13 and 14 will address natural supramolecular precursors and structures in more detail. We therefore will not dwell on them in this section. Selected biomolecular hosts are shown in **Table 11.6.**

All biological systems rely at some level on supramolecular structure to transfer information, conduct metabolism, build structures, and interface with the environment. Supramolecular hosts in the form of enzymes, DNA, heme, ionophores, neuroreceptors, and chemical receptors accommodate diverse substrates like sugars, intercalation materials, transition metals ions, alkali metal ions, acetylcholine, and hormones, respectively. In the case of heme, once iron is

TABLE 11.6	*Biological Supramolecular Structure Precursors and More*	
Biological supramolecule	**Comments**	**Examples**
Enzyme pockets	70–75% Substrates and cofactors are anions. Substrates include phosphites, phosphates, sulfates, chloride	

Generic rendition of an enzyme pocket is shown. Many cooperative intermolecular and intramolecular (bond breaking) reactions occur in such pockets.

Ionophores	Membrane transport of cations. Lipophilic carriers of cations Valinomycin K$^+$/H$^+$ Nonactin and enniatins have less selectivity for cations than valinomycin and nonactin.	

The ionophore nonactin is depicted. Its action is similar to that of valinomycin. The ionophore is specific to K$^+$.

Coenzymes	NAD$^+$ and NADH Electron transfer in oxidoreductase enzymes Molecule is all covalent bonded. Action is by intermolecular attractions.	

Nicotinamide adenine dinucleotide (NAD$^+$) is a coenzyme that provides the vehicle to carry electrons. It therefore is a major participant in redox reactions. NAD$^+$, an oxidizing agent, is reduced to NADH during metabolic processes. NAD is a coenzyme that is able to transfer electrons in the form of hydrides.

(*continued*)

TABLE 11.6 (CONTD.)	*Biological Supramolecular Structure Precursors and More*	
Biological supramolecule	**Comments**	**Examples**
Natural chelation agents	Synthesized by bacteria (*Streptomyces*) Used as antibiotic	
		Tetracycline is a natural chelation agent that is produced by the bacterium Streptomyces. As an antibody, its function is based on the ability to inhibit the action of ribosomes in procaryotic bacteria by binding with t-RNA. In other words, it has the capability to inhibit protein synthesis. Is it the host or is t-RNA the host?
Porphyrins	Chlorophyll Porphyrins Tetrapyrroles Vitamin B_{12}	
		Left: A porphyrin ring guest coordinates an iron cation. In unoxygenated form, the ground state Fe cation has a +2 charge and exists in the Fe(II) state. It is paramagnetic with four unpaired electrons. In this configuration, the heme complex has a domed shape [7]. In the bound state, electron transfer occurs from Fe(II) to the oxygen to form an Fe(III) species. The system is diamagnetic and the porphyrin ring is flattened out. This example illustrates an intricate relationship between the host (the porphyrin ring), the guest (the iron cation), and a substrate (the oxygen molecule). The binding of O_2 is influenced by the allosteric effect in that as soon as one O_2 is bound, the affinity for oxygenation increases until all sites in hemoglobin are filled. Right: vitamin B_{12} is depicted. Cobalt(III) is coordinated by six dative bonds to anchor the core of the vitamin.

TABLE 11.6 (CONTD.)	*Biological Supramolecular Structure Precursors and More*	
Biological supramolecule	**Comments**	**Examples**
	Chlorophyll-*a*	

Chlorophyll-a is a conjugated π–system that is characterized by low-energy π–π* electron transition capability [7]. Perhaps it is one of the most amazing of the supramolecules; it captures photons and begins the process of making energy that drives most forms of life on this planet. The magnesium tetrapyrrole complex is anchored by the Mg^{2+} cation. The long aliphatic (phytyl) chain anchors the molecule into a phospholipid membrane of the chloroplast. Light harvesting antennae produce an electron on excitation that is transferred away from the excited state center to a chemical reduction center. The hole left behind is utilized to oxidize water into oxygen. The superstructure consisting of chlorophylls, electron transfer moieties, and all other infrastructure must be considered to be the premier accomplishment of natural supramolecular chemistry—all existing at the nanoscale nonetheless.

incorporated during a "primary level host–guest synthesis process," it becomes the guest of a quaternary protein structure to become the integral functional part of hemoglobin. All of this is in place so that hemoglobin can perform the vital function as host to molecular oxygen.

Molecular recognition is the foundation of biological materials and of life itself. Life began with the earliest forms of recognition: Two molecules combined to form a lower energy, lower entropy but restless state. Molecular recognition operated at the simplest level: the binding and transport of metal cations by simple molecules and the self-assembly of lipid bilayers. For us, it now operates at the highest level, where we rely on molecular recognition to guide our senses, our neural networks, and our chemical receptors to cope with the pressures and stresses of the outside world. Adrenaline, acetylcholine, dopamine, vitamins, and many more all have the appropriate receptors in our bodies to help us survive— where other supramolecules like DNA can make sure that we continue.

11.2 DESIGN AND SYNTHESIS OF SELECTED SUPRAMOLECULAR SPECIES

"The Chemist as Engineer and Architect" is a title that should be applied to this section. Although chemists have been designers of chemicals since making the scene in ancient times, and especially in the glory days of organic and polymer chemistry and later for pharmacological synthesis, this field offers as much as or more in terms of creativity, engineering, and design. The supramolecular chemist is interested in systems of great complexity, adding molecule after molecule to make species that can be quite large. Supramolecular chemistry, the chemistry of self-assembly and modification of nanomaterials, offers new sets of challenges as its golden age is upon us with no perceivable limit in sight.

The transformation of materials from precursors and substrates into hierarchical nanostructures is reviewed in **Figure 11.13.** Precursor molecules, synthesized by traditional chemical methods, are groomed to become hosts (and guests) and then undergo transformation into supramolecules. Once formed, supramolecules react, again primarily via intermolecular interactions, to form integrated structures. Although we are not emphasizing applications in the text, some are listed to provide perspective.

Supermolecules are large "molecules" that are held together by covalent, ionic, or metallic bonds; others provide alternative definitions. Supermolecules are largely homogeneous (e.g., they are made of repeating units of the same molecule or atom) and are not in the strictest sense considered to be hosts or guests. Obviously, carbon nanotubes, gold surfaces, and latex beads are members of this class of materials. Supermolecules therefore logically serve as substrates for other chemical species. Substrates are activated chemically and bonds are formed with surfaces via chemisorption (covalent) or physisorption processes. Therefore, supermolecules are not supramolecules, but they serve an important role as substrates that are able to anchor interesting species. Many types of substrates are nanomaterials.

The major components involved in a supramolecular design process are summarized in **Figure 11.14.** A simple but rather obvious rule applies to supramolecular synthesis: "The larger the system is, the more complicated its synthesis becomes" and the more aspects need to be considered. We start once again with precursors—more specifically, the selection of precursors that have the capability to serve as a host species in a potential supramolecular synthesis. Most precursor species are already available for commercial consumption—obviously, the best-case scenario.

The design process involves the following key elements: (1) selection of precursor molecules (starting materials); (2) analysis of thermodynamic and kinetic variables (this includes consideration of solvent effects, intermolecular interactions, and reaction schemes); (3) selection or design of receptors and substrates (another way to say "host–guest"); (4) design of the target hierarchical structure; and (5) process termination (passivation) schemes. Integral to every step of the process are: (a) the synthesis conditions, (b) purification steps (if required), and (c) characterization of products (and, of course, abiding by all safety protocols!). One is compelled to ask what the difference is between this and any other usual form of chemical synthesis. The major difference is stated in (3): the host–guest relationship.

Jean-Marie Lehn, one of the founders of the field, provides the following strategy [4]:

Precursors → receptor synthesis → host–guest reaction → supermolecule

Lehn uses the term supermolecule (not by our definition). There is more:

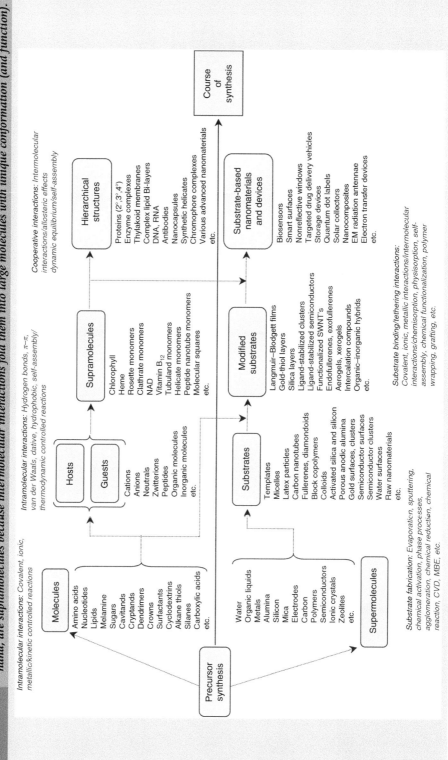

Fig. 11.13

The synthesis process is tracked through a material perspective. Starting with precursors, supramolecular synthesis proceeds through a succession of more complex interactions until an end product is obtained—whether a device or a natural product. Supermolecules, according to our definition in this text, are held together with covalent, ionic, or metallic bonds and are not perceived as host species, although good arguments can be made for the opposite case (e.g., carbon nanotubes and polymer wrapping). Substrates in general can be considered to be supermolecules. Even in nature, supramolecules rely on supermolecules for support, like the phosphate–deoxiribose backbone in DNA. Are peptides supermolecules? Are they precursors? According to our definition, they are precursors (or substrates). Proteins, on the other hand, are supramolecules because intermolecular interactions fold them into large molecules with unique conformation (and function).

Intramolecular interactions: Covalent, ionic, metallic/kinetic controlled reactions

Intramolecular interactions: Hydrogen bonds, π–π, van der Waals, dative, hydrophobic, self-assembly/thermodynamic controlled reactions

Cooperative interactions: Intermolecular interactions/allosteric effects dynamic equilibrium/self-assembly

Substrate binding/tethering interactions: Covalent, ionic, metallic interactions/intermolecular interactions/chemisorption, physisorption, self-assembly, chemical functionalization, polymer wrapping, grafting, etc.

Substrate fabrication: Evaporation, sputtering, chemical activation, phase processes, agglomeration, chemical reduction, chemical reaction, CVD, MBE, etc.

Precursor synthesis

Molecules
Amino acids
Nucleotides
Lipids
Melamine
Sugars
Cavitands
Cryptands
Dendrimers
Crowns
Surfactants
Cyclodextrins
Alkane thiols
Silanes
Carboxylic acids
etc.

Supermolecules
Water
Organic liquids
Metals
Alumina
Silicon
Mica
Electrodes
Carbon
Polymers
Semiconductors
Ionic crystals
Zeolites
etc.

Hosts

Guests
Cations
Anions
Neutrals
Zwitterions
Peptides
Organic molecules
Inorganic molecules
etc.

Substrates
Templates
Micelles
Latex particles
Carbon nanotubes
Fullerenes, diamondoids
Block copolymers
Colloids
Activated silica and silicon
Porous anodic alumina
Gold surfaces, clusters
Semiconductor surfaces
Semiconductor clusters
Water surfaces
Raw nanomaterials
etc.

Supramolecules
Chlorophyll
Heme
Rosette monomers
Clathrate monomers
NAD
Vitamin B_{12}
Tubuland monomers
Helicate monomers
Peptide nanotube monomers
Molecular squares
etc.

Modified substrates
Langmuir–Blodgett films
Gold-thiol layers
Silica layers
Ligand-stabilized clusters
Ligand-stabilized semiconductors
Functionalized SWNT's
Endofullerenes, exofullerenes
Aerogels, xerogels
Intercalation compounds
Organic–inorganic hybrids
etc.

Hierarchical structures
Proteins (2°, 3°, 4°)
Enzyme complexes
Thylakoid membranes
Complex lipid Bi-layers
DNA, RNA
Antibodies
Nanocapsules
Synthetic helicates
Chromophore complexes
Various advanced nanomaterials
etc.

Substrate-based nanomaterials and devices
Biosensors
Smart surfaces
Nonreflective windows
Targeted drug delivery vehicles
Storage devices
Quantum dot labels
Solar collectors
Nanocomposites
EM radiation antennae
Electron transfer devices
etc.

Course of synthesis

Fig. 11.14

One form of a supramolecular design process is summarized. Preorganization implies that the host molecule does not undergo significant conformational changes upon accommodation of a guest species. Complementarity is achieved after preorganizational deliberations are accomplished. Complementarity simply implies a good stable fit between guest and host. Obviously, thermodynamic and kinetic considerations permeate all of supramolecular chemistry. There are many factors that affect the thermodynamics and kinetics of reactions and a few are summarized in the figure. The solvent is another crucial variable in these reactions. When all is in place, activation takes place and the reaction concludes with the formation of a host-guest complex. With due foresight and consideration, hierarchical structures consisting of supramolecular subunits self-assemble into a functioning nanomaterial.

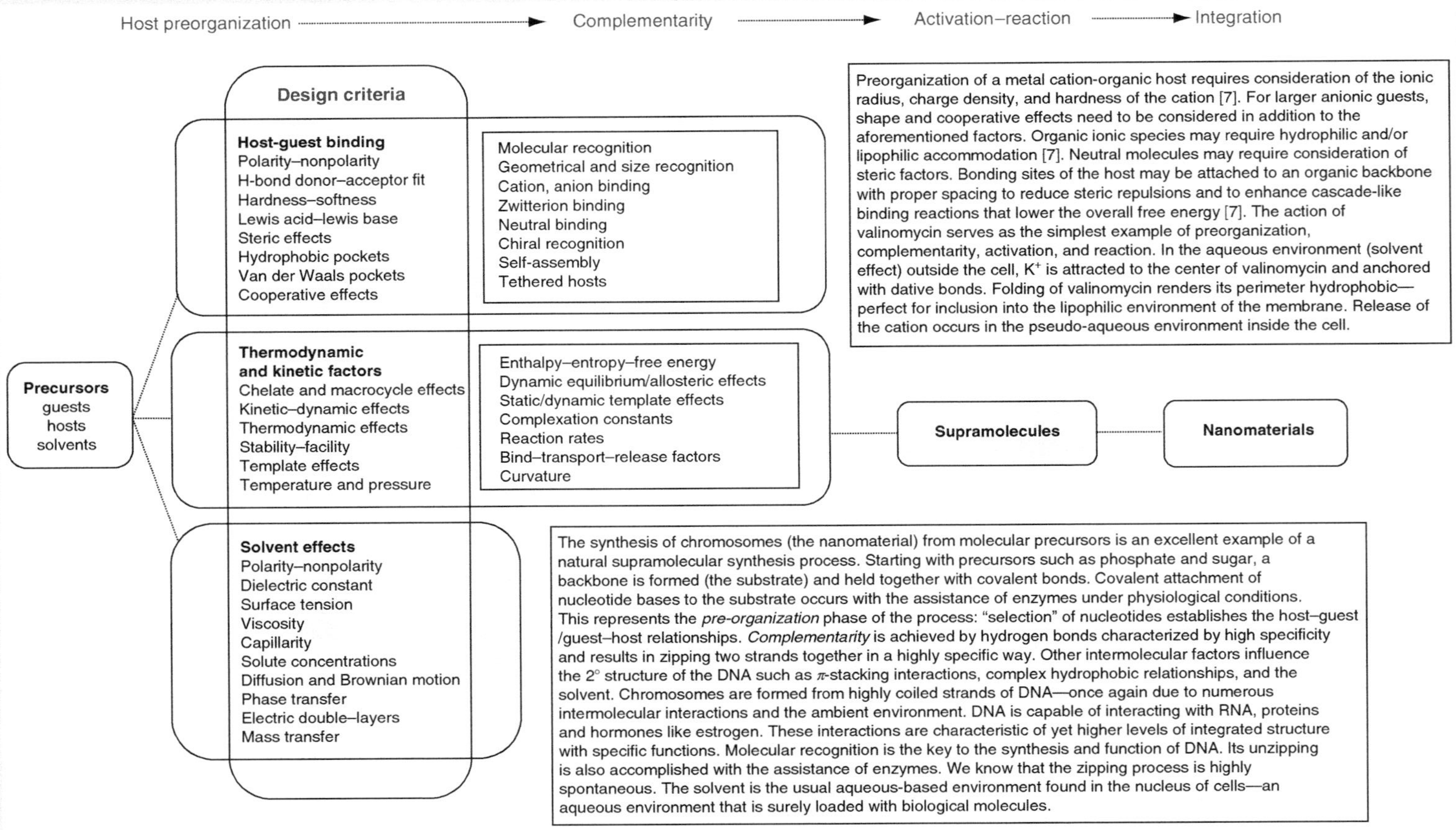

Host preorganization ───▶ Complementarity ───▶ Activation–reaction ───▶ Integration

Precursors
guests
hosts
solvents

Design criteria

Host–guest binding
Polarity–nonpolarity
H-bond donor–acceptor fit
Hardness–softness
Lewis acid–Lewis base
Steric effects
Hydrophobic pockets
Van der Waals pockets
Cooperative effects

Molecular recognition
Geometrical and size recognition
Cation, anion binding
Zwitterion binding
Neutral binding
Chiral recognition
Self-assembly
Tethered hosts

Thermodynamic and kinetic factors
Chelate and macrocycle effects
Kinetic–dynamic effects
Thermodynamic effects
Stability–facility
Template effects
Temperature and pressure

Enthalpy–entropy–free energy
Dynamic equilibrium/allosteric effects
Static/dynamic template effects
Complexation constants
Reaction rates
Bind–transport–release factors
Curvature

Solvent effects
Polarity–nonpolarity
Dielectric constant
Surface tension
Viscosity
Capillarity
Solute concentrations
Diffusion and Brownian motion
Phase transfer
Electric double-layers
Mass transfer

Supramolecules

Nanomaterials

Preorganization of a metal cation-organic host requires consideration of the ionic radius, charge density, and hardness of the cation [7]. For larger anionic guests, shape and cooperative effects need to be considered in addition to the aforementioned factors. Organic ionic species may require hydrophilic and/or lipophilic accommodation [7]. Neutral molecules may require consideration of steric factors. Bonding sites of the host may be attached to an organic backbone with proper spacing to reduce steric repulsions and to enhance cascade-like binding reactions that lower the overall free energy [7]. The action of valinomycin serves as the simplest example of preorganization, complementarity, activation, and reaction. In the aqueous environment (solvent effect) outside the cell, K⁺ is attracted to the center of valinomycin and anchored with dative bonds. Folding of valinomycin renders its perimeter hydrophobic—perfect for inclusion into the lipophilic environment of the membrane. Release of the cation occurs in the pseudo-aqueous environment inside the cell.

The synthesis of chromosomes (the nanomaterial) from molecular precursors is an excellent example of a natural supramolecular synthesis process. Starting with precursors such as phosphate and sugar, a backbone is formed (the substrate) and held together with covalent bonds. Covalent attachment of nucleotide bases to the substrate occurs with the assistance of enzymes under physiological conditions. This represents the *pre-organization* phase of the process: "selection" of nucleotides establishes the host–guest /guest–host relationships. *Complementarity* is achieved by hydrogen bonds characterized by high specificity and results in zipping two strands together in a highly specific way. Other intermolecular factors influence the 2° structure of the DNA such as π-stacking interactions, complex hydrophobic relationships, and the solvent. Chromosomes are formed from highly coiled strands of DNA—once again due to numerous intermolecular interactions and the ambient environment. These interactions are characteristic of yet higher levels of integrated structure with specific functions. Molecular recognition is the key to the synthesis and function of DNA. Its unzipping is also accomplished with the assistance of enzymes. We know that the zipping process is highly spontaneous. The solvent is the usual aqueous-based environment found in the nucleus of cells—an aqueous environment that is surely loaded with biological molecules.

Supermolecule → recognition, transformation, translocation → polymolecular self-assembly/self-organization + functional components → molecular and supramolecular devices

Jonathan W. Steed and Jerry L. Atwood [7] place emphasis on the following parameters:

Chelate–macrocycle effects + preorganization–complementarity + thermodynamic–kinetic selectivity + intermolecular–supramolecular interactions → HOST DESIGN

Not all of these categories are independent parameters. Accordingly, host *pre-organization* is perhaps the most important aspect of the design process—the host–guest relationship: Is there *complementarity* between the host and the guest? Do electronic and steric interactions, physical size, molecule spacing, and potential conformational changes of the host match those of the guest? Electronic compatibility includes polarity and charge, hardness–softness, H-bond alignment, and Lewis acid–base complementarity, to name just a few [7].

Successful *preorganization* results in hosts that exhibit little conformational changes after binding with a guest [7]. This is an important factor in host–guest design because the most stable complexes are achieved with preorganized hosts [7]. Interestingly, dramatic (and dynamic) conformational changes occur routinely in enzymatic hosts during and after binding. The conformational changes serve to strain bonds in substrates once drawn into the active pocket of an enzyme. It is much easier, for example, to break strong covalent bonds that are artificially lengthened or bent in this way. The coordination of substrates with one water molecule is a kinetic effect that facilitates the dissociation of substrate bonds by providing a sources of hydrolysis of polarity required for orientation. In an ideal synthetic world, we wish our supramolecular structures that comprise the bulk of the nanomaterial to be static once formed, unless the purpose is that of sensing. In sensors, allowing for guests to "check out" ensures that the sensor is able to continue with its purpose.

11.2.1 Thermodynamic and Kinetic Effects

Thermodynamic and kinetic factors are major players on the supramolecular design stage. The biological domain offers remarkable and numerous examples of selectivity. It also offers examples of nonselectivity (e.g., the oxidation–reduction of many kinds of substrates by oxidoreductases implies there is some flexibility in choice of substrate). One of the prime directives of supramolecular chemistry is selectivity because stable structures are preferred to those that accommodate more than one species. Obviously it would be problematic to build a uniform structure from a pool of similar but structurally varied materials. Therefore, analysis of binding constants and kinetic parameters of potential guests beforehand is a prudent path to take.

Selectivity and stability of the resulting host–guest complex are interrelated. The "right guest" usually tends to form the most stable H–G complex. Chelate effects, macrocycle effects, intermolecular interactions, solvent effects, concentration profiles, and reaction rates all influence the thermodynamics and kinetics of a potential H–G couple. Selectivity by itself is not enough, however.

The guest molecule must also form a thermodynamically stable complex with the host—in some instances, one that is able to release the guest if necessary (as in living things). For synthetic purposes, irreversible complexes are usually desired except in the case of sensors where reversible binding is preferred [7]. The magnitude of the binding constant (a.k.a. complexation, association, or formation constant) reflects the degree of thermodynamic stability of a host–guest complex:

$$H + G \leftrightarrow HG \tag{11.2}$$

$$K_{f,1} = \frac{[HG]}{[H][G]} \tag{11.3}$$

where K_f is the equilibrium constant, and H and G represent the concentrations of the host and guest, respectively. Selectivity with regard to the equilibrium constants of two or more potential guest molecules is the ratio of the respective binding constants [7]:

$$Selectivity = \frac{K_{G,1}}{K_{G,2}} \tag{11.4}$$

The proper choice of guest (lock and key), preparation of the host (preorganization), and complementarity (host–guest fit) all contribute to a successful supramolecular design process [7,8].

The Chelate and Macrocycle Effects Revisited.   Binding constants are estimated by experiment and by simulation. Molecular dynamic modeling in particular has proven to be extremely useful in describing H–G energy and structures. Numerous experimental methods, some direct and some indirect, have been employed to determine binding constants and other H–G-related parameters. Calorimetry, solubility, titration, potentiometry, spectroscopy, x-ray crystallography, and the vast array of nanotools discussed in chapter 3 have all provided useful information concerning substrate selection and binding.

Although binding constants have a favorable enthalpy due to the chelate or macrocycle effects, the entropy of their formation is usually unfavorable [8]. The association constant of a 1:1 (monodentate) H–G reaction was shown in equation (11.3). For metal binding, chelates with a higher number of binding sites demonstrate dynamic molecular recognition as the binding constant changes with the addition of another guest:

$$G + G \leftrightarrow HG_2; \quad K_{f,2} = \frac{[HG_2]}{[HG][G]} \tag{11.5}$$

and the overall reaction (with equation 11.2):

$$H + 2G \leftrightarrow HG_2; \quad \beta_2 = \frac{[HG_2]}{[HG][G]^2} \tag{11.6}$$

With $\beta_1 = K_{f,1}$, β_n overall is [8]

$$\beta_n = \prod_1^n K_{f,n} \qquad (11.7)$$

The macrocycle effect is similar to the chelate effect. Macrocycles are more stable than their respective open chain analogs [8]. Increased stability of macrocycle complexes is because; (1) macrocyclic hosts are less solvated (therefore require less energy to desolvate); (2) macrocycles are rigid to begin with and therefore potential ΔS upon complexation is smaller than for a flexible system; and (3) macrocycle complexes are relatively inert chemically compared to their acyclic analogs [8]. Although having one more binding site in a single host is better in terms of complex stability, larger cycles do not necessarily translate into more stable complexes [7]. Formation constants range from 10^2 to 10^4 for podates (single chain complexing agents) and 10^4 to 10^6 for corates (cyclic complexing agents) due to the macrocyclic effect. The *macrobicyclic effect* demonstrates even higher K_f. A macrobicycle host is characterized by two macrocyclic rings linked at the poles and is represented by the cryptands. The formation constant macrobicycle complexes like cryptates range from 10^6 to 10^{12} [7].

Recall from earlier discussions that chelates and templates exert a kinetic effect on processes. The ability to bring reactants into close proximity results in species that may not ordinarily form. This is a kinetic effect. Kinetic studies are based on reaction rates, concentration, and temporal coordinates.

A study by Nachtigall, Lazzarotto, and Nome provides an excellent example of a complex binding study [38]. The host is calix[4]arene (**Fig. 11.15**).

EXAMPLE 11.1   *Coordination Thermodynamics and the Chelate Effect*

Calculate ΔG and the formation constants K_f (and log K_f) for the following coordination reactions at 298 K and discuss the reasons for the differences. M = metal cation.

(a) $M^{2+}_{(aq)} + 4NH_{3(aq)} \rightarrow [M(NH_3)_4]^{2+}_{(aq)}$ $\Delta H° = -53$ kJ·mol^{-1}, $\Delta S° = -42$ J·mol^{-1}·K^{-1}

(b) $M^{2+}_{(aq)} + 4CH_3-NH_{2(aq)} \rightarrow [M(CH_3-NH_2)_4]^{2+}_{(aq)}$ $\Delta H° = -57$ kJ·mol^{-1}, $\Delta S° = -67$ J·mol^{-1}·K^{-1}

(c) $M^{2+}_{(aq)} + 2(NH_2-CH_2-CH_2-NH_2)_{(aq)} \rightarrow [M(NH_2-CH_2-NH_2)_2]^{2+}_{(aq)}$
$\Delta H° = -56$ kJ·mol^{-1}, $\Delta S° = 10$ J·mol^{-1}·K^{-1}

Solution

In all cases, use $\Delta G° = \Delta H° - T\Delta S°$ *and* $\Delta G° = -RT \ln K_f$ *(where log* $K_f = \ln K_f/2.303$*)*

(a) $\Delta G = -53$ kJ·mol^{-1} $- (298 K)(-0.042$ kJ·mol^{-1}·K$^{-1}) = -41$ kJ·mol^{-1}
$\ln K_f = \Delta G/[-RT] = -41$ kJ·mol^{-1}/$[-0.00834$ kJ·mol^{-1}·K$^{-1} \times 298 K] = 16.5$
log $K_f = \ln K_f/2.303 = 7.2$; $K_f = 1.5 \times 10^7$ M^{-1}

(b) $\Delta G = -57$ kJ·mol^{-1} $- (298 K)(-0.067$ kJ·mol^{-1}·K$^{-1}) = -37$ kJ·mol^{-1}
$\ln K_f = \Delta G/[-RT] = -37$ kJ·mol^{-1}/$[-0.00834$ kJ·mol^{-1}·K$^{-1} \times 298 K] = 15$
log $K_f = \ln K_f/2.303 = 6.5$; $K_f = 3.1 \times 10^6$ M^{-1}

(c) $\Delta G = -56$ kJ·mol^{-1} $- (298 K)(+0.010$ kJ·mol^{-1}·K$^{-1}) = -59$ kJ·mol^{-1}
$\ln K_f = \Delta G/[-RT] = -59$ kJ·mol^{-1}/$[-0.00834$ kJ·mol^{-1}·K$^{-1} \times 298 K] = 23.8$
log $K_f = \ln K_f/2.303 = 10.3$; $K_f = 2.3 \times 10^{10}$ M^{-1}

The formation constant is higher for multidentate ligand in (c) by several orders of magnitude. Notice that the entropic consideration for (c) is positive—a definite plus if spontaneity is desired in a reaction.

FIG. 11.15

FIG. 11.15 *Calixarene host species are shown. Left: A p-t-butylcalix[4]arene molecule binds a sodium cation in its lower tier structure forming an endo-calix complex. Middle: Bonding of toluene guest is shown in the upper tier binding complex. Bonding is accomplished primarily with van der Waals interactions. Right: Binding of an aliphatic amine to form an exo-calix complex is shown. In the experiment of Nachtigall et al. [38], a calixarene without para-substituted groups was used.*

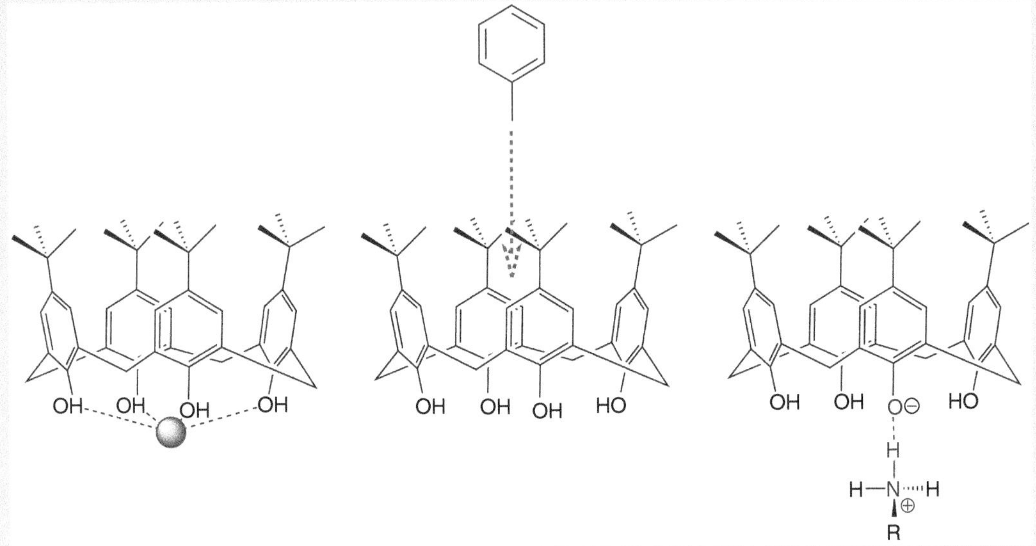

Calixarenes are members of the family of cyclophanes because they possess bridged aromatic groups [7]. They are formally known as substituted [1,1,1,1] metacyclophanes. Macrocyclic calixarenes are synthesized from linked phenol groups to form a four-membered macrocycle. *p*-Methylcalix[4]arene was synthesized in 1956 by R. F. Hunter and B. T. Hayes. Calixarenes are able to bind cations, anions, and neutral molecules.

Calixarenes offer more than one means to complex guest species. If the guest is a monovalent cation, the lower rim (acidic) hydroxyl groups are able to collectively bind, with dative bonds, the cation by the hydroxyl groups. This is referred to as the *endo* mode of binding. On the other hand, the binding of aliphatic amine complexes was found to proceed via an *exo* mode of binding in which the bound guest is outside the lower rim phenolic substituents—akin to a perching mode mentioned earlier [38]. The complex is called an *exo*-calix-ammonium complex. The major intermolecular interaction responsible for binding is the hydrogen bond between N^+–$H \cdots \cdots O^-$–*Phe*. The calixarene is also capable of binding toluene and other similar guests within the hydrophobic upper rim region of the molecule to form an inclusion compound. In this case, van der Waals forces and steric and hydrophobic interactions need to be considered during the molecular design process.

Nachtigall et al. studied the binding between calix[4]arenes (without substituted *para* groups) and aliphatic amines by conductance, spectroscopic, and NMR methods [38]. Titrations were carried out in acetonitrile by hexylamine at constant temperature (298.0 ± 0.1°C). UV absorbance (310 nm) versus

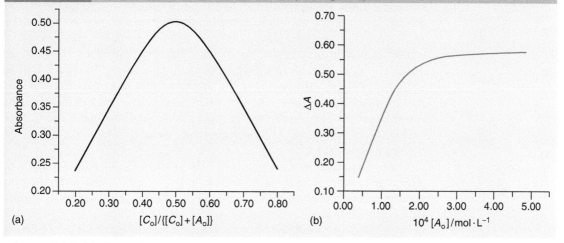

Source: F. F. Nachtigall et al., *Journal of the Brazilian Chemical Society*, 13, 295–299, 2002.

$[C_o]/\{[C_o] + [A_o]\}$ indicated 1:1 stoichiometry, where $[C_o]$ is the total concentration of the calixarene and $[A_o]$ is the total concentration of the amine. $[C_o] + [A_o]$ = 3.0 × 10⁻⁴ M (**Fig. 11.16a**). The concentration relationships between one-on-one host–guest complexes are often portrayed by Job plots (**Fig. 11.16a**). The concentration of the complex [C] is plotted against the ratio of the host concentration to the total concentration of host and guest: $[H]/\{[H] + [G]\}$. From the figure it is obvious that, in a 1:1 ideal circumstance, the maximum in complex concentration is achieved when each component of host and guest has the same concentration—an idealistic case that does not consider higher order complexation and numerous other factors [7].

The equilibrium between the calixarene and the amine conformed to the form of an acid–base reaction:

$$\text{Calix-OH} + \text{NR}_3 \rightarrow \text{calix-O}^- + \text{HNR}_3^+ \qquad (11.8)$$

and K_p, the equilibrium constant, is equal to

$$K_p = \frac{[C^-][A^+]}{[C][A]} \qquad (11.9)$$

Rewriting using total concentrations,

$$0 = \left(1 - \frac{1}{K_p}\right)\left[C^-\right]^2 - \left\langle [C_o] + [A_o] \right]\left[C^-\right] + [C_o][A_o] \right\rangle \qquad (11.10)$$

The concentrations are related to the Beer–Lambert law by

$$\Delta Abs = \varepsilon \left(\frac{[C_o]+[A_o]-\sqrt{([C_o]+[A_o])^2 - 4[C_o][A_o]\left(1-\frac{1}{K_p}\right)}}{2\left(1-\frac{1}{K_p}\right)} \right) \qquad (11.11)$$

with ε the molar absorptivity equal to 3954 mol·L^{-1} and $K_p = 18$ for the proton transfer reaction. The pK_a value for the hexylamine reaction was determined to be 18.26. The pK_a for phenol is higher at 26.6 due to the lack of stabilization by intramolecular hydrogen bonding in the calixarene. The plot of absorbance versus $[A_o]$ of the calixarene–hexylamine complex is shown in **Figure 11.16b.** The value of K_p was extracted by application of nonlinear regression analysis. At lower concentrations, the K_p form (the dissociation into ions) of the apparent two-step process was dominant. At higher concentrations, the K_a form (the acid–base reaction) assumed more importance—also the part that involved the hydrogen bond [38].

Other titrant species were applied in the study, and differences in K_p were attributed to differences in the basicity of the amines. Conductimetric titration measurements and NMR titration analysis (chemical shifts of the ^{1}H signal) showed similarly shaped curves during the course of the titrations. Fluorescence, calorimetric, and potentiometric titrations are other means to measure binding constants and other parameters. Extraction methods, in which there is a partition of a guest between two immiscible phases, are a means by which macrocycle selectivity can be determined [7].

Thermodynamic Control in Natural Systems.   The self-assembly of a double helix consisting of adenine and uracil nucleic acids will serve us as a perfect example of the concept of dynamic equilibrium [39]. Dynamic equilibrium is defined as a dynamic state of balance between forces—a characteristic of many metabolic processes. After some back and forth, dynamic equilibrium will ultimately drive the process to favor the thermodynamic product [26]. In **Figure 11.17**, the free energy of complexation is tracked along the course of nucleation and propagation of two complementary strands of RNA (comprising only of adenine and uracil nucleic acid bases). Although ΔH is negative from the onset (nucleation), ΔG is positive due to the unfavorable $T\Delta S$ term, and it becomes clear from the graphs that a minimal number of base pairs are needed before a double helix can be formed by self-assembly. After four base pairs are zipped together, ΔG becomes negative and the reaction proceeds spontaneously. During propagation, the entropic factor plateaus and self-assembly fall into place to zip together the remainder of the complementary base pairs [26,39]. The bottom line is that assembling larger supramolecular assemblies becomes easier as more noncovalent bonds are involved in the process [26,39]. The activation energy required to overcome the entropic penalty of the initial step is probably overcome by participation of an enzyme.

The tobacco mosaic virus (TMV) shown earlier illustrates another example of thermodynamic influence over a self-assembly process [26]. Recall that the virus is a cylindrical structure about 300 nm in length with a diameter of 18 nm. It is

FIG. 11.17

The free energy of nucleation and propagation of RNA zipping (consisting only of adenine and uracil) is tracked in the figure. Two components are contributing to the overall free energy: enthalpy and entropy. Initially, ΔG is positive due to the unfavorable $T\Delta S$ term. This makes sense because zipping two strands together contributes inherent entropy hits to the overall free energy. From the graph it is clear that a minimal number of bases have to be zipped together before the enthalpy term dominates the process. The entropic component plateaus as expected because each increment provides the same level on negentropy. This phenomenon is an example of dynamic equilibrium. Once a complex is established, the affinity for establishing more complexes increases and it is driven by increasingly favorable enthalpic factors. In the very broadest sense, this process resembles an allosteric effect (a phenomenon characteristic of enzymatic systems) in that ligands (the nucleotides) are able to affect the progress of complexation and exhibit a degree of cooperativity.

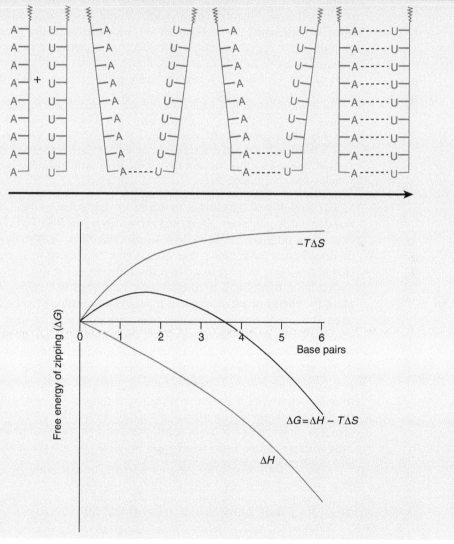

Source: M. E. Craig, D. M. Crothers, and P. Doty, *Journal of Molecular Biology*, 62, 383–401, 1971. With permission.

made of 2130 identical protein subunits wrapped around a single strand of RNA in a helical fashion. The RNA contains ca. 6400 nucleotide bases. During the assembly progresses of the virus, the rush to completion gains momentum as more units are self-assembled into the body of the virus.

During the synthesis of this structure, several nanoscale interactions are evident: (1) preorganization, complementarity, and molecular recognition; (2) thermodynamic control and the dynamic equilibrium effect; (3) self-assembly of a hierarchical structure; (4) template effects (of the RNA and the preformed protein wedges); (5) self-correction (due to thermodynamic factors); and (6) efficiency (due to high selectivity). In the first step, a wedge-shaped protein monomer is synthesized. The monomers self-assemble (by molecular recognition) to form disc-shaped subunits. Once a disc is formed, it binds to a single strand of RNA that induces a conformational change in the disc's geometry into that of a helical structure [26]. The RNA strand serves as a templating mechanism around which more discs are added until the entire virus is formed. As more units are added, there is a rush to completion, a cascading effect of sorts, until the virus is fully assembled and termination of the process occurs. Trent H. Galow of the University of Edinburgh summarizes the process:

> Equilibrium means reversibility. Intermediates that do not represent the thermodynamic minima can progress in a forward or reverse direction—relatively easily to overcome the activation barrier. Once the lowest energy aggregate is formed, it is hard to overcome the activation barrier for the reverse stage. Equilibrium continues to the right-hand-side until complete. Inherently, a self-correcting process.

Self-assembly of the virus is an example of dynamic molecular recognition and dynamic equilibrium, illustrated previously by the RNA and TMV examples. Once a few building blocks are in place and the initial threshold overcome, the thermodynamic conditions that exist at the beginning of the process have changed in favor of more assembly. The function of the disc, an expression of molecular recognition, is to interact with a strand of RNA. Once the RNA is in place, the energy of self-assembly is reduced and addition of more discs is facilitated; this is akin to a cascade effect (one with purpose) that gains momentum with each step. If the virus decomposes, the individual components are capable of reassembly into a TMV with all its natural function restored [26].

The example of the TMV involves thousands of molecules, scores of proteins, and 2130 protein discs, but only one strand of RNA. These molecular recognition and self-assembly processes occur under physiological conditions. Backing up the TMV synthesis is a central command system that orchestrates this process. The simplest of life forms exhibits mastery of chemical processes that we only wish we can replicate in a beaker; however, we are getting closer, step by step, with each breakthrough.

11.2.2 Basic Design Parameters: The Host, the Guest, and the Solvent

Precursor Molecules. Precursor molecules are defined as rather small molecules that require making and breaking of covalent bonds during their synthesis. The

FIG. 11.18	*Condensation reaction between two amino acids to form a small dipeptide is depicted. Covalent bonds are broken and formed in this reaction. In living things, enzymes mediate such processes under conditions that are considered to be rather mild. Peptides should be considered as supermolecules held together by covalent bonds with relatively similar composition.*

precursor molecule or molecules will form the basis of the next step: designing the receptor and substrate. Proteins, for example, are the result of a condensation reaction between two amino acid precursor molecules **(Fig. 11.18)**. In polymer chemistry, precursor molecules are considered to be monomers. The backbones of polymers are analogous to those of proteins in that covalent bonds hold them together. The interactions between polymeric chains, however, can be covalent in nature, but many are considered to be intermolecular. Interactions between proteins (or within one protein) can be intermolecular or covalent (the disulfide linkage). Are proteins supermolecules or supramolecular molecules?

Once proteins are folded and provide a specific function, do they then become supramolecules? How about a compromise? Let us call peptides superprecursors. It is easier to define synthetic supramolecular precursors. All synthetic precursor molecules consist of covalent bonds that are made by hard chemistry techniques (under kinetic control conditions). The synthesis of the first crown ether shown earlier is a typical example.

Receptors and Substrates. Molecular receptors (hosts) and substrates (guests) are generally classified as organic chemicals (precursor molecules) held together by strong covalent bonds [4]. The key to designing receptor (and substrate) molecules is the ability to convert organic molecules into ones that have potential for molecular recognition [4]. This paradigm does not apply to all precursor molecules. For example, the precursor molecules that combine to form the rosette (melamine and cyanuric acid) already possess a high degree of molecular recognition via their complementary H-bonding structure. But we get the point nonetheless. There are many possible kinds of substrates (guests): cationic, anionic, zwitterionic, or neutral species. These species originate from inorganic, organic, or biological sources.

Another challenge faced by designers of drugs is estimation of binding constants of drugs to receptors [40]. Molecular recognition mechanisms bring the parties together, but what happens afterwards is also important. Partitioning free energy of binding or association according to the type of bonding (hydrogen bonds, $\pi-\pi$ interactions, salt bridges, or the hydrophobic effect) is one approach, but that would ignore the cooperative effect that intermolecular attractions are known to exhibit. The strength of the hydrophobic effect, a measure that depends significantly on the type of method applied, must be measured in the context of the medium in which it occurs and not in isolation [40]. This brings us to solvent effects.

Solvent Effects.   Macromolecular interactions with water are one of the most important factors to consider when designing a supramolecular system. Structure, function, specificity, stability, and the dynamical potential of supramolecules are all influenced by solvation with water if taking place in aqueous media. For example, proteins rely on water for their three-dimensional structure [41]. Water also plays a vital role in protein–nucleic acid recognition by mediating bridging hydrogen bonds between functional groups [42]. Water acts as a lubricant that facilitates hydrogen bonding. Proteins have *conformational flexibility* and in solution exhibit a high degree of hydration states (e.g., the existence of ordered water in the form of a hydration shell that is denser than bulk water and facilitates proton transfer [41]. The energetic price tag for

---

EXAMPLE 11.2   *A Simple Demonstration of a Solvent Effect*

The dipole of α-alanine ethylester ($\mu = 2.09$) is in line with a potassium cation at a distance of 0.5 nm. What is the energy of the interaction in water ($\varepsilon_r = 78.85$)? in methanol ($\varepsilon_r = 32.63$)? in trimethyl amine ($\varepsilon_r = 2.44$) at 298 K? The

Solution

$$\text{Use } E_{Ion-Dipole} = \frac{ze\mu\cos\theta}{4\pi\varepsilon_o\varepsilon_r r^2}$$

Solution:
Permittivity of free space, $\varepsilon_o = 8.854 \times 10^{-12}$ $C^2 \cdot J^{-1} \cdot m^{-1}$, $e = 1.6022 \times 10^{-19}$ C

$$E_{Ion-Dipole}(water) = \left[\frac{(+1)(1.6022\times10^{-19}\,C)(2.09\times3.336\times10^{-30}\,C\cdot m)\cos0}{4\pi(8.854\times10^{-12}\,C^2\cdot J^{-1}\cdot m^{-1})(78.85)(0.5\times10^{-9}\,m)^2}\right]\left(\frac{6.022\times10^{23}}{mol}\right)\left(\frac{1kJ}{1000J}\right) = -0.31\,kJ\cdot mol^{-1}$$

Reducing all nonvariable components to 24.18 (the variable component is the dielectric constant of the solution):

$$E_{Ion-Dipole}(methanol) = \frac{24.18}{32.63} = -0.74\,kJ\cdot mol^{-1}$$

$$E_{Ion-Dipole}(trimethylamine) = \frac{24.18}{2.44} = -9.91\,kJ\cdot mol^{-1}$$

The solvent matters. If you are designing a host species that is trying to attract an ion with a dipolar appendage, then choice of solvent is very relevant.

hydrating a new surface or removing water from a surface is between 3.6 and 63 kJ · mol^{-1} · Å^{-2} [42].

Hydrophobic effects influence the three-dimensional folding of proteins. Entropic considerations arise when a hydrophobic surface is in contact with polar water molecules. Quasi-crystalline organization of water occurs on the surface of the hydrophobic entity, but H-bonding saturation (four bonds) with other water molecules does not occur and this phenomenon results in water that is denser than it is in its natural state. As a result, slight negative entropy is obtained when water molecules are released from the pocket into the bulk aqueous solution as each H_2O is bonded tetrahedrally to other water molecules.

Solvent effects were covered in some detail in chapter 10 but we will elaborate upon this important parameter further in this section. The solvent plays a role in the recognition process—particularly with regard to the degree of solvation. Going back to our discussion of hydration spheres in chapter 10, removing the hydration sphere around a cation would have a positive (unfavorable) enthalpy component, although entropy would increase as a result (favorable). When designing a host–guest relation that takes place within a solvent, a priori inspection of the solvation properties of potential guest molecules and the thermodynamic trade-offs between the host and guest must be evaluated.

In addition to the solvation interactions and the electrostatic factors discussed in example 11.2, other solvent evaluation criteria that weigh into the design process include the fact that the hydrophobic effect (from chapter 10) affects the formation (complexation) constant of H–G reactions. If a H–G complex is rendered insoluble in the parent solvent, then the tendency to bind to the guest is enhanced. The binding of anion guests depends strongly on the type of solvent. For example, binding constants of small inorganic anions are better in carbon tetrachloride than it is in water [7].

The design of supramolecular systems is not trivial and requires detailed analysis of precursors, hosts and guests, and the complexes themselves once formed. Consideration of thermodynamic and kinetic effects as well as the contribution of the solvent, all in concert, is a formidable task that may require the assistance of a computer program. Nature, of course, does this on a routine basis—the result of hundreds of millions of years of trial and error. Perhaps we can devise a combinatorial chemistry scenario that is able to accelerate the replication (molecular recognition) process, at least to the level of the simplest primitive, most rudimentary organism.

11.3   EXTENDED SUPRAMOLECULAR STRUCTURES

Most chemical systems are fabricated by mixed bottom-up methods. Mixed methods involve several kinds of materials, intra- and intermolecular interactions, solvents, and reaction conditions. Each step requires meticulous forethought and design. We show one remarkable example: the formation of monolayers of gold clusters, molecular squares, and surface tethered groups.

11.3.1   *Golden Molecular Squares*

The synthesis of molecular squares is straightforward (**Fig. 11.19**) [43]. It is based on the transition metal–ligand interaction and it has recently been applied to construct metal macrocycles and cages [43]. Molecular squares from osmium tetroxide, olefin, and bispyridyl ligands come together to form an electrically neutral octahedral coordination center around osmium, making them amenable for hydrogen bonding without interference from a charged species. Mixtures contained small to large macrocyclic osmate esters (COEs). It was found that higher ordered structures were obtained when aromatic diols were added to the mix. Apparently, π-stacking and hydrogen bonding self-assembly occur between the COEs and the diols. The osmium(VI) is capable of directing the synthesis of COEs and higher structures by coordination, chemical reaction, hydrogen bonding, and aryl stacking. Hydrogen bonds form between the alcohol and the available osmium oxygens. Aryl stacking was achieved between 24 individual components. The size of the diol plays a role in the size of the supramolecule [43]. This strategy is an interesting route to synthesize integrated supramolecular structures.

Lahav et al. in 1999 developed ordered arrays of supramolecular structures for sensing, electronic and photochemical purposes [44]. Fujita et al. in 1994 synthesized molecular squares with the goal of fabricating new types of host–guest cavities [45]. The squares consist of bipyridine units linked by transition metal ions such Pd^{2+} and Pt^{2+} [45]. Au-molecular square structures are depicted in **Figure 11.20** and illustrate a multistep fabrication process involving numerous kinds of chemistry—a trait very typical of modern nanoscale engineering and chemistry. Precursors such as 3-aminopropyltriethoxysilane are synthesized according to traditional organic chemical routes and serve as the precursor molecules. An indium–tin-oxide (ITO) supermolecular surface is activated to generate hydroxyl groups. The ethoxy group on the silane is exchanged by a surface hydroxyl and the silane is anchored to the surface. The monolayer is further stabilized by lateral interactions between adjacent silanes. The amino-terminus is made positive by a subsequent protonation step. Prefabricated supermolecular gold colloids of ca. 12 nm diameter are introduced to the positively charged surface and allowed to equilibrate. The gold colloids were made by a reduction of gold salts in a citrate solution. Molecular squares are then introduced onto the Au colloid modified surface. Au and cross-linking molecules are alternatively added until the desired number of layers is achieved. Amazingly, all the reaction sequences occur at room temperature under mild conditions. This type of synthesis illustrates the utility and practicality of self-assembly processes.

The typical plasmon absorption of Au colloids of that size is $\lambda = 518$ nm, but another absorption was noticed at 650 nm, suggesting the presence of aggregates due to interparticle coupling [44]. The molecular squares carry positive charges and were thus able to cross-link with the Au nanoparticles by "multi-site" ion pairing [44]. The ratio of Au:Pd was 30:1; the ratio of Au:Cd was 100:1. As a sensor, the cross-link molecular squares act as π-receptors for π-donor substances such as dialkoxybenzene derivatives. This is an excellent example of a self-assembly process following covalent interactions to synthesize

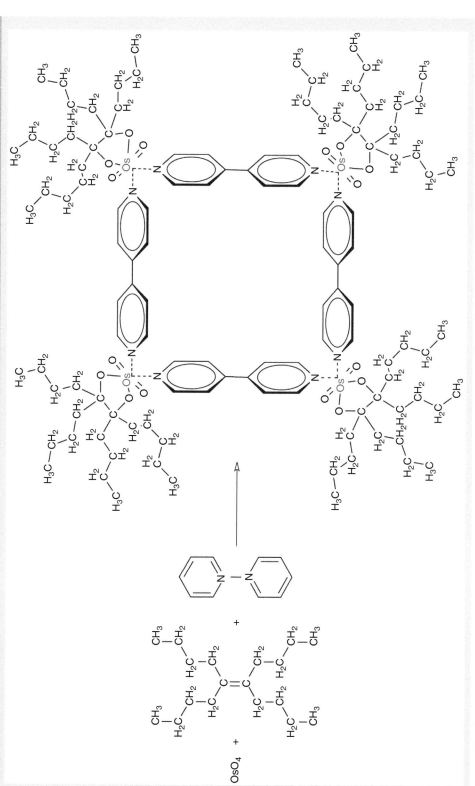

FIG. 11.19 Molecular square supramolecular complex anchored by osmium(VI) is depicted. The repeating units are capable of forming hexagonal and larger structures.

Fɪɢ. 11.20   *Synthesis and structure of a multilayer golden molecular square complex is depicted [44]. The figure shows a process in which covalent bonding, electrostatic interactions, and π-processes are in effect to form the structure. Fujita et al. developed a series of cyclophane-like molecular square precursors that consist of bipyridines linked by metal ions (e.g., molecular squares) [45] (Fig. 11.8o and p). Gold colloid (represented by gold spheres) and molecular square (represented by blue squares) were assembled into an integrated structure. The assembly was formed by noncovalent cross-linking of 12-nm Au colloids with molecular square transition metal complexes. An activated ITO surface was reacted with 3-aminopropyltriethoxysilane by covalent interactions. The amine groups yielded a positively charged surface. The chemical-sensing layers were fabricated by alternating applications of citrate stabilized Au colloids molecular squares. The molecular squares served to cross-link*

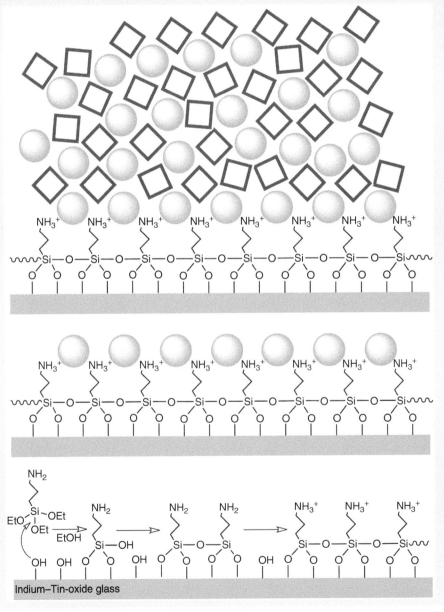

Source: M. Lahav et al., *Chemical Communications*, 1937–1938, 1999.

FIG. 11.20 (CONTD.) *the Au nanoparticles. Molecular squares are a new generation of hosts that contain cavities suitable for binding guests. For example, the squares can be used to sense (complexation) anions. An ITO substrate is first activated to generate accessible hydroxyl groups. A reactive silane (3-aminopropyl-triethoxysilane) was applied to the surface to generate a positively charged monolayer. Citrate-stabilized Au colloids (12 nm) were then applied to the monolayer. Palladium squares [(en)Pd(4,4'-bipy)]4(NO₃)₈ or cadmium squares [Cd(4,4'-bipy)]4(NO₃)₈ adhered to the gold colloids and fabrication of the "first receptor layer" was completed. Alternating steps of gold colloid and square addition led to the structure depicted at the top of the figure. The plasmon band absorption of gold colloids of this size (12 nm) is 518 nm. Interestingly, an additional plasmon absorption that was significantly red shifted (to 650 nm) was detected [44]. The shifted peak is attributed to interparticle plasmon coupling—proof that particles are in contact with each other via the molecular squares. The absorption at 650 nm increased as the number of layers was increased. Electrical conductivity through the layer was also enhanced as more layers were added.*

a supramolecular structure. It also serves as an excellent example of a potential sensor application based on π-interactions.

11.3.2 Synthesis of Benzocoronene Complexes

Benzocoronenes are a versatile class of nanographitic precursors that are relatively easy to modify chemically and are able to form aryl stacks by self-assembly. Benzocoronene materials have promising applications in electronics. The fabrication of highly ordered two- or three-dimensional polymolecular architectures consisting of benzocoronene discotic liquid crystals involves intramolecular, intermolecular, and interfacial forces. Replacement of the R-groups is possible with a variety of species. Benzocoronenes are graphite analog precursor molecules synthesized by conventional organic chemistry by making and breaking covalent bonds (**Fig. 11.21**). Chemical modification of benzocoronenes is accomplished by organic chemical means as well. Ethylene oxide residues can be attached to the modifiable regions of the molecule to alter properties (**Fig. 11.21**). Once synthesized, the precursors self-assemble to form columnar structures called liquid crystals (**Fig. 11.22**).

The structure of liquid crystals (LCs) is somewhere between that of a crystalline solid and a liquid [46]. There are two kinds of liquid crystals: thermotropic (in which phase transitions are observed as a function of temperature) and lyotropic (in which phase transitions occur as a function of concentration). Liquid crystals were discovered in 1888 by the Austrian botanist Friedrich Reinitzer after he noticed that cholesteryl benzoate exhibited two melting points. At 145°C it melted into an opaque liquid and at 179°C, it melted but this time into a clear liquid. The term liquid crystal was credited to Georges Friedel in 1922.

LCs are found in two phases: *smectic* and *nematic*. Smectic phases are characterized by positional order (along one direction) and orientational order (along the layer normal or tilted away from the layer normal). Nematic phases exhibit no positional order but are characterized by long-range orientational order; all

FIG. 11.21   *A planar hexa-peri-benzocoronene is depicted. The diameter of the molecule is 0.15 nm. Left: Synthesis of hexabenzocoronene from an alkene is depicted. Right: Chemical modification of R-groups yields many kinds of hexabenzocoronenes.*

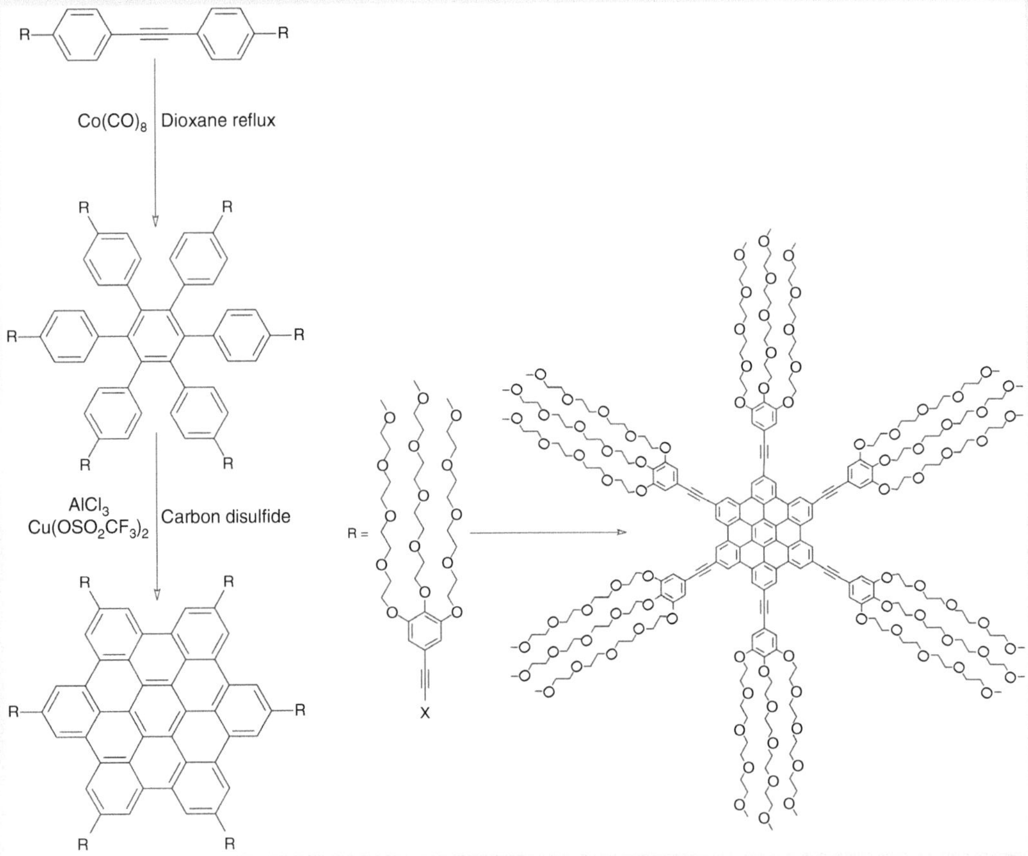

Source: Redrawn with permission from Professor Klaus Müllen, Max-Planck Institut für Polymer Forschung.

crystals are oriented ($a = b \ll c$) in the same direction within their respective domains. Nematic LCs can be aligned by external magnetic fields.

Benzocoronenes form uniaxial discotic (nematic) columnar liquid crystals. The columns can be further organized into rectangular or hexagonal arrays. The structure following mild heating is shown in **Figure 11.22** [47,48]. A benzocoronene species with $-C_{12}H_{25}$ substituted constituents showed LC phase at 106.5 and 124.4°C. A single LC phase was demonstrated at ~400°C [49]. In another study, hexa-peri-hexabenzocoronene (also a discotic liquid crystal) was combined with perylene dye to form thin films with vertically separated perylene–hexabenzocoronene domains with extended interfacial surface area [50]. The structure was applied to a photodiode device in which a photovoltaic response on the order of 34% efficiency was achieved at 490 nm [50]. The photovoltaic response was due to charge transfer between vertically aligned dye and the hexabenzocoronene π-systems. Over 50 different cores have been created from which about 3000 discotic LCs have been developed [51].

Fig. 11.22	*A schematic illustrating the transformations of benzocoronenes. Left: Self-assembly in water; middle: herringbone structure following self-assembly; right: liquid crystal columnar discotic structure.*

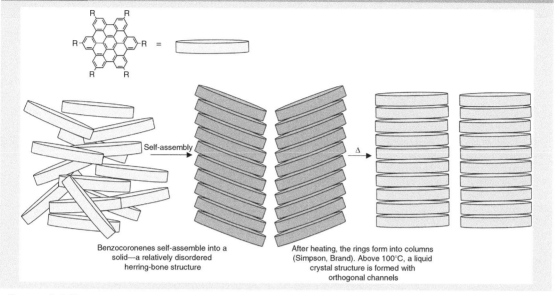

Benzocoronenes self-assemble into a solid—a relatively disordered herring-bone structure

After heating, the rings form into columns (Simpson, Brand). Above 100°C, a liquid crystal structure is formed with orthogonal channels

Source: C. D. Simpson et al., *Journal of Materials Chemistry,* 14, 494–504, 2004; A. M. van der Craats et al., *Advanced Materials,* 10, 36–38, 1998. With permission.

Graphitic nanotube analogs can be made from chemically modified benzocoronenes. Mixing R-groups (**Fig. 11.23**) to create an amphiphilic benzocoronene results in helical self-assembly in THF or THF–water mixtures [52]. Self-assembly in THF results in a tighter helix.

The aspect ratio of the tubes was ca. 1000 and the diameter of the opening was 14 nm. The wall was 3 nm in thickness and consists of a helical array of π-stacked graphene molecules. The exterior and interior of the molecules were covered

Fig. 11.23	*Asymmetrically modified benzocoronenes self-assemble into helical structures that resemble derivatized carbon nanotubes.*

Source: J. P. Hill et al., *Science,* 304, 1481–1483, 2004. With permission.

with hydrophilic triethylene glycol residues. The tubes were electrically conductive [52]—and all by self-assembly of a precursor molecule.

11.3.3 *Helical Supramolecular Polymers*

We provide one more example of a self-assembled supramolecular complex. This one is a DNA mimic. Bifunctional ureido-*s*-triazines with penta(enthylene oxide) side chains were self-assembled in water to form helical columns due to cooperative hydrophobic stacking of hydrogen bonded pairs—just like in DNA (**Fig. 11.24**) [53]. The monofunctional version was not able to form helical columns. Stacking occurred when the concentration exceeded 5×10^{-6} *M*, and the process was monitored by fluorescence spectroscopy [53]. Brunsveld et al. claim that the stacking configuration enhances the formation of a hydrophobic

FIG. 11.24

Ureido-s-triazines modified with ethylene oxide side chains are depicted in the figure. These materials appear to be nucleotide mimics and it is no wonder that they are able to form helical structures in water. The structures dimerize via four-fold hydrogen bonding and are able to stack aromatic cores by aryl–aryl intermolecular interactions. The core is protected from the solvent by the creation of a hydrophobic interior—all by self-assembly. The column radius was found by small angle neutron scattering (SANS) to be 0.16 nm. Interestingly, columns formed by the molecule on the left in the figure had discs that were able to rotate. There was no disc rotation in columns formed by the entity on the right due to the hexamethylene spacer between dimers. This experiment demonstrated the formation of highly ordered aggregates by cooperative intermolecular interactions: hydrophobicity, π–π stacking, and hydrogen bonding—the same three that are responsible for DNA's conformation.

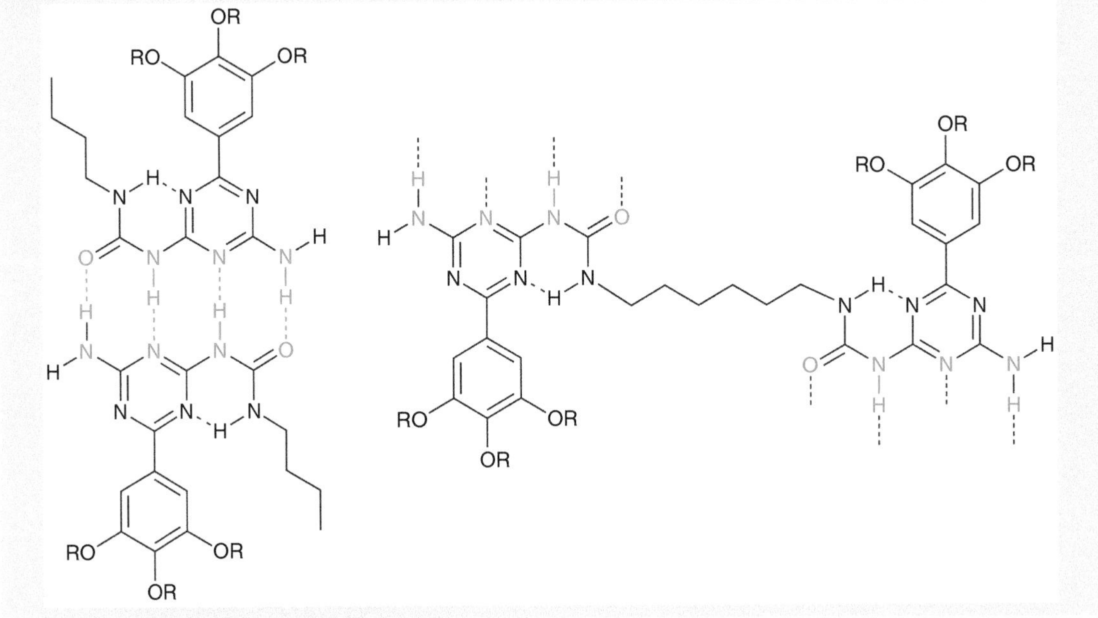

Source: L. Brunsveld et al., *Proceedings of the National Academy of Science USA*, 99, 4977–498, 2002. With permission.

"microenvironment" that promotes intermolecular hydrogen bonding to occur at higher concentrations due to shielding from water. Addition of side chains skews the helicity of the structures. Specially designed precursors allow for control over the design of chiral nanotubes—nanoscience in action! The R-groups are either long chain aliphatics or chainlike derivatives of ethylene oxide.

References

1. N. J. Turro, Supramolecular organic photochemistry: Control of covalent bond formation through non-covalent supramolecular interactions and magnetic effects, *Proceedings of the National Academy of Science USA*, 99, 4805–4809 (2002).
2. N. J. Turro, From molecular chemistry to supramolecular chemistry to superdupermolecular chemistry. Controlling covalent bond formation through non-covalent and magnetic interactions, *Chemical Communications*, 20, 2279–2292 (2002).
3. F. M. Menger, Supramolecular chemistry and self-assembly, *Proceedings of the National Academy of Science USA*, 99, 4818–4822 (2002).
4. J.-M. Lehn, *Supramolecular chemistry: Concepts and perspectives*, VCH Verlagsgesellscahft GmbH, Weinheim (1995).
5. J.-M. Lehn, Organic chemistry: Its language and its state of the art, *Proceedings Centenary Geneva Conference*, M. V. Kisakürek, VHCA, Basel; VCH, Weinheim (1993).
6. J.-M. Lehn, Cryptates: Inclusion complexes of macrocyclic receptor molecules, *Pure Applied Chemistry*, 50, 871–892 (1978).
7. J. W. Steed and J. L. Atwood, *Supramolecular chemistry*, John Wiley & Sons, Ltd., Chichester (2000).
8. P. D. Beer, P. A. Gale, and D. K. Smith, *Supramolecular chemistry*, Oxford Chemistry Primers, Oxford Science Publications, Oxford University Press, Oxford (2003).
9. J. G. E. M. Fraaijie, Soft matter. Presentation ISM-Series, University of Leiden, wwwchem.leidenuniv.nl/scm/Presentations/ISM/ISM09/ISM09.pdf (viewed 2007).
10. I. W. Hamley, *Introduction to soft matter*, John Wiley & Sons, Ltd., Chichester (2000).
11. J. H. Morowitz, B. Heinz, and D. W. Deamer, The chemical logic of a minimum protocell, *Origins of Life and Evolution of Biospheres*, 18, 281–287 (1988).
12. A. Pappelis and S. W. Fox, Domain protolife, *Journal of Biological Science*, 20, 129–132 (1995).
13. S. L. Miller, A production of amino acids under possible primitive earth conditions, *Science*, 117, 528–529 (1953).
14. D. Ring, Y. Wolman, N. Friedmann, and S. L. Miller, Prebiotic synthesis of hydrophobic and protein amino acids, *Proceedings of the National Academy of Science USA*, 69, 765–768 (1972).
15. B. Commoner, Unraveling the DNA myth: The spurious foundation of genetic engineering, *Harper's Magazine*, February, 39–47 (2002).
16. B. Commoner, Biochemical, biological and atmospheric evolution, *Proceedings of the National Academy of Science USA*, 53, 1183–1194 (1965).
17. S. B. Prusiner, The prion diseases: One protein, two shapes, *Scientific American*, 272, 48–57 (1995).
18. J.-M. Lehn, *Supramolecular chemistry: Concepts and perspectives*, Weinheim, Germany: VCH (1995).
19. J.-M. Lehn, *Organic chemistry: Its language and its state of the art*, Proceedings Centenary Geneva Conference, M. V. Kisakürek, VHCA, Basel; VCH, Weinheim (1993).
20. J.-M. Lehn, Cryptates: Inclusion complexes of macrocyclic receptor molecules, *Pure Applied Chemistry*, 50, 871–892 (1978).
21. J. Rebek, Jr., "Some got away, but others didn't...," *Journal of Organic Chemistry*, 69, 2651–2660 (2004).

22. C. J. Pedersen, Cyclic polyethers and their complexes with metal salts, *Journal of the American Chemical Society*, 89, 7017–7036 (1967).

23. C. J. Pedersen, Cyclic polyethers and their complexes with metal salts, *Journal of the American Chemical Society*, 89, 2495–2496 (1967).

24. Molecular recognition, Wikipedia, en.wikipedia.org/wiki/Molecular_recognition (2007).

25. S. Shinkai, M. Ikeda, A. Sugasaki, and M. Takeuchi, Positive allosteric systems designed on dynamic supramolecular scaffolds: Toward switching and amplification of guest affinity and selectivity, *Accounts of Chemical Research*, 34, 494–503 (2001).

26. T. H. Galow, Non-covalent synthesis, University of Edinburgh. Presentation, www.chem.ed.ac.uk/teaching/chem4-5/resources/course_d.pdf (viewed 2007).

27. P. Bhattacharyya, W. Epstein, and S. Silver, Valinomycin-induced uptake of potassium in membrane vesicles from *Escherichia coli*, *Proceedings of the National Academy of Science USA*, 68, 1488–1492 (1971).

28. H. Brockmann and G. Schmidt-Kastner, Valinomycin 1, XXVII, Mitteil Über Antibiotica aus Actinomycetes, *Chemische Berichte*, 88, 57–61 (1955).

29. Z. Syed, Y. Ishida, K. Taylor, D. A. Kimbrell, and W. S. Leal, Pheromone reception in fruit flies expressing a moth's odorant receptor, *Proceedings of the National Academy of Science USA*, 103, 16538–16543 (2006).

30. F. Gräter, W. Xu, and H. Grubmuller, Pheromone discrimination by the pheromone-binding protein of *Bombyx mori*, *Structure*, 14, 1577–1586 (2006).

31. J. C. Love, L. A. Estroff, J. K. Kriebel, R. G. Nuzzo, and G. M. Whitesides, Self-assembled monolayers of thiolates on metals as a form of nanotechnology, *Chemistry Review*, 105, 1103–1169 (2005).

32. Z. Tang, D. Shangguan, K. Wang, H. Sui, K. Sefah, P. Mallikratchy, H. W. Chen, Y. Li, and W. Tan, Selection of aptamers for molecular recognition and characterization of cancer cells, *Analytical Chemistry*, 79, 4900–4907 (2007).

33. D. H. J. Bunka and P. G. Stockley, Aptamers come of age—at last, *National Review of Microbiology*, 4, 588–596 (2006).

34. E. Smyth, RNA in therapy, Horizon Symposia, Connecting Science to Life, Nature Publishing Group, nature.com/horizon/rna/background/therapy.html (2003).

35. B. A. Sullenger and E. Gilboa, Emerging clinical applications of RNA, *Nature*, 418, 252–258 (2002).

36. C. P. Poole and F. J. Owens, *Introduction to nanotechnology*, Wiley-Interscience, John Wiley & Sons, Inc., Hoboken, NJ (2003).

37. J. N. Israelachvili, *Intermolecular & surface forces: With applications to colloidal and biological systems*, 2nd ed., Academic Press, San Diego (1992).

38. F. F. Nachtigall, M. Lazzarotto, and F. Nome, Interaction of calix arene and aliphatic amines: A combined NMR, spectrophotometric and conductimetric inverstigation, *Journal of the Brazilian Chemical Society*, 13, 295–299 (2002).

39. M. E. Craig, D. M. Crothers, and P. Doty, Relaxation kinetics of dimer formation by self-complementary oligonucleotides, *Journal of Molecular Biology*, 62, 383–392 (1971).

40. D. H. Williams and B. Bardsley, Estimating binding constants—The hydrophobic effect and cooperativity, *Perspectives in Drug Discovery and Design*, 17, 43–59 (2004).

41. B. Barbiellini, Water interaction with proteins, Northeastern University, stardec. ascc.neu.edu/~bba/RES/PROT/WATER.htm

42. J. Barciszewski, J. Jurczak, S. Porowski, T. Specht, and V. A. Erdmann, The role of water structure in conformational changes of nucleic acids in ambient and high pressure conditions, *European Journal of Biochemistry*, 260, 293–307 (1999).

43. Y. L. Cho, H. Uh, S.-Y Chang, H.-Y. Chang, M.-G. Choi, I. Shin, and K.-S. Jeong, A double-walled hexagonal supermolecule assemble by guest binding, *Journal of the American Chemical Society*, 123, 1258–1259 (2001).

44. M. Lahav, R. Gabai, A. N. Shipway, and I. Willner, Au–colloid—"Molecular square" superstructures: Novel electrochemical sensing interfaces, *Chemical Communications*, 1937–1938 (1999).
45. M. Fujita, F. Ibukuro, H. Hagihara, and K. Ogura, Quantitative self-assembly of a catenane from two preformed molecular rings, *Nature*, 367, 720–723 (1994).
46. P. J. Collings and M. Hird, *Introduction to liquid crystals*, Taylor & Francis, Bristol, PA (1997).
47. C. D. Simpson, J. Wu, M. D. Watson, and K. J. Mullen, From graphite molecules to columnar superstructures—An exercise in nanoscience, *Journal of Materials Chemistry*, 14, 494–504 (2004).
48. A. M. van der Craats, J. M. Warman, K. Müllen, Y. Geerts, and J. D. Brand, Rapid charge transport along self-assembling graphitic nanowires, *Advanced Materials*, 10, 36–38 (1998).
49. S. Ito, M. Wehmeier, J. D. Brand, C. Kubel, J. P. Rabe, and K. Müllen, Synthesis and self-assembly of functionalized hexa-peri-hexabenzocoronenes, *Chemistry A European Journal*, 6, 4327-4342 (2000).
50. L. Schmidt-Mende, A. Fechtenkötter, K. Müllen, E. Moons, R. H. Friend, and J. D. MacKenzie; Self-organized discotic liquid crystals for high-efficiency organic photovoltaics, *Science*, 293, 1119–1122 (2001).
51. S. Kumar, Self-organization of disc-like molecules: Chemical aspects, *Chemistry Society Reviews*, 35, 83–109 (2006).
52. J. P. Hill, W. Jin, A. Kosaka, T. Fukushima, H. Ichihara, T. Shimomura, K. Ito, T. Hashizume, N. Ishii, and T. Aida, Self-assembled hexa-peri-hexabenzocoronene graphitic nanotube, *Science*, 304, 1481–1483 (2004).
53. L. Brunsveld, J. A. J. M. Vekemans, J. H. K. K. Hirschberg, R. P. Sijbesma, and E. W. Meijer, Hierarchical formation of helical supramolecular polymers via stacking of hydrogen-bonded pairs in water, *Proceedings of the National Academy of Science USA*, 99, 4977-4982 (2002).

Problems

11.1 In your spare time, create an academic tree that relates all the disciplines and subdisciplines of chemistry. Create one with a timeline as well.

11.2 Define in your own words the concepts of thermodynamic and kinetic control. Are they one and the same (a part of a continuum) or are they distinctly different phenomena?

11.3 Design a supramolecule from the bottom up. Consider solvent effects, substrates, and precursors. Will your system provide a function?

11.4 Give the IUPAC name for valinomycin.

11.5 Name all of the types of natural or synthetic examples of self-assembly that you witnessed before becoming a chemist, biologist, or physicist. Is the formation of the gritty layer on the surface of a bathtub after one takes a bath an example of self-assembly?

11.6 Calculate the thickness of a layer of 2 mL of a large *n*-alkane spread out over the surface of water. How much water is covered, assuming a tightly packed layer?

11.7 Summarize the differences between hard and soft matter in one paragraph.

11.8 Define the terms host–guest, complementarity, and preorganization.

11.9 What is molecular recognition? Define the terms lock and key, dynamic molecular recognition, positive allosteric effect, and negative allosteric effect.

11.10 Describe the mechanism of a dehydrogenase enzyme (e.g., alcohol dehydrogenase). Is the enzyme able to oxidize and reduce substrates? Under what conditions is it able to do so? What are these types of enzymes called?

11.11 Would you consider valinomycin a molecule with a specific function or simply a molecule that reacts in a stochastic manner?

11.12 Is there a difference between a supramolecule and a supermolecule?

11.13 Summarize the relative bond strengths of all the types of intermolecular interactions in this chapter.

11.14 Is there an entropic factor involved in this reaction?

$$M(H_2O)_6^{2+} + 6NH_3 \rightarrow [M(NH_3)_6]^{2+} + 6H_2O$$

How about this reaction?

$$M(H_2O)_6^{2+} + 3 [H_2N-CH_2CH_2-NH_2]$$
$$\rightarrow [M(H_2N-CH_2-CH_2-NH_2)_3]^{2+} + 6H_2O$$

Which one of the two is favored and why?

11.15 a. We know there is a difference in reactivity between linear bidentate (and greater) chelating agents and monodentate ligands like NH_3. Is there a difference between linear polydentate chelates (left) and macrocycle chelates (right) containing the same amount of ligand moieties? What are those differences and which kinds are entropically favored in a reaction? Why?

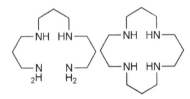

b. Is there a limit on the number of ligands a macrocycle can have before it encounters unfavorable energetics? Why or why not?

11.16 Donating and accepting electron pairs are important solvent effects. Which of the following are good donors? Good acceptors? Good donors and acceptors? Neither?

(a) $R-\overset{\overset{\displaystyle O}{\|}}{S}-R$

(b) H_2O

(c) $R-OH$

(d) $R-C\equiv N$

(e)

(f)

11.17 Research: Look into *Gutmann donor* and *acceptor numbers*. Create a table of molecules that lists donor number, acceptor number, and the dielectric constant of each. What trends do you see?

11.18 A recent study showed that stability constants of Ag(I) and Cu(II) metal cations with selected macrocycles were higher in perchlorate than in nitrate solutions. Stability constants were also higher in DMF rather than DMSO solutions. Can you explain these results? (A. Sil et al., *Supramolecular Chemistry*, 15, 451–457, 2003.)

11.19 Cation complexes (Ca^{2+}, NH_4^+, K^+, or Na^+) with crown ring size of 18 atoms seem to demonstrate the highest stability constants. Why do you think this is the case?

11.20 Give examples of a *chelate effect*, a *macrocycle effect*, and a *macrobicyclic effect*. Which kind generally demonstrates the highest binding constants?

11.21 What is the difference between the *thermodynamic template effect* and the *kinetic template effect*?

11.22 Alkali metals (Li^+, Na^+, K^+, Rb^+, and Cs^+) are bound by a variety of hosts in the following order of decreasing $-\Delta G$ (J. W. Steed et al., *Supramolecular Chemistry*, Wiley, New York, 2000): spherands > cryptaspherands > cryptands > hemispherands > corands > podands > solvents. Explain this trend.

11.23 Metal chemical hardness versus softness is also a factor in supramolecular design, as are hard acids and hard bases.

Devise a scenario that involves these parameters.

11.24 What is the *synergic effect*? (J. W. Steed et al., *Supramolecular Chemistry*, Wiley, New York, 2000.)

11.25 What is *chiral recognition*? Is it important in biological systems? Name some important chiral biological molecules and their corresponding receptors.

11.26 What is an *amphiphilic receptor*? List a few examples.

11.27 What is a *π-acid ligand*?

11.28 Anion coordination chemistry is important in biology. Because anions are larger than cations, what kinds of receptors are able to accommodate such species? Name a few examples. Why do anions have higher solvation ΔG (more spontaneous) than cations of similar size?

11.29 Anions assume a variety of shapes: spherical, linear, planar, tetrahedral, and octahedral. Give examples of each. Also, provide an appropriate receptor for each.

11.30 Design a nanoscale molecular light conversion device. What components are required to transmit a signal?

11.31 Design a nanoscale electrochemical sensor. What components are required to transmit a signal?

11.32 What kinds of molecules would serve well as molecular wires? Molecular rectifiers? Molecular switches? Molecular machines?

11.33 Design a problem that involves a bipyridine species as the chromophore. Use the Beer–Lambert law. Find a species that absorbs and then luminesces. Calculate the quantum efficiency.

11.34 What host species can act as enzyme mimics? (Consult J. W. Steed et al., *Supramolecular Chemistry*, Wiley, New York, 2000.)

11.35 With classmates, discuss the types and structures of liquid crystals. Design a system that performs a function.

11.36 Much of what we have covered previously seems to be more related to organic or biochemistry. Although important, we also need to ask some nano-oriented questions. One good question to ask concerns the size of supramolecules. In general, what happens to solubility as size is increased?

11.37 Many enzymes are extremely large. How do they coexist with their immediate surroundings?

CHEMICAL SYNTHESIS AND MODIFICATION OF NANOMATERIALS

If a metal is downsized to a critical dimension on the nanoscale, the classical physical laws are no longer valid to describe its physical properties but instead, quantum mechanical rules must be applied. Particles of that size are called nanoclusters, nanoparticles or, related to their special physical behaviour, quantum dots. Quantum dots are a fundamental part of nanoscience and will play a decisive role in future nanotechnology. They will influence our life dramatically, since applications from information technology up to diagnosis and therapy in medicine on a molecular level will become possible.

GÜNTER SCHMID, UNIVERSITY OF ESSEN (2006)

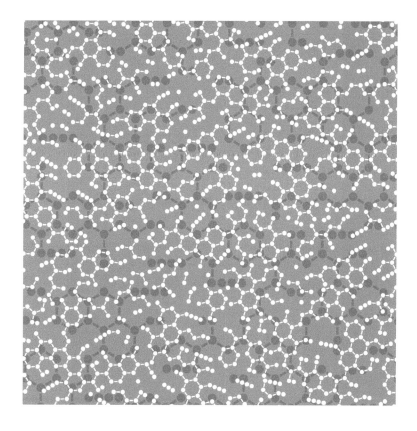

FIG. 12.0

Professor Günter Schmid of the University of Essen (non Essen–Duisburg) has made major contributions to modern cluster chemistry and catalysis. The synthesis of Au-55 quantum dots, their physical and chemical properties, and applications have been described in great detail by Schmid.

Source: Image courtesy of Professor Dr. Günter Schmid of the University of Essen in Germany. With permission.

THREADS

Supramolecular chemistry revolves around intermolecular interactions and host–guest relationships. The precursors to supramolecules are manufactured by mostly traditional means: kinetically controlled reactions involving the making and breaking of covalent bonds. We have shown how dative bonds play a major role in biological and supramolecular complexes—in both their synthesis and function. In this chapter, we continue with intermolecular interactions but also review chemical reactions that are more traditional in their makeup that involve breaking old bonds and making new ones. Polymer chemistry is reviewed in the context of nanomaterial synthesis. Polymers are projected to make a significant contribution to the science of drug delivery by providing substrates and capsules for active components. Self-assembled monolayers have been in existence for several decades but are proving to be a significant material for sensors and electronic and optical devices, and they are proving to be facilitative in terms of their scale-up potential. With this chapter we round out the chemistry division of the text; the two chapters of the biological division remain. Our progress through perspectives, tools, properties and phenomena, chemical synthesis, and the biological domain is nearly completed.

12.0 CHEMISTRY AND CHEMICAL MODIFICATION

Supramolecular chemistry is the chemistry of the molecule and the intermolecular bond. Chemical modification of precursor molecules, supramolecules, and nanomaterials in general proceeds with an arsenal of techniques, including covalent, ionic, and metallic interactions and with all the intermolecular forms we have delved into in previous chapters. In many ways, this chapter is focused more on synthesis than on describing types of molecules. The bottom-up method of fabricating nanomaterials is a core ingredient of every process.

12.0.1 Types of Synthesis Processes

Chemical synthesis strategies are numerous and varied. We assemble another catalog to describe them and classify them. Although there is much overlap, we attempt to place each kind of synthesis method within a convenient perspective. One convenient means of drawing boundaries is to place a synthesis method in one of the major classifications of nanomaterials: zero-, one-, and two-dimensional systems. This makes some sense, but we all know that several types of synthesis methods cross the boundaries of material geometry. For example, the process of forming a silica layer on a surface is also relevant for zero-, one-, and two-dimensional materials or on anything that has a surface, even one with a nano-surface. A special section is devoted to template methods of synthesis. Template methods offer a kinetically controlled environment that is able to alter the course of reactions that may otherwise not happen. In the last section, we cover the enormous domain of polymer chemistry and its relation to nanoscale materials. **Table 12.1** lists a few major chemical synthetic methods. Once again boundaries are placed for convenience; for example, we are aware that many template methods involve the sol–gel process.

12.0.2 Introduction to Molecular Self-Assembly

In the broadest sense, molecular self-assembly (MSA) implies an effortless construction of an object—a fabrication that simply, quickly, and automatically "falls into place." Molecular self-assembly has a similar meaning but it occurs at the level of the nanoscale with molecules. Molecules spontaneously fall into place without significant energy or direction from an outside source. MSA is influenced by chemical reactions, electrostatic interactions, and physical phenomena such as capillarity [3]. Just like with any reaction, minimization of chemical potential is the overall driver. With nanomaterials and surfaces, the driving force behind self-assembly of monolayers is to reduce the high energy associated with nanoparticles and reduce the high energy of surfaces to form a low-energy stable structure.

There are two general kinds of molecular self-assembly. *Intermolecular self-assembly* is the spontaneous aggregation (or reaction) of molecules. The molecules may collectively react with a substrate via covalent bonds (e.g., chemisorption) or by means of intermolecular forces (e.g., physisorption). *Intramolecular self-assembly* occurs within large molecules or complexes like proteins. Proteins are able to assume their 2° structure (folding to make helices or pleated sheets) due to the interactions of hydrogen bonds. The 3° structure of proteins (higher level of folding) is due to covalent interactions in the form of disulfide linkages and intermolecular interactions like hydrogen bonding, and/or hydrophobic interactions in the presence of water. Protein quaternary structure is due to proteins self-assembling into complex structures—mainly in the form of

TABLE 12.1	*Types of Chemical Synthesis*
Type of synthesis	**Description**
Template synthesis	Template synthesis involves a solid (or liquid) material that is capable of directing the geometrical development of a nanomaterial during its chemical or physical fabrication of nanomaterials to resemble its own image (or negative image). If pores or channels are small enough, like they are inside carbon nanotubes or zeolites, the template is capable of exerting kinetic confinement effects during a chemical process.
Sol–gel methods	Sol–gel methods have been in existence for several decades. It is a process whereby monodisperse particles can be generated in large quantities. Most sol–gel systems are based on silicate chemistry but others exist as well, such as the formation of monodisperse hematite (α-Fe_2O_3) from ferric hydroxides/ferric oxyhydroxides [1]. Sol–gel chemistry is exploited to form silica nanoparticles as drug delivery vehicles.
Metal reduction	Reduction of metals, particularly gold salts like hydrogen tetrachloroaurate [$HAuCl_4$], by organic bases such as sodium citrate is a very old procedure. As the solution becomes saturated, Au nanoparticles nucleate and start to precipitate. Control of solution parameters such as pH, concentration of reducing agent, and potential stabilizing ligands lead to control over particle size [2].
Molecular self-assembly	Molecular self-assembly is the spontaneous assembly of precursors to form metastable or stable nanoparticles. Molecular self-assembly is a process that is driven by favorable free energy to form zero-, one-, and two-dimensional nanomaterials.
Atomic layer deposition	ALD is a precise means of applying layers of select materials onto any surface that is capable of chemical modification. The method was discussed in some detail in chapter 4. Formation of covalent bonds and intermolecular bonds occurs during ALD deposition. ALD is based on one-to-one stoichiometry that is highly specific (e.g., any reactants that are allowed to react with a substrate react completely and all excess reactants are purged before the next step in the process).
Emulsion polymerization	Emulsion polymerization is a method that involves an emulsifier with a monomer species and a surfactant. This method is used to form latex particles that have applications as a vehicle for targeted drug delivery. In addition, the solution may contain comonomers, initiators, stabilizers, buffering agents, and preservatives.
Block copolymerization	Block copolymers are formed from two or more monomeric species. The net polymer is composed of blocks that contain one kind of polymer that is linked covalently or otherwise to another block made of another kind of polymer. In the majority of cases, a patterned array is formed that permeates the structure of the polymer—at least to the extent of forming domains with good order.
Electrodeposition electroless deposition	Electrodeposition is a means to apply thin metallic layers on conducting surfaces. Electroless deposition techniques are applied to form metallic layers on the surfaces of any material that is capable of chemical modification. Formation of three-dimensional nanomaterials is possible by this technique.

globulin proteins like hemoglobin. These interactions are more along the lines of glorified intermolecular self-assembly. Self-assembly is a subset of the general class of chemical modification. If breaking old covalent bonds and making new ones happens under mild conditions quickly, effortlessly, and spontaneously, it should be labeled as a self-assembly process. For example, the nucleophilic substitution of ester-modified silanes by surface hydroxyl groups occurs rather spontaneously, to completion, and with relatively small equilibrium constant—surely a bona fide self-assembly episode.

Molecular self-assembly is a characteristic of supramolecular reactions. Natural processes, in which energetic considerations are always at a premium, readily exploit such reactions to accomplish many metabolic and physiological tasks. Although nature is quite efficient, any error in the transfer of information, the code itself, or any environmental perturbation of the process could lead to defective products that can become detrimental to the organism. For example, it is known that certain amyloid proteins in Alzheimer's patients have flawed structure due to improper folding of the protein and that those forms tend to self-assemble. The anomalous proteins (prions) are rendered insoluble and tend to agglomerate (self-assemble) into structures called plaques in the brain. What happened to these proteins during the self-assembly (intramolecular) folding process that resulted in such devastating consequences to the organism?

According to George Whitesides et al. [4]:

> Molecular self-assembly is the spontaneous association of molecules under equilibrium conditions into stable, structurally well-defined aggregates joined by non-covalent bonds. Molecular self-assembly is ubiquitous in biological systems and underlies the formation of a wide variety of complex biological structures .... Self-assembly is also emerging as a new strategy in chemical synthesis, with the potential to generate non-biological structures with dimensions of 1 to 100 nm.

and

> Molecular synthesis is a technology that chemists use to make molecules by forming covalent bonds between atoms. Molecular self-assembly is a process in which molecules (or parts of molecules) spontaneously form ordered aggregates and involves no human intervention: the interactions involved usually are non-covalent. In molecular self-assembly, the molecular structure determines the structure of the assembly. Synthesis makes molecules: self-assembly makes ordered ensembles of molecules (or ordered forms of macromolecules). The structures generated in molecular self-assembly are usually in equilibrium states (or at least in metastable states).

George Whitesides was mentioned in the introduction and is one of the leaders in the field of molecular self-assembly. Interestingly, self-assembly is not the exclusive domain of nanomaterials. Micromaterials and larger materials are also able to fall into place under the proper conditions [4]. We discussed in chapter 8, "Nanothermodynamics," how self-assembly processes are characteristic of dissipative systems that are fueled from outside sources (the Bénard cell). The concept of self-assembly actually can be extended to include galaxies! However, within the frame of this text and the prevailing definition, let us adhere to molecular processes that require little input of energy or management.

12.0.3 Introduction to Chemical Functionalization

Nanomaterials tend to aggregate, react chemically, or decompose if not treated properly. Nanomaterials are metastable materials that need to find ways to address the inherently unhappy state of high surface energy. There are many means to stabilize nanosurfaces and render them chemically receptive to further modification (e.g., attachment of ligands with good leaving groups or of ligands that can be easily displaced by ones that make stronger interaction with the nanoparticle). All kinds of nanomaterials are capable of undergoing chemical functionalization. Attaching molecular recognition moieties to the surface of ligand-stabilized quantum dots, adding polymer linking groups to the surface of single-walled carbon nanotubes, and attaching photo-sensing charge transfer dye molecules to titanium dioxide nanoparticles to initiate the exchange of solar energy to PV devices are all means of modifying nanomaterial substrates for further chemical processing.

There are many ways to chemically modify nanomaterials. Chemical modification methods involve addition, subtraction, or exchange of chemical groups on the surface of nanomaterials or internal restructuring of the molecule itself. One objective of chemical modification is to enhance the chemical behavior of nanomaterials without altering (at least drastically) the physical properties of the underlying molecule. Nanocomposite physical properties such as mechanical strength and electrical and thermal conductivity benefit composites immensely if reinforced with carbon nanotubes. However, pristine tubes are ineffective in the matrix of the polymer without chemical modification of the surface (e.g., not much is able to stick to the surface of carbon nanotubes). Bonding chemically labile species to the surface of nanotubes enhances cross-linking between nanotubes and with the materials of the polymer matrix. The degree of cross-linking affects the mechanical properties of the composite but also the intrinsic physical properties of the carbon nanotube. This trade-off needs to be considered during the design phase of the nanocomposite—an issue for the mechanical and chemical engineers to address. The terms *functionalization, modification,* and *derivatization* are used interchangeably in this chapter.

12.0.4 *The Nano Perspective*

Bottom-up methods will assume more importance in nanotechnology as the years go by. Obstacles such as the apparent lack of long-range order will diminish as better precursors are developed, techniques are refined, and atomically smooth substrates become commonplace. Formation of undesirable side products becomes less important as supramolecular chemistry and site-specific chemistry become better understood. Whole-scale manufacturing of nanomaterials by bottom-up techniques at the nanoscale is not a reality at this time, although its upside is enormous (in particular, its high throughput and inexpensive capital needs). However, there are many examples of well-developed (and profitable) manufacturing processes to make nanoparticles of all sorts from the bottom up. Latex particles that have nanometer to micron dimensions have been around for decades. In the computer industry, bottom-up techniques are just starting to make their presence known. Industrial fabrication of multiwalled carbon nanotubes from the bottom up by flame techniques for use as polymer fillers is acquiring an increasingly bigger share of the global market.

Efforts to copy nature (biomimetics) are ongoing as more and more of nature's secrets are unraveled. The key to success must lie in the ability to fabricate incredibly complex materials at body temperature and neutral *p*H—easier said than done.

12.1 SELF-ASSEMBLY REVISITED

We have been directly and indirectly alluding to the concept of molecular self-assembly in this chapter and previous chapters. MSA occurs between and among mobile molecules and between molecules and surfaces. The surface can be flat (planar) or with high curvature (colloids, cavities, nanowires, edges, etc.). Molecular self-assembly is the method of choice of bottom-up chemical synthesis of nanostructure materials for many reasons—not all of them fully realized at this stage of its development. Some forms of MSA require high specificity, as in the case of supramolecular recognition, and others not so much, such as in the case of monolayers and metal surfaces. Self-assembly implies that materials come together in a rather spontaneous way without the need for excess energy. The term self-assembly also implies that chemical reactions are reversible (e.g., relatively small but positive reaction equilibrium constants, K_{rxn}).

In addition to intermolecular and intramolecular MSA, there is also static and dynamic MSA. *Static self-assembly* is the embodiment of the equilibrium condition (e.g., in that $\Delta G = 0$) and the process operates under the conditions of thermodynamic equilibrium. *Dynamic self-assembly* processes exist away from equilibrium conditions, dissipate energy, and stop existing when the steady-state flow of energy ceases. What kind of system resembles a dynamic self-assembly process? The answer is that living organisms are full of processes that resemble dynamic self-assembly.

We have now encountered a small dilemma. All along we have been stating that nature relies on self-assembly processes to accomplish its multiple purposes and that these processes are spontaneous, directed by molecular recognition, and occur under mild conditions. These are all true but the bottom line is that all self-assembly processes in living things are solar powered. This implies that MSA in living systems is a dynamic process and not a static one per se. Recall that the general state of thermodynamic equilibrium is only achieved when the organism ceases to exist—the condition of cellular death. Are there examples of static equilibrium in natural systems? In the beaker, molecular self-assembly occurs at moderate temperatures. If the system achieves equilibrium, it is called static MSA. There is no dilemma because the energy required to drive it is supplied by a few integral portions of $k_B T$. After all, very few self-assembly processes are conducted in a refrigerator. For our synthetic purposes, we wish that reactions terminate (achieve thermodynamic equilibrium or kinetic stability) so that we can use the products to accomplish other tasks. MSA for us implies equilibrium; for nature, it implies nonequilibrium.

The formation of macromolecules and macromolecular systems from inorganic starting materials that become integrated into living things is the biological foundation for self-assembly. As with most biological processes, a change in state is effected from that of a disorganized system to one that inherently demonstrates a higher level or organization. Entropy therefore must be accounted for in any self-assembly process. Driving forces for self-assembly include chemisorption of the potential layer to the substrate, surface effects, hydrophobic–hydrophilic factors, intermolecular forces, and capillary forces [3].

Self-assembly processes tend to produce products with relatively short-range order. Long-range order, although extremely desired, breaks down due to the

FIG. 12.1

Image of an n-alkane thiol (derived from tetradecane) with a 2° amine capping (terminal) group is shown in the figure. The thiol group (yellow) chemisorbs to gold surfaces to form a monolayer. The alkane tails of the SAM interact with each other via lateral van der Waals forces. The capping group (the amine, blue) undergoes numerous organic reactions to form chemically modified monolayers.

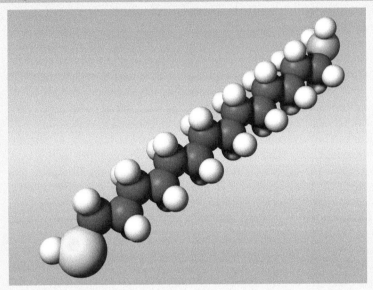

occurrence of any number of possible aberrations during the synthesis process or perturbations transferred to the process by surface defects. The goal of achieving long-range order is one of the holy grails of molecular self-assembly research. Interactions that result in forming labile bonds and molecular recognition leading to selectivity are key ingredients of MSA.

A generic molecule that is suitable for self-assembly on a surface consists of a surface-active head group, a cylindrical body that usually is an aliphatic chain, and a capping (or terminal) group (**Fig. 12.1**). The head group is either an electron-rich moiety like sulfur that is willing to donate electrons to a transition metal surface or one that contains labile ester groups willing to exchange with surface hydroxyls. The body of the typical surface-active molecule is usually a long-chain alkane that experiences lateral van der Waals interactions to form a tightly bound monolayer once bound to the surface. The capping group is usually capable of further derivatization with the purpose of adding additional layers or highly specific host molecules (**Table 12.2**).

Self-Assembled Monolayers.   Self-assembled monolayers (SAMs) form a distinct class of molecular self-assembled materials. In the spirit of this chapter, we shall continue with the premise that intermolecular attractions and bonds form the core of the structures of interest, although a strong case can be made for covalent "self-assembly" interactions as well. Yes, a self-assembled monolayer is a coating on a substrate by a single layer of molecules that fall together spontaneously, but covalent bonds can be and are formed between the monolayer entity and an

TABLE 12.2	*Types of Self-Assembly Methods and Materials*	
Generic type of SAM	**Description**	**Applications substrates**
Langmuir–Blodgett films on liquid surfaces	*A generic alkyl acid is depicted: Y = functional group, usually* $-CH_3$, $-OH$, *or* $-CO_2H$	Hydrophobic or hydrophilic liquid substrates Electrostatic interactions: Charge–dipole Dipole–dipole
Self-assembled monolayers on solid surfaces	Thiols *Sulfur-containing SAMs bond well to metals. Left top: n-alkane thiol; top right: two-tailed thiol; bottom: a disulfide is depicted. Disulfide linkages play a major role in determining protein structure*	Neat, structured Au, Ag, Cu surfaces Covalent *d–d* interactions ~180 $kJ \cdot mol^{-1}$
	 Phosphonates and fatty acids. R = –H, –CH_3, or other hydrocarbon groups	Glass, SiO_2, Al_2O_3, AgO, or other polar surfaces SiO_2, Al_2O_3, TiO_2 Ionic/electrostatic interactions ~50 $kJ \cdot mol^{-1}$

Other SAM ligand systems on thin films, bulk substrates, nanomaterials [5] R–OH on iron oxides, Si–H, Si
$R-CO_2^-$ and $R-CO_2H$ on alumina, iron oxides, Ni, Ti, and TiO_2
$R-NH_2$ on FeS_2, mica, stainless steel, CdSe
R–CN on Ag, Au
R–SH on Ag, AgS, Au, CdTe, CdSe, CdS, Cu, Ge, Hg, InP, Ir, Ni, PbS, Pt, Ru and Zn, ZnS
R–S–S–R′ on Ag, Au, CdS, Pd
R–SeH on Ag, Au, CdS, CdSe
$P-R_3$, $P(Ph)_3$ on Au, FeS_2, CdS, CdSe, CdTe
$R_3-P=O$ on Co, CdS, CdSe, CdTe
$R-PO_4^-$ on Al_2O_3, Nb_2O_5, Ta_2O_3, TiO_2
$RHC=CH_2$ on Si
RCCH on Si(111):H

(*continued*)

TABLE 12.2 (CONTD.)	*Types of Self-Assembly Methods and Materials*	
Generic type of SAM	**Description**	**Applications substrates**
Silanization		Hydroxylated surfaces (in general) and carbon Covalent Si–O ~440 kJ·mol⁻¹ Covalent Si–C ~300 kJ·mol⁻¹

$$ \text{Covalent Si–O } \sim 440 \text{ kJ·mol}^{-1} $$

Alkylsilanes. X = halogen, ester, hydroxide, amine (1°, 2°) or other reactive group; R = –H, –CH₃, or hydrocarbon moiety

activated surface. Yes, the interactions involved in spreading a monolayer of stearic acid on the surface of water is due to weak interactions, such as dipole–dipole, van der Waals, and hydrophobic effects, but many kinds of self-assembled monolayers form strong covalent (chemisorbed) bonds with the substrate. In other words, it is perhaps better to classify the self-assembly process as occurring between a superprecursor (the monolayer moiety) and a supermolecule (the metal) that results in the formation of a stable layer.

Regardless where we draw the line, the application of a self-assembled monolayer onto a surface is accomplished by liquid-phase and gas-phase methods (e.g., by atomic layer deposition [ALD]). The self-assembled monolayer technique is capable of forming layer upon layer until a desired hierarchical structure or thickness is obtained. Since this is a process with more thermodynamic than kinetic character (once again, where do we draw such lines?), the driving force to form SAMs is the reduction of the overall chemical potential [3]. Monolayers of special molecules can be formed on surfaces of bulk materials composed of metals, metal oxides, semiconductors, ceramics, polymers, and various liquids. The layers can be organic or inorganic. Covalent bonds are generally formed between the layer and the substrate as well as at the terminal (solution) end, the reactive site at which an additional layer can be added.

Substrates. Substrate preparation is accomplished by well-established techniques [5]: physical vapor deposition (PVD), evaporation (thermal and electron beam), and electroless and electrodeposition, to name a few. Silicon wafer surfaces (in the natural oxidized form) are commonly used as substrates. A 1- to 5-nm layer of Ti, Cr, or Ni (or Ta) is applied to provide adhesion for noble metals. The thickness of Au films after deposition ranges from 10 to 200 nm. Such metal films form polycrystalline domains (on the order of 10 nm–1 μm) usually of the (111) orientation for *fcc* metals. Grain size and surface structure can be modified by manipulation of temperature, composition, deposition rate, annealing, chemical treatment, and mechanical methods. Cleaved mica substrates provide "pseudosingle crystal surfaces." Terraced Au(111) layers by epitaxial growth on mica(100) surface are prepared by thermal evaporation of gold (@ 0.1–0.2 nm · s^{-1}) onto mica surface at 400–650°C. Interestingly, self-assembled layers on liquid mercury have shown high levels of organization [5]. Liquids tend to form "perfect" surfaces, whereas solid surfaces are never perfect unless it is a single crystal material.

SAM Techniques.   Several techniques take advantage of SAM technology. Metal-coordinating pyridines serve as gate molecules to link electrodes. The link is formed by SAM mechanism on the gold surfaces of either electrode. Nano-imprinting methods are able to produce patterned surfaces by application of hexadecanethiol onto a gold layer. Silicon is coated with a thin Ti layer upon which Au is applied. Microcontact printing (μCP) technique is able to form patterns of self-assembled monolayers with the alkanethiol SAM as the ink (**Fig. 12.2**). A mask is first prepared by optical, x-ray, or e-beam lithography and its pattern is subsequently transferred to stamp made of silicon elastomer (poly[dimethylsiloxane], or PDMS). The patterned PDMS is then dipped into the ink (the *n*-alkanethiol) and the substrate is stamped accordingly [6], thus transferring the ink to the substrate. Etching and applying another kind of thiol (e.g., one terminating in carboxylic acid groups that also reacts with the metal surface) yields a surface that is completely covered in SAMs but that has patterns. The regions coated with thiol are hydrophobic and thereby resistant to wet chemical etching.

12.1.1 Langmuir–Blodgett Films

In 1891, Agnes Pockels, an "uncredentialed" physicist, submitted a letter to Lord Rayleigh. The letter summarized work she had accomplished in her kitchen sink between times she washed dishes and conducted other household chores. She did not know at the time that her work would lay down the foundation for a new field of chemistry and that the "trough" she used to study the effect of substances on water would essentially form the basis for the trough used by Langmuir and Blodgett in their landmark discovery. The great Ostwald in 1932 paid tribute to her work [7]: "Every colleague who is now engaged on surface layer or film research will recognize that the foundation for the quantitative method in this field … has been laid by observations fifty years ago."

The definition of a Langmuir–Blodgett (LB) film, according to KSV Instruments, Ltd. is

> Lipids, polymers or other water insoluble atoms or molecules can form ultrathin and organized monolayers at the air/water interface, i.e., Langmuir films. These films can be deposited on solid substrates to form highly organized regular multilayer stacks called Langmuir–Blodgett films. The LB films are prepared by successively dipping a solid substrate up and down through the monolayer at a constant molecular density or surface pressure. In this way multilayer structures of hundreds of a few nanometer thick monolayers can be produced.

Langmuir–Blodgett films are two-dimensional nanostructured materials. They are molecular monolayers that have length and width but no apparent thickness.

FIG. 12.2   *Microcontact printing (µCP) developed by the Whitesides group of Harvard University. This technique is an inexpensive process (except for the single top-bottom lithographic step involved in making the mask) with high potential throughput. After application of the first monolayer, the wettability of the surface (one way or another) is altered. This allows for specific treatment of the remaining surface with other types of SAMs. The patterned surface is characterized by domains with different chemical and physical properties. The sample surface depicted in the figure is a hypothetical pattern and is not grounded in any form of reality.*

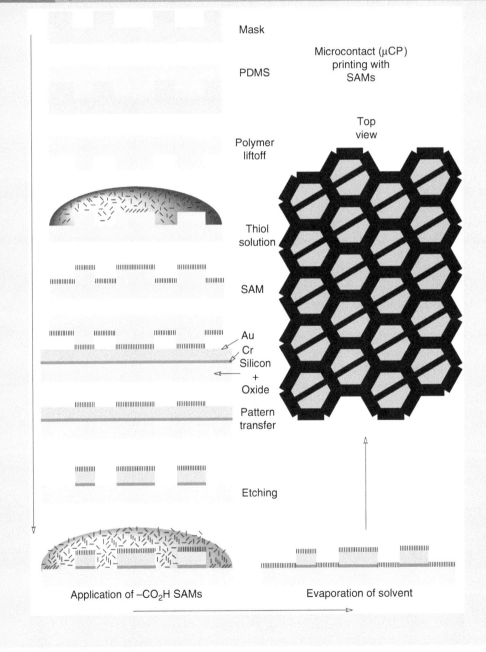

If linking groups are attached to the caps, more layers can be applied to form contiguous robust films. Additional layers can be added and held by simple hydrophobic–hydrophilic interactions but without the integrity afforded by covalent linkage between layers. In many ways, LB films are a chemist's dream because they are easy to prepare; are reproducible; possess excellent molecular organization, orientation, and structural properties; and are easy to characterize by a multitude of analytical techniques. Molecules that are not soluble in the liquid medium (usually water) qualify as candidates for LB application. Gibbs layers, on the other hand, are formed from molecules that are soluble in the liquid medium. Micelles and similar structures are members of the family of Gibbs materials.

Langmuir–Blodgett films were discussed briefly in chapter 4, "Fabrication Methods." The substrate is lowered normal to the surface of the liquid phase. Langmuir–Schaefer deposition, a related technique, involves lowering a solid substrate in a horizontal orientation to the liquid phase. Self-assembled/organized deposition of amphiphilic molecular layers at a liquid–air interface (or solid–liquid–air) forms the basis for the Langmuir–Blodgett film technique. The formation of thin organic films is accomplished by transferring monolayers of amphiphiles (surfactants) from the surface of water, where they are entrained, to a solid substrate. The observations of Benjamin Franklin in 1774 and Lord Rayleigh later about the spread of an oil monolayer at an air–water interface—in this case, that of a pond—have already been mentioned.

What is the importance of monolayers? To begin with, monolayers have the ability to modify the surface energy of a solid or liquid. A surface that was previously hydrophilic (e.g., hydrated alumina) can be rendered hydrophobic after the application of a long-chain carboxylated hydrocarbon such as stearic acid. The surface tension of solutions is also altered by application of a monolayer, as astutely observed by Benjamin Franklin. Surfactants have an affinity for interfaces and the reduction in surface tension results in the reduction of the overall free energy of the interfacial system. Monolayers are able to alter the optical properties of a surface. The best example of this is the invisible glass developed by Katherine Blodgett that consisted of 44 monolayers. Monolayers provide a means of electrical communication between a conducting substrate such as gold and sensing moieties tethered to capping groups. Because the thickness of the layer can be controlled and *preorganized* molecular arrangement is achievable, Langmuir–Blodgett films support a wide variety of molecular, electronic, and bioelectronic devices. Such films have already achieved widespread commercial success and therefore must be considered as one of the first applications of nanoscience to make an impact.

All of us chemists have estimated Avogadro's number by determining the molecular cross-section of stearic acid on the surface of water. It is incredibly ironic how thoughts of nanotechnology never entered our heads during our freshman laboratory endeavors. All we were interested in that lab was to determine Avogadro's number, the thickness of the stearic acid monolayer, calculating the molecular volume, and passing the class (right?). The spread of stearic acid on the surface of water is a pure example of MSA. It involves only intermolecular interactions; each stearic acid molecule retains its chemical identity and a condition of static MSA in equilibrium is attainted. The polar group of stearic acid interacts with polar water molecules via hydrogen bonding. The linear long-chain aliphatic components of stearic acid attract each other by van der Waals interactions to provide an evenly spaced, hexagonally distributed contiguous layer.

The Langmuir–Blodgett Technique.   The Langmuir–Blodgett technique and a schematic of the mechanism are shown in **Figure 12.3a** and **b.** KSV Instruments, Ltd. Series 5007 instrument and an LB-substrate, are illustrated in **Figure 12.4a** and **b,** respectively [8]. A flat planar material with an appropriately treated surface (e.g., clean and chemically compatible with the chemistry of the monolayer) is thrust vertically downward into the liquid phase supporting the monolayer. Following equilibration, the plunger is withdrawn slowly as the monolayer adheres to the surface. Meanwhile, a moving barrier applies pressure laterally on the film to ensure maximum contact among the amphiphiles in the monolayer. The LB technique is a very simple technique that is capable of fabricating two-dimensional nanomaterials. Film thickness is linearly proportional to the number of LB steps [9].

As molecules are squeezed together in the trough, the incipient monolayer undergoes phase changes that range from the gaseous (the highly dilute phase)

FIG. 12.3

(a) *Successive layers of amphiphiles are added to a surface. A solid substrate is dipped into a liquid (usually water) that has a monolayer of a selected amphiphile on its surface. In the figure, the solid surface is an activated silica that is decorated with hydroxyl groups. The surface therefore is hydrophilic and attracts the hydrophilic moiety of the amphiphile (the spherical structures of the molecule). For the second application, the reverse process occurs; specifically, it is now the hydrophobic parts of the amphiphile that interact with the hydrophobic parts of the first monolayer. The process continues in this way until enough layers are built to satisfy the specification. (b) Graphical depiction of the Langmuir–Blodgett apparatus and operation. The SAM molecules are introduced onto the surface of the liquid (usually water) with a syringe. Lateral pressure is applied to the monolayer from the barrier plunger apparatus. It is amazing that Agnes Pockels conducted the first LB experiments in her kitchen sink.*

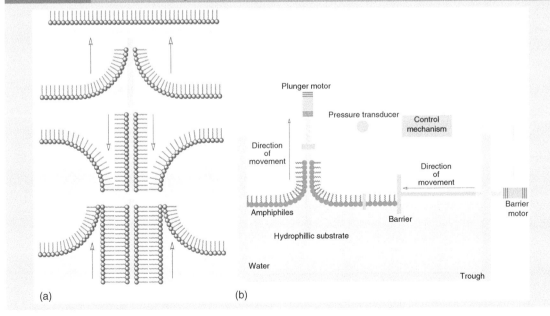

(a)   (b)

FIG. 12.4	*(a) Image of a KSV Langmuir–Blodgett 5007 Series instrument provide by KSV Instruments of Finland. Compare the components in this image with the schematic shown in Figure 12.3. (Image courtesy of Tapani Viitala of KSV Instruments, Inc.) (b) Image of a substrate undergoing the LB procedure is shown. The surface is that of a silicon wafer with a native oxide coating. The type of surface and its rate of entry and withdrawal affect the quality of the final SAM on the surface.*

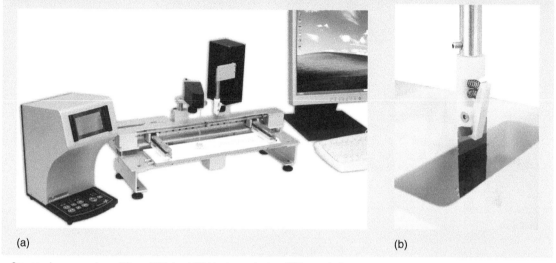

(a)   (b)

Source: Image courtesy of Tapani Viitala of KSV Instruments, Inc. With permission.

to liquid phases (expanded to condensed) and finally into the solid phase. A generic pressure–area isotherm profile is shown in **Figure 12.5.** Many factors affect the quality of LB films:

- Most LB substrates are made of hydrophilic materials like quartz, glass, silica, and metals. The computer industry employs silicon wafers covered with a thin layer of silicon oxide. Substrates with extremely high surface energy like neat silver and gold make monolayer transfer problematic. Metal surfaces are also easily contaminated [3].
- The chemical nature and structure of the amphiphile also influence the character of the LB process and the resulting monolayer. Careful selection of chain length and polar head strength is important. Head groups that interact strongly with the water may not be suitable for LB processing. If the strength of the head group is on the weaker side, then it may be difficult to form an LB film due to lack of interaction with the substrate [3]. For C_{16} amphiphiles, the most stable films are formed by surfactants with head groups that terminate (in general) with alcohols, esters, carboxylic acids, cyanides, and amides. Weaker films result with ethers and peroxides, and very weak LB films are generated if halogens are employed. Strong soluble end groups include the class of sulfate-based anions and ammonium (1, 2, and 3°) cations [10].

Fig. 12.5

A typical pressure–area isotherm is shown in the figure. Upon compression, the system proceeds from a gaseous (highly dilute) phase depicted on the far right through a liquid–gas phase and then into transitions between expanded liquid and condensed liquid phases—all distinct regions on the graph. When the solid phase is achieved, the distance between molecules is minimized and the molecular area equals the area of the molecule. If the monolayer is compressed beyond the solid phase boundary, the film collapses into a three-dimensional structure. The surface tension of water is also shown as a function of surface coverage—it decreases as molecules pack the surface. The monolayer provides the requisite interactions with undercoordinated water molecules at the surface.

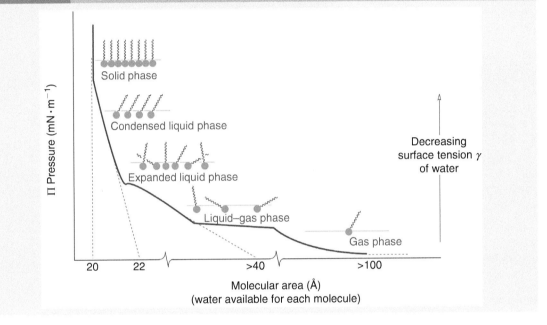

- The kind of solvent (mostly water), its *p*H, temperature, and the relative concentration of amphiphiles all influence the LB process to some degree.
- The cleanliness of the substrate surface, the rate of harvesting, and the rate of compression also influence the course of an LB process.

Zwitterionic Phospholipids. A recent study by L. Cristofolini et al. utilized LB layers of zwitterionic phospholipids as a method to immobilize DNA. An LB film consisting of dipalmitoylphosphatidylcholine (DPPC) (**Fig. 12.6**) was formed in a KSV Instruments 5000 Langmuir trough. All monolayers were formed on a DNA subphase with *p*H 6.0 containing 5 mM $CaCl_2$ and 10 mM NaCl. The DNA covered between 20% and 30% of the surface area. Compression was accomplished at a rate of 10 mm·min^{-1} [11]. The purpose of the study was to find a way to transport DNA into mammalian cells by nonviral gene therapy. The authors found that the immobilized DNA retains its double-helix structure when stabilized by the zwitterionic lipid monolayer [11]. An incredible representation of the tiered structure is shown in **Figure 12.7**.

| FIG. 12.6 | *Dipalmitoylphosphatidylcholine (DPPC), a zwitterionic phospholipid, is a two-tailed SAM molecule.* |

Diplalmitoylphosphatidylcholine (DPPC)

PET–Gold Clusters. Langmuir–Blodgett monolayers of nanoparticles of phenyle-thylthiolate (PET)-passivated gold were fabricated at various Π (surface pressures) [12]. The diameter of PET–gold nanoparticles $[Au_{38}(Ph-C_2S)_{24}]$ was approximately 1.6 nm [13]. A monolayer on water was prepared by addition of the PET–Au onto the surface of water in an LB trough and then deposited on an interdigitated electrode array. Measurement of interparticle spacing was accomplished with transmission electron microscopy (TEM) and scanning tunneling microscopy (STM). The surface pressure in the trough is related to the interparticle distance; π–π stacking of the phenyl groups of neighboring particles occurred during the self-assembly of

| FIG. 12.7 | *The DPPC monolayer–DNA interaction is depicted. The model is based on x-ray reflectivity data. The formation of the zwitterionic DPPC–DNA–Ca^{2+} system (the calcium cations are represented by the red spheres) at the air–water interface demonstrates the versatility of the LB technique. Ca^{2+} is a necessary ingredient for this structure to form. Because DNA did not denature in this configuration, its utility as a method for DNA immobilization (hence transport) offers inroads to developing advanced gene-therapy vehicles.* |

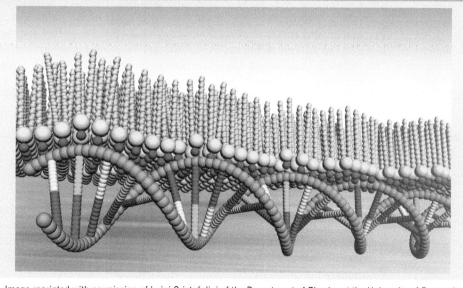

Source: Image reprinted with permission of Luigi Cristofolini of the Department of Physics at the University of Parma, Italy.

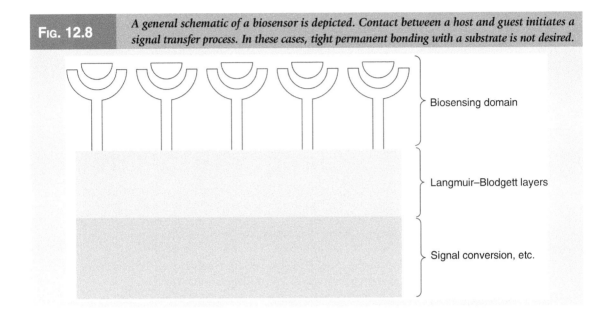

FIG. 12.8   *A general schematic of a biosensor is depicted. Contact between a host and guest initiates a signal transfer process. In these cases, tight permanent bonding with a substrate is not desired.*

Biosensing domain

Langmuir–Blodgett layers

Signal conversion, etc.

the LB film. Current-potential profiles indicated that interparticle coupling occurred through the aromatic ring overlap of neighboring particles.

Electron transfer occurred through bonds and through space via van der Waals contacts between interdigitated ligands. The authors concluded that the ligand van der Waals interactions are enhanced by the π–π stacking of the phenyl groups from adjacent particles. By way of computer modeling, they found that the edge-to-edge separation of particles was 1.1 nm when the phenyl groups completely overlapped [12].

Biosensors with LB Films.   A biosensor is a molecular sensing element coupled to a transducer (**Fig. 12.8**). A biosensor operates at a higher level of molecular recognition with emphasis on the "cognitive" part of the term—the ability to recognize and record a molecular recognition episode. In this way, sensing is reduced to a single step, unlike multiple steps involved in traditional analytical techniques. Designing a biosensor is similar to supramolecular design and is about identification of the target molecule (the guest), selection of a suitable molecular host species, tethering strategies to fix biosensing elements, and, of course, the associated electronics that are required. The design of such sensors involves chemists, biochemists, engineers, and electronics specialists. The detection limit and resolution of such devices should be better that 1 nmol.

12.1.2   *Gold–Thiol Monolayers*

The formation of self-assembled monolayers on metals is not considered to be supramolecular chemistry, technically speaking. It is the reaction between an organic molecule (not considered generally to be a supramolecule) and an inorganic metal surface (the supermolecule?) that is often made of gold. Although no ionic, dipole–dipole, and hydrogen bonding interactions take place, a combination of dative bonding (chemisorption) and van der Waals forces, as well as hydrophobic interactions, is in effect. The chemisorption of a monolayer upon a metal substrate implies the formation of stronger bonds (e.g., bonds between electron-donating moieties like sulfides and disulfides to an electronegative metal like Au). One of the driving forces for chemisorption (and physisorption) on metal or metal oxide surfaces is reducing the surface excess energy (chapter 6). Surface metal atoms in particular are undersaturated with regard to bonding and are willing to seek electrons from other sources for stabilization—especially from electron-rich elements like sulfur.

The synthesis of Au–thiol SAMs is relatively straightforward. The result is a monolayer system that is one of the most fundamental, flexible, and useful in nanotechnology; gold is the perfect substrate due to its high conductivity and relative chemical inertness, and that sulfur loves gold. The ability to chemically modify the tail group of the attached moieties allows for additional layers and specific chemical functionalization in later steps. According to George Whitesides et al. of Harvard, SAMs (1) are easy to prepare; (2) conform to the topography of kind of shape; (3) are able to modulate information between the environment and its substrate metal structure via electrical I–V response, surface plasmon resonance, and refractive index; and (4) are the nanostructured component of macroscopic "interfacial phenomena" like surface wetting, adhesion, and friction [5].

The tethering of alkylthiolate monolayers (ca. 1–3 nm) on gold occurs by dative bonding between sulfur and gold. The strong affinity of S for Au is one factor that plays a role in the stability (and hence popularity) of Au–thiol systems, but intermolecular van der Waals forces are also important; they are responsible for the lateral packing of the monolayer. Lateral packing provides uniformity and robustness to the layer. The strength of the lateral interaction, dependent on chain length, was found to be ca. 4–8 kJ · mol^{-1} per methylene group [14], but 6–8 kJ · mol^{-1}, according to others [15]. The chemisorption energy of dimethlydisulfide adsorption was found by the TPD method to be 117 kJ · mol^{-1} [16]; others report a larger bond strength of ~180 kJ · mol^{-1} [15]. The bond between S and Au is formed by a chemisorption mechanism. The generic reaction to form such a monolayer is as follows:

$$2R(CH_2)_nSH + 2Au^\circ \rightarrow 2R(CH_2)_nS^- + 2Au^{1+} + H_{2(g)} \tag{12.1}$$

The capping R-group represents species that are either active (available for further chemistry) or terminal (allowing no further chemistry). The hydrophobic aliphatic tail is optimally 2–3 nm in length and tilts at an angle of 30° from the normal to the gold surface. This tilt angle maximizes the interchain van der Waals attractions of the hydrocarbon chain and hence lateral continuity. The Au(111) surface provides an optimal substrate and directs the thiols to be laid out in a hexagonal pattern (**Fig. 12.9**).

Metal surfaces are characterized by high surface energy. In other words, if a droplet of water is placed on the surface, the contact angle is expected to be very small. Chemical modification of the gold surface by an alkane–thiol monolayer (depending on the chemical nature of the capping group) is able to drastically alter the wettability characteristics of the gold surface. If the capping group is a terminal methyl group, the contact angle of water on a hexadecane thiolate SAM-modified surface was found to be 115°. If, instead, a hydrophilic substituent is the capping group, such as [HS(CH_2)_{16}–OH], the contact angle is less than 10°. Special surfaces can be fabricated to suit the wettability requirements with mixed species of monolayer [17].

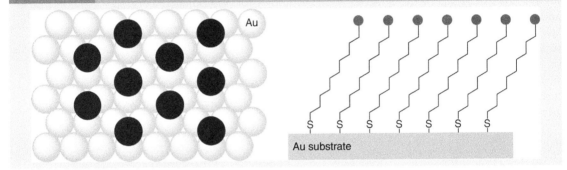

FIG. 12.9

Left: alkanethiol distribution on Au(111) surface is depicted. The strength of the Au–S bond is relatively strong (ca. 184 kJ·mol^{-1}) and the intermolecular attraction of the alkane chains of the thiol groups is approximately 6–8 kJ·mol^{-1}. Right: idealized rendition of thiol layer on gold substrate. Red dots represent tethered functional groups at the end of the thiol aliphatic chains. Alkane chains with 12 or more methylene units form dense, well-ordered monolayers. The sulfur of the thiol attaches to the three-fold pocket (pinning site) formed by adjacent gold atoms. Distance between pinning sites is ca. 5 Å. This provides an area equal to 21.4 Å^2 with which to accommodate a chemisorbed molecule. The van der Waals radius of the thiol is only 2.3 Å (area = 17 Å^2); therefore, it cannot completely fill the space provided by the gold interstitial surface area. The chain therefore tilts 30° to enhance van der Waals interactions between and among the monolayer molecules.

Source: A. Ulman, *An Introduction to Ultrathin Organic Films: From Langmuir–Blodgett to Self-Assembly*, Academic Press, Boston (1991). With permission.

Since their inception, numerous forms of thiol-based monolayer systems have been developed, tested, and applied. A typical SAM system consists of the following components:

- Substrate made of smooth silicon, glass, mica, plastic, or another metal
- Interstitial metal layer that provides better adhesion between the substrate and the noble metal
- Noble metal layer, usually an *fcc* metal with a (111) crystal face, deposited via thermal evaporation, RF sputtering, electrodeposition, physical vapor deposition, or electroless deposition
- Organo-sulfur compound that consists of (a) the head group (the sulfur, which has an affinity for metal), (b) the *n*-alkane shaft $(-CH_2)-_n$, and (c) the tail group (ranging in reactivity from inert to very reactive). The organosulfur compound is applied via solution or vapor deposition methods

The energetics of thiol layers formed of linear *n*-alkanethiols, particularly on Au surface, is determined by temperature programmed desorption (TPD) **(Fig. 12.10)** [18]. Two forces act in concert to construct a monolayer on Au and two peaks are apparent in the spectrum of a TPD experiment [18]. A lower energy peak is a physisorption peak that was found to increase linearly with increasing chain length, indicating a van der Waals type relationship. A higher temperature peak did not vary with alkane chain length and is due to the chemisorbed S–Au interaction. A third peak at higher energy appeared when

FIG. 12.10

A temperature programmed desorption profile of hexanethiol layer deposited at 208°C after 2460 seconds. The lower peak represents physisorbed component of the monolayer. The higher energy peak represents the chemisorbed component. Chain length only affected the position of the lower energy peak.

Source: Image redrawn with permission from Giacinto Scoles, Department of Chemistry, Princeton University.

chain length exceeded 14 methylene groups [18]. The overall relationships are, however, more complex.

12.1.3   Organosilanes

The chemisorption of alkysilanes to an activated (hydroxylated) silicon surface results in robust monolayers for a couple of reasons: (1) the covalent attachment to the surface and (2) lateral cross-linking with other silica groups as the monolayer grows (**Fig. 12.11**). The cross-linking is accomplished through a

FIG. 12.11

A generalized version of acid catalyzed esterification—a nucleophilic addition–elimination reaction. The carboxylic ends of long-chain hydrocarbons interact with hydroxyl groups anchored to the surface of a substrate.

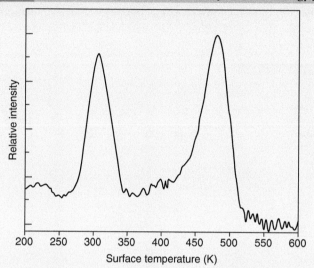

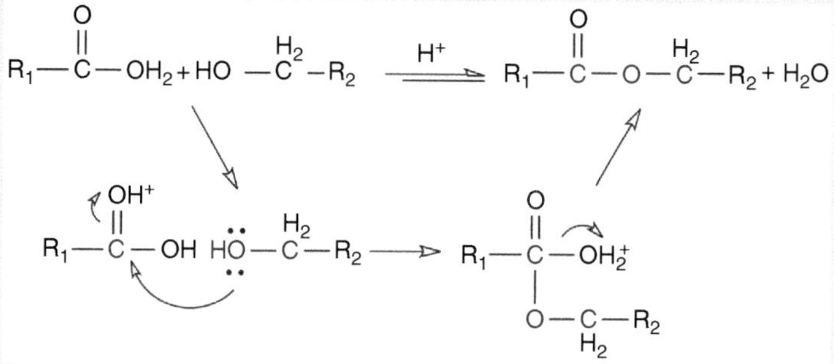

FIG. 12.12

Counterclockwise from top left. Hydroxyl-terminated silane esters are added to an activated silicon surface to form the first monolayer. In the next step, 3-aminopropyl-dimethyl-ethoxysilane is reacted with the first monomer by an ester substitution reaction. The red dot represents the amine group. A thin layer of styrene monomers is applied by spin-coating onto the surface of the bilayer. The system is exposed to ultraviolet radiation, thereby initiating cross-linking of the polymer to form a thin robust coating.

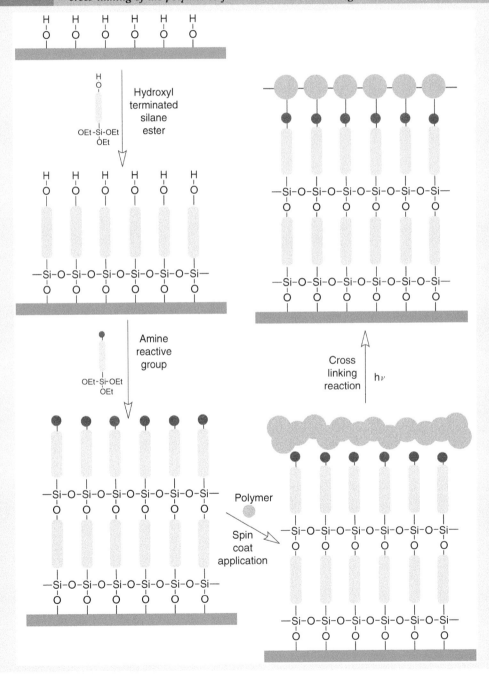

–Si–O–Si– covalent network. The reaction of the silyl ester groups proceeds in a manner similar to a standard esterification reaction between a carboxylic ester and a hydroxyl group but with more facility (**Fig. 12.12**):

$$K_{Eq} = \frac{[R_1COOCR_2][H_2O]}{[R_1COOH][R_2COH]} = 3.38 \tag{12.2}$$

These kinds of reactions, like the one for acetic acid + ethanol in the presence of an acid catalyst, have a relatively low equilibrium constant and therefore are reversible.

12.2 SYNTHESIS AND CHEMICAL MODIFICATION OF NANOMATERIALS

We touched upon several bottom-up chemical synthesis methods in chapter 4. Here we give more detailed examples of the synthesis and chemical modification of selected nanomaterials from the zero-, one-, and two-dimensional categories.

12.2.1 *Synthesis and Modification of Zero-Dimensional Materials*

Colloids, clusters, and quantum dots are members of the extended family of one-dimensional nanomaterials. We dedicate a special section to Au_{55} clusters—a truly versatile and important nanomaterial. The development of Au_{55}, its synthesis, and ligand decoration schemes are attributed to the work of Günter Schmid and his colleagues at University of Essen (now Essen-Duisberg) [19–27].

Au-55 Clusters. Synthesis of uniformly small particles proceeds best from the bottom up. One of the most popular methods to form metal clusters and colloids is through the reduction of metal cations. Common reducing agents include the base complements of organic acids such as sodium citrate, reducing alcohols, Na_2S, borohydrides [B_2H_6], sodium borohydride [$NaBH_4$], and even hydrogen gas. We understand why small clusters wish to agglomerate to form larger clusters via Ostwald ripening. In order to fabricate metal clusters of predetermined size (and we assume that nanoscale clusters are desired rather than colloids), special steps need to be taken. First of all, addition of a potential ligand species to the reaction mixture is required. The ligand serves to bind reduced metals and thereby modulate the growth of the embryonic clusters. Depending on relative concentrations of reactants (the metal salt, reducing agent and ligand), growth of clusters that are monodisperse with desired dimensions is possible. The generic process for synthesis of nanoclusters is summarized as

Reduction of metal cation → agglomeration prevention → ligand stabilization or ligand exchange → extraction from solvent → further surface modification

The synthesis of a triphenylphosphine ligand-stabilized gold cluster is shown in **Figure 12.13**.

Gold has a rich history. It has been part of our culture for over 5,000 years. With its malleability, corrosion resistance, and inherent luster, it is no wonder it has contributed to civilization's course—both positively and negatively. Gold leaf, for example, can be beaten into a layer that is 50 nm in thickness. Its inclusion into beautiful glasses is well documented. The chemistry of gold was problematic due its chemical inertness. With aqua regia and cyanide solutions the only means to dissolve gold, the evolution of its chemistry was limited in scope. Michael Faraday, of course, was well known for his chemical manipulation of gold species. Faraday reduced tetrachloroaurate [$AuCl_4^-$] with white phosphorus (the reducing agent) to generate colored solutions of gold colloids [19]. Ostwald was the first to bring attention to the property dependence of nanoclusters on their surface and, by that postulate, predicted that nanoparticles possess remarkable properties [19].

Gold colloid synthesis is straightforward. Of all the procedures available today, a method called the "Turkevitch route" is very popular [19,28]:

$$HAuCl_4 + (C_6H_5O_7)Na_3 \rightarrow Au^\circ + \text{oxidized products} \qquad (12.3)$$

FIG. 12.13

Reaction scheme of formation of Au_{55} ligand-stabilized cluster is depicted. At the top left, a solution containing dissolved metal cations is shown. The cations are converted into gold atoms after addition of a reducing agent like citrate. Once formed, the atoms nucleate and grow into aggregates that eventually stop at the cluster phase (depending on the reaction conditions). Ligands attach to the vertices of the Au_{55} cluster; there are 12 vertices in this structure. Not shown are the counter-anions, the chloride atoms of $Au_{55}[P(Ph)_3]_{12}Cl_6$.

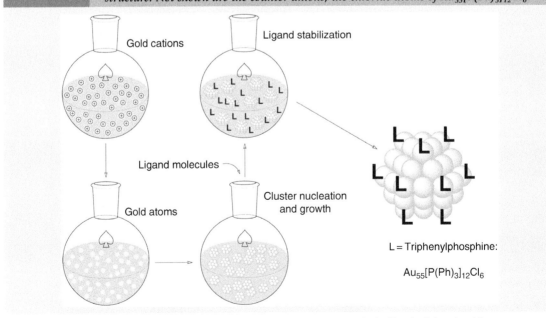

L = Triphenylphosphine:

$Au_{55}[P(Ph)_3]_{12}Cl_6$

Source: Image redrawn with permission of Prof. Günter Schmid, Institut für Anorganische Chemie, University of Essen.

Approximately 5×10^{-6} mol of $HAuCl_4$ is dissolved in 19 mL of deionized water and heated to boiling; 1 mL of 0.5% sodium citrate is added with constant stirring for 30 min. The solution undergoes color changes from yellow to clear, gray, purple, deep purple, and finally to ruby-red. Water is added to maintain the level of solution to 20 mL. The "Brust route" is similar but employs sodium borohydride as the reducing agent:

$$HAuCl_4 + [CH_3(CH_2)_7]_4NBr \text{ (TOAB)} + toluene + NaBH_4 \rightarrow Au^\circ \quad (12.4)$$

This technique utilizes an emulsion layer made of water and toluene; 4.0×10^{-3} mol of tetraoctylammonium bromide (TOAB, surfactant, the phase transfer catalyst and the stabilizing ligand) is added to 80 mL of water and then 9.0×10^{-4} mol of $HAuCl_4$ in 30 mL water is added to the TOAB solution and stirred vigorously for 10 min. The aqueous phase is clear and the organic phase is orange. Sodium borohydride is added dropwise to the mixture and the color changes from orange to white to purple to dark red. To make sure that the product clusters are monodisperse with regard to size, the solution is stirred for an additional 24 h. The organic phase is washed with sulfuric acid to neutralize the solution. TOAB is not considered to be a strong ligand and will readily undergo ligand exchange with stronger ligands like thiols that bind covalently to the gold clusters [29].

Synthesis of gold clusters from $HAuCl_4$ by a sonolysis technique is used to create gold ribbons 20–30 nm in diameter. Glucose and hydroxyl radicals, formed at the cavities formed by sonication, serve as the reducing agents. When cyclodextrin is used, only spherical particles are synthesized [29]. The size of the clusters, their distribution, and stabilization are controlled by the following factors: reductant concentration, temperature, pH, stirring rate, rate of addition of reducing agent, and chemical composition of the solution.

G. Schmid et al. were the first to describe the physical properties of Au_{55} clusters [20]. Gold-55 (Au_{55}) is a magic-number cuboctahedral cluster. Stable forms of Au_{55} are synthesized by

$$AuCl[P(C_6H_5)_3] + B_2H_6 \rightarrow Au_{55}[P(C_6H_5)_3]_{12}Cl_6 + H_3B-P(C_6H_5)_3 \quad (12.5)$$

Gaseous diborane is passed through a warm 150-mL solution of benzene containing 3.94 g of $AuCl[P(C_6H_5)_3]$. Diborane is the best reducing agent, but it also acts as a Lewis acid that binds phosphines. This process limits the amount of free ligand available at any time during the course of the reaction. Excess ligand concentration leads to smaller complexes and clusters, an undesirable outcome [20]. The temperature is raised to 50°C. After 40 min, the colorless solution turns dark brown. Upon cooling, a dark precipitate settles to the bottom of a now colorless solution. The precipitate is filtered, rinsed with dichloromethane, and filtered again through Celite to remove unwanted solids (e.g., colloidal gold). The product, $Au_{55}[P(C_6H_5)_3]_{12}Cl_6$, is re-precipitated slowly in 250 mL pentane to ensure that phosphine ligands saturate the Au_{55} cluster. The overall yield of the process is 29% [20]. The cluster is 2.1 nm in diameter (the cluster is 1.4 nm). An exact stoichiometric relationship between reactants and products has not been derived for this reaction [20].

The Au_{55} product is a dark brown powder that is soluble in dichloromethane and pyridine and insoluble in petroleum ether, benzene, and alcohols [20]. In air, the ligand-stabilized cluster decomposes to solid gold (agglomeration) and in solution reverts back to its precursor state. Mössbauer spectroscopy reveals the presence of four kinds of gold atoms: 13 central atoms, 24 uncoordinated peripheral atoms, 12 atoms coordinated to phosphine ligands, and 6 atoms coordinated to chlorine [20]. Three simulated renditions of $Au_{55}[P(Ph)_3]_{12}Cl_6$ are shown at the top of **Figure 12.14**. On the bottom, a molecular model of a gold cluster attached to DNA is shown.

The triphenylphosphine ligands of the cluster are labile and undergo ligand exchange readily in phase transfer reactions:

$Au_{55}[P(Ph)_3]_{12}Cl_6$ (in dichloromethane) + $P(Ph)_2(C_6H_4SO_3)Na$ (in water)

$\rightarrow Au_{55}[P(Ph)_2(C_6H_4SO_3)Na]_{12}Cl_6$

$\rightarrow Au_{55}[P(Ph)_2(C_6H_4SO_3)]_{12}(Cl_6)^{12-} + 12Na^+$ (12.6)

A thiol derivative of an organo-silsesquioxanes (OSS) is one of the most intriguing ligands synthesized by G. Schmid et al. of the University of Essen in Germany [27]. Its synthesis is shown in **Figure 12.15**. Several changes in chemical reactivity and physical properties occur in substituted Au_{55} clusters. The

FIG. 12.14 *Left: various molecular models of $Au_{55}[P(Ph)_3]_{12}Cl_6$ are depicted. The last image shows the electron density, especially that of the phenyl rings. Right: molecular model of the interaction of Au_{55} clusters with DNA. The strong interaction of gold clusters with the major groove of DNA makes them extremely toxic to living things.*

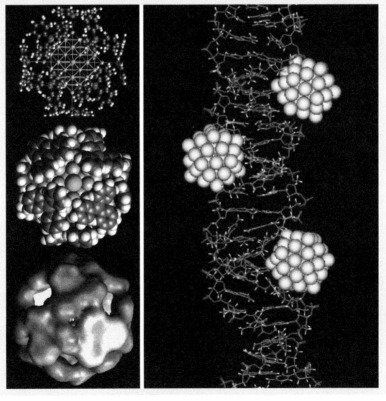

Source: Images courtesy of Prof. Günter Schmid, Institut für Anorganische Chemie, University of Essen. With permission.

Fig. 12.15 Synthesis of T_8-OSS-SH is shown. The ultimate Au-T_8-OSS-SH complex is soluble in pentane, which is not possible for the triphenylphosphine derivative that is soluble only in dichloromethane type solvents. In addition to providing an impressive visual, the utility of these ligand-stabilized clusters in two-dimensional arrays shows promise.

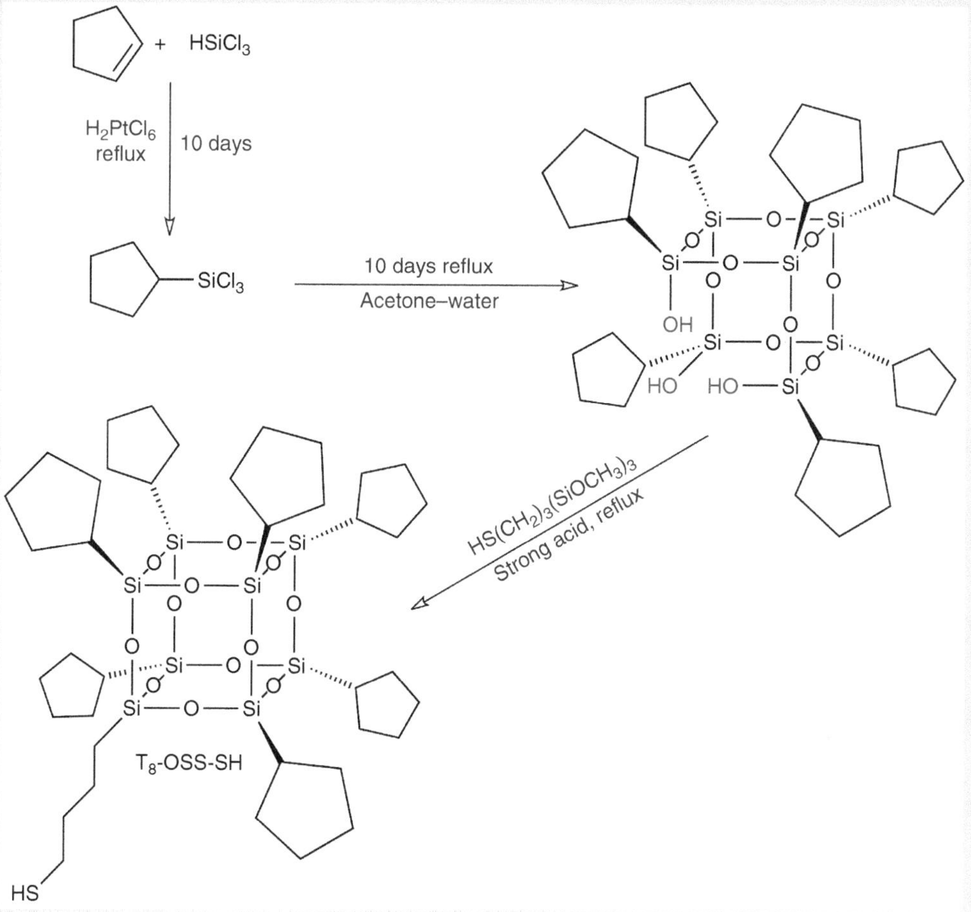

Source: Image redrawn with permission of Prof. Günter Schmid, Institut für Anorganische Chemie, University of Essen.

ligand-stabilized cluster increases in dimension from 2.1 to 4.4 nm upon exchange of the ligands. It is soluble in pentane as well as dichloromethane and is considerably more stable (due to its strong Au–S covalent bond) than its phosphine-substituted counterpart. An increase in the activation energy of electron tunneling increased from 0.16 eV for Au$_{55}$[P(Ph)$_3$]$_{12}$Cl$_6$ to 0.26 eV for the T_8-OSS-SH ligand system [27]. A ligand exchange chemical reaction scheme is shown in **Figure 12.16.** Ligand exchange occurs according to the following scheme:

$$Au_{55}[P(Ph)_2(C_6H_4SO_3)Na]_{12}Cl_6 + 12\ T_8\text{-OSS-SH}$$
$$\rightarrow Au_{55}[T_8\text{-OSS-SH}]_{12}Cl_6 + 12\ PPh_3 \qquad (12.7)$$

However, when Au$_{55}$[P(Ph)$_2$(C$_6$H$_4$SO$_3$)Na]$_{12}$Cl$_6$ ligand is exposed to T_8-OSS-SH, no ligand exchange takes place (equation 12.7), possibly due to phase transfer kinetic factors.

FIG. 12.16   *Ligand exchange reactions with Au$_{55}$[P(Ph)$_3$]$_{12}$Cl$_6$ are graphically depicted. Both the closo-dodecaborate and the OSS ligands are able to displace the triphenylphosphine. The utility of the thiol group in binding gold-based materials is once again demonstrated. [Au$_{55}$(B$_{12}$H$_{11}$SH)$_{12}$]$^{24-}$ is soluble in water and [Au$_{55}$(T$_8$–OSS–SH)$_{12}$] is soluble in pentane. Au$_{55}$[P(Ph)$_3$]$_{12}$Cl$_6$ is soluble in dichloromethane: three ligand-stabilized clusters with three radically different solubility properties.*

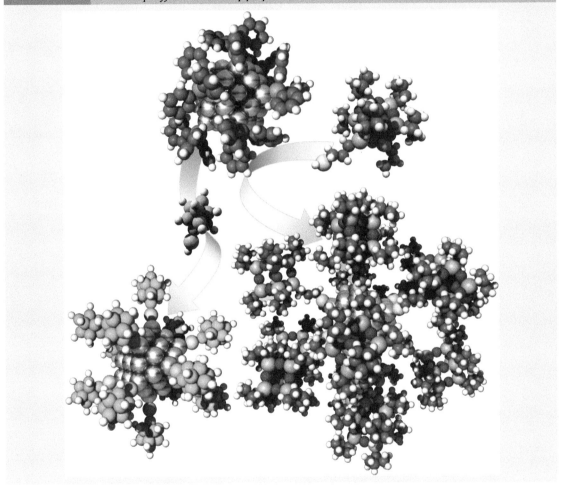

Source: Image redrawn with permission of Prof. Günter Schmid, Institut für Anorganische Chemie, University of Essen.

We represent one more interesting ligand consisting of 12 boron atoms [23]. The exchange of the phosphine occurs thusly in dichloromethane:

$$Au_{55}[P(Ph)_3]_{12}Cl_6 + Na_2[B_{12}H_{11}SH] \rightarrow Au_{55}[(B_{12}H_{11}SH)Na_2]_{12}Cl_6 \quad (12.8)$$

Na$^+$ can be exchanged by (octyl)$_4$N$^+$, thereby rendering the complex water soluble. The boron cluster is depicted in **Figure 12.16.** Both the OSS and boron ligand systems increase the size of the cluster and thereby increase the spacing of interdigitated arrays of two-dimensional ligand-stabilized quantum dots relative to the Au$_{55}$ phosphine ligand-stabilized cluster.

FIG. 12.17	*A colorized TEM image of an array of ligand-stabilized $Au_{55}[P(Ph)_3]_{12}Cl_6$ clusters. Self-assembly in this case resulted in a square planar array of ligand-stabilized gold nanoparticles.*

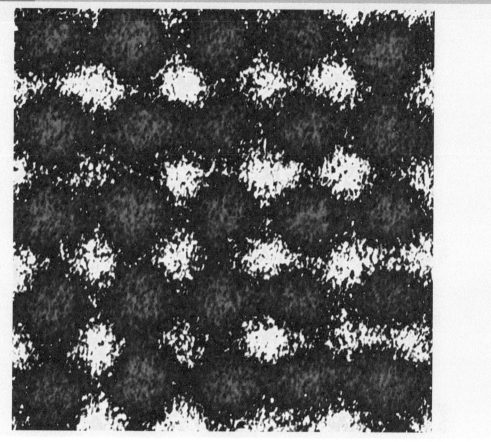

Source: Image courtesy of Prof. Günter Schmid, Institut für Anorganische Chemie, University of Essen. With permission.

Au_{55} clusters can be formed into arrays depending on the type of ligand and the concentration of clusters. After ligand-stabilized clusters are spread on a two-dimensional surface such as copper, configurations ranging from liquid (disordered) to solid de-interdigitated (similar in structure to a two-dimensional hexagonal close-packed structure) to "interdigitated" [30,31]. **Figure 12.17** shows a TEM image of ligand-stabilized Au_{55} clusters formed into a square array. The number of ligand systems that are able to interconnect quantum dot clusters is enormous. The clusters can be linked by covalent systems (e.g., sulfide linkages) or by intermolecular interactions such as π–π interactions, van der Waals, and electrostatic ones. A few interdigitated systems and other linking systems are shown in **Figure 12.18**. The size of cuboctahedron metal clusters (*sans* ligands) also affects the interlayer spacing of arrays: $[Au_{55}(PPh_3)_{12}Cl_6]$ (1.4 nm), $[Pt_{309}O_{30}]$ (1.8 nm), $[Pd_{561}(Phen_{36}O_{200})]$ (2.5 nm), $[Pd_{1415}(Phen_{60}O_{\sim1100})]$ (3.0 nm), and $[Pd_{2057}(Phen_{86}O_{\sim1600})]$ (3.6 nm).

FIG. 12.18 *Ligands interlinking clusters do so by several mechanisms ranging from covalent bonding (–R–S–R–) or disulfide (–S–R–S–S–R–S–) linkages that are covalent to aryl group interdigitation, and scores of other intermolecular interactions.*

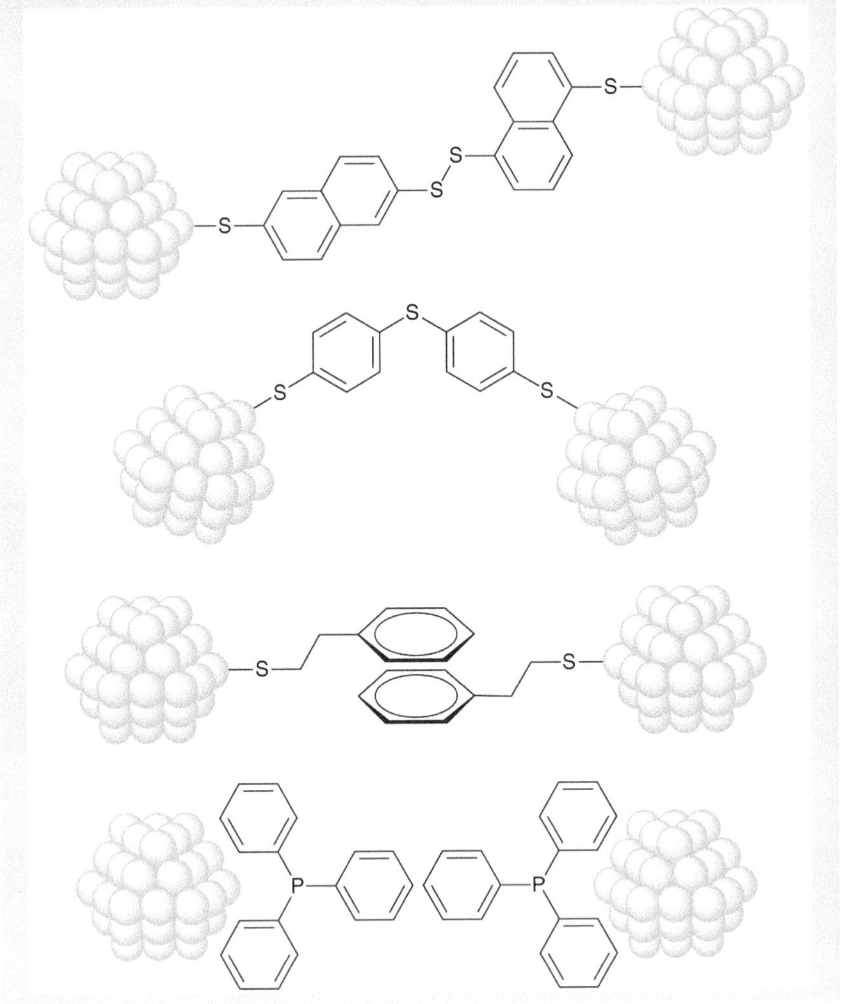

Semiconductor Quantum Dot Synthesis. There are several challenges facing the synthesis of quantum dots: (1) synthesis of dots that are monodisperse with regard to size, shape, and orientation and (2) fabrication of those dots into a uniform array. The inverse micelle-emulsion method of fabrication of semiconductor quantum dots has become increasingly popular. (Please take note that there are numerous methods to fabricate semiconductor *q*-dots from the top down that are not discussed in this chemical synthesis chapter.)

Semiconductors can be efficiently synthesized in reverse (inverse) micelles (emulsions of inverse micelles). Reverse micelles are formed in nonpolar solutions in which the hydrocarbon tails of the amphiphile project outward from the micelle into the nonpolar medium. Naturally, the polar groups of the

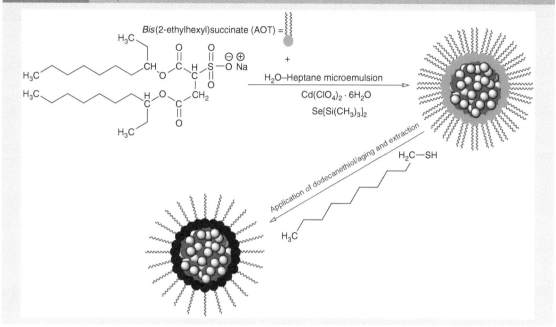

FIG. 12.19 *Formation of CdSe semiconductor quantum dots by an emulsion procedure. A very versatile surfactant known as NaAOT (aerosol-OT) is often used in these reactions. Reactants are sequestered within the micellar structures and react when mixed.*

amphiphiles project into the micelles structure. An example of a CdSe quantum dots formed in inverse micelles from sodium *bis*(2-ethylhexyl)sulfosuccinate (a.k.a. NaAOT), $Cd(ClO_4)_2 \cdot 2H_2O$, and $Se[Si(CH_3)_3]_2$ ([*bis*(trimethylsilyl)Se) is shown in **Figure 12.19**. Emulsions can be created by vigorously mixing water with a nonpolar solvent like heptane.

Inverse micelles form in solutions in which the majority liquid component is nonpolar and the minority component is made of water. The polar core of the micelle is called the "water pool" and is characterized by the water-surfactant molar ratio W:

$$W = \frac{[H_2O]}{[S]} \tag{12.9}$$

where S is the concentration of the surfactant [32,33]. Large Ws are characteristic of emulsions while reverse micelles have $W < 15$. The water pool radius, R_{Water}, is calculated from

$$R_{Water} = \frac{3V_{aq}[H_2O]}{\sigma[S]} \tag{12.10}$$

where V_{aq} is the volume of the water molecules and σ is the area of head of the polar group. Because Na(AOT) is used quite often, its R_{Water} is defined by the empirical relation [33]:

$$R_{Water} = 1.5W \tag{12.11}$$

In other words, the water pool radius is a linear function of the water content. Larger micelles can be achieved by increasing the water-surfactant ratio [3,34].

One group of reverse micelles containing S^{2-} anions is mixed into a solution that contains reverse micelles containing Cd^{2+} cations. Following mixing, micelles are able to interact to form CdS semiconductor clusters. Thiol capping of the CdS dots occurs by disrupting the micelles with thiol capping moieties [35]. The resultant ligand-stabilized CdS quantum dots are separated by centrifugation or precipitation.

M. L. Steigerwald, A. P. Alivisatos, and coworkers applied inverse micelles in the synthesis of CdSe semiconductor quantum dots (**Fig. 12.19**) [36]. Mixed semiconductors can also be synthesized with the reverse micelle solutions of the same W: S^{2-} and Te^{2-} with appropriate Na(AOT) surfactant are mixed with metal cationic solutions like $Cd(AOT)_2$, $Zn(AOT)_2$, $Mn(AOT)_2$, or Ag(AOT) [35]. The subsequent syntheses are controlled by predetermined ratios of reactants:

$$x_a = \frac{\left[M_1^{2+}\right]+\left[M_2^{2+}\right]}{\left[X_1^{2-}\right]}; \quad x_b = \frac{\left[M_1^{2+}\right]+\left[M_2^{2+}\right]}{\left[X_2^{2-}\right]}$$

$$x_c = \frac{\left[M_3^{2+}\right]+\left[M_4^{+}\right]}{\left[X_1^{2-}\right]}; \quad x_d = \frac{\left[M_3^{2+}\right]+\left[M_4^{+}\right]}{\left[X_2^{2-}\right]} \tag{12.12}$$

Following mixing, several procedures are available for further processing of the reaction. Semiconductor cluster centers are extracted from the micelles by addition of a binding group with high (and selective) affinity for the clusters. Groups such as dodecanethiol are able to coat the semiconductor and are separated by a subsequent phase transfer reaction. The original surfactant is liberated from the clusters by addition of ethanol. The combination of dodecanethiol addition and extraction of the clusters can be accomplished by different procedures: (1) immediate addition–immediate extraction; (2) delayed addition (2 days)–immediate extraction; (3) immediate addition–delayed extraction; and (4) 90-min delay in addition of dodecanethiol [33]. Thermodynamic equilibrium, water pool radius, and other factors are allowed to exert influence over the size and quality of the semiconductor quantum dot clusters [33].

Aging (time before addition of dodecanethiol), time of extraction, oxidation, cation ratios, and cation–anion ratios, in addition to the solvent and surfactant, all affect the size and crystallinity of quantum dots. CdTe are made under conditions of excess Cd in order to prevent oxidation of Te: $x = [Cd^{2+}]/[Te^{2+}] = 2$ [33]. Procedure (1) resulted in dots 2.6–3.4 nm in diameter while procedure (2) allowed for larger particles: 3.4–4.1 nm [33]. The difference was readily observable from fluorescence spectra. CdS by procedure (1) showed no fluorescence. By procedure (2) and with an excess of CdS ($x = [Cd^{2+}]/[S^{2-}] = 2$), size-dependent, but otherwise weak, fluorescence was observed (e.g., red shift with increasing particle size). In other words, the exciton peak was visible when aging of the particle was allowed to proceed [33]. However, strong fluorescent peaks were observed in CdS dots when the cation–anion ratio favored the anion. The addition of excess S prevents the vacancies (e.g., $x = [Cd^{2+}]/[S^{2-}] = 1/2$) from forming in the crystal. Although aging enhances the crystallinity of the CdS dots, elimination of S vacancies by adding excess S^{2-} proved to enhance the peak intensity [33].

Triangular nanocrystals of CdS are formed from $Cd(AOT)_2$/isooctane/H_2O mixture with $W = 30$ [33]. Addition of H_2S and N_2 generates cadmium and sulfur coprecipitates. Dodecanethiol is added after 2 days. TEM analysis and computer simulation showed that the particles are pyramidal in shape. Alloyed semiconductor crystals of the form $Cd_{1-y}Zn_yS$ are made by procedure (1). The solid is made from precursors containing excess sulfur: $x = \{[Cd^{2+}] + [Zn^{2+}]\}/[S^{2-}] = 1/2$ [33].

M. Guglielmi et al. synthesized CdS semiconductor dots from cadmium acetate $[Cd(CO_2CCH_3) \cdot 2H_2O]$, thioacetamide $[CH_3CSNH_2]$, and the surfactant 3-mercaptopropyl-trimethoxysilane (MPTMS) in methanol and resulted in particles on the order of 8 nm in diameter [M. Guglielmi et al., *Journal of Sol–Gel Science Technology*, 8, 1017, 1992]. Other CdS dots formed by mixing cadmium dichlorate $[Cd(ClO_4)_2]$ with $[Na(PO_3)_6]$ complex with H_2S produced nanoparticles of differing size. The size was dependent on the pH of the starting solution. At pH 9.8, the CdS nanoparticle fluoresced blue color ($\lambda \sim 430$ nm); at pH 9.0, the color of the solution was turquoise ($\lambda \sim 480$ nm); and at pH 8.1, the solution appeared greenish ($\lambda \sim 500$ nm) [A. Henglein et al., *Journal of the American Chemical Society*, 109, 5649, 1987].

Of course, as with gold clusters, monolayers and more complex structures are possible with semiconductor quantum dots. AgS dots on a TEM grid showed hexagonal packing characterized by relatively large domains (>100 μm). Chain interdigitation, chain length, chain packing, and particle interaction all contribute to the formation of structures with higher order. AgS quantum dots decorated with alkyl chains—$(CH_2)_x$—have formed into *fcc* "supracrystals" with $x = 8$ [33]. **Figure 12.20** shows a two-dimensional hexagonal packed layer of a generic quantum dots.

Gold colloids were also be formed in inverse micelles [37]. The majority solvent used was *n*-heptane and the surfactant was NaAOT. Micelles consisting of $HAuCl_4$ and $N_2H_4 \cdot H_2SO_4$ were mixed to form gold colloids ranging in size from 30 to 100 nm (with aggregates of 200–400 nm). The size of the gold colloids increased as the $[H_2O]/[AOT]$ ratio was increased [37]. In another microemulsion synthesis, NaAOT-*Span80*-isooctane (aqueous $HAuCl_4$ and $N_2H_5 \cdot OH$) were mixed:

$$4HAuCl_4 + 3N_2H_5 \cdot OH \rightarrow 4Au^o + 16HCl + 3N_2 + 3H_2O \qquad (12.13)$$

The non-ionic cosurfactant *Span80* acted as a stabilizer and structural modifier for the gold colloids. W was 8.0 and spherical gold nanoparticles 3.5–8.6 nm were achieved [37]. A small amount of rod-like and cube-shaped particles were also synthesized (95 nm in length) [37]. When W approached 20, particle size increased in the range of 10–40 nm and rods and cubes became more prevalent. When cosurfactants NaAOT and tetra(ethylene glycol)dodecylether were used together with isooctane as the oil phase, pure *fcc* gold up to 80 nm in size was produced. The morphology (spherical, rod-like, or trigonal systems) of the gold crystals was dependent on experimental conditions [37].

Aqueous Incipient Wetness Impregnation. This method is used heavily in the field of catalyst preparation. Although the resulting catalyst particles are not considered to be quantum dots, they certainly are small enough to be included in that special family of materials. AIWI requires a support material that aids in

Hexagonal array of semiconductor quantum dots decorates a surface. Differing surface arrangements can be achieved depending on the size of the dot and its ligands (or if there are mixed ligand systems).

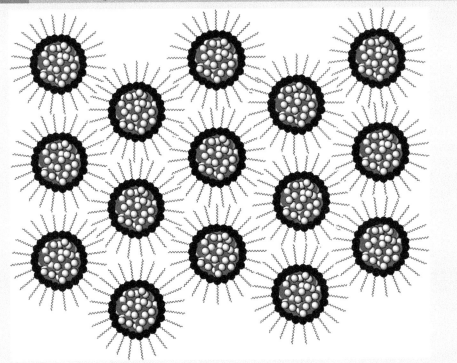

size control of the catalyst particles. The support is often in the form of fumed alumina ($85-115$ $m^2 \cdot g^{-1}$). One procedure is reviewed next to provide sufficient perspective about the procedure.

Fe–Mo catalyst for the formation of single-walled carbon nanotubes is prepared by mixing $Fe_2(SO_4)_3 \cdot 5H_2O$, $(NH_4)_6Mo_7O_{24} \cdot 4H_2O$ with fumed Al_2O_3 in deionized water at $80°C$ for 1 h with application of ultrasonication [38]. The mixture is dried overnight at $100°C$ under a nitrogen blanket. The resulting powder is then ground with mortar and pestle and calcined at $900°C$ for 10 min. After cooling, the catalyst-support powder is reground into a fine dispersion. The ratio of starting materials results in 2.7 mg of metal in a 6:1 Fe:Mo molar ratio on ~50 g of alumina support material [38]. Nanoparticles on the order of 1 nm to a few nanometers are produced by this method. The powder is then placed into a quartz boat in a reduction atmosphere to produce metal catalyst particles. CVD is then allowed to proceed and single-walled carbon nanotubes (SWNTs) are formed. Very simple.

Nb_2O_5–TiO_2 catalysts are prepared by AIWI of TiO_2 with aqueous solutions of niobium oxalate–oxalic acid. Nb loading of $0.48-8.5$ wt.% was prepared by this process. Time of flight secondary ion mass spectrometry (ToF SIMS) experiments on the catalyst complex revealed the presence of monomeric, dimeric, and trimeric Nb_xO_y clusters, even at low loading fractions, and that the degree of aggregation of the niobia clusters increased with higher loading levels [39].

Once again, a very simple, low-cost, straightforward chemical procedure takes center stage in the arena of nanoscience. The power of nano!

Sol–Gel Synthesis. One of the oldest forms of nanotechnology comes in the form of sol–gel synthesis. Colloid chemists have developed this technology dating back numerous decades (or more), developing one of the simplest, inexpensive, low-temperature, and most effective bottom-up wet chemical synthesis of nanomaterials that are highly pure and monodisperse in size. Inorganic metal oxides as well as inorganic–organic hybrid materials are synthesized routinely by sol–gel methods. The *sol* is the homogeneous solution of molecular reactant precursors that are concerted into an infinite molecular weight three-dimensional polymer: the *gel* that forms an elastic-solid fill material with the same volume as the liquid [40]. Mixtures of precursors (different metals, oxides, and even organics) lead to binary, ternary, and higher order systems [40]. Colloid size, structure, and morphology are dependent on the solution pH, temperature, composition, concentrations, and the solvent. Transformation from the sol into the gel, for example, is stimulated by a change in the pH of the solution [40].

Sol–gel synthesis proceeds by hydrolysis and condensation of silicate precursors like triethylorthosilicate (TEOS) (**Fig. 12.21**). Hydrolysis is the process of breaking bonds by the action of H_2O. Condensation is the formation of a bond with the simultaneous release of a water molecule (similar to the synthesis of a peptide bond). Hydrolysis of a silicon precursor proceeds:

$$Si(OR)_4 + yH_2O \rightarrow Si(OR)_{4-y}(OH)_y + yROH \qquad (12.14)$$

where R is an ethyl group in TEOS. Subsequent condensation is illustrated by:

$$2Si(OR)_{4-y}(OH)_y \rightarrow [(RO)_{4-y}(OH)_{y-1}]-Si-O-Si-[(OR)_{4-y}(OH)_{y-1}] + H_2O \quad (12.15)$$

FIG. 12.21 One of the most celebrated workhorses of sol–gel synthesis is tetraethylorthosilicate (TEOS). In water (or alcohol–water mixtures), TEOS undergoes dimerization and eventual polymerization. The main processes involved in sol–gel synthesis techniques are hydrolysis and condensation.

Triethylorthosilicate (TEOS)

Hydrolysis, condensation, gelification, aging

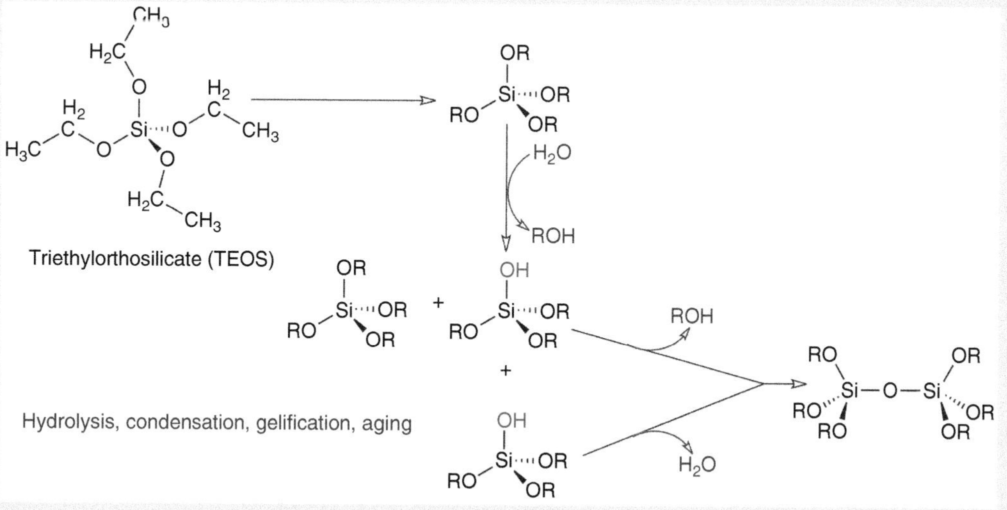

Although we have sol–gel synthesis in the section of zero-dimensional materials, its versatility is easily expanded to include one-dimensional, two-dimensional, and higher order structures and morphologies. Different metal groups and combinations of metals like aluminum, potassium, titanium, and others result in colloidal (and eventually ceramic) materials of great diversity, properties, and function.

Colloidal materials abound around us. Porous rocks, clays, mist, smoke, blood, opal, pearl, bones, milk, elgin, ice cream, and other foods are some of the natural materials that are made of colloidal materials [41]. Synthetic colloidal materials include paint, toothpaste, photographic silver iodide, polyurethane foam, and a multitude of other industrial materials. Typically, colloids range in size from 1 nm to 1 μm [41]. Because of high surface-to-volume ratios, colloids actually exist somewhere between physical phases—a separate and distinct colloidal phase. Colloidal particles are therefore sensitive to external forces such as electric fields and to intermolecular forces that we discussed earlier. Colloidal particles undergo Brownian motion, balancing attractive and repulsive forces by the interplay of electrostatic and steric factors. Reversible agglomeration is known as *flocculation* and irreversible agglomeration is called *coalescence* [41]. Agglomeration is prevented by the application of charge and steric stabilization—topics we discussed briefly in chapter 6.

Colloidal systems consist of a dispersed phase within a continuous phase. The colloidal system we are most interested in is that of a solid (dispersed) phase dispersed within an aqueous (continuous) phase. Solid–liquid dispersions contain a high solid content. Types of colloidal system dispersions (with examples in parentheses) include of liquid–gas (liquid aerosols), solid–gas (solid aerosols), gas–liquid (foams, froths), liquid–liquid (emulsions), gas–solid (solid foams), liquid–solid (ice cream), and solid–solid (opal). Colloids are characterized by *rheological* techniques (flow behavior) in which traits like thickening, shear thinning, and viscoelasticity are prominent. There are many ways to characterize particle size and shape (discussed briefly in chapter 3), such as light-scattering, small-angle x-ray scattering (SAXS) and small angle neutron scattering (SANS) analysis. Sedimentation (not discussed in chapter 3) relies on the following relationship:

$$\left(\rho_s - \rho_l\right)\frac{4}{3}R^3 g = 6\pi \upsilon R \eta \tag{12.15}$$

where

ρ_s and ρ_l are the densities of the solid particles and the liquid, respectively
υ is the sedimentation velocity
g is the acceleration constant of gravity
R is the radius of the colloidal particle
η is the viscosity of the medium

Metal nanoparticles are often synthesized on inorganic supports. Sol–gel processes, in this way, contribute to the synthesis of metal zero-dimensional materials. The inorganic support serves to isolate metal precursors before reduction and thereby prevent agglomeration of the metal [42]. The sol–gel product also serves in the capacity of a template. Vondrova et al. made use of a cyanogel that consisted of coordination polymers in which two metal centers are bridged by a cyanid ligand: $PdCl_4^{2-}$ and $Co(CN)_6^{3-}$. Upon heating, the cyanide groups are able to reduce the metals. In their experiments, silica gel was prepared from $[SiO_2/Na_2O]$ and aqueous solutions of Na_2PdCl_4 and $K_3Co(CN)_6$ were added to the solution and allowed to form a gel. The co-gel was dried at 600°C and the cyanogel phase was reduced to metal [42]. The Pd/Co alloy metal particles had an average diameter of 7 nm [42].

Sol–gel synthesis demonstrates versatility and facility at low cost. It is truly and purely chemistry. Sol–gel methods are revisited in section 12.3.

12.2.2 Synthesis and Modification of One-Dimensional Materials

There are many techniques available from which to form one-dimensional materials, better known as nanowires. Obviously, it will not be possible to cover all the methods in this section. Selected methods are summarized in **Table 12.3.**

A broad class of substances can be synthesized in the form of one-dimensional nanomaterials (nanotubes, nanocylinders, whiskers, fibers, nanorods, and nanowires). The wires can be electrically conducting, semiconducting, or insulating. Metals, ionic compounds, ceramics, and organic materials can all be made into nanowires. Recently, biological materials such as DNA have been shown to be nanowires with great utility, versatility, and durability. The ultimate nanowire is a chain of atoms. We have already discussed carbon nanotubes in chapter 9 and will not present any topics related to them in this section.

Conductive nanowires 100 nm in diameter were fabricated by self-assembly of amyloid protein fiber biological templates (Step 1) and selective metal

TABLE 12.3	Bottom-Up Chemical Nanowire Synthesis Methods	
Method	**Description**	**Examples**
Template-based methods	Electrodeposition	Gold nanocylinders in porous alumina materials. Electrodeposition is valid for all metals; some require nonaqueous solvents
	Electroless autocatalytic deposition	Electroless Au, Ag, Pd, Co, Ni, Cu
	Colloid, solution impregnation	Sol–gel or solution mixtures to make metals, semiconductors, ceramics, and other materials
	Microemulsions	Semiconductor rods and wires formed in micellular-based cavities
	Block copolymers	Formed in emulsions to form ordered materials with cavities, channels, and other template forms
	Biological materials	DNA and other biological materials with periodic structure serve as templates
Chemical vapor deposition	Decomposition of precursor materials	Carbon nanotubes Ge nanowires by diffusion of Ge_2H_6 vapor into mesoporous silica + heat [3,43]

deposition of Ag or Au (Step 2) [44]. In another study, block copolymers were first assembled to form within "lithographically defined" channels on a silicon substrate [45]. The block copolymers, 40 nm across, were introduced into the channels and self-assembled into long cylindrical molds after heat treatment at 230°C. Aqueous solutions of metal ions then were made to infiltrate the lines. The polymer was removed by application of oxygen plasma etching leaving the metal nanowire. The wires were 10 nm in diameter and over 50 μm in length. Buriak's group fabricated 25 parallel Pt wires. Jillian Buriak, the leader of the research team, stated:

> We've figured out a way to use molecules that will self-assemble to form the lines that can be used as wires. Then we use those molecules as templates and fill them up with metal, and then we have the wires that we want. You use the molecules to do the hard work for you.

Brown University researchers have found a straightforward way to synthesize Fe–Pt nanorods and nanowires by a technique that offers diameter, length, and composition control [46]. In their solution-based process, varying the ratio of the solvent (octadecene) and surfactant (oleylamine), nanowires ranging from 20 to 200 nm have been produced. Adding more surfactant resulted in longer nanowires, whereas more solvent yielded shorter rods. A 1:1 ratio produced 20-nm rods.

Solution synthesis of germanium nanowires was accomplished at 1 atm and less than 400°C [47]. The precursor molecule germanium 2,6,-dibutylphenoxide $[Ge(DBP)_2]$ was dissolved in oleylamine surfactant and then immediately transferred into a 1-octadecene solution at 300°C. The resulting nanowires were single crystal of cubic phase and coated with oleylamine. The length of the wires ranged from 0.1 to 10 μm. A self-seeding mechanism and an aggregation mechanism are proposed to explain the growth process.

Organic nanowires are also produced by bottom-up methods. P. Nickels et al. showed that polyaniline (PAN) nanowires can be grown within a DNA template [47]. Polyaniline was grown on DNA templates immobilized on a surface by (1) oxidative polymerization with ammonium persulfate; (2)enzymatic oxidation with hydrogen peroxide (horseradish peroxidase); or (3) photo-oxidation using a Ru complex as the photo-oxidant [48].

Ultralong $Cd(OH)_2$ nanowires were synthesized according to a hydrothermal methods from $Cd(CH_3COO)_2 \cdot 2H_2O$ and $C_6H_{12}N_4$ aqueous solution at 95°C without the use of templates [49]. Wires with aspect ratio of several thousands were formed. The formation mechanism is attributed to the oriented attachment of small particles. Transformation into semiconductor material occurred by calcination at 350°C for 3 h [49].

Electrodeposition. DC electrodeposition of metals into templates is an effective method to form nanowires that are monodisperse with regard to diameter [50–63]. The length of the nanowire is dependent on the duration of the electrodeposition process. Both metal and semiconductor nanowires can be formed by electrodeposition. Electrodeposition was discussed briefly in chapter 4. The optical properties of nanogold rods synthesized in anodic porous alumina membranes are dependent on the size, shape, and orientation within the template material [50–63]. The optical response of nanogold–alumina membrane composites was discussed in chapter 5.

Electroless Methods. Electroless deposition plating is the autocatalytic, continuous, and chemical reduction of metal ions onto a substrate. The substrate does not have to be conducting. The required components of an electroless plating solution include an aqueous solution of metal ions of interest, a catalyst (usually a minute amount of the metal in reduced form), reducing agents, complexing agents (help monitor *p*H and control free metal ion concentration), and solution stabilizers (catalytic inhibitors to prevent out of control reactions that result in poorly structured products) operating in a specific range of *p*H, metal ion concentration, and temperature. No electrical current is required in this form of deposition—hence the name electroless. Whether forming wires or plating on a two-dimensional surface, the final metal layer or wire conforms to the contour of the template or the surface.

A gold electroless plating solution, for example, consists of: 0.01 M gold chloride–hydrochloride trihydrate; 0.014 M sodium potassium tartarate; 0.013 dimethylamine borate; 400 mg·L^{-1} sodium cynanide; *p*H = 13.0 and temperature 60°C. An example of the electroless deposition of copper is

$$Cu^{2+} + EDTA^{4-} \rightarrow Cu(EDTA)^{2-} \tag{12.16}$$

$$2H_2C{=}O + 4OH^- \rightarrow 2HCO_2^- + 2H_2O + H_2 + 2e^- \tag{12.17}$$

$$Cu^{2+} + 2e^- \rightarrow Cu^o \tag{12.18}$$

$$Cu(EDTA)^{2-} + 2H_2C{=}O + 4OH^-$$
$$\rightarrow Cu^o + 2H_2O + H_2 + 2HCO_2^- + EDTA^{4-} \tag{12.19}$$

12.2.3 *Synthesis and Modification of Two-Dimensional Materials*

George Whitesides' rule for designing surfaces includes the following considerations, starting with the SAM [5]: (1) in general, functionalized alkane–thiols will form ordered SAMs with the terminal capping group pointed away from the substrate; (2) rule number 1 applies best for *n*-alkane thiols with small cap groups (e.g., –OH, –CN, –CO$_2$H, etc.); and (3) rule number 1 does not take into account steric, molecular free volume, lateral segregation (multicomponent assemblies), and neighbor and surface interactions. He goes on to say, "The development of guidelines for designing SAMs that predict the organization and composition of the monolayer and incorporate all these elements remains to be worked out."

In general, although we are getting better at supramolecular design, it is exceedingly more difficult to predict the physical and chemical properties of SAMs and we must rely on semi-empirical studies for at least a little while longer. However, modification of SAMs occurs by both covalent and intermolecular mechanisms. There is no limit to the variety of derivatization that can occur to receptive monolayers. Monolayer surfaces, especially, require modification if future design involves biological or biochemical applications [5].

Chemical Modification of SAMs. The chemistry of covalent chemical modification is influenced strongly by nucleophilic substitution reactions—e.g., basic organic chemistry. Bromine is a good leaving group and is used in many chemical modification procedures in nucleophilic substitution reactions.

A basic S$_N$2 nucleophilic substitution is shown. Bromine is a good leaving group in the presence of nucleophilic moieties. In SAM systems, due consideration is always allotted to designing the chemistry of the end group.

A basic nucleophilic substitution reaction is shown in **Figure 12.22**. If such a group is a terminal capping group on a SAM molecule, then the case for chemical modification is made possible. The strength of leaving groups depends on several parameters: size of the atom, electronegativity, electric charge, steric properties, and the solvent, to name a few. For example, F is not a good leaving group, whereas Br is a good leaving group. Inspection of some of the most basic SAM chemical procedures reveals that –OH is a good nucleophile and that –Cl and various alkoxides serve as good leaving groups. The conjugate bases of strong acids are good leaving groups. The sulfonate ion is an excellent leaving group and is employed in many kinds of substitution-chemical modification reactions of monolayer end groups.

The solvent, just like in supramolecular chemical reactions, is also an important consideration in chemical modification reactions and plays an important role in the rate of displacement reactions [64]. For example, the displacement of iodine in methyl iodide by an acetate anion takes place 10 million times faster in dimethylformamide (DMF) than it does in methanol [64]. If an S$_N$2 nucleophilic displacement reaction involves ions, then polar solvents are required. *Hydroxylic solvents*, for example, promote hydrogen bonding of reactants and products. Hydroxylic solvents include water and alcohols. *Polar aprotic solvents* include acetonitrile, DMF, dimethyl sulfoxide, and acetone. Nonpolar solvents consist of the family of low molecular weight alkanes, toluene, benzene, and other familiar solvents. The free energy of activation $\Delta G^\ddagger$ of reaction of nucleophiles with methyl iodide varies significantly from solvent to solvent. For example, the free energy barrier of CN$^-$ + CH$_3$I substitution is 59 kJ · mol^{-1} in DMF but as high as 92 kJ · mol^{-1} in methanol [64].

The other generic kind of chemical modification takes advantage of intermolecular interactions. A sample of such a reaction was illustrated by the golden molecular squares in chapter 11. Chemical modification by grafting, wrapping, associating, hydrogen bonding, electrostatic charge interactions, and all the other intermolecular interactions discussed so far occurs with high frequency in the literature. It all depends on what exactly the goal of the chemical treatment is: robust inert layers? biosensors? optical response? electrical conduction? The list goes on and on.

If the terminal group of the SAM molecule is a methyl group (not a good leaving group at all), then for all practical purposes the chemical portion of the synthesis is completed, unless one applies external stimuli by photo-excitation or an electrochemical potential [5]. The surface of the SAMs capped with methyl groups is hydrophobic; the surface is a low-energy surface (large contact angle) and is relatively nonreactive, if that is the place one wishes to stop.

More often than not, additional layers are desired to improve robustness and enhance properties. The most popular example of this process is the generation of additional layers by adding silane moieties to the previous layer. Our example of ALD shown in chapter 4 is analogous to this philosophy of layer building. During each step, a hydroxyl group acts as a nucleophile with silicon that has good leaving groups attached—for example, the $Si-O(Me)_3$ or $-SiCl_3$ type groups. Cross-linking of layers occurs via condensation reaction. The end result is a thick layer (dependent on the number of cycles employed) with lateral robustness.

Four advantages lend themselves to SAM modification following monolayer deposition: (1) application of commonly used procedures; (2) ligand incorporation into SAMs after monolayer deposition impossible a priori; (3) derivatization of many SAM samples by many kinds of ligands; (4) preservation of the molecular, short-, and long-range order of the parent monolayer; and (5) small amount of ligand (e.g., 10^{-9} mol) [5]. Disadvantages of SAM films are related to the surface coverage; the extent of the parent and modified structure is relatively unknown and heterogeneous coverage may be produced during chemical modification [5].

Table 12.4 lists several strategies of SAM chemical modification [5]. In addition to the chemical natures of the terminal group and the potential modifying moieties, the SAM itself is able to affect the chemical reaction by steric and geometric factors, restricting accessibility to functional groups, and modifying the nature of the solvent near the SAM–solution interface. The characteristics of the solution at the surface may be quite different from that of the bulk [5]. The SAM is a chemical entity, albeit a complex one, and as such exhibits properties that are based in its structure and composition. The organization of the chains (e.g., level of crystallinity or disorder), the surface density, orientation of functional groups and their distance from the substrate, and lateral steric effects (and van der Waals interactions) all contribute to the chemical and physical character of the monolayer.

Electron Beam Writing.   Enhancement of an electron beam writing procedure was accomplished by attachment of silica-based SAMs capped with nitro-phenol [65,66]. Monolayers were formed on activated silicon surface substrate by application of nitrophenoxy-propyltrimethyloxysilane (NPPTMS). The terminal nitro groups were converted into amines by exposure to an electron beam (5–6 keV). The nitrophenyl group is easy to reduce chemically to form amines but that would limit the scope of selectivity over surface patterning. How would one selectively mark off areas that require modification of the nitrophenyl groups by such a broad-base chemical treatment? Application of the e-beam solves this problem quite well (**Fig. 12.23**). Modification of the amine groups by altering the *p*H renders the amine groups (now in the ammonium state) receptive to reaction with citrate-stabilized gold nanoparticles. Gold in this form carries inherent negative charges. The process is similar to the formation of expanded monolayers of the golden molecular squares [67].

Covalent Chemical Modification.   Covalent chemical modification of the distal layers of fluorinated self-assembled monolayers (F-SAMs) $[CF_3(CF_2)_7(CH_2)_2-S-Au]$

TABLE 12.4	*Functional Groups and Chemical Modification*	
Type of modification	**Mechanism**	**Examples Terminal group + ligand**
Covalent reactions	Nucleophilic substitution, esterification, acylation, nucleophilic addition, cyclo-addition, metal-catalyzed olefin cross-metathesis	Maleimide + thiol (R = peptides, carbohydrates) Disulfide + thiol (R = DNA) Hydroxyl + siloxane Halogen + phosphine (lone pair donors) Amine + thiocyanate Vinyl group + acrylamide (Ru-catalyzed) Acetyl group + azide (of alkylthiolates) → triazole Alkyne + azide Ester + azide (formation of amide)
	Surface activation to form a reactive intermediate that is able to react with the ligand (e.g., formation of amide linkages by interchain anhydride intermediate) *Activation* by external stimuli (electric fields, photoexcitation) transforms unreactive groups into active ones.	Dehydrated carboxylic acid (via TFAA)[‡] + amine → amide Activated carboxylic acid (via NHS)[*] + amine → amide Hydroquinone (via electrochemical oxidation) → quinone + diene
	Covalent bond breaking reactions of terminal cap group. In these reactions, there is release of a molecule bound to a surface group into solution.	Propionic ester modified quinones release ester and lactone during electrochemical reduction, resulting in intramolecular cyclization
	Surface-initiated polymerization is able to graft polymers to a SAM surface by (1) covalent linking of preformed polymer chains to reactive SAM terminal groups or (2) direct growth of the polymer from the terminal group.	(1) Poly(acrylic acid)/poly(ethylene glycol) (2) Poly(acrylonitrile) (by photoinitiated radical method) Block copolymers
Intermolecular interactions	*Nonspecific adsorption* of molecules from the solution phase by van der Waals forces, electrostatic interactions, combinations thereof, and hydrophobic effects	Surfactants, polymers, polyelectrolytes, proteins, organic dyes, and colloidal particles Nonspecific adsorption implies no control (or relatively little control) over orientation and thickness
	Fusion of vesicles: adsorption of lipid vesicles results in supported bilayers, single layers, or hybrid layers, the chemical nature of which depends on the chemistry of the functional group.	Supported bilayers of phospholipid vesicles yield layers that are defective. Hybrid bilayers serve as dielectric barriers with few pinholes. The combination of SAM + bilayer = good model for cell membranes.
	Selective deposition is a function of hydrophobicity and charge density. Electrostatic charge on SAM aids orientation of adsorbate molecules. Due to selectivity, the method is applicable to patterning techniques.	pH control of adsorption of polyelectrolytes, polyallylamine (PAA), and polyethyleneimine (PEI) onto carboxylic acid- or oligoethylene glycol (OEG)-terminated SAMs favors PAA on OEG at pH 4.8, while PEI deposits primarily on carboxylic acid groups. Positively charged surfaces adsorb antibodies. *Cytochrome c* adsorbs onto positive SAM surfaces with favorable electron-transfer orientation without losing its structure. Negatively charged surfaces do not produce the same effect.

TABLE 12.4 (CONTD.)	*Functional Groups and Chemical Modification*	
Type of modification	**Mechanism**	**Examples Terminal group + ligand**
	Modifications by designed *molecular recognition*—the domain of supramolecular chemistry This kind of modification is reversible. Due to high selectivity, the precision of the method allows placement of ligands close together.	Use of hydrogen bonding networks, dative bonds between metals and ligands, electrostatic interactions, and/or hydrophobic interactions Concerted, cooperative host–guest relationships stabilize surface complexes. Reactions have small equilibrium constants and are therefore reversible (e.g., dissociation of adsorbates by addition of excess ligand).

Source: J. C. Love et al., *Chemical Reviews*, 105, 1103–1169 (2005).

can be attained by bombardment with low energy (>100 *e*V) polyatomic ions [68]. The projectile ion was $CH_2Br_2^+$ with an m/z ratio equal to 172. Detection of the species CH_2F^+, $CHBrF^+$, and CF_2Br^+ was proof that the F-SAM underwent chemical modification [68].

Corrosion Prevention. In an attempt to improve the corrosion resistivity of aluminum, a self-assembled monolayer of hexadecanedioic acid (HDDA) was applied to the surface of the metal (**Fig. 12.24**) [69]. To enhance the layer further, cross-linking within the HDDA layer was attempted by application of octyltrichlorosilane (OTS). Use of perfluorinated carboxylic acid as the base layer reveals that OTS is able to displace the 1° organic SAM and from Si–O (siloxane) linkages to the metal surface [69]. Carboxylic acids do not form the strongest links with metal surfaces and during any chemical modification, such considerations need to be included in the design phase of the monolayer experiment. If the result of chemical modification is the decomposition of the SAM, then another strategy needs to be considered.

12.3   TEMPLATE SYNTHESIS

Template synthesis is the fabrication of nanomaterials within porous materials and interstitial spaces. We focus primarily on porous materials formed by inorganic and organic–inorganic hybrid materials. According to the IUPAC definition (as before), there exist three classifications of porous materials: macroporous ($d > 50$ nm), mesoporous ($2 < d < 50$ nm), and microporous ($d < 2$ nm). An example of each kind of template is presented. Porosity ε is defined as the porous or void fraction of a material (as before). In more specific terms, it is also defined as the fraction of void volume available to an adsorbate or analyte. Obviously, this omits porous spaces that are inaccessible to adsorbates. The various kinds of porous materials are displayed in **Figure 12.25.**

Chemistry in Confined Spaces. Chemists are used to working within the framework of nonconfined systems such as the beaker and the flask. But how are

FIG. 12.23

The conversion of the nitro group on the benzene ring into amines is accomplished with an electron beam. By this way, specific sectors are created on an otherwise homogenous monolayer coating. Further chemical modification is facilitated with the amine terminus of the molecule. Such patterning is possible by direct writing with the electron beam. Adjustments in pH convert the amine into an ammonium moiety that is receptive to electrostatic interaction with citrate-stabilized gold colloids that have negative charges. The latter part of this procedure was used effectively to produce the golden molecular squares encountered in chapter 11 [65–67].

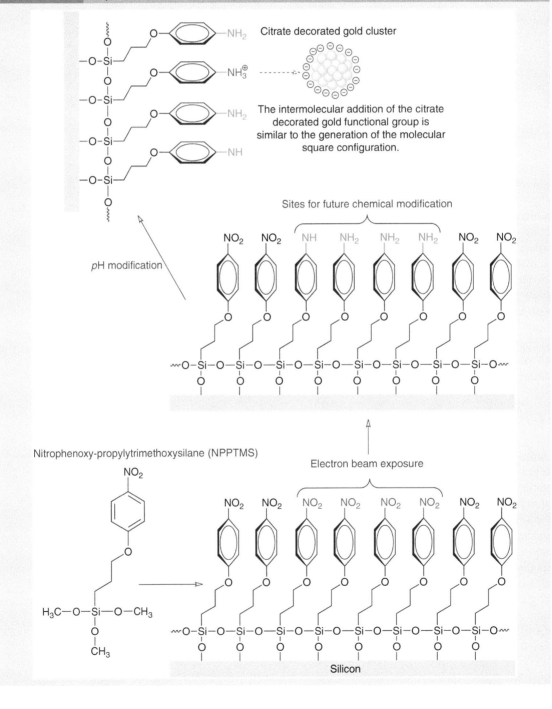

FIG. 12.24	*Hexadecanedioic acid (HDDA) was applied to the surface of the metal [69]. To enhance the layer further, cross-linking of the HDDA later was attempted by application of octyltrichlorosilane (OTS).*

Hexadecanedioic acid (HDDA)

Octyltrichlorosilane (OTS)

Source: L. D. Seger, J. P. Rasimas, R. Pesce-Rodriguez, and R. Fifer, Chemical modification and attempted polymerization of self-assembled monolayers of hexadecanedioic acid at aluminum surfaces, Army Research Lab, www.stormingmedia.us/21/2171/A217133.html (1997). With permission.

chemical reactions affected if that chemistry is accomplished in a space similar in scale to that of the process itself? Molecularly confined spaces are defined loosely as three-dimensional spaces less than 1 nm in diameter. In essence, we are talking about nanosized chemical reactors.

In one example, the pyrolysis of surface-linked 1,3-diphenylpropane (DPP) in MCM-41 zeolite was found to exhibit accelerated rates and alterations in product selectivity—both kinetic effects [70]. S. Polarz reported that the formation of electrochromic oxides like MoO_3 is limited to first-order kinetics in confined spaces and that the kinetics of transformation of precursors is also altered [71]. In another study, researchers found that chirality of cations can be controlled in confined spaces [72]. R. Anwander showed that surface-confined metal species are amenable to small-ligand chemistry of alkoxide metal centers. These species are not found in traditional solutions due to agglomeration [73]. Encapsulation and surface grafting contributed to nonagglomerated metal–ligand species as a result of steric effects—pathways originating from steric and highly unsaturated metal centers [73].

Diffusion Behavior in Porous Membranes. Diffusion of gases and liquids through porous materials is a major field of study in nanoscience. Purification of natural gas to remove toxic compounds and noncombustibles is a major undertaking by today's energy suppliers. Membranes that offer selectivity along with high throughput are highly sought after. Unfortunately, the two characteristics reside on opposite ends of the spectrum (e.g., you can have one but not the other). There are a few kinds of diffusion processes through porous membranes. The general categories are (**Fig. 12.26**): (1) bulk diffusion (nonselective, no separation); (2) Knudsen flow diffusion in which the mean free path of the molecule is on the order of the pore diameter; and (3) molecular or restricted diffusion (when pore size is on the order of molecular size). The Knudsen number K_n (the ratio between the mean free path and the system size) indicates which diffusion domain is applicable:

Various kinds of porous materials are displayed. Top left: the "ink bottle" pore cavity is quite common and serves as an excellent nanosized beaker within which studies of chemical confinement are conducted. A structure is considered to be a pore if its width is less than its depth. Top middle: interstitial spaces are formed with hard spheres. The spaces are exploited in templating processes such as nanosphere lithography. Top right: interstitial spaces are found between cylinders formed by a sol–gel process. Such spaces are also found in single-walled carbon nanotube bundles. The interspatial cavity is defined by the van der Waals gap between and among bundles. Bottom left: enclosed cavities are not useful in the templating process due to their inaccessibility. These kinds of cavities, however, are able to affect the physical properties (e.g., thermal conductivity) of the material. Bottom middle: zeolites exhibit highly ordered structures that come in the form of channels or three-dimensional interconnected porous networks. Bottom right: the pore channels of porous alumina can be made to transverse the entire thickness of the membrane or be left capped at one end.

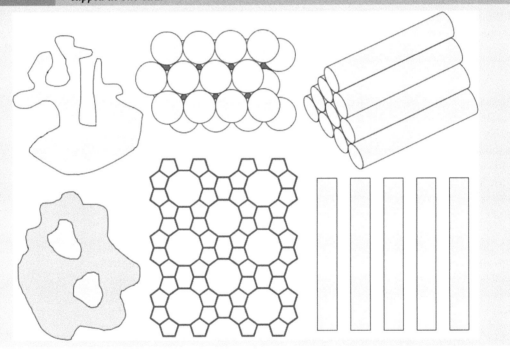

$$K_n = \frac{\lambda}{d} \tag{12.20}$$

where λ is the molecular mean free path and d is the pore diameter. When $K_n \gg \lambda$, the condition of bulk diffusion prevails. Pores greater than 100 nm in diameter fall into this category ($D > 10^{-4}$ m$^2 \cdot$ s^{-1}). When $K_n \sim \lambda$, Knudsen flow predominates and diffusion behavior is described by Fick's law of diffusion. Pore diameter lies between 100 and 1 nm ($10^{-8} < D < 10^{-4}$ m$^2 \cdot$ s^{-1}). When $K_n \ll \lambda$, molecular or restricted flow (configurational diffusion and surface migration) is in effect. Pore diameter in the molecular flow is less than 1 nm ($10^{-8} < D < 10^{-16}$ m$^2 \cdot$ s^{-1}).

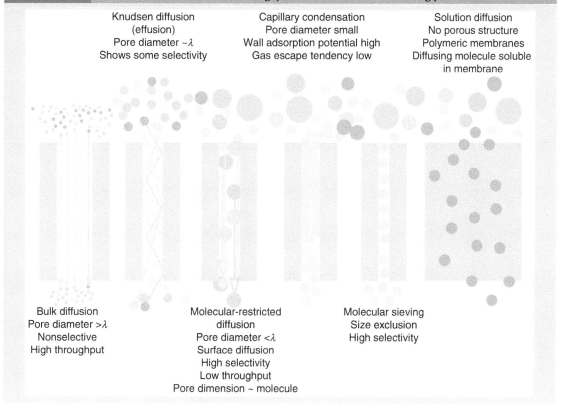

FIG. 12.26 — *Diffusion through porous materials is important to many industrial processes. Various kinds of diffusion are depicted. Pressure differentials are important in bulk diffusion (far left) but lose their importance as pore size is diminished. In general, selectivity increases as pore size diminishes. However, throughput diminishes with decreasing pore diameter.*

Knudsen diffusion
(effusion)
Pore diameter $\sim\lambda$
Shows some selectivity

Capillary condensation
Pore diameter small
Wall adsorption potential high
Gas escape tendency low

Solution diffusion
No porous structure
Polymeric membranes
Diffusing molecule soluble
in membrane

Bulk diffusion
Pore diameter $>\lambda$
Nonselective
High throughput

Molecular-restricted
diffusion
Pore diameter $<\lambda$
Surface diffusion
High selectivity
Low throughput
Pore dimension $\sim$ molecule

Molecular sieving
Size exclusion
High selectivity

Four basic "selective" mechanisms of gas transport (mass transport) through porous materials are *Knudsen diffusion* (more akin to effusion), surface diffusion, capillary condensation, and molecular sieving [74]. The pore diameter under Knudsen diffusion conditions is on the order of the mean free path of the molecule. Mesoporous systems exhibit Knudsen flow. In other words, as molecules make their way through the pore channels, collisions with the wall are likely. Their rate of transport through the pore channel is inversely proportional to their molecular mass. The separation factor between two molecular species is the ratio of the squares of their respective masses. For optimal selectivity, the pore diameter should be less than the mean free path λ of the molecule. Therefore, pore size, surface chemistry, and the chemical nature of the molecule affect mass transport through the pore channels. *Surface diffusion* involves interaction of molecules with the pore surface; polar molecules interact with polar surfaces; and nonpolar molecules interact with nonpolar surfaces (van der Waals forces) (**Fig. 12.27**). If the orifice is small enough to promote overlap of potential fields within the pore, then the case of *capillary condensation* results as molecules transit the pore channel. The propensity of escape is dependent on the molecule's ability to overcome the adsorption potential. Lastly, molecular sieving occurs

Chemical modification of the pore walls of porous materials affects both physical and chemical properties. Additional layers (or just the presence of one layer if the pore diameter is small enough) constrict the orifice and thereby enhance the size-exclusion ability of the material. Chemical modification of the surface allows for separation based on intermolecular chemical interactions.

when the pore diameter is so small that size exclusion keeps large molecules out of the channel. Small molecules like CO_2 and H_2 that have kinetic diameter near 3 Å qualify for passage through very small pores.

We will not dwell much more on diffusion through porous materials but will make mention that chemical modification of pore channels is able to impact both the physical properties of the pore channel (e.g., the size and flow) and the chemical selectivity of potential substrates (e.g., traction with specific substrates that affect the kinetics of the flow processes). **Figure 12.27** illustrates the chemical modification of two surfaces of porous alumina formed by an anodize process. The shape, size, charge, and chemical affinity of the diffusing molecule play more important roles in smaller pore channels (also confinement effects). Gas permeation and separation, dialysis, electrodialysis, osmosis, and reverse osmosis are all membrane-based phenomena but detailed discussion beyond what was already presented is beyond the scope of this section.

The Periodic Minimal Surface. A minimal surface is a surface of a structure that results in the smallest possible area [75]. An example of a two-dimensional minimal surface is afforded by a film of soap formed within a wire frame [75].

The surface tension γ of the film of soap is proportional to the surface area (e.g., the as-formed film "is consistent with the shape of the frame and with the requirement that the mean curvature of the film be zero at all points" [75]. Of course with zeolites, we are interested in the *triply periodic minimal surface* (TPMS). These surfaces abound in nature, especially in the silicates, lyotropic colloids, detergent films, and lipid bilayers—all materials that are formed by self-assembly processes and that, interestingly, can serve the function of a template in synthetic processes to form nanomaterials. The skeleton of a sea urchin exhibits this phenomenon at the interface between calcite crystals and its soft-matter structure [76,77]. Atoms in zeolites lie on minimal surfaces [78,79].

Up until the discovery of template-based surfactant synthesized zeolites, minimal surfaces were difficult to quantify. It was found that these new zeolitic materials possessed structures that resemble liquid–liquid minimal surfaces. The synthetic zeolites have predetermined, tunable, and well-defined pore size and shape; excellent thermal and hydrolytic stability; and high order over macroscopic scales [75]. The correspondence between structure determined experimentally and theoretical calculations is excellent; the cubic mesophase of the H_2O–cetyltrimethylammonium bromide (CTAB) is the same as those predicted by calculation and the same as those found in the zeolite from which it was synthesized [80]. Ordered graphite foams from fullerenes and crystalline metal oxides also possess triply periodic minimal surfaces. We can never drift too far off the topic of energy and how it is always minimized in the structures that we study in nanoscience—or any other science, for that matter.

12.3.1 Macroporous Template Materials

There are many kinds of macroporous materials. Electrochemical etching of aluminum (anodizing) is capable of forming alumina structures with pores greater than 50 nm up to 200+ nm in diameter. In general, anodically formed porous aluminas are classified as mesoporous systems and will be discussed later in this section. Electrochemical etching of silicon, chemical etching of glass, ion track etching of polymers, and excimer laser micromachining are techniques capable of producing one-dimensional (pores oriented in one direction) macroporous materials [81]. Microemulsion chemistry and colloid sintering processes are capable of generating porous structures that range in two or three dimensions. Sacrificial scaffolds of crystalline materials have increasingly been used as templates to form porous materials [81].

The most popular method of forming macroscopic porous materials is from templates consisting of close-packed spherical polymer beads (a.k.a. latex beads) by an infiltration-sintering process. The process to form them is as follows: (1) synthesis of polymeric latex beads by emulsion polymerization; (2) packing of the beads into arrays by several techniques, such as evaporation, filtration, settling, and dip-coating; (3) infiltration of the interstitial volume with silicate, zirconia, titania, or aluminate sol–gel precursor solutions by capillary action or vacuum induction of the solution or by infiltration by solutions containing metal–anion components; and (4) treatment at high temperature (~500°C or more) to remove polymeric components and simultaneously sinter to form ceramic matrix. Filling scaffolds by electrochemical deposition and chemical vapor deposition are also effective methods of creating porous materials.

Close-packed open structures of 320–360 nm have been formed in a single-step process. The void spaces in stacked latex spheres, for example, were formed by vacuum-assisted induction (packing) that were covered with the metal alkoxide precursors. The assembly was calcined at 575°C [82]. Approximately 500-nm thick films that contain three-dimensional arrays of macropores titania, zirconia, or alumina microcrystalline material were formed.

Hybrid macroporous materials are also fabricated by colloidal crystal templating methods that yield porous thiol–metal oxide structures. [83]. Examples of hybrid macroporous materials include thiol-functionalized titania or zirconia with propylsiloxane, ethylsulfonate, or propylsulfonate linkages. The templating materials in this case are made of polymethyl methacrylate (PMMA). The purpose of using macro- rather than meso- or microporous materials is to improve the mass transport of a potential analyte through the porous material. The colloidal crystal system has built-in "synthetic flexibility" in that the selectivity, activity, and throughput characteristics of an adsorbent can be tailor made to fit an application. Hybrid macroporous materials consisting of zirconia and titania were chosen because those materials offer greater flexibility for chemical functionalization than do the silica-based substrates. Three types each of titania and zirconia macroporous structures are discussed and properties are displayed in **Table 12.5** for use in removal of heavy metals from solution by ion adsorption. An example of a template method that uses latex spheres is shown in **Figure 12.28a**.

12.3.2 Mesoporous Template Materials

Anodic Aluminum Oxide Membranes. Anodization of gate (refractory) metals leads to another class of mesoporous system. In anodizing, a metal is made the anode and oxidation of that metal in an electrolytic solution occurs. Anodizing aluminum has been an industry practice dating back to the first quarter of the twentieth century. Following the introduction of aluminum as a major industrial material, NPL of Teddington, United Kingdom, filed a British patent (no. 290,901) in 1926 that later was acquired by Alcoa in the United States in the 1930s. Anodizing is a process that creates an insulating oxide layer on a conductive metal anode, usually aluminum, in an electrolytic solution, usually a dilute polyprotic acid. By providing hexagonally packed pore channels that are simple to fabricate and the ability to manipulate pore diameter and length during and after anodizing, the anodic alumina oxide (AAO) offers a perfect template for nanoscale material synthesis. The utility of the AAO as a template is widespread.

TABLE 12.5	Hybrid Inorganic–Organic Macroporous Structure Size and Surface Area		
Macropore material	**Pore size (nm)**	**Template size (nm)**	**Surface area ($m^2 \cdot g^{-1}$)**
TiO_2–O_3Si–Pr–SH	265	300	37
TiO_2–O_3S–Et–SH	270	300	12
TiO_2–O_3S–Pr–SH	275	300	12
ZrO_2–O_3Si–Pr–SH	415	480	18
ZrO_2–O_3S–Et–SH	450	480	20
ZrO_2–O_3S–Pr–SH	460	480	14

Metals and alloys, polymers, semiconductors, carbon nanotubes, and ceramics are just a few examples of the major classes of materials that can be fabricated or inserted into the pore channels of AAOs. The presence of hydroxyl groups on the surface of AAOs facilitates a wide range of chemical applications.

Robust porous oxides are formed on metals known as gate metals. These include aluminum, titanium, tantalum, and niobium. Zirconium and vanadium have also been anodized successfully. We limit our discussion, however, to

Fig. 12.28

Macroscopic and mesoscopic materials are compared. Mesoscopic materials have pores that are "officially" under 50 nm. (a) Macroporous template synthesis: porous ceramic (titania or silicate) ordered arrays formed from spherical latex templates. The structure on the bottom is interconnected with channels similar to those found in zeolites. The size of the spheres can be several hundred nanometers.

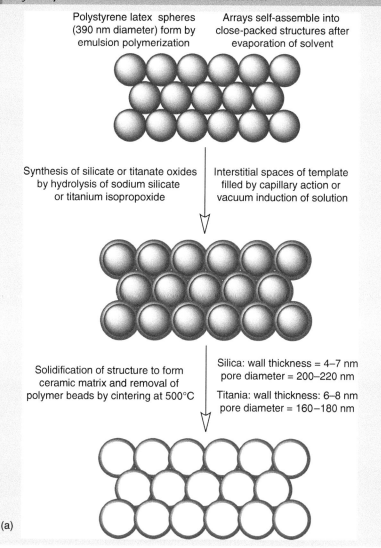

Polystyrene latex spheres (390 nm diameter) form by emulsion polymerization

Arrays self-assemble into close-packed structures after evaporation of solvent

Synthesis of silicate or titanate oxides by hydrolysis of sodium silicate or titanium isopropoxide

Interstitial spaces of template filled by capillary action or vacuum induction of solution

Solidification of structure to form ceramic matrix and removal of polymer beads by cintering at 500°C

Silica: wall thickness = 4–7 nm
pore diameter = 200–220 nm

Titania: wall thickness: 6–8 nm
pore diameter = 160–180 nm

(a)

(continued)

FIG. 12.28 (CONTD.)

(b) Mesoporous template synthesis: porous aluminas are able to transcend all boundaries with regard to classification. In general, most anodically formed porous structures have pores in the range of 20–50 nm. Commercially available anodic films have pores that are much larger. The structure of anodically formed porous aluminum oxide is shown again, this time compared to a barrier layer oxide. The porous layer is formed on top of the barrier layer. The barrier layer thickness is the same throughout the process. The scalloped metal surface is shown in the figure. The size of the scallop depends on the applied voltage. By means of a double-anodize method, a perfect honeycombed structure can be formed. During the first anodization, usually consisting of an extended time period (e.g., 24 h), equilibrium positioning of the pore channels is achieved. All the pore channels have the same diameter with the same barrier layer thickness. After dissolution of the primary layer, re-anodization produces a new film that sprouts from the preexisting scalloped structure—already with the equilibrium size and distribution.

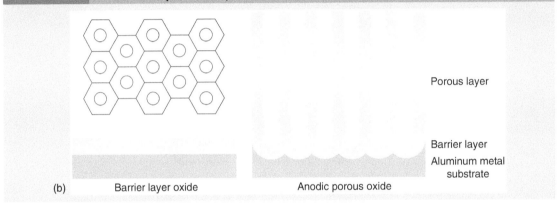

Porous layer

Barrier layer
Aluminum metal substrate

(b)   Barrier layer oxide   Anodic porous oxide

oxides formed exclusively on aluminum. Three general types of oxides are grown (**Fig. 12.28b**). First, nonporous barrier layer oxides are formed when aluminum is anodized in weak acids such as tartaric or boric acids. In these acid solutions, the as-formed oxide is completely insoluble and a thin layer (a few nanometers) is formed. Secondly, irregular nonporous oxides are formed in strong monoprotic acids such HCl and HNO_3, acids in which the as-formed oxides are highly soluble. The third category of porous AAOs is the most important with regard to template synthesis. To achieve a structured, uniform porous anodic oxide, aluminum metal must be anodized in a polyprotic acid in which the aluminum oxide is slightly soluble. Polyprotic acid anions are also comparable in size to the aluminate anion and this factor may play a role in the oxide's solubility and its complex electrochemistry during formation. Oxide thickness over 300 μm has been achieved on aluminum.

Anodize Methods.   Porous alumina membranes comprise a remarkable class of nanoscale template materials. Fabrication is fairly straightforward. Equipment and supplies considered to be nothing more than "low tech" are required. Apparatus and materials include a DC power supply, aluminum substrate plate stock, electrode contacts, ammeter and voltmeter, temperature-controlled bath, an agitator system, and a solution of a dilute polyprotic acid (usually phosphoric, chromic,

TABLE 12.6	*Anodize Parameters for Common Polyprotic Electrolytes*					
% Electrolyte (w/w)	Temp. (°C)	Potential (V)	Duration (h)	Thickness (μm)	Pore diameter (nm)	Pore density (N · cm⁻²)
4% H_3PO_4	0	130	8	50	200	1.3×10^8
1% H_2CO_4	−5	90	2	40	120	2.2×10^9
1% H_2CO_4	0	70	3	30	86	4.3×10^9
1% H_2CO_4	0	40	6	40	60	1.2×10^{10}
4% H_2CO_4	2	30	10	25	52	1.4×10^{10}
10% H_2SO_4	0	20	4	20	32	3.7×10^{10}
10% H_2SO_4	5	15	6	15	22	8.0×10^{10}
15% H_2SO_4	8	10	3	5	10	3.8×10^{11}
20% H_2SO_4	20	2	1	<1	~5	1.5×10^{12}

Source: G. L. Hornyak, *Characterization and optical theory of nanometal/porous alumina composite membranes*, Ph.D. dissertation, Colorado State University (1997).

sulfuric, or oxalic acids). Basic anodize formulations are listed in **Table 12.6.** The porosity ε for all types of membranes ranges from 20 to 30%. Porosity can be increased or decreased following anodizing and membrane detachment.

Control of pore diameter under voltage control conditions is accomplished by variation of applied voltage. The range in pore diameter varies from a few to several hundreds of nanometers. The thickness of the membrane is a function of the duration of anodizing. Thickness can be varied from a few nanometers to greater than 300 μm. Pore channel aspect ratio ($\mathcal{A}_{pc} = \ell/d$) ranges from less than one to greater than several thousand. An atomic force microscopy (AFM) image of a porous aluminum oxide made in oxalic acid is shown in chapter 1, **Figure 1.4.**

Two general strategies are applied during anodization. In the first, anodic potential is held constant for the duration of film growth while the current is allowed to seek its steady-state level. This configuration leads to pore channels that have the same diameter throughout the thickness of the membrane. In the second type, the applied current is held constant and the voltage is allowed to wander. In this case, pore channels form in conical configurations. **Figure 12.29** shows the current density time transient under voltage control on anodize time for two types of acid solutions: oxide slightly soluble (porous type, dashed red line) and oxide insoluble (barrier type, solid blue line). Initially, both transients appear to exhibit the same behavior due to the initial surge in current as the system strives to achieve the voltage according to Ohm's law:

$$V = I\Omega \tag{12.21}$$

Resistance Ω in the anodic circuit is provided by the newly formed barrier layer. At the onset, no oxide (insulating) coating is present and the current achieves its maximum value under specific voltage, electrolyte, and temperature conditions. However, as the oxide layer begins to form, a rapid decay in current occurs in both electrolytes. In the boric acid medium, the current decays exponentially without recovery. This results in a nonporous and thin barrier oxide

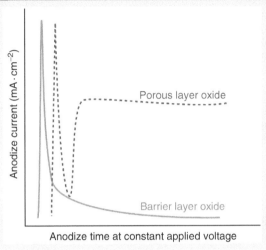

Anodize current–time transients under voltage control conditions are shown for nonporous barrier oxides (blue) and porous oxides (dashed red). The plateau represents the steady-state current once the balance between electrochemical dissolution and field-assisted growth has been established. The current is proportional to the thickness of the barrier layer.

layer. In polyprotic acids, the current reemerges and achieves a steady-state level for the duration for the anodize process. A porous oxide layer is grown under these conditions.

A porous layer is formed due to a complex interplay between the chemical dissolution process and the electric field-assisted growth. This behavior is responsible for the steady-state plateau region of **Figure 12.29.** The steady-sate current is a function of the barrier layer thickness.

As-formed membranes are complex in composition and contain a wide range of amorphous, hydrated, and crystalline alumina species in addition to incorporated electrolyte. Three layers contribute to the as-formed membrane: the aluminum substrate, the barrier layer, and the porous layer. The barrier layer allows anodization to occur and is the place where chemical dissolution and electric field-assisted growth of the oxide exist in the steady state. The barrier layer thickness is a function of applied anodic voltage. The pore diameter, as a result, also depends on the applied potential. The barrier layer behaves much like an ohmic device. Ohm's law describes the electronic behavior of the barrier layer oxide:

$$V = \rho d i_+ \tag{12.22}$$

where ρ is the specific resistance of the oxide material during anodizing in terms of ohm centimeters, d is the thickness of the barrier layer in centimeters, and i_+ is the ionic current density at the anode in amperes per square centimeter.

Pore formation and spacing are explained by phenomenological reasoning. Aluminum metal plates of 99.999% purity are used to manufacture the best membranes. The plate is electropolished to form a shiny, smooth, and reflective surface. Electropolishing is similar to anodizing except that chemical dissolution of the as-formed oxide is favored over film growth: high temperature, high current, very high or very low *p*H conditions. Unless formed from single crystal aluminum, the surface is saturated with crevices, pits, ridges, and other irregularities. These areas are regions where current is focused, akin to a reversed lightning rod. As the potential is applied, a competition for available current density is initiated. An oxide layer is formed and dissolved, and pockets of lower resistance outcompete areas where thicker oxide layers were formed. Scalloping of the aluminum surface occurs as the current density vector increases in strength, favoring their survival over bigger or smaller cells. The cells with the appropriate barrier layer thickness survive and pack two-dimensionally into a hexagonal array. In membranes formed by this process, branched pore channels are commonplace as pore cells migrate to find equilibrium positions in the hexagonal array as a function of anodize time. Surface defects and imperfections prevent long-range order from happening in AAOs, although domains on the order of microns are possible.

Highly Ordered Membranes. Fabrication of highly ordered membranes involves a two-step anodize process [84]. In the first step, a porous membrane is formed according to the scheme described previously. After 24 h of anodizing, all scallops are the same size and equally distributed on the aluminum surface. At this time, the oxide is dissolved in acid, leaving an aluminum plate that is pretemplated for another round of anodizing. With the scallops preformed and in the equilibrium state, anodizing is resumed, but this time the oxide layer grows directly from the template scalloped layer. The result is an ordered array of nanochannels with no branching and no heptagonal or pentagonal channel lattices (e.g., hexagonal cells are prefered in the "equilibrium" structure). Hideki Masuda of the Tokyo Metropolitan University is credited with its development [H. Asoh, K. Nishio, M. Nakao, T. Tamamura, H. Masuda conditions for fabrication of ideally ordered anodic porous alumina using pretextured Al, *J. Electro. Chem. Soc.*, 148, B152–B156 (2001)].

Membrane Detachment. There are several ways to remove the AAO membrane from its aluminum metal substrate. The most efficient means is to apply the procedure of stepwise voltage reduction. We know that pore size is a function of applied potential. If we were to reduce the potential stepwise, then a network of smaller and smaller pore channels would form in the barrier layer. The detachment process current–time transient is similar in form to **Figure 12.29** except that it contains a series of steps (**Fig. 12.30**), each with less current intensity as voltage is turned down sequentially. The barrier layer is a highly permeable ionic membrane during anodizing. If an instantaneous 10% voltage reduction is applied, the current undergoes depletion and then recovers to the new steady-state condition dictated by the newly applied potential. Essentially, a new set of pores, proportionally smaller than the parent pore, are formed in the barrier layer. This process is repeated until the pore diameter is reduced along with a concomitant thinning of the barrier layer.

When 1 V is achieved (pore diameter $\approx$1.5 nm, barrier layer thickness $\approx$1 nm), the process is stopped and the aluminum and oxide still attached to the metal plate is placed in an acid detachment solution (25% H_2SO_4 for sulfuric

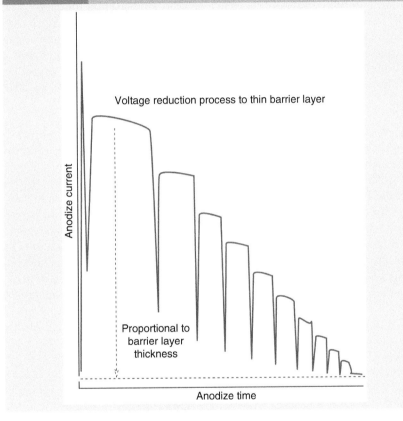

FIG. 12.30   *The voltage reduction process is shown in this figure. Because the thickness of the barrier layer and the pore diameter are dependent on the anodic voltage, decreases in voltage result in thinner barrier layers and smaller pore channels. After several steps of reduction, a fractal network of smaller pore channels is formed. This process in effect increases the exposure of alumina at the metal interface to dissolution by acids. The result is detachment of the bulk oxide film from the aluminum metal substrate.*

acid-formed films at room temperature, conditions that favor dissolution). The dissolution reaction is as follows:

$$2Al_2O_3 + 12H^+ \rightarrow 4Al^{3+} + 6H_2O + 3H_2(g) \qquad (12.23)$$

$$2Al^\circ + 6H^+ \rightarrow 2Al^{3+} + 3H_2(g) \qquad (12.24)$$

Hydrogen gas is produced between the alumina film and the metal substrate. Bubbles coalesce underneath the film indicating detachment and, after rinsing and drying, the film essentially pops off intact. Another method to remove the film that is practical for thin aluminum metal substrates is the application of saturated mercuric chloride. The mercuric chloride only attacks aluminum metal and not the oxide, and an amalgam between aluminum and mercury is formed.

In other words, treatment of the aluminum metal substrate with mercuric chloride removes the aluminum and leaves an intact barrier layer (unlike in the voltage reduction procedure). This is a good method to use if the barrier layer is deliberately left intact. In this way, the membrane is closed at one end of the pore channel—a perfect nanobeaker.

The AAO membrane has some ideal chemical, physical, thermal, and optical properties because it is made from alumina and, after drying, the membrane is insulating. Each cell could contain capacitors, resistors, conductors, or semiconductors and be insulated from its neighbors. Alumina is relatively inert chemically. Although it is an amphoteric material (able to dissolve in both acids and bases), at neutral pH, the material is inert to most chemical dissolution reactions. Optically, the membranes are highly transparent. Just like with colloids and quantum dots, light scattering (which causes diffusivity) does not occur until pore diameter is larger than 25 nm. That is why membranes made in sulfuric acid (with pore diameter <25 nm) are perfectly clear. Membranes with pores in the 50- to 100-nm range appear clear but slightly diffuse. Inclusion organic oxalates may contribute to the diffusivity in addition to the larger pore size. Membranes made in phosphoric acid, with pore diameter greater than 150 nm, appear opaque, grayish to white. AAOs can withstand thermal stress until phase changes occur. As we noted earlier, many phases of alumina, along with incorporated electrolyte, make up the matrix. At temperatures above 600°C, alumina undergoes a phase change into γ-Al$_2$O$_3$. At temperatures greater than 800°C, sulfates decompose into SO$_2$ and SO$_3$ gases and the matrix undergoes catastrophic loss of pore structure. At temperatures higher than 900°C, alumina is converted into γ-Al$_2$O$_{3-}$.

Chemical Derivatization of Anodic Surfaces. Chemical modification of the surfaces of the pore channels is quite straightforward. The alumina surface is a natural source of hydroxyl groups that are able to act as nucleophiles in ester exchange reactions. Alkylphosphonic acids and diethylbutylphosphonate are used to change the wetting character of the surface from hydrophilic to hydrophobic [85]. Silanes, carboxylates, phosphonates, and numerous other species are able to react with the pore channel walls of AAO membranes.

Template Synthesis with Porous Aluminas. Template synthesis with AAO hosts results in nanomaterials in the pore channels. This method is a general approach that allows for the synthesis of many kinds of nanomaterials by many types of synthetic methods [56,86,87]. Monodisperse materials are fabricated in highly ordered membranes. Two forms of synthetic strategies are employed: one in which the AAO membrane is detached from the aluminum metal substrate and another where the membrane is left on the aluminum with the barrier intact.

Synthesis with intact membranes requires thinning of the barrier layer by acid dissolution or by the voltage reduction method described previously so that electrical conductivity and ionic transport are facilitated through the barrier layer. The barrier layer, once thinned, allows for AC electrodeposition of metal and semiconductor materials. In another application, one that is gaining importance as a component in flat panel displays, carbon nanotubes are synthesized from Co, Fe, or Ni nanoparticles that are electrodeposited at the base of the pore channels by the AC method. The process comprises four steps following the initial anodizing process: barrier layer thinning, AC electrodeposition of metal catalyst nanoparticles, chemical vapor decomposition of carbon feedstock gas, and, lastly, the growth of carbon nanotubes.

Electroless deposition procedures to form honeycomb replicas or negative replicas of AAOs also require AC deposition to place a catalyst at the pore basin. In electroless deposition, a conductive material is deposited onto another material, conductive or nonconductive, by the autocatalytic reduction of metal ions in a chemical solution without using electrodes. H. Masuda of the Tokyo Metropolitan University has taken this form of plating to high levels in nanostructured material fabrication.

Electroplating metals and semiconductors into the pore channels of AAO straightforward. Following detachment and removal of the barrier layer, the freestanding membrane is coated with a layer of silver, usually applied by RF sputter procedure. The silver performs the function as a cathode and is easily removed following electroplating with HNO_3 treatment. Formation of gold nanostructures is depicted in **Figure 12.31.** Multitiered structures are also possible in concentric geometry (**Fig. 4.15,** if viewed from the top).

Alumina → derivatized walls → polymer coating → metal nanowire

Hybrid Inorganic–Organic Mesostructured Materials. Organically functionalized mesostructured materials were formed by the co-condensation of TEOS and organoalkylsilane with a neutral alkyl amine surfactant. Organosilane species included octyltriethoxysilane, mercaptopropyltrimethoxysilane, phenyltriethoxysilane, butyltrimethoxysilane, and propyltrimethoxysilane—all relatively long-chained alkylsilanes. At least 3-methylene units were required to interact with the hydrophobic core of the structure-directing micelle. This allowed for successful incorporation of the silane into the pore walls of the mesostructured material [88].

Most ordered mesoporous solids consisting of metal oxides, metals, and carbon have amorphous pore walls. Hierarchical ordered mesoporous solids with molecular-scale pore surface periodicity, however, certainly are not amorphous. Inagaki et al. managed to synthesize, with the aid of surfactant mediation, an ordered benzene–silica hybrid that contains hexagonally distributed mesopores with lattice constant equal to 5.25 nm and structural period of 0.76-nm spacing along the channel axis [89].

The tube was "self-assembled" by a combination of alternating hydrophilic silica and hydrophobic benzene layers with structural directing between the precursor molecules and the surfactants. The synthesis is as follows. An organosilane monomer $[(C_2H_5O)_3Si-C_6H_6-Si(C_2H_5O)_3]$ is heated in a basic solution of surfactant template material. Interactions between and among the surfactant, the organosilane, silica, and benzene organize to direct the self-assembly process to form into a tube. Average diameters range from 3 to 5 nm, and the diameter can be controlled by selection of the surfactant moiety. As the material was developed, sulfonation of the material was accomplished successfully without disruption of the mesoporous structure or periodicity [89] that proved to be stable at 500°C in air. Mesoporous structure synthesis can also be accomplished with emulsion chemistry [90].

The study of chemistry in confined spaces continues with the addition of each new kind of template. The overlap between "supramolecular chemistry in confined spaces" and cage-like structures continues as new insights into chemistry

FIG. 12.31 *Synthesis of gold nanorods by template synthesis/electrodeposition method. Top left: porous alumina was removed from alumina metal substrate by the process of voltage reduction in 25% oxalic acid. Pore channel diameter after anodizing is between 10 and 50 nm. Middle left: a layer of Ag is applied by RF plasma sputter process. The Ag layer is the cathode in all electrodeposition steps. Bottom left: Ag posts are electroplated into pore channels to provide platform deeper into the oxide. Bottom right: Au is plated on top of the Ag posts. The aspect ratio of the Au cylinders is controlled by the pore diameter and the time of electrodeposition. Middle right: the Ag is removed with a nitric acid wash. Top right: if required, the alumina is removed by reaction in NaOH solution. The red color of the gold cylinders resembles the actual color of the composite membrane*

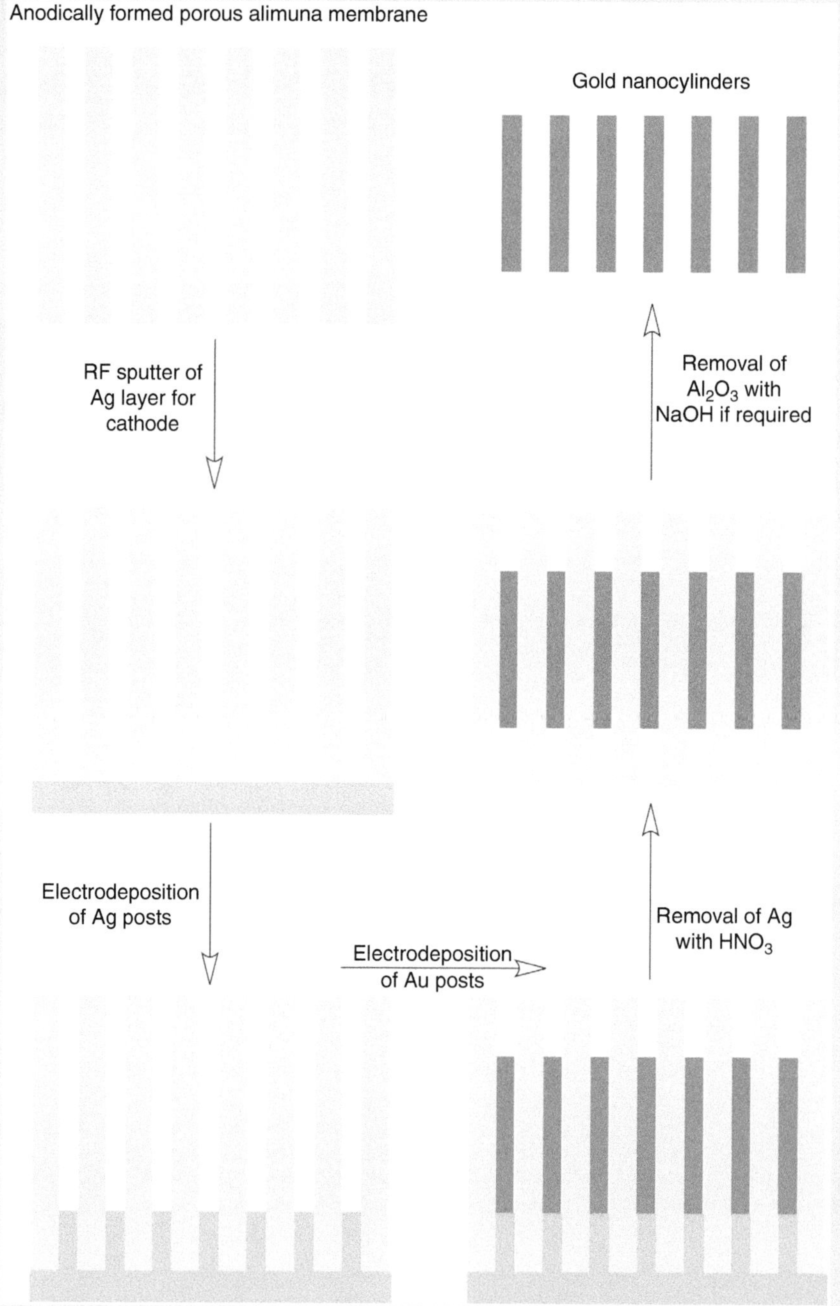

Anodically formed porous alimuna membrane

Gold nanocylinders

RF sputter of
Ag layer for
cathode

Removal of
Al$_2$O$_3$ with
NaOH if required

Electrodeposition
of Ag posts

Removal of Ag
with HNO$_3$

Electrodeposition
of Au posts

and kinetics continue to be uncovered [91]. These nanometer-ordered "minimal surface" structures continue to amaze.

12.3.3 *Microporous Template Materials*

Microporous templates have the smallest pore cavities and channels (<2 nm). The best known of these are the zeolites. Although zeolite pore channel size spans the microporous domain and into the meso range, we shall confine their discussion to this section (another boundary). Axel F. Cronstedt, the Swedish chemist who discovered nickel in 1751, coined the word zeolite (from the Greek *zein*, "to boil," and *lithos*, "stone") for a microporous natural material, after observing boiling chips in water. There are approximately 50 naturally occurring microporous zeolites and hundreds of synthetic forms that are classified as micro- or mesoporous. Zeolites are considered to be of a class of materials known as aluminosilicates that are microporous materials with a three-dimensional structure consisting of an interconnected network of pores. Si and Al atoms exist in tetrahedral crystal structure coordinated by oxygen atoms. Scolecites, natrolites, phillipsites, and analcime are some naturally occurring zeolites.

Mobil crystalline materials (MCMs) are synthetic zeolites. MCM zeolites are characterized by pore diameter less than 2 nm but as large as 50 nm, depending on synthesis conditions. The largest pores found in naturally occurring zeolites are 1–1.2 nm. The "liquid crystal template system" of Kresge et al. showed that pore size can be controlled by intercalation of layered aluminosilicates between ordered surfactant micelle species [92]. A calcination process follows chemical synthesis to form the zeolite. Uses of zeolites include molecular sieves, adsorption, shape-selective catalysis, membrane separations, nonlinear optics, and detergents.

12.3.4 *Other Interesting Template Materials*

Aerogels and Xerogels. An aerogel is an extremely low-density (porosity >90%) and high internal surface area (>1000 $m^2 \cdot g^{-1}$) microporous structure derived from a sol–gel process in which the liquid component is replaced by gas. Xerogels generally have porosity that is much lower (and hence higher density) [3]. During the sol–gel process and following a period of aging, the system is exposed to a pressure and temperature above that of the supercritical point of the solvent—a process called supercritical drying. This technique results in removal of the liquid vehicle without causing structural damage to the solid matrix of the material. Supercritical conditions are employed to mitigate the effect of capillary forces that would cause the collapse of the gel network otherwise. (Note: see Laplace's equation that involves the radius of curvature, contact angle, and surface energy.) S. S. Kistler, the first person to make aerogels, stated in 1931 [132]:

> Obviously if one wishes to produce an aerogel, he must replace the liquid with air by some means in which the surface of the liquid is never permitted to recede with the gel. If a liquid is held under the pressure always greater than the vapor pressure, and the temperature is raised, it will be transformed at the critical temperature into a gas without two phases having been present at any time.

The critical conditions for water are 374°C and 22 MPa; for carbon dioxide, they are 31°C and 7 MPa.

Aerogels are made from silicas, aluminum oxides, and carbons by a sol–gel process involving hydrolysis, condensation, and gelification. The synthesis is relatively expensive and requires long times for gelation—especially in the extreme low-density materials. Kistler reacted 1 mol of Na_2SiO_3 with 2 mol of HCl to form 1 mol of a silicon oxide complex $[SiO_2 \cdot H_2O]$ and 2 mol of NaCl. He treated the aged sol in an autoclave held at 374°C @ 221 bar. Kistler extracted the water with ethanol and thereby lessened the required temperature parameter from 374 to 243°C @ 64 bar. He produced aerogels with density ranging from 0.02 to 0.2 g·cm^{-3}. The system was extended to other inorganic gels, various combinations of precursors, and to organic systems.

12.4 POLYMER CHEMISTRY AND NANOCOMPOSITES

Polymers (from the Greek *polymers,* "having many parts," from *poly,* "many," and *meros,* "part") are materials that consist of a large number of similar units (monomers) that are bonded together. Swedish chemist J. J. Berzelius coined the word *polymer* in 1833, although his original intent was to describe materials with the same empirical formula that had different molecular weights. Polymers, a direct by-product of modern chemical engineering, have made a significant impact on our technological development by means of their versatility. Polymers are made of various mixtures of carbon and hydrogen, nitrogen, fluorine, or silicon. Molecular weights of bulk polymers are in the thousands to millions. Colloidal polymeric materials and nanocomposites are two areas that have received much attention recently.

12.4.1 Introduction to Polymer Chemistry

Polymeric materials exhibit many forms. Inorganic polymers include various strains of clay, sand, and fibers. Adhesives, fibers, coatings, plastics, and rubber

---

EXAMPLE 12.1   *Degree of Polymerization of Polyethylene*

Calculate "*n*" for a polyethylene polymer that is 180 nm in length. Assume that the polyethylene is not coiled and that all the bonds are in one plane. The C–C bond length is 0.154 nm.

Solution:
The monomeric repeating unit of polyethylene is $-(C_2H_4)-n$

The C–C–C bond angle is 109.5° for a tetrahedral structure. The actual linear bond length is $\sin[(1/2)\,109.5°] \times$ *C–C bond length* = 0.1258 nm
  Since every monomeric component has two bonds,
  $n = 180/[(0.1258) \times 2] = 714$ *monomers*

constitute the class of synthetic organic polymers. Well-known biological polymers include polysaccharides, proteins, and naturally occurring gum rubber. Polymers are generally formed by addition (adding to a double bond) or condensation (e.g., elimination of water) reactions. Epoxies (cross-linked) and polyurethanes (urethane linkages) are members of other classes of polymers. A *copolymer* is a mixture of two polymers. A *block copolymer* consists of repeating units of blocks of two types of polymers. The degree of polymerization, represented usually by n, indicates the number of monomer units that have bonded to generate the polymer. The value of n ranges from a few hundred to several thousand. Molar mass is related to the degree of polymerization—roughly equal to the average molecular weight of the constituents times the number of monomeric units.

Types of Polymers.   There are many kinds of monomers and hence many kinds of polymers. A few are listed in **Figure 12.32.**

Bonding in Polymers.   Because polymers are complex structures containing large molecules, one can surmise that many possible types of forces exist between and among polymer chains. One does not expect polymers to be highly ordered but they do have considerable crystalline character. A stretched rubber band is the classic example of how some level of order is imparted to a polymer as it is elongated, freezing the components momentarily into a lattice. Although the order may not transcend the entire polymer (long-range order), ordered domains, called crystallites, exist within the structure of a polymer.

Some types of polymer chains are attached to each other by strong covalently bonded substituents. This is called cross-linking. Weak intermolecular forces discussed previously abound in polymers. Weak dispersion forces in poly(ethylene) and van der Waals forces in poly(propylene) exist in polymers and arise from synchronous attractions of electrons between large molecules, of which there is an abundance in polymers [93]. Medium to weak dipole–dipole forces are found in poly(vinylchloride) and poly(methylmethacrylate). Much stronger hydrogen bonding also plays a major role in the structure of condensation polymers such as polyamides and proteins and in cellulose. Strong electrostatic intermolecular forces are found in ionomeric compounds. Every type of inter-molecular force is represented in polymers.

12.4.2 Polymer Synthesis

There are two general strategies available to fabricate polymeric materials. The first method is by the process of *addition*. There are three stages in the addition process: initiation, propagation, and termination. Addition processes involve addition to the double bond of an alkene monomer. Initiators can be free radicals, cations, anions, or organometallic. Organometallic catalysts called Ziegler catalysts consist of an organometallic (e.g., R_3Al) and a metal halide (e.g., $TiCl_4$). Because radicals are highly reactive, rapid chain reactions abound in the propagation step. All reactions are terminated when radical products combine with other radicals in the mixture. The second general mechanism is the *condensation* process. The term condensation implies that water is eliminated as bonds are formed between two monomers. Other small molecules, however, can also be eliminated, such as hydrochloric acid, methanol, and carbon dioxide.

Fig. 12.32	*Some of the most common polymers are shown in the figure. All can be applied to make nanomaterials.*

Polymerization method	Starting material (monomer)	Monomer structure
Free radical 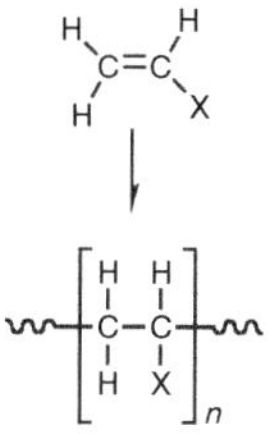	Poly(ethylene) PET	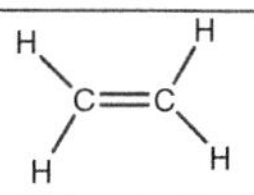
	Poly(propylene)	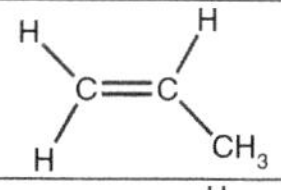
	Poly(vinylchloride) PVC	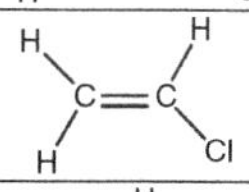
	Poly(styrene)	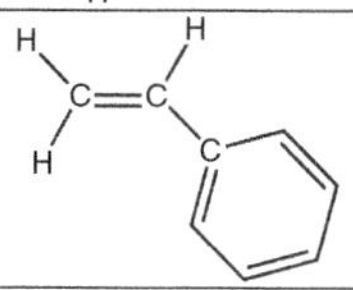
	Poly(acrylonitrile) Orlon	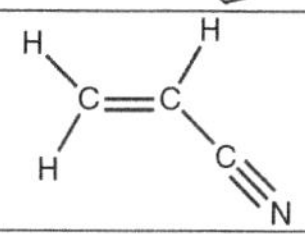
	Poly(tetrafluoroethane) Teflon	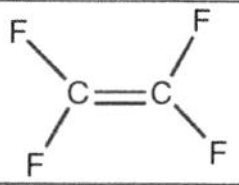
Hydrolysis	Poly(vinylalcohol) PVA	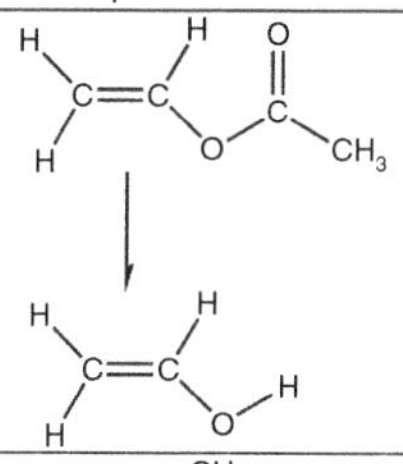
Anionic	Poly(methylmethacrylate) PMMA (and free radical)	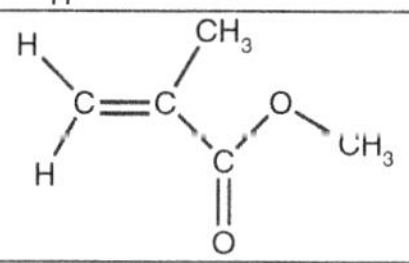
Cationic	Butyl rubber	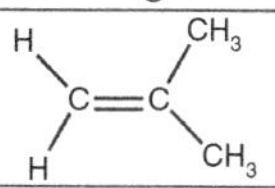
Condensation	Poly(butylenesuccinate) PBS	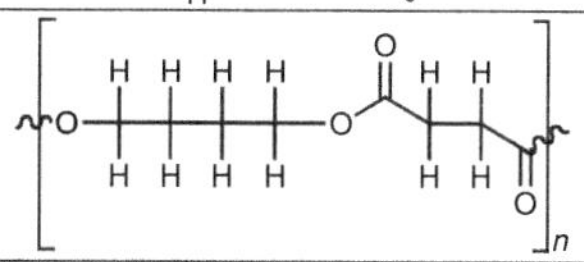
	Polyesters Combination of dicarboxylic acid or diester+diol	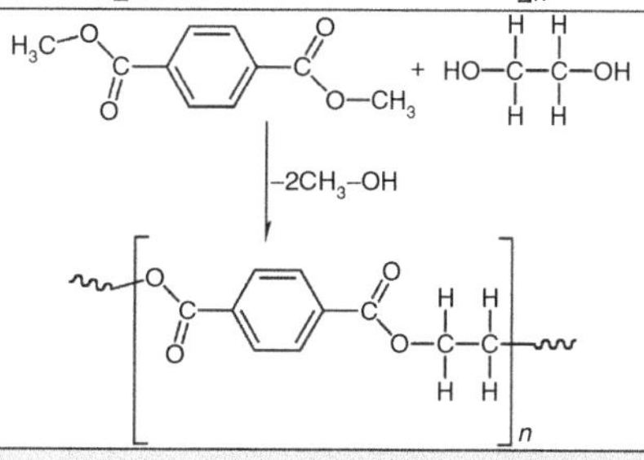

A free radical polymerization reaction is depicted. Following initiation, the polymer propagates until reactants are depleted by termination. The thermal decomposition of organic peroxides (e.g., R = benzoylperoxide) is one means of generating the radical initiator required for the polymerization of the ethylene monomer (Step 1). Step 2 involves consecutive addition of monomeric units until chain lengths of several hundreds to thousands are achieved. The radical process propagates in this way until reaction with another radical is achieved, thus terminating the polymerization (Step 3).

FIG. 12.33

Step 1: Initiation

Step 2: Propagation

Step 3: Termination

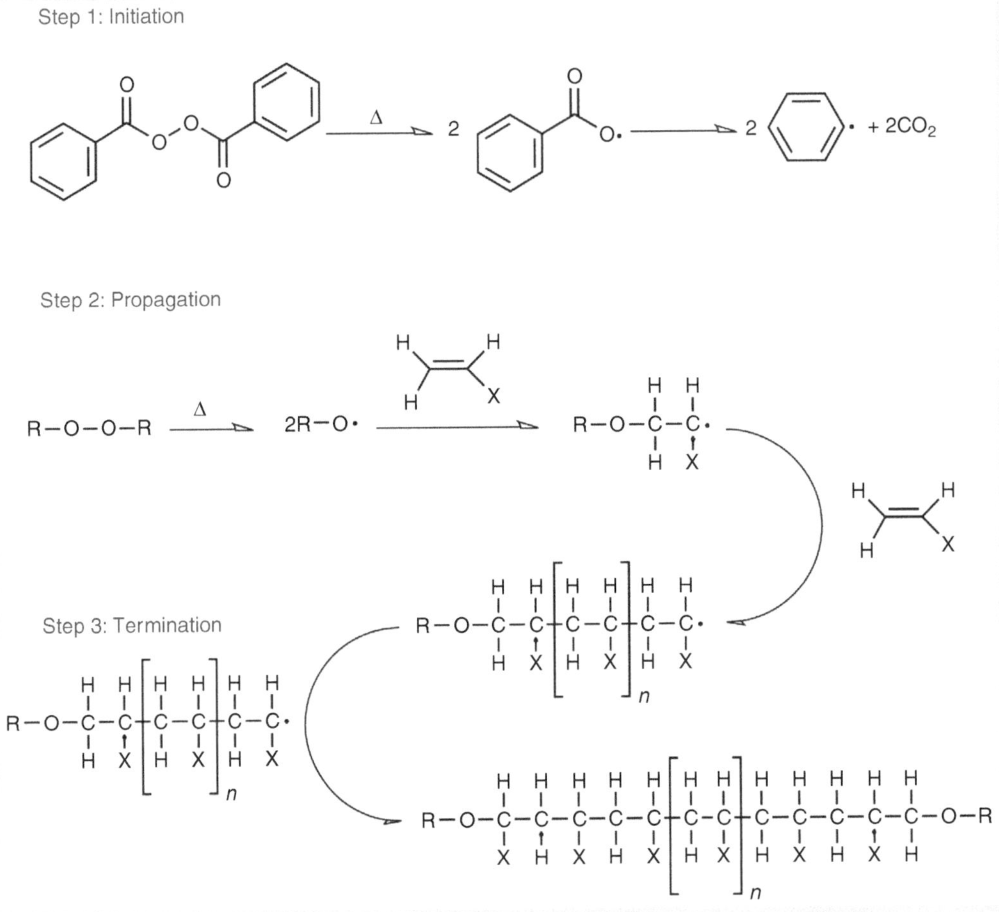

The *radical chain reaction process* is one way to form addition polymers (**Fig. 12.33**). The thermal decomposition of organic peroxides (e.g., R = benzoyl peroxide) is one means of generating the radical initiator required for the polymerization of the ethylene monomer (Step 1). Step 2 involves consecutive addition of monomeric units until chain lengths of several hundreds to thousands are achieved. The radical process propagates in this way until reaction with another radical is achieved, thus terminating the polymerization (Step 3).

For such addition polymer processes, if "X" is a hydrogen, methyl, chlorine, nitrile, or benzene moiety, then polyethylene, polypropylene, polyvinyl chloride, polyacrylonitrile, or polystyrene is formed, respectively. If all four constituents

FIG. 12.34	*An example of the anionic polymerization of ethylene is shown below. A nucleophilic reagent attacks a carbon member of the double bond to form a carbanion. A powerful base such as t-butyl-lithium can serve as the anionic initiator.*

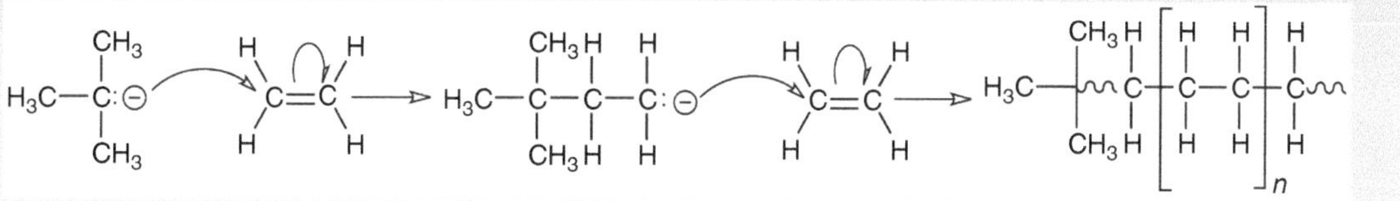

of the ethylene core are substituted with fluorine, then Teflon, poly(tetrafluoro-ethane), is the product of the polymerization process. There are many more possible types of polymers based on the free radical polymerization of the ethylene class of alkene monomers.

Anionic or cationic molecules are also able to initiate addition polymerization. An example of the anionic polymerization of ethylene is shown in **Figure 12.34.** A nucleophilic reagent attacks a carbon member of the double bond to form a carbanion. A powerful base such as *t*-butyl lithium can serve as the anionic initiator. Since primary carbanions are more stable than tertiary carbanions, the reaction proceeds to add more ethylene and thus propagates the polymer reaction to form poly(ethylene).

In cationic addition, an electrophilic carbocation or other cationic initiator attacks the double bond and creates cationic intermediates during the propagation step of the polymerization process.

Condensation reactions are commonly used to form polyamides like nylon and polyesters. Nylon-66 is formed by condensation between adipic acid and 1,6-hexamethyldiamine. Condensation reactions are responsible for the generation of the peptide linkage between the carboxylate and the amine of two amino acids to form a polypeptide. Poly(butylenesuccinate), or PBS, is formed by a condensation reaction in which water is eliminated (**Fig. 12.35**).

Polymers that are formed in this way, with water as a by-product, may be susceptible to future hydrolysis and subsequent degradation.

12.4.3 Block Copolymers

A copolymer is a macromolecular species that contains two or more monomers. Block copolymers are characterized by relatively long, ordered sequences of repeating units made of immiscible blocks of polymers. Block copolymers can be highly ordered nanostructures (e.g., high-density magnetic storage) or nano-structured without long-range order (e.g., active component in filters) or can form individual nanoscale materials such as vesicles (e.g., vehicles for drug delivery) [94]. Block copolymers are able to form a unique class of templates from soft-matter materials. Because they are able to function as a template, block copolymers are used in lithographic pattern transfer techniques. For polymers to be successful as nanotemplates, they must be ordered into molecular-scale assemblies [95]. Synthesis of diblock polymers requires pure reagents and strict control of reaction conditions. This is because there are no termination

Fig. 12.35	*Condensation reactions are commonly used to form polyamides like nylon and polyesters. Nylon-66 is formed by condensation between adipic acid and 1,6-hexamethyldiamine. Condensation reactions are responsible for the generation of the peptide linkage between the carboxylate and the amine of two amino acids to form a polypeptide. Poly(butylenesuccinate), or PBS, is formed by a condensation reaction in which water is eliminated.*

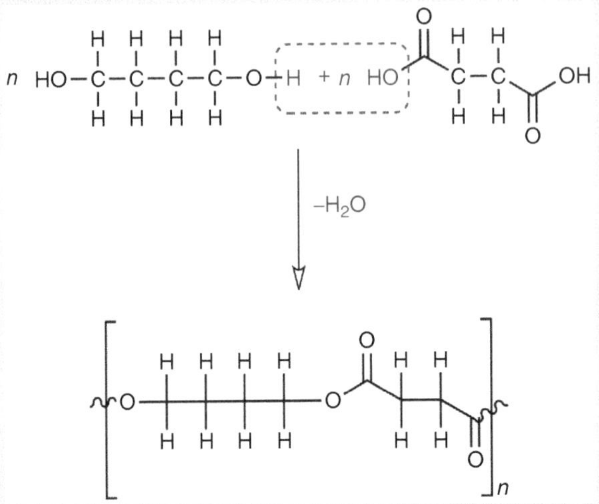

steps involved in the process and all processes (self-assembly, etc.) occur simultaneously [95].

An **A–B** diblock copolymer has the following sequence of monomer **A** and monomer **B** units:

A–A–A–A–A–A–A–A–A–A–A–A–A–B–B–B–B–B–B–B–B–B–B–B

An **A–B–C** triblock copolymer has the following sequence:

A–A–A–A–A–A–A–A–B–B–B–B–B–B–B–B–C–C–C–C–C–C–C–C

Other forms like gradient, graft, and star copolymers also exist. A–B block copolymers are useful templates in preparing spheres and cylinders [94,95]. A–B–C triblock copolymers are applied in synthesis of more complex nanostructures. Polymer chains, before assembly, tend to be on the order of 10 nm in size ($\sim 10^5$ g·mol^{-1})—for example, the radius of gyration, R_g [94,95]. Size and spacing of periodic structures can be controlled by altering the length of the block copolymer subunits or by swelling domains by adding more A or B (**Fig. 12.36**).

An important property of block copolymers is phase separation. There are two types: (1) macrophase separation and (2) microphase separation. Macrophase separation, as the name implies, is undesirable. At the nanoscale, however, microphase separation leads to new properties. Microphase separation occurs with polymers that are dissimilar (immiscible). The size of the microphases is modulated by the balance between enthalpic (larger the better) and entropic (smaller the better) considerations. In diblock systems, laminar structures are favored if **A** and **B** have $f_A/f_B = 50/50$; the gyroid form at $f_A/f_B \sim 40/60$; hexagonal cylinders between 40/60 and 20/80; and *bcc* structures at $f_A/f_B \sim 10/90$. Note that the ratios can be reversed.

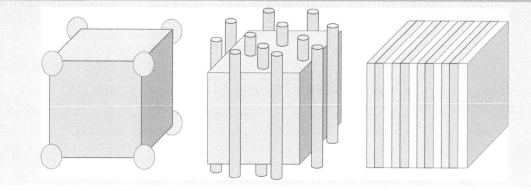

FIG. 12.36	*Block copolymer morphology. Left: A–B (spherical, bcc); middle: A–B (cylindrical, hcp); right: A–B–C (lamellar). A third common form, cubic gyroid, is not pictured. A–B lamellar forms result if the packing fraction f is on the order of 50%. For cylindrical block copolymers with f_A = 32%, cylindrical structure is favored. If the temperature is raised, the structure transforms into the lamellar phase depicted in the right (except that it is a diblock copolymer and not the triblock system depicted in the figure).*

Sources: J. A. Elliott, *Mixtures of polymers,* Presentation, University of Cambridge, www.cus.cam.ac.uk/~jae1001/teaching/materials/M6_Lecture_4.pdf, 2004, and N. P. Balsara and H. Han, in *The chemistry of nanostructured materials,* P. Yang, ed., World Scientific Publishing Co., Ltd., Hackensack, NJ, 2003. With permission.

Block copolymers combine the physical properties of different polymers (homopolymers) and are excellent precursor materials for synthesizing a vast array of nanomaterials [94]. The morphology and properties of the product nanostructured polymeric material depend on polymer component molecular weight, distribution, and fraction present in the block copolymer. Diblock copolymers self-assemble into two phases after prolonged heating above the respective glass transition temperatures. For example, magnetic nanowires can be formed by biblock copolymer polystyrene (PS) and PMMA [96]. Block copolymers of styrene (f = 70%) and methyl methacrylate nanomers (MMA) yielded cylindrical PMMA of 20–40 nm. The diblock copolymer was applied by spin-coating with annealing at 170°C for 24 h to yield a film 100 nm in thickness. The PMMA phase was then decomposed by application of UV light (253.7 nm). Deposition of Co was accomplished by electrodeposition to form the nanowires.

Cobalt nanowires with 14 nm diameter were formed in diblock copolymer template consisting of PS and poly(butadiene) (PB) [94,97]. In this case, a thin film of PS + PB block copolymer was applied by spin-coating technique onto a SiN surface. After annealing, an array of 5-nm PB spheres formed in the PS matrix. Upon contact with ozone, the PB was removed, leaving an array of holes. Reactive ion etching (RIE) was applied to transfer the pattern to the SiN substrate, after which templating synthesis of nanomaterials could proceed by various methods.

Nanoscale vesicles similar to those found in cells can be formed from block copolymers. Poly(ethyleneoxide) (PEO) and poly(ethylethylene) combine to form biological-type vesicles in water. The hydrophobic PEE spontaneously forms the core of the center while the hydrophilic PEO extends into the aqueous medium [98]. Vesicles made of block copolymer are more durable and less permeable to water than those formed from phospholipids that are used in drug delivery [98]. Individual polymers are able to assemble into several forms depending on reaction conditions and polymer constitution. The hydrophobic core can be a single strand that folds in on itself and cross-links between cofolded sections of the strand. Vesicles with projections (appendages consisting of the hydrophilic component of the diblock polymer) from the surface of the hydrophobic core come in three varieties: (1) *hairy nanospheres* are made if the appendage length is less than the diameter of the sphere; (2) *star polymers* result if the appendage is much greater than the sphere diameter; and (3) *brush polymers* occur if short appendages are formed on a flat surface [99].

12.4.4 Emulsion Polymerization

Polymers are very important to nanoscience and nanotechnology. Block copolymers are used in template synthesis among other applications and biodegradable polymeric nanoparticles are becoming increasingly important in biomedical applications. The field of nanocomposites is increasingly gaining momentum. A useful and versatile way to form nanoparticles that have functionality is by emulsion polymerization. Couvreur et al. in 1979 introduced this method of polymerization to design nanoparticles with biodegradable polymers for the in vivo delivery of drugs [100]. Common monomers used in emulsion polymerization include esters like the methacrylates, alkyl-cyanoacrylates, butadiene, styrene, vinyl acetate, acrylic, and methacrylic acids [101]. Copolymers can also be grown by this method.

Latex (from the Latin *latex*, meaning "liquid, fluid") is a fluid that contains a dispersion of polymer particles in water. Latex solutions are traditionally associated with paints and coatings. Several natural latex solutions exist; natural rubber latex is a raw material used for the synthesis of rubber polymers. In 1835, white, milky fluid from plants was designated as latex. Latex is formed from polymerization of monomeric emulsions. Many kinds of polymers can be synthesized by emulsion polymer techniques: poly(vinyl acetate), poly(styrene), poly(chlorophene), and poly(alkyl cyanoacrylate) [102]. Polymerization to form latex particles occurs within dispersed particles suspended in an aqueous medium that are stabilized by surfactants. Latex particles are high molecular weight, spherical, and traditionally in the 100-nm diameter range, but they can be as large as 5 μm in diameter.

Emulsion polymerization involves a monomeric entity, a surfactant, and an aqueous phase that contains an initiator. The technique is a relatively simple process that requires mechanical stirring and temperature control (the beauty and elegance of bottom-up methods!). At the onset, an emulsion of monomer that is slightly soluble in the water medium is introduced [101]. A surfactant (a.k.a. emulsifying agent) is responsible for forming micelles that in turn are responsible for solubilizing monomers [101]. The emulsifying agent is responsible for the overall stability of the monomer, the embryonic polymer, and the polymer product. The polymerization reaction—initiation and the early stages of propagation—occurs inside the monomer-containing micelles.

Initiators form free radicals that seek out monomers for reaction as the polymer grows in size within the ever expanding micelles. Both initiation and propagation take place inside the micelles that provide safe haven for the slightly

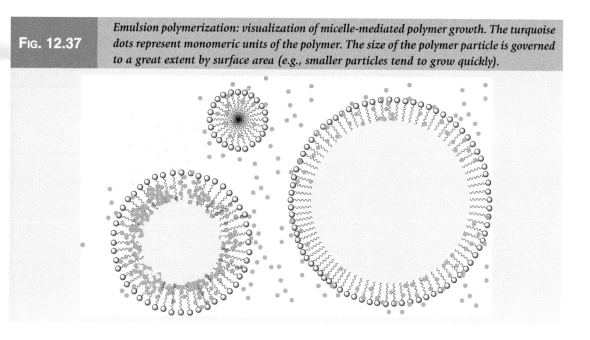

FIG. 12.37 *Emulsion polymerization: visualization of micelle-mediated polymer growth. The turquoise dots represent monomeric units of the polymer. The size of the polymer particle is governed to a great extent by surface area (e.g., smaller particles tend to grow quickly).*

soluble monomer molecules. **Figure 12.37** shows the stages of growth of the polymer inside a micelle by emulsion polymerization. The mechanism proceeds as follows. Not much polymerization occurs in the large droplets (monomer-swollen latex particles) due to the high surface area advantage of smaller micelles. Thus, by statistical factors, droplet size is maximized at some predetermined level.

Polymeric cationic surfactants (a.k.a. polysoaps) are used in polymerization of styrene to form latex microspheres. It was found that the polymerization properties and the eventual properties of the latex balls depend sensitively on the characteristics of the emulsifier and the charge on the initiator [103]. With cationic initiators, for example, all emulsifiers guide the formation of stable monodisperse latexes [103]. In another study, it was found that critical micelle concentration (CMC; usually 10^{18} micelles $\cdot$ mL^{-1}) and average polystyrene particle size increased with decreasing alkyl chain length of the emulsifying agent [104]. As you might recall, the CMC is the equilibrium concentration of an amphiphilic component at which the formation of micelles is initiated. Anionic surfactants include alkyl benzene, sodium vinyl, alpha olefin, and alcohol sulfonates. Non-ionic surfactants (insensitive to pH) include alkyl polysaccharides and amines ethoxylates, alcohol, and alkylphenol, to name just a few. A common initiator used in emulsion polymerization is persulfate (a sulfate peroxide), a water-soluble species that decomposes at low temperatures (50°C) into sulfate radicals.

Poly(Alkylcyanoacrylate) Drug Delivery Vehicles. Poly(alkylcyanoacrylates) (PACCAs) were employed as polymers in the early 1980s. However, the corresponding monomers have been used since at least 1966 because of their excellent adhesive properties resulting from the bonds of high strength they are able to form with most polar substrates, including living tissues and skin. Therefore, the monomers have been used extensively as tissue adhesives for the closure of skin wounds and as surgical glue. More recently, one application of these biodegradable polymers is as nanoparticulate drug carriers. This area of research that arose in the 1980s has gained increasing interest in therapeutics, especially for cancer trea'ents. Simultaneous cellular resistance to multiple drugs represents a major problem in cancer chemotherapy. The resistance mechanism can have different origins: either directly linked to a specific mechanism developed by the tumor tissue or connected to the more general problem of distribution of a drug towards its targeted tissue. One of the most common mechanisms of cellular resistance to chemotherapeutic drugs has been attributed to an active drug efflux from resistance cells linked to the presence of transmembrane P-glycoprotein. With resistant cells, upon entering the cell, the drug binds to P-glycoprotein and is pumped out of the cell. PACA nanoparticles have been used to overcome multidrug resistance at the cellular level by co-encapsulation of doxorubicin (an anticancer drug) and cyclosporin A (a chemosensitizing agent). The extremely promising research in this field now makes PACA nanoparticles the most promising polymer colloidal drug delivery system and they are currently in Phase II clinical trials in the treatment of resistant cancers [131].

The synthesis of a PACA is shown in **Figure 12.38.** An anionic polymerization scheme is depicted. There are many ways to synthesize methacrylate polymers. H. Eerikäinen et al. synthesized methacrylic polymers and drugs by an aerosol flow reactor method [105]. PACA nanoparticles have applications in the treatment of intracellular infections, nonresistant and resistant cancers, delivery of oligonucleotides, peptides, proteins, and vaccines.

One of the most exciting applications of PACA nanoparticles is in the treatment of cancers. Intravenously administered drugs are distributed throughout the body as a function of the physicochemical properties of the molecule. The nonspecificity of the delivery of the drug means that the same pharmacologically active concentration of drug that reaches the tumor also affects the rest of the body. Colloidal drug carriers such as PACA nanoparticles offer a way to address this problem in cancer therapy. PACA nanoparticles are taken up by the liver (macrophages of the mononuclear phagocyte system (MPS), spleen, and, to a lesser extent, by bone marrow after intravenous injection [106]. The uptake of the PACA nanoparticles by the MPS is a nonspecific foreign body response by the host that is based on adsorption of blood proteins and complement activation [100,106]. Polymerization to form PACA follows an anionic mechanism initiated by nucleophiles such as OH^-, CH_3O^-, and CH_3COO^- leading to the formation of nanoparticles of low molecular mass due to rapid polymerization. In an aqueous medium, anionic polymerization can be controlled to produce higher molecular weight nanoparticles.

The pH of the solution plays an important part in the reaction; the formation of PACA nanoparticles above pH of 3 is not possible, most likely due to aggregation. Lowering the pH with a mineral acid such as HCl and adjusting the concentration of the anionic polymerization inhibitor (SO_2) in the monomer [106], along with using stabilizers (dextran-70, dextran-40, dextran-10, polyoxamer-18, -184, -237, etc.) and surfactants (polysorbate-20, -40, or -80), can be used to control particle size and molecular mass of the nanoparticles [107]. The concentration of the monomer and the speed of stirring are also important parameters in nanoparticle formation. The size of the nanoparticles varies from 50 to 300 nm [107–109].

Poly(alkylcyanoacrylate) (PACA) can be used as tissue adhesives—drug carriers that are able to overcome multidrug resistance. Three different mechanisms for polymerization exist: free radical, anionic, and zwitterionic. Free radical mechanism is determined by high activation energy (125 kJ·mol⁻¹). The polymerization is very slow and the reaction rate depends strongly on the temperature and the quantity of radicals. Anionic and zwitterionic techniques are more rapid and easier to handle and are better suited for biomedical applications. Propagation of polymerization occurs after formation of carbanions that are able to react with another monomer molecule leading to the formation of living polymer chains. Termination is due to the presence of action that leads to that end of the polymerization. Anionic chain terminators include O_2, CO_2, H_2O, and HCl.

Source: R. D'Sa, *Poly(alkylcyanoacrylates): Biodegradable Polymeric Nanoparticles for Biomedical Applications*, U. of Colorado at Denver, 2004. With permisssion.

A generalization of this procedure involves: 100 μL of the monomer is dispersed in a pH 2.5 solution (adjusted with HCl) containing 10 mL of a 1% solution of dextran-70 [110]. The solution is allowed to polymerize for 3–4 h under strong magnetic stirring. The addition of drugs to this mixture will entrap them in the nanoparticle [111,112]. A generalized synthesis is shown in **Figure 12.39.**

Biodegradation of PACA Nanoparticles. In order for a polymeric drug delivery system to be suitable in vivo, the material must be biocompatible and biodegradable or easily excreted (e.g., by the kidneys). In nature, enzymes secreted by fungi, bacteria, and the digestive ingredients that reside within mammalian stomachs are able to effect biodegradation reactions. Many kinds of synthetic biodegradable polymers have been invented. The mechanism for PACA degradation depends on the local environment. Two main mechanisms exist and are shown in **Figure 12.40.** The first and predominant mechanism involves

FIG. 12.39

Emulsion polymerization is a very popular approach used to synthesize polymer colloids. Introduced in 1979 to design nanoparticles with biodegradable polymers for the in vivo delivery of drugs, polymerization is initiated by the hydroxyl ions of water and elongation of the polymer chain occurs according to an anion polymerization mechanism. To produce higher molecular mass as well as stable nanoparticles, polymerization must be carried out in an acidic medium (pH 1.0–3.5). After dispersing the monomer in an aqueous acidic medium containing surfactant and stabilizer, polymerization is continued for 3–4 h by increasing the pH of the medium to obtain the desired products. Particle size and molecular mass of nanoparticles depend upon the type and concentration of the stabilizer and/or surfactant used. The size and molecular mass of nanoparticles depend upon the pH of the polymerization medium, but nanoparticle production is not possible above a pH of 3.0, probably due to the aggregation and stepwise molecular mass increase at lower pH. Other factors that influence the formation of nanoparticles include the concentration of monomer and the speed of stirring.

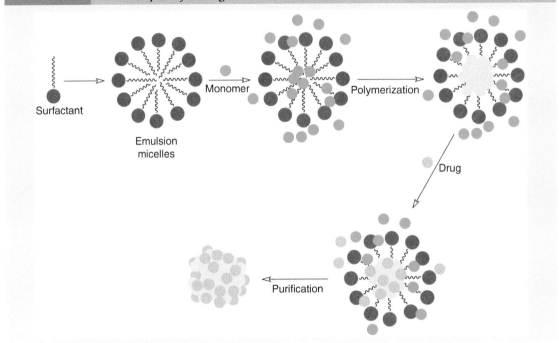

Source: R. D'Sa, *Poly(alkylcyanoacrylates): Biodegradable Polymeric Nanoparticles for Biomedical Applications*, U. of Colorado at Denver, 2004. With permisssion.

hydrolysis of the ester bond of the alkyl side chain to form alkyl alcohol and poly(cyanoacrylic acid) [113–115]. The key ingredient in the degradation process is that selected chemical bonds must be labile to hydrolysis. For example, base-catalyzed hydrolysis of polyester is initiated by the nucleophilic attack on a carbonyl group by a hydroxyl group. Both degradation products are water soluble and are therefore easily excreted by the kidneys. The esterases found in serum, lysosymes, and pancreatic juice [116,117] catalyze this reaction in vivo within a few hours, depending on the length of the alkyl chain [114,118,119].

FIG. 12.40 *Enzymatic degradation catalyzed by esterases leading to the production of alkylalcohol and poly(cyanoacrylic acids) is shown. One of the degradation mechanisms described in the literature consists of the hydrolysis of the ester bond of the alkyl side chain of the polymer. Degradation products consist of an alkylalcohol and poly(cyanoacrylic acid), which are soluble in water and can be eliminated in vivo via kidney filtration. This degradation has been shown to be catalyzed by esterases from serum, lysosomes, and pancreatic juice and is believed to occur as the major degradation pathway in vivo. According to this mechanism, nanoparticles are usually degraded within a couple of hours depending on the alkyl side chain length of the PACA forming the nanospheres.*

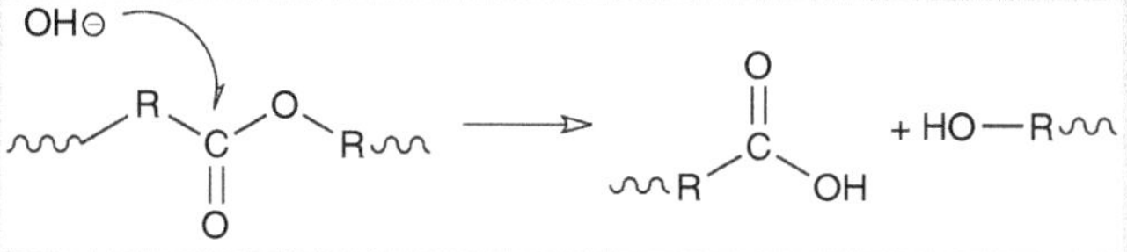

Source: R. D'Sa, *Poly(alkylcyanoacrylates): Biodegradable Polymeric Nanoparticles for Biomedical Applications*, U. of Colorado at Denver, 2004. With permisssion.

The second mechanism is an unzipping depolymerization followed by an immediate repolymerization to give a lower molecular weight polymer (oligomers). This base-catalyzed reaction [120] is thought to be initiated by amino acids of proteins. Since this reaction takes place very fast it is difficult to observe in complex biological systems and therefore has never been described. The alkylcyanoacrylate moieties are recovered, and this process continues until the polymer is consumed (**Fig. 12.41**).

12.4.5 Nanocomposites

A composite (from the Latin *compositus*, "to put together" (*com*) + *ponere*, "to place") is a material that is made up of various components. In other words, it is not a homogeneous material. The term is often applied to construction materials. Graphite fiber epoxy composites developed in the late 1970s revolutionized airframe and wing designs. Nanogold deposited in the pores of porous alumina membranes is a composite material.

There are many kinds of composites that contain nanomaterials. Concrete, for one, is an age-old bulk material composite that contains aggregate and

FIG. 12.41 *Unzipping depolymerization–repolymerization mechanism producing oligomers of PACA [110]. The alkylcyanoacrylate moieties are recovered, and this process continues until the polymer is consumed.*

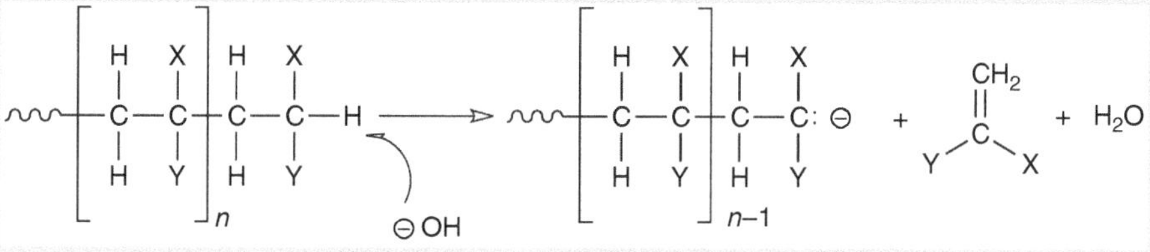

cement binder (10–100 μm). Cement is formed by grinding clay and calcined limestone into a fine powder. The performance of concrete has recently been shown to improve dramatically with the addition of SiO_2, fly ash, TiO_2, and Fe_2O_3 nanoparticles [121,122]. Addition of nanoparticles enhances the mechanical properties of concrete. Nanoparticles are able to occupy voids between cement grains, accelerate hydration and improve bonding between aggregates, improve toughness, shear, tensile and flexural strength, and enhance crack arrest. Experts are predicting that the concrete industry will be dramatically impacted by such nanotechnological enhancements.

Nature has developed complicated structures with nanostructured materials that exhibit incredible mechanical strength and hardness. For example, biominerals (calcite, aragontie, hydroxyapatite) interact with biomacromolecules (proteins, polysaccharides, proteoglycans) to form structures with hierarchical long-range order [123,124]. If you look inside an abalone shell you will see a pearly iridescent layer called nacre (mother of pearl), nature's incredible armor. Although the components are brittle (but deformable), the shell exhibits a robust structure [125]. Thin layers (5–20 nm) of elastic organic biopolymeric materials separate hexagonal platelets of aragonite ($CaCO_3$). The dimensions of the aragonite platelets are 10–20 μm by 200–500 nm thick. The lamellar composite has a 2-fold advantage in strength and 1000-fold advantage in toughness over the individual material components [125]. The iridescence is due to interference colors produced by the interaction of the 500-nm thick platelets with light. Nature puts this nanocomposite to use in the form of the mussel. Attempts to mimic this structure are ongoing [126]. Synthetic efforts that try to mimic the bricklike ultrastructure are in abundant supply.

Carbon Nanotube–Polymer Composites. Nanocomposite based on carbon nanotube reinforced polymers shows great potential. We now discuss the science of composites that utilize chemically modified carbon nanotubes for structural reinforcement and electrical, optical, and thermal enhancement. From chapter 9 we learned about several ways to chemically modify carbon nanotubes [127,128].

There are two generalized approaches for incorporating carbon nanotubes into a polymer matrix. One is to use aligned arrays, usually of the multi-walled carbon nanotube (MWNT) type, grown by chemical vapor deposition (CVD) that are millimeters in height [128]. The array can be impregnated with polymers in order to preserve the alignment. Secondly, use of loose, unaligned nanotubes of high purity, primarily of the SWNT type, can be applied towards multifunctional polymer composites, membranes, coatings, and fiber fabrication [128].

Polymer grafting discussed earlier is an effective way to place carbon nanotubes within a polymeric matrix. In a reversal of roles, researchers have grafted MWNTs onto silicon carbide cloth by chemical vapor deposition growth to form a SiC–CNT polymer nanocomposite [129]. The cloth-nanotube product can be infiltrated with a high-temperature polymer (epoxy) and then stacked and cured in a furnace [129]. The carbon nanotubes act like Velcro fasteners and afford strength to the nanocomposite in what is known as the *through thickness*, the direction normal to the plane of the cloth. The CNT-treated cloth shows fourfold improvement in fracture tests, fivefold improvement in energy dissipation by structural damping and structural energy, and a threefold increase in dimensional stability when compared to control materials [129]. Thermal and electrical conductivity performance was also significantly enhanced [129].

Single-walled carbon nanotubes have also been successfully incorporated within polymer matrices (Nylon-6,6, formed from two six-carbon precursors) to form nanocomposites by a process called interfacial in situ polymerization [130]. Interfacial polymerization of the polyamide Nylon-6,6 takes place, as the name indicates, between the interfacial layer of an aqueous phase and an organic (toluene) phase. Within the organic phase, the condensation monomer precursor is adipoyl chloride and within the aqueous phase the complementary precursor is 1,6-hexamethylene diamine, both in equimolar proportions to produce the neat nylon polymer.

Carboxylic acid-functionalized SWNTs are subjected to the same polymerization procedure as described before. Purified neat SWNTs, purified functionalized SWNTs, and NaDDBS-stabilized (sodium dodecylbenzene sulfonate surfactant) SWNTs were tested. The only SWNT species that showed good dispersion in both the solvent and the composite were the ones that were chemically functionalized.

References

1. T. Sugimoto, K. Sakata, and A. Muramatsu, Formation mechanism of monodisperse pseudocubic a-Fe$_2$O$_3$ particles from condensed ferric hydroxide gel, *Journal of Colloid Interface Science*, 159, 372–382 (1993).
2. V. R. Reddy, Gold nanoparticles: Synthesis and applications, *Synlett*, 11, 1791–1792 (2006).
3. G. Cao, *Nanostructures & nanomaterials: Synthesis, properties & applications*, Imperial College Press, London (2004).
4. G. M. Whitesides and M. Boncheva, Beyond molecules: Self-assembly of mesoscopic and nanoscopic components, *Proceedings of the National Academy of Science*, 99, 4769–4774 (2002)
5. J. C. Love, L. A. Estroff, J. K. Kriebel, R. G. Buzzo, and G. M. Whitesides, Self-assembled monolayers of thiolates on metals as a form of nanotechnology, *Chemical Reviews*, 105, 1103–1169 (2005).
6. J. L. Wilbur, A. Kumar, H. A. Biebuyck, E. Kim, and G. M. Whitesides, Microcontact printing of self-assembled monolayers: Applications in nanofabrication, *Nanotechnology*, 7, 452–457 (1996).
7. C. H. Giles, *The origins of surface film balance: Studies in the early history of surface chemistry*, Part 3, Society of Chemical Industry, January, 43–53 (1971).
8. Tapani Viitali, KSV Instruments, Inc., Finland.
9. N. Tillman, A. Ulman, and T. L. Penner, Layered metal phosphates and phosphonates: From crystals to monolayers, *Langmuir*, 5, 101–111 (1989).
10. N. K. Adam, *The physics and chemistry of surfaces*, 3rd ed., Oxford University Press, London (1941).
11. L. Cristofolini, T. Berzina, S. Erokhina, O. Konovalov, and V. Erokhin, Structural study of the DNA dipalmitoylphosphatidylcholine complex at the air–water interface, *Biomacromolecules*, 8, 2270–2275 (2007).
12. S. Pradhan, D. Ghosh, L.-P. Xu, and S. Chen, Interparticle charge transfer mediated by π–π stacking of aromatic moieties, JACS Communications, *Journal of the American Chemical Society*, 129, 10622–10623 (2007).

13. R. L. Donkers, Y. Song, and R. W. Murray, Substituent effects on the exchange dynamics of ligands on 1.6 nm diameter gold nanoparticles, *Langmuir,* 20, 4703–4707 (2004).

14. L. Salem, Attractive forces between long saturated chains at short distances, *Journal of Chemical Physics,* 37, 2100–2113 (1962).

15. A. Ulman, *An introduction to ultrathin organic films: From Langmuir–Blodgett to self-assembly,* Academic Press, Boston (1991).

16. R. G. Nuzzo, B. R. Zegarski, and L. H. L. Dubois, Fundamental studies of the chemisorption of organosulfur compounds on gold(111)—Implications for molecular self-assembly on gold surfaces, *Journal of the American Chemical Society,* 109, 733–740 (1987).

17. I. Engquist, Self-assembled monolayers, Department of Physics, Linköping University, Sweden, www.ifm.liu.se/applphys/ftir/sams.html, (1996).

18. G. Scoles, Physisorption and chemisorption of alkanethiols and alkylsulfides on Au(111), Princeton University, www.princeton.edu/~gscoles/theses/sean/chapter4.html (2006).

19. G. Schmid and B. Corain, Nanoparticulated gold: Synthesis, structures, electronics and reactivities, *European Journal of Inorganic Chemistry,* 3081–3098 (2003).

20. G. Schmid, R. Pfeil, R. Böse, F. Banderman, S. Meyer, G. H. M. Calis, and J. W. A. van der Velden, $Au_{55}[P(C_6H_5)_3]_{12}Cl_6^-$ ein Goldcluster ungewöhnlicher Größe, *Chemische Berichte,* 114, 3634 (1981).

21. G. Schmid, Metal nanoclusters as quantum dots. In *Encyclopedia of nanoscience and nanotechnology,* H. S. Nalwa, ed., American Scientists Publisher, 5, 387–398 (2004).

22. G. Schmid, E. Emmrich, J.-P. Majoral, and A.-M. Caminade, The behavior of Au_{55} nanoclusters on and in thiol-terminated dendrimer monolayers, *Small,* 1, 73–75 (2005).

23. G. Schmid, R. Pugin, W. Meyer-Zaika, and U. Simon, Clusters on clusters: Closododecaborate as a ligand for Au55 clusters, *European Journal of Inorganic Chemistry,* 2051–2055 (1999).

24. G. Hornyak, M. Kröll, R. Pugin, T. Sawitowski, G. Schmid, J.-O. Bovin, G. Karsson, H. Hofmeister, and S. Hopfe, Gold clusters and colloids in alumina nanotubes, *Chemistry: A European Journal,* 3, 1951–1956 (1997).

25. G. L. Hornyak, S. Peschel, T. Sawitowski, and G. Schmid, TEM, STEM and AFM as tools to study clusters and colloids, *Micron,* 29, 183–190 (1998).

26. G. Schmid and G. L. Hornyak, Metal clusters: New perspectives in future nanoelectronics, *Current Opinions in Solar State and Materials Science,* 2, 204, (1997).

27. G. Schmid, R. Pugin, J.-O. Malm, and J.-O. Bovin, Silsesquioxanes as ligands for gold clusters, *European Journal of Inorganic Chemistry,* 1998, 813–817 (1998).

28. J. Turkevitch, P. C. Stevenson, and J. Hillier, Nucleation and growth process in the synthesis of colloidal gold, *Discussions of the Faraday Society,* 11, 55 (1951).

29. Colloidal gold, en.wikipedia.org/wiki/Colloidal_gold, October (2007).

30. A. Badia, S. Singh, L. Demers, L. Cuccia, G. R. Brown, and R. B. Lennox, Self-assembled monolayers on gold monolayers, *Chemistry: A European Journal,* 2, 359–363 (1996).

31. N. Sandhyarani, T. Pradeep, J. Chakrabarti, M. Yousuf, and H. K. Sahu, Distinct liquid phase in metal-cluster superlattice solids, *Physics Reviews B,* 62, R739 (2000).

32. M. P. Pileni, ed., *Reactivity in reverse micelles,* Elsevier, Amsterdam (1989).

33. M. P. Pileni, Semiconductor nanocrystals. In *Nanoscale materials in chemistry,* K. J. Klabunde, ed., Wiley-Interscience, New York (2001).

34. S. M. Emin, C. D. Dushkin, S. Nakabayashi, and E. Adachi, Growth kinetics of CdSe nanoparticles synthesized in reverse micelles using *bis*(trimethylsilyl) selenium precursor, *Central European Journal of Chemistry,* 5, 590–604 (2007).

35. A. N. Shipway, E. Katz, and I. Willner, Nanoparticle arrays on surfaces for electronic, optical and sensor applications, *ChemPhysChem,* 1, 18–52 (2000).

36. M. L. Steigerwald, A. P. Alivisatos, J. M. Gibson, T. D. Harris, R. Kortan, A. J. Muller, A. M. Thayer, T. M. Duncan, and L. E. Brus, Surface derivatization and isolation of semiconductor cluster molecules, *Journal of the American Chemical Society*, 110, 3046–3050 (1988).

37. D. Ganguli and M. Ganguli, *Inorganic particle synthesis via macro- and microemulsions: A micrometer to nanometer landscape*, Kluwer Academic Press/Plenum Publishers (Springer), New York (2003).

38. G. L. Hornyak, L. Grigorian, A. C. Dillon, P. A. Parilla, and M. J. Heben, A temperature window for chemical vapor deposition growth of single walled carbon nanotubes, *Journal of Physical Chemistry B*, 106, 2821–2825 (2002).

39. S. B. Bukallah, M. Houalla, and D. M. Hercules, Characterization of supported Nb catalysts by ToF-SIMS, *Surface Interface Analysis*, 29, 818–822 (2000).

40. R. Kelsall, I. Hamley, and M. Geoghegan, *Nanoscale science and technology*, John Wiley & Sons, Ltd., Chichester (2005).

41. I. W. Hamley, *Introduction to soft matter*, John Wiley & Sons, Ltd., Chichester (2000).

42. M. Vondrova, T. Klimczuk, V. L. Miller, B. W. Kirby, N. Yao, R. J. Cava, and A. B. Bocarsly, Supported superparamagnetic Pd/Co alloy nanoparticles prepared from silica/cyanogel Co-Gel, *Chemical Materials*, 17, 6216–6218 (2005).

43. R. Leon, D. Margolese, G. Stucky, and P. M. Petroff, Nanocrystalline Ge filaments in the pores of a mesosilicate, *Physics Review B*, 52, R2285–R2288 (1995).

44. T. Scheibel, R. Parthasarathy, G. Sawicki, X.-M. Lin, H. Jaeger, and S. L. Lindquist, Conducting nanowires built by controlled self-assembly of amyloid fibers and selective metal deposition, *Proceedings of the National Academy of Science*, 100, 4527–4532 (2003).

45. J. Chai, D. Wang, X. Fang, and J. M. Buriak, Assembly of aligned linear metallic patterns on silicon, *Nature Nanotechnology*, 2, 500–506 (2007).

46. C. Wang, Y. Hou, J. Kim, and S. Sun, A general strategy for synthesizing FePt nanowires and nanorods, *Angewandte Chemie International Edition*, 46, 6333–6335 (2007).

47. H. Gerung, T. J. Boyle, L. J. Tribby, S. D. Bunge, C. J. Brinker, and S. M. Han, Solution synthesis of germanium nanowires using a Ge^{2+} alkoxide precursor, *Journal of the American Chemical Society*, 128, 5244–5250 (2006).

48. P. Nickels, W. U. Dittmer, S. Beyer, J. P. Kotthaus, and F. C. Simmel, Polyaniline nanowire synthesis templated by DNA, *Nanotechnology*, 15, 1524–1529 (2004).

49. M. Ye, H. Zhong, W. Zheng, R. Li, and Y. Li, Ultralong cadmium hydroxide nanowires: Synthesis, characterization, and transformation in CdO nanostrands, *Langmuir*, 23, 9064–9068 (2007).

51. C. A. Foss, Jr., G. L. Hornyak, J. A. Stockert, and C. R. Martin, Optical properties of composite membranes containing arrays of nanoscopic gold cylinders, *Journal of Physical Chemistry*, 96, 7497–7499 (1992).

52. C. A. Foss, Jr., M. J. Tierney, and C. R. Martin, Template-synthesis of infrared-transparent metal microcylinders: Comparison of optical properties with the predictions of effective medium theory, *Journal of Physical Chemistry*, 96, 9001–9007 (1992).

53. C. J. Brumlik, C. R. Martin, and K. Tokuda, Microhole array electrodes based on microporous alumina membranes, *Analytical Chemistry*, 64, 1201–1203 (1992).

54. C. A. Foss, Jr., G. L. Hornyak, J. A. Stockert, and C. R. Martin, Template synthesis and optical properties of small metal particle composite materials: Effects of particle shape and orientation on plasmon resonance maxima, *Proceedings of Material Research Society Symposium on Nanometals*, Boston, MA, 286, 431–436 (1993).

55. C. A. Foss, Jr., G. L. Hornyak, J. A. Stockert, and C. R. Martin, Optically transparent nanometal composite membranes, *Advanced Materials*, 5, 135–136 (1993).

56. C. R. Martin, Nanomaterials—A membrane-based synthetic approach, *Science*, 266, 1961–1966 (1994).

57. C. A. Foss, Jr., G. L. Hornyak, J. A. Stockert, and C. R. Martin, Template-synthesized nanoscopic gold particles: Optical spectra and the effects of particle size and shape, *Journal of Physical Chemistry*, 98, 2963–2971 (1994).

58. G. L. Hornyak, K. L. N. Phani, D. L. Kunkel, V. P. Menon, and C. R. Martin, Fabrication, characterization and optical theory of aluminum nanometal/nanoporous membrane thin film composites, *Proceedings of the 2nd International Conference on Nanostructured Materials*, H.-E. Schaefer, R. Würschum, H. Gleiter, and T. Tsakalokos, eds., 6, 839–842 (1995).

59. G. L. Hornyak and C. R. Martin, Optical properties of a family of Au-nanoparticle-containing alumina membranes in which the nanoparticle shape is varied from needle-like (prolate) to spheroid, to pancake-like (oblate), *Thin Solid Films*, 300, 84–88 (1997).

60. G. L. Hornyak, C. J. Patrissi, C. R. Martin, J-C.Valmalette, J. Dutta, and H. Hofmann, Dynamical Maxwell-Garnett optical modeling of nanogold-porous alumina composites: Mie and kappa influence on absorption maxima, *Nanostructured Materials*, 9, 575–578 (1997).

61. G. L. Hornyak, C. J. Patrissi, and C. R. Martin, Fabrication, characterization and optical properties of gold-nanoparticle/porous-alumina composites: The non-scattering Maxwell-Garnett limit, *Journal of Physical Chemistry B*, 101, 1548–1555 (1997).

62. J. C. Hulteen, C. J. Patrissi, D. L. Miner, E. R. Crosthwait, E. B. Oberhauser, and C. R. Martin, Changes in the shape and optical properties of gold nanoparticles contained within alumina membranes due to low temperature annealing, *Journal of Physical Chemistry B*, 101, 7727–7731 (1997).

63. G. L. Hornyak, *Characterization and optical theory of nanometal/porous alumina composite membranes*, Ph.D. dissertation, Colorado State University (1997).

64. A. Streitwieser and C. H. Heathcock, *Introduction to organic chemistry*, Macmillan Publishing Co., New York (1985).

65. N. Tillman, A. Ulman, J. S. Schildkraut, and T. L. Penner, Incorporation of phenoxy groups in self-assembled monolayers of trichlorosilane derivatives: Effects on film thickness, *Journal of the American Chemical Society*, 110, 6136–6144 (1988).

66. J. A. Preece, The micro–nano device, presentation, University of Birmingham (2004).

67. T. Zhu, X. Fu, T. Mu, J. Wang, and Z. Liu, pH-dependent adsorption of gold nanoparticles on aminothiophenol-modified gold substrates, *Langmuir*, 15, 5197–5199 (1999).

68. N. Wade, T. Pradeep, J. Shen, and R. G. Cooks, Covalent chemical modification of self- assembled fluorocarbon monolayers by low-energy $CH_2Br_2^+$ ions: A combined ion/surface scattering and x-ray photoelectron spectroscopic investigation, *Rapid Communications in Mass Spectrometry*, 13, 986–993 (1999).

69. L. D. Seger, J. P. Rasimas, R. Pesce-Rodriguez, and R. Fifer, Chemical modification and attempted polymerization of self-assembled monolayers of hexadecanedioic acid at aluminum surfaces, Army Research Lab, www.stormingmedia.us/21/2171/A217133.html (1997).

70. A. C. Buchanan, III, M. K. Kidder, P. F. Britt, Z. Zhang, and S. Dai, Effects of pore confinement on the pyrolysis of 1,3-diphenylpropane in mesoporous silica, Oak Ridge National Laboratory, www.ornl.gov/~webworks/cppr/y2001/pres/115857.pdf (2001).

71. S. Polarz, Chemistry in confined spaces, presentation, ERA Chemistry, Workshop Mainz Bochum, www.erachemistry.net/index/file/158 (2005).

72. J. Sivaguru, H. Saito, M. R. Solomon, L. S. Kaanumalle, T. Poon, S. Jockusch, W. Adam, V. Ramanurthy, Y. Inoue, and N. J. Turro, The control of chirality by cations

in confined spaces: Photooxidation of enecarbamates inside zeolite supercages, *Photochemistry and Photobiology*, 86, 123–131 (2006).

73. R. Anwander, SOMC@PMS: Surface organometallic chemistry at periodic mesoporous silica, *Chemical Materials*, 13, 4419–4438 (2001).

74. Carbon nanotube membranes and adsorbants, private communication, M. J. Heben, National Renewable Energy Laboratory (2001).

75. Periodic minimal surfaces, University of Cambridge, Jacek Klinowski Group, www-klinowski.ch.cam.ac.uk/jkhome.htm (2002).

76. G. Donnay and D. L. Pawson, X-ray diffraction studies of echinoderm plates, *Science*, 166, 1147–1150 (1969).

77. H. U. Nissen, Crystal orientation and plate structure in echinoid skeletal units, *Science*, 166, 1150–1152 (1969).

78. S. Andersson, S. T. Hyde, K. Larsson, and S. Lidin, Minimal surfaces and structures: From inorganic and metal crystals to cell membranes and bio-polymers, *Chemical Reviews*, 88, 221–242 (1988).

79. V. Alfredsson, M. W. Anderson, T. Ohsuna, O. Terasaki, M. Jacob, and M. Bojrup, Cubosome description of the inorganic mesoporous structure MCM-48, *Chemical Materials*, 9, 2066–2070 (1997).

80. J. Charvolin and J. F. Sadoc, Cubic phases as structures of disclinations, *Colloid Polymer Science*, 268, 190–195 (1990).

81. Y. Xia, Y. Lu, K. Kamata, B. Gates, and Y. Yin, Macroporous materials containing three dimensionally periodic structures. In *The chemistry of nanostructured materials*, P. Yang, ed., 69–100, World Scientific Publishing Co. Pte. Ltd., Singapore (2003).

82. B. T. Holland, C. F. Blanford, and A. Stein, Synthesis of macroporous minerals with highly ordered three-dimensional arrays of spherical voids, *Science*, 281, 538–540 (1998).

83. R. C. Schroden, M. Al-Daius, S. Sokolov, B. J. Melde, J. C. Lytle, A. Stein, M. C. Carbajo, J. T. Fernandez, and E. E. Rodriguez, Hybrid macroporous materials for heavy metal ion adsorption, *Journal of Materials Chemistry*, 12, 3261–3267 (2002).

84. H. Asoh, K. Nishio, M. Nakao, T. Tamamura, and H. Masuda, Conditions for fabrication of ideally ordered anodic porous alumina using pretextured Al, *Journal of the Electrochemical Society*, 148, B152–B156 (2001).

85. P. G. Mingalyov, M. V. Buchnev, and G. V. Llisichkin, Chemical modification of alumina and silica with alkylphosphonic acids and their esters, *Russian Chemical Bulletin*, 50, 1693–1695 (200).

86. V. M. Cepak, J. C. Hulteen, G. Che, K. B. Jirage, B. B. Lakshmi, E. R. Fisher, and C. R. Martin, Fabrication and characterization of concentric tubular composite micro- and nanostructures using the template synthesis method, *Journal of Materials Research*, 13, 3070–80 (1998).

87. G. Che, B. B. Lakshmi, C. R. Martin, E. R. Fisher, and R. A. Ruoff, Chemical vapor deposition based synthesis of carbon nanotubules and nanofibers using a template method, *Chemical Matererials*, 10, 260–267 (1998).

88. L. Mercier and T. J. Pinnavaia, Direct synthesis of hybrid organic–inorganic nano-porous silica by a neutral amine assembly route: Structure–function control by stoichiometric incorporation of organosiloxane molecules, *Chemical Materials*, 12, 188–196 (2000).

89. S. Inagaki, S. Guan, T. Ohsuna, and O. Terasaki, An ordered mesoporous organosilica hybrid material with a crystal-like wall structure, *Nature*, 416, 304–307 (2002).

90. M. Yates, K. Ott, E. Birhbaum, and T. McCleskey, Hydrothermal synthesis of molecular sieve fibers: Using microemulsions to control crystal morphology, *Angewandte Chemie International Edition*, 41, 476–478 (2003).

91. N. J. Turro, Molecular structure as a blueprint for supramolecular chemistry in confined spaces, *PNAS*, 102, 10766–10770 (2005).

92. C. T. Kresge, M. E. Leonowicz, W. J. Roth, J. C. Vartuli, and J. S. Beck, Ordered mesoporous molecular sieves synthesized by a liquid crystal template mechanism, *Nature*, 359, 710–712 (1992).

93. J. D. Roberts, R. Stewart, and M. J. Casserio, *Organic chemistry*, W. A. Benjamin, Inc., New York (1971).

94. N. P. Balsara and H. Han, Block copolymers in nanotechnology. In *The chemistry of nanostructured materials*, P. Yang, ed., World Scientific Publishing Co., Ltd., Hackensack, NJ (2003).

95. P. Yang, ed., *The chemistry of nanostructured materials*, World Scientific Publishing Co., Pte., Ltd., Singapore (2003).

96. V. V. Warke, M. G. Bakker, D. E. Nikles, J. Mays, and P. Britt, Block copolymer nanolithography for the preparation of patterned magnetic recording media, poster, MINT Center, University of Alabama, Fall (2005).

97. A. Urbas, R. Sharp, Y. Fink, E. L. Thomas, M. Xenidou, and L. J. Fetters, Tunable block-copolymer–homopolymer photonic crystals, *Advanced Materials*, 12, 812–814 (2000).

98. B. M. Discher, Y. Won, D. S. Ege, J. C. Lee, F. S. Bates, D. E. Discher, and D. A. Hammer, Tough vesicles made from diblock copolymers, *Science*, 284, 1143–1146 (1999).

99. C. P. Poole and F. J. Owens, *Introduction to nanotechnology*, Wiley-Interscience, John Wiley & Sons, Inc., Hoboken, NJ (2003).

100. P. Couvreur, B. Kante, M. Roland, P. Guiot, P. Bauduin, and P. Speiser, Polycyanoacrylate nanocapsules as potential lysosomotropic carriers: Preparation, morphological and sorptive properties, *Journal of Pharmacy and Pharmacology*, 31, 331–332 (1979).

101. H. Y. Erbil, *Surface chemistry of solid and liquid interfaces*, Blackwell Publishing, Oxford (2006).

102. M. E. Karaman, L. Meagher, and R. M. Pashley, Surface chemistry of emulsion polymerization, *Langmuir*, 9, 1220–1227 (1993).

103. D. Cochin and A. Lascheqsky, Emulsion polymerization of styrene using conventional, polymerizable, and polymeric surfactants. A comparative study, *Macromolecules*, 30, 2278–2287 (1997).

104. S. Demharter, W. Richtering, and R. Mülhaupt, Emulsion polymerization of styrene in the presence of carbohydrate-based amphiphiles, *Polymer Bulletin*, 34, 271–277 (1995).

105. H. Eerikäinen, *Preparation of nanoparticles consisting of methacrylic polymers and drugs by an aerosol flow reactor method*, VTT Publications 563, Finland (2005).

106. F. Lescure, C. Zimmer, D. Roy, and P. Couvreur, Optimization of polycyanoacrylate nanoparticle preparation: Influence of sulfur dioxide and pH on nanoparticle characteristics, *Journal of Colloid Interface Science*, 154, 77–86 (1992).

107. S. J. Douglas, L. Illum, and S. S. Davis, Particle size and size distribution of poly(butyl 2-cyanoacrylate) nanoparticles. II. Influence of stabilizers, *Journal of Colloid Interface Science*, 103, 154–163 (1985).

108. M. J. Alonso, A. Sanchez, D. Torres, B. Seijo, and J. L. Vila-Jato, Joint effect of monomer and stabilizer concentrations on the physico-chemical characteristics of poly(butyl-2-cyanoacrylate) nanoparticles, *Journal of Microencapsululation*, 7, 517–526 (1990).

109. B. Seijo, E. Fattal, L. Roblot-Treupel, and P. Couvreur, Design of nanoparticles of less than 50 nm diameter: Preparation, characterization and drug loading, *International Journal of Pharmacy*, 62, 1–7 (1990).

110. C. Vauthier, C. Dubernet, E. Fattal, H. Pinto-Alphandary, and P. Couvreur, Poly(alkylcyanoacrylates) as biodegradable materials for biomedical applications, Advances in Drug Delivery Reviews, 55, 519–48 (2003).

111. K. S. Soppimath, T. M. Aminabhavi, A. R. Kulkarni, and W. E. Rudzinski, Biodegradable polymeric nanoparticles as drug delivery devices, *Journal of Control Release*, 70, 1–20 (2000).

112. C. Vauthier and P. Couvreur, Degradation of poly(alkylcyanoacrylates). In *Miscellaneous biopolymers and biodegradation of synthetic polymers handbook of biopolymers*, 9, J. P. Matsumara and A. Steinbuchel, eds., Wiley–VHC, New York (2002).

113. V. Lenaerts, P. Couvreur, D. Christiaens-Leyh, E. Joiris, M. Roland, B. Rollman, and P. Speiser, Degradation of poly(isobutyl cyanoacrylate) nanoparticles, *Biomaterials*, 5, 65–68 (1984).

114. L. Vansnick, P. Couvreur, D. Christiaens-Ley, and M. Roland, Molecular weights of free and drug-loaded nanoparticles, *Pharmacy Research*, 1, 36–41 (1985).

115. K. Langer, E. Seegmüller, A. Zimmer, and J. Kreuter, Characterization of polybutylcyanoacrylate nanoparticles: Quantification of PBCA polymer and dextran, *International Journal of Pharmceutics*, 110, 21–27 (1994).

116. R. H. Müller, C. Lherm, J. Herbort, and P. Couvreur, In vitro model for the degradation of alkylcyanoacrylate nanoparticles, *Biomaterials*, 11, 590–595 (1990).

117. D. Scherer, J. R. Robinson, and J. Kreuter, Influence of enzymes on the stability of polybutylcyanoacrylate nanoparticles, *International Journal of Pharmaceutics*, 101, 165–168 (1994).

118. F. Leonard, R. K. Kulkarni, G. Brandes, J. Nelson, and J. J. Cameron, Synthesis and degradation of poly(alkyl-cyanoacrylates), *Journal of Applied Polymer Science*, 10, 259–272 (1996).

119. C. Lherm, R. Muller, F. Puisieux, and P. Couvreur, II. Cytotoxicity of cyanoacrylate nanoparticles with different alkyl chain length, *International Journal of Pharmaceutics*, 84, 13–22 (1992).

120. B. Ryan and G. McCann, Novel sub-ceiling temperature rapid depolymerization—Repolymerization reactions of cyanoacrylate polymers, *Macromolecular Rapid Communications*, 17, 217–227 (1996).

121. K. Sobolev and M. Ferrada-Gutiérez, Nanotechnology of concrete, *Nano News* press release, The NanoTechnology Group, Inc. (2005).

122. K. Sobolev and M. Ferrada-Gutiérez, How nanotechnology can change the concrete world: Part 2, *American Ceramic Society Bulletin*, 11, 16–19 (2005).

123. G. Fu, S. Valiyaveettil, B. Wopenka, and D. E. Morse, $CaCO_3$ biomineralization: Acidic 8-kDa proteins isolated from aragoniteic abalone shell nacre can specifically modify calcite crystal morphology, *Biomacromolecules*, 6, 1289–1298 (2005).

124. X. Li, W.-C. Chang, Y. J. Chao, R. Wang, and M. Chang, Nanoscale structural and mechanical characterization of a natural nanocomposite material: The shell of red abalone, *Nano Letters*, 4, 613–617 (2004).

125. M. Berger, Nature's bottom-up nanofabrication of armor, NanoWerk, LLC, www.nanowerk.com/spotlight/spotid=870.php (2006).

126. P. Podsiadlo, Z. Liu, D. Paterson, P. B. Messersmith, and N. A. Kotov, Fusion of seashell nacre and marine bioadhesive analogs: High-strength nanocomposite by layer-by-layer assembly of clay and L-3,4-dihydroxyphenylalanine polymer, *Advanced Materials*, 19, 949–955 (2007).

127. P. Liu, Modifications of carbon nanotubes with polymers, *European Polymer Journal*, 41, 2693–2703 (2005).

128. D. B. Geohegen and H. M. Christen, Functional nanomaterials—Capabilities, presentation, Center for Nanoscale Materials Sciences, Oak Ridge National Laboratory, Tennessee (2005).

129. V. P. Veedu, A. Cao, X. Li, K. Ma, C. Soldano, S. Kar, P. M. Ajayan, and M. H. Ghasemi-Nejhad, Multifunctional composites using reinforced laminae with carbon-nanotube forests, *Nature Materials*, 5, 457–462 (2006).

130. R. Haggenmueller, F. Du, J. E. Fischer, and K. I. Winey, Interfacial in situ polymerization of single wall carbon nanotube/nylon 6,6 nanocomposites, *Polymer*, 47, 2381–2388 (2006).

131. R. D'Sa, *Poly(alkylcyanoacrylates): Biodegradable polymeric nanoparticles for biomedical applications*, U. of Colorado at Denver (2004).

132. S. S. Kistler, Coherent expanded aerogels and jellies, *Nature*, 127, 741–744 (1931).

Problems

12.1 What are the distinctions between intermolecular self-assembly and intramolecular self-assembly?

12.2 Describe the construction of a protein from amino acids all the way from 1 through 4° structure. Is all of it a self-assembly process or are other mechanisms involved?

12.3 List the generic forms of chemical functionalization.

12.4 Intermolecular interactions have advantages in nanosynthesis over traditional covalent procedures. Make a list.

12.5 What are so-called "full-shell clusters"? Why do they have a special stability? (Question courtesy of Günter Schmid, University of Essen.)

12.6 Why is a shell of ligand necessary to synthesize and to isolate metal nanoparticles? (Question courtesy of Günter Schmid, University of Essen.)

12.7 Porous alumina membranes are often employed by template synthesis methods. Ag nanopedestals are fabricated in the pore channels of anodic alumina by an electrodeposition process. How many coulombs of charge are required to form a nanocylinder of Ag 300 nm in length inside pores of a membrane that is 30% porous? The area of the alumina is 3.14 cm². Use:

$$l_{cyl} = \left(\frac{CM_{Ag}}{\mathcal{F}A\varepsilon\rho} \right) \times 10^7$$

where
  C is coulombs
  M_{Ag} is the atomic mass of Ag

$\mathcal{A}$ is the area of the alumina film
ε is the porosity as before
$\mathcal{F}$ is the Faraday constant
ρ is the density of the Ag

From the preceding, are we able to calculate the diameter of the Ag cylinders? If the pore diameter is 50 nm, what is the aspect ratio of the Ag cylinders?

12.8 Pick a monomer and a polymer process (radical, anion, etc.) and lay out a polymerization process: initiation/growth and termination. Discuss the properties of your polymer.

12.9 The degree of polymerization δ is represented by:

$$\delta = \frac{Molecular\ weight\ of\ polymer}{Molecular\ weight\ of\ monomer}$$

For a polymer of poly(methylmethacrylate) (PMMA) that has 30,000 amu, what is the degree of polymerization?

12.10 What is the difference between a block copolymer and a regular copolymer?

12.11 Estimate the length of *n*-dodecane thiol, a molecule often used in self-assembly on gold surfaces. Two formulas are useful: (1) the root-mean-square length (left) and the extended length formula (right) (J. F. Shackelford, *Introduction to Materials Science for Engineers*, 4th ed., Prentice Hall, Englewood Cliffs, NJ, 1998):

$$\bar{L} = l\sqrt{m} \quad \text{and} \quad L_{ext} = ml\sin\frac{109.5°}{2}$$

where *l* is the length of a single bond and *m* is the number of bonds.

12.12 Estimate the thickness of a monolayer of olive oil (*triolein*; look it up) from 5 mL that is spread over 0.5 acres. Assume that the overall volume of the monolayer material remains the same (5 mL).

12.13 A tradition in general chemistry laboratory: Calculate the number of stearic acid molecules required to cover a watchglass. What preliminary conditions must be in place before the student can estimate Avogadro's number? How would you design such an experiment?

12.14 In an early experiment, Irving Langmuir determined that long chain fatty acid molecules have the same area (20 Å^2) per molecule regardless of the number of carbon atoms and that the polar end was stuck in the water (the subphase). What other important conclusion did Langmuir glean from this experiment?

12.15 Surface pressure π (a "two-dimensional" pressure) in a Langmuir trough is related to the difference between the surface tension of water (γ_o) and the surface tension of the subphase γ that is covered by the amphiphiles. Recall that in this scenario pressure is applied by the barrier laterally in the trough. The consequence of this action is to squeeze the monolayer together and form a tight assembly:

$$\Pi = \gamma_o - \gamma$$

If Π typically has a value of 10 mN·m^{-1}, after compression, what is the value of γ?

Relating the lateral two-dimensional pressure to the internal pressure of a bulk material in terms of newtons per square meter (e.g., the monolayer) requires the following relation (P. C. Hiemenz and R. Rajagopalan, *Principles of Colloid and Surface Chemistry*, 3rd ed., Marcel Dekker, New York, 1997.):

$$P = \frac{\Pi}{\Phi}$$

where Φ is the thickness of the monolayer. Assuming that the monolayer is 2 nm thick, what is the value of P?

Akin to the ideal gas laws ($P \propto V^{-1}$), is there an analogous inverse relationship between π and area (is $\pi \propto A^{-1}$)? Under what conditions is the preceding equation valid? Under what conditions do you expect it to break down?

12.16 When designing a biosensor using the LB technique, what fundamental considerations must be in place during the design process?

12.17 Research cantilevers on <veeco.com>. Calculate the vibrational frequencies of cantilevers of mass ___, dimensions, _____, made of: (1) silicon, (2) steel, (3) silica glass, (4) quartz, (5) carbon nanotube.

12.18 From the preceding examples, calculate the changed vibrational frequency if a single molecule of a protein with a mass of (1) 700, (2) 1200, and (3) 3000 Da is attached to the tip of the cantilever.

12.19 Design a simple circuit to stimulate and monitor resonant vibration of the preceding cantilevers.

12.20 For a cantilever of dimensions $500 \times 100 \times 0.5$ μm made of single crystal silicon, with an antibody of 14,000 Da attached to its tip, which binds to an antigen of 43,000 Da, calculate the rate of vibration with and without the attached antigen.

12.21 Use Stoney's equation to determine the relative deflections produced for a silicon cantilever of length 500 μm and thickness 0.5 μm versus a silicon nitride cantilever of length 600 μm and thickness 0.65 μm.

Section 5

Natural and Bionanoscience

Natural Nanomaterials

Nature's R&D with photonic systems that manipulate the flow of light and colour has been ongoing for at least 500 million years. She has developed methods and protocols for creating every manner of animal and plant appearance. These optical systems are self-assembled and have precisely specified optical functionalities. They serve diverse biological purposes and have been produced as a result of complex evolutionary selection pressures and developmental constraints. When it comes to photonics, nature truly is an experienced practitioner.

Where better, then, to look for inspiration…?

P. Vukusic, University of Exeter

FIG. 13.0	*(a) Professor Peter Vukusic of the University of Exeter, School of Physics. (b) Tiles on a butterfly wing illustrate the complex hierarchy and beauty of natural materials interacting with light on the nanoscale.*

(a)   (b)

Source: Image supplied courtesy of P. Vukusic, University of Exeter. Used with permission.

THREADS

The chemistry and physics underlying nanoscience have been covered in whirlwind fashion in previous chapters. We now give an overview of naturally occurring materials. In this chapter we describe examples of interesting and remarkable natural nanomaterials, focusing on their structure and function and the role the nanoscale plays in macroscopic expression.

This chapter marks the first formal incursion into the biological division, although we have certainly touched upon related topics throughout the text. Many natural nanomaterials comprise inorganic materials; we discuss a few selected examples of those as well. Following this chapter, we delve in more detail into biochemical nanoscience in chapter 14 and take a look behind the structure and function of biochemical machinery from the top down as well as from the bottom up.

13.0 NATURAL NANOMATERIALS

In this chapter we give examples of natural nanomaterials and the nanoscale forces and effects that give rise to their macroscale properties. What do we mean by natural nanomaterials? Every material can be described at the nanoscale, but nanomaterials are materials that are remarkable because of their inherent nanostructure. We define natural nanomaterials as materials that occur in natural environments, without artificial modification or processing. This includes the biological domain as well as the materials found around it.

The chemical identity and properties of a substance depend upon its *molecular* structure. The nanostructure of biological material is due to its *supramolecular* level—arrangements of tens to hundreds of molecules into shapes and forms that can give the material striking properties through interactions with light, water, and other materials.

13.0.1 Nanomaterials All around Us

Nanomaterials abound in living systems and in inorganic rocks and minerals. The materials highlighted in this chapter are a small sample of the natural nanomaterials around us.

13.0.2 Aesthetic and Practical Value of Natural Nanomaterials

Natural materials are rich in texture and color created by specific nanoscale structures. The geometric patterns on seashells, the patterns and colors on butterfly wings, and the shape and texture of bird feathers are a few examples. The iridescence of feathers and insect bodies is also due to layered structures with nanoscale spacing. The striking and remarkable effects of these natural nanomaterials have inspired artists, decorators and engineers, and scientists.

13.0.3 Learning from Natural Nanomaterials

The effects of nanoscale properties are often striking and easily apparent to the casual observer. For this reason natural nanomaterials have attracted scientific curiosity. This interest has led to a greater understanding of nanoscale phenomena. By studying how nanomaterials produce the effects so readily observable on the macroscale, we can learn much about the effects that nanoscale structures have on the behavior of materials in general. Natural nanomaterials are a source of guidance for chemists, physicists, and engineers to create sophisticated synthetic materials with unique functions.

The physical origins of the remarkable properties of many biological materials are due to a complex, often hierarchical structure that can provide a model for designing radically improved artificial materials for engineering and medicine. These nano-inspired materials possess outstanding levels of adaptivity, multifunctionality, and mechanical performance.

Artificial materials based on principles of design from nature are called *biomimetic* materials. We will cover biomimetic materials and their applications in a later chapter on nanotechnology. In this chapter we will focus on the natural nanomaterials themselves and on the principles of nanoscience that we can learn from them.

13.0.4 The Nano Perspective

Natural nanomaterials provide an inspiring way of introduction to nanoscience. Their beauty and intriguing properties arouse curiosity and reward investigation. It can be enlightening to realize that common functional materials that we take for granted, such as paper, cloth, and clay, depend as much on their physical nanostructure as on their chemistry to give them their useful properties. In this chapter we call attention to aesthetic and human interest as we discuss natural nanomaterials.

Naturally occurring nanostructures are rich in patterns and properties adapted through natural selection over evolutionary time scales. Science has long recognized the value of natural chemical compounds; the emerging nanoscience is now examining the properties and potential uses of natural nanostructures.

13.1   INORGANIC NATURAL NANOMATERIALS

A great variety of natural nanomaterials are formed by living organisms—by complex biochemical life processes. But there are also many natural nanomaterials that form spontaneously by much simpler chemical and physical processes. Among them are a number of simple self-assembling minerals and crystalline oxides of metals. Many complex natural inorganic substances, such as clays, carbonaceous soots, and natural inorganic thin films, exhibit properties based on their supramolecular nanoscale structure, and hence fall into our definition of natural nanomaterials. In this section we look at *inorganic natural nanomaterials*.

13.1.1   *Minerals*

Many minerals have complex nanostructures that give rise to optical, mechanical, absorbent, and catalytic properties. *Clays* are complex minerals composed of nanoscale particles derived from the weathering of silicate rocks containing oxides of the element silicon. Silicon can form many types of oxides, or silicate groups, that constitute anions to combine with different metals. Most of the outer crusts of the Earth and Moon are made up of silicate minerals.

Silicon.   Silicon is to rocks what carbon is to organic material. It is abundant, combines readily with oxygen and hydrogen, and its tetravalent bond structure leads to a great variety of molecular geometries. Silicon, located in the periodic table below carbon, is a much larger atom, and has a complete stable octet shell of quantum shell $n = 2$ electrons in addition to four valence electrons in its $n = 3$ outer quantum shell. Whereas carbon compounds form thin flexible films in water or else dissolve in it, silicon compounds are denser and less soluble and more readily form three-dimensional structures. Silicon compounds tend to form crystals that may incorporate or exclude water molecules. In the organic world of carbon, water is a partner. In the inorganic world of silicon, water is incorporated into the silicate crystalline structure. For example, when water is incorporated into the crystal lattice of silicates, nanoscale dislocations are formed.

The mechanical and optical properties of two minerals with the same chemical composition depend on the supermolecular arrangement of material at the nanoscale. Silica minerals are excellent examples of materials of the same or very similar chemical composition that have dramatically different properties. Clear quartz crystal has a strikingly different appearance compared with opal, even though both are composed of silica; the difference is due not

FIG. 13.1   *Silica.*

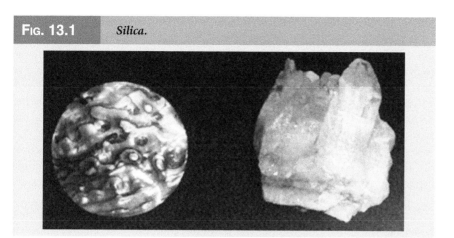

FIG. 13.1   *Silica.*

to chemical composition, or to crystal structure, but rather to nanoscale structures (**Fig. 13.1**).

Minerals were originally classified by crystalline structure and color, even as their chemical compositions were still being discovered. Silicate minerals crystallize with distinctive geometries based on the structure of their silicate ion group. For example, *tectosilicates*, or "framework silicates," have a three-dimensional crystal structure composed of silicate tetrahedra with SiO_2 in a 1:2 ratio. Tectosilicates comprise nearly 75% of the crust of the Earth. With the exception of the quartz group, tectosilicates are *aluminosilicates*, containing another abundant element, aluminum. Some of the interesting nanoscale-derived properties of silicates result from their interaction with metals and water.

Zeolites. *Zeolites* are hydrated aluminosilicate minerals made of tetrahedral building blocks of (AlO_4) and (SiO_4) linked by rings. These units form a rigid, three-dimensional crystalline structure with a network of interconnected tunnels and cages. More than 40 naturally occurring zeolites are known. An example is the mineral natrolite $[Na_2Al_2Si_3O_{10} \cdot 2H_2O]$. Zeolites differ in the number of aluminosilicate building blocks in the rings, the ratio of silicon to aluminum, and in how the building blocks are arranged. The rings are composed of six to a dozen or more tetrahedrally coordinated silicon (and/or aluminum) atoms linked by shared oxygen atoms.

The larger the aluminosilicate ring is, the larger is the size of the tunnels and cages within its superstructure that open into pores on the surface of the mineral. These tunnels pass completely through the crystal like a honeycomb enabling water, for example, to filter freely. In a given form of zeolite, the pore and channel sizes are nearly uniform, allowing the crystal to act as a nanoscale filter, or *molecular sieve*.

Molecular sieves are materials that can sort molecules based on their size and chemical or electronic affinity. The size of molecular or ionic species that can enter the pores of a zeolite is controlled by the diameters of the tunnels (e.g., size exclusion). The types of molecules that are allowed to pass through the pores are influenced by their electrical charges and chemical interaction with the sieve matrix (e.g., the chemistry of the surface).

The aluminosilicate crystalline structure is a negatively charged framework with pores that hold cations such as sodium, potassium, calcium, and magnesium (Na^+, K^+, Ca^{++}, Mg^{2+}), among others. These cations readily exchange with other alkali and alkaline earth ions, making zeolites natural *ion exchange systems*.

The crystal matrix of regularly spaced, precise, nanosized pores makes the zeolites natural molecular sieves. Due to the strong electron charge inside the pores (e.g., chemically active surfaces) and high surface energy, zeolites are able to act as catalysts. The internal surface of the pores is surrounded by the electron orbitals of the (AlO_4) and (SiO_4). The strongly electronegative environment is capable of rearranging the bonds of molecules trapped inside or passing through the openings in the matrix.

The size and electronic charge in the pores is also affected by the cation that is held in the pore bonding site. The pores in one form of zeolite are approximately 4 Å (0.4 nm) across when occupied by Na^+. With a larger K^+ ion, the pore opening is reduced to approximately 0.3 nm. A Ca^{2+} ion, with two charges, exchanges with two singly charged ions on the zeolite matrix. Thus, the pore opening increases to approximately 0.5 nm when Ca^{2+} is exchanged for sodium or potassium.

The nanoscale architecture of the silicate and alumina in zeolites gives them their remarkable properties. Because of their powerful properties as filters and catalysts, zeolites are invaluable scientific tools. Their practical usefulness makes them the basis for a valuable and important industry. Novel types of zeolites have been synthesized in laboratories since 1948, and new natural forms are still being discovered.

13.1.2   Clays

When aluminosilicates are weathered by the action of water containing dissolved carbonic acid or other chemicals, they break down into particles less than 2 μm in diameter to form clays. *Clays* are hydrous aluminum silicates rich in silicon and aluminum oxides and hydroxides. Clays have distinctive properties compared to other finely divided mineral particles due to their small particle size, layered structure, "stickiness" (self-adhesion), and plasticity, which is due to their ability to incorporate variable amounts of water into their structure. The layered crystalline structure of many clays (similar to a deck of cards) makes them shrink and swell as water is absorbed and removed between the layers.

Moderate drying by the sun removes water from clays and results in a hard material that is relatively strong and resistant to reabsorption of water. When clays are heated to high temperatures ("fired"), they lose water and form permanent bonds between particles, transforming them from a soft pliable substance to a hard durable material. From prehistoric times, clays have been used for art and ceremonial objects, as a building material, and as one of the earliest media for holding written records.

Montmorillonite Nanoclay.   Clay comes in many different colors and exhibits differing properties depending on its proportions of silicon, aluminum, and other elements and their molecular combinations with oxygen and water. One especially important class of clays are the *montmorillonites* [1,2]. Montmorillonite is formed by the weathering of rocks that have a relatively low silica content. It is named after Montmorillon in France, where it is found in abundance. It is also an important soil constituent in many regions of North America. It is the main constituent of the volcanic ash weathering product, *bentonite*.

Montmorillonite's crystal structure has two tetrahedral sheets sandwiching a central octahedral sheet. The microscopic particles in montmorillonite clays are plate shaped with an average diameter of approximately 1 μm or less. In natural montmorillonite, particle size ranges from 50 to 500 nm. Its chemical composition is hydrated sodium calcium aluminum magnesium silicate hydroxide— $[(Na,Ca)_{0.33}(Al,Mg)_2Si_4O_{10}(OH)_2 \cdot nH_2O]$.

The sodium and calcium may be replaced by potassium, iron, and other cations, in different relative abundances, with resulting variations in physical properties. The structure unit of montmorillonite consists of two silica tetrahedral sheets with alumina octahedral sheets in between, as shown in **Figure 13.2**.

Like other clays, montmorillonite swells with the addition of water, but its volume changes with water content to an unusual extent. The degree of expansion depends on the exchangeable cation substituted in the molecule. With sodium as the predominant cation, the clay can swell to several times its original volume with the uptake of water. Montmorillonite swelling is a significant factor in soil stability and has to be taken into account in building roads, dams, and buildings where it is a major soil constituent. The expansion properties of sodium montmorillonite make it useful as a nonexplosive agent for splitting rock in stone quarries and for the demolition of concrete structures where the use of explosive charges is undesirable.

The absorbant properties of montmorillonite have led to its use in traditional medicines and cosmetics. Modern uses include anticaking additives for animal foods and other commodities, drilling muds for lubricating and expanding fissures in rocks during oil exploration and production, ion exchange agents for chemical processing, catalyst supports, binding agents for toxins and microorganisms, and water purification. Montmorillonites catalyze the self-assembly of certain organic molecules. Montmorillonite and other clays are truly versatile nanomaterials with many emerging uses and newly appreciated properties.

FIG. 13.2   *Montmorillonite.*

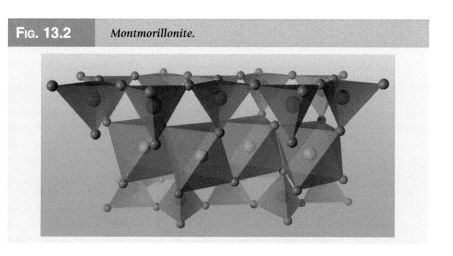

Montmorillonite is now commonly referred to as *nanoclay* in the emerging natural nanomaterials market, where new applications are being discovered each year. Commercial production may involve grinding to reduce particle size, sorting for uniform particle size, aggregation of particles into porous granules for easy handling, and chemical treatment to yield a desired mix of cations.

The remarkable properties of montmorillonite and other clays are due to their crystal structure and its particle size. These properties are a result not just of its chemical structure, but of the interaction of the clay nanoparticles with water and with each other. Although the diameter of natural montmorillonite clay particles varies, the thickness is on the order of nanometers. The large surface area of the clay particles enhances interaction with water and solutes. If the crystals were not divided into nanoparticles, the montmorillonite molecules would take the same form as other harder minerals with a similar crystalline symmetry. Instead, the exchange of cations produces disorder in the crystals, preventing the growth of large crystals and resulting in nanoparticles [3].

Catalyst for Life.   Montmorillonite and other clays may have acted as catalysts in the evolution of life from prebiotic materials. During the 1990s, scientists discovered that clays catalyze the chemical reactions needed to construct RNA from *nucleotides* [4,5]. This discovery was bolstered by a recent finding that clays can catalyze the transformation of lipid micelles into vesicles. In previous chapters we discussed how lipids in water spontaneously form micelles, making a region surrounded by a single lipid layer. In living organisms, the membranes of vesicles and cells are composed of lipid bilayers.

Under certain conditions fatty acid micelles spontaneously form into vesicles with lipid bilayer membranes, similar to those in living cells, but the transition is very slow without a catalyst or nucleation site for the membranes [6,7]. In recent years, it was found that clays act as nucelation sites for the self-assembly of micelles into vesicles. In 2004 a research group at the Howard Hughes Medical Institute (HHMI) showed that small amounts of montmorillonite greatly accelerated the formation of vesicles from fatty acid micelles [8].

In addition to catalyzing RNA synthesis, montmorillonite catalyzes the formation of vesicles that may have been precursors to the evolution of the first living cells. The HHMI group was also able to induce vesicles to grow and to split into separate vesicles under laboratory conditions. By using montmorillonite with RNA attached, they demonstrated that some of the RNA was contained in the newly formed vesicles and remained there without leaking out. The researchers succeeded in inducing the vesicles to grow by incorporating additional micelles. When large vesicles were extruded through small pores, they divided into smaller vesicles containing RNA, with structure and properties similar to the precursors. They described these processes as "a proof of principle that growth and division is possible in a purely physical–chemical system," which is "vaguely analogous to biological cell division."

There are many more complicated processes necessary to get from vesicle replication with RNA to a living cell, but nanoclays are beginning to provide clues about their role in the prebiotic world [9,10]—a perfect example of one nanomaterial interacting with others.

13.1.3   *Natural Carbon Nanoparticles*

With its covalent bonding, carbon readily forms complex two- and three-dimensional structures with nanoscale features under a variety of conditions. Diamond and graphite are two naturally occurring forms of carbon that differ in bonding between the carbon atoms, and as a result have different crystal structures and physical properties.

Diamond Nanoparticles.   Diamonds are networks of carbon atoms each linked by four bonds to their neighbors in a strongly bonded crystalline lattice. On the surface of the lattice, outer bonds are terminated by hydrogen atoms. The smallest diamond structure possible is the adamantane molecule (chapter 9), with a lattice of 10 carbon atoms linked in a tetrahedral lattice, each capped by a hydrogen atom. Large gemstone diamonds are formed only at very high pressures and temperatures, since these are the only conditions under which the diamond crystalline phase of carbon is more stable than graphite or amorphous carbon. The only natural conditions on Earth sufficient for the formation of large diamonds are more than 120 km (75 miles) deep within the Earth's mantle, where the high pressures and temperatures (40 kbar and 900°C) can compress carbon into the compact structure of diamond.

Until the emergence of nanoscience, it was generally taken for granted that diamond material could only be formed under extreme conditions of pressure and heat that existed uniquely deep within the Earth's molten depths. This is true of large, single crystalline diamonds, but very small nanodiamonds or even thin films of diamond can be formed under less rigorous conditions. Once scientists became aware of and interested in the structure of minerals and sediments on the nanoscale, they began to discover many occurrences of nanosized diamonds. The quest to understand how these nanodiamonds could have been formed has led to fascinating insights into the Earth's history.

Suitable conditions for formation of very small diamond crystals and films may occur close to and even on the surface of Earth. Nanoscale diamond particles may occur naturally as the result of rapid combustion of organic material or in meteorite impacts. Industrial researchers have developed methods to grow extremely thin diamond films at very low pressures, using gas phase plasmas and ion deposition, where the usual phase rules that govern materials in equilibrium do not apply. These are the types of conditions in flames, lightning discharges, radioactive decays, explosions, and high-energy impacts. It is interesting that ancient beliefs in the Indus civilization attributed the formation of diamonds to lightning strikes on rocks.

Nanodiamonds as Indicators of Impact Events.   Micro- and nanosized diamonds found in recent years in nonigneous minerals may have originated from meteorite impacts, radioactive decay, or other processes not yet understood. Some tiny diamonds are strongly associated with known meteorite impacts. Some of the most dramatic examples are microscopic diamonds found in Meteor Crater, Arizona, formed when an iron-rich meteor crashed 50,000 years ago. The diamonds may have been formed from carbon in the impacted soil or from graphitic particles in the meteorite itself. Some of these diamonds have a unique hexagonal crystal structure only found in impact craters, which was named lonsdaleite after British mineralogist Dame Kathleen Lonsdale.

Canadian scientists have found diamond particles 3–5 nm across embedded in a layer of sediment 65 million years old that is linked by other evidence to the time when many scientists believe a giant meteor slammed into Earth and caused the mass extinction of dinosaurs and many other Cretaceous animal and plant species [11]. The nanodiamonds, along with iridium and other indicators in sediments, form a worldwide marker layer from which the nature and date of the impact have been inferred. Other research on the diamonds from this impact dust layer, found in areas of Montana and New Mexico, has shown that their isotope abundance is characteristic of diamonds formed from carbon on Earth rather than carbon particles formed in space and embedded in meteorites. This work further supports the hypothesis that these nanodiamonds were formed in the impact.

Recently, nanodiamonds have become the primary evidence for the existence of a massive meteoric impact 12,900 years ago. Scientists had suspected some sort of catastrophic event caused at least four major changes found in the fossil record. The first was a mini ice age, called the Younger Dryas cold phase, which started about 12,900 years ago and lasted for about 1,000 years. The second is the sudden disappearance of the Clovis culture, the earliest known human inhabitants of North America, in the same time period. The third is the disappearance of the large animals such as mastodons from North America. The fourth is the disruption of human settlements and disappearance of primitive Stone Age cultures in Europe during the same time period. The nanodiamond dust found over a wide area was the key piece of evidence tying other clues together. This is an example of how nanoscience has led to surprising breakthroughs in diverse fields of study. With new awareness of the existence of nanomaterials and the tools to identify them, scientists have obtained new insights into the history of the Earth.

Nanodiamonds from Radiation.   Another unusual source of natural nanodiamonds on Earth is the action of radioactive isotopes in carbon-rich minerals. Scientists examining uranium-rich coal from Russia have discovered diamonds ranging in size from 2 to 40 nm and containing a few thousand to a few million carbon atoms. The diamonds appear to have formed when uranium atoms in the rock underwent fission, breaking up into high-energy ions. These fission fragments tear through the carbon structure, breaking the bonds and heating up a microscopic area around the fission track. As the carbon atoms move back together to form new bonds, they sometimes reconfigure into a diamond lattice structure [12].

Nanodiamonds from Petroleum Deposits.   An even more surprising source of nanosized diamond particles is petroleum. These *diamondoids* had first shown up as deposits in crude oil equipment, and have now been shown to be present in petroleum deposits under the Gulf of Mexico. They are extremely small nanoparticles, typically containing up to 11 *adamantane* units (10 carbon atoms arranged in a diamond lattice). Although scientists have been able to synthesize similar diamondoid particles in the laboratory, they have not been able to make particles larger than four adamantane lattice cages. Some of the oil-based diamondoids have unique structures, such as one consisting of six adamantane units arranged in a disk, named cyclohexamantane [13,14].

It is not clear how the diamondoids found in oil are formed. One possibility is that clay minerals catalyze reactions between methane and other hydrocarbon precursors, adding carbon to adamantane seed crystals. One possibility is that these nanodiamonds were formed by the giant meteor impact in the Gulf of Mexico some 65 million years ago and migrated into the oil-bearing strata along with other organic material. This hypothesis may be supported in that the worldwide distribution of diamondoids in oil follows a pattern consistent with known meteor impact events. These particles are on the boundary between chemical molecules and nanoparticles, where many interesting and useful properties arise. Researchers are studying their use as building blocks for nanomachines, as platforms for antiviral drugs, and in other applications.

Diamond Nanoparticles of Unknown Origin. The sources of some micro- and nanodiamonds are still a mystery. Tiny diamond grains 20–80 µm in size have been found in metamorphic rocks (gneisses) from southwestern Norway and Asia. These are not the igneous rocks from the molten magma that is the source of gem diamonds. The metamorphic rocks have not yet been linked to any known meteor impacts. Diamonds that were weathered out of igneous rock might have been deposited in sediments that were the precursor of the metamorphic gneisses where they are now found. But sedimentary rocks underwent pressure and heat that transformed them into their present metamorphic phase. Diamonds are unlikely to have survived this transformation; instead, they would have been oxidized or converted to graphite. Although diamond is the hardest form of carbon and graphite is one of the softest minerals known, graphite is actually more stable than diamond [15–17].

The study of nanodiamonds is still very new, with many unknowns waiting to be resolved. Nanoscience has increased the awareness and attention given to small-scale particles in geoscience just as in other fields. As the basic scientific understanding of nanoscience grows, more knowledge and tools for solving mysteries in geochemistry, mineralogy, and archeology will be made available.

Graphite Nanoscience. Graphite is black and soft; it can be considered to be the purest form of coal. Diamond and petroleum deposits may be converted over geological time into graphite, as the most stable form of pure carbon. A single particle of carbon with a graphitic structure no longer has the properties associated with graphite, since these properties depend on the interaction between different nanolayers. Nevertheless, nanoparticles with basic graphitic structure do exist in the natural world.

The most familiar application of graphite is in the common pencil lead. Graphite powder is valued for its lubricating properties. Both applications depend on the nanostructure of the flat sheets of linked carbon atoms. For a long time, it was believed that the sheets were loosely coupled and could easily slice over each other, giving graphite its useful properties. However, it has been shown that in a vacuum (such as in space), graphite is a poor lubricant. This led to the discovery that the lubrication is due to the absorbed fluids between the layers, such as air and water. Recent studies suggest that an effect called *superlubricity* can also account for graphite's lubricating properties.

When a large number of crystallographic defects binds these planes together, graphite loses its lubrication properties and becomes what is known as *pyrolytic carbon*. This material is useful for blood-contacting implants such as prosthetic heart valves because it does not cause blood clotting. It is also highly diamagnetic; thus it will float in midair above a strong magnet.

Graphite forms *intercalation* compounds with some metals and small molecules. In these compounds, the host molecule or atom gets "sandwiched" between the graphite layers, resulting in compounds with variable stoichiometry. An example of an intercalation compound is potassium graphite, denoted by the formula KC_8.

Amorphous Carbon Nanoparticles. Amorphous carbon is a term applied to coal, soot, and other carbon compounds that are neither graphite nor diamond. However, these materials are not truly amorphous; they are polycrystalline or nanocrystalline materials of graphite or diamond within an amorphous carbon matrix.

Soot, or "black carbon," produced from burning organic material is ubiquitous in the environment. Soot and smoke contain a variety of carbon nanoparticles, which can serve as both an adsorbent and a catalytic site for chemical reactions. Soot from combustion where the amount of oxygen is limited has been found to contain significant amounts of carbon buckyballs.

13.1.4 *Nanoparticles from Space*

Nanoparticles with unusual properties can form in the exotic conditions that exist in interplanetary and interstellar space. Meteorites sometimes contain microscopic particles of diamond or other complex nanostructures. Diamonds could be formed by collisions between asteroids or with other planets, but they could also be formed in deep space. Some nanodiamonds embedded in meteorites contain an isotopic mixture of xenon not found in the solar system. It is believed that such diamonds were formed in collapsing stars as they died in supernova explosions. If this hypothesis is correct, stardust may contain tiny diamonds.

Astronomers studying interstellar space have determined that nanoparticles make up a large part of interstellar dust. Optical measurements indicate that carbonaceous particles ranging from soot to diamond exist in disks of gas and interstellar clouds. These particles may have formed directly from condensation of gases or from collision fragments. In the low density of space, exotic crystal structures may gradually coalesce into nanoparticles with unusual structures and properties. Astronomers have evidence that nanodust particles in space act as catalysts in formation of compounds in clouds of gas and dust, leading to building up of raw materials that may play a part in formation of life after the clouds coalesce into planets [18].

13.2 Nanomaterials from the Animal Kingdom

In this section we look at natural nanomaterials formed by animals. We start with organic structural fibers and matrices common to most forms of life. In this section we trace the progression of mineralization from shells to teeth and bones, and look at the polymers that make up scales, hair, and feathers. In later sections we will explore other directions that plants and insects took in producing their own unique nanomaterials.

13.2.1 Building Blocks of Biomaterials

In this section we look at some protein and biopolymer building blocks that illustrate some important principles and examples of nanostructure architecture in animals, plants, and single-celled organisms.

Multicellular organisms synthesize a protein matrix to hold cells together in a structured body. These *extracellular matrices* provide the scaffolding, support, and strength to tissues for all living organisms. Organisms arrange protein, mineral, and polymer building blocks from the bottom up, linking them in complex scaffolds and networks. They make an exquisite variety of composite materials whose organization on the nanoscale gives them strengths and features that far surpass the properties of their components. These *natural nanomaterials* have evolved far beyond the basic function of holding cells together: Natural nano-materials protect, support, and serve an amazing variety of functions in the bodies of organisms.

We now look at some important structural proteins and some examples of the protein nanomachinery that controls the building of natural nanomaterials. Long, straight chain *collagen* fibers intertwine to provide strength; elastic *proteo-glycans* and *elastins* give resilience; adhesive *glycoproteins* such as *fibronectin* and *laminin* bind the matrix and cells. These and other structural proteins are key building blocks for the nanostructured materials we will examine in this chapter. In addition to the building blocks, we will show a few examples of the proteins that act as nanomachines to guide and control the assembly of nanostructures. Nonprotein polymers are also important in natural structural materials. Three such polymers we will discuss are *cellulose, chitin,* and *keratin*. We encounter a few other interesting examples along the way. Cellulose is produced by plants, chitin is produced mainly by shellfish and insects, and keratin is produced mainly by vertebrates.

Most organisms combine protein and polymer building elements with inor-ganic minerals such as *carbonates, phosphates,* and *silicates,* ordering their nano-structure to create strong and rigid materials with unique optical and electrical properties. Some general principles of structural organization should emerge from this overview of natural nanomaterials. One recurring theme will be that of *ordered fibrous polymers embedded in a mixture of nonfibrous substances* to give a *composite* with superior properties to the single components. Where high strength and rigidity are required, nature generally selects *fiber-forming macromolecules,* usually ones that aggregate into *near-crystalline arrays*. Where elasticity and resil-ience are important, *highly branched polymers* such as *elastin* meet the requirement. In many natural materials from different sources and different compositions, we will find a *hierarchy of fiber assemblies* from the molecular level up to the macroscopic, with *interlocking linkages* to spread loads. An important theme is the importance of *calcium carbonates* and *phosphates* as components of skeletal structure, as well as serving as buffers or sinks for carbon dioxide and regulators of cellular metabolism.

A good introduction to fibers in connective tissues begins with the *collagens*, made up of *tropocollagen* building blocks arranged in a variety of near-crystalline aggregations into fibrils in a wide variety of organisms and tissues. The straight chain fibrils of collagen are supplemented by more branched polymers to provide elasticity and resilience. We encounter more details about collagens and similar polymer units, such as chitin and keratin, in many settings as we traverse through the nanoscale architecture of biomaterials in this and later sections [19].

Single-Celled Organisms, Plants, and Animals.   Plants and animals have radically divergent life patterns and equally different body architectures and materials. The most important plant structural materials are nanostructured cellulose and lignin, familiar as paper, cotton, and wood. The highly important functional and structural nanodesign of plant leaves, flowers, seeds, and roots is beyond the scope of this section.

Before proceeding, we devote this section to the biological materials made by multicellular animals: nanostructures combining proteins and organic polymers with mineral crystals of carbonate and phosphate. These important nanostructures include *exoskeletons* of sponges, corals, and crustaceans (seashells); eggshells; cartilage; scales; teeth; bone; turtle shells; claws; hair; fingernails; feathers; and other tissues in which carbonate and phosphate crystals are layered in with organic polymers and proteins to provide increased hardness, rigidity, and strength. In the vertebrates, mineralized carbonates and phosphates are developed into *endoskeletons* of cartilage and bone.

Mineralized Biological Nanomaterials.   Tissues that contain solidified inorganic salts are said to be *mineralized*. Mineralized tissues, such as shells and bones, are widespread among multicellular animals; sponges, corals, and sea urchins make mineralized tissues to support their cells. The origin of multicellular animals appears to be tied closely to carbonate-based chemistry. The most primitive mineralized materials made by multicellular animals include crystals of *calcium carbonate*, $CaCO_3$, as a component of their structure.

The first traces of biomineralization in the geological record are found in rocks dated to the *Ediacaran* era (which is when the first fossil evidence of multicellular animal life appears)—immediately following a period of high carbon dioxide content in the Earth's oceans. The Ediacaran was followed by the Cambrian epoch 545–490 million years ago. The "Cambrian explosion" was the era in which primitive life forms diverged into the major classes that make up life on Earth. The carbonate ion plays a central role in their metabolism [20].

Carbonate Chemistry.   Let us look at the chemistry that is the foundation for the building of carbonate nanostructures. The carbonate ion, CO_3^{-2}, is formed when carbon dioxide, CO_2, dissolves in water:

$$CO_2 + H_2O \rightarrow H_2CO_3 \rightarrow H^+ + HCO_3^- \rightarrow 2H^+ + CO_3^{2-} \qquad (13.1)$$

The neutral gas, carbon dioxide, readily dissolves in water where it is in equilibrium with the bicarbonate ion and the carbonate ion. This equilibrium buffers the acidity of the solution. Metal ions such as sodium, potassium, magnesium, calcium, and zinc can associate with the hydrated carbonate ion. The lighter and more active metals, sodium and potassium, form strongly hydroscopic and highly soluble salts, whereas magnesium and calcium and heavier metals are less soluble and tend to precipitate out of solution to form mineral crystals. Higher organisms use dissolved carbon dioxide as a source of carbon for both photosynthetic reduction of carbon into carbohydrates and for mineralization, as well as a means of controlling *p*H. This *sequestration* of carbon dioxide through photosynthesis and mineralization is an important factor through which life regulates the amount of carbon dioxide in the oceans and atmosphere.

Crystalline Structures of Carbonates.   Calcium carbonate is the most abundant form of mineral carbonate, both in geological sediments and in biomaterials. Pure calcium carbonate can crystallize in several different forms, with the molecules stacked in different arrangements to form different crystal shapes and symmetries. The most common crystalline form of calcium carbonate is *calcite*, in which the carbonate ion groups lie in a single plane pointing in the same direction, leading to a *trigonal symmetry*. Another common carbonate mineral is *aragonite*, in which the carbonate ions lie in two planes, with adjacent carbonate groups pointing in opposite directions, giving an *orthorhombic* symmetry. Either of these forms of calcium carbonate can form by precipitation from solution depending on the concentration, temperature, *p*H, and presence of other salts. The formation of both can be catalyzed by protein enzymes.

Cellular Chemistry of Carbonate.   In living cells, carbonate precipitation is catalyzed by specific protein enzymes, which determine the crystalline form of the mineralized carbonate. The family of enzymes called *carbonic anhydrases* catalyzes the conversion of dissolved carbon dioxide into carbonate ions. In bacteria these enzymes get rid of excess carbon dioxide and produce carbonate salt, which serves as a buffer in the cell cytoplasm, helping to regulate the *p*H in the cell by reacting with excess acid. In sponges and other multicellular organisms, this carbonate continues to serve its roles in regulating cell chemistry, but it can also be precipitated to form a mineralized skeleton. Special proteins bind to the carbonate and direct its precipitation in either calcite or aragonite form.

13.2.2   Shells

By exacting analysis of the nanostructure of seashells, scientists have found that the mantle of tissue next to the growth surface of the shell lays down layers of chitin coated on both sides by special proteins. The chitin determines the size and position of the calcium crystals, and the proteins control the deposition of calcium in specific crystalline forms for the plates or prisms.

Nanostructure of Shell Growth.   In corals and shellfish, the development of carbonate-based skeletons has evolved as a support and protection for the organism. In animals that build carbonate exoskeletons, the carbonate minerals typically are precipitated as crystals, or *prisms*, embedded in layers or matrices of proteinaceous and polysaccharide material. In mollusks, the layer of the shell closest to the living organism consists of very smooth pearly nacre, whose mineral crystals are in the form of flattened aragonite platelets, while the thick body of the shell is built of calcite prisms.

The exoskeleton or shell is grown by a layer of cells that first lays down a coating of protein supported by polysaccharide polymers such as chitin. The proteins act like a nano-assembly mechanism to control the growth of the calcium carbonate crystals. The calcium and carbonate ions are also secreted by the mantle, but it is the tailored environment created by the organic matrix that determines how the crystals grow. The honeycomb-like matrix of protein and chitin remains as an encapsulating layer around each crystal. This envelope serves as a relatively flexible, crack-deflecting matrix for the mineral particles. The strong covalent bonding of the protein and chitin complements the ionic bonding of the brittle mineral to give the finished material a combination of toughness and hardness. The size of each crystal is typically around 100 nm.

Strength and Toughness Based on Nanostructure.   As a result of the nanostructure biocomposite of polygonal carbonate crystals embedded in a polymer matrix, the nacre of mollusk shells has extraordinary physical properties. Nacre consists of vertical stacks of flat polygonal aragonite crystal tablets (or platelets). The argonite is arranged in parallel growth layers called *lamellae*. Each lamellae layer is built of several thousand layers of aragonite plates stacked in an interlocking offset pattern somewhat like a brick wall. The plates are mortared to each other by thin layers of protein and chitin glue. Platelets are typically hexagonal, with thicknesses between 0.3 and 1.5 µm and diameters ranging from 5 to 20 µm. Each plate is encased in a protein and chitin coating a few tens of nanometers thick. The matrix has pores that allow mineral bridges to connect the plates to each other. This structure makes nacre 3000 times as fracture resistant as crystalline mineral aragonite.

Methods for Investigating Shell Nanostructure.   To investigate the microscopic structure of calcium carbonate crystals in snail shells and the interactions between proteins and the crystals on the nanoscale, researchers use atomic force microscopy (AFM). The AFM is guided and controlled by scanning electron microscopy (SEM), light microscopy (LM), and biochemical methods. For studies of nanoscale interactions of proteins with crystal faces, extracts of proteins taken from shells are mixed with solutions of calcium carbonate and calcium chloride.

Calcium Control by Protein Nanomachinery.   A number of calcium-controlling proteins have been identified. Some of these proteins have active sites that orient calcium carbonate into seed crystals of either calcite or aragonite, a process called *crystal nucleation*. Other shell-building proteins bind onto specific faces and edges of calcium carbonate crystals, where they may promote or inhibit either calcite or aragonite formation. The inhibitory proteins may prevent the growth of the crystal in some directions while allowing growth in preferred directions. The inhibitory proteins have been shown to inhibit the dissolution of crystals once they are formed. Some of the calcium-controlling proteins found in shells have active sites that closely resemble portions of proteins found in milk whey, which are associated with the deposition of calcium in the bones of mammals. The complex mixture of calcium-building proteins is specific to each organism, resulting in growth of different shell types, patterns, and strengths.

The polysaccharide chitin is secreted by the mantle cells and acts as a scaffold to which the protein molecules are attached. Some proteins have been shown to act as adhesives between argonite tablets (the large protein lustrin A, found in abalone nacre). Other proteins, such as perlucin, nucleate new layers in calcium carbonate crystals. Other proteins, whose function is not yet fully understood, such as perlustrin, are similar to insulin- and insulin-like growth factor-binding proteins found in higher organisms. Structures for the calcium-controlling proteins from seashells are still being fully determined but their amino acid sequences are related to proteins that stimulate proliferation and differentiation of cells involved in building bone in vertebrates, such as fibroblasts, bone-marrow cells, and osteoblasts. Recently, the full structure has been worked out for ovocleidin-17, a protein similar to perlucin, which plays a similar role to build calcium layers in chicken eggs. Nature is very conservative in preserving useful nanotools once they are evolved, as evidenced by the close relationships between the calcium-controlling machinery across many organisms. Yet, as we find when we learn about other mineralized tissues, nature is infinitely inventive in the structures that organisms build to meet the needs of survival in their environments [21–28] (**Fig. 13.3**).

13.2.3 *Exoskeletons*

The mollusks adapted to a defensive lifestyle of building a hard protective shell into which the organism could retreat. Predators in turn evolved first teeth and claws, and then jaws, to crush and penetrate the shells, leading to an eons-long arms race in which the nanostructure of the shell was optimized. Other marine animals that originated during the Cambrian took on a more mobile and flexible form of mineralized exoskeleton. Crustaceans, such as crabs, lobsters, and shrimp, have a tough, flexible, chitin-rich exoskeleton divided into sections connected by flexible joints—ideal for supporting movement.

The crustacean's modular exoskeleton is divided into three parts: head, thorax, and abdomen—a body architecture that characterizes insects and vertebrates

FIG. 13.3   *Shells.*

as well. In addition to mobility, the lighter, more flexible exoskeleton of the crustaceans allows them to shed or *molt* their exoskeletons as they grow and then grow new ones. The thin exoskeleton is not only easy to break out of compared to shells, but also involves less of an investment in energy and nutrients to rebuild. The crustacean exoskeleton is also an excellent instructive example of how strength and flexibility are related to the nanostructure of a material with a minimal investment in material weight, volume, and components. Crustacean exoskeletons show a strong anisotropy in their properties, resulting from their hierarchical organization from the nanometer to the millimeter scale [29–31].

Crustacean exoskeletons have four layers, all relatively rich in proteins compared to mollusk shells. The outer layer is a soft chitin-free coating of lipids, proteins, and calcium salts, which seals and protects the exoskeleton. The exocuticle is a matrix of chitin and protein with embedded mineral particles, mainly calcite. The endocuticle is also a mineralized matrix of chitin and protein, but with less calcification and polymer cross-linking than in the exocuticle. The membranous layer is also composed of protein and chitin, but is not mineralized. The thicknesses of these layers vary in different types of crustaceans. They also vary on different parts of the body (claws, abdomen, eye covering, etc.) and depend on the stage of development and molting. In a typical crab, the epicuticle is on the order of 10 μm, the exocuticle 30 μm, the endocuticle several hundred micrometers, and the membranous layer 20–30 μm [32].

The crustacean exoskeleton is an outstanding example of a nanostructured natural material optimized to trade off high strength, light weight, and economy of fabrication. The protein and chitin in crustacean exoskeletons form fibrils arranged in ordered, three-dimensional, twisted honeycomb patterns. This structure, sometimes referred to as "twisted plywood," is formed by the helical stacking of plates composed of crystalline chitin–protein fibers. The stacked planes are gradually rotated about their normal axis, thereby creating complex structures that appear as arches when cut in cross-sections. This structure is also called the Bouligand pattern [29], named for the scientist who first deduced it from electron micrograph observations.

However, these twisted plywood planes are not simple arrays of parallel chitin–protein fibers. Interconnected fibers form a honeycomb-like network filled with pores in which calcite mineral particles are distributed. The different layers of the exoskeleton differ in density and hardness.

The finely woven twisted plywood structure of the crustacean exocuticle has a high stiffness (8.5–9.5 GPa). The hardness increases within the exocuticle between the surface region (130 MPa) and the region close to the interface to the endocuticle (270 MPa). The endocuticle is a much more coarsely twisted plywood structure, in which both the stiffness (3–4.5 GPa) and hardness (30–55 MPa) are much smaller than in the exocuticle. The different layers are important in absorbing mechanical loads on the exoskeleton and in allowing for staged growth [33].

Preventing Propogation of Crack Failure.   The cuticles of spiny and slipper lobsters have a pattern of surface beads, or *tubercles*, and a system of pits that resist failure by blunting any cracks that start to form in the shell. The layered structure of the shell dissipates forces horizontally, while the tubercle and pit systems increase the surface area available to dissipate forces and interrupt the growth of cracks through the material. These systems may represent evolutionary solutions to predation, particularly by predators that strike, crush, or bite holes, rather than those that engulf.

13.2.4 *Endoskeletons*

Thus far we have discussed animals that develop hard exoskeletons to protect and support their bodies. Another group of animals began to build an internal structure for support. This strategy disencumbered the body of a restrictive heavy coating, and gave the advantages of mobility, unrestricted growth, and less investment in material for a given body size. In some of the simplest multicellular animals, the sponges, the extracellular matrix contains both mineral crystals and *collagen*. Collagen acts as a structural material in many animals, including *arthropods* (crustaceans and insects), where it plays a secondary role to the exoskeleton.

Cartilage.   In *chordates* and *vertebrates*, strands of collagen are combined with other structural polymers to form an *endoskeleton* of *cartilage*.

The nanostructure of cartilage is important in giving it the strength to support the animal body and in providing a framework for mineralization that eventually leads to the development of *bone* in later descended vertebrates. Cartilage tissue can also be found in invertebrates such as horseshoe crabs, marine snails, and cephalopods, where its nanostructure is optimized to support muscular movement and attachment of tissue to the shell. Although cartilage is found in invertebrates, no invertebrate cartilage mineralizes in vivo. Mineralization of cartilage is a vertebrate innovation.

Cartilage forms a dense connective tissue that can withstand large shear and other stresses. The nanostructure of cartilage is very complex. A very important difference between cartilage and the materials of mollusk and insect exoskeletons is that cartilage has living cells embedded within the tissue, so it grows from within rather than from a boundary layer of cells. This difference opens the possibility of many more nano and micro architectures. This also enables cartilage to grow much more rapidly than if growth could occur only at the surface, with important implications for the life cycle of the organism.

Types of Cartilage.   In vertebrates there are three main types of cartilage adapted to different functions. Each type differs in arrangement and concentration of collagen fibers in relation to the matrix and in the nanostructure of the fibrils themselves. For a given type of cartilage, the orientation of the fibers varies depending upon the loads that the cartilage bears during development.

> *Hyaline cartilage* is a translucent cartilage found in growing tissue and in embryos, serving as a center of ossification or bone growth. The name hyaline comes from the Greek word *hyalos*, meaning glass. This refers to the translucent matrix or ground substance. Hyaline cartilage also forms the gristly *articular* cartilage lining bones in joints and is present inside bones as well. In hyaline cartilage, collagen molecules are arranged in cross-striated fibers, 15–45 nm in diameter.

Elastic cartilage is a dense yellow tissue that supports stiff organs such as the ear, nose, and larynx. Elastic cartilage is stiffer and more elastic than hyaline cartilage due to elastin bundles distributed in the matrix and attached to the collagen fibers.

Fibrocartilage is a white cartilage with high tensile strength, with larger and more densely concentrated collagen fiber bundles than in other types of cartilage. Fibrocartilage forms the tough tissue forming the discs separating vertebrae and connecting tendons or ligaments to bones.

Matrix. The matrix interacts with the fibers of cartilage to give its elastic properties. The matrix contains *proteoglycans*, which are large molecules with a protein backbone and *glycosaminoglycan* side chains. These molecules fill the spaces between the collagen fibers. The proteoglycans are hydroscopic, holding water that can be expelled and reabsorbed in response to loads, both compression and stretching. This gives cartilage resistance to compression and elastic resilience (ability to spring back into shape after load). An important example of a proteoglycan is *aggrecan*, whose glycosaminoglycan side chains contain *chondroitin sulfate* and *keratan sulfate*. Aggrecan is found in the tough gristly cartilage that supports bone joints. The matrix also contains the precursor building block molecules of collagen, *tropocollagen*.

Collagen: Self-Assembled Hierarchical Structure. The fibrous collagen strands in the matrix of cartilage have a hierarchical, self-assembling architecture. Tropocollagen building blocks bundle into *fibrils*, and collagen *fibrils* bundle into *fibers*.

Tropocollagen molecules have a typical molecular weight of about 300,000, but there are many variants of different chain lengths. Each tropocollagen building block is composed of three polypeptide strands, each of which is a left-handed helix. Three left-handed helices self-assemble into a right-handed triple helix coil about 300 nm long and 1.5 nm in diameter. The triple coil is not bound by chemical bonds, but is a *cooperative quaternary structure* stabilized by hydrogen bonding.

The bonding of tropocollagen subunits into self-assembled fibrils is controlled by a mixture of strong and weak intermolecular forces. There is some covalent cross-linking between tropocollagen units along the 300-nm long triple helices, and a variable amount of covalent cross-linking between neighboring tropocollagen helices in the fibrils. The result is a large number of different nanostructure combinations, giving rise to many different types of collagen found in tissues. These spontaneously bound structures with their large number of possible arrangements can have different mechanical and structural properties.

Quaisi-Crystalline or Near-Crystalline Structure. Unlike the highly ordered and regularly consistent bonding between molecules in a crystal, the aggregations between proteins are more variable, yet not random. This type of organization of matter is called *quaisi-crystalline, near-crystalline, or cooperative structure*. It is characteristic of very large molecules with both covalent and ionic sites for inter- and intramolecular interactions, both bondings and repulsions. In biomaterials it allows for great variety in nanostructure to produce materials with a great range of properties.

Types of Collagen.   There are some 24 different classified types of collagen. In *hyaline* cartilage, collagen units are arranged in cross-striated fibers, 15–45 nm in diameter, called type II collagen. Type II collagen has tropocollagen units stacked in parallel with regularly staggered ends, which give this type of cartilage a characteristic banded appearance under high magnification under the electron microscope. Type II cartilage is the main type found in fully developed cartilage. Type I and type III collagens are less strong and are found in bone and its precursors [34–36].

Buehler has shown in a mathematical model that the force required to separate two fibers of collagen bonded along their length is dependent on the length of the fibers, and that there is an optimal length depending on the structure of the fibers [35,36].

Collagen and Cartilage as Precursors of Bone.   Strong fibrous collagen molecules are not only the main structural protein in cartilage, but are also the main organic framework for bone and other supportive tissue in vertebrates. Cartilage is the precursor of bone in the evolutionary and embryonic development of vertebrates, forming the skeleton of the earliest types of vertebrates, of which lampreys, rays, and sharks are modern representatives [37].

The cartilage in sharks and rays consists primarily of a mesh of collagen fibers embedded in a gelatin-like matrix, along with a scattering of cartilage-generating cells called *chondrocytes*. (Sometimes the cartilage is surrounded by a thin layer of mineralized tissue that gives it a little extra stiffness.) The result is softness and flexibility.

Bone.   Bone differs from shell in having a phosphorous mineral component that largely replaces calcium carbonate. The first step towards development of bone was the development of proteins, which fixed phosphorous as phosphate, so that the calcium mineralization was supplemented by the calcium phosphate mineral *hydroxyapatite*—$Ca_{10}(PO_4)_6(OH)_2$.

Types of Bone.   Bone is formed with differing densities, microstructures, and nanostructures that give each type of bone its structural features. Each type developed to serve a different function in different animals.

Aspidine.   The earliest appearance of bony tissue was in jawless fishes with bony mouth plates (similar to those of the hagfishes) and bony armor plating on the exterior of their bodies. These plates, with a characteristic structure called aspidine, had a hierarchical structure and composition like bone. Aspidine was composed mostly of hydroxyapatite; it had a characteristic microstructure, with three distinct layers. Each of the microstructure layers had a different nanostructure, with different densities and orientations of calcium phosphate crystals. The outer layer is a solid, highly mineralized array of dentine tubercles similar to the patterns on some shark teeth, underlain by a porous mineralized, honeycomb. These early animals probably had an endoskeleton made of cartilage, with a small percentage of mineralization, like the toothed but boneless jawed fishes.

Teeth.   In the *sharks* and other *jawed fishes* with cartilage endoskeletons, hydroxyapatite serves mainly to provide the strong *enamel* and *dentin* layers of teeth.

Enamel is the hardest and most highly mineralized tissue in vertebrates, consisting mostly of flattened, needle-like crystallites of apatite supported by a small amount of protein, which adds toughness and resists cracking. *Dentin* is a softer, less mineralized, and honeycombed material that supports the enamel shell from within and below, much like the porous aspidine supported the hard outer layer.

Tooth enamel contains more than 90% mineral crystals, of which about 95% is phosphate. The rest of the mineral content is calcium carbonate with traces of magnesium and other alkali metals. Traces of fluoride anion are found in shark and other teeth, adding to hardness and making them less soluble in contact with corrosive agents such as acids and enzymes.

Unlike dentin and bone, tooth enamel does not contain collagen. Instead, it has two unique classes of proteins called *amelogenins* and *enamelins*. While the role of these proteins is not fully understood, it is believed that they serve as a framework in the growth of enamel layers, and they may add some strength by resisting crack propagation.

Because of its high mineral and low protein content, enamel is relatively brittle compared to dentin. For sharks this problem is solved by shedding teeth continually and growing new ones. Like seashells, teeth have a composite nanostructure, which gives them more strength than their simple mineral constituents. Unlike shells, teeth and other bones have a complex inner structure in which living cells continually replenish the mineral and structural protein content.

Nanostructure of Bone Growth.   In growth and development from embryo to adult in vertebrates, bone is formed by mineralization of a cartilage scaffolding (**Fig. 13.4**). The evidence given by research is that the mineralization of tissue is associated primarily with the type I collagen fibril, that the fibril is formed first and then mineralized, and that mineralization replaces water within the fibril

Fig. 13.4   *Collagen.*

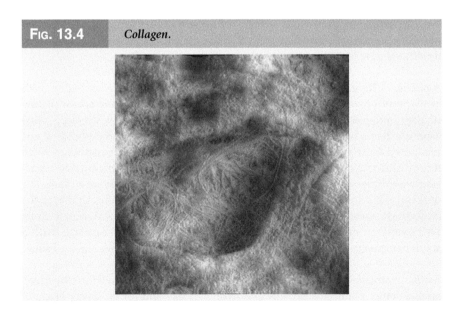

with mineral. The collagen fibril therefore plays an important role in mineralization, providing the aqueous compartment in which bone minerals grow. Collagen remains in the bone and contributes to its elasticity and fracture resistance [38].

Nanostructure of Collagen Mineralization Sites.   In bone, the collagen triple helices lie in a parallel, staggered array, with gaps of 40 nm between the ends of the tropocollagen subunits. These spaces seem to serve as nucleation sites for the deposition of long, hard, fine crystals of the mineral component, mainly calcium hydroxyapatite, with some calcium phosphate.

Cortical and Trabecular Bone.   Compact, or *cortical*, bone forms the hard dense outer shell of large bones; spongy or *trabecular* bone fills the interior spaces of bones.

Woven and Laminar Bone.   Cortical bone is typically either woven or lamellar (layered). *Woven bone* has a small number of randomly oriented collagen fibers; it grows quickly without a preexisting cartilage scaffolding, but is relatively weak. *Lamellar bone* is stronger, formed of numerous stacked layers, or *lamellae*, and filled with collagen fibers. Woven bone is found mainly in areas of regrowth after a break. Most woven bone is gradually replaced by slower growing lamellar bone, on a scaffolding of calcified hyaline cartilage through a process known as *bony substitution.*

In laminar bone, the collagen fibers run parallel to other fibers in the same layer. The fibers run in opposite directions in alternating layers, assisting in the bone's ability to resist torsion forces. Plate-shaped carbonate apatite crystals are arranged in layers across the fibers; like the fibers, the orientations of these crystals are different in alternate layers. At the next higher level of organization in laminar bone, the laminar layers are grouped into cylindrical structures [39].

Mechanical Properties of Bone.   Bone has two major characteristics: rigidity to support weight and toughness, the ability to withstand breaking. Bone has compressive strength some two times its torsional or tensional strength. This selectiveness is due to the arrangement of the apatite crystal and the collagen fibers. The apatite is arranged in small crystalline plates about $60 \times 30 \times 8$ nm in size found within and around collagen fibrils. The plates overlap and are aligned parallel to the long axis of the fibril. Also, the plates within each fibril have the same angular orientation, which differs from the orientation in adjacent fibrils. As compression is applied, the matrix material is squeezed together as the collagen deforms to absorb the force. The plates slide towards each other until they start to touch. This initial movement absorbs some of the applied force and prevents the bone from breaking. As the plates touch, frictional forces take over. The plates can no longer slide and the compression force is resisted. What happens when the forces are too great for the crystalline plates? They fracture, but because the plates are so small, large cracks cannot form. Also, because the plates of adjacent fibrils are not oriented at the same angle any small crack that does form cannot propagate easily from one plate to the next.

Piezoelectric Properties of Bone.   Like many minerals, hydroxyapatite is piezoelectric; the ionic components of the crystals generate a small electric field in response to stress. Repeated stress, such as weight-bearing exercise or bone healing, results in the bone thickening at the points of maximum stress. It has been hypothesized that this is a result of bone's piezoelectric properties. This would imply a response of the collagen and matrix to the electric fields, which is still an area of active research. We will discuss some aspects of the response of bone to electromagnetic fields in the chapter on biomedical nanotechnology.

Micromechanical Models of Bone.   A number of micromechanical mathematical models for the Young's modulus of bone have been developed to account for the strength and anisotropy of the material [40]. The models incorporate the platelet-like geometry of the load-bearing crystals, the alternating thin and thick lamellae, and the orientations of the crystal platelets in the lamellae. The thin and thick lamellae are modeled as orthotropic composite layers made up of thin rectangular apatite platelets within a collagen matrix, and classical orthotropic elasticity theory is used to calculate the Young's modulus of the lamellae. Bone is viewed as an assembly of such orthotropic lamellae bent into cylindrical structures, and having a constant, alternating angle between successive lamellae.

Bony Exoskeletons.   The shells of turtles, armadillos, and some other vertebrates are outgrowths of the animal's skin (*dermus*) and endoskeleton. The individual plates are covered by scales called *scutes*, made of the tough protein keratin, like claws and fingernails. Underneath the scute is a layer of dermal tissue and calcified shell, or carapace, which is formed by fusion of vertebrae and ribs during development.

Unlike seashells and arthropods, turtle shells have living cells, blood vessels, and nerves, including a large number of cells on the calcareous shell surface and scattered throughout its interior. Bone cells that cover the surface and are dispersed throughout the shell secrete protein and mineral so that the bone can grow and reshape continuously.

13.2.5   *Skin and Its Extensions*

Keratin.   *Keratin* is a polypeptide-based material with a complex hierarchical structure that is the major component of skin, scales, claws, beaks, feathers, horns, hair, and other hard, tough outgrowths of the dermal layer that covers the bodies of vertebrates. The properties that make structural proteins like keratins useful depend on their supermolecular aggregation. These depend on the properties of the individual polypeptide strands, which depend in turn on their amino acid composition and sequence.

Relation of Molecular Biology and Biochemistry to Nanostructure.   The α-helix and β-pleated sheet motifs, and disulfide bridges, are crucial to the conformations of globular, functional proteins like enzymes, many of which operate semi-independently, but they take on a completely dominant role in the architecture and aggregation of keratins, just as in collagens.

Hair, wool, and related fibers all have keratins as their principal constituent. Their different physical properties are due to differences in their protein arrangement at the nanoscale level. The polypeptide chains of keratin are folded, but when stretched, the chain unfolds. The unfolding is reversible, giving the fibers their elastic properties (the stretched form of keratin is called β-keratin and the folded form is called α-keratin). Keratins can have different degrees of folding depending upon the side chains attached to the polypeptide. We will see more details of polypeptides in chapter 14, "Biomolecular Nanoscience".

Glycine and Alanine.   Keratins contain a high proportion of the smallest of the 20 amino acids, glycine, whose "side group" is a single hydrogen atom, as well as the next smallest, alanine, with a small and uncharged methyl group. In the case of β-sheets, this allows sterically unhindered hydrogen bonding between the amino and carboxyl groups of peptide bonds on adjacent protein chains, facilitating their close alignment and strong binding. Fibrous keratin molecules can twist around each other to form helical intermediate filaments.

Limited interior space is the reason why the triple helix of the (unrelated) structural protein collagen, found in skin, cartilage, and bone, likewise has a high percentage of glycine. The connective tissue protein elastin also has a high percentage of both glycine and alanine. Silk fibroin can have these two as 75–80% of the total, with 10–15% serine, and the rest having bulky side groups. The chains are antiparallel, with an alternating C–N orientation. A preponderance of amino acids with small, unreactive side groups is characteristic of structural proteins, for which H-bonded close packing is more important than chemical specificity.

Disulfide Bridges.   In addition to intra- and intermolecular hydrogen bonds, keratins have large amounts of the sulfur-containing amino acid cysteine, required for the disulfide bridges that confer additional strength and rigidity by permanent, thermally stable cross-linking—a role sulfur bridges also play in vulcanized rubber. Human hair is approximately 14% cysteine. The pungent smells of burning hair and rubber are due to the sulfur compounds formed. Extensive disulfide bonding contributes to the insolubility of keratins, except in dissociating or reducing agents such as urea.

The more flexible and elastic keratins of hair have fewer interchain disulfide bridges than the keratins in mammalian fingernails, hooves, and claws (homologous structures), which are harder and more like their analogs in other vertebrate classes. Hair and other keratins consist of α-helically coiled single protein strands (with regular intrachain H-bonding), which are then further twisted into superhelical ropes that may be further coiled. The keratins of reptiles and birds have β-pleated sheets twisted together, then stabilized and hardened by disulfide bridges.

Feathers.   Feathers, produced by birds and their extinct relatives, are keratin structures, among the most complex structural organs found in vertebrates. The keratins in feathers, beaks, and claws—and the claws, scales, and shells of reptiles—are composed of protein strands hydrogen-bonded into β-pleated sheets, which are then further twisted and cross-linked by disulfide bridges into structures even tougher than the keratins of mammalian hair, horns, and hoof. The complex micro- and nanostructure of feathers controls the flow of air and heat around the surface of the animal, yielding exceptionally high performance in aerodynamics and insulation. An interesting aspect of the nanoscience of feathers is the aerodynamic interaction on the nanoscale between the complex surface of the finely divided feather structure and the surrounding air.

Feathers are attractive for humans for their beauty of form and color. A nanoscale phenomenon called the *Dyck texture* is what causes the colors blue and green in most parrots. This is due to an optical effect caused by the nanostructure of the feather itself, rather than pigment, or the *Tyndall effect,* as was previously believed (dispersion of shorter light wavelengths).

Hair.   Hair, produced by mammals in many forms, has varied complex hierarchical keratin structures, which give hair it curliness and resilience. The nanostructure of the keratin fiber arrangement in hair has been studied using scanning microbeam small-angle x-ray scattering (SAXS). Scanning microbeam SAXS patterns of hair fibers can reveal the differences in patterns between the inner and outer sides of the curvature in curly hair. In very curly Merino and Romny wool, different types of cortices exist; the macroscopic curl of the hair fibers arises from inhomogeneous nanoscale distribution of two types of cortices. The same effect is observed for human hair [41].

Wools.   Wool has two qualities that distinguish it from hair or fur: It has scales that overlap like shingles on a roof and it is crimped; in some fleeces the wool fibers have more than 20 bends per inch. Because of its economic importance, wool fiber keratins were one of the first proteins whose folding behavior was understood. X-ray crystallography was used to reveal differences in regular spacings of the protein structure between relaxed and stretched wool fibers [42]. The molecular conformation and the microstructure of wool fibers can also be studied by newer techniques, such as Raman spectroscopy and atomic force microscopy. Raman spectroscopy shows absorption by symmetrical vibrations, so it gives information about the C–C skeletal vibration region, and the S–S and C–S bond vibration regions or keratins. Raman spectroscopy experiments can reveal the secondary structural transformation from α-helical to β-pleated sheets, which takes place in wool fiber stretching. The stretching mechanism of wool fibers can be divided into two different mechanisms at different levels of the nanoscale structural hierarchy: (1) slippage of the folds of the polypeptide chain on the protein molecular level and (2) disordered conformation of fibrils and breaking of some portion of the S–S bonds linking keratin fibers at the supramolecular level [43].

13.2.6   Summary

In this section we have seen how the nanoscale architecture of natural biomaterials determines their properties. Polymers and proteins with strong covalent bonds provide tensile strength. Hydrogen bonding and cross-linking in proteins and complex polymers provide resilience and flexibility. And ionic bonded mineral crystals provide load-bearing and compression strength. Combinations of these materials in a hierarchy of organization from the molecular to nano- to micro- to macroscale create optimized material properties for different requirements. These materials provide examples of important principles of structure on the nanoscale.

An important theme is how proteins orchestrate the architecture of natural biomaterials on the nanoscale, sometimes with the aid of polymer scaffolds. Mineral crystals, micelles, the bilayer membranes of cells, and polymer scaffolds are self-assembling: They organize into predetermined shapes and networks based on the molecular structure. The proteins, by contrast, are synthesized by a bottom-up, serial mechanism and they perform their catalytic control of the building of biomaterials in a bottom-up fashion, intervening in the crystallization process of mineralization and in the extension of polymer meshes. The result is an intricate highly functional nanostructure.

In addition to strength and elasticity, the nanostructure of biomaterials gives rise to optical, electrical, hydrodynamic, and aerodynamic properties—for example, in diatoms, nautilus shell, and feathers. We will provide examples in the following sections on plant fibers, butterfly wings, lotus leaves, gecko feet, and others.

13.3 NANOMATERIALS DERIVED FROM CELL WALLS

Plants and some bacteria differ from animals in having a cell wall surrounding their cellular membranes. In bacteria and most plants the cell wall is made of cellulose or a similar carbohydrate material. These cell walls have an intricate nanostructure. When the organisms die, their cellular cytoplasm and membranes are dissolved away, leaving the cell wall as a nanomaterial residue.

13.3.1 Paper

Paper is the most common familiar substance made from the cellulose residue of digested reeds, stems, wood, or fibers such as cotton. The oldest form of paper was made by the ancient Egyptians from the papyrus reed that grows on the Nile River, from which our word "paper" derives. Although paper is an artificial material, it consists of natural *cellulose* that has been purified and processed. Its properties that we value, such as absorbency and strength, depend on the nanostructure of the cellulose fibers. The microstructure and nanostructure of the fibers allows them to entangle and link by hydrogen bonding to form strong mats. The process of papermaking harnesses the natural self-assembling properties of the fibers. The nanostructure of the pores made by the plant cells makes paper fibers absorbent to water and inks, and makes it possible to hold in fillers such as clays and titanium dioxide to impart sheen and smooth finishes.

Ancient papermaking consisted essentially of gathering, pulping, and pressing the natural materials. Today, processing of natural cellulotic materials to make various types of papers is an extremely sophisticated technological field involving chemistry, mechanical engineering, and nanotechnology. Many types of fiber and fillers are added to modern papers. The appreciation of how nanoscience can be applied to paper is relatively new and is covered in a later section under nanotechnology.

At Louisiana Tech University, the potential for applying nanoscience to improved papermaking is being developed in a pioneering research and education program [44]. Canadian companies are also rapidly developing ways to improve paper through the application of nanoscience [45]. Similar developments are taking place in Europe, Israel, and the Far East.

13.3.2   *Cotton*

Cotton is a natural plant material whose nanostructure gives it highly valued properties. Cleaned cotton fibers consist of almost pure *cellulose*, the same material from which paper is made; however, the nanostructure of cotton makes it unique. Cotton has been prized for centuries for its absorbent wicking properties, which make it comfortable in warm climates, make excellent bandages and wound dressings, and allow cotton fabrics to be easily dyed with colorful pigments. These characteristics are due to the nanostructure of the fibers as produced by the cotton plant (genus *Gossypium*) in its seed pod.

The high strength, durability, and absorbency of cotton are due to the nanoscale arrangement of the fibers. Each fiber is made up of 20–30 or more concentric layers of cellulose, called *lamenae*, coiled in a spiral winding to form a series of natural springs. When the cotton boll is opened, the fibers dry into flat, twisted, ribbon-like shapes and become kinked together and interlocked. This interlocked form is ideal for spinning into a fine yarn.

Natural cotton fibers are covered with a cuticle layer of waxes and pectin, which is removed by cleaning. The outer wall of the fiber is composed of crystalline cellulosic fibrils. Inside this wall are three layers of closely packed spiraling parallel fibrils. The direction of the spiral is reversed in the middle layer relative to the two layers that enclose it, adding strength and elasticity to the fiber. The center of the fiber is the lumen, which contains the cell contents in the immature fiber and is empty when the cotton boll is fully ripe and open. The cross-section of the fiber is kidney or bean shaped when dry and may swell to a round shape when it absorbs moisture. The fibrils swell with absorption of moisture—more so in the immature fiber because it has less crystalline cellulose.

Inside the spiraling fibrils is a level of supramolecular nanostructure that optimizes the performance of the fiber. The cellulose molecules in cotton fibrils are organized into parallel formations called crystallites. Each elementary fibril is made up of 36 cellulose chains, which make up one basic crystalline unit of cotton cellulose. In the fiber, elementary fibrils are gathered into microfibrils and larger aggregates called macrofibrils. The diameters of elementary fibrils measure 3.5–10 nm wide, the microfibrils 10–40 nm, and the macrofibrils 60–300 nm. The exact arrangement of fibrils determines the properties of the fiber (**Fig. 13.5**).

The cellulose in the fibrils is laid down in a series of concentric growth rings, the lamellae, whose number and properties are affected by temperature fluctuations and moisture in the plant during fiber development. Up to 50 lamellae have been reported in fibers from some plants, but between 25 and 40 lamellae are more commonly found in a mature fiber. Different growing conditions, climate, and variety of plant can yield different qualities of fiber.

Various forms of microscopy have been used to elucidate the morphology and internal structure of cotton. With transmission electron microscopy (TEM), fibrils on the surface of cotton fibers and in thin transverse sections can be

FIG. 13.5   *Cotton.*

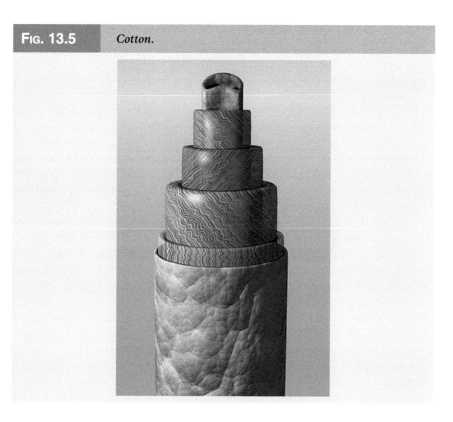

resolved, but at the cost of pretreatments that may alter the natural state or introduce artifacts.

The scanning probe microscope (SPM) technique is nondestructive, and permits examining cotton fibers in ambient conditions in the presence of natural humidity. This technique uses a probe tip and measures the force changes between the tip and the sample as the tip is scanned. Information about the nanomechanical properties of the fibers can be obtained by operating SPM in force mode. In this mode, the force as a function of distance between the tip and the sample is recorded. The AFM and the scanning tunneling microscope (STM) have also been used to investigate the surface structure of cotton fibers [46,47].

In the following sections we will look at some natural nanomaterials synthesized by single-celled organisms, yeasts, fungi, and plants. Their nanomaterials are made from organic carbon polymers and from silicon.

13.3.3 *Bacterial Fibers*

Cellulosic nanomaterials are produced by single-celled organisms such as algae and bacteria as well as by plants. Some of these organisms produce useful materials that have familiar uses in many cultures, as well as promising new uses. In this section we show some examples.

Certain strains of *Acetobacter xylinum* bacteria produce fine cellulose fibrils with a diameter of 3–8 nm and a length of 50–80 nm. Traditionally, this bacterium is used in the Philippines to ferment coconut milk, producing a jelly-like food base known as *nata de coco* (Spanish, "cream of coconut"). This jelly is an ingredient used in Asian desserts. It can be prepared as a jelly, an agar, or a pudding with gelatin features. Nata de coco has also been used as a sunblock ingredient and moisturizer in lotions and cosmetics.

These bacterial fibers form an entangled mass, making a three-dimensional nanonetwork with extremely small pores. The cellulose produced by these bacteria has very different characteristics from most other naturally occurring cellulose. Bacterial cellulose fibrils are relatively insoluble because of their high molecular weight, stiffness, and extensive hydrogen bonding. The stiffness of the gels led to traditional use as a food additive when prepared as emulsions of 99% water and 1% cellulose.

Acetobacter xylinum bacterial cellulose can be pressed and dried into sheets with remarkable mechanical strength; the Young's modulus is as high as >15 GPa across the plane of the sheet. The high Young's modulus has been attributed to the unique supermolecular nanostructure in which fibrils of biological origin are preserved and bound tightly by hydrogen bonds. A team of scientists and engineers led by Sony in Japan have studied the mechanical properties of cellulose sheets and used the material in a commercial speakerphone. Because of their strength, stability, and optical transparency, these fibers have also been used as reinforcement material in flexible displays for televisions, personal computers, and portable phones developed in Japan by NTT, Mitsubishi, and others.

Researchers led by Dr. Mari Tabuchi in Japan have found new uses for this unique natural nanomaterial. Recognizing the potential of its polymer strength, insolubility, and nanoporiosity, they have used it to produce high-performance nanopore filters for a nanoscale gene detection system. The filters made from bacterial cellulose resisted up to 2 MPa of pressure without any clogging, whereas conventional membranes were capable of less than 0.5 MPa. The bacterial gel also gives improved performance as a separation medium for macromolecules and for light amplification in the system's photodetectors.

Pulps made from bacterial cellulose give a strong paper and are used for reinforcing conventional pulp papers and enabling paper-making from some fibrous materials, *Acetobacter xylinum*, cyanobacteria, and algae [48–50].

Biofilms.   Bacteria can attach to surfaces where they grow in colonies. They secrete a complex matrix, which holds the colony in place, called a *biofilm*. Common examples of biofilms are the slippery coating in a stale puddle of water or in a vase of flowers, and dental plaque on teeth. Biofilms are important for health, sanitation, and their clogging of equipment in water. In this section we look at the nanostructure of the biofilm polymer matrix and how it is secreted.

Bacteria and other microorganisms that form biofilms synthesize an extracellular *polysaccharide* matrix, sometimes called EPS for *extracellular polymeric substance*. Polysaccharides are polymers made up of sugar units (or *saccharides*) bonded together. These large molecules are amorphous, adhesive, insoluble, and resistant to attack by enzymes that would normally attack sugars and starches. They make an ideal protective home for colonies of bacteria living in water.

This polymer matrix protects the cells in the biofilm. Channels within the matrix help distribute nutrients and signaling molecules. The tough biofilm matrix may persist after the bacteria die and may even become fossilized. Fossilized biofilm matrices are among the earliest evidence for primitive life on Earth.

How are bacteria and fungi able to stick to surfaces and form such strong films, even in flowing water? Cells use special protein molecules and hairlike nanostructures called *pili* (Latin for "hairs") to attach to surfaces and to each other. Each pilus (Latin for "hair" [singular]) is about 10 nm in diameter and may be 100+ nm in length. Pili are composed of proteins with binding sites specialized for receptors on the surfaces of other bacteria or on environmental surfaces. After attachment with the pili, cells begin to secrete the organized biofilm matrix. Once an attachment to a surface has been made, it is easier for other cells to attach to the first ones, and the biofilm can grow to macroscopic dimensions.

Cell cultures growing in water often form a surface biofilm called a *pellicle*. Complex pellicles can be formed on the surface of fermenting food preparations or on any liquid in which organic matter is decomposing. Cut flowers left in a water vase generate pellicles that are easy to harvest and study under the microscope. Some bacteria form stable pellicles with nanopores and other interesting properties.

13.3.4 Diatoms

One class of microorganisms, the *diatoms*, have cell walls composed of a silica skeleton with an overlay of pectin. The rigid silica cell walls support the pectin just like the cellulose walls of bacteria and most plants. But the silica is a permanent nondegradable material that accumulates in sediments after the cell dies. Like other microorganisms, diatoms are very abundant in the oceans. There can be anywhere from thousands to hundreds of millions of phytoplankton cells in 1 L of seawater, depending on nutrients, temperature, and light [51,52].

Sediments composed largely of diatom skeletons are found in some places and are known as *diatomaceous earths*. The fossil record of unique species preserved in the sediments makes diatoms a useful tool in dating sediments for modern ecological and evolutionary researchers.

The silica skeletons of diatoms, called *frustules*, have a complex micro- and nanostructure. There are more than 200 genera of diatoms; hundreds of thousand of species have been identified, with a great variety of intricate skeletal shapes and patterns. Diatom skeletons range in size from 1 or 2 μm to hundreds of microns. Their intricate substructures and pores are on the nanometer to micrometer scale. Pores can vary from 20 to 200 nm in diameter in different species. Some species have a finer network of secondary pores within the primary pores (**Fig. 13.6**).

The detailed structure of diatom frustules is below the resolving power of ordinary light microscopes. Nanoscale features of diatoms can be characterized using SEM, AFM, EDAX, and secondary ion mass spectrometry (SIMS).

Adhesive Nanofibers from Diatoms. Some species of diatom extrude adhesive nanofibers, which are used to anchor the cells in biofilm colonies. Researchers in Australia have examined the adhesive and mechanical properties of these

FIG. 13.6   *Diatoms.*

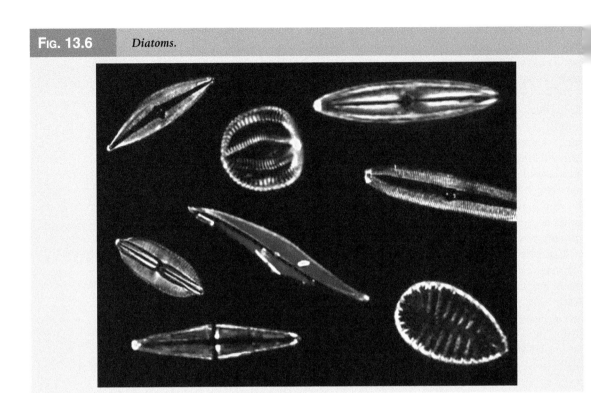

fibers using atomic force microscopy on the live diatom cells. The resulting force curves have a regular sawtooth pattern, the characteristic fingerprint of modular proteins. The protein fibers do not lose their elasticity even when stretched and relaxed for up to 600 repetitions. The high force required for bond rupture, high extensibility (~1.2 μm), and the accurate and rapid refolding upon relaxation together provide strong and flexible properties ideally suited for the cell-substratum adhesion of fouling diatoms. Studies of natural nanofibers like these allow us to understand the mechanism responsible for the strength of adhesion, and may one day lead to improved man-made adhesives and fibers [53].

13.3.5   Lotus Flower

The sacred lotus, *Nelumbo nucifera*, is an aquatic perennial, flowering plant. The plant roots itself to the bottom of ponds and rivers, with the leaves floating on the surface. Supported by thick stems, the lotus flowers rise several centimeters above the leaves. The sacred lotus is indigenous to a wide area, extending from eastern Asia to the Middle East. The plants were introduced to Europe in the eighteenth century and are now found worldwide, where they are typically raised in botanical green houses. *N. nucifera* flowers are used to supplement food as a garnish and add flavor to a variety of dishes. The petals are also used to make lotus tea and are considered an herbal remedy.

The sacred lotus is very important in practices of the Asian religions, Hinduism and Buddhism. It is revered due to the striking white flower that remains free of dirt even after having been submerged in muddy river water. It is a symbol of purity. From Bhagavad Gita 5.10, "One who performs his duty without attachment, surrendering the results unto the Supreme Lord, is unaffected by sinful action, as the lotus leaf is untouched by water."

The quote from the Bhagavad Gita illustrates a remarkable property of the lotus flower: its ability to easily shed water and dirt. This confers several survival benefits. By preventing plant pathogenic microorganisms (e.g., bacteria and virus) from gaining a foothold on the plant surface, the lotus can avoid some natural infections. Also, clean plant surfaces have been associated with decreased plant temperatures. Rates of diffusion through the stomata (pores in the plant epidermis) of water and respiratory gases are higher.

Apparently, surface adhesion to dirt is reduced by a combination of a waxy surface coating and microtopography of the plant surface. Closer examination reveals that the flower petals are essentially nonwettable.

The wettability of a surface is related to the *contact angle* (CA) between a droplet of liquid and the horizontal surface upon which it rests. The CA value results from the forces that develop between the interfaces of liquid, gas, and solid phases of the system. These *surface tension* forces are modified by the physical and chemical properties of the materials (liquid, gas, or solid) that comprise the system. If the angle is between 0° and 90°, the surface is considered wettable, with the former value denoting complete wettability. The result is that the liquid will spread out to form a monomolecular thin film on the horizontal surface.

If the contact angle of the system is between 90° and 180°, the surface is non-wettable; that is, the liquid drop is in contact with the surface at a single point and exhibits low interfacial tension. Although this state is difficult to achieve in real plant systems, they can get close. This condition is called *superhydrophobicity*. Because the CA between the liquid and the surface is very high (e.g., water drops on a waxed surface), the drops bead up and, if the surface is not normal to gravity, will roll off. Much as a snowball collects more snow as it rolls, the drops collect dust and dirt and carry them away, leaving the petal surface clean.

Wilhelm Barthlott and Christoph Neinhuis first noted that the combination of surface topography and chemistry of leaves and petals of the sacred lotus exhibited superhydrophobicity. The lotus cuticle consists of a combination of an insoluble polyester polymer called *cutin* impregnated with and covered by lipids called *epicuticular wax crystalloids*. The waxes consist of long-chain hydrocarbons that self-assemble to form the nanometer-sized crystalloids. These appear to be partially responsible for the surface hydrophobicity.

The surface of the lotus flower has an additional adaptation that increases the hydrophobicity effect. Scanning electronmicrographs show a very bumpy surface, consisting of a series of micrometer-sized projections called *epidermal papillae* or *trichomes*. Some plants have a combination of a waxy cuticle and low trichomes, which creates a relatively smooth surface. The droplets slide rather than roll over the surface. Dirt and dust are not picked up as efficiently. The sacred lotus has a very rough surface produced by underlying epidermal cells, which push upwards to create numerous trichomes. The spaces between the trichomes trap air and the hydrophobic epicuticular waxes exclude water. This reduces the contact area between the water drops and decreases wettability. As a result, water droplets ride on the tips of the projections and a cushion of air.

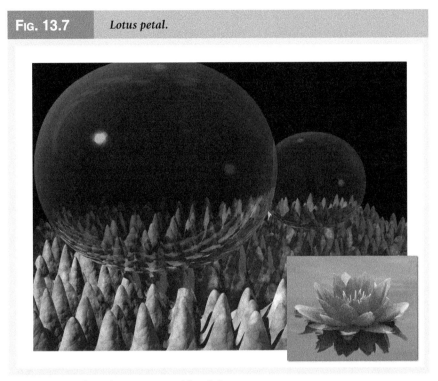

Fig. 13.7   *Lotus petal.*

Source: Lotus flower image courtesy of Gary B. Lawrence.

However, the wettability of dirt particles is increased and they adhere to the water droplet. The drops roll, trapping the particles and effectively cleaning the petal surface (**Fig. 13.7**).

The sacred lotus is an example of a surface that, due to the physical and chemical conditions at the micro- and nanometer scale, is able to produce a remarkable self-cleaning effect, which aids the survival of the plant. The lotus effect can be observed in other organisms, such as tulips and water cress; in the wings of butterflies and chitinous coverings; and in the feet of the gecko [54–57].

13.4   NANOMATERIALS IN INSECTS

In this section we present a discussion on the nanostructure of some insects, including the truly remarkable and colorful nature of butterfly wings, and how their properties are due to their nanostructure.

13.4.1   *Chitin*

Chitin is a hard translucent natural material made by plants, insects, and other animals. Chitin is a major component of insect bodies and wings. It is also the main component of the exoskeletons of arthropods such as the crustaceans (crab, lobster, and shrimp) and of the beaks of cephalopods (squid and octopus).

Like cellulose, chitin is one of many naturally occurring polymers synthesized by organisms to give structure, strength, and protection to cells and bodies. Chitin and cellulose are closely related in molecular structure: Both are long, unbranched chains of glucose derivatives. Chitin is a long-chain *polysaccharide*, a polymer of *β-glucose* sugar units. The chemical formula of chitin is $(C_8H_{13}O_5N)_n$. Chitin differs chemically from cellulose by having one hydroxyl group on each glucose monomer substituted with an *acetylamine* group. The nitrogen in the amine group leads to stronger hydrogen bonding between adjacent polymer chains, making the chitin–polymer matrix stronger than cellulose.

The polysaccharide chitin polymer is translucent, pliable, strong, and resilient. In arthropods such as insects, the chitin exoskeleton may be made tougher by being embedded in a hardened proteinaceous matrix. The nanostructure of chitin polymer is organized into layers and interlocking structures whose growth is guided by special chitin-binding matrix proteins. These matrix proteins have chitin-binding domains that bind to saccharides and catalyze their polymerization into polysaccharide chitins. The amino acid sequences determine the folding of the proteins, which in turn determines the nanostructure of the chitin polymer material.

The larval stage of insects frequently has body walls with less protein than the adults. The strength given to the polysaccharide by intertwined protein chains can be seen by comparing the body walls of a caterpillar and a beetle. This is an example of supramolecular strengthening of materials on the nanoscale, as we saw in detail earlier when we discussed shell and bone. Next we will look at some interesting optical properties that are due to nanoscale structure in insect wings [58].

13.4.2 Chitin Structures in Insect Wings

Different types of organisms have different proteins that build structures adapted for the organism's environmental niche. A rich variety of structures have evolved over time and have been selected to perform functions that aid in species survival. Insects provide many examples of chitin nanostructures. By studying these nanostructures we can learn how insects do some of the remarkable and strange things of which they are capable. We may even be able to adapt these properties to artificial nanomaterials for new and useful applications.

Some insects such as the cicada (*Pflatoda claripennis*) and termite (family Rhinotermitidae) have ordered hexagonal packed array structures on their wings made of chitin. The spacings of the arrays range from 200 to 1000 nm. The structures tend to have a rounded shape at the apex and protrude some 150–350 nm out from the surface plane. Wing structures with spacings at the lower end of the range may be adapted to serve as an optimized antireflective coating (a kind of natural "stealth technology"). These close-spaced arrays may also act as a self-cleaning coating (another example of the *lotus effect* described in the previous section).

Structures with spacing at the upper end of the range might provide mechanical strength to bear loads imposed in flight. Arrays with these larger spacings might control the flow of air over the wing to improve the aerodynamic efficiency of the insect. The research team of Watson and Watson has demonstrated the multipurpose design of natural structures such as these in its studies [59].

13.4.3   Butterfly Wings

A butterfly's head and chest are covered with plates of hard chitin, while the abdomen is covered with soft chitin. The wings of a butterfly are a chitinous membrane. The brightly colored wings of some Lepidoptera species of butterfly, such as *Morpho rhetenor*, and moths, such as *hawkmoths*, are a consequence of the nanotopography of the wing's surface and its interaction with light [60]. The resulting colors are quite striking and visible at great distances, reportedly up to half a mile. There have been a number of hypotheses for this coloration. The most common explanation for why such a feature evolved is that the brilliant colors are used for communication and mate selection and have an obvious survival benefit for the species.

The wings also exhibit *iridescence*, which is defined as a shift in the color of an object when observed from different positions. This characteristic is easily demonstrated in optical storage media, such as compact discs (CDs). The very fine submicrometer scale repeating rows of pits, which encode digital information on the CD's surface, cause the light falling on them to be diffracted into the range of colors that comprise the visible spectrum. Each frequency of light is visible at a particular angle with respect to the surface of the disc. In effect, the CD acts as a diffraction grating that exhibits iridescence. Natural diffraction gratings appear to be reasonably common and are observed in a diversity of land and sea organisms, such as crustaceans (the seed shrimp, *Cyridinid ostracod*), the scarab family Scarabaeidae, and other insects.

13.4.4   Color and Structure

The color of some materials is determined by a *pigment*—a chemical that absorbs certain frequencies of light and reflects others. This type of color is called *chemical color*. *Iridescence* or *physical color*, on the other hand, is a result of the interaction of light with the physical structure of the surface. The structures that interact with wavelengths of visible light are nanoscale in size. When light strikes a transparent surface some of the light is reflected. A few of the light rays penetrate into the material, reflect off the bottom surface, and pass upwards to join the light rays reflected from the upper surface.

If the two rays are *in phase* they combine to form a bright color. This is called *constructive interference*. If the rays are out of phase, the rays cancel and the light dims. This is *destructive interference*. The color, intensity, and angles of iridescence are dependent on the thickness and the refractive index of the substrate, and on the incident angle and frequency of the light striking the surface (**Fig 13.8**).

Natural iridescence is produced by several mechanisms. The most famous is that of the opal, the sedimentary mineral prized for its gem-like qualities. Here the iridescence is produced by packed silica spheres in the nanometer range. In opals that produce strong iridescence, the spheres are uniform in size and are arranged in layers. This provides the appropriate conditions for interference. Another example is the iridescence of diatoms, which also behave as photonic crystals due to their nanoscale structure (**Fig. 13.6**).

Butterflies and moths produce iridescence in their own unique way. A closer look at the wing surface shows rows of scale arranged much like tiles on a roof

FIG. 13.8	*Interference.*

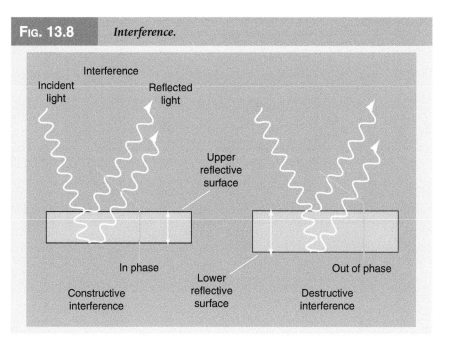

(**Fig. 13.9**). Each scale has smaller structures on its surface (**Fig. 13.10**). There are raised ribs that run the length of the scale and are spaced evenly part. There are even smaller, uniformly spaced "microribs" that interconnect the larger ribs (**Fig. 13.11**).

At a microscopic size the surface of the scale has a very intricate and highly ordered structure. The size of the spacing of the ribs and spars is comparable to the wavelength of light; about 400–700 nm. The spaces between the ribs and spars form natural photonic crystals or diffraction gratings that can generate constructive and destructive interference. A *photonic crystal* is a periodic nanostructure

FIG. 13.9	*Butterfly wing tiles.*

FIG. 13.10   *Butterfly wing scales.*

that can modify the passage of light rays. The refractive indices of the materials and the cavities restrict the frequencies of light that can propagate well. A particular system excludes the propagation of a particular set of frequencies; this is called a *photonic band gap*. The gaps will appear dark and represent the frequencies where the light rays are out of phase and undergo destructive interference.

Morpho Butterfly.   The architecture of the surface of the scales varies between species. The male *Morpho rhetenor*, with its dazzling blue wings (capable of reflecting 80% of the incident light at the blue wavelengths), has a series of very small longitudinal ribs on the wing scale that in cross-section look like evergreen trees.

FIG. 13.11   *Butterfly wing microribs.*

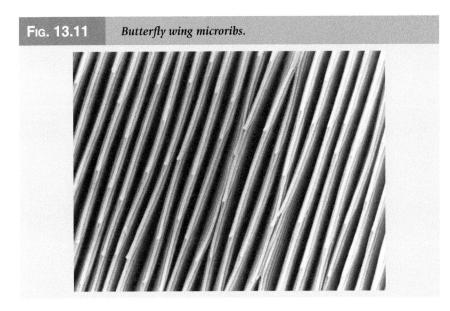

FIG. 13.12 *Nanoscale cross-section of microribs.*

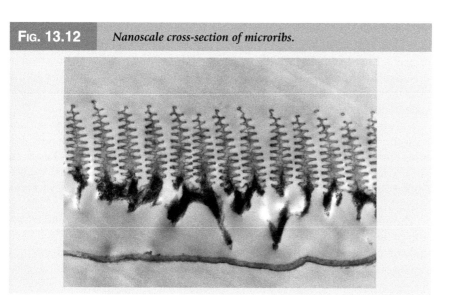

These are called *setae*. The ridges are made of a *cuticle* layer, composed of chitin polymerized from glucose (**Fig. 13.12**).

The lowest "branches" are the longest and they get smaller towards the tip of the setae. Each branch is transparent and separated from the next by an air-filled space. This alternating cuticle/air structure, with a refractive index ratio of 1.5:1, creates the right condition for interference to occur (**Fig. 13.13**). As light falls on

FIG. 13.13 *Nanoscale schematic of microribs.*

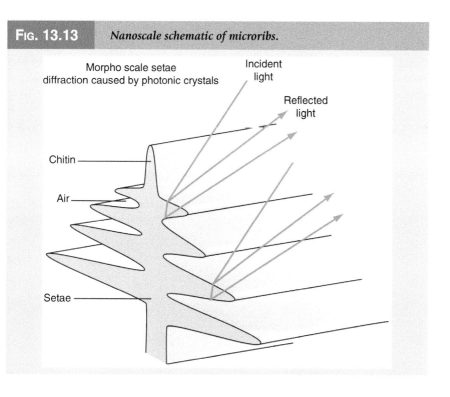

the setae it first strikes the upper branches and, as noted before, some of the light passes through the cuticle and some is reflected. Because the lower branches project further out than the higher ones, they also transmit, refract, and interfere with light. The cumulative effect is to produce constructive interference in the blue wavelengths and generate a strong blue color. This is the characteristic appearance of the wings of the *Morpho* butterfly.

In some moth species the nanotopography is different. Rather than surface projections, the series of transparent ribs, microribs, and the smaller nanostructures enclose subsurface cavities. This creates a spongy volume that has been referred to as the "pepper-pot structure." Again, these structures have a size range similar to that of the wavelength light and can generate interference. In many ways, this arrangement is similar to the iridescent opal.

The basic theme of objects that show iridescence is that diffraction and refraction of light is caused by the physical surface itself, not necessarily the chemical makeup of the underlying substrate. This characteristic has been utilized by a variety of systems, with varying architectures, to generate the brilliant and shifting colors associated with spectacular iridescence.

13.5   GECKO FEET: ADHESIVE NANOSTRUCTURES

Many members of the family Gekkonidae, commonly called the gecko lizard, have the ability to cling to virtually any surface and at any orientation, even upside down on a glass surface (**Fig. 13.14**). Aristotle observed this in the fourth century B.C. This remarkable ability is independent of the surface type. They can adhere equally well to smooth or rough surfaces. The survival benefits are evident. A number of models had been proposed, ranging from suction cups to sticky adhesives. These prior models had significant problems. Gecko feet do not exude any sticky material and observations of the anatomy of the feet also showed no suction-like features, even at microscopic size. The feet have to adhere well, but also release easily. This gives the gecko the ability to move rapidly over any surface. It was difficult to explain this seemingly paradoxical

| FIG. 13.14 | *Gecko on glass.* |

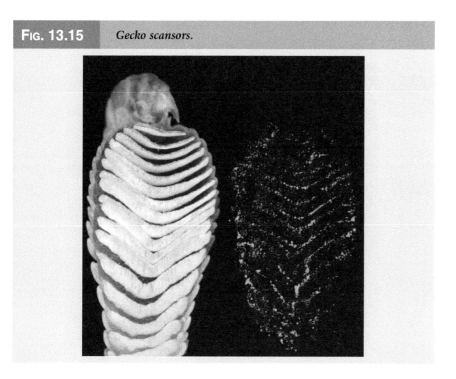

FIG. 13.15   *Gecko scansors.*

situation. Iterative rounds of hypothesis testing have evolved the working model of how the gecko feet work to the present form, which focuses on the micro- and nanotopography of the gecko feet [61].

13.5.1   Gecko Feet

The gecko foot has a series of subdigital transverse ridges called *scansors* (**Fig. 13.15**), which contain numerous, 100-μm long projections called *setae* (**Fig. 13.16**). The setae are formed from epidermal extensions that were originally thought to assist in skin shedding but have evolved in some species to provide an adhesive function. There are approximately half a million of these setae on the feet of the Tokay gecko. Each seta is further subdivided into about a thousand 200-nm wide projections called *spatulae*. Each spatula has a flattened end, much like a spoon (hence the name spatula). As a result, the total surface area of the gecko feet is enormous and this provides a clue about the mechanism of the gecko's remarkable clinging abilities [62].

13.5.2   Mechanism of Adhesion

Although surfaces may appear to be smooth, at molecular dimensions these same surfaces are very rough; steep molecular mountain ranges rise up between deep valleys. So even when two objects appear to be closely touching, in reality only the rough tips touch and the surface area of contact is relatively small. The spatulae of the gecko's foot are able to project into the valleys between the mountains and increase the surface contact area tremendously. The normally

Fig. 13.16   *Gecko setae.*

minute forces that do not come into play when the contact area is low now dominate. The force is called van der Waals attraction. Since van der Waals forces are physical rather than chemical, the gecko feet stick to hydrophobic and hydrophilic surfaces with the same force.

Van der Waals forces arise when the molecules have an uneven distribution of electrons across their surfaces. This creates a charge separation, which results in a molecule where one end is more negative than the other end. This is called a dipole. As unlike charges attract, the negatively charged area of the dipole is attracted to nearby positive charges and so generates a slight force. The force is typically very short range, within 2 nm or so. In order for strong adhesion to occur, you need a very close contact between the molecules of the two surfaces and a very large surface area. The gecko's spatulae are flexible enough to essentially mold themselves into the molecular nooks and crannies of any surface; this provides intimate contact and large surface area. The result is strong adhesion. Studies of the mechanics of adhesion of a single seta showed that it binds with a force of 200 μN. The theoretical maximum force that a single Tokay gecko with its 6.5 million setae can apply is on the order of 130 kg. In reality, not all of the setae can adhere simultaneously, so the practical maximum force is considerably lower [63].

13.5.3 Attachment and Release of Grip

How do the geckos apply and release their grip? The ability of the gecko to attach and detach its foot when it wants to is related to the manner in which the lizard applies its foot to a surface. The gecko places its toes on to a surface by first placing the base of the toe (the proximal end) to the surface. It then rolls the toe forward, towards the far (distal) end, incorporating more surface area, until the entire toe pad is in contact with the substrate. This begins adhesion, but the maximal force is reached when the toe is pulled slightly backward, about 5 μm (a force is applied parallel to the substrate surface). The movement changes the angle of the seta to the surface to below criticality, which in this case is about 30°. This apparently increases the contact area of the spatulae and surface to the practical maximum [64].

To release, the sequence of events is reversed. The distal end of the toe lifts and curls upward, increasing the contact angle to above 30°, at which point the spatulae detach. The ability of the gecko to rapidly move (up to 1 m sec^{-1}) across the surface is therefore a combination of the morphology of the foot, cycles of toe and foot movements, and the nanotopology of the scansors.

13.5.4 Self-Cleaning

Another interesting property of the gecko foot is its ability to stay clean even as it sticks to dirty walls. It appears to exhibit differential or competitive adhesion to dirt compared to a surface. Hansen and Autumn dusted the feet of the Tokay gecko with ceramic microspheres. These spheres essentially attenuated the van der Waals forces normally present between the spatulae and the wall. After the gecko had been walking over the surface for a while the spheres were shed. This suggested that the spheres were more likely to stick to the wall than the feet. The investigators have analyzed the interaction of the spatulae and the microspheres and concluded that about 25–60 spatulae were needed for the feet to cling to the spheres. Electronmicrography showed that far fewer fibers on average adhered to the spheres and the spheres were more likely to adhere to the wall. This is a form of self-cleaning that is unlike that of the lotus flower, but is nonetheless a result of the topography of the surfaces at the molecular level.

13.6 MORE NATURAL FIBERS

13.6.1 Spider Silk

One of the more interesting examples of naturally engineered biomaterial is the spider's web that serves a vital role in the survival of its engineer. Its main function is predation; prey items are caught in the sticky component of the web and available to the spider for a leisurely meal. The web itself needs to survive a variety of assaults. It must withstand the constant buffeting by the elements and the force of impact of the insect. It is able to stretch and not break, but also exhibits great strength. Spider silks are comparable in strength to the best synthetic materials made by man. Depending on how the measures are taken (or whom you talk to), the strength of spider silk is some five times that of steel of the same weight. Also, one author noted that a 450-g strand of spider silk can stretch around the world. What is it about the structure and composition of the silk that allows it to exhibit such remarkable properties?

When discussing the qualities of spider silk (or any fibrous structure) it is necessary to evaluate how well that material can withstand stresses. Young's modulus of elasticity is a quantitative measure of that characteristic. It is defined as the amount of stretch a material can exhibit for a given stress or a measure of the stiffness of the material. Simply, it is the ratio between the stress applied and the resulting strain exhibited by the material. The stress is defined as the force (F) applied over the cross-sectional area (A) of the material and has the dimensions of pressure, F/A or newtons per square meter. This is called a pascal (Pa, or megapascal, MPa). The strain is the change in length over the original length. The value is dimensionless and is represented as a percentage, $\Delta L/L_{original}$. The slope of the curve is a measure of the stiffness of the material: the steeper the slope is, the stiffer the material is. As the stress increases, the material will stretch until it breaks; this is the maximum height of the curve and is called the tensile strength of the material. The area under that curve is the total energy required to break the material, as force × distance = work (energy). The curve shown here is linear. Most materials exhibit a nonlinear curve, which is characteristic of a particular material. Generally, Young's modulus varies over the course of the applied stress (**Fig. 13.17**).

Spider Silk Structure. The major superclass protein of spider silk is fibroin. From analysis of the gene and amino acid sequences the structure of the fibroin has been well characterized. Depending upon the degree of hydration and the tension on the fiber, the fibroin chains can be loosely folded (**Fig. 13.18a**) or aligned in a quasi-crystalline matrix (**Fig. 13.18b**). The primary amino acid sequences of the fibroin show several distinct repeated motifs. These motifs are responsible for the secondary structures found in the silk proteins. β-pleated sheet is a type of secondary structure formed by sequences of 6–12 alanine (Ala or polyAla) and glycine/alanine (Gly/Ala) residues. As the protein folds back on itself, each fold forms hydrogen bonds between the N–H and C=O of adjacent folds to create a structure called a β-pleated sheet [65].

Fibrous regions are formed by glycine-rich and glycine-/proline-repeat sequences. In some cases the amino acids hydrogen bond to form β-spiral and "nanospring" like structures. There is also some folding, although not to the extent seen in the β-pleated sheets. These regions are noncrystalloid, unlike the

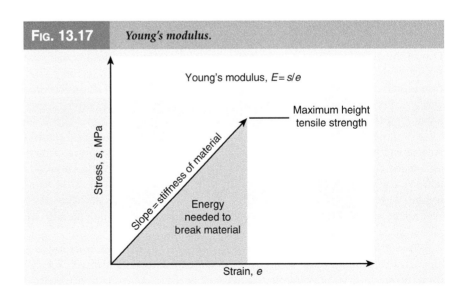

Fɪɢ. 13.17   *Young's modulus.*

FIG. 13.18 *Nanostructure of spider silk fiber.*

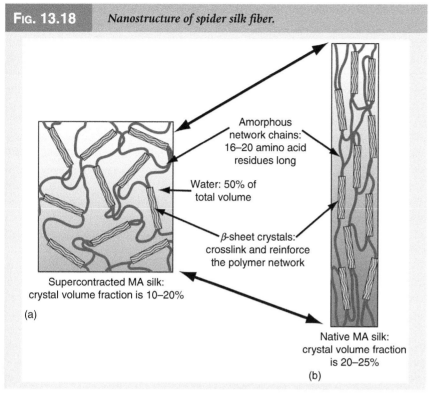

Amorphous
network chains:
16–20 amino acid
residues long

Water: 50% of
total volume

β-sheet crystals:
crosslink and reinforce
the polymer network

Supercontracted MA silk:
crystal volume fraction is 10–20%

(a)

Native MA silk:
crystal volume fraction
is 20–25%

(b)

Source: Image courtesy of J. M. Gosline et al. (1999) *Journal Exp. Biol.*, 202, 3295–3303.

β-pleated sheet regions. As a result the fibers tend to be more elastic with lower tensile strength. Local β-pleated sheets can crystallize, forming stable regions that cross-link the extensible, noncrystalline fibers. This results in spider silks with both high stiffness and extensibility.

Changes in Young's modulus are not constant as a smoothly increasing force is applied to the silk fiber. Rather the slope of the curve can change abruptly, resulting in some discontinuities in the curve. These are called yield points. This is likely due to the sudden breaking of hydrogen bonds in both the fiber and β-pleated sheet regions of the protein, with the resulting change in the shape of the silk protein. The stress/strain curve behaves differently after this yield point when compared to the curve before, reflecting the change in the architecture of the silk fiber [66].

The silk is formed by polymerization of silk proteins. These proteins are stored in a liquid form in the glands located in the abdomen of the spider. Across the various arachnid species seven different gland and silk types have been identified. Not all species produce all seven of the silks. At most a spider may produce up to six types. The glands are connected via spinning ducts to one to four pairs of spinnerettes, which contain anywhere from 2 to 50,000 small tubes. The fibroin liquid is extruded through the tubes to produce the silk fibers.

In general, the concentrated liquid protein, called the spinning dope, is stored in the glands. The proteins in the dope are not aggregated into the typical fibrous spider silk, but are in a globular state. The material is quite viscous at this time, as high protein concentration (50% protein/water) causes the long polymers to become entangled. As the dope passes through the spinning duct, the pH of the dope drops from 6.9 to 6.3. This triggers a conformational change in the proteins and they start to untangle. The α-helical and β-pleated sheet structures develop as the side chains of the exposed amino acids form hydrogen bonds. The ejection of the dope through the duct increases shearing forces in the fluid, which align the long fibers, decrease the viscosity, and facilitate polymerization of the fibroin into spider silk [67].

Types of Spider Silk.   The preceding process is a generalized description of silk formation. It varies depending on the type of silk being made. The production of the varieties of spider silk is dependent not only on the protein structure of the silk, but also on how the silk was produced, the diet of the spider, and the environmental condiments at the time of spinning. There are several varieties of spider silk, each adapted to particular functions in various parts of the web (**Fig. 13.19**) [68].

Dragline silk contains two subclasses of fibroin called major ampullate spidroin 1 and 2 (MaSP1 and 2, >several 200–300 kDa) that are produced by the major ampullate gland. This type of silk is rich in PolyAla sequences and Gly/Ala repeats and has a high tensile strength and elasticity. This is the first silk the spider makes when building a web. The dragline is extruded, the wind catches it, and it attaches (by chance) to another anchor point. The spider then moves back and forth across the line and produces more silk to reinforce it. The dragline eventually becomes part of the web frame. The fiber needs to be tough as it serves to support the bulk of the web and provides safety lines for the spider.

Minor ampullate glands produce another subclass of fibroin called the minor ampullate spidroin 1 and 2 (MiSP 1 and 2). These polymerize to form smaller

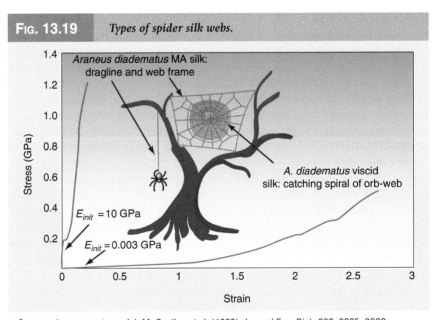

FIG. 13.19   *Types of spider silk webs.*

Araneus diadematus MA silk: dragline and web frame

A. diadematus viscid silk: catching spiral of orb-web

E_{init} = 10 GPa

E_{init} = 0.003 GPa

Stress (GPa)

Strain

Source: Image courtesy of J. M. Gosline et al. (1999) *Journal Exp. Biol.*, 202, 3295–3303.

TABLE 13.1	*Spider Gland and Silk Types*
Glands	**Silks**
Major ampullate	Dragline and web frame
Minor ampullate	Web radii and capture
Flagelliform	Spiral capture and viscid
Aggregate	Web glue
Pyriform	Web attachment discs
Aciniform	Dragline and wrapping
Coronatae	Adhesive threads

diameter fibers than dragline silk. Although these proteins show fewer β-pleated sheet structures, they nonetheless have a comparable tensile strength to dragline silk, but lower elasticity. This is probably due to differences in protein cross-linking. The silks are used to reinforce the web with radial fibers and are also involved in producing the capture spiral. The flagelliform, aggregate, and aciniform glands produce web silk protein variants involved in the production of capture silks, web glues, and cocoon wrapping, respectively (summarized in **Table 13.1**).

From any perspective, the spider silk is a remarkable material. It is clear that a synthetic and easily manufactured version of spider silk would have great commercial implications. Researchers have attempted to produce such items with limited success. The complex chemical structure and the many different environmental conditions under which the silk can be produced suggest synthesis will be a difficult task. What the spider does so easily is not so easily mimicked [69].

13.6.2 Sponge Fibers

Sponges are a primitive form of multicellular life. Their cells secrete an intercellular matrix that supports and holds the cells together. Some sponges secrete a skeletal matrix that is high in silica content. The secretion of silica grains or *phytoliths* is not uncommon in sponges, sea invertebrates, and some grasses, but the optical properties of the skeleton of the deep-sea sponge *Euplectella*—also known as the glass sponge—are unusual. This sponge grows long fibers in its exoskeleton that guide light in a manner similar to man-made fiber-optic cables. But these fibers, made up of three layers of material, are tougher than artificial glass optical fibers [65].

Joanna Aizenberg and colleagues at Bell Labs in the United States and Tel Aviv University in Israel have been studying the *Euplectella* sponge and have found that the skeleton of the sponge is constructed from amorphous hydrated silica. The fibers are made of a network of *spicules*, composed of silica spheres between 50 and 200 nm in diameter assembled together to yield the final 100-μm fiber. In optical properties, the sponge fibers resemble the fibers used in telecom networks, with a high-refractive-index core and low-index cladding. When Aizenberg coupled light into them, she found that they acted like single- or few-mode waveguides.

The natural fiber has advantages over its man-made counterparts. It is tougher than brittle commercial fiber. The layers contained within the spicules' structure are connected by organic molecules that provide an effective crack-arresting mechanism. The fibers can be bent to a much tighter radius than man-made fibers.

The *spicules* grow in an ambient environment, rather than the high-temperature furnaces needed to make optical fiber. The natural, bottom-up growth of the fibers allows for the structure to become precisely doped with impurities that improve the fiber's performance.

The sea sponge fibers are multimode when surrounded by air or seawater and single mode when embedded in a material. Multimode fiber can carry multiple light waves over shorter distances, and single-mode fiber can carry a single light wave over longer distances. The free-standing spicules are multimode because the refractive index contrast is greater between the spicule shell and air than between its core and shell. This allows light to fill the whole fiber rather than just the core.

The fibers have a complex structure that incorporates a hierarchal set of elements that provide strength similar to the trusses in bridges or high-rise buildings. They are being studied to learn principles of optical and structural design that may improve artificial fibers.

13.7 SUMMARY

In this brief overview of natural nanomaterials, some general principles of structural organization have been repeated in each type of material surveyed. A recurring theme is that natural materials are built from the bottom up, layer by layer, with incorporation of many heterogeneous atoms and molecules into the structure.

The bottom-up structures are not random, but highly ordered. They are, however, highly heterogeneous: typically, *fibrous polymers are embedded in a mixture of nonfibrous substances* to give a *composite* with superior properties to the single components. Where high strength and rigidity are required, we found *fiber-forming macromolecules*, aggregated into *ordered arrays*. Where elasticity and resilience are important, *highly branched polymers* such as *elastin* were found to meet the requirement.

In many natural materials, we found a *hierarchy of fiber assemblies* from the molecular level up to the macroscopic, with *interlocking linkages* to spread loads. In many different types of material, fibers are embedded in *layers with alternating orientations*. Another important theme is the importance of *calcium carbonates* and *phosphates* as components of skeletal structure as well as serving as buffers or sinks for carbon dioxide and regulators of cellular metabolism. A final important theme is the occurrence of sulfur linkages and silica compounds to meet unique optical and structural requirements.

References

1. J. K. Mitchell, *Fundamentals of soil behavior*, 3rd ed., John Wiley & Sons, New York, (2005).
2. K. Katti and D. Katti, Effect of clay–water interactions on swelling in montmorillonite clay. In *ASCE 16th engineering mechanics conference: Techniques for*

experimental analysis, instrumentation, and mechanics, University of Washington, Seattle, 1–9 (2003).

3. A. Viani, A. Gualtieri, and G. Artioli, The nature of disorder in montmorillonite by simulation of x-ray powder patterns, *American Mineralogist*, 87, 966–975 (2002); see web page at: webmineral.com/data/Montmorillonite.shtml

4. G. W. Beall and M. Goss, Self-assembly of organic molecules on montmorillonite, *Applied Clay Science*, 27, 179–186 (2004).

5. G. Ertem and J. P. Ferris, Template-directed synthesis using the neterogeneous templates produced by montmorillonite catalysis. A possible bridge between the prebiotic and RNA worlds, *Journal of the American Chemical Society*, 119, 7197–7201 (1997).

6. M. Franchi, E. Bramanti, L. M. Bonzi, P. L. Orioli, C. Vettori, and E. Gallori, Clay–nucleic acid complexes: Characteristics and implications for the preservation of genetic material in primeval habitats, Origins of Life and Evolution of the Biosphere, 29, 297–315 (1999).

7. J. Leng, S. U. Egelhaaf, and M. E. Cates, Kinetics of the micelle-to-vesicle transition: Aqueous lecithin–bile salt mixtures, *Biophysics Journal*, 85, 1624–1646 (2003).

8. M. M. Hanczyc, S. M. Fujikawa, and J. W. Szostak, Experimental models of primitive cellular compartments: Encapsulation, growth, and division, *Science*, 302, 618–622 (2003).

9. M. M. Hanczyc and J. W. Szostak, Replicating vesicles as models of primitive cell growth and division, *Current Opinions in Chemical Biology*, 8, 660–664 (2004).

10. L. E. Orgel, Prebiotic chemistry and the origin of the RNA world, *Critical Reviews in Biochemistry and Molecular Biology*, 39, 99–123 (2004).

11. D. B. Carlisle and D. R. Braman, Nanometre-size diamonds in the Cretaceous—Tertiary boundary clay of Alberta, *Nature*, 352, 708–709 (1991).

12. R. Monastersky, Radioactive alchemy: Diamonds from coal, *Science News*, 149, 133 (1996).

13. J. E. P. Dahl et al., Isolation and structural proof of the large diamond molecule, cyclohexamantane ($C_{26}H_{30}$), *Angewandte Chemie International Edition*, 42, 2040–2044 (2003).

14. J. E. P. Dahl, S. G. Liu, and R. M. K. Carlson, Isolation and structure of higher diamondoids, nanometer-sized diamond molecules, *Science*, 299, 96–99 (2002).

15. R. Monastersky, Microscopic diamonds crack geologic mold, *Science News*, 148, 22 (1995).

16. N. V. Sobolev and V. S. Shatsky, Diamond inclusions in garnets from metamorphic rocks: A new environment for diamond formation, *Nature*, 343, 742–745 (1990).

17. L. F. Dobrzhinetskaya, E. A. Eide, R. B. Larsen, B. A. Sturt, R. G. Trønnes, D. C. Smith, W. R. Taylor, and T. V. Posukhova, Microdiamond in high-grade metamorphic rocks of the Western Gneiss region, Norway, *Geology*, 23, 597–600 (1995).

18. T. Henning, H. Mutschke, S. Schlemmer, and D. Gerlich, Nanoparticles in space and the laboratory, NASA Laboratory Astrophysics Workshop, held May 1–3 2002 at NASA Ames Research Center, Moffett Field, CA 94035-100. Edited by F. Salama, NASA Conference Proceedings NASA/CP-2002-21186, 175 (2002).

19. R. Har-El and M. L. Tanzer, Extracellular matrix 3: Evolution of the extracellular matrix in invertebrates, *FASEB Journal*, 7, 1115–1123 (1993).

20. D. J. Jackson, L. Macis, J. Reitner, B. M. Degnan, and G. Wörheide, Sponge paleogenomics reveals an ancient role for carbonic anhydrase in skeletogenesis, *Science*, 316, 1302 (2007).

21. S. Blank, M. Arnoldi, S. Khoshnavaz, L. Treccani, M. Kuntz, K. Mann, G. Grathwohl, and M. Fritz, The nacre protein perlucin nucleates growth of calcium carbonate crystals, *Journal of Microscopy*, 212, 280–291 (2003).

22. K. Mann, F. Siedler, L. Treccani, F. Heinemann, and M. Fritz, Perlinhibin, a cysteine-, histidine-, and arginine-rich miniprotein from abalone (*Haliotis laevigata*) nacre,

inhibits in vitro calcium carbonate crystallization, *Biophysics Journal*, 93, 1246–1254 (2007).

23. J. P. Reyes-Grajeda, A. Moreno, and A. Romero, Crystal structure of ovocleidin-17, a major protein of the calcified *Gallus gallus* eggshell: Implications in the calcite mineral growth pattern, *Journal of Biological Chemistry*, 279, 40876–40881 (2004).

24. D. R. Lide, *CRC handbook of chemistry and physics*, 88th ed., CRC Press, Boca Raton, FL (2007).

25. J. L. Katz, Mechanics of hard tissue. In *The biomedical engineering handbook*, 3rd ed., CRC Press, Boca Raton, FL, Section VI: Biomechanics, chap. 47 (2006).

26. J. D. Bronzino, ed., *The biomedical engineering handbook*, 3rd ed., CRC Press, Boca Raton, FL (2006).

27. Y. Bar-Cohen, *Biomimetics: Biologically inspired technologies*, CRC Press, Boca Raton, FL (2005).

28. T. Vo-Dinh, *Nanotechnology in biology and medicine: Methods, devices, and applications*, CRC Press, Boca Raton, FL (2007).

29. D. M. Skinner, S. S. Kumari, and J. J. O'Brien, Proteins of the crustacean exoskeleton, *American Zoologist*, 32, 470–484 (1992).

30. D. Raabe, C. Sachs, and P. Romano, The crustacean exoskeleton as an example of a structurally and mechanically graded biological nanocomposite material, *Acta Materialia*, 53, 4281–4292 (2005).

31. P. Romano, H. Fabritius, and D. Raabe, The exoskeleton of the lobster *Homarus americanus* as an example of a smart anisotropic biological material, *Acta Biomaterialia*, 3, 301–309 (2007).

32. J. J. O'Brien, S. Kumari, and D. M. Skinner, Proteins of crustacean exoskeletons: I. Similarities and differences among proteins of the four exoskeletal layers of four brachyurans, *Biology Bulletin*, 181, 427–441 (1991).

33. S. F. Tarsitano, K. L. Lavalli, F. Horne, and E. Spanier, The constructional properties of the exoskeleton of homarid, palinurid, and scyllarid lobsters, *Hydrobiologia*, 557, 9–20 (2005).

34. R. E. Burgeson, New collagens, new concepts, *Annual Review of Cell Biology*, 4, 551–577 (1988).

35. M. J. Buehler, Nature designs tough collagen: Explaining the nanostructure of collagen fibrils, *PNAS USA*, 103, 12285–12290 (2006).

36. M. J. Buehler and S. Y. Wong, Entropic elasticity controls nanomechanics of single tropocollagen molecules, *Biophysics Journal*, 93, 37–43 (2007).

37. B. K. Hall, Consideration of the neural crest and its skeletal derivatives in the context of novelty/innovation, *Journal of Experimental Zoology* (Mol. Dev. Evol.), 304B, 548–557 (2005).

38. D. Toroian, J. E. Lim, and P. A. Price, The size exclusion characteristics of type I collagen, *Journal of Biological Chemistry*, 282, 22437–22447 (2007).

39. H. D. Wagner and S. Weiner, On the relationship between the microstructure of bone and its mechanical stiffness, *Journal of Biomechanics*, 25, 1311–1320 (1992).

40. J. M. Crolet, M. Racila, R. Mahraoui, and A. Meunier, A new numerical concept for modeling hydroxyapatite in human cortical bone, *Computer Methods in Biomechanics and Biomedical Engineering*, 8, 139–143 (2005).

41. Y. Kajiuraa, S. Watanabea, T. Itoua, K. Nakamuraa, A. Iidab, K. Inouec, N. Yagic, Y. Shinoharad, and Y. Amemiya, Structural analysis of human hair single fibres by scanning microbeam SAXS, *Journal of Structural Biology*, 155, 438–444 (2006).

42. W. T. Astbury and H. J. Woods, The x-ray interpretation of the structure and elastic properties of hair keratin, *Nature*, 126, 913–914 (1930).

43. H. Liu and W. Yu, Microstructural transformation of wool during stretching with tensile curves, *Journal of Applied Polymer Science*, 104, 816–822 (2007).

44. See Websites for Nano Pulp and Paper Llc, http://www.nanopulpandpaper.com/, and Better Paper Technologies Llc, http://www.bptech.cc/

45. M. Koepenick, Paper innovation ramps up with nanochemistry, *Pulp and Paper Canada*, 102, 17–19 (2001).

46. J. M. Maxwell, S. G. Gordon, and M. G. Huson, Internal structure of mature and immature cotton fibers revealed by scanning probe microscopy, *Textile Research Journal*, 73, 1005–1012 (2003).

47. A. A. Baker, W. Helbert, J. Sugiyama, and M. J. Miles, Surface structure of native cellulose microcrystals by AFM, *Applied Physics*, A 66, S559–S563 (1998).

48. R. M. Brown, Jr., Algae as tools in studying the biosynthesis of cellulose, nature's most abundant macromolecule. In *Experimental phycology. Cell walls and surfaces, reproduction, photosynthesis*, W. Wiessner, D. G. Robinson, and R. C. Starr, eds., pp. 20–39, Springer-Verlag, Berlin (1990).

49. M. Tabuchi, Nanobiotech versus synthetic nanotech? *Nature Biotechnology*, 25, 389–390 (2007).

50. M. Tabuchi, F. Tomita, K. Kobayashi, S. Miki, T. Saijo, A. Shibata, and Y. Baba, Self-regulated bionanostructured membranes for MEMS, MEMS 2006 Istanbul, *19th IEEE International Conference on Micro Electro Mechanical Systems*, 514–517 (2006).

51. J. Bradbury, Nature's nanotechnologists: Unveiling the secrets of diatoms, *PLoS Biology*, 2, e306 (2004).

52. N. D. Crosbie and M. J. Furnas, Picocyanobacteria and nano-/microphytoplankton growth, *Aquatic Microbial Ecology*, 24, 209–224 (2001).

53. T. Dugdale, R. Dagastine, T. Chiovitti, P. Mulvaney, and R. Wetherbee, Single adhesive nano-fibers from a live diatom have the signature fingerprint of modular proteins, *Biophysics Journal*, 89, 4252–4260 (2005).

54. A. Summers, Biomechanics: Secrets of the sacred lotus, *Natural History*, 406, www.naturalhistorymag.com/master.html?http://www.naturalhistorymag.com/0406/0406_biomechanics.html (2006).

55. Nelumbo nucifera, Pacific Island Ecosystems at Risk (PIER) www.hear.org/pier/species/nelumbo_nucifera.htm (2007).

56. A. Agrawal, Wettability, nonwettability and contact angle hysteresis, MIT, Non-Newtonian Fluid Dynamics Research Group: web.mit.edu/nnf/education/wettability/wetting.html (2004).

57. W. Barthlott and C. Neinhuis, Purity of the sacred lotus, or escape from contamination in biological surfaces, *Planta*, 202, 1–8 (1997).

58. Z. Shen and Z. M. Jacobs-Lorena, Evolution of chitin-binding proteins in invertebrates, *Journal of Molecular Evolution*, 48, 341–347 (1999).

59. G. S. Watson and J. A. Watson, Natural nano-structures on insects—Possible functions of ordered arrays characterized by atomic force microscopy, *Applied Surface Science*, 235, 139–144 (2004) (8th European Vacuum Conference and 2nd Annual Conference of the German Vacuum Society).

60. A. R. Parker and H. E. Townley, Biomimetics of photonic nanostructures, *Nature Nanotechnology*, 2, 347–353 (2007).

61. K. Autumn, Properties, principles, and parameters of the gecko adhesive system. In *Biological adhesives*, A. Smith and J. Callow, eds., pp. 225–255, Springer-Verlag, Berlin, (2006).

62. K. Autumn, K., M. Sitti, A. Peattie, W. Hansen, S. Sponberg, Y. A. Liang, T. Kenny, R. Fearing, J. Israelachvili, and R. J. Full, Evidence for van der Waals adhesion in gecko setae, PNAS USA, 99, 12252–12256 (2002).

63. K. Autumn, A. Dittmore, D. Santos, M. Spenko, and M. Cutkosky, Frictional adhesion: A new angle on gecko attachment, *Journal of Experimental Biology*, 209, 3569–3579 (2006).

64. K. Autumn, Y. A. Liang, S. T. Hsieh, W. Zesch, W.-P. Chan, W. T. Kenny, R. Fearing, and R. J. Full, Adhesive force of a single gecko foot-hair, *Nature*, 405, 681–685 (2000).

65. J. M. Gosline, P. A. Guerette, C. S. Ortlepp, and K. N. Savage, The structure of spider silk, *Journal of Experimental Biology*, 202, 3295–3303 (1999).

66. V. Sundar, A. D. Yablon, J. L. Grazul, J. Aizenberg, and M. Ilan, Fibre-optical features of a glass sponge, *Nature*, 424, 899 (2003).

67. M. Beals, L. Gross, and S. Harrell, *Spider silk: Stress–strain curves and Young's modulus*, www.tiem.utk.edu/~mbeals/spider.html (1999).

68. X. Hu, K. Vasanthavada, K. Kohler, S. McNary, A. M. F. Moore, and C. A. Vierra, Molecular mechanisms of spider silk, *Cellular and Molecular Life Sciences*, 63, 1986–1999 (2006).

69. M. A. Garrido, M. Elices, C. Viney, and J. Pérez–Rigueiro, Active control of spider silk strength: Comparison of drag line spun on vertical and horizontal surfaces, *Polymer*, 43, 1537–1540 (2002).

Problems

13.1 The setae of geckos adhere to hydrophobic as well as hydrophilic surfaces by van der Waals forces. This implies that the size and shape of the tips of the setae are responsible for adhesion and chemistry—a truly nano phenomenon! Assume that each seta is capable of 200 µN of adhesion force and that there are 14,400 setae per square millimeter. What area of setae is required to lift a 180-lb human up from the floor?

13.2 Each seta consists of approximately 500 spatulae—the actual nanostructured material. How many spatulae are there per square millimeter? What is the average force load capability per spatulae?

If you were to use carbon nanotubes instead of spatulae, what would be the density per square millimeter (assume SWNTs with diameter = 1 nm)? If W (adhesion energy for van der Waals surfaces) equals 50 mJ·m^{-2}, what is the lifting capability of such a synthetic adhesive?

13.3 In nature, as body mass is increased, not only does the number of setae increase, but also the structural components responsible for adhesion (the spatulae) get smaller (reduced diameter). This is evidenced in the progression in biological mass from a small beetle to a fly to a spider to a gecko. (H. Gao and H. Yao, Proceedings of the National Academy of Science, 101, 7851–7856, 2004.) Why do you think this trend occurs as it does?

13.4 The average mass of a diatom is approximately 60 pg. What is the maximum and minimum diatom biomass in a liter of ocean water, based on the population densities cited in this chapter?

13.5 Look up the properties of silkworm silk and compare them to those of spider silk. What are some of the difficulties in artificially making a fiber that compares to spider silk? Discuss and support your answers with references.

13.6 Would you expect a calcite crystal to be isotropic in Young's modulus? Why? Look up the values for Young's modulus for calcite, argonite, and rock salt (NaCl) and explain in terms of the crystal structure.

13.7 Go to the following Web pages for the interesting exercises using CD diffraction gratings to determine wavelength of incident light or measure the groove spacing:
<http://maxwell.physics.mun.ca/mpl/Physics1051_MM_F05/lab6.pdf>
<http://chem.lapeer.org/PhysicsDocs/GroovyCD.html>

13.8 Aragonite is a polymorph of calcite, which means that it has the same chemistry as calcite but it has a different structure and, more importantly, different symmetry and crystal shapes. Aragonite's more compact structure is composed of triangular carbonate ion groups (CO_3), with a carbon at the center of the triangle and the three oxygens at each corner. Unlike in calcite, the carbonate ions do not lie in a single plane pointing in the same direction. Instead, they lie in two planes that point in opposite directions, thus destroying the trigonal symmetry that is characteristic of calcite's structure. To illustrate the symmetries of calcite and aragonite, draw or construct an equilateral triangle, a threefold rotation with three mirror planes that cross in the center. Now join two of these triangles together at their bases and you have a diamond-shaped figure with the symmetry of a twofold rotation with one mirror plane in the middle. This is the effect of the two carbonate planes with opposite orientations on the symmetry of this structure. Aragonite has an orthorhombic symmetry (2/m 2/m 2/m) instead of calcite's "higher" trigonal (bar 3 2/m) symmetry. A very rare mineral called vaterite is also a polymorph of aragonite and calcite—making them all trimorphs. Vaterite has a hexagonal symmetry (6/m 2/m 2/m). Draw or make models of each of these crystals. Draw or model the smallest unit from which you can construct the complete crystal in each case.

13.9 The sodium ion has a radius of ___. The potassium ion radius is ___, calcium ___, magnesium ___, and barium ___. The bond length of the Al–O bond in an Al–O–Al linkage is ___. The bond length of the Si–O–Si bond is ___. The bond length of the Al–O–SI bonds is ___. What size zeolite unit ring would be needed to give a pore size of at least 0.4 nm with each cation? Hint: Look up bond lengths in CRC Handbook of Chemistry and Physics.

13.10 An interesting set of exercises for learning about biofilms, titled "A Manual of Biofilm-Related Exercises," is available from the American Society of Microbiology at: <http://www.personal.psu.edu/faculty/j/e/jel5/biofilms/>.

13.11 What is the difference between single- and multimode optical fiber conductance of light?

BIOMOLECULAR NANOSCIENCE

Some have said to me that "sequencing the human genome will diminish humanity by taking the mystery out of life." Nothing could be further from the truth. The complexities and wonder of how the inanimate chemicals that are our genetic code give rise to the imponderables of the human spirit should keep poets and philosophers inspired for millennia.

J. CRAIG VENTER, PH.D., CEO
Celera Genomics, Inc.

*C*hapter 14

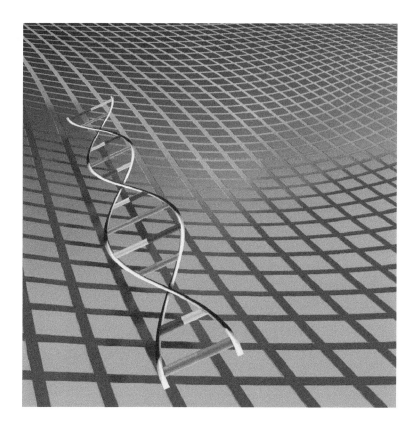

FIG. 14.0	*DNA strands.*

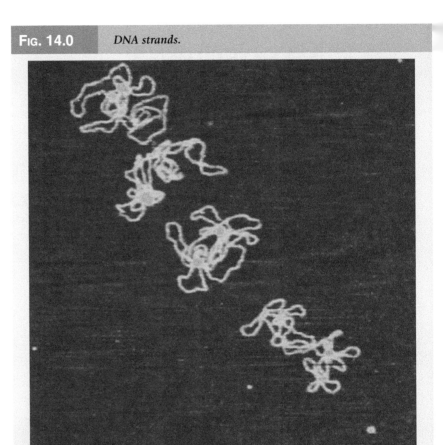

Source: Image courtesy of Veeco Instruments, Inc.

THREADS

Chapter 14 presents a short course in biochemistry and cellular biology, with emphasis on the nano perspective. Most physicists and engineers never took a formal course in biology. Because nanotechnology is highly interdisciplinary, knowledge of biochemical and cellular processes is essential. This chapter will hopefully bridge that gap. Chapter 13 introduced natural nanomaterials. Before that we studied the chemistry and physics of nanomaterials. With all that now in perspective, we look at living things and what makes them tick in a different light—emanating from the nanoscale.

This chapter introduces the organization of life as related to nanoscale properties of matter. From the intramolecular forces in water to the self-organizing forces of giant macromolecules such as proteins, life depends upon and is an emergent property of the energetics of matter on the nanoscale. Understanding how nanoscale forces orchestrate life will be exciting work for future scientists. How biological systems operate at the nanoscale and how nanoscale machinery integrates into larger systems are an important part of the broader discipline of nanoscience. Biological nanosystems, the product of millions of years of evolutionary development, will reward the nanoscientist with rich and intriguing examples of nanostructures, nanomachines, and principles of operation for nanosystems.

The nanoscience of living systems displays finely honed design, efficiency of material and energetics,

THREADS (CONTD.)

stability, and tolerance of errors and interferences that are the product of billions of years of exploration of the possibilities of structure and engine building on the molecular and cellular scale. The natural nanomachinery of living organisms has been rigorously optimized through natural experimentation and competitive selection to evolve into the elegant and finely tuned systems that our science has only begun to understand. Advances in understanding the basis of living systems will in turn be greatly empowered by the current progress in the fields of nanoscience and nanotechnology. Exposure to the challenge of biosystems on the nanoscale is part of a multidisciplinary approach that will inspire lateral thinking in approaching problems on the nanoscale. This is the motivation for a detailed component in the basics of life science on the nanoscale as a part of this treatise.

14.0 INTRODUCTION TO BIOMOLECULAR NANOSCIENCE

Life can be considered to be nanoscience in the wild native state. Living organisms are proof that self-replicating molecular nanoengines exist and function, performing work; building material structures of great strength and subtlety; reproducing and repairing themselves; and processing information to identify and control sources of materials and energy, ward off threats, and carry out survival strategies. In this chapter we will explore some of the nanoscale phenomena that underlie life processes.

14.0.1 Definitions: Biomolecular Nanoscience

Biomolecular nanoscience is the study of living systems on the nanoscale level. Biomolecular nanoscience studies how phenomena at the nanoscale underlie higher levels of organization in living things. Biomolecular nanoscience is an integrated transdisciplinary approach to the science of life. Biomolecular nanoscience includes aspects of biophysics, molecular biology, protein structure, and biochemistry; it aims to unify the branches of physics, biology, and biochemistry that deal with phenomena at the nanoscale and large-scale implications.

The large-scale aspects of life include how a person, an organism, or an organ is able to perform its functions without failure in the face of external threats and internal complications. These functions include metabolism, growth, development, reproduction, heredity, evolution of species, and intelligent behavior. Biomolecular nanotechnology provides a foundation upon which such phenomena can be related to underlying forces, structures, and organizations.

Research in nanoscience has opened new windows into the causes and mechanisms of previously mysterious biological phenomena, as shown by examples in the previous chapter such as the role of clay in catalyzing the synthesis of biomolecules, the relation of bone strength to nanostructure, the coloring of butterfly wings, and adhesion of gecko feet. There is every reason to expect that further investigations into biomolecular nanoscience will yield even more important insights.

14.0.2 *Historical Origins*

Modern nanoscience arose from a bottom-up approach, beginning with the physical aspects of manipulating atoms, as well as a top-down approach, working from the analysis of existing natural structures and substances in finer and finer detail down to the nano level. Biomolecular nanoscience is top down in its historical development, since whole organisms were studied long before their small components.

The story of how scientists and doctors, working over many centuries, explored ever finer details down to the nanoscale is a fascinating one. Since the first microscopes, there has been a drive to examine and study biological structures at an ever smaller level, seeking to understand the structure and basis of life. The earliest application of microscopy was the observation of microbiological specimens. Physical phenomena were applied to early experiments with life (e.g., the understanding of electricity was advanced by Galvani's study of frog legs). The study of x-rays was linked to their application for anatomical imaging from their first discovery. The study of proteins and molecular biology by physical chemists using x-ray diffraction laid the basis for understanding how proteins act as nanomachines. And now it is no wonder that nanoscience also includes life and living things.

Biochemistry in the nineteenth century made important progress in analyzing small molecules and elucidating important energy and material cycles in living systems. However, traditional chemistry dealt only with freely mixing combinations of molecules in bulk. In the highly structured organization of matter that exists in cells and at their boundaries, models of molecular interaction based on freely circulating and randomly mixed chemical molecules are no longer an adequate description. Great progress was made in the twentieth century in understanding macromolecules, catalysts, and the orchestration of molecular processing in the cell. The physical chemistry of catalysts, surface reactions, and polymers was useful in describing parts of the picture, but the biological environment proved even more complex—for example, in the processes by which proteins orchestrate cell division. In the twenty-first century, biomolecular nanotechnology develops on the foundations of the traditional sciences of biophysics, biochemistry, and molecular and cellular biology to meet the challenges of describing and understanding the nanoscale behind the life.

14.0.3 *Biomolecular Nanoscience: Roots in Traditional Science*

Nanoscience depends on traditional sciences of physics, biology, and chemistry. In this section we give an historical overview leading from bulk properties of matter and whole organisms to the understanding of matter on the nanoscale, as applied to biomolecular science.

Physics.   The "integration of everything" allows us to move from one discipline to another. Understanding light and electricity from the perspective of physics enabled the application of spectroscopy chemistry and its insights into the structure of atoms and molecules. In the same way, understanding of how properties change with interactions at the nanoscale is opening surprising new insights and applications.

Chemistry.   Classical chemistry, with strong input from physics, had begun to unravel matter at the molecular scale in the nineteenth century. By the middle of the twentieth century, chemists manipulated matter on the atomic scale, but largely only in bulk phase. By extremely subtle and intricate methods, chemists influenced the molecular structure by combining external influences such as heat, light, and electricity. By introducing specific reagents together in controlled conditions, they produced interactions that changed the organization of bonds and geometric connections (or stereochemistry) of molecules. This was accomplished in traditional chemistry with multiple trillions of molecules, all reacting at the same time and in the same way in a test tube with external control of bulk reaction conditions such as temperature and pressure.

When Smalley, Kroto, Curl, and coworkers discovered the structure of carbon buckyballs in the 1980s, they helped shift perspectives towards molecules as engineered structures that are produced directly in high yields by electronic vacuum processes and other engineering approaches. This viewpoint began to supplement and even supplant the predominant attitude that complex chemical molecules were destined to be the product of elegant, elaborate, carefully planned synthetic reactions involving many steps and much laboratory time by highly skilled chemists with intimate knowledge and subtle repertoires of reactions and source materials. This trend was already underway with polymer, catalyst, and ion complex ligand chemistry, including progress in the synthesis of conductive polymers.

In the meantime, life scientists and molecular biologists were working out how cell membranes, DNA, RNA, and protein structures brought about carefully controlled chemical and physical interactions in the cell. Biochemists and biotechnologists were thinking of the molecular components of cells as machinery performing tasks on the molecular level. Protein and membrane structures became as much a focus as enzyme and metabolite kinetics [1–6].

Molecular Biology.   Armed with the concepts of chemistry, biochemists and molecular biologists began thinking of living structures as networks of chemical reactions [7,8]. With the discovery of DNA and the mechanism of transcription by Watson and Crick (and Franklin), foreshadowed by the discovery of helical protein structures by Pauling, molecular biologists began to think in terms of three-dimensional machinery that operates on the subcellular level [9–11]. In the meantime, the mechanism of cell-wall penetration by bacteriophages was elucidated, along with rotary flagellation engines propelling cilia, electrochemical and stereochemical mechanisms operated gates in cell membranes, DNA replicating and repairing machinery, cell division machinery, and the whole mechanism of life based on molecular mass transport and information exchange.

Medical scientists such as A. Gilman, J. P. Nolan, and Leonard A. Herzenberg began to map the complex network of molecular signaling pathways that control the lives of cells. In the process, they called on the work of physicists and engineers to develop new tools and techniques such as flow cytometry to separate cells based on differences in genome expression [12–15], statistical sampling microscopy to resolve details below the limits of optical microscopes [16], and statistical analysis of complex networks with many cross-talk interactions to identify relationships between pathways [17]. The knowledge of molecular control pathways casts new light on the understanding of evolution and development, including insights into physical and chemical constraints that limit and define what would otherwise be thought to be random selection of evolutionary pathways [18,19].

Engineering and Biotechnology.   A number of initiatives accelerated the energetic exploration of nanobiology and nanomedicine [20,21]. Biological and microelectronic materials continued to be integrated into biochips and biosensors [22–27]. Groups at Los Alamos and Sandia National Laboratories made progress towards self-assembly of nanostructures to contain cells; the goal was to create a nanostructured environment in which cells would remain viable without external buffers or fluidic architecture. This work opened the proverbial door for cell-based sensors based on immobilized cells [28]. Protective implant capsules were synthesized for transplanted cells to deliver insulin or other complex substrates. Pores in the capsule are small enough to allow delivery of insulin while not the entry of rejection antibodies [29]. Other researchers developed ways to harness the replication machinery of DNA and synthesize useful proteins in bulk for medical applications. To obtain useful throughput quantity, the proteins are then purified robotically followed by automated analysis [30]. DNA of higher animals can also be modified; for example, human antibodies are being produced in the milk of genetically modified goats and cows [31,32].

DNA protein replicating machinery operating outside the cell has also been developed. This simplified the process of separating and purifying end products and opened the way for bulk production of proteins similar to conventional chemical processes. For example, at the Harvard Institute of Proteomics, at the company DNANO, cell-free extracts containing DNA, RNA, and substrates are currently being employed to synthesize proteins for biochemical and immunological products [33].

14.0.4   *The Nano Perspective*

In this section we see how the advancement of nanoscience benefited from insights and techniques shared with physics, chemistry, and cellular and molecular biology. Biomolecular nanoscience was inspired by naturally occurring molecular machinery in the cell, and molecular and genetic biologists found it increasingly useful to employ concepts and systematic analysis of molecular machinery in common with nanoscience. Thus, biomolecular nanoscience can be considered as the pursuit of nanoscience with relevance and application to biological science.

Biomolecular nanoscience draws upon engineering concepts and methods to contribute new models for understanding life based on complex chemical, physical, and structural machinery, with subtle control systems that involve detailed and highly specific interactions at the molecular level. This interdisciplinary approach, *biomolecular nanoscience*, provides a perspective from which we model and understand living processes at the level of molecular machinery.

14.1   MATERIAL BASIS OF LIFE

In this section we give an overview of the development of tools and methods for studying matter, starting historically with bulk properties of matter and whole organisms, and leading to matter on the nanoscale, as applied to biomolecular science.

Biomimetics and Bionanotechnology.   Biomimicry is defined as the study of forms and functions in nature in order to apply them to engineering and manufacturing practice. The imitation of nature is as old as invention itself, but the understanding of biological systems and materials on the molecular scale has opened an immensely rich resource for designers and engineers [34]. Lessons learned from biology are being applied to nanotechnology and, in turn, are applied to biotechnology [35]. A compelling counterexample comes to mind: the elucidation of the structure of carbon nanospheres by the engineering designs of Buckminster Fuller led researchers to the discovery of the correct structure, at least in part, based on dodecahedron geometry [36].

To understand the domain of biomolecular nanoscience and to be able to use its tools, it is important to have a grasp of what is being measured, manipulated, and constructed and to have some appreciation for the marvelous and complex natural structures of life. This is true for the individual building blocks as well as the network of their interactions. These components and network pathways are the guiding architecture of cells and organisms for their development and metabolic life cycles.

14.1.1   *Molecular Building Blocks—From the Bottom Up*

In the following section we start with the smallest building blocks of matter that play a direct role in living systems and describe how they are combined into progressively larger components, up to the level of the living biological cell. We introduce and define some fundamental terms, such as DNA and RNA—important building blocks and mechanisms that we explore in more detail in later sections. The chemical building blocks that are significant in life processes, leaving aside the important aspect of electron and charge transfer, are the atomic species: C, H, O, N, P, and S—all elements that are coincidentally important to organic chemists. The simplest molecules of greatest importance are H_2O, CO_2, N_2, and O_2.

Water.   Although we have discussed the importance of *water*, its structure, and chemical properties in previous sections, we reinforce its significance. Its ability to make hydrogen bonds (**Fig. 14.1**); dissociate into OH^- and H^+ (pH); and solvate essential minerals like Na^+, Cl^-, Ca^{+2}, K^+, Mg^{2+}, Fe^{3+}, Fe^{2+}, I^-, and numerous other metal cationic and anionic species cannot be understood. The interaction

FIG. 14.1

Water structure—hydrogen bonds are continuously forming and rearranging. The two hydrogen and one oxygen atoms that make up the water molecule can associate into aggregates with other water molecules by forming hydrogen bonds. The hydrogen bonds between water molecules lead to exclusion of the hydrophobic organic portions of large biological molecules such as lipids and proteins. This exclusionary force drives the formation of micelles and double layer membranes. It is the main force determining the folding and shaping of proteins. Water molecules solvate with any hydrophilic portion of a protein or fatty acid through hydrogen bonds. The hydrophobic portion of the molecule is squeezed into a shape determined by the attraction of water for the hydrophilic portions.

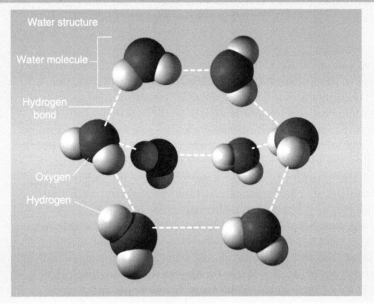

Water structure

Water molecule

Hydrogen bond

Oxygen

Hydrogen

Source: Image by Anil Rao.

of hydrophilic and hydrophobic species by water forms the basis for structural order in cellular systems. It is no wonder that NASA is always looking for water on other planets.

Elements.   *Carbon* is equally essential. The ability of carbon to form bonds with itself, hydrogen, oxygen, nitrogen, sulfur, and phosphate groups makes it the key building unit for life. The presence of heterocyclic atoms (nitrogen, oxygen, sulfur) in aromatic rings alters the electronic properties of the electron bond system and forms the basis for the macrocycle effect discussed in chapters 10 and 11. *Oxygen* is a key element that, in combination with hydrogen, gives rise to the unique properties of water; oxygen also modifies the properties of carbon compounds. Oxygen serves as a source of unpaired electrons that are available for dative binding of metals and is the central ingredient in aerobic respiration. *Nitrogen* is found in nucleotide purine and pyrimidine bases such as adenine, guanine, thymine, and cytosine (and uracil). It is a participant in the formation of peptide bonds in proteins. *Phosphorous* is a component of the backbone of DNA and RNA and of ADP and ATP (adenosine di- and triphosphate). *Sulfur* plays a major role in protein folding—a process that affords proteins their functionality.

Biological Molecules. Life depends on a variety of organic and inorganic compounds: food (metabolites including sugars, starches, fats, vitamins, and minerals); sensing and signaling materials (hormones, neurotransmitters, enzymes, regulator compounds such as nitric oxide); structural components (collagen fibers, shells, bones); chemical transport structures (hemoglobin); digestive entities (enzymes, stomach acids), and energy catalysts (ATP, chlorophyll). In **Table 14.1** we review a few essential biological molecules and their roles.

Amino Acids. A set of extremely important building blocks for life are the *amino acids*—organic compounds of carbon, hydrogen, oxygen, and nitrogen (see **Table 14.1**). The carboxylic oxygen component possesses acidic hydrogens and the amine nitrogen component is basic: Amino acids are *zwitterions* and, given the right solvent conditions, a positive charge from the acid part of the molecule can transfer to the base portion, leaving the two parts of the same molecule with opposite charges.

Amino acids in which the acidic carboxylic group and the basic amine group are attached to the same carbon atom are called *alpha amino acids*. Opposite electron affinities pulling on the same carbon make these alpha carbon atoms hungry for electrons, and thus they are very reactive. Alpha carbons readily form bonds to electron-rich nitrogen atoms in amino groups, linking amino acids together to form long chains.

In these chains, the amine group on one molecule links to the alpha carbon of a neighbor; in biochemistry this link is termed as a *peptide bond*. Peptide linkages are covalent metastable chemical bonds that are strong enough to make stable chains, but weak enough to be broken by catalysts in water so that chains can be rearranged under mild conditions. Out of all amino acids that are chemically possible, only alpha amino acids, with the carboxyl and amino groups located on the same carbon at the end of the molecule, form peptide linkages found in living organisms. Twenty specific alpha amino acids are the building blocks of proteins. Amino acids are precursor molecules with subnanometer dimensions. When they are assembled to form a peptide or a protein, at that stage, they become nanomaterials.

Proteins. Proteins are the primary building blocks for the molecular nanostructures and nanomachines that perform the work of cellular life and are bona fide nanomaterials. Proteins are long organic chains of chemically linked amino acid molecules (polyamides). The connecting links of the chain are the peptide bonds between alpha carbons and nitrogen atoms in an amine group. The backbone of the chain thus consists of carbon and nitrogen with bonds to hydrogen, oxygen, and organic groups and sometimes to phosphorous and sulfur. The minimum chain size for a protein to have a structural biological function seems to be about 40 amino acid groups. Shorter chains are referred to as peptides or polypeptides.

Peptides interact with proteins to serve as messengers and perform other roles in cells. The protein chains can be several thousand amino acid units in length that fold onto themselves to form stable shapes held together by intermolecular

TABLE 14.1   *Biological Molecules*

Biomolecule	Structure	Comments
Amino acids	Glycine $H_2N-CH\overset{\overset{O}{\|\|}}{C}-OH$ with H below Tryptophan $H_2N-CH\overset{\overset{O}{\|\|}}{C}-OH$ with CH_2 and indole ring	Glycine is the smallest amino acid: <0.5 nm, MW = 75 Tryptophan is the largest natural amino acid in proteins; <0.7 nm, MW = 246 All amino acids that make up life have the following basic structure: The **R** represents different groups in each amino acid; the amine group (red) and the carboxylic acid group (blue) are both attached to alpha carbon atoms that react with the nitrogen in other amino acid molecules to form peptide bonds. $H_2N-\overset{\overset{H}{\|}}{\underset{\underset{R}{\|}}{C}}-\overset{\overset{O}{\|\|}}{C}-OH$
Carbohydrates	α-Furanose (furanose ring structure) D-Glucose $OHC-\cdots-CH_2OH$ (with OH OH H OH above and H H OH H below)	Carbohydrates (from the Greek *carbo*, "sugar," and *hydra*, "water") are aldehydes and ketones with hydroxyl groups. Carbohydrates function in the storage and transport of energy (starch, glycogen) and in structure (polysaccharides, cellulose, chitin). Monosaccharides include glucose, fructose, ribose, and deoxyribose. The size of simple sugars is on the order of amino acids. Polysaccharides can get relatively large. Glycogen is ~150 nm.
Lipids, fatty acids, and phospholipids	Myristoleic acid $CH_3(CH_2)_3CH=CH(CH_2)_7COOH$ Oleic acid $CH_3(CH_2)_7CH=CH(CH_2)_7COOH$ Cardiolipin (Diphosphatidylglycerol) (phospholipid structure with OH, H_2C, CH, CH_2, O=P–OH, HO–P=O groups)	Many fatty acids are long-chain aliphatics with a carboxylic acid group at the terminus. Unsaturated fatty acids have no double bonds. Phospholipids contain a phosphate group and constitute a major component of cell membranes.
Nucleotide bases	Adenine guanine (adenine ring structure with NH_2) (guanine ring structure with O, NH, NH_2)	The four base nucleotides of DNA. The sizes and molecular weights are on the order of the amino acids. Thymine and adenine specifically complex together by hydrogen bonding, as do cytosine and guanine.

TABLE 14.1 (CONTD.)	Biological Molecules	
Biomolecule	**Structure**	**Comments**
	Cytosine thymine 	
Adenosine triphosphate	Adenosine-5′-triphosphate 	ATP serves the role of storing, transporting, and releasing energy to drive reactions in cells. All cells use ATP as their primary energy currency. Energy is stored in the phosphate–phosphate group bonds. These bonds are formed with energy consumed in respiration in the mitochondria and photosynthesis in plants, with the aid of special nanostructures that harvest and redirect electron energy. MW = 499; size = 0.95 nm
Proteins	1° Primary structure is the amino acid backbone 2° Determined by intramolecular hydrogen bonding. Two kinds: α-helix or β-sheet 3° α-helix + β-sheet; disulfide linkages 4° Quaternary structure consists of more than one protein (for hemoglobin)	Peptides formed from condensation reaction of amino acids to form peptide bonds Size ranges from 2 to 50 nm or greater; MW ranges from 6000 to >1,000,000 The simplest protein (peptide) is insulin. Hemoglobin, enzymes, fibrinogen, albumin, collagen
Cofactors Coenzymes Vitamins	Fat soluble: vitamins A, D, E, and K Water soluble: thiamin, riboflavin, nicotinic acid, pantothenic acid, ascorbic acid, folic acid, B_{12}, biotin NAD, FAD	The class of organic substances that function in trace amounts. Compounds that are essential to sustain the life are called cofactors. If an organism cannot synthesize a cofactor, it must obtain it from its environment.
Steroids	β-Estradiol (estrogen) Testosterone 	Includes cholesterol, vitamin D, and many hormones. Vitamin A (retinol) is an isoprenoid derivative, as is vitamin E (tocopherol). Terpenoids can be found in all classes of living things, and are the largest group of natural organic compounds. The flavors, scents, and colors of many plants are due to terpenoids, as are antibacterial and biologically active substances such as menthol, camphor, and the cannabinoids produced by the *Cannabis* plant. Steroid hormones include testosterone, progesterone, estradiol (an estrogen), cortisol, and others. Vitamin D is a steroid hormone precursor (cholecalciferol, also called calcitriol).

forces such as hydrogen bonding between water and hydrophilic portions of the chain (**Fig. 14.2**). The affinities of hydrophobic organic portions of the chain for each other and their repulsion by the surrounding aqueous environment play a strong role in the folding of proteins. Protein shape is highly correlated with their purported function: controlling access to chemically active reaction sites on the chains. Proteins serve as structural units but also act as catalysts and transport engines [37–43].

Nucleotides.   Nucleotides are made up of three portions: a heterocyclic base, a sugar, and one or more phosphate groups. In the cell they have important roles in energy production, metabolism, and signaling. Nucleotides are linked in polymeric chains of nucleic acids anchored to a phosphate-sugar backbone to form nucleic acids such as DNA and RNA—encoding and transcribing genetic information in living organisms. Nucleic acids are the nanomachinery of DNA and RNA.

FIG. 14.2

Structure of proteins—the basic building blocks for life. Proteins are chains of amino acids: 20 common acids called "residues" linked by carbon nitrogen bonds, "peptide bonds." Chain length can range from 40 to several thousand, commonly several hundred. Chains fold: folding stays in place by intramolecular forces weaker than peptide bonds. Primary structure = residue sequence. Secondary structure = folding: folds are of different types classified as: α helix and β sheet. Tertiary structure = shape of a single protein or "conformation." Quaternary structure = fitting together of several proteins to form a complex.

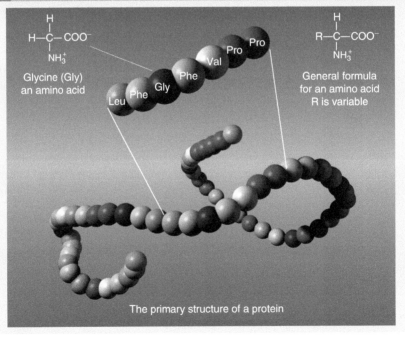

The primary structure of a protein

Source: Image by Anil Rao.

14.1.2   Cells and Organized Structures

Bottom Up Meets Top Down.   In order to understand how proteins, DNA, and RNA function in living systems, we need to put them into the special context of increasing levels of organization. Starting with the organization of DNA into genes and then of folded proteins, structures such as pores and prions, the simplest life forms like viruses, phages, bacteria, and cells all show hierarchical development. Therefore, we now switch to a more traditional top-down approach and introduce a description of cellular anatomy. However, to maintain the nanobiology context, we place more emphasis on size and scale than is done in most traditional approaches.

Cells and Internal Structure.   A brief overview of cellular biology is presented [44,45]. A typical cell in the human body is on the order of 10 µm (10,000 nm) in diameter—the size of a lymphocyte or red blood cell. A yeast cell is typically about 2–3 µm; a bacterium such as *Escherichia coli* is only 1/10 of that size and a flu virus is about 100 nm across. The human egg cell is about 35 µm in diameter. A typical cell is shown in **Figure 14.3.**

Living forms are classified into three categories based on their cell types. The oldest and simplest types of cells, the *archea* and the *bacteria,* are single celled life forms; although simple in structure, they are highly evolved. The more complex cells are called the *eukaryotes* and include all multicelled animals and plants as well as single-celled organisms that have a *cellular nucleus.* All cells have external *membranes* separating their *cytoplasm* from the environment. Most highly developed *eukaryotic* cells have a nucleus containing genetic material. Cells have distinct *organelles* such as *mitochondria* and *ribosomes, microtubules,* and a *cytoskeleton.*

The simpler and generally smaller cells characteristic of *prokaryotes* do not have a separate nucleus but do have DNA. Some bacteria form *capsules* that protect and anchor the cell and contain *spores* that store DNA in an inactive state. When conditions for normal cell growth are unfavorable, the spores are able to regenerate a new bacterial cell when the surrounding conditions change for the better. Bacteria have a *cell wall* made of *peptidoglycan,* a polymer consisting of sugars and amino acids that forms a mesh-like layer surrounding the outside of the plasma membrane. A typical bacterial cell is shown in **Figure 14.4.** Note the cell wall, the absence of a nucleus, and the generally less complex internal structure.

An *organelle* is a descriptive term for any discrete structure found inside a cell. Originally, organelles were identified and named because they showed up as visible regions and bodies under the microscope. Today, organelles are defined in terms of their roles in the material, energy, and information networks of the cell. There are many types of organelles with specialized functions, particularly in the eukaryotic cells of higher organisms. An organelle is to the cell what an organ is to the human body.

We focus especially on some of the molecular nanomachinery that performs essential functions in the cell such as maintaining the structure, sending and receiving signals to other cells, and synthesizing the building blocks used by the cell. Organelles are depicted inside the cell in **Figures 14.3** and **14.4.** Ribosomes are ~30 nm in size; mitochondria are cigar-shaped ellipsoids $500 \times 900 \times 300$ nm; the chloroplast is ~4 µm, lysosomes are ~700 nm, and vacuoles are ~10 mm—all,

FIG. 14.3

A typical eukaryote cell (animal kingdom). Cells are tiny chemical factories, using protein nanomachinery to produce and deliver materials according to instructions encoded in the DNA. In this figure of a typical animal cell we see the nucleus holding the DNA, surrounded by the nuclear envelope and communicating with the rest of the cell through the nuclear pores. The DNA produces RNA which interacts with the ribosomes to synthesize proteins. Ribosomes are concentrated in the nucleolus, but some are attached to sites on the rough endoplasmic reticulum and some float freely in the cytoplasm. The proteins are sorted and tagged for their destinations in the Golgi apparatus. From there they are transported in transport vesicles to their destinations in the cell machinery, either internally, in the cell membrane, or outside the cell as a secretion (either through a cell pore or a secretory vesicle). Mitochrondia produce energy to drive the processes. Lysosomes break down proteins and other substances that are no longer needed. Centrioles hold disassembled scaffolding units for erecting the apparatus that will be used when the cell divides. The membranes of the smooth endoplasmic reticulum serve as sites where various internal enzymatic reactions needed for the cell functions take place.

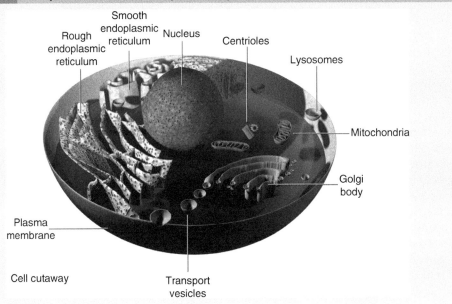

Source: Image by Anil Rao.

without any doubt, considered to be nanomaterials. In **Figure 14.3,** in a typical animal cell, the *nucleus* is central and prominent. The nucleus sequesters DNA and is surrounded by a *nuclear envelope*. Communication with the rest of the cell is conducted through pores in the nuclear envelope. DNA in the nucleus produces the template for *m*-RNA (messenger RNA) that in turn translates the blueprint to the ribosomal protein synthesis machinery.

Ribosomes are concentrated in the *nucleolus* (**Fig. 14.3**); some are attached to sites on the *rough endoplasmic reticulum* and some float freely in the cytoplasm. Proteins are sorted and tagged for their destinations in the *Golgi apparatus* and transported by vesicles to destinations in the cell, either internally, in the cell

FIG. 14.4 *A typical bacteria cell. In this figure of a typical bacterial (prokaryotes) cell, the plasma membrane is protected by a cell wall and capsule. There is no nucleus, and the DNA floats as a plasmid in the cytoplasm, along with ribosomes. A flagellum extends through the cell membrane and wall and is capable of rotary movement. Pili are extensions of the capsule and help attach the cell to its neighbors or to a stationary anchor point.*

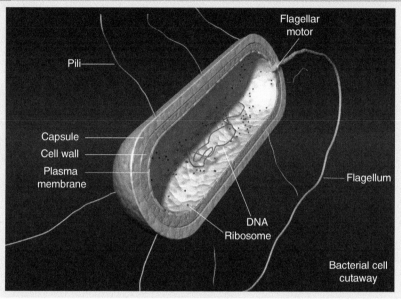

Source: Image by Anil Rao.

membrane or outside the cell via secretion (through a cell pore or a secretory vesicle). *Mitochondria* produce energy to drive cellular processes. *Lysosomes* break down proteins and other substances that are no longer needed. *Centrioles* hold disassembled scaffolding units for erecting apparatus for cell division. The membranes of the *smooth endoplasmic reticulum* serve as sites where various internal enzymatic reactions needed for cell functions take place.

We have made reference to cell membranes throughout the text. All cells have an outer *cellular membrane* (also called the *plasma membrane*) composed of a double layer of lipid molecules with a hydrophilic head and hydrophobic tail (**Fig. 14.5**). These amphiphilic molecules form a *lipid bilayer*, with the hydrophobic lipid tails lying together inward between the outer surfaces formed by the hydrophilic heads. This membrane is interspersed with proteins, carbohydrates, and a number of other large molecules (such as cholesterol) that are held in place by a combination of hydrophilic and lipophilic intermolecular forces.

The *cytoskeleton* is a macromolecular scaffolding or nanoskeleton that gives structure and support to the cytoplasm within cells. It is not present in most prokaryotes, but is an essential structure in all eukaryotic cells (**Fig. 14.6a**). The cytoskeleton maintains the shape and the internal organization of the cell but is not static or rigid; it is able to reconfigure within the cytoplasm depending on prevailing conditions. The cytoskeleton is composed of three types of protein

FIG. 14.5
FIG. 14.5   *Phospholipid bilayer of the cell membrane. Note the inclusions and attachments, including peripheral, integral, and transmembrane proteins.*

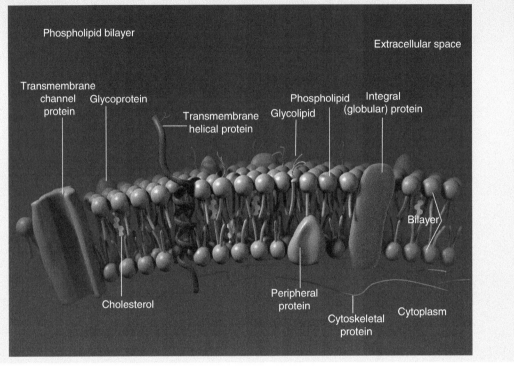

Source: Image by Anil Rao.

FIG. 14.6   *Cells showing endoplasmic reticulum. (a) Atomic force micrograph of cellular endoplasmic reticulum.*

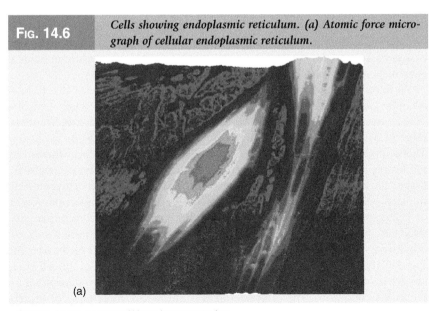

(a)

Source: Image courtesy of Veeco Instruments, Inc.

**FIG. 14.6
(CONTD.)**   *(b) Actin filament. (c) Actin and Cadherin filaments support adhesion between cells at the cell membranes.*

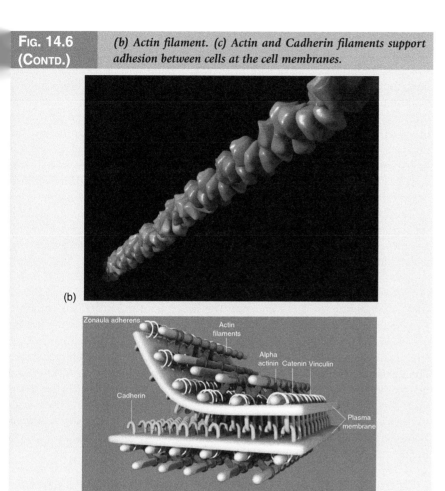

(b)

(c)

Source: Image by Anil Rao.

filaments, each organized in a different way and serving a different function: *actin filaments* (or microfilaments), *intermediate filaments,* and *microtubules.*

Actin filaments or microfilaments are composed of two chains of actin protein intertwined in an interlocking spiral to form the 7-nm diameter filaments (**Fig. 14.6b**). Actin filaments are linked by smaller protein units in a framework next to the cell membrane that supports the membrane and its junctions to other cells and the extracellular matrix (**Fig. 14.6c**). Actin filaments are very important in cell migration processes within an organism's body (*cytokinesis*) and anchor *myosin* for muscle contraction. Actin filaments are also distributed within the interior of many types of cells, where they control streaming of components suspended in the cytoplasm. The filamentous system is a remarkable example of a hierarchical nanostructured material.

Intermediate filaments consist of several different types of cytoskeleton elements that are components of supporting frameworks within the interior of cells. Intermediate filaments are generally larger (8–11 nm in diameter) and more stable than actin filaments. There are four types of intermediate filaments, classified according to their compositions and functions. *Vimentins* provide structural support to muscle and other cells. *Keratins* form a scaffold that supports the nucleus of *epithelial* (skin) cells. *Lamins* make up the nuclear lamina, a support framework within the nuclear envelope. Lastly, *neurofilaments* strengthen the long *axons* of *neurons*. Microtubles are hollow cylinders about 25 nm in diameter, made up of *protofilaments* composed of α- and β-tubulin proteins. The functions of different types of microtubules include: (1) intracellular transport of organelles like mitochondria and vesicles, (2) mitotic spindles in cell division, (3) inner core skeleton of cilia and flagella, and (4) scaffolds for synthesis of the cell wall in plants. Thus, the microtubule elements of the cytoskeleton act as tramways for movement of vesicles and organelles within the cell, and for separation of chromosomes during cell division. Portions of the cytoskeleton are anchoring points for cell movement mechanisms such as flagella and cilia, which have microtubule inner core skeletons.

Details of the Nucleus.   The nucleus of eukaryotic cells has a double membrane enclosure holding the DNA material. The structure of the nucleus is complex; we will describe some of the main components in this section and in the section on DNA. The nucleus is enclosed in a *nuclear envelope* (or nuclear membrane), which separates the contents of the nucleus from the rest of the cell cytoplasm. The nuclear envelope is a double membrane composed of two separate lipid bilayer membranes. The *perinuclear space* between the two membranes that make up the nuclear envelope is about 20–40 nm wide. The outer membrane is continuous with the rough endoplasmic reticulum. The nuclear envelope contains pores that regulate the exchange of proteins, RNA, and other substances between the nucleus and the cytoplasm.

A *chromosome* is a single long chain molecule of DNA ca. 2 nm in diameter. In eukaryotic cells, the chromosome is packaged in the cell nucleus and includes a surrounding of special proteins, the lamins, which package and protect the DNA strand. In bacteria and other prokaryotic cells without nuclei, the chromosome is bare; it is attached to the cell membrane or floats in the cytoplasm as a closed loop of DNA called a plasmid. In eukaryotes, DNA undergoes several levels of folding before it is called a chromosome [101]. The 2-nm strands are much too long to fit into a several micron diameter cell. The first coiling is the double-stranded DNA. The second coiling event involves a 140-base pair sequence that coils around a histone protein to form a bead (histone-associated DNA is called *chromatin*) to form structures called *nucleosomes*, the diameter of which is 11 nm. These nucleosomes stack on top with linker regions in between (for flexibility). At this stage the protochromosome is 30 nm in diameter. Series of nucleosomes form a condensed region of the chromosome that is 300 nm in diameter that, in turn, condense to form 700-nm structures per strand. Two chromosomes then add to make a structure that is 1400 nm wide and clearly visible with an optical microscope.

The *nucleolus* is a dense body within the nucleus in which ribosomes, RNA, and proteins are concentrated; the ribosomes are assembled from their constituent protein and RNA components and protein synthesis takes place here. The nucleolus is not separated from the rest of the nucleus by a membrane and is a dynamic structure. The cell nucleus is surrounded by a double membrane called the nuclear envelope, which is supported by a framework of lamin protein fibers, called the nuclear lamina. The nuclear envelope separates the nuclear contents, the DNA, and nucleoli from the rest of the cytoplasm. The chromosomes are attached to the lamina; during mitosis, the lamina disintegrates prior to formation of the spindle fibers along which the chromosomes separate. The lamina connects with the rough endoplasmic reticulum adjacent to the nuclear envelope.

Plasmids.   A plasmid is a circular strand of DNA separate from the chromosomal DNA that is capable of autonomous replication, generally at a higher rate than a chromosome found in the nucleus. Plasmids are important in bacteria where they are associated with rapid genetic adaptation and exchange of genetic material with other bacteria. In bacterial cells there may be little difference between a large plasmid and a small chromosome. Some plasmids are also found in eukaryotes. Mitochondria and chloroplasts resemble plasmids in structure. It is thought that eukaryotic organisms arose from ancient cells engulfing others to form symbiotic relationships.

Endoplasmic Reticulum.   The endoplasmic reticulum is a network of tubules, vesicles, and sacs found in all eukaryotic cells. This interconnected network supports specialized functions in the cell, including protein synthesis, calcium sequestration, steroid production, storage and production of glycogen, and insertion of membrane proteins. Structure and composition of the membranes making up the endoplasmic reticulum are similar to those of the plasma membrane.

The *rough endoplasmic reticulum* is located near the nucleus. It holds ribosomes and other apparatus used in protein synthesis and transport. The *smooth endoplasmic reticulum* is located closer to the periphery of the cell. The membrane network of the smooth endoplasmic reticulum provides surface area for enzyme activity and storage and processing of enzyme products. In highly specialized cells, the smooth endoplasmic reticulum is adapted for specialized functions. In muscle cells, its vesicles and tubules store calcium, which is released by *calcium pumps* during the contraction process.

Lysosomes.   Lysosomes are organelles surrounded by globular membranes within the cytoplasm. Lysosomes digest excess or damaged organelles and proteins, food particles, and engulfed viruses or bacteria. The membrane surrounding a lysosome prevents its digestive enzymes from destroying the cell. Lysosome structure resembles the lipid bilayer self-assembled structures discussed earlier.

Mitochondria and Chloroplasts.   Mitochondria and chloroplasts are organelles (plasmids) within cells that have their own DNA and carry out special functions such as energy production and photosynthesis, respectively. They represent extremely evolved *symbionts* that have become highly integrated components of higher cells; they carry out essential molecular processing functions for the host cell but they retain their own DNA and replication cycle, and depend on the host for most essential life functions other than their specialized contribution. In many cells mitochondrial DNA codes for some of the many proteins that form the ribosomes. Thus, the cell nucleus and the mitochondria are mutually dependent for the production of proteins [46,47]. One only gets mitochondria, for example, from one's mother.

Ribosomes.   A ribosome is a special type of organelle involved directly in the synthesis of proteins. Ribosomes contain the active ingredient of protein replication (ribosomal RNA or *r*-RNA) along with the associated proteins. Ribosomes are active in the nucleolus, bound to the endoplasmic reticulum, and floating freely in the cytoplasm, depending on the type and destination of the proteins they synthesize. The molecular recognition system involved in protein synthesis, starting with DNA, is indeed one of the most remarkable examples of living chemistry.

14.1.3 Viruses

Existing on the boundary between living and nonliving structures, viruses consist of a nucleic acid (DNA or RNA) genetic core enclosed in a protein coat. The virus is able to hijack the operation of specific host cells once the viral particle gains entrance. The protein coat, called the *capsid*, protects the core from attack by antibodies and provides a mechanism for penetrating the wall of the host cell—the simplest forms of survival. The core may also contain one or more protein enzymes used to aid replication in the cell, but the virus uses the molecular machinery of the host for most of its reproductive functions [48]. Outside the host cell, the virus particle with its protein coat is a dormant particle called a *virion* that may form crystals like a nonliving molecular material.

Viruses are of particular relevance for nanoscience because of their small size and the simplicity of their molecular machinery. A typical virus particle is between 10 and 300 nm in diameter. Many are roughly spherical in shape, but some are filaments with length up to 1400 nm. The diameter of the capsid is only about 80 nm. Most viruses are too small to view with a light microscope, but some are as large as or larger than the smallest bacteria and can be seen under high optical magnification.

14.1.4 Prions

Another type of subcellular disease factor, whose existence was controversial when first reported, is the prion [49]. Prions are misfolded proteins that appear to be able to catalyze misfolding of related proteins in living systems, resulting in dysfunction in cells and organisms. Like viruses, their existence was first inferred by the ability of tissue extracts to transmit disease in the absence of any known identifiable disease agent such as microbes or viruses. Their action is more biological in nature than mere toxins due to their ability to reproduce copies of themselves, leading to plaques (especially in the brain) and to disrupt normal metabolic functions. Thus, they are sometimes referred to as "quasi-bionts" like the viruses, but they do not have the structure of RNA or a protein coating like virions.

Prions were discovered after exhaustive research into the causes and cures for the devastating condition known as "mad cow disease" or BSE (bovine spongiform encephalitis). BSE is related to a similar disease in sheep and is transmitted to humans by consumption of contaminated meat products. Prions seem to be extremely refractory in that they are able to survive the digestive system and make their way to the brain, where they produce their damaging effects (a compelling reason to cook food thoroughly). All known prions are believed to infect and propagate by formation of an amyloid fold—polymerization of the protein into a fiber with a core consisting of tightly packed β-sheets [50].

14.1.5   Toxins and Disruptive Nanoparticles

Traditionally, toxins are defined as molecules that disrupt the operation of cells to cause damage or death by chemical action. Chemically inert particles, if sufficiently small, also are able to disrupt the molecular machinery of cells. Nonliving particles such as asbestos fibers and soot particles that lodge into the fine structures of the lung and some other tissues thereby have the capability to disrupt processes of cell growth and division, causing cancers and cell death. Immune cells are able to absorb and address many types and sizes of contaminating nonliving particles as well as bacteria [51]. Classical toxins range from heavy metal atoms that tie up and precipitate cytoplasm to complex biomolecules such as enzymes and proteins that disrupt and disorganize specific cellular metabolism pathways [52]. Toxic inert particles include asbestos, silica dust, metal oxides, and soot (although some particles may be chemically active as well and microscopically disruptive at the same time).

In general, toxins are nonliving chemical substances that interface with the complex nanomachinery of living systems and, in the worst case, cause blockage or disruption. Molecules that are toxic to one type of organism or cell may be harmless or useful to others. Many living organisms produce highly specific toxins as a defense mechanism. In a later section we explore how the immune systems of some organisms, including humans, produce extremely sophisticated toxins to protect the organism against harmful invaders. We also examine some concerns about possible toxic effects of new types of nanoparticles that are entering the environment as a result of nanotechnology.

14.1.6   Completing the Circle from Top
Down to Bottom Up

When we started our exploration of the building blocks of life, we went from the simplest chemical constituents up to the complex structure of DNA and RNA. We are able to describe these constituents in terms of chemistry and the physical interactions between molecules up to the nanoscale. Above this level of organization, we begin to encounter structures on the order of hundreds of nanometers and higher. Above this point it becomes quite cumbersome to describe the components and their interactions at the detailed level of atoms and molecules; new levels of organization begin to emerge. Aggregates of molecules form units that act as identifiable separate entities and maintain their structures and identities. To the extent that such aggregates are stable and act as units, it is consistent with reality to consider them as objects on a higher emergent level of organization. Thus, we continue our description by starting with observable features under the microscope, such as cells, and work our way back down to molecular components in the context of cellular anatomy and functions.

Levels of Organization in Living Systems.   Although it would be theoretically possible to describe the behavior of a complete living organism in terms of its chemical and molecular interactions, it is more efficient to employ larger aggregated units of organization as shorthand in order to simplify the description. This higher level of description not only is more efficient, but also makes for a better fit to our human ability to grasp complexity and to understand, model, plan, and manipulate. This description is analogous to grouping of lines of computer programming code into functions and subroutines; it helps us to deal with the complexity of large systems. We put assemblages of molecules into an aggregated conceptual box and use higher levels and features to describe living structure and behavior. In studying nanoscience or any complex system, there is always the possibility of an element of chaotic behavior in which new features emerge unexpectedly. This can be true even when we try to carefully control experiments, designs, and fabrications. Thus, the study of living systems is instructive for nanoengineers engaged in any complex design [53].

Emergent Properties: Complex Systems, Chaos, and Catastrophe Theory.   We introduced emergent property theory briefly in chapter 8, "Nanothermodynamics." The emergence of new features and structures from apparently chaotic behavior is a widespread but constructive phenomenon. Stable, robust structures and performance can be based in underlying chaotic and random processes. This is the domain of chaos theory, fuzzy logic, and catastrophe theory, practical applications of which include complex robotic controls; simulated annealing for exploring optimal designs of circuits, polymers, and proteins; neural networks; signal processing; and track correlation. It is noteworthy that many early insights into the nature of chaos theory owe their origins to the study of natural living systems, such as ecological population dynamics and morphogenesis and embryonic development [54–58]. An important aspect of biological nanoscience is the insight that understanding of processes at the nanoscale can provide in revealing how large numbers of very small components can organize to produce effective behaviors without centralized direction or control.

Lower Size Limits for Life.   What is the size of the smallest aggregate of molecules that is able to support all of the functions of life? As we have seen with viruses and prions, there is a continuum between living and nonliving aggregates of matter. Biomolecular nanoscience, aided by ever more powerful microscopy, continues to discover and investigate new naturally occurring life forms and life-related particles, as well as nonliving particles that can interact with living metabolisms, down to sizes in the nanoscale. With microscopy tools able to resolve and study structures at the nanoscale, an understanding is emerging of a rich complexity of life components and disease agents at the subcellular level.

    The generally accepted lower limit for cellular life forms such as bacteria is 200 nm, which happens to be the lower limit of resolution of a standard light microscope. But noncellular quasi-bionts such as viruses have been found in sizes from 10 to 300 nm in diameter.

In 1999, a workshop convened by the U.S. National Academy of Sciences estimated that "the minimum diameter of a spherical cell compatible with a system of genome expression and a biochemistry of contemporary character would appear to lie somewhere between 200 and 300 nm, probably closer to the latter" [59]. For noncellular life forms such as viruses [60]:

> The lower size limits are seen in some RNA viruses like the Qß virus (which contains only three genes), and in certain animal viruses (e.g., poliovirus) that are in the range of 25 to 50 nm in diameter, while most others are in the range of 100 to 200 nm or even larger.

Symbiotic organelles or bacteria are also commonly found in the 200-nm range and are sometimes smaller. These include bacteria that can only grow inside cells of other organisms, intracellular organelles (e.g., mitochondria or chloroplasts), and the enigmatic nanobacteria smaller than 200 nm that have been reportedly found in kidney stones [61–63]. Thus, at the smallest and lowest number of components, organic units of matter can no longer have an independent existence; they can only carry out life functions by acting symbiotically with other life forms. It is somewhere on the boundary of independent living cells and dependent symbionts that the boundary between life and nonliving matter lies.

In addition to asking how small an organism can be in physical volume or number of molecules, it is of interest from the nano- and molecular biology perspective to ask how small the set of DNA machinery that programs the cell—the genome and the genes that it encodes—can be.

Upper and Lower Size Limits for Genomes in Cells.   From a nanoscience and nanotechnology point of view, it is useful to learn how the simplest organisms are structured because they can provide a starting point for designing artificial nanomachinery for biochemical functions. For these reasons as well as fundamental scientific interest, it is of interest to ask, "What is the smallest number of genes needed to construct a viable organism?" [64].

Advances in the ability to create new types of organisms, either through modification of existing genomes or synthesizing DNA from components, have led to the new field of synthetic biology [65]. The emerging field of synthetic biology leads to a view of the cells as an assemblage of parts put together to build an organism with desired characteristics. This program has been approached theoretically and experimentally in the laboratories of Craig Ventnor, Francis Collins, and elsewhere [66,67]. To put the question in context, consider that the organism with the largest genome found thus far is a microscopic protozoan, *Amoeba dubia*, with a genome of 670 billion base pairs (670 Gb). A related microbe, *Amoeba proteus*, has a mere 290 billion base pairs, but that number is 100 times larger than the human genome. The 3-Gb-long human genome is thought to encode fewer than 30,000 genes, perhaps as few as 23,000 that are required to program the human body.

The free-living organism with the smallest genome found thus far is one of the most abundant proteobacteria in the world's oceans, *Pelagibacter ubique*, with a DNA sequence length of only 1.3 Mb. *P. ubique* is suited for a robust independent existence, with complete biosynthetic pathways for all 20 amino acids and all but a few cofactors [68]. Parasitic or symbiotic bacteria may have even smaller genomes because they depend on their host for essential life functions [69].

The mitochondria organelles in the cells of higher organisms may have originated as an independent symbiotic microbe that invaded host cells (or were engulfed) and gradually coadapted to integrate fully with the host. There is much genetic evidence for symbionts adapting to their host and becoming simpler and smaller in the process, notably in bacteria adapted to live inside the cells of insects [70,71]. The smallest such examples yet found have apparently undergone extreme genome reduction in adapting to intracellular life within their hosts and have many similarities to mitochondria. An alternative but less widely accepted explanation that has been posited is that extreme endosymbionts represent a stage in the evolution of an organelle that split off from the host nuclear genome [72].

Even smaller hypothetical nanobacteria may fragment into nongrowing entities that appear considerably smaller than the true, viable organisms, and these fragments may come together at a later time to form a viable organism. Some plant viruses exhibit just such a pattern. Each particle package separates RNA, and sometimes three separate particles are needed to establish an infection. While this strategy is known for a few RNA viruses, there are as yet no examples among the bacteria or other prokaryotes [60,73]. Other requirements for small cell size include: (1) reduction of the average size of proteins, (2) an RNA-world approach in which a single type of molecule accomplishes both catalytic and genetic functions, and (3) the use of overlapping genes and genes on complementary strands [59,60].

Genome Sizes in Viruses.   Viruses, which need no genes for constructing and maintaining cellular structure and depend entirely on host cells for reproduction, may have as few as three genes. Most viruses encode from 3 to 400 genes. The largest known viral genome is found in the *mimivirus*: approximately 1.2 megabases that encode about 1000 genes. The discovery of this virus blurs the boundary between higher living organisms and viruses [74].

Continuing the Search for Life at the Nanoscale.   The search at the nanoscale for new types of minimally sized life forms invites theoretical questions of extraterrestrial life and in studies of the origins of life on Earth. This search is also important for disease agents, including toxins, catalytic particles, and life forms. Determining their roles in human and animal disease remains an active and open area of investigation, with many open questions. Some evidence has been reported to implicate anobacteria or other nano life forms as agents in deposits of calcium as arterial plaque in blood vessels, kidney stones, and joints, as well as in tooth decay. However, alternative explanations have been presented, so the matter remains controversial [75,76].

14.2   CELLULAR MEMBRANES AND SIGNALING SYSTEMS

In the following section, we look into aspects of the life cycle of cells and how cells communicate with their surroundings.

All organisms consist of cells that multiply through cell division. An adult human being has approximately 100,000 billion cells, all originating from a single cell, the fertilized egg cell. In adults there is also an enormous number of continuously dividing cells replacing those dying. Before a cell can divide it has to grow in size, duplicate its chromosomes and separate the chromosomes for exact distribution between the two daughter cells. These different processes are coordinated in the cell cycle. [77]

We discussed cell size briefly in chapter 5 and continue from another perspective here. Cells grow to a certain size and may divide to form new cells. In each type of tissue or in each type of single-celled organism, the optimal cell size is determined by the processes by which nutrients, oxygen, and waste products are brought into and out of the cell. Size matters because of the costs of transport; hence, the ratio of cell volume to surface area is a critical factor. The optimal size of a cell is related to its rate of metabolism, its shape (ratio of volume to surface), and the particular mechanism for transporting metabolic raw materials and products across the cell—as well as the particular specialized function that the cell is programmed to perform.

Cells in complex multicellular organisms organize and function as tissues and organs. Tissues in the human body have a very specific and intricately organized nanostructure, adapted to enhance their performance in load bearing; stress transport; fluid transport; resistance to penetration by hostile pathogens; and protection, balance, and containment of vital substances such as water, oxygen, salts, and nutrients. Organs have complex and specialized functions, which are interdependent and regulated by chemical and electrochemical signals in an interplay of nerve and chemical communications networks. All of the communication of cells with the surrounding fluid environment must involve the cell membrane. For organs to develop, function, and be maintained, cells must recognize and respond appropriately to their neighbors and environment.

14.2.1   *Cell Membrane Function*

Cells are contained and maintained by their membranes, both external and internal to the cytoplasm. The nucleus has its own membrane, as do other specialized structures such as mitochondria, organelles, etc. The external cell membrane contains specialized structures with sophisticated mechanisms for regulating what passes between the inside of the cell and the external environment. Each of these structures is a functional biological nanomachine, built of biological molecules such as proteins, glycoproteins, peptides, etc. [78,79].

Specialized ion channels and membrane domains act as gateways to control molecular events on the nanoscale that govern electrostatic and material balances, interactions with the external world, and communication with other cells of the same and different organs. Molecular biology, biophysics, and molecular genetics techniques for the study of ion channels are active areas of research on the nanoscale domain to understand topics such as the molecular origin of voltage dependence in ion channels, functioning of calcium-activated K^+ channels of large conductance, gap junction tunnels, the process of exocytosis, and other special aspects of cell membrane nano-anatomy [80]. More strongly ionic species— such as cations of sodium, calcium, and potassium ions; anions of chlorine; and most proteins—have low solubility and affinity for the lipid layer, which is a natural barrier between the cytoplasm and the exterior. In general, hydrophilic species can only pass across the lipid barrier by active expenditure of energy facilitated at special gateway sites (e.g., valinomycin).

Figure 14.5 showed a section of cell membrane composed of the phospholipid bilayer with several types of inclusions interspersed into the bilayer. The bilayer itself is composed of many pairs of phospholipids, with their hydrophilic tails forming a lipid-rich region sandwiched between the hydrophilic heads, which form two outer layers facing the external and internal environments. The outer environment will typically be filled with intercellular fluids that bathe the cells. The inner environment is the cell's cytoplasm. Both are generally based in aqueous media.

The total thickness of the lipid bilayer is only about 5–10 nm or less, including the water of hydration attached to the phosphate groups depending on proteins and lipids embedded within the bilayer. The core region is very hydrophobic and the phosphate surfaces are very hydrophilic: The effective concentration of water changes from nearly 0 inside the membrane (between the two layers of the lipid) to a concentration of around 2 M in the phosphate layers [81,83].

Cholesterol is an essential component of all cell membranes. The cholesterol molecule is amphiphilic, with a hydrophilic alcohol group coupled to a lipophilic steroid and hydrocarbon chain. With the alcohol anchoring the cholesterol molecule to the phosphate head layer of the membrane, the lipophilic portion inserts itself between the lipid chains, where it enhances membrane flexibility—somewhat like a lubricant. Cholesterol molecules also group together to form lipid rafts in the membrane that reduce the permeability of the membrane to hydrogen ions (acid), sodium, and other ions. Glycoproteins and glycolipids attach to the membrane and project out from the surface with specific markers that are recognized as tags to identify the cell, with hydrophilic sugar residues attached to the part of the polypeptide exposed at the surface of the cell. While the most common function of these projections is to allow recognition of the cell by other cells, enzymes, and antibodies, some types of "projecting molecules" serve as anchors to attach the cell to its surroundings or to other cells, in specialized types of cells. They play an important role in attaching cells to each other to construct tissues and organs.

Proteins that are embedded in the cell membrane are anchored in place by filaments of cytoskeletal protein (on the interior of the cell) or tethered to an extracellular matrix (composed of collagen and other polymers, which form scaffoldings between cells). Some embedded proteins and lipids are free to float through the membrane. Some types of proteins are embedded in the membrane to perform material transport functions. The attachment locations of these proteins can be *peripheral, integral,* or *transmembrane.* Peripheral proteins are inserted in or attached only to the phosphate surface or only to one half of the lipid bilayer. Integral proteins are embedded through both halves of the bilayer. Transmembrane proteins are integral proteins that cross through the membrane to make connection between the inner and outer environments of the cell.

Peripheral membrane proteins are associated with membranes but do not penetrate the hydrophobic core of the membrane. Peripheral membrane proteins typically have only weak electrostatic and temporary attachment to the membrane. They are often attached to integral proteins instead of or in addition to the phosphate layer, serving in a secondary attachment role in the process associated with the integral protein. Because they have weak electrostatic attachments, peripheral proteins can be separated from the membrane by salt solutions, whereas integral proteins can be disrupted from their membranes only with strong detergents [84,85].

14.2.2 Ion Pumps, Ion Channels, and Maintenance of the Cellular Environment

Membrane pores act as ion channels that help to control the concentration of ions inside the cell. This difference in ion concentration establishes an electronic potential difference across the cell membrane. Ion channels are present in all types of living cells. The controlled passage of ions through the potential gradient of the channels plays an important role in cell physiology for nerve and muscle cells.

Each type of cell can have many different kinds of ion channels controlling different types of ions. The most important ions for cell physiology are sodium, potassium, calcium, and chlorine (Na^+, K^+, Ca^{2+}, and Cl^-). A channel that can open and close is called a gated channel. A cell may have several different types of channels controlling the same ion in different ways in response to different conditions. Several types of mechanisms control gated channels. A channel can switch between an activated or open state to an inactivated or closed state in response to: (1) voltage changes (voltage-gated channels), (2) activation of receptors (receptor-gated channels), and (3) specific ions and chemicals (ligand-gated channels). In the sections that follow we will discuss some of the characteristics of ion channels and the functions that they enable in nerve and muscle cells.

14.2.3 Transmission of Neural Impulses: Action Potential and K Channel

Key Advances and Tools.   Since the 1700s, people have known that nerve impulses are somehow associated with electricity. Electricity was first known as the buildup of static charge on pieces of amber. In 1771, Luigi Galvani discovered that frog legs can be made to contract by application of electric charge. This led to the direct experimentation of the effects of electricity on muscle and nerve tissue. By the mid-nineteenth century, Helmholtz measured the speed of propagation of the *action potential* (the membrane potential) along frog neurons. Helmholtz found that the propagation velocity was about 40 m per second—slower than would be the case if the action potential were due simply to electrical conduction through a wire. Thus, the action of nerves had to involve a more complex biochemical process. The term *synapse* was coined by Sherrington for these junctions in 1897.

Microscopic techniques revealed much about the structure of synapses in the following years, but it would be another 60 years before their mechanism of operation could be unraveled. This depended upon first understanding how electrical impulses were carried along the membranes of the nerve fibers, the *axons*, whose long processes stretched out from the cell body to connect with other neurons. At this time it was understood that the membrane potential was maintained by a difference in ion concentration gradients across the cell membrane. Using the microelectrode technique, it was determined that the potential gradient in nerve cells was due to an imbalance between sodium and potassium ions on the inside and outside of the membrane. The sodium and potassium gradients were in opposing directions. When the ion gates opened and the concentration gradient collapsed, the result was an electrical propagation along the nerve fiber. Somehow, the potential would have to be restored after each such discharge. The voltage clamp was an advance on the microelectrode, adding a second reference electrode and controlling the potential inside the cell with a feedback circuit [85].

In a famous series of experiments published in 1952, Alan Hodgkin and Andrew Huxley used the voltage clamp method to separate the components of the membrane currents due to potassium (K^+) and sodium (Na^+) ions. They were able to separate the components of the membrane conductance into three factors: Na^+ conductance, K^+ conductance, and Na^+ inactivation. They found that the Na^+ conductance is activated in response to depolarization but closed without repolarization. They concluded from this that there must be two control gates for the Na^+ channels. The K^+ channels, working in the reverse direction from the Na^+ channels, were activated by polarization but did not have a deactivation response. The K^+ channels work to restore the membrane potential to its resting level after a discharge. Since the K^+ channel gates work continuously to maintain the resting potential, they require only an activation mechanism. With advances in amplifiers, it is possible to measure extremely small currents corresponding to the passage of single ions through an ion channel [86].

Mathematical Description of the Cell Membrane.   In the previous section we described in qualitative terms how the balance of chemical forces results in an electrical potential difference across the lipid bilayer membrane. In this section we will show the same concept in more precise mathematical terms. Molecules that are free to move, as in a gas or a solution, tend to diffuse from a region of higher concentration to a region of lower concentration (a *concentration gradient*). Because it requires expenditure of energy to concentrate a substance in solution or gas, we can describe the tendency to diffuse away from concentrated areas as a *force*.

If concentrated salt water is separated by such a membrane from pure water, the water will diffuse into the region of higher salt concentration until the salinity is equal on both sides of the membrane. This phenomenon is called *osmosis* (from the Greek *endo*, "inside" + *smos*, "push or thrust") and is observed when water is absorbed into plant tissues. It represents a real force because it can do the work of raising large amounts of water from roots of trees up to leaves, far in excess of the work that could be done by capillary action alone.

The Nernst Equation.   The force associated with concentration diffusion is measured by the work done by the change in volume as water diffuses across a membrane. If a force in the form of pressure is applied to one side of the membrane, water is forced to cross the membrane from the side of high salt concentration to the other side. This is the basis for water purification by reverse osmosis. The equation analogous to the ideal gas law for the work due to differences in concentrations is

$$W_c = 2.303\,RT\,\log\frac{[A]_{out}}{[A]_{in}} \tag{14.1}$$

where W_c is the work, or energy; $[A_{out}]$ is the concentration in one region of solution (by convention, outside the cell); and $[A_{in}]$ is the concentration a different region of the solution (inside the cell). The concentration is expressed in terms of moles.

If, as in the cell, we have a membrane separating two unequal concentrations of ions, such as potasium (K^+) and chlorine (Cl^-), and the membrane is differentially permeable to the positive and negative ions, there is an electrical potential difference across the membrane if the concentrations of the two species differ on either side. The work required to separate the two ions is given by

$$W_E = \mathcal{F}\mathcal{E} \tag{14.2}$$

where W_E is the work or energy; $\mathcal{F}$ is Faraday's constant, a measure of the quantity of electrical charge per mole of substance; and $\mathcal{E}$ is the electrical potential difference across the membrane. Note that this work is independent of the distance between the charges and only depends on the potential difference, no matter over what distance the potential exists, independent of the membrane thickness.

When the chemical forces are in balance with the electrical forces, the system is at equilibrium; there is no net movement of ions across the barrier. The chemical force tending to move, say K^+ into the cell, will be equal to the electrical force tending to move it outside (or vice versa), and the work given by equation (14.1) will equal the work in equation (14.2), so

$$W_c = 2.303\,RT\,\log\frac{\left[K^+\right]_{out}}{\left[K^+\right]_{in}} \tag{14.3}$$

It follows that the membrane potential, $\mathcal{E}$, is given by

$$E = 2.303\,\frac{RT}{F}\,\log\frac{\left[K^+\right]_{out}}{\left[K^+\right]_{in}} \tag{14.4}$$

This is the *Nernst equation,* first derived by Walther Nernst in 1888. The membrane potential E is also called the *Nernst potential* or the *diffusion potential.*

For any given ion species, such as K^+, Cl^-, etc. and given type of membrane, such as a squid neuron, the Nernst potential is the *equilibrium potential* at which there is no net flux for that ion across the membrane. This *membrane potential* is of fundamental importance to electrical activity in all cells and to the nanoscience of living things.

At 18°C, the constant $2.3\,RT/\mathcal{F} = 58$ mV. For the giant squid axon, the ratio for concentration of potassium outside versus inside the cell is 1 to 20 [87], and the membrane potential is −75 mV. This is the equilibrium potential that the membrane tends to assume when membrane is permeable to potassium (i.e., when the potassium ion channel gates are open).

The Membrane Equivalent Circuit.   A convenient way to represent the membrane potential is with an equivalent electronic circuit, in which the lipid membrane is represented by a capacitance element. The membrane potential due to each type of ion is represented by a battery connected across the membrane capacitor. Each battery has the voltage and polarity corresponding to the ion for which it is an analogue. For each type of ion, the ion gate is represented by a variable resistance in series with the battery. Ohm's law for electrical currents relates resistance and potential (voltage) to the current as

$$E = IR \qquad (14.5)$$

where E is the electrical potential in volts (V), I is the current in amperes (A), and R is the resistance in ohms. The reciprocal of resistance is conductance (g) measured in siemens (S).

Conductance is an analogue for membrane permeability (P).

Current flowing across the giant axon membrane may be represented by the sum of conductive components (which we identify as ion channels) and capacitance (the cell membrane). The electrical properties can be represented by the circuit diagram in **Figure 14.7,** the classic *membrane equivalent circuit.*

The Hodgkin–Huxley Model Equations.   Hodgkin and Huxley [88,89] discovered that the conductivity of different ions was dependent upon the potential difference across the cell membrane (V_m) and the equilibrium potentials ($\mathcal{E}$) of the ions. They assumed the membrane potential was not equal to the equilibrium potential when there is a net flow of ions proportional to the difference between the membrane potential and the equilibrium potential. They represented the observed conductivity across the membrane as the sum of currents through the conductive elements of the equivalent circuit, or:

$$I_{ion} = g_{ion}\left(V_m - E_{ion}\right) \qquad (14.6)$$

In the simple form of the Hodgkin–Huxley model represented by the preceding equivalent circuit diagram, there are only three types of channel: a sodium channel with conductivity g_{Na}, a potassium channel with conductivity g_K, and a weaker chlorine channel with conductivity g_{Cl}.

From their voltage clamp measurements, Hodgkin and Huxley found that g_{Na} and g_K change over time as well as with voltage, but the conductances of the

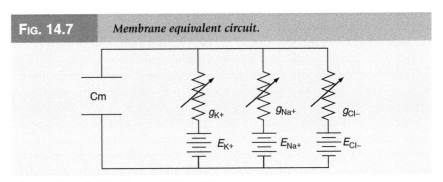

Fɪɢ. 14.7   *Membrane equivalent circuit.*

Source: Image by Mike Hammer, Lightspeed Llp.

other ions are constant. When the axonal membrane was depolarized, they observed a transient increase in Na^+ conductance and a slower noninactivating increase in K^+ conductance. In cells, conductance is caused by opening of individual ion channels, but each channel is either open or closed. In the Hodgkin–Huxley model, a probability function x (known as a gate variable) may be constructed to define what fraction of the channels are open as a function of the membrane voltage and time.

In the original simplified model, the time dependence of the conductance is represented by an activation coefficient x, which represents the probability of a gate in the channel being open. The conductance for a time-dependent channel can thus be written in terms of its activation coefficient x ($0 \leq x \leq 1$) and the maximum conductance $g_{ion,max}$ that is possible when all channel gates are open:

$$g_{ion} = g_{ion,max}(x) \tag{14.7}$$

where x is given by the equation:

$$\frac{dx}{dt} = \alpha_x x(1-x) - \beta_x x \tag{14.8}$$

where α_x and β_x are rate coefficients that are nonlinear functions of voltage (units are t^{-1}).

The activation coefficient can be raised to a higher power to represent the case in which a larger number of gates are present in a single ion channel. Additionally, each channel may have multiple activation coefficients to describe activation followed by, for example, voltage-dependent inactivation. With these refinements, the current is given by an equation with the following general form:

$$I_{ion} = g_{ion,max} x y (V_m - E_{ion}) \tag{14.9}$$

where x is the activation coefficient and y is the inactivation coefficient. This is the form of the equation for the Na^+ conductance responsible for the depolarization during the action potential because it inactivates without repolarization. Because the Na^+ conductance opened in response to depolarization but closed without repolarization, Hodgkin and Huxley reasoned that there must be two control gates. One is activated when a threshold depolarization is achieved; the second closes more slowly. The two gates are represented by two variables: m is the activation coefficient (substituted for x), and h is the inactivation coefficient (substituted for y). The time dependencies of the two gates are described by the differential equations:

$$\frac{dm}{dt} = \alpha_m x(1-m) - \beta_x m \tag{14.10}$$

$$\frac{dh}{dt} = \alpha_h x(1-h) - \beta_x h \tag{14.11}$$

where α and β are rate constants that are time-invariant functions of voltage. The Na^+ conductance may then be written:

$$g_{Na} = g_{Na,max}\left(m^3 h\right) \tag{14.12}$$

where $g_{Na,max}$ is the maximum Na$^+$ conductance.

Expressions for rate constants were found by fitting curves to experimental data. When this was first done, the data from squid giant axons had to be numerically integrated by hand-cranked numerical calculators, an enormous effort described in Huxley's later paper [88].

The Na$^+$ current is described by

$$I_{Na} = g_{Na}\left(V_m - E_{Na}\right) \tag{14.13}$$

$$I_{Na} = g_{Na,max}\left(m^3 h\right)\left(V_m - E_{Na}\right) \tag{14.14}$$

where E_{Na} is the Na$^+$ reversal potential. The potassium current may be described similarly, although, instead of voltage-operated activation and inactivation gates, one need only include an activation gate.

The Hodgkin–Huxley model explained the mechanisms for initiation and propagation of action potentials in neurons and has remained the basis for computational modeling of nerve function. Hodgkin and Huxley received the Nobel Prize for physiology or medicine in 1963 in recognition of this accomplishment and related work on understanding neuronal mechanisms.

Potassium Channel: An Example of a Nano Ion Pump Engine.   As an example and to give an appreciation of how ion channels work from a nanoscience point of view, we illustrate one type of gated channel for which the structure has been determined. Potassium channels are very important and common in many different types of cells, from bacterial to vertebrate nerve and muscle cells. There are a large number of different types of potassium ion (K$^+$) channels, but one of the most important is the voltage-gated family that opens and closes in response to changes in the membrane potential. This type of potassium channel is especially important for its role in the action potential of nerve cells. Voltage-gated potassium channels are found in almost all organisms including bacteria, where they maintain the ionic balance of the cell.

The molecular model shows a K$^+$channel from the bacterium *Streptomyces lividans* (**Fig. 14.8**). The channel is formed from a transmembrane protein group with a central pore through which the potassium ion is gated.

Functional Components: Filter and Gate Subunits.   The K$^+$ channel consists of two major functional parts: the filter pore, which selectively allows potassium ions to pass, and the gate, which opens and closes the channel based on the membrane potential. The pore part of the channel both selectively filters and conducts ions through the membrane. The voltage-sensing gate portion of the channel changes its conformation in response to changes in transmembrane voltage, thus closing the pore. The filter region forms the extracellular end of the pore. The voltage activation gate lies at the opposite end of the channel, at its cytoplasmic mouth.

How the Subunits Are Constructed.   The potassium channel is made up of four identical protein chains that link together to encircle the central pore space that

| FIG. 14.8 | *Potassium ion channel.* |

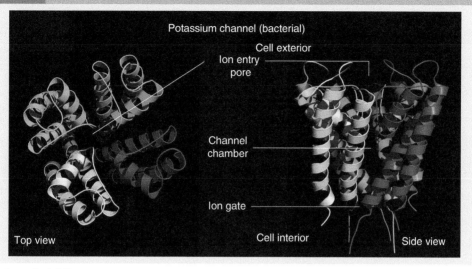

serves as the ion conduction pathway. The two functional domains of the channel, the filter and the gate, are formed by the opposite ends of the four grouped protein chains. Each protein chain is composed on six alpha helices. The first four helices of each chain form the voltage gate, which is thus composed of 16 alpha helices in all. The last two transmembrane helices of each chain, along with a loop of protein chain that circles back into the membrane (called a re-entrant P loop), are embedded inside the lipid portion of the membrane to form the pore walls. The P loop includes a short helix and a region of polypeptide chain that forms the ion selective filter.

The amino acid chain sequence of this potassium channel transmembrane protein is similar to all known K^+ channels, particularly in the pore region. X-ray analysis with data to 3.2 Å has shown that the four identical protein subunits create an inverted teepee, or cone, cradling the selectivity filter of the pore in its outer end. The narrow selectivity filter is only 12 Å (0.12 nm) long, whereas the remainder of the pore is wider and lined with hydrophobic amino acids. A large, water-filled cavity and helix dipoles are positioned so as to overcome electrostatic destabilization of K^+ ions in the pore in the interior of the membrane bilayer. Carbonyl oxygen atoms attached to the protein chain line the selectivity filter. The selectivity filter contains two K^+ ions about 7.5 Å apart that provide electrostatic repulsive forces to overcome attractive forces between K^+ ions and the polarizable covalent bonds in the protein chain making up the selectivity filter.

How It Works. The ion conduction pathway can switch between two main functional states: open and closed. When the cell membrane is polarized so that the interior of the cell is at a negative voltage relative to the exterior, the K^+ channels remain closed. When the membrane is depolarized, these channels open rapidly (in less than 1 ms), allowing ions to flow passively down their electrochemical gradients at near diffusion rates (up to one hundred million ions per second). K^+ channels have a selective permeability for potassium of at least 10,000 times greater than for sodium.

Comparing the x-ray structures of K^+ channels crystallized in a closed conformation to similar channels crystallized in an open conformation suggests that bending or kinking of the inner pore-lining helices plays a key role in pore gating. Studies of the potassium channel have identified two segments that contain several charged protein residues, and these charged residue regions apparently undergo conformational changes in response to the potential difference across the membrane; therefore, they could contribute to the gating mechanism.

Recent research is revealing that voltage-gated K^+ channels all have very similar pore structures and permeability mechanisms, although they use diverse mechanisms of gating (the processes by which the pore opens and closes). Certain features of potassium channels indicate a multi-ion conduction mechanism involving single-file queuing of ions inside a long, narrow pore. Because of these properties, K^+ channels are classified as "long pore channels." The pores of all the K^+ channels are blocked by tetraethylammonium (TEA) ions, making this and related compounds strong toxins.

Ball and Chain Mechanism: An Example of Nanomachinery in the Ion Gate.　In many voltage-gated potassium channels, a protein ball and chain functions to open and close the ion gate, like a cork bottle stopper on a string. The amino-terminal domain of the protein chain, on the inside of the cell where the gate is located, forms the ball and chain. The ball region swings freely through the cytoplasm on its protein chain tether. The ball is the right size and shape to plug the pore when it swings in from the cytoplasm, thus deactivating the channel. The tether is long enough for the ball to reach the inside surface of the plasma membrane, where the positively charged, hydrophobic ball can be attracted and attached to anionic phospholipids of the lipid bilayer. A change in the membrane potential is apparently sufficient to dislodge the ball from the membrane and into the plugged position [90].

14.2.4　Synapses and Neurotransmitters

We have seen how the ion gating mechanism works for the propagation of the action potential along nerve cells. In the 1950s, Katz, Fatt, and Exxles demonstrated that ion channels were also fundamental to signal transmission across the *synapses*, the junctions between nerve cells. In the synapses, the ion channels are not gated by changes in membrane voltages as were the action potential gates. Instead, the synaptic ion channels were found to be regulated by small molecules, ligands like *acetylcholine*. In the 1960s and 1970s, working with large amounts of tissue to isolate very small amounts of ligands, researchers discovered many other small molecules and peptides that gate synapse channels, including glutamate, GABA, glycine, serotonin, dopamine, and norepinephrine. These became important in understanding and regulating overall brain and nerve activity.

Synapses.　The junction between a neuron and another cell is called a synapse. Most commonly we think of synapses as being between two neurons, but synapses also connect nerve endings to muscles and sensory organs. Synapses connecting a nerve to muscle fibers are also called *neuromuscular junctions* or *myoneural junctions*.

Neurons.   To make the function of the synapse clear requires an overview of the structure and function of *nerve cells*, the *neurons*. Neurons are the specialized cells that carry messages in the body. Neurons carry out specialized roles as motor neurons, sensory neurons, and highly specialized types of neurons found in the brain. A typical neuron has four main morphological regions: the *cell body*, *dendrites*, *axon*, and *presynaptic terminals*. The *cell body* is similar to that of other cells, with a nucleus and cytoplasm. The cell body is also referred to by the neurological terms *perikaryon* and *soma*. It is the base station that maintains the metabolic functions for the more specialized appendages, the *dendrites* and *axon*, collectively called *nerve processes*.

When action potentials travel down the axon to the presynaptic terminal, they initiate the release of neurotransmitters into the gap between the cell junctions (called the *synaptic cleft*). The action potential triggers the opening of calcium ion (Ca^{2+}) channels in the axon membrane, allowing calcium to enter the cell. The calcium ions in turn cause the opening of small vesicles bound to the membrane, containing neurotransmitters. The neurotransmitters are released into the narrow cleft and bind to ion channel receptors on the receiving cell membrane. Typically, the receptors are ligand-gated ion channels, which open in response to the neurotransmitter to change the membrane potential of the receiving cell (**Fig. 14.9**).

The potential change in the receiving cell may cause excitation or inhibition. Excitation usually is associated with the flow of positive ions into the cell and inhibition by negative ion flow. If enough excitation signals are received, they trigger an action potential in the dendrite of the receiving neuron, thus propagating the signal to the next cell. Synapses may release neuroinhibitors, which have the effect of lowering the activity of the receiving cell.

The receptors in the synapses are sensitive to any compound recognized by the gate in their ion channels. Toxins, anesthetics, and drugs that act on the nervous system may mimic the neurotransmitters, block the receptors, or instigate other effects to interfere with normal nerve function.

14.2.5   *Hormones and Regulation of Cell Growth and Metabolism*

In the previous section we saw how ion channels in the cell membrane work and how changes in electrical potential trigger the release of chemical neurotransmitters in the nervous system. Now we take a look at a more generalized system of intercellular communication: the regulation of growth and metabolism by chemical messengers acting in a less localized region than the synapses. As we go through some highlights of the system of hormone regulation, it is important to remember that the same fundamental mechanisms of communication through receptors embedded in the cell membrane are at work throughout all cells of the organism.

Endocrine System.   Historically, the first cell signaling chemicals to be recognized were those associated with the *ductless glands* of the human body, which were

| **FIG. 14.9** | *Neuronal synapse.* |

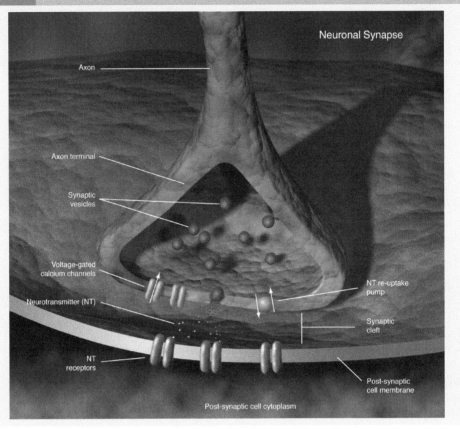

Source: Image by Anil Rao.

classified anatomically as the *endocrine glands*, in contrast to the *exocrine* glands, which had obvious visible ducts to deliver their secretions into the blood or digestive tract, like the sweat, salivary, and bile glands. The *endocrine system* is an anatomical and physiological classification for nine ductless glands that secrete important *hormones* [91,92].

Hormones are chemical regulators that are carried throughout the body, where they are recognized by specific cell receptors to regulate *growth, metabolism*, and *sexual development*. Over 100 different hormones have been identified and classified in humans. In insects, hormones control the growth and transition between different phases of development. Recent advances are leading to the realization that almost every organ of the body releases some type of chemical signaling compound. Thus, the endocrine system represents the most dramatic example of the phenomenon of cell signaling, which plays a much more generalized role in the body.

The Control of Growth. The growth of a cell or tissue is now recognized to be a complex process, controlled by a network of interactions among genes, metabolism, nutrition, and hormones. Different aspects of growth include developmental patterns, growth rate, body and tissue size, aging, and cancer. These interactions are under the control of *chemical, electrical,* and *physical* influences to which the individual cells respond. Because the cells are programmed to respond in specific ways to specific chemicals, we may think of these chemical messengers as *signals* that carry information to the cells. The signals are received and recognized at specific receptors.

Cell signaling requires recognition of the hormone signal by the receiving cell, which involves interaction of the hormone with a protein receptor. The *receptor protein* may be in the cell membrane or internal to the cell. Typically, the hormone fits into a steric structure in the receptor protein and produces a conformal or electronic change in the state of the protein. This process is similar to the way in which gating locations in cell membrane proteins interact with ions and other ligands, as we saw with the potassium ion channel.

Any regulatory interaction where the result causes the originating step to moderate is called a *negative feedback* loop. Negative feedback tends to stabilize a process by regulating the rate of the driving step. With negative feedback loops, any deviation from the normal or optimal control *set point* gives rise to a countering action that tends to return the system to the control state. In contrast, *positive feedback* leads to a runaway process that grows or speeds up to a breaking point or discontinuity, after which the cycle starts over [58].

Cellular Hormone Production Factory: Pancreas.   An important example of a hormone system is the *insulin* and *glucogen* system, which regulates metabolism throughout the body. The complex of cells that interact to synthesize insulin and related hormones is an illustrative example of how the cellular signaling system interacts to control hormone production. Insulin up-regulates the metabolism of glucose, and glucogen down-regulates glucose metabolism in cells throughout the body. Insulin is produced in the *pancreas* gland by specialized *beta cells* located in regions of the pancreas called the *islets of Langerhans*.

14.3   DNA, RNA, AND PROTEIN SYNTHESIS

A fundamental aspect of the structure of living organisms is the pair of special molecules that control protein formation. This molecular machinery generates proteins from a physical code stored in a polymer chain by means of an elegant and beautiful nanoprocessing engine that is inspiring and instructive for the design of molecular computers and actuators. Proteins are synthesized in living organisms by a catalytic chain building process directed by two special types of long chain molecular structures, called nucleic acids (or nucleosides) because they were first isolated in the cell nucleus [44,45,93].

14.3.1   DNA and RNA Function and Structure

The two types of nucleic acid play separate roles in protein synthesis: DNA (deoxyribonucleic acid) stores the information used to make a protein. DNA is the template that makes RNA (ribonucleic acid); the latter actually encodes the

FIG. 14.10   *The DNA double helix structure and the bases from which it is formed.*

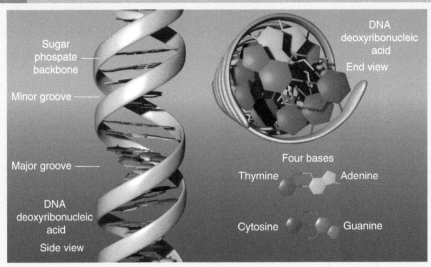

Source: Image by Anil Rao.

amino acid sequence and makes the protein chains with the aid of enabling catalysts.

DNA consists of pairs of nucleotides (also called bases) arranged in a double helix (like a ladder twisted into a spiral) (**Fig. 14.10**). Each nucleotide has three parts: a five-carbon deoxyribose sugar ring, a phosphate group composed of phosphorous and oxygen, and a nitrogen-containing base. The DNA nucleotide bases may be one of four compounds: adenine, cytosine, guanine, or thymine (**Fig. 14.10**). The four bases that are part of nucleotides in DNA are represented in code sequences by the letters A, C, G, and T. The sequences of these bases in a DNA molecular strand comprise the genetic code. The three-dimensional chemical structure of the bases is critical for allowing DNA to form helical chains and to self-replicate. Adenine can form two hydrogen bonds with thymine while cytosine can form three hydrogen bonds with guanine. A–T and C–G are called complementary pairs. This coupling property of the bases is the mechanism for the bonding process in DNA self-replication.

Bases in the DNA strand pair up with complementary bases to form a chain of RNA. RNA has the three bases—adenine, cytosine, and guanine—in common with DNA, but in RNA uracil replaces the DNA thymine. Each RNA nucleotide also has three parts: a five-carbon ribose sugar ring (instead of the deoxyribose in DNA), a phosphate group, and one of the nitrogen-containing bases (**Fig. 14.11**).

The RNA chain is usually arranged in a single helix. The bases in the RNA in turn pair up with amino acids to form a protein chain. The amino acids are chemically attracted onto the bases of the RNA strand and then linked by an enzyme into a protein polymer chain. Three bases in sequence on the RNA chain match three portions of each amino acid so as to absorb it in place during the chain-building process (**Fig. 14.12**).

| **FIG. 14.11** | *Structure of RNA.* |

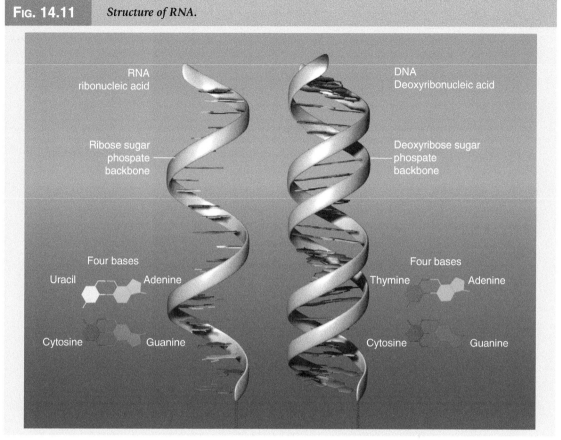

Source: Image by Anil Rao.

Thus, sets of three bases are said to code for each amino acid and are referred to as "codons." For example, the codon GCU uniquely attracts and binds to the amino acid alanine. The codon UAG does not match any of the 20 amino acids used in protein building, and causes the enzyme-linking process to stop. Recall from the discussion of amino acids that each amino acid has a base (amino) and acid (carboxyl) location on the molecule. These opposite chemical functional groups are aligned in neighboring amino acids by the RNA replicating process, ready to be joined by a polymerase catalyst. **Figure 14.12** illustrates how DNA synthesizes RNA, which in turn synthesizes proteins.

14.3.2 DNA Replication

DNA is replicated with the aid of enzymes called DNA polymerases. The DNA polymerases copy a template DNA chain by polymerizing nucleotides to form a DNA chain that is complementary to the original. Polymerases extend the DNA strand by adding nucleotides to the end of the growing chain. Different types of DNA polymerases carry out repair of damaged DNA strands. Polymerization of

FIG. 14.12 *DNA transcribes RNA which produces proteins DNA codes for synthesis of proteins in the cell by producing complementary RNA (messenger RNA, mRNA). Transfer RNA (tRNA) catalyzes the alignment of polypeptide chain protein constituents and their peptide bonding.*

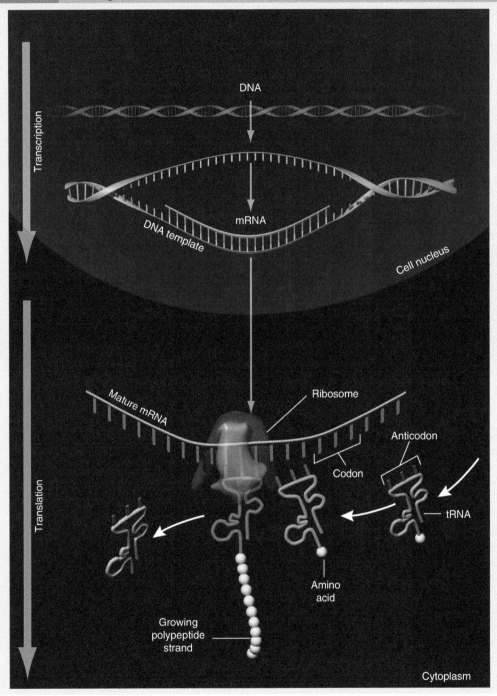

Source: Image by Anil Rao.

the DNA chain occurs by joining the hydroxyl group on a carbon of the sugar to the phosphate of an adjacent nucleotide. Other enzymes catalyze the building of RNA chains from DNA and the synthesis of proteins from RNA (**Fig 14.12**).

14.3.3 *DNA as a Genetic Information Storage Material*

DNA sequences preserve the information on how to program the synthesis of proteins. This information is not changed by the replication process, so it is available for reuse as many times as the living organism needs it, so long as the DNA strand is not damaged or altered. DNA is copied when organisms reproduce, so this information is passed down to subsequent generations. DNA is the material that encodes the genetic information of all living things. The units of information that encode for particular proteins or phenotypic traits are called genes.

DNA and Genes.   A specific sequence of bases in the DNA chain will encode a sequence on the RNA chain that generates a protein. Most proteins are common to a large family of organisms and only a few are unique to a single species of life. The region on a chain of DNA that encodes a protein or any feature of inheritance is called a gene. Some proteins regulate the activity of others, and the complex mechanisms of the cell include regulation of the protein replication process itself, so the DNA and RNA by themselves do not provide a simple direct map to the phenotype of the organism.

The genome, defined as the total number of base pairs in the DNA of a particular type of cell, does not relate in a predictable fashion to the number of genes. DNA in many cases appears to have long sequences of base pairs whose function is not fully known, many of which are highly repetitive. Each gene codes for the synthesis of a protein indirectly by coding for an RNA codon sequence.

Exon and Intron.   The portion of the DNA that codes for RNA that synthesizes a protein is called an exon. Some regions of the genome consist of long sequences in the DNA called introns, which are transcribed into RNA but not translated into protein. These may serve to separate the genes or regulate their transcription [94,95].

Genotype and Phenotype.   Historically, starting with Mendel in the nineteenth century, genes were identified as the unit that carried any inherited characteristic, such as the color of a flower or shape of a leaf, beak, or wing. Genes were identified by the ability of observable features to be inherited. Mendel originated the concept of genes by counting the relative abundance of observable features in the offspring of crossed varieties of peas. His ideas were found to have universal significance long before the physical basis of the gene was understood.

When biologists began to understand the inner structure of the cell and the process of cell division, the gene was associated with the chromosomes located in the cell nucleus of higher cells. Once the role of DNA in encoding genes was understood, it was realized that the gene is not a simple physical entity. A gene can include segments of DNA that do not code for a protein directly but instead have sequences that bind other regulatory molecules involved in the protein synthesis [93].

The composition and configuration of the compounds making up an organism—from its molecular composition to the arrangement of its proteins and cells to the shapes and interconnections of its organs and appendages—are specific for each organism and are called its *phenotype*. The DNA sequence that encodes the genes and controls the development of the phenotype is called the *genotype*. It is of great theoretical and practical importance to be able to determine which genotypes give rise to the synthesis of particular products. By comparing the genotypes of different organisms and individuals we can understand relationships and the causes of inherited diseases and disorders.

DNA Sequencing to Determine Genotypes and Characterize Genes. A large body of knowledge has been developed by biologists and geneticists, starting late in the nineteenth century, on genetic inheritance based on observations of deformities and modification of the shapes and number of chromosomes in the cell—structures that can be made visible with staining and preparation under the microscope. By the middle of the twentieth century biologists had established that the number of chromosomes in each cell and the amount of nucleic acid were an invariant characteristic for normal members of any species, thus indicating strongly that genetic inheritance was carried by nucleic acid in the chromosomes. Some plants and animals have large chromosomes, making them useful subjects for experimental study. Higher animals such as mammals have similarities in their chromosomes—a state that leads to generalizations and understanding of the effects of chromosome abnormalities. Since chromosomes have specific shapes and numbers at certain stages of cell development, genetic biologists were able to map various inherited characteristics onto specific regions of the chromosomes, especially for species where chromosomes were easily observable, such as fruit flies, squid, and a number of plants. When the molecular mechanisms of DNA replication began to be worked out, it began to be possible to envisage correlating knowledge about genetic inheritance and chromosomes with detailed molecular sequences at various places on the DNA strand and, ultimately, with complete genotypes.

Genotypes can be determined in principle by analyzing the sequence of bases in the strands of DNA in the cell. Since there are millions of nucleotides in the genome of even the simplest organism, the analysis of their sequence is an enormous and very difficult accomplishment (the human genome contains about 3 billion base pairs, notated as 3 gigabases [3 Gb]). To perform a chemical analysis of each of the millions of nucleotides while keeping track of their sequence in the DNA chain requires complex and sophisticated strategies if the task is to be made feasible in even the simplest organisms.

The first methods developed to sequence DNA were termination sequencing developed by Frederick Sanger and Howard Chadwell and chemical sequencing developed by Allan Maxam and Walter Gilbert, both during the 1970s. Because of the difficulty and labor involved in performing sequencing, only short portions of genomes were sequenced at first. Since the award of the Nobel Prize in 1980 to Sanger and Gilbert, intensive and large-scale efforts have been made to extend these first steps. Worldwide coordinated projects were started to achieve rapid and inexpensive sequencing of complete genomes for any organism or individual, culminating in the first nearly complete draft sequences of the human genome in 2001 and 2002, spurred by the U.S. government-sponsored Human Genome Project [93–98].

The development of genome sequencing was a unique large-scale development involving many cooperative and competitive efforts, analogous for biology to space exploration programs or to the high-energy physics accelerator projects for discovery of elemental particles. A number of important steps marked the rapid progress in ability to decipher genome sequences. The first sequencing of a complete organism was carried out in 1983 by a team of scientists at the U.K. Medical Research Council for the Epstein–Barr virus, whose genome had approximately 170,000 base pairs (170 kb). The same year Mullis and coworkers developed a way to amplify replication of strands of DNA by repetitive cycling, which could be readily automated. This development, called variously the polymerase chain reaction (PCR), DNA thermocycling, or DNA amplification, made it feasible to obtain workable amounts of DNA from very small amounts of starting material and revolutionized DNA sequencing and cloning [99].

Initially, DNA sequencing reactions are only capable of analyzing the sequence of a few thousand nucleotides at a time. PCR and computer-assisted sequence matching programs developed in the 1990s made it possible to determine the sequence for much larger pieces of DNA. Since genomes typically contain millions of billions of base pairs, it was necessary to break up DNA into segments in order to sequence it. One had to somehow keep track of where the breaks in the DNA strand were made in order to reconstruct the entire sequence. One approach to this problem is called shotgun sequencing. In the shotgun sequencer method, the DNA is broken up and the fragments amplified by replicating them. Then the fragments are sequenced individually and, using computer matching programs, they are compared for overlaps to try to deduce the complete sequence of the original DNA.

Quite apart from progress in determining the sequence of entire genomes, DNA sequencing of shorter characteristic portions of the genome was very useful in genetic and forensic applications. Polymerase chain reaction (PCR), computer matching techniques, and other advances have pushed DNA sequencing from a small-scale, labor-intensive technology capable of reading several kilobase pairs of sequence per day into an automated, high-throughput, multibillion dollar industry. Nanotechnology has enabled sequencing technologies to be developed based on advances in electroanalytical chemistry and photosensors on the nanoscale that do not involve either electrophoresis or Sanger sequencing chemistries. For example, sequencing by synthesis (SBS) involves multiple parallel microsequencing addition events occurring on a surface where data from each round are detected by imaging [100].

By understanding and manipulating the synthetic molecular nanomachinery of protein synthesis we can produce useful protein products and modify the characteristics of organisms—for example, to eliminate genetic defects or synthesize desirable hormones, drugs, foods, or fuels in bioreactors, crops, or livestock. The ability to decipher and manipulate DNA sequences leads to the possibility of artificially making identical copies of individual organisms (called *cloning*) and inserting genes into bacteria, plants, or animals to alter their phenotypes (genetic engineering, making transgenic or chimeric organisms, performing gene surgery).

The protein products of DNA and RNA synthesis are the nanomachinery of the cell, determining the development and maintenance of the organism. Every aspect of shape, size, metabolism, and behavior of the organism is based on the structure and function of the proteins encoded by its DNA, and how they are expressed and energized.

14.3.4   RNA and DNA Nanoengines: Viruses and Phages

Now that we have given highlights of the functions of biomolecular nanoma-chinery, we will look at a very simple example of how DNA and protein struc-tures work together as nanomachines. This example is easy to visualize because it is not confined inside the complex environment of the cell. In section 14.1.3 we introduced viruses in the survey of the sizes of organisms in relation to their biomolecular structure. Viruses fall below the minimum size possible to have all of the molecular machinery necessary for an independent reproduction cycle, but they can reproduce by invading and hijacking the machinery of cells. Now that we have surveyed how DNA and RNA work to replicate copies of them-selves and to reproduce proteins, it is possible to understand how viruses can take over this machinery and alter it to make copies of viral DNA and protein coats.

In some viruses, called phages (from the Greek word meaning "to eat"), the protein coat has an active function: It attaches to the cell wall of a bacterial host and drives the viral DNA or RNA core through the wall into the body of the cell where it can replicate. Thus, we see a step in the continuum between nonliving molecular machinery and living organisms (**Fig. 14.13**).

FIG. 14.13	*Bacteriophage. Bacteriophages use protein nanomachinery to insert DNA into a host bacterium, where the phage DNA takes over the protein replication process of the bacteria to produce more bacteriophages.*

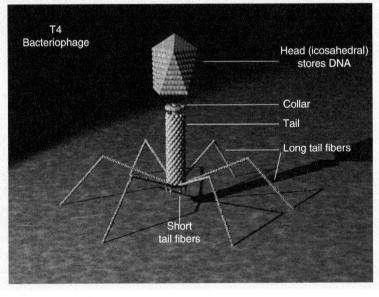

Source: Image by Anil Rao.

14.3.5   *The Role of the Protein Environment*

To understand how the nanomachinery of the cell works, it is necessary to take account of the role of the water and protein environment in which the process operates. As emphasized in earlier sections, water, hydration, and hydrogen bonding play a key role in biological processes at the nano level. As one example, we illustrate the role of "zinc sticky fingers" in shepherding the RNA transcription process (**Fig. 14.14**). The "sticky fingers" in this case are proteins bound to a zinc ion known as transcription factor IIIA. These finger-shaped proteins surround and stabilize the RNA backbone helix during the transcription process. The complex role of zinc finger proteins took many years of experimentation and separations to elucidate before it was isolated and its structure determined. We conclude this chapter with this example to remind potential entrants into the world of biomolecular nanoscience and nanotechnology that biological systems are extremely subtle and complex, with many surprises. In an introduction of this type there is space and time for only a brief introduction to some of them. Hopefully, this introduction will spark further curiosity about and appreciation for the amazing possibilities of biomolecular nanoscience.

| **Fig. 14.14** | *Transcription actor IIIA zinc "Sticky Fingers" protein. Transcription factor IIIA is a protein that binds to RNA, shaping the rRNA backbone during the process of transcription.* |

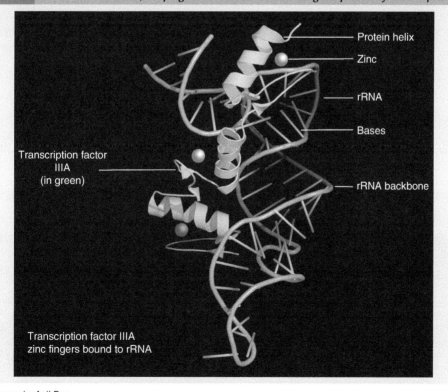

Source: Image by Anil Rao.

14.4   Concluding Remarks

14.4.1   *Emerging Concepts and Developments*

This chapter introduced life sciences from a nanoscience viewpoint. Approaching life as an emergent property of the energetics of matter on the nanoscale gives an integrated viewpoint for studying the self-organizing properties of macromolecules such as proteins, RNA, and DNA.

Biological nanosystems must be seen as the product of a long and complex process of evolutionary development, but understanding how they work should reveal intelligent short cuts for remedying disorders, exploiting biological machinery to produce new products, and designing soft nanomachines for energy, computation, and communication. Approaching the complex world of bioscience from the viewpoint of nanoscience is one way of introducing a field that is far too broad and intricate to master in an introductory survey. Since natural biology is very rich with nanoscale phenomena, it is not surprising that nanoscience intersects in many areas with cellular and molecular biology. As the field matures it is possible that nanoscience will contribute to the understanding of larger scale phenomena and unraveling of the complexities of signaling and immune systems.

Future Possibilities.   Biomolecular nanoscience is a novel way of looking at the world of living systems. It should be a productive endeavor leading to new insights and knowledge about biology and life, as well as guiding the development of new types of artificial nanoengines. Biomolecular nanoscience is an integrated approach to the complex phenomena of living systems. As such, it lays a foundation for new approaches to biotechnology and biomedicine. We will explore some aspects of the possibilities for these fields in chapters of the companion volume, *Introduction to Nanotechnology*.

References

1. J.-M. Lehn, Perspectives in supramolecular chemistry—From molecular recognition towards molecular information processing and self-organization, *Angewandte Chemie International Edition*, 29, 1304–1319 (2003).
2. F. Vogtle, *Fascinating molecules in organic chemistry*, John Wiley & Sons, Inc., New York, (1992).
3. T. Sekine, T. Niori, J. Watanabe, T. Furukawa, S. W. Choi, and H. Takezoe, Spontaneous helix formation in smectic liquid crystals comprising achiral molecules, *Journal of Materials Chemistry*, 7, 1307–1309 (1997).
4. H. Aldersey-Williams, *The most beautiful molecule: Discovery of the buckyball*, John Wiley & Sons, Inc., New York, (1997).
5. M. Meyyappan and L. Kelly, *Carbon nanotubes: Science and applications*, CRC Press, Boca Raton, FL, (2005).
6. S. Reich, C. Thomsen, and J. Maultzsch, *Carbon nanotubes*, John Wiley & Sons, Inc., New York, (2004).
7. A. Ullmann, ed., *Origins of molecular biology: A tribute to Jacques Monod*, ASM Press, Washington, D.C. (2003).
8. J. Cairns, J. D. Watson, and G. S. Stent, eds., *Phage and the origins of molecular biology*, CSHL, Woodbury, NY (2000).

9. R. B. Corey and H. R. Branson, The structure of proteins: Two hydrogen-bonded helical configurations of the polypeptide chain, by Linus Pauling, *Proceedings of the National Academy of Science, USA*, 37, 205–211 (1951). Available on JSTOR at www.jstor.org/view/00278424/ap001041/00a00030/0

10. L. Fredholm, The discovery of the molecular structure of DNA—The double helix: A scientific breakthrough, on The Official Web Site of the Nobel Foundation, at: http://nobelprize.org/educational_games/medicine/dna_double_helix/readmore.html

11. J. D. Watson and F. H. C. Crick, A structure for deoxyribose nucleic acid, *Nature*, 171, 737 (1953), Commentary by T. Zinnen on National Health Museum web site: www.accessexcellence.org/RC/AB/BC/casestudy2.htm

12. Honoring the lifetime achievements of Mack J. Fulwyler, special issue, *Cytometry*, A67, 53–179 (2005).

13. R. Kondratas, The history of the cell sorter, Smithsonian Institution Videohistory Collection, www.si.edu/archives/ihd/videocatalog/9554.htm

14. M. R. Loken, D. R. Parks, and L. A. Herzenberg, Two-color immunofluorescence using a fluorescence-activated cell sorter, *Journal of Histochemistry and Cytochemistry*, 25, 899–907 (1977).

15. M. B. Hale and G. P. Nolan, Phospho-specific flow cytometry: Intersection of immunology and biochemistry at the single-cell level, *Current Opinions in Molecular Therapy*, 8, 215–224 (2006).

16. A. J. Westphal et al., Kinetics of size changes of individual *Bacillus thuringiensis* spores in response to changes in relative humidity, *Proceedings of the National Academy of Science, USA*, 100, 3461–3466 (2003).

17. K. Sachs, O. Perez, D. Pe'er, D. A. Lauffenburger, and G. P. Nolan, Causal protein-signaling networks derived from multiparameter single-cell data, *Science*, 308, 523–529 (2005).

18. S. B. Carroll, *Endless forms most beautiful: The new science of Evo Devo and the making of the animal kingdom*, W. W. Norton & Co., New York (2005).

19. M. J. Denton, C. J. Marshall, and M. Legge, The protein folds as platonic forms: New support for the pre-Darwinian conception of evolution by natural law, *Journal of Theoretical Biology*, 219, 325–342 (2002).

20. E. D. Wolf, ed., Advanced submicron research and technology development at the national submicron facility, *Proceedings of the IEEE*, 71, 589–600 (1983). Available at IEEE archives website at: ieeexplore.ieee.org/iel5/5/31325/01456911.pdf?arnumber=1456911

21. NSF Nanobiotechnology Center, center website at: www.nbtc.cornell.edu/

22. R. E. Jenkins and S. R. Pennington, Arrays for protein expression profiling: towards a viable alternative to two dimensional gel electrophoresis, *Proteomics*, 1, 13–29 (2001).

23. M. J. Heller and A. Guttman, *Integrated microfabricated biodevices*, Marcel Dekker, New York (2002).

24. M. Schena, *Microarray analysis*, John Wiley & Sons, Inc., New York (2002).

25. J. S. Albala and I. Humphery-Smith, *Protein arrays, biochips and proteomics*, Marcel Dekker, New York (2003).

26. N. H. Malsch, ed., *Biomedical nanotechnology*, CRC Press, Boca Raton, FL (2005).

27. A. Guiseppi-Elie and T. Vo-Dinh, eds., *Biochips handbook*, CRC Press, Boca Raton, FL (2007).

28. H. K. Baca, C. Ashley, E. Carnes, D. Lopez, J. Flemming, D. Dunphy, S. Singh, Z. Chen, N. Liu, H. Fan, G. P. Lopez, S. M Brozik, M. Werner-Washburne, and C. J. Brinker, Cell-directed assembly of lipid–silica nanostructures providing extended cell viability, *Science*, 313, 337–341 (2006).

29. B. Gimi, T. Leong, Z. Gu, M. Yang, D. Artemov, Z. M. Bhujwalla, and D. H. Gracias, Self-assembled three-dimensional radio frequency (RF) shielded containers for cell encapsulation, *Biomedical Microdevices*, 7, 341–345 (2005).

30. P. Braun, Y. Hu, B. Shen, A. Halleck, M. Koundinya, E. Harlow, and J. LaBaer, Proteome-scale purification of human proteins from bacteria, *Proceedings of the National Academy of Science, USA*, 99, 2654–2659 (2002).

31. P. H. C. van Berkel, M. M. Welling, M. Geerts, H. A. van Veen, B. Ravensbergen, M. Salaheddine, E. K. J. Pauwels, F. Pieper, J. H. Nuijens, and P. H. Nibbering, Large scale production of recombinant human lactoferrin in the milk of transgenic cows, *Nature Biotechnology*, 20, 484–487 (2002).

32. E. Behboodi et al., Viable transgenic goats derived from skin cells, *Transgenic Research*, BW2118, 1–10 (2004).

33. T. V. Murthy, W. Wu, Q. Q. Qiu, Z. Shi, J. LaBaer, and L. Brizuela, Bacterial cell-free system for high-throughput protein expression and a comparative analysis of *Escherichia coli* cell-free and whole cell expression systems, *Protein Expression and Purification*, 36, 217–225 (2004).

34. J. Benyus, *Biomimicry: Innovation inspired by nature*, William Morrow and Co., New York (1997).

35. D. S. Goodsell, *Bionanotechnology: Lessons from nature*, Wiley-Liss, Inc., Hoboken, NJ (2004).

36. R. E. Smalley, Discovering the fullerenes, *Reviews of Modern Physics*, 69, 723–730 (1997).

37. C. Branden and J. Tooze, *Introduction to protein structure*, 2nd ed., Routledge, Abington, OX, UK (1999).

38. A. M. Lesk, *Introduction to protein architecture: The structural biology of proteins*, Oxford University Press, Oxford, UK (2001).

39. D. Whitford, *Proteins: Structure and function*, John Wiley & Sons, Inc., New York (2005).

40. A. Fersht, *Structure and mechanism in protein science: A guide to enzyme catalysis and protein folding*, 3rd ed., W. H. Freeman, New York (1998).

41. G. C. Howard and W. E. Brown, *Modern protein chemistry: Practical aspects*, CRC Press, Boca Raton, FL (2001).

42. G. A. Petsko and D. Ringe, *Protein structure and function*, New Science Press, London (2004).

43. H. Rehm, *Protein biochemistry and proteomics*, Elsevier, Burlington, MA (2006).

44. G. Karp, *Cell and molecular biology*, John Wiley & Sons, Inc., New York (2005).

45. H. Lodish, A. Berk, P. Matsudaira, C. A. Kaiser, M. Krieger, M. P. Scott, and L. Zipursky, J. Darnell, *Molecular cell biology*, 5th ed., W. H. Freeman, New York (2003).

46. J. Bereiter-Hahn, Behavior of mitochondria in the living cell, *International Review of Cytology*, 122, 1–63 (1990).

47. L. Grohmann, A. Brennicke, and W. Schuster, The mitochondrial gene encoding ribosomal protein S12 has been translocated to the nuclear genome in Oenothera, *Nucleic Acids Research*, 20, 5641–5646 (1992).

48. N. J. Dimmock, A. J. Easton, and K. N. Leppard, *Introduction to modern virology*, 6th ed., Blackwell Publishing, Inc., Malden, MA (2006).

49. J. Collinge, Prion diseases of humans and animals: Their causes and molecular basis, *Annual Review of Neuroscience*, 24, 519–550 (2001).

50. K. M. Pan, M. Baldwin, J. Nguyen, M. Gasset, A. Serban, D. Groth, I. Mehlhorn, Z. Huang, R. J. Fletterick, F. E. Cohen, and S. B. Prusiner, Conversion of alpha-helices into beta-sheets features in the formation of the scrapie prion proteins, *Proceedings of the National Academy of Science, USA*, 90, 10962–10966 (1993).

51. G. Oberdörster et al., Principles for characterizing the potential human health effects from exposure to nanomaterials: Elements of a screening strategy, *Particle and Fiber Toxicology*, 2, 8 (2005).

52. C. D. Klaassen, J. B. Watkins, and L. J. Casarett, eds., *Casarett & Doull's essentials of toxicology*, McGraw–Hill, New York (2001).

53. B. Onaral and J. P. Cammarota, *Complexity, scaling, and fractals in biomedical signals, Biomedical engineering handbook,* Chap. 58, CRC Press, Boca Raton, FL (2000).

54. R. Thom, *Structural stability and morphogenesis,* 2nd rev ed., Addison Wesley, Boston (1989).

55. G. Nicolis and I. Prigogine, *Self-organization in nonequilibrium systems: From dissipative structures to order through fluctuations,* John Wiley & Sons, Inc., New York (1977).

56. E. Ott, *Chaos in dynamical systems,* Cambridge University Press, Cambridge, UK (2002).

57. Z. Li, W. A. Halang, and G. Chen, eds., *Integration of fuzzy logic and chaos theory: Studies in fuzziness and soft computing,* Springer–Verlag, Berlin, (2006).

58. W. A. Halang and A. D. Stoyenko, *Constructing predictable real time systems,* Kluwer Academic Publishers, Dordrecht, the Netherlands (1991).

59. C. de Duve and M. J. Osborn (panel 1 moderators), Constraints on size of a minimal free-living cell. In *Size limits of very small microorganisms: Proceedings of a workshop* (1999), National Academy of Science, USA (2000), www.nap.edu/html/ssb_html/ NANO/nanopanel1.shtml

60. K. Nealson (panel 2 moderator), Is there a relationship between minimum cell size and environment? In *Size limits of very small microorganisms: Proceedings of a workshop* (1999), National Academy of Science, USA (2000), www7.nationalacademies. org/ssb/nanopanel2.html

61. E. O. Kajander and N. Çiftçioglu, *Nanobacteria:* An alternative mechanism for pathogenic intra- and extracellular calcification and stone formation, *Proceedings of the National Academy of Science, USA,* 95, 8274–8279 (1998).

62. J. T. Hjelle, M. A. Miller-Hjelle, I. R. Poxton, E. O. Kajander, N. Ciftcioglu, M. L. Jones, R. C. Caughey, R. Brown, P. D. Millikin, and F. S. Darras, Endotoxin and nanobacteria in polycystic kidney disease, *Kidney International,* 57, 2360–2374 (2000).

63. J. O. Cisar, D.-Q. Xu, J. Thompson, W. Swaim, L. Hu, and D. J. Kopecko, An alternative interpretation of nanobacteria-induced biomineralization, *Proceedings of the National Academy of Science, USA,* 97, 11511–11515 (2000).

64. A. Mushegian, The minimal genome concept, *Current Opinion in Genetics & Development,* 9, 709–714 (1999).

65. D. Ferber, Synthetic biology: Microbes made to order, *Science,* 303 158–161 (2004).

66. J. I. Glass, N. Assad-Garcia, N. Alperovich, S. Yooseph, M. R. Lewis, M. Maruf, C. A. Hutchison, III, H. O. Smith, and J. C. Venter, Essential genes of a minimal bacterium, *Proceedings of the National Academy of Science, USA,* 103, 425–430 (2006).

67. H. O. Smith, C. A. Hutchison, III, C. Pfannkoch, and J. C. Venter, Generating a synthetic genome by whole genome assembly: X174 bacteriophage from synthetic oligonucleotides, *Proceedings of the National Academy of Science, USA,* 100, 15440–15445 (2003).

68. S. J. Giovannoni, H. J. Tripp, S. Givan, M. Podar, K. L. Vergin, D. Baptista, L. Bibbs, J. Eads, T. H. Richardson, M. Noordewier, M. S. Rappé, J. M. Short, J. C. Carrington, and E. J. Mathur, Genome streamlining in a cosmopolitan oceanic bacterium, *Science,* 309, 1242–1245 (2005).

69. V. Pérez-Brocal, R. Gil, S. Ramos, A. Lamelas, M. Postigo, J. M. Michelena, F. J. Silva, A. Moya, and A. Latorre, A small microbial genome: The end of a long symbiotic relationship? *Science,* 314, 312–313 (2006).

70. S. G. E. Andersson, Perspectives genetics: The bacterial world gets smaller, *Science,* 314, 259 (2006).

71. A. Nakabachi, A. Yamashita, H. Toh, H. Ishikawa, H. E. Dunbar, N. A. Moran, and M. Hattori, The 160-kilobase genome of the bacterial endosymbiont *Carsonella, Science,* 314, 267 (2006).

72. S. G. E. Andersson, A. Zomorodipour, J. O. Andersson, T. Sicheritz-Pontén, U. C. M. Alsmark, R. M. Podowski, A. K. Näslund, A.-S. Eriksson, H. H. Winkler,

and C. G. Kurland, The genome sequence of *Rickettsia prowazekii* and the origin of mitochondria, *Nature*, 396, 133–140 (1998).

73. J. Maniloff, Nanobacteria: Size limits and evidence, *Science*, 276, 1773–1776 (1997).

74. B. La Scola, S. Audic, C. Robert, L. Jungang, X. de Lamballerie, M. Drancourt, R. Birtles, J.-M. Claverie, and D. Raoult, A giant virus in amoebae, *Science*, 28, 2033 (2003).

75. J. Hogan, Are nanobacteria alive or just strange crystals? *New Scientist*, 182, 6–7 (2004).

76. M. R. Taylor, *Dark life: Martian nanobacteria, rock-eating cave bugs, and other extreme organisms of inner Earth and outer space*, Scribner, New York (1999).

77. Key regulators in the cell cycle, Nobel Assembly press release: Summary of award of Nobel Prize in physiology or medicine to Leland H. Hartwell, R. Timothy (Tim) Hunt, and Paul M. Nurst (2001).

78. G. B. Childs, Membrane structure (2003) University of Texas Medical Branch web page at: cellbio.utmb.edu/cellbio/memembrane.htm

79. Y. Yawata, *Cell membrane: The red blood cell as a model*, Wiley-VCH Verlag GmbH & Co., Weinheim, Germany (2003).

80. R. Latorre and J. C. Sáez, eds., *From ion channels to cell-to-cell conversations*, Springer-Verlag, Berlin (1997).

81. B. Hille, *Ion channels of excitable membranes*, 3rd ed., Sinaner Associates, Sunderland, MA (2001).

82. D. Marsh, Polarity and permeation profiles in lipid membranes, *Proceedings of the National Academy of Science, USA*, 98, 7777–7782 (2001).

83. D. Marsh, Membrane water-penetration profiles from spin labels, *European Biophysics Journal*, 31, 559–562 (2002).

84. J. Nagle and S. Tristram-Nagle, Structure of lipid bilayers, *Biochimica Biophysica Acta*, 1469, 159–195 (2000).

85. F. Goñi, Non-permanent proteins in membranes: When proteins come as visitors, *Molecular Membrane Biology*, 19, 237–245 (2002).

86. E. R. Kandel, J. H. Schwartz, and T. M. Jessell, *Principles of neural science*, 4th ed., McGraw-Hill, New York (2000).

87. G. M. Shepherd, *Neurobiology*, 3rd ed., Oxford University Press, Oxford, UK (1994).

88. A. Huxley, From overshoot to voltage clamp, *Trends in Neurosciences*, 25, 553–558 (2002).

89. A. L. Hodgkin and A. F. Huxley, A quantitative description of membrane current and its application to conduction and excitation in nerve, *Journal of Physiology (London)*, 117, 500–544 (1952).

90. D. A. Doyle, J. M. Cabral, R. A. Pfuetzner, A. Kuo, J. M. Gulbis, S. L. Cohen, B. T. Chait, and R. MacKinnon, The structure of the potassium channel: Molecular basis of K+ conduction and selectivity, *Science*, 280, 69–77 (1998).

91. C. A. Janeway, Jr., P. Travers, M. Walport, and M. J. Shlomchik, *Immunobiology*, 5th ed., Garland Publishing, Inc., New York (2001).

92. L. Du Pasquier and G. W. Litman, *Origin and evolution of the vertebrate immune system*, Springer-Verlag, Berlin (2000).

93. J. D. Watson, T. A. Baker, S. P. Bell, A. Gann, M. Levine, and R. Losick, *Molecular biology of the gene*, 5th ed., Benjamin Cummings Publishing Company, San Francisco, CA (2003).

94. P. J. Beurton, R. Falk, and H.-J. Rheinberger, *The concept of the gene in development and evolution (Cambridge studies in philosophy and biology)*, Cambridge University Press, Cambridge, UK (2003).

95. P. Portin, The concept of the gene: Short history and present status, *The Quarterly Review of Biology*, 68, 173–223 (1993).

96. J. Davis, *Mapping the code: The human genome project and the choices of modern science*, John Wiley & Sons, Inc., New York (1991).

97. M. A. Palladino, *Understanding the human genome project*, 2nd ed., Benjamin Cummings Publishing Company, San Francisco, CA (2005).

98. Human Genome Program Report, *Genomics and its impact on science and society*, U.S. Department of Energy, Washington, D.C. (2003).

99. P. Rabinow, *Making PCR: A story of biotechnology*, University of Chicago Press, Chicago, IL (1996).

100. K. Mitchelson, ed., *New high-throughput technologies for DNA sequencing and genomics 2*, Elsevier, Amsterdam, the Netherlands (2007).

101. C. P. Poole and F. J. Owens, *Introduction to nanotechnology*, Wiley Interscience, Hoboken, NJ (2003).

Problems

14.1 There are on the order of one hundred trillion cells (1×10^{14}) in a human body. Starting with a single fertilized egg cell, how many cell divisions are required to result in a full human body?

14.2 Cells—think nano:

a. Why are there so many cells in a mammalian body?

b. In general, why are cells the sizes that they are?

c. What advantages does a multicellular organism have compared to a single-celled organism?

d. What are the advantages as building blocks of a cell over a virus or macrophage?

14.3 Oxygen and nutrients must be transported into cells from their surfaces. Likewise, waste products must be transported out. For three spherical cells with radii of 1000, 2000, and 5000 nm, calculate their surface areas, volumes, and ratios of surface area to volume. What would you expect the relationship to be between cell volume and sustainable rate of metabolic activity?

14.4 Refering to Problem 14.1, and using the typical size for a human cell cited in Section 14.1.2 of this chapter, calculate the approximate total cellular surface area in a human body.

14.5 In a typical healthy human digestive tract, the number of bacteria is an order of magnitude higher than the number of cells in the body. How many bacteria are in a typical human digestive tract? Assuming that these bacteria are the size of *E. coli* (see Section 14.1.2), what is their total volume and surface area?

14.6 What is the difference between DNA and RNA? In the context of DNA and RNA, what is meant by "complementary?" Which of the following are complementary to each other?

a. Sugar and phosphate

b. Deoxyribose and ribose

c. Thymine and cytosine

d. Adenine and quinine

14.7 For the DNA segment below:

a. What is the RNA segment that you would expect to be transcribed corresponding to it?

b. Does it matter whether you transcribe from left to right or right to left to get the RNA?

c. For the protein translated by this sequence, does the direction of transcription matter?

d. What is a sequence of amino acids in the polypeptide chain that it would transcribe?

(If the direction matters, work out both peptides for both directions.)

DNA Segment: TAGTGCAAAGCTCAGCAT

14.8 Describe the structure of a cell membrane. Include polysaccharide chains, cross-linking, peptide chains, lipo-poly-saccharides, lipid, proteins, and meso-somes in your discussion.

14.9 Ribosomes, ca. 18 nm in size, consist of two subunits (50S and 30S). Explain their role in protein synthesis. Why is this structure such a perfect nanomaterial? Relate your answer to size, structure, and function.

14.10 What is the function of the *cristae* found in mitochondria? Discuss the morphology and structure of mitochondria. What nanocomponent(s) is (are) exerting its (their) effects in mitochondria?

14.11 Explain the structure of the endoplasmic reticulum and the *cisternae*. What nanocomponent(s) is (are) exerting its (their) effects in the smooth ER? The rough ER?

14.12 Perhaps the pinnacle of molecular recognition is illustrated by the immune globulins—for example, the antibodies. These classes of proteins are not only able to recognize and neutralize antigenic materials that have invaded the body, but are also able to do so in a dynamic, versatile way. Antigens are considered to be divalent; antibodies are considered to be polyvalent. Explain what this means and relate your understanding to the overall solubility of antibody–antigen complexes.

14.13 Polarity, pH, solubility, and structure all play a role in amino acid chemistry. There are basically four kinds of amino acids: those with nonpolar R groups, uncharged polar R groups, positively charged basic R groups, and negatively charged acidic groups. In an ion-exchange column at high pH, what do you predict the order of elution would be for the following amino acids: glycine, phenylalanine, arginine, and glutamic acid? Which ones migrate towards the anode? The cathode? In an electrophoretic gel—also at high pH.

14.14 Polypeptides normally do not exist in linear form (extended). The α-helix is a form often adopted. What is the length in nanometers of a polypeptide chain that has 80 amino acids in the α-helix form? In the open linear extended form? Hint: The C–N–C–N–C backbone consists of single bonds with alternating carbons and nitrogens. The peptide bond containing the α-carbon exists in a plane. Determine the repeating unit, the pitch, and the rise—also bond lengths and angles—and calculate the lengths. Any biochemistry text will contain the relevant information with images.

14.15 Compare the structures of the α-helix and the β-sheet. Why the difference?

14.16 The behavior of proteins in solution is quite important to metabolism. List five characteristics of proteins that affect their solubility.

14.17 What is the length of all the unraveled human DNA in a human cell if set end to end? What assumptions did you use to make your calculation?

Index

C

E

F